ENVIRONMENTAL OXIDANTS

Volume

28

in the Wiley Series in

Advances in Environmental
Science and Technology

JEROME O. NRIAGU, Series Editor

ENVIRONMENTAL OXIDANTS

Edited by
Jerome O. Nriagu

and

Milagros S. Simmons

Department of Environmental and Industrial Health,
School of Public Health, The University of Michigan,
Ann Arbor, Michigan

A WILEY-INTERSCIENCE PUBLICATION

JOHN WILEY & SONS, INC.

New York · Chichester · Brisbane · Toronto · Singapore

Library of Congress Cataloging in Publication Data:

Environmental oxidants / edited by Jerome O. Nriagu and Milagros S.
    Simmons.
        p.   cm.—(Advances in environmental science and technology :
    v. 28)
    "A Wiley-Interscience publication."
    Includes index.
    ISBN 0-471-57928-9 (cloth : acid-free paper)
    1. Oxidizing agents—U.S. Environmental aspects. 2. Oxidizing agents—
Health aspects. 3. Sewage—Purification—Ozonization. I. Nriagu,
Jerome O. II. Simmons, Mila S. III. Series.
TD196.095E54 1994
628'.01'541393—dc20   93-4970
                                                                    CIP

Printed in the United States of America

10 9 8 7 6 5 4 3 2

CONTRIBUTORS

NEIL ALEXIS, The Gage Research Institute, 223 College Street, Toronto, Ontario, Canada M5T 1R4

JOHN AUSTIN, Meteorological Office, London Road, Bracknell, Berkshire, RG12 2SZ, United Kingdom

OGUZ K. BASKURT, Department of Physiology, Akdeniz University, Medical Faculty, Morfoloji Binasi, Dumlupinar Bulvari, 07070 Antalya, Turkey

PAUL J. A. BORM, Department of Health Risk Analysis and Toxicology, University of Limburg, P. O. Box 616, 6200 MD Maastricht, The Netherlands

ALDO BRUCCOLERI, Department of Chemistry, University of Calgary, 2500 University Drive NW, Calgary, Alberta, Canada T2N 1N4

CHRISTOPH BRÜHL, Division of Atmospheric Chemistry, Max-Planck-Institute for Chemistry, P. O. Box 3060, 55020 Mainz, Germany

JANUSZ Z. BYCZKOWSKI, ManTech Environmental Technology, Inc., P. O. Box 31009, Dayton, Ohio 45437-0009

I. COLBECK, Department of Chemistry and Biological Chemistry, University of Essex, Wivenhoe Park, Colchester CO4 3SQ, United Kingdom

PAUL J. CRUTZEN, Division of Atmospheric Chemistry, Max-Planck-Institute for Chemistry, P. O. Box 3060, 55020 Mainz, Germany

SIMON H.R. DAVIES, Department of Civil and Environmental Engineering, Michigan State University, A349 Engineering Building, East Lansing, Michigan 48824-1226

WOLF DIETER GROSSMANN, Department of Applied Landscape Ecology, Environmental Research Center Leipzig, Permoserstr. 15, D-04318 Leipzig, Germany

WILLIAM E. HOGSETT, U.S. Environmental Protection Agency, 200 SW 35th Street, Corvallis, Oregon 97333

HITOSHI HORI, Department of Biological Science and Technology, Faculty of Engineering, The University of Tokushima, 2-1 Minamijosanjima, Tokushima 770, Japan

YVONNE M. W. JANSSEN, Department of Health Risk Analysis and Toxicology, University of Limburg, P. O. Box 616, 6200 MD Maastricht, The Netherlands

ERIC JOOS, Electricité de France, Direction des Etudes et Recherches, 6, quai Watier, 78400 Chatou, France

ARUN P. KULKARNI, Toxicology Program, College of Public Health MDC-56, University of South Florida, 13201 Bruce B. Downs Blvd., Tampa, Florida 33612

COOPER H. LANGFORD, Department of Chemistry, University of Calgary, 2500 University Drive NW, Calgary, Alberta, Canada T2N 1N4

RICHARD A. LARSON, Institute of Environmental Studies, University of Illinois, 1101 W. Peabody Drive, Urbana, Illinois 61801

E. HENRY LEE, ManTech Environmental Technology, Inc., 200 Southwest 35th Street, Corvallis, Oregon 97333

JOS LELIEVELD, Air Quality Department, Wageningen University, 6700 EV Wageningen, The Netherlands

GUISEPPE LEPORE, Department of Chemistry, University of Calgary, 2500 University Drive NW, Calgary, Alberta, Canada T2N 1N4

A. R. MACKENZIE, Department of Chemistry, Centre for Atmospheric Science, University of Cambridge, Lensfield Road, Cambridge CB2 1EW, United Kingdom

KAREN A. MARLEY, Institute of Environmental Studies, University of Illinois, 1101 W. Peabody Drive, Urbana, Illinois 61801

JOANNE P. MARSH, Department of Pathology, Medical Alumni Building, University of Vermont, Burlington, Vermont 05405

L. ROBBIN MARTIN, Mechanics and Materials Technology Center, The Aerospace Corporation, E1 Segundo, California 90245-4691

SUSAN J. MASTEN, Department of Civil and Environmental Engineering, Michigan State University, A349 Engineering Building, East Lansing, Michigan 48824-1226

BROOKE T. MOSSMAN, Department of Pathology, Medical Alumni Building, University of Vermont, Burlington, Vermont 05405

ARUN B. MUKHERJEE, Department of Limnology and Environmental Protection, University of Helsinki, SF-00710 Helsinki, Finland

MOHAMMAD G. MUSTAFA, Department of Environmental Health Sciences, School of Public Health, University of California, Los Angeles, Los Angeles, California 90024-1772

HIDEKO NAGASAWA, Pharmaceutical Institute, School of Medicine, Keio University, 35 Shinanomachi, Shinjuku, Tokyo 160, Japan

CHRISTIAN SEIGNEUR, ENSR Consulting and Engineering, 1320 Harbor Bay Parkway, Alameda, California 94501

FRANCES SILVERMAN, The Gage Research Institute, 223 College Street, Toronto, Ontario, Canada M5T 1R4

SUSAN M. TARLO, The Gage Research Institute, 223 College Street, Toronto, Ontario, Canada M5T 1R4

HIROSHI TERADA, Faculty of Pharmaceutical Sciences, The University of Tokushima, Shomachi-1, Tokushima 770, Japan

ANNE M. THOMPSON, Atmospheric Chemistry and Dynamics Branch, NASA/Goddard Space Flight Center, Code 916, Greenbelt, Maryland 20771

DAVID T. TINGEY, U.S. Environmental Protection Agency, 200 SW 35th Street, Corvallis, Oregon 97333

SEMA YAVUZER, Department of Physiology, Ankara University, Medical Faculty, Ankara, Turkey

CHARLES F. YOCUM, Departments of Biology and Chemistry, The University of Michigan, Ann Arbor, Michigan 48109-1048

INTRODUCTION
TO THE SERIES

The deterioration of environmental quality, which began when mankind first congregated into villages, has existed as a serious problem since the industrial revolution. In the second half of the twentieth century, under the ever-increasing impacts of exponentially growing population and of industrializing society, environmental contamination of the air, water, soil, and food has become a threat to the continued existence of many plant and animal communities of various ecosystems and may ultimately threaten the very survival of the human race. Understandably, many scientific, industrial, and governmental communities have recently committed large resources of money and human power to the problems of environmental pollution and pollution abatement by effective control measures.

Advances in Environmental Science and Technology deals with creative reviews and critical assessments of all studies pertaining to the quality of the environment and to the technology of its conservation. The volumes published in the series are expected to service several objectives: (1) stimulate interdisciplinary cooperation and understanding among the environmental scientists; (2) provide the scientists with a periodic overview of environmental developments that are of general concern or that are of relevance to their own work or interests; (3) provide the graduate student with a critical assessment of past accomplishment, which may help stimulate him or her toward the career opportunities in this vital area; and (4) provide the research manager and the legislative or administrative official with an assured awareness of newly developing research work on the critical pollutants and with the background information important to their responsibility.

As the skills and techniques of many scientific disciplines are brought to bear on the fundamental and applied aspects of the environmental issues, there is a heightened need to draw together the numerous threads and to present a coherent picture of the various research endeavors. This need and the recent tremendous growth in the field of environmental studies have clearly made some editorial adjustments necessary. Apart from the changes in style and format, each future volume in the series will focus on one particular theme or timely topic, starting with Volume 12. The author(s) of each pertinent section will be expected to critically review the literature and the most important recent developments in the particular field; to critically evaluate new concepts, methods, and data; and to focus attention on important unresolved or controversial questions and on probable future trends. Monographs embodying the results of unusually extensive and well-rounded investigations will also be published in the series. The net result of the new editorial policy should be more integrative and comprehensive volumes on key environmental issues and pollutants. Indeed, the development of realistic standards of environmental quality for many pollutants often entails such a holistic treatment.

JEROME O. NRIAGU, Series Editor

PREFACE

Environmental oxidants such as ozone and the oxides of nitrogen, sulfur, and carbon have attracted public attention in relation to the smog episodes and their adverse effects on human and ecosystem health. Other oxidants, especially the oxygen species such as molecular oxygen, hydroxyl radical, singlet oxygen, superoxide anion, and hydrogen peroxide, can trigger reactions resulting in the destruction of the ozone layer and in climatic changes. At the systemic and cellular levels, oxygen radicals are believed to be involved in liver injury associated with alcohol consumption and in DNA damage and mutational changes while oxidative stress has been implicated in arteriosclerosis, xenobiotic toxicity, and ischemia/reperfusion injury. In the recent past, considerable attention has been devoted to the use of oxidants and other processes that generate the oxygenated species in hypoxic cell-specific anticancer drugs and for the treatment of contaminated waters. Because of the multifarious effects of oxidants from the global down to the cellular levels, the production and loss of oxidants associated with human activities must be of some concern.

This book provides a broad overview of the environmental chemistry and toxicology of oxidants and their role in pollution control. The topics covered include the evolution, production, distribution, and fate of oxidants in the atmosphere, hydrosphere, and biosphere; the influence of human activities on oxidative processes in the atmosphere; oxidative stress at the cellular, systemic, and ecosystem levels; and the use of oxidants in wastewater treatment processes. This book does not pretend to be an encyclopedia encompassing all aspects of the environmental oxidants; this would have been a Herculean task indeed. Rather it provides the reader with critical reviews of current research on selected key topics within the broad field that transcends several scientific disciplines.

Because of the multidisciplinary flavor of the chapters and the multinational makeup of the authors, this book should be of some interest to anyone concerned about air pollution and its effects on our health and the world we live in.

JEROME O. NRIAGU
MILAGROS S. SIMMONS

Ann Arbor, Michigan
February 1994

CONTENTS

ADVANCES IN ENVIRONMENTAL SCIENCE AND TECHNOLOGY
Jerome O. Nriagu, Series Editor

1

PHOTOSYNTHETIC OXYGEN EVOLUTION

Charles F. Yocum

Departments of Biology and Chemistry, University of Michigan, Ann Arbor, Michigan 48109–1048

Environmental Oxidants, Edited by Jerome O. Nriagu and Milagros S. Simmons.
ISBN 0–471–57928–9 © 1994 John Wiley & Sons, Inc.

1. INTRODUCTION

Green plants, algae, and cyanobacteria convert H_2O into O_2 in a chlorophyll-catalyzed photoreaction. A complex multisubunit enzyme system called photosystem II (PSII) catalyzes H_2O oxidation and donates the resulting electrons to the subsequent reactions that ultimately reduce CO_2 to sugars. Photosynthesis in the biosphere has been estimated to account for an annual global production of fixed carbon amounting to as much as 10^{10} to 10^{11} tons (Kamen, 1963); about the same amount of O_2 would be liberated into the biosphere as a by-product. The last two decades have seen substantial advances in understanding all aspects of photosynthesis. Readers interested in topics related to photosynthetic electron transfer beyond photosystem II and the O_2-evolving reaction will find the review by Andreasson and Vanngard (1988) to be a useful source of information. In addition, there are a number of recent technical reviews on the components of photosystem II and the O_2-evolving reaction. Those by Babcock (1987), Brudvig et al. (1989), Dismukes (1986), Hansson and Wydryzinski (1990), Ghanotakis and Yocum (1990), Rutherford et al. (1992), and Van Gorkom (1985) will provide the interested reader with additional information, and where appropriate are cited again in the relevant sections that follow. As the reader will see, research on the mechanism of H_2O oxidation and the structure of PSII has occupied the talents of a large number of investigators, ranging from biophysical chemists to molecular biologists. In this review I will briefly discuss selected aspects of the chemistry of H_2O oxidation that pertain to the biological reaction and then discuss evidence for the probable mechanism of the reaction that occurs in photosystem II. The functions of essential inorganic ion components of the reaction will also be discussed, as will proposed roles of extrinsic and intrinsic polypeptides that make up the enzyme system.

2. WATER OXIDATION CHEMISTRY AND EXPERIMENTAL OBSERVATIONS ON THE MECHANISM OF BIOLOGICAL OXYGEN EVOLUTION

The overall reaction for oxidation of water to oxygen,

$$2H_2O \longrightarrow O_2 + 4H^+ + 4e^- \qquad E_{m,7} = +0.815 \text{ V} \qquad (1)$$

requires a very high redox potential in comparison to most other biological electron transfer processes. Alternative mechanisms for the reaction could involve either sequential 1-electron oxidation steps or concerted multielectron reactions. A hypothetical mechanism invoking sequential 1-electron oxidation reactions presents thermodynamic difficulties. If any of the individual steps in a sequential mechanism possessed a potential higher or lower than 4-electron potential cited in Equation (1), then other 1-electron oxidation steps in the mechanism would of necessity require adjustments in their individual potentials. A major thermodynamic barrier to sequential, 1-electron oxidation mechanisms is raised if one considers the redox potential for 1-electron oxidation of H_2O to form the hydroxyl radical (Koppenol, 1978):

$$H_2O \longrightarrow OH\cdot + H^+ + e^- \qquad E_{m,7} = +1.8 \text{ V} \qquad (2)$$

Modification of this very high redox potential to a value near the overall 4-electron potential for H_2O oxidation would be facilitated by changing the pH or by assuming that H_2O binding to PSII occurs with an exceptionally high affinity. For example, an alkaline shift of about 15 pH units would achieve the desired alteration in redox potential. Such changes lie outside the realm of probability for a biological system, and therefore proposals that biological H_2O oxidation involves concerted multielectron oxidation/reduction reactions are favored, and are supported by experimental evidence.

Investigations of the mechanism of photosynthetic H_2O oxidation undertaken by Kok et al. (1970) and Joliot et al. (1969) employed sensitive electrodes to detect O_2, and utilized an illumination protocol whereby samples were subjected to flashes of light of sufficiently short duration (a few microseconds) so that one electron was transferred on every photoexcitation of PSII. The results of this work revealed that O_2 was released in a burst from PSII after a sequence of three flashes, and again after each fourth flash excitation event. In order to account for the flash-number dependence of the O_2 yield from PSII, Kok et al. proposed the existence of a linear series of discrete oxidation states, denoted S-states ($S_i = 0, 1, 2, 3$, and 4), associated with the O_2-evolving reaction in PSII. The S-states were proposed to operate by means of a cycle in which the most-oxidizing species (S_4^{4+}) would decay spontaneously to yield O_2 and S_0:

$$S_0^0 \longrightarrow S_1^{+1} \longrightarrow S_2^{+2} \longrightarrow S_3^{+3} \longrightarrow S_4^{+4} \longrightarrow S_0^{+0} \qquad (3)$$
$$2H_2O \longrightarrow O_2 + 4H^+$$

The experimental observation that O_2 appeared after the third, rather than the fourth, light flash in previously dark-adapted material was accommodated to the Kok model by proposing that S_1 is the dominant stable species in the resting enzyme system (Joliot and Kok, 1975). Each S-state system is isolated from other O_2-evolving centers; accumulated oxidizing equivalents are not shared among centers. With the exception of S_4, higher S-states decay to S_1 with half-times of about 30 sec. The period-4 oscillation in O_2 yield eventually damps to roughly equivalent yields on each flash, a phenomenon attributed to causes that include photochemical "misses" (a particular center is not excited by the flash) and "double hits" (a center recovers from illumination and is reexcited by the same light pulse, provided it is of sufficient duration). Reducing agents such as NH_2OH and H_2O_2 will react with the S_1 state to form a new state, S_{-1}, which is characterized by the appearance of an O_2 burst on the fifth, rather than the third, flash in a series (Bouges, 1971). Lastly, as will be described in detail in a later section, four manganese atoms ligated to PSII are an essential component of the redox active site of H_2O oxidation; it is widely believed that oxidation of these manganese atoms correlates with S-state advancement.

A number of experimental approaches have been utilized to probe the interaction of H_2O with the S-states, the goal being to identify the operation sequential or concerted mechanisms for the 4-electron oxidation of the substrate. Measurements of proton release from PSII after a train of short light flashes revealed a stoichiometry of 1, 0, 1, and $2H^+$ associated with S-state advancement from S_0 to S_4 (Saphon and Crofts, 1977). One interpretation of this result would be that protons are liberated in the course of sequential oxidation of H_2O during S-state advancement:

$$2H_2O + S_0^0 \longrightarrow S_1^{+1} \longrightarrow S_2^{+2} \longrightarrow S_3^{+3} \longrightarrow S_4^{+4} \longrightarrow S_0^0 + O_2$$

$$ \downarrow \downarrow \downarrow$$

$$ 1H^+ 1H^+ 2H^+ (4)$$

These proton release stoichiometries have been reported by a number of investigators, but the most recent experiments, designed to account for possible interfering structural properties associated with the photosynthetic membranes examined, have revealed noninteger stoichiometries of H^+ release on the $S_1 \rightarrow S_2$ and $S_3 \rightarrow S_4$ transitions. Based on this observation it is now proposed that protons are released from the protein matrix near the water-splitting apparatus, instead of directly from the active site of the reaction itself (Jahns et al., 1991). If this is so, then the operation of a concerted multielectron mechanism for H_2O oxidation by PSII would seem to be likely possibility:

$$S_0^0 \longrightarrow S_1^{+1} \longrightarrow S_2^{+2} \longrightarrow S_3^{+3} \longrightarrow S_4^{+4} + 2H_2O \longrightarrow S_0^0 + O_2$$

$$\downarrow \downarrow \downarrow \downarrow \downarrow \downarrow$$

$$4H^+ 3H^+ <3H^+ <2H^+ 4H^+$$

$$\downarrow \downarrow \downarrow \downarrow$$

$$1H^+ 0.5H^+ 1H^+ 1.5H^+ (5)$$

Support for a concerted 4-electron oxidation mechanism like that shown in Equation (5) has been obtained in experiments that employed H_2O containing isotopically labeled oxygen (^{18}O) in combination with sensitive mass spectroscopy to detect oxidation products from photosynthetic H_2O oxidation catalyzed by short light flashes (Radmer and Ollinger, 1986; Bader et al., 1987). The results of these experiments showed that the oxygen isotope in H_2O added to a sample in S_3 appeared in the O_2 released after the enzyme system had been advanced through S_4. The central finding of these experiments, that the final isotopic composition of O_2 produced by the photosynthetic decomposition of H_2O can be determined at the level of S_3, has been interpreted to indicate that intermediate oxidation products of water are not formed in association with the S-state advancements $S_0 \rightarrow S_3$.

Other evidence in support of a concerted 4-electron oxidation reaction is to be found in experiments where manganese oxidation state changes driven by exposure of PSII to short light flashes were detected by optical spectroscopy (Dekker et al., 1984a, b, c). These positive going absorption changes centered at about 300 nm have been assigned to ligand-to-metal charge-transfer bands (Dekker, 1992), and are proposed to specifically reflect $Mn^{3+} \rightarrow Mn^{4+}$ oxidation state transitions. The absorbance changes exhibit a dark stability consistent with S-state lifetimes, accumulate up to the S_3 state, and are quenched rapidly after the light flash that catalyzes the $S_3 \rightarrow S_4 \rightarrow S_0$ transition. Debate over the amplitude and spectrum of the absorbance change accompanying the $S_0 \rightarrow S_1$ transition (see, for example, Lavergne, 1986, 1987) has unfortunately obscured the importance of the fact that the overall pattern of absorbance changes is remarkably consistent in indicating advancements in a charge-accumulating system that resets itself after formation of S_4. Combined with the O-isotope experiments, such data point to the concerted 4-electron oxidation model for H_2O oxidation in PSII. There are alternative views that favor sequential, multielectron mechanisms that might reasonably involve formation of a peroxyl intermediate as a means of bypassing 1-electron oxidation of H_2O. Arguments for sequential mechanisms may be found in a recent publication from the laboratory of Renger (Messinger et al., 1991) and in a review article by Vanngard et al. (1992).

3. ELECTRON TRANSFER REACTIONS ASSOCIATED WITH ADVANCEMENT OF THE S-STATES

Chlorophyll *a* catalyzed photochemistry in PSII is ultimately responsible for triggering a series of reactions that result in S-state advancement and H_2O oxidation. These reactions are represented by the linear sequence of electron carriers shown in Equation (6). The estimated redox potential of the various components is given in parentheses beneath the electron transfer scheme:

$$S^N \longrightarrow Y_Z \longrightarrow P680 \longrightarrow \text{Pheophytin } a \longrightarrow Q_A \longrightarrow Q_B$$

$$(+0.8\,V) \qquad (+1.0\,V) \qquad (+1.2\,V) \qquad\qquad (-0.6\,V) \qquad\qquad (0\,V) \qquad (+0.06\,V)$$

$$(6)$$

In Equation (6), S^N represents the S-states, Y_Z is a redox-active tyrosine residue on one of the constituent polypeptides of the enzyme system, P680 is a redox-active chlorophyll *a*, and pheophytin *a* is a chlorophyll *a* molecule lacking the Mg^{2+} atom normally ligated in the center of the pigment ring system. Two quinones, Q_A and Q_B, serve in sequence as the terminal electron acceptors for photochemically generated electrons. These species are plastoquinones, a form of quinone with an isoprenoid side-chain that is unique to O_2-evolving photosynthetic systems. The quinone called Q_A is very tightly bound to the electron transfer system, whereas the second quinone (Q_B) is more loosely associated with its binding site.

Light absorption by P680 initiates a series of very rapid electron transfer reactions within the ensemble of PSII-associated chromophores, as shown in Equation (7) on the next page:

$$S^N \longrightarrow Y_Z \longrightarrow P680 \longrightarrow \text{Pheophytin } a \longrightarrow Q_A \longrightarrow Q_B$$

$$\downarrow \text{hv}$$

$$S^N \longrightarrow Y_Z \longrightarrow P680+ \longrightarrow \text{Pheophytin } a^- \longrightarrow Q_A \longrightarrow Q_B$$

$$\downarrow \text{dark}$$

$$S^N \longrightarrow Y_Z^+ \longrightarrow P680 \longrightarrow \text{Pheophytin } a \longrightarrow Q_A^- \longrightarrow Q_B \qquad (7)$$

$$\downarrow \text{dark}$$

$$S^{N+1} \longrightarrow Y_Z \longrightarrow P680 \longrightarrow \text{Pheophytin } a \longrightarrow Q_A \longrightarrow Q_B^-$$

Repetitions of this cycle of events create advancements in the S-states, while the doubly reduced form of Q_B ($Q_B H_2$) is released from its binding site and replaced by an oxidized plastoquinone on every other cycle of redox activity. The kinetics of the reactions shown in Equation (7) are summarized in Table 1; the half-times shown are estimates provided by Rutherford (1988). The time scale for electron transfer ranges from a few picoseconds for primary photochemistry to about $100 \, \mu\text{sec}$ for Q_B reduction; the slowest event shown, S_4^{4+} decay, occurs in about 1 msec. The polypeptide structure of PSII would seem to feature an organization wherein chromophore binding sites are sufficiently insulated from one another so as to prevent various reactive intermediates from undergoing rapid charge recombination reactions with one another. Table 1 presents the estimated half-times for some of the charge recombination reactions that would prevent S-state advancement. As can be seen, these recombination reactions are estimated to occur on time scales that are substantially slower than those for the reactions that produce S-state advances.

In view of the extreme positive and negative redox potentials of the electron carriers in PSII and their existence as free radical species during the efficient operation of PSII, it is not surprising that the system can be damaged in the course of photocatalyzed electron transfer. Substantial evidence has accumulated for the existence of a process called photoinhibition that inactivates PSII. Recovery from such damage requires protein synthesis, indicating that polypeptide constituents of the enzyme system have been altered or destroyed. Proposed

Table 1   Electron Transfer Kinetics in PSII

Forward Electron Transfer Reaction	Half-time	Recombination Reaction	Half-time
P680/Pheo $a \rightarrow$ P680$^+$/Pheo a^-	< 10 psec	—	—
Pheo a^-/$Q_A \rightarrow$ P680$^+$/Pheo a/Q_A^-	250 psec	Q_A^-/P680$^+ \rightarrow Q_A$/P680	150 μsec
Y_Z/P680$^+$/$Q_A^- \rightarrow Y_Z^+$/P680/Q_A^-	50–250 nsec	Y_Z^+/$Q_A^- \rightarrow Y_Z$/Q_A	20 msec
Y_Z^+/Q_A^-/$Q_B \rightarrow Y_Z^+$/Q_A/Q_B-	100 μsec	Y_Z^+/$Q_B- \rightarrow Y_Z$/Q_B	400 msec
S^N/Y_Z^+/$Q_B- \rightarrow S^{N+1}$/Y_Z/Q_B-	30 μsec–1 msec	S^{N+1}/$Q_B- \rightarrow S^N$/Q_B	30 sec

mechanisms for such damage include toxic effects of accumulated quinone semi-quinone or tyrosine radicals. Evidence also exists that points to triplet oxygen as a causative agent for photoinhibition under some conditions. It is probable that all of the reactive species mentioned contribute to loss of PSII activity. For further details on photoinhibition, the reader is referred to a recent article by Barber and Andersson (1992), and the references contained therein.

4.   THE ROLES OF Mn, Ca^{2+}, AND Cl$^-$ IN S-STATE CATALYZED H$_2$O OXIDATION

An extensive literature documents proposed functions for Mn, Ca^{2+}, and Cl$^-$ in H$_2$O oxidation (Amesz, 1983; Babcock, 1987; Brudvig et al., 1989; Coleman and Govindjee, 1987; Debus, 1992; Dismukes, 1986; Yocum, 1991), and research in this area is particularly intensive at the present time. Among available techniques, optical spectroscopy, magnetic resonance spectroscopy (electron paramagnetic resonance, or EPR, and nuclear magnetic resonance, or NMR), and X-ray absorption techniques (XANES, or X-ray absorption near-edge structure, and EXAFS, or extended X-ray absorption fine structure) have emerged as prominent methods that can be successfully applied as probes of manganese in the active site of H$_2$O oxidation. As will be seen, paramagnetic forms of manganese in the S$_2$ state are readily detected by EPR, while NMR has proven useful in probing the reduction and extraction of manganese from the active site of O$_2$ evolution; XANES and EXAFS spectroscopy can be used to address questions of both the structural organization and oxidation states of manganese in PSII. When applied to purified PSII preparations that afford high concentrations of the enzyme, these biophysical methods have proven useful in providing data leading to working models for the active site of H$_2$O oxidation. The prevalent view at present is that manganese is the redox-active species responsible for H$_2$O oxidation. The roles of Ca^{2+} and Cl$^-$ are less clearly defined; both ions are essential activators of the redox reactions that result in H$_2$O oxidation.

4.1.   Manganese

High rates of O$_2$ evolution activity by PSII correlate closely with the presence of four manganese atoms per reaction center in the enzyme (Cheniae, 1980; Yocum et al., 1981). In dark-adapted PSII, manganese is not EPR- detectable at room temperature, nor does it give rise to strong enhancements of H$_2$O proton spin-lattice relaxation rates (Sharp and Yocum, 1981). Although the metals resist extraction during the rigorous purification procedures used to extract PSII from photosynthetic membranes, some relatively specific yet gentle treatments will remove manganese in the form of Mn^{2+} to produce an inhibition of steady-state O$_2$ evolution. The most widely used of these treatments are exposure to the buffer Tris at high concentrations (0.8 M, pH 8) in the light (Cheniae and Martin, 1978), or dark exposure to low concentrations (1−5 mM) of reductants such as NH$_2$OH

(Cheniae and Martin, 1971). The light requirement for Tris extraction has been demonstrated to arise from a specific binding of the amine buffer to the S_2 state (Frasch and Cheniae, 1980). For NH_2OH, illumination prevents inhibition, a result that suggests either that higher S-states are immune to attack by the reductant (Sharp and Yocum, 1981) or that reduction of manganese is rapidly reversed by reoxidation of the metal during illumination (Ghanotakis et al., 1984c). In darkness, about $3Mn^{2+}$ per reaction center are extracted by NH_2OH; about $2Mn^{2+}$ per reaction center are released by Tris in the light (Yocum et al., 1981). For inhibitory reductants like NH_2OH, it is likely that loss of manganese proceeds through a series of steps involving the formation of a reduced S-state termed S_{-1}, and the subsequent formation of a metastable state containing several atoms of Mn^{2+} (Beck and Brudvig, 1987, 1988) that collapses to release the reduced metals from the enzyme system.

Under the appropriate conditions, PSII samples from which manganese has been extracted by Tris or NH_2OH can be reactivated by reinsertion of the metal into its binding sites. Because this process depends on light, it is called photoactivation. Optimal photoactivation requires the presence of Mn^{2+}, Ca^{2+}, and Cl^- (Tamura and Cheniae, 1988; Miyao-Tokutomi and Inoue, 1991). A careful analysis of the conditions leading to photoactivation have revealed that at least two quanta must be absorbed by a reaction center to create an intermediate, metastable state from which complete reactivation can be achieved by continuous illumination. The reader is referred to the mechanism proposed by Tamura and Cheniae (1988) for further details.

As noted above, magnetic resonance probes have been extensively applied to PSII in attempts to unravel the structure and oxidation states of the manganese atoms that catalyze H_2O oxidation. Although Mn^{2+} bound to relatively immobile sites such as those that would exist in a large membrane system would ordinarily provide efficient relaxation pathway(s) for H_2O protons in an NMR experiment, such relaxation enhancements are difficult to detect in active PSII samples. While inhibitory reduction of manganese in PSII by NH_2OH has been shown to create substantial enhancements of H_2O proton spin-lattice relaxation rates (Sharp and Yocum, 1981), the intrinsic relaxivity of active preparations is near that observed for the suspending buffer. This finding indicates that in the S_1 state, manganese atoms in the PSII active site are in slow exchange with bulk solvent. At the same time, Srinivasan and Sharp (1986a, b) have detected small relaxation enhancements that show an oscillatory pattern that can be modeled in terms of S-state cycling as well as the relaxation efficiencies of higher oxidation states of manganese.

The utility of low-temperature EPR as a probe of PSII-associated manganese was demonstrated by the discovery that the S_2 state was paramagnetic (Dismukes and Siderer, 1981). Illumination of PSII under conditions that ensure a 1-electron oxidation of the S_1 state (200 K, or the presence of inhibitors that prevent reoxidation of Q_A^-) produced an EPR-detectable $g = 2$, $S = 1/2$ "multiline" signal spanning about 1600 G with about 16 lines. A number of properties of this S_2-associated signal have now been established (see reviews by Brudvig et al., 1989; Dismukes, 1986). It is widely agreed that the multiline signal arises from two or more manganese atoms in a mixed valence state (i.e., Mn^{3+}/Mn^{4+}); studies of synthetic multinuclear manganese complexes indicate that a minimum of two manganese atoms in differing oxidation states will produce a multiline-type signal (see Pecoraro, 1988, for a review). Of the possible combinations of four manganese atoms (for example, dimer-dimer, monomer-trimer, or tetramer; Pecoraro, 1988), a tetranuclear manganese cluster has been advanced as the species that seems to provide a combination of properties that can account for the origin of the multiline EPR signal (Brudvig and Crabtree, 1986).

A second EPR signal forms in PSII under certain conditions, such as lowered illumination temperature (160 K) or addition of species (NH_3, F^-) that are known to displace Cl^- from the enzyme (Casey and Sauer, 1984; Zimmermann and Rutherford, 1984; Beck and Brudvig, 1986). The EPR signal is centered around $g = 4.1$ with a spin of 3/2 or 5/2 (de Paula et al., 1986a; Haddy et al., 1992). If samples of PSII illuminated at a temperature that promotes formation of the $g = 4.1$ species ($\ll \sim 190$ K) are subsequently warmed and then reannealed, the multiline species replaces the $g = 4.1$ signal. The observation of two separate EPR signals from the S_2 state has occasioned proposals that different manganese populations might constitute the molecular origin of the signals. Recently, however, Britt and his associates (Kim et al., 1990, 1992) have provided evidence that both signals very likely originate from the same ensemble of metal atoms, and these data are taken to support the existence of a tetranuclear manganese cluster as the origin of both the multiline and $g = 4.1$ EPR signals associated with the S_2 state.

In terms of manganese oxidation states that would generate a paramagnetic species within a four-atom ensemble of the metal, the oxidation state $3Mn^{4+}/Mn^{3+}$ is a reasonable choice for the EPR-active S_2 state. The existence of this oxidation state is supported by the results of X-ray absorption near-edge structure (XANES) spectroscopy, which can provide information on manganese oxidation states (Cole et al., 1987; Guiles et al., 1990). From these investigations, S_1 is proposed to contain $2Mn^{4+}/2Mn^{3+}$. The identity of the valencies of manganese in other S-states is unsettled. The S_3 state is not paramagnetic, which could be taken as evidence that the remaining Mn^{3+} in S_2 has been oxidized. Evidence also exists that an organic radical can be generated during S_3 formation, but with low quantum yield and only under conditions where the O_2-evolving complex has been modified so as to inhibit or slow normal S-state advancement (Boussac et al., 1989, 1990a). Lastly, it has been proposed that S_0 contains Mn^{2+}, based on the observation of oxidation of S_0 to S_1 in the dark (Styring and Rutherford, 1987) by a dark-stable tyrosine radical called Y_D^+ that is a ubiquitous constituent of PSII.

If the current assignments of manganese oxidation states in S_1 and S_2 are correct, and if the metal were to be the only species oxidized during S-state advancement, then a tentative scheme for manganese oxidation states in each

S-state could be written as:

$$(S_0) \longrightarrow (S_1) \longrightarrow (S_2) \longrightarrow (S_3) \longrightarrow (S_4)$$
$$Mn^{2+}/Mn^{3+}/2Mn^{4+} \longrightarrow 2Mn^{3+}/2Mn^{4+} \longrightarrow Mn^{3+}/3Mn^{4+} \longrightarrow 4Mn^{4+} \longrightarrow 3Mn^{4+}/Mn^{5+}$$

$$(8)$$

Such a model predicts that on S-state transitions up to the formation of S_3 only two manganese atoms undergo oxidation, and leaves open the possibility that the $2Mn^{4+}$ species present in S_0 are redox-inert throughout much of the S-state cycle.

Structural information on the arrangement of the manganese atoms in PSII is derived principally from spectroscopic investigation employing X-ray absorption fine structure (EXAFS), a technique whose application to PSII was pioneered by M. P. Klein and coworkers and subsequently utilized by several other research groups. The principal features of the manganese array revealed by this technique are Mn—O or —N ligands at 1.8–1.9 Å, a Mn—Mn distance at 2.7 Å, and a feature at 3.3 Å ascribed either to a Mn—Mn interaction (George et al., 1989; Guiles et al., 1990; Penner-Hahn et al., 1990) or to another X-ray scatterer, perhaps Ca^{2+} (M. P. Klein, personal communication). These distances can be arranged into geometries consistent with pairs of metals arranged in a dimer-of-dimers configuration, as a monomer-trimer, or as a tetranuclear cluster. The chemical nature of the ligands to manganese cannot be determined with certainty from EXAFS; mu—oxo bridges, carboxylate oxygens, and imidazole nitrogens are all likely candidates. Although manganese prefers an ionic ligation environment, which would argue for a predominance of carboxylate ligands or mu—oxo bridges between metal atoms, Britt et al. (1990) have detected the presence of at least one nitrogen ligand associated with manganese in the S_2 state.

A complete understanding of the structure and function of the ensemble of manganese atoms that constitute the redox-active core of photosynthetic H_2O oxidation requires substantial additional experimentation; only general conclusions can be drawn at the present time. The four atoms of the metal that are required to form the redox-active site are ligated to PSII in oxidation states that are very likely higher than $+2$ in the S_1 state. It can be demonstrated that at least two oxidizing equivalents are stored in the form of oxidation state advances in manganese $(S_0 \rightarrow S_1 \rightarrow S_2)$, and optical absorbance changes ascribed to $Mn^{3+} \rightarrow Mn^{4+}$ transitions implicate the metal as the site of storage of additional oxidizing equivalents during S-state cycling. Equation (8) can be interpreted to indicate that only two of the four manganese atoms undergo oxidation reduction in the course of H_2O oxidation, leaving $2Mn^{4+}$ in a redox-silent form throughout the catalytic cycle, a proposal that has been advanced by Kretschmann et al. (1991). Ligands to the manganese atoms appear to arise from oxygen and nitrogen associated with amino acid residues; mu—oxo bridges between manganese atoms are also very likely to form part of the overall ligand environment that produces the PSII-bound form of the metals.

4.2. Chloride

Chloride has been demonstrated to be an essential cofactor in the occurrence of H_2O oxidation (Bové et al., 1963). Localization of the action of the anion within the S-state mechanism was established by Izawa and his colleagues (1969). Other anions can substitute for Cl^-; the general order of effectiveness, based on the steady-state rates of activity observed, is $Cl^- > Br^- > I^- \sim NO_3^-$ (Kelley and Izawa, 1978). Utilization of EPR to detect manganese oxidation state advancement indicates that in the absence of Cl^-, one oxidizing equivalent can be abstracted from S_1 to form an unusual form of S_2 characterized by an abnormally long lifetime (Ono et al., 1986). The multiline signal is not observed in Cl^--deficient S_2, but addition of the anion after illumination catalyzes formation of the signal in the dark. The effect of Cl^- depletion has also been examined by a technique called thermoluminescence. This is a method for analyzing the properties of charge-separated states by detecting light emission from samples of PSII that have been illuminated (to produce a charge separation), frozen, and then warmed gradually to stimulate light emission as charge recombination occurs. The result of such measurements indicates that the redox potential of Cl^--depleted S_2 is decreased by some 60 to 80 mV (Vass et al., 1987) relative to a Cl^--sufficient control.

Chloride also affects that ability of reductants to attack and reduce the manganese cluster in PSII material from which extrinsic proteins have been removed, a treatment that exposes bound manganese to reagents that cannot gain access to the metal in the intact enzyme. In this structurally modified form of PSII, Cl^- addition slows the rate at which p-phenylenediamines inactivate O_2 evolution by reducing manganese that is then released as Mn^{2+} from the enzyme (Tamura et al., 1986; Mei and Yocum, 1990). Similarly, Cl^- slows the rate of S_{-1} formation by N-methylated derivatives of NH_2OH (Beck and Brudvig, 1988) in PSII material that has not been perturbed, and the anion can also produce a weak protective effect against NH_2OH inhibition of O_2 evolution; a further examination of this latter phenomenon showed that the Cl^- effect was abolished by removal of extrinsic polypeptides (Mei and Yocum, 1990). Amines such as Tris have been suggested to act as manganese reductants, and this phenomenon is Cl^--sensitive as well (Rickert et al., 1991).

Two hypotheses regarding the molecular mechanism of Cl^- action of PSII have been advanced. The first of these proposes that the anion acts as a counterion to neutralize positive charges in the vicinity of the active site of H_2O oxidation (Homann, 1988a). Based on analogies with the activating effect of halides on some enzymes, this proposal would envision anions as regulating the overall charge of the environment in which manganese catalyzes water oxidation. Proponents of this hypothesis have demonstrated convincingly that anions (including Cl^-) can regulate the stability and structure of PSII in the vicinity of the O_2-evolving reaction (Homann, 1988b). This stabilization effect is clearly detected in facilitating retention of certain essential extrinsic proteins by PSII under extremes of ionic strength and pH, and is very likely the origin of the halide-induced attenuation of NH_2OH inhibition of H_2O oxidation. A careful analysis of the stabilizing effectiveness of such anions produces an order that can be related to so-called lyotropic effects, wherein an empirically determined series of anions will promote stability (or denaturation) of proteins.

The second hypothesis states that Cl^- is ligated directly to manganese atoms in PSII (Sandusky and Yocum, 1984, 1986). This proposal stems from a number of experimental observations showing that Lewis bases (primary amines, F^-, OH^-, and carboxylic acids) produce reversible inhibitions of H_2O oxidation. For amines, binding to the S_2 state is the origin of the inhibition, and it has been proposed that these species bind to manganese (Velthuys, 1975). An examination of amine-induced inhibition of PSII showed that the potency of amine inhibition correlated with the pK_a of the amine, so it was proposed that a Lewis base–Lewis acid interaction (Angelicci, 1973) was responsible for inhibition (Ghanotakis et al., 1983). Since manganese in higher oxidation states is a strong acid, amines were proposed to inhibit O_2 evolution by direct ligation to manganese.

The kinetics of the interaction between Cl^- and inhibitory Lewis bases is competitive (Sandusky and Yocum, 1984, 1986). Only inhibition by NH_3 can be shown to occur at a second, Cl^--insensitive binding site (Beck et al., 1986). On account of the ability of series of metal-ligating species to compete with Cl^- for its site of action in PSII, it was proposed that Cl^- ligates to manganese and in doing so facilitates the redox activity of the metal. One prediction of this model, that NH_3 can also bind to manganese at a Cl^--insensitive site, has been confirmed spectroscopically; bound NH_3 modifies the appearance of the multiline signal (Beck and Brudvig, 1986) and electron spin-echo envelope modulation has definitively shown that the modification arises from bound NH_3 (Britt et al., 1989). It has further been demonstrated that Cl^- displacement from PSII by amines and F^- causes formation of the $g = 4.1$ form of the multiline signal (Casey and Sauer, 1984). Spectroscopic investigations by EPR have failed to detect the presence of Cl^- as a direct ligand to manganese (Yachandra, et al., 1986). More recent X-ray studies have indicated that EXAFS may not be sufficiently sensitive to detect the anion (Penner-Hahn et al., 1990), but a recent report of Br^- EXAFS experiments on samples where that anion replaced Cl^- suggests that Br^- is, in fact, bound to manganese (Yachandra et al., 1991).

It is probable that Cl^- can exert dual effects on PSII, regulating both the reactivity of the manganese atoms at the active site of H_2O oxidation and the overall structure and stability of polypeptides that make up part of the O_2-evolving complex. Further research is required to sort out the connections between these potential modes of action of the anion. Among the unsettled questions is that of the stoichiometry of the anion in PSII. A range of values is cited in the literature, from 1 tightly bound species, through 4–5, up to as many as 40–50 molecules of Cl^- per reaction center. The lower stoichiometries would favor a direct role in manganese redox chemistry, while much larger values would be indicative of a more general effect on the protein structure of the enzyme system.

4.3. Calcium

Calcium is required for H$_2$O oxidation, and the role of this metal in PSII is the subject of extensive research. Extraction of extrinsic 23- and 17-kDa proteins (to be discussed in detail in Section 5) from PSII preparations depresses O$_2$ evolution activity (Akerlund et al., 1982); this inhibition can be reversed by addition of relatively high concentrations of Ca^{2+} (Ghanotakis et al., 1984a; Miyao and Murata, 1984 a). As a consequence of this observation, it has been discovered that one role of the 23-kDa protein is to sequester Ca^{2+} and prevent its release from PSII (Ghanotakis et al., 1984b; Miyao and Murata, 1986; Waggoner and Yocum, 1987). Investigations of Ca^{2+} stoichiometries in PSII have produced a variety of results. Initial analyses found very high levels of the metal (Ghanotakis et al., 1984a). Later work on PSII membrane preparations from spinach found two atoms of the metal per reaction center (Cammarata and Cheniae, 1987); release of one atom by acidification to pH 3 with citrate suppresses activity (Ono and Inoue, 1988). In wheat, three atoms of Ca^{2+} were detected (Cammarata and Cheniae, 1987). Other studies (Shen et al., 1988a,b; Enami et al., 1989b) in cyanobacteria and in rice and spinach PSII preparations found < 2 atoms of Ca^{2+} per reaction center and cast doubt on the necessity for the metal in O$_2$ evolution activity. Preparations with lower Ca^{2+} content and/or the absence of a Ca^{2+} effect on activity exhibit lowered O$_2$ evolution rates and lose manganese during Ca^{2+} extraction, phenomena not reported by other workers.

Determination of the metal-ion specificity of the Ca^{2+} site in PSII revealed that, of the metals tested, only Sr^{2+} restored activity (Ghanotakis et al., 1984a) at low levels due to slowed turnover of the S-states (Boussac and Rutherford, 1988a). Replacement of Ca^{2+} by Sr^{2+} produces a multiline signal with narrowed linewidths (Boussac and Rutherford, 1988b). Lanthanides compete for the Ca^{2+} site in PSII (Ghanotakis et al., 1985); without added Ca^{2+} the trivalent metals destroy activity and release extrinsic proteins and Mn^{2+}. Techniques have now been developed to reversibly bind lanthanides to PSII (Bakou et al., 1992). In this manganese-retaining, lanthanide-substituted form of PSII, illumination is incapable of catalyzing manganese oxidation. Among other metals tested for an interaction with the PSII Ca^{2+} site, only Cd^{2+} has so far been shown to be a freely reversible competitive inhibitor of activity [K_i = 0.3 mM (Waggoner and Yocum, 1990)] that blocks multiline signal formation (Ono and Inoue, 1989). Monovalent metals (Na$^+$, K$^+$, and Cs$^+$) are very weak competitive inhibitors of Ca^{2+}-dependent activity (Waggoner et al., 1989). In cyanobacteria, Na$^+$ is reported to substitute for Ca^{2+} (Becker and Brand, 1985), and PSII preparations from thermophilic cyanobacteria appear not to require either Ca^{2+} or Cl$^-$ for activity (Pauly et al., 1990).

The effect of Ca^{2+} depletion on S-state advancement has been extensively investigated by EPR. Initial findings indicated that in the absence of the metal, S$_2$ formation was blocked (Blough and Sauer, 1984; Toyoshima et al., 1984; de Paula et al., 1986b; Ghanotakis et al., 1887). This finding and the observation that Ca^{2+} depletion slows Y$_Z^+$ decay have been repeated with purified PSII preparations

(Kalosaka et al., 1990). However, other results have indicated that the effect of Ca^{2+} depletion on S_2 formation is not so strightforward. Extraction of Ca^{2+} in high salt per 30 mM EGTA [ethylene glycol bis(β-aminoethyl ether)-N,N,N',N'-tetraacetate] and subsequent dialysis to rebind extrinsic proteins (all in the light) produces a modified S_2 multiline signal (Boussac et al., 1989, 1990a). Similar results with EDTA (ethylenediamine tetraacetate) and citrate acidification to remove Ca^{2+} have also been reported (Ono and Inoue, 1990a, b). Significantly, it has also been shown that in the absence of Ca^{2+}, the low-temperature limit at which S_2 forms (ca. 200 K) is increased to about 270 K (Ono and Inoue, 1990a). An exploration of the diverse factors surrounding Ca^{2+} depletion by chelators and temperature optima indicate that modification of the multiline signal is due to the presence of the chelator (Boussac et al., 1990a). A major consequence of Ca^{2+} depletion/illumination in the presence of chelators has been the discovery that turnovers subsequent to S_2 formation produce a new EPR-detectable species that is proposed to be a histidine radical (Boussac et al., 1990b). Further research is required to resolve the discrepancies surrounding the role of Ca^{2+} in manganese oxidation. The temperature optimum for S_2 formation and the extent of Ca^{2+} depletion in samples exhibiting multiline signals need to be carefully reexamined; additional characterizations of the organic radical seen in Ca^{2+}-depleted PSII should also prove interesting.

Data on Ca^{2+} function in PSII suggest that the metal-binding site is similar to Ca^{2+} sites in other proteins. The specificity of the site for select divalent species (Ca^{2+}, Sr^{2+}, Cd^{2+}), as well as monovalent and trivalent metals, is seen in other proteins (Martin, 1984; Levine and Williams, 1982; Einspahr and Bugg, 1984). Where such sites have been examined, oxy-ligation predominates; Ca^{2+} release from PSII by acidification (Ono and Inoue, 1988) is interpreted to be indicative of Asp or Glu ligands to Ca^{2+}. The Sr^{2+}-altered multiline signal may be diagnostic of a structural modification of the manganese ligand environment by the larger metal; Ca^{2+} binding is central to the structural modification or stabilization of many proteins. Lastly, estimates of Ca^{2+} affinity in 23 and 17-kDa- depleted PSII provide a useful estimate of how this Ca^{2+} site compares with those found in other proteins. Initial K_M values for Ca^{2+} in PSII preparations depleted of extrinsic proteins were unusually high [>2 mM (Ghanotakis et al., 1984a)]. More recent measurements provide a range of affinities (K_M values) from about 0.001 to 0.050 mM (Kalosaka et al., 1990; Waggoner and Yocum, 1990). These estimates are well within the range of (K_D) values found in a variety of other Ca^{2+}-binding proteins. The PSII K_M values were obtained in the steady state; single-turnover experiments suggest that S_3 has an appreciably lower Ca^{2+} affinity than do other S-states (Boussac and Rutherford, 1988c), and it has also been shown that manganese extraction abolishes Ca^{2+} binding to PSII (Tamura and Cheniae, 1988). Thus, Ca^{2+}-binding affinity in PSII would seem to be governed by the presence of manganese as well as by the oxidation state(s) of this metal.

Recent experiments have begun to address the possible structural effects of Ca^{2+} binding in PSII. An examination of manganese reduction by NH_2OH in preparations depleted of Ca^{2+} by acidification (the extrinsic 23- and 17-kDa species are retained) suggests that the metal acts as a structural "gatekeeper" to exclude H$_2$O from its site of oxidation (Tso et al., 1991). Our own work on this topic (Mei and Yocum, 1991) utilized extrinsic polypeptide-depleted PSII preparations where Ca^{2+} can equilibrate freely with its binding site and reductants of varying size have free access to PSII manganese (Ghanotakis et al., 1984c). Under such conditions, Ca^{2+} did not slow the rate of manganese reduction by NH$_2$OH. However, the rate of inhibition of O$_2$ evolution was slowed, and appreciable amounts of EPR-detectable Mn^{2+} remain associated with Ca^{2+}-supplemented PSII membranes after NH$_2$OH treatment. These data have been interpreted to indicate that Ca^{2+} binding stabilizes the protein matrix that makes up the ligand environment surrounding manganese in PSII.

It has also been suggested (V. L. Pecoraro, personal communication; Rutherford, 1989) that Ca^{2+} might function in PSII as the binding site for substrate H$_2$O. These speculations are based on the observed ligation properties of the metal in several Ca^{2+}-binding proteins. In most complexes of Ca^{2+}, the metal accepts six to seven ligands. In a number of proteins, the coordination environment includes several protein-associated oxygen ligands (for example, from Glu or Asp), but in addition, one to three molecules of H$_2$O are usually present to complete the coordination sphere.

A brief summary of current information on the inorganic ion cofactors required for O$_2$ evolution is given in Table 2. On the basis of the evidence at hand, manganese atoms constitute the active site of H$_2$O oxidation, although doubts have been raised as to whether each oxidation event in the S-state cycle is associated with manganese oxidation. Calcium may function to promote formation of an essential structural conformation of the active site of the enzyme system, although the nature of this structure is unclear. Likewise, there is no clear answer to the question of why perturbations that seem to arise from Ca^{2+} extraction affect manganese-centered redox reactions associated with formation of S$_2$ and S$_3$. Chloride removal from PSII appears to directly affect the capacity of PSII to oxidize manganese, although as with Ca^{2+}, the exact nature of Cl$^-$ participation in manganese redox activity is unclear. Two effects, which may be interrelated, exist for the anion. One of these is a direct action on the redox activity of manganese, while the second is a stabilizing effect on the protein matrix. It is evident that oxidation of manganese in photoactivation and S-state cycling and

Table 2 Summary of Inorganic Ion Functions in PSII

No. of atoms:	Mn	Ca^{2+}	Cl$^-$
Active Site:	4	2–3	> 5 (?)
Proposed function:	Oxidation of H$_2$O	Structural stabilization	Mn redox activity; stabilization of PSII proteins
Surrogate ions:	None	Sr^{2+}	Br$^-$ > I$^-$ > NO$_3^-$

the reduction of manganese (by added reagents) to S_{-1} (or even more reduced forms of the enzyme) are strongly influenced by Ca^{2+} and Cl^-.

5. POLYPEPTIDES OF PHOTOSYSTEM II AND THE O_2-EVOLVING COMPLEX

The past decade has witnessed remarkable progress in the purification and structural analysis of PSII. Mechanical disruption of chloroplast thylakoid membranes followed by purification with a phase partitioning method (Akerlund et al., 1982), or solubilization of membranes with nonionic detergents such as Triton X-100 or digitonin (Berthold et al., 1981; Kuwabara and Murata, 1982) to produce material free of photosystem I, cytochrome complexes and ATP synthase have yielded preparations that are enriched in O_2 evolution activity and contain as many as 20 different polypeptides. Genes encoding these proteins reside in either the chloroplast or the nucleus; nuclear-encoded proteins are synthesized in precursor form, transported across the thylakoid envelope, and processed to produce the mature protein, which is in many cases assembled into multisubunit enzyme systems (Keegstra et al., 1989). Table 3 lists the polypeptides of PSII according to molecular masses estimated from mobilities on SDS-polyacrylamide gel electrophoresis. Actual molecular weights, derived from amino acid analyses and from amino acid sequences derived from DNA sequences, differ in most cases; Table 3 presents the values most often found in the

Table 3   Selected Properties of PSII Polypeptides

M.W. (kDa)	Coding Site[a]	Extrinsic, Intrinsic	Proposed Function
47	C	I	Chl a-binding; antenna
43	C	I	Chl a-binding; antenna
33	N	E	Mn stabilization
34 (D2)	C	I	Reaction center; Y_D, Q_A, P680, Pheo a
32 (D1)	C	I	Reaction center; Y_Z, Q_B, P680, Pheo a
28	N	I	Chl-binding; extraction affects quinone reduction activity
23	N	E	Part of ion concentration structure; required for Ca^{2+} retention
22	N	I	?
17	N	E	Part of ion concentration structure; facilitates Cl^- retention
10	N	E/I	?
9	C	I	Cytochrome b559 subunit
4.5	C	I	Cytochrome b559 subunit
< 5	C	I	Several small species of unknown function

[a]C, chloroplast; N, nucleus.

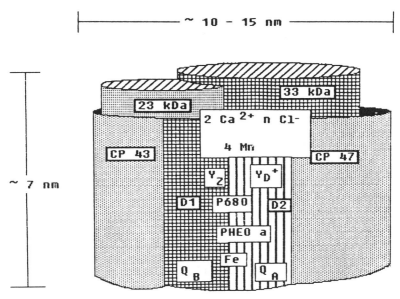

Figure 1. Schematic representation of the polypeptide structure of PSII based on image reconstruction investigations reported by Haag et al. (1990). The extrinsic 17-kDa polypeptide and several intrinsic species have been omitted. The location of chromophores within the polypeptide matrix shown is for illustrative purposes only; actual ligation sites are unknown.

literature, which are derived from estimates based on mobility of the species in SDS (sodium dodecyl sulfate)-polyacrylamide gel electrophoresis.

The fully competent O_2-evolving form of PSII comprises both integral membrane proteins (ranging in size from 47 to 3 kDa) and extrinsic species (33, 23, and 17 kDa). Boekema and his colleagues (Dekker et al., 1990; Haag et al., 1990) have examined purified PSII preparations from spinach using image reconstruction techniques. To orient the reader with respect to the polypeptide structure of PSII, a model of the enzyme, derived from image construction experiments (Haag et al., 1990), is presented in Figure 1. The polypeptides shown, as well as the species listed in Table 3, appear to be present in single copies in each PSII structure (De Vitry et al., 1987; Murata et al., 1984), and in several cases have been removed from PSII and then reconstituted, or have been deleted by genetic manipulations. In many cases, one can assign specific lesions in the electron transfer activity of the enzyme system to the absence of one or more of these polypeptides; from such experiments, a general picture of their contributions to the structure and/or function of the enzyme is emerging.

5.1. Extrinsic PSII Proteins

In eukaryotes, all three PSII extrinsic proteins (33, 23, and 17 kDa) are nuclear-encoded. In prokaryotes (cyanobacteria), the 23- and 17-kDa species appear to be absent. The 17- and 23-kDa species have been deleted by mutations in the green alga *Chlamydomonas*; the resulting mutants are photoautotrophic, but grow more slowly than the wild type (Mayfield et al., 1987b). Biochemical studies on spinach PSII preparations have shown that the 23- and 17-kDa species can be easily removed by elevated ionic strength (optimal extraction occurs at 1–2 M NaCl). The extracted preparations show depressed rates of O_2 evolution that are partially restored upon rebinding of the 23- and 17-kDa polypeptides (Akerlund et al., 1982). As already noted in Section 4.3, the lesion created by extraction of the smaller extrinsic proteins is within the S-state cycle itself, and arises from loss of Ca^{2+} and Cl^-. Results of extraction/reconstitution experiments indicate that the 23-kDa polypeptide plays a role in Ca^{2+} retention (Miyao and Murata, 1986; Waggoner and Yocum, 1987), whereas the 17-kDa polypeptide seems to facilitate Cl^- retention in the O_2-evolving complex (Miyao and Murata, 1985).

Removal of 23- and 17-kDa species from PSII also modifies the structure of the O_2-evolving complex so as to remove a barrier between manganese-binding sites and the external medium. Intact PSII preparations are not sensitive to inhibition by bulky, manganese specific reductants such as hydroquinone or *p*-phenylenediamines; after extraction of the 23- and 17-kDa polypeptides, these same reductants produce Mn^{2+} coincident with loss of O_2 evolution activity (Ghanotakis et al., 1984c; Tamura et al., 1986). A similar diffusion barrier arising from the 23- and 17-kDa polypeptides is detected in Ca^{2+}-depleted PSII preparations reconstituted with the extrinsic proteins, where externally added Ca^{2+} requires long (40–60 min) incubation periods for optimal reactivation of O_2 evolution activity (Ghanotakis et al., 1984b; Ono and Inoue, 1988).

The 33-kDa extrinsic protein is more difficult to remove from PSII. Although extraction of manganese with Tris also removes this polypeptide, NH_2OH-catalyzed extraction of manganese seems not to perturb binding of the protein (Murata et al., 1983). Exposure of PSII to either 1 M $CaCl_2$ or 2.5 M urea/200 mM NaCl will extract the polypeptide without manganese loss, and lowered rates of O_2 evolution can be obtained in this system provided Ca^{2+} and Cl^- are added (Ono and Inoue, 1983, 1984; Miyao and Murata, 1984b); even more activity is recovered upon rebinding of the 33-kDa species (Ono and Inoue, 1984). Preparations depleted of the 33-kDa protein slowly lose manganese (Ono and Inoue, 1983), so it appears that the principal role of this species is to stabilize manganese binding to PSII. On the basis of the results just cited, the protein seems unlikely to have a function in the direct ligation of manganese, a supposition borne out by the finding (Burnap and Sherman, 1991) that deletion of the 33-kDa polypeptide by mutagenesis in a cyanobacterium produces transformed cells capable of slow autotrophic growth. Curiously, a similar mutation in *Chlamydomonas* results in loss of O_2 evolution activity (Mayfield et al., 1987a).

The mode of binding of extrinsic proteins to PSII has been investigated in some detail. Cross-linking experiments show that the 33-kDa protein interacts with a site on CP47, a 47-kDa Chl *a*-binding protein of the intrinsic portion of PSII (Enami et al., 1989a). Epitope mapping experiments (Frankel and Bricker, 1987) show that removal of the 33-kDa protein exposes sites on CP47. Binding of the 23-kDa protein requires the presence of the PSII-associated 33-kDa protein, and binding of the 17-kDa protein in turn requires the presence of the 23-kDa polypeptide (Miyao and Murata, 1989). Estimates of the binding affinities of these proteins have been carried out; the respective numbers are 12, 2.7, and 3.8 nM (Miyao and Murata, 1989). Although the presence of a functional manganese center has not been considered to be essential for extrinsic protein binding to PSII, Kavelaki and Ghanotakis (1991) have shown that manganese is necessary for rebinding of the 23- and 17-kDa proteins to a manganese-depleted preparation retaining the 33-kDa protein. Other binding interactions among the extrinsic polypeptides and intrinsic components of PSII may be revealed in the future; separated from the intrinsic proteins of PSII, none of the extrinsic proteins can be shown to bind to one another (Miyao and Murata, 1989).

Characterization of the three extrinsic PSII proteins has progressed to a point where it is clear that play an important role in regulating the structure and stability of the O$_2$-evolving enzyme system. Since none of these polypeptides has been demonstrated to have a metal-binding capacity, and since O$_2$ evolution can proceed in their absence, it would be difficult to envision any of these proteins playing a role in manganese ligation in which one or another of the species contributed amino acid ligands to the metals. It has been suggested (Wales et al., 1989), on the basis of a possible homology between the 33-kDa protein and a mammalian Ca^{2+}-binding protein, that this largest PSII extrinsic protein may play a role in binding the metal. If, as a group, the extrinsic species are not critical for ligation of metals in the active site of PSII, then their binding to the intrinsic proteins of the enzyme may in some way regulate the organization and/or structure of the polypeptides that do contribute to manganese (or Ca^{2+}) ligation in the O$_2$-evolving complex.

5.2. Intrinsic Proteins of PSII

Preparations of PSII from which the extrinsic proteins and manganese have been depleted contain a number of intrinsic proteins, which are listed in Table 3. The 9- and 4.5-kDa polypeptides ligate the heme of a cytochrome, called b559, which is tightly bound to PSII but whose function remains uncertain. Several of the larger proteins bind Chl *a* (47, 43, 34, 32, and 28 kDa). Drawing on analogies with the crystal structure of the reaction center of *Rhodopseudomones Viridis*, Michel and Deisenhofer (1986) and Trebst (1986) proposed that the 34- and 32-kDa proteins of PSII (also called D2 and D1 after their diffuse appearance on Coomassie Blue-stained SDS-polyacrylamide gels) contained the requisite components (P680, Pheo *a*, quinones, and tyrosine radicals) of the photochemical reaction center. These predictions have been verified by site-directed mutagenesis experiments for the tyrosine radicals (Debus et al., 1988a, b; Vermaas et al., 1988), and by the biochemical isolation of a heterodimer complex of D1 and D2, along with cytochrome b559, which exhibits the photochemically catalyzed formation of the EPR-detectable triplet form of the reaction center Chl *a*, P680 (Nanba and Satoh, 1987; Okamura et al., 1987).

The X-ray crystal structures of bacterial reaction centers (Deisenhofer et al., 1984, 1985; Yeates et al., 1988) have provided insights into the possible structure of the D1/D2 polypeptide complex of PSII, as well as the contrasting properties of the two reaction center systems. The pigment content of D1/D2 is higher [minimally six Chl a and two Pheo a versus four and two, respectively, for bacteria (Dekker et al., 1989; Kobayashi et al., 1990)]. While bacterial preparations can be crystallized with bound quinones, PSII reaction centers lose plastoquinones during purification, and for this reason, stable, long-lived charge separations have not been achieved in isolated D1/D1/b559 preparations. Estimates of chromophore stoichiometries in the PSII reaction center rest on cytochrome b559 and pheophytin contents. This is a controversial area; if two b559 hemes are assumed to be present, then the PSII reaction center contains as many as 12 Chl a. Alternative determination methods that assume that PSII pheophytin content is identical to that found in the reaction center of photosynthetic bacteria (two per reaction center) result in fractional heme stoichiometries (Nanba and Satoh, 1987).

The observations that are now available leave no doubt that the D1 and D2 polypeptides of PSII ligate reaction center chromophores; at the same time, the limited photoactivity of this material would suggest that the PSII reaction center in its fully functional form requires more polypeptides than are found in the analogous bacterial reaction center system. Characterizations of PSII preparations containing 47- and 43-kDa Chl a-binding proteins (CP47 and CP43 in Figure 1) in addition to D1/D2/b559 have detected the presence of the dark-stable tyrosine radical (Y_D^+), light-induced Y_Z^+ formation, and bound quinone (Satoh, 1983). Biochemical extraction of the 43-kDa protein destabilizes Y_D^+, and Q_A is also extracted (Petersen et al., 1990). Deletion of this protein by mutagenesis, however, does not cause any detectable loss of Q_A binding (Rogner et al., 1991). Other intrinsic proteins (28 and 22 kDa) are observed in highly purified PSII preparations (Ghanotakis et al., 1987). The 22-kDa protein is of uncertain function; extraction does not markedly perturb O_2 evolution activity, for example. The 28-kDa protein, which binds Chl a and perhaps some Chl b, can also be extracted from PSII without loss of O_2 evolution activity, but the Q_B site seems to be altered by this treatment; herbicides no longer inhibit activity, and an anionic electron acceptor [Fe $(CN)_6^{3-}$] shows a remarkable increase in ability to catalyze electron transfer activity (Bowlby et al., 1989).

Further experimentation will undoubtedly clarify structure/function relationships among various PSII intrinsic proteins. At the present time, it is clear that the D1 and D2 polypeptides form a structure capable of binding the chromophores required for photochemical activity. The larger Chl a-binding species (43 and 47 kDa) most likely function as antenna systems to funnel excitation energy to P680, and also function as structural elements to form more effective binding environments for the quinone electron acceptors and promote longer lifetimes for intermediates (for example, tyrosine radicals) involved in charge separation. The function of cytochrome b559 remains a mystery, as do the roles of very small (< 5 kDa) intrinsic species that are found in association with purified PSII preparations. The structure of PSII will ultimately be solved by X-ray crystallography; several groups have now obtained crystals of various forms of the biochemically resolved enzyme, and hopefully these crystals will be developed to a stage where they are suitable for structural analysis.

5.3. Which PSII Polypeptides Ligate Metal Ion Cofactors of the O_2-Evolving Complex?

While spectroscopic investigations have proven useful in elucidating the redox states and certain structural properties of the manganese atoms that catalyze H_2O oxidation, little is known about the identities of the polypeptides that participate in metal atom ligation. The absence of data to implicate the extrinsic proteins in metal binding, coupled with identification of primary electron transfer reactions in the D1 and D2 species, has focused attention on these latter proteins as potential sites of manganese and Ca^{2+} binding. There are experimental data to support this belief. Measurements of the microwave power saturation properties of Y_Z^+, for example, show that the radical is strongly relaxed by manganese (Yocum and Babcock, 1981). Manganese–Y_D^+ interactions have also been detected (Styring and Rutherford, 1987). Amino acid sequences of the proteins, derived from DNA sequences, have been subjected to hydropathy plotting routines (Trebst, 1986). The resulting models for the organization of D1 and D2 in the membrane show that the tyrosine radicals are appropriately positioned on the correct side of the membrane, and that amino acid residues (Glu, Asp, and His, for example) that could serve as ligands to either manganese or Ca^{2+} are also present in lumenal domains of both proteins.

Directed mutagenesis experiments utilizing cyanobacteria are now in progress in a number of laboratories to attempt to identify putative metal ligands on D1 and D2. The challenges to this approach are formidable. Both polypeptides are highly conserved among species ($> 90\%$ conservation of amino acid sequence similarities), so the possibility that genetic alterations to a particular amino acid will disrupt the entire protein complex is always present. An additional challenge arises from the lack of high-resolution structural information; in changing a particular amino acid residue, one cannot ensure that a subtle effect on the structure of the enzyme is the cause of a defect produced by mutagenesis. Nevertheless, site-directed mutagenesis of carboxy amino acids on both D1 and D2 have been shown to modify or abolish sites that appear to contribute to manganese binding, and to modify O_2 evolution activity. Residues in proximity to tyrosine 161 (Y_Z) on the D1 protein have become favorite targets for genetic alteration; Nixon and Diner (1992) have recently reported that substitution of Ser for Glu in a lumenal-side position (170) of D1 abolishes the manganese-binding environment as monitored by photooxidation of Mn^{2+}. Glutamate 169 on D2 has been modified in the laboratory of Vermaas (Vermaas et al., 1990), and some of the substitutions destabilize manganese binding at the active site of H_2O oxidation.

Given the difficulties inherent in targeting amino acid residues for mutagenesis and then analyzing the consequences of these modifications, any successful results from site-directed mutation experiments must be viewed with optimism. The large number of potential targets of mutagenesis and the relatively small number of investigators working in the area indicate that progress will be slow. At the same time, the emerging results are encouraging and additional useful findings should appear in the near future.

6.   CONCLUDING REMARKS

Measured against the information available a decade ago, the present view of PSII is much clearer. At the same time, new information makes clear the remaining depths of ignorance surrounding a number of critical aspects of the structure and function of the enzyme. The areas that have drawn the most attention, and which will probably see more advances in the near future, are the structure of the enzyme, the roles of the inorganic cofactors, and the identification of metal ion-binding sites on the constituent polypeptides of PSII. Details of the actual mechanism of H_2O oxidation (substrate binding and steps in the catalytic mechanism) will probably remain obscure for the immediate future, given the short lifetime of the S_4 state ($\sim$ 1 msec) and the resulting difficulties that would be associated with trapping and probing this state. Another question deserving attention is that of how H^+ release by PSII is coupled to formation of the proton gradient that drives ATP synthesis in intact photosynthetic membranes. The accumulating evidence that protons are released from the proton matrix of PSII rather than from H_2O as a consequence of S-state advancement could lead to a productive reexamination of details of the interaction between PSII electron transfer and the chemiosmotic proton gradient used for ATP synthesis.

In assessing the current situation and the future prospects for a better understanding of PSII, it is worth noting one of the many aphorisms of Efraim Racker, quoted by Kornberg (1989): "Don't waste clean thinking on dirty enzyme." With purification of the enzyme in its fully active as well as its inactive, simplified forms under control, there are many routes open to a better understanding of its function. Perhaps the major challenge in the future will be to avoid wasting dirty thoughts on a clean enzyme.

ACKNOWLEDGMENTS

The author gratefully acknowledges support from the Cellular Biochemistry Program of the National Science Foundation and from the USDA Competitive Research Grants Office for research on the structure and function of photosystem II.

REFERENCES

Akerlund, H. E., Jansson, C., and Andersson, B. (1982). Reconstitution of photosynthetic water splitting in inside-out thylakoid vesicles and identification of a participating polypeptide. *Biochim. Biophys. Acta* **681**, 1–10.

Amesz, J. (1983). The role of manganese in photosynthetic oxygen evolution. *Biochim. Biophys. Acta* **726**, 1–12.

Andreasson, L. E., and Vanngard, T. (1988). Electron transport in photosystems I and II. *Annu. Rev. Plant Physiol. Plant Mol. Biol.* **39**, 379–411.

Angelicci, R. J. (1973). Stability of coordination compounds. In G. Eichhorn (Ed.), *Inorganic Biochemistry*. Elsevier, Amsterdam, pp. 63–101.

Babcock, G. T. (1987). The photosynthetic oxygen-evolving process. In J. Amesz (Ed.), *Photosynthesis*. Elsevier, Amsterdam, pp. 125–158.

Bader, K. P., Thibault, P., and Schmid, G. H. (1987). A study of the properties of the S_3 state by mass spectrometry in the filamentous cyanobacterium *Oscillatoria chalybea. Biochim. Biophys. Acta* **893**, 564–571.

Bakou, A., Buser, C., Dandulakis, G., Brudvig, G., and Ghanotakis, D. F. (1992). Calcium binding site(s) of photosystem II as probed by lanthanides. *Biochim. Biophys. Acta* **1099**, 124–131.

Barber, J., and Andersson, B. (1992). Too much of a good thing: Light can be bad for photosynthesis. *Trends Biochem. Sci.* **17**, 61–71.

Beck, W. F., and Brudvig, G. W. (1986). Binding of amines to the O_2-evolving center of photosystem II. *Biochemistry* **25**, 6749–6756.

Beck, W. F., and Brudvig, G. W. (1987). Reactions of hydroxylamine with the electron-donor side of photosystem II. *Biochemistry* **26**, 8285–8295.

Beck, W. F., and Brudvig, G. W. (1988). Resolution of the paradox of ammonia and hydroxylamine as substrate analogs in the water-oxidation reaction catalyzed by photosystem II. *J. Am. Chem. Soc.* **110**, 1517–1523.

Beck, W. F., de Paula, J. C., and Brudvig, G. W. (1986). Ammonia binds to the manganese site of the O_2-evolving complex of photosystem II in the S_2 state. *J. Am. Chem. Soc.* **108**, 4018–4022.

Becker, D. W., and Brand, J. J. (1985). *Anacystis nidulans* demonstrates a photosystem II cation requirement satisfied only by Ca^{2+} or Na^+. *Plant Physiol.* **79**, 522–558.

Berthold, D. A., Babcock, G. T., and Yocum, C. F. (1981). A highly resolved, oxygen-evolving photosystem II preparation from spinach thylakoid membranes: EPR and electron transport properties. *FEBS Lett.* **134**, 231–234.

Blough, N. V., and Sauer, K. (1984). The effects of mono- and divalent salts on the O_2-evolution activity and low temperature multiline EPR spectrum of photosystem II preparations from spinach. *Biochim. Biophys. Acta* **767**, 377–381.

Bouges, B. (1971). Action de faibles concentrations d'hydroxylamine sur emission d'oxygene des algues *Chlorella* et des chloroplasts d'épinards. *Biochim. Biophys. Acta* **234**, 103–112.

Boussac, A., and Rutherford, A. W. (1988a). S-state formation after Ca^{2+} depletion in the photosystem II oxygen-evolving complex. *Chem. Scr.* **28A**, 123–126.

Boussac, A., and Rutherford, A. W. (1988b). Nature of the inhibition of the oxygen-evolving enzyme of photosystem II induced by NaCl washing and reversed by the addition of Ca^{2+} and Sr^{2+}. *Biochemistry* **27**, 3476–3483.

Boussac, A., and Rutherford, A. W. (1988c). Ca^{2+} binding to the oxygen evolving enzyme varies with the redox state of the Mn cluster. *FEBS Lett.* **236**, 432–436.

Boussac, A., Zimmermann, J. L., and Rutherford, A. W. (1989). EPR signals from modified charge accumulation states of the oxygen evolving enzyme in Ca^{2+} deficient photosystem II. *Biochemistry* **28**, 8984–8989.

Boussac, A., Zimmermann, J. L., and Rutherford, A. W. (1990a). Factors influencing the formation of modified S_2 EPR signal and the S_3 EPR signal in Ca^{2+}-depleted photosystem II. *FEBS Lett.* **277**, 69–74.

Boussac, A., Zimmermann, J. L., Rutherford, A. W., and Lavergne, J. (1990b). Histidine oxidation in the oxygen-evolving photosystem-II enzyme. *Nature (London)* **347**, 303–306.

Bové, J. M., Bové, C., Whatley, F. R., and Arnon, D. I. (1963). Chloride requirement for oxygen evolution in photosynthesis. *Z. Naturforsch., B: Anorg. Chem., Org. Chem., Bichem., Biophys., Biol.* **18B**, 683–688.

Bowlby, N. R., Ghanotakis, D. F., Yocum, C. F., Petersen, J., and Babcock, G. T. (1989). The functional unit of oxygen-evolving activity: Implications for the structure of photosystem II. In S. E. Stevens and D. Bryant (Eds.), *Light Energy Transduction in Photosynthesis: Higher Plant and Bacterial Models*. American Society of Plant Physiologists, Rockville, MD, pp. 215–226.

Britt, R. D., Zimmermann, J. -L., Sauer, K., and Klein, M. P. (1989). Ammonia binds to the catalytic Mn of the oxygen evolving complex of photosystem II: Evidence by electron spin echo envelope modulation spectroscopy. *J. Am. Chem. Soc.* **111**, 3522–3532.

Britt, R. D., DeRose, V. J., Yachandra, V. K., Kim, D., Sauer, K., and Klein, M. P. (1990). Pulsed EPR studies of the manganese center of the oxygen-evolving complex of photosystem II. In M. Baltscheffsky (Ed.), *Current Research in Photosynthesis*. Kluwer Academic Publishers, Dordrecht, The Netherlands, pp. 769–772.

Brudvig, G. W., and Crabtree, R. H. (1986). Mechanism for photosynthetic O_2 production. *Proc. Natl. Acad. Sci. U.S.A.* **83**, 4586–4588.

Brudvig, G. W., Beck, W. F., and de Paula, J. C. (1989). Mechanism of photosynthetic water oxidation. *Annu. Rev. Biophys. Biophys. Chem.* **18**, 25–46.

Burnap, R., and Sherman, L. (1991). Deletion mutagenesis in *Synechocyctis* sp. PCC6803 indicates that the Mn-stabilizing protein photosystem II is not essential for O_2 evolution. *Biochemistry* **30**, 440–446.

Cammarata, K., and Cheniae, G. M. (1987). Studies on 17, 24 kD-depleted photosystem II membranes. I. Evidence for high and low affinity calcium site in 17, 24-kD depleted PSII membranes from wheat versus spinach. *Plant Physiol.* **84**, 857–895.

Casey, J., and Sauer, K. (1984). EPR detection of a cryogenically photogenerated intermediate in photosynthetic oxygen evolution. *Biochim. Biophys. Acta* **767**, 21–28.

Cheniae, G. M. (1980). Manganese binding sites and presumed manganese proteins in chloroplasts. In A. San Pietro (Ed.), *Methods in Enzymology*. Academic Press, New York, Vol. 69, pp. 349–363.

Cheniae, G. M., and Martin, I. F. (1971). Effects of hydroxylamine on photosystem II. I. Factors affecting decay of oxygen evolution. *Plant Physiol.* **47**, 568–575.

Cheniae, G. M., and Martin, I. F. (1978). Studies on the mechanism of Tris-induced inactivation of oxygen evolution. *Biochim. Biophys. Acta* **502**, 321–344.

Cole, J., Yachandra, V. K., Guiles, R. D., McDermott, A., Britt, R. D., Dexheimer, S. L., Sauer, K., and Klein, M. P. (1987). Comparison of the structure of the Mn-complex in the S_1 and S_2 states of the photosynthetic O_2-evolving complex. An X-ray absorption edge spectroscopy study. *Biochemistry* **26**, 5974–5981.

Coleman, W. J., and Govindjee. (1987). A model for the mechanism of chloride activation of oxygen evolution in photosystem II. *Photosynth. Res.* **13**, 199–223.

Debus, R. J. (1992). The manganese and calcium ions of photosynthetic oxygen evolution. *Biochim. Biophys. Acta* **1102**, 269–352.

Debus, R. J., Barry, B. A., Babcock, G. T., and McIntosh, L. (1988a). Site-directed mutagenesis identifies a tyrosine radical involved in the photosynthetic oxygen evolving system. *Proc. Natl. Acad. Sci. U.S.A.* **85**, 427–430.

Debus, R. J., Barry, B. A., Sithole, I., Babcock, G. T., and McIntosh, L. (1988b). Directed mutagenesis indicates that the donor to P_{680}^+ in photosystem II is tyrosine-161 of the D1 polypeptide. *Biochemistry* **27**, 9071–9074.

Deisenhofer, J., Epp, O., Miki, K., Huber, R., and Michel, H. (1984). X-ray structure analysis of a membrane protein complex: Electron density map at 3 A resolution and a model of the chromophores of the photosynthetic reaction center from *Rhodopseudomonas viridis*. *J. Mol. Biol.* **180**, 385–398.

Deisenhofer, J., Epp, O., Miki, K., Huber, R., and Michel, H. (1985). Structure of the protein subunits of the photosynthetic reaction centre of *Rhodopseudomonas viridis* at 3 A resolution. *Nature (London)*, **318**, 618–624.

Dekker, J. P. (1992). Optical studies on the oxygen-evolving complex of photosystem II. In V. L. Pecoraro (Ed.), *Manganese Redox Enzymes*. VCH Publishers, New York, pp. 85–103.

Dekker, J. P., Van Gorkom, H. J., Brok, M., and Ouwehand, L. (1984a). Optical characterization of photosystem II donors. *Biochim. Biophys. Acta* **764**, 301–309.

Dekker, J. P., Van Gorkom, H. J., Wensink, J., and Ouwehand, L. (1984b). Absorbance difference spectra of the successive redox states of the oxygen-evolving apparatus of photosynthesis. *Biochim. Biophys. Acta* **767**, 1–9.

Dekker, J. P., Plijter, J. J., Ouwehand, L., and Van Gorkom, H. J. (1984c). Kinetics of manganese redox transitions in the oxygen-evolving apparatus of photosynthesis. *Biochim. Biophys. Acta* **767**, 176–179.

Dekker, J. P., Bowlby, N. R., and Yocum, C. F. (1989). Chlorophyll and cytochrome b559 content of the photochemical reaction center of photosystem II. *FEBS Lett.* **254**, 150–154.

Dekker, J. P., Betts, S. D., Yocum, C. F., and Boekema, E. J. (1990). Characterization by electron microscopy of isolated particles and two-dimensional crystals of the CP47-D1-D2-cytochrome b559 complex of photosystem II. *Biochemistry* **29**, 3220–3231.

de Paula, J. C., Beck, W. F., and Brudvig, G. W. (1986a). Magnetic properties of manganese in the photosynthetic O_2-evolving complex. 2. Evidence for a manganese tetramer. *J. Am. Chem. Soc.* **108**, 4002–4009.

de Paula, J. C., Li, P. M., Miller, A. F., Wu, B. W., and Brudvig, G. W. (1986b). Effect of the 17- and 23-kilodalton polypeptides, calcium and chloride on electron transfer in photosystem II. *Biochemistry* **25**, 6487–6494.

De Vitry, C., Diner, B. A., and Lemoine, Y. (1987). Chemical composition of photosystem II reaction centers (PSII); phosphorylation of PSII polypeptides. In J. Biggins (Ed.), *Progress in Photosynthesis Research*. Nijhoff, Dordrecht, The Netherlands, pp. 105–108.

Dismukes, G. C. (1986). The metal centers of the photosynthetic oxygen evolving complex. *Photochem. Photobiol.* **43**, 99–115.

Dismukes, G. C., and Siderer, Y. (1981). Intermediates of a polynuclear manganese cluster involved in photosynthetic water oxidation. *Proc. Natl. Acad. Sci. U.S.A.* **78**, 274–278.

Einspahr, H., and Bugg, C. E. (1984). Crystal structure studies of calcium complexes and implications for biological systems. *Met. Ions Biol. Syst.* **17**, 51–97.

Enami, I., Miyaoka, T., Mochizuki, Y., Shen, J. -R., Satoh, K., and Katoh, S. (1989a). Nearest neighbor relationships among constituent proteins of oxygen-evolving photosystem II membranes: Binding and function of the extrinsic 33 kDa protein. *Biochim. Biophys. Acta* **973**, 35–40.

Enami, I., Kamino, K., Shen, J. R., Satoh, K., and Katoh, S. (1989b). Isolation and characterization of photosystem II complexes which lack light-harvesting chlorophyll a/b proteins but retain three extrinsic proteins related to oxygen evolution from spinach. *Biochim. Biophys. Acta* **977**, 33–39.

Frankel, L. K., and Bricker, T. M. (1987). Use of a monoclonal antibody in structural investigations of the 49-kDA polypeptide of photosystem II. *Arch. Biochem. Biophys.* **256**, 295–301.

Frasch, W. D., and Cheniae, G. M. (1980). Flash inactivation of oxygen evolution. Identification of S_2 as the target of inactivation by tris. *Plant Physiol.* **65**, 735–745.

George, G. N., Prince, R. N., and Cramer, S. P. (1989). The manganese site of the photosynthetic water-splitting enzyme. *Science* **243**, 789–791.

Ghanotakis, D. F., and Yocum, C. F. (1990). Photosystem II and the oxygen-evolving complex. *Annu. Rev. Plant Physiol. Plant Mol. Biol.* **41**, 255–276.

Ghanotakis, D. F., O'Malley, P. J., Babcock, G. T., and Yocum, C. F. (1983). Structure and inhibition of components on the oxidizing side of photosystem II. In Y. Inoue, A. R. Crofts, Govindjee, N. Murata, G. Renger, and K. Satoh (Eds.), *The Oxygen Evolving System of Photosynthesis*. Academic Press, Tokyo, pp. 91–102.

Ghanotakis, D. F., Babcock, G. T., and Yocum, C. F. (1984a). Calcium reconstitutes high rates of oxygen evolution in polypeptide depleted photosystem II preparations. *FEBS Lett.* **167**, 127–130.

Ghanotakis, D. F., Topper, J. N., Babcock, G. T., and Yocum, C. F. (1984b). Water-soluble 17 and 23 kDa polypeptides restore oxygen evolution activity by creating a high-affinity binding site for Ca^{2+} on the oxidizing side of photosystem II. *FEBS Lett.* **170**, 169–173.

Ghanotakis, D. F., Topper, J. N., and Yocum, C. F. (1984c). Structural organization of the oxidizing side of photosystem II. II. Exogenous reductants reduce and destroy the Mn complex in photosystem II membranes depleted of the 17 and 23 kDa polypeptides. *Biochim. Biophys. Acta* **767**, 524–531.

Ghanotakis, D. F., Babcock, G. T., and Yocum, C. F. (1985). Structure of the oxygen-evolving complex of photosystem II: Calcium and lanthanum compete for sites on the oxidizing side of photosystem II which control the binding of water-soluble polypeptides and regulate the activity of the manganese complex. *Biochim. Biophys. Acta* **809**, 173–180.

Ghanotakis, D. F., Demetriou, D. M., and Yocum, C. F. (1987). Isolation and characterization of an oxygen-evolving photosystem II reaction center core preparation and a 28 kDa chla-binding protein. *Biochim. Biophys. Acta* **891**, 15–21.

Guiles, R. D., Yachandra, V. K., McDermott, A. E., DeRose, V. J., Zimmermann, J. L., Sauer, K., and Klein, M. P. (1990). Structures and oxidation states of Mn in several S-states of photosystem II detected by X-ray absorption spectroscopy. In M. Baltscheffsky. (Ed.), *Current Research in Photosynthesis*. Kluwer Academic Publishers, Dordrecht, The Netherlands, pp. 789–792.

Haag, E., Irrgang, K. D., Boekema, E. J., and Renger, G. (1990). Functional and Structural analysis of photosystem II core complexes from spinach with high oxygen evolution capacity. *Eur. J. Biochem.* **189**, 47–53.

Haddy, A., Dunham, W. R., Sands, R. H., and Aasa, R. (1992). Multifrequency EPR investigations into the origin of the S_2-state signal at $g = 4.1$ of the O_2-evolving complex. *Biochim. Biophys. Acta* **1099**, 25–34.

Hansson, O., and Wydrzynski, T. (1990). Current perceptions of photosystem II. *Photosynth. Res.* **23**, 131–162.

Homann, P. (1988a). Structural effects of Cl^- and other anions on the water oxidizing complex of chloroplast photosystem II. *Plant Physiol.* **88**, 194–199.

Homann, P. (1988b). The calcium and chloride requirement of photosynthetic water oxidation: Effects of pH. *Biochim. Biophys. Acta* **934**, 1–13.

Izawa, S., Heath, R. L., and Hind, G. (1969). The role of chloride ion in photosynthesis. III. The effect of artificial electron donors upon electron transport. *Biochim. Biophys. Acta* **180**, 388–398.

Jahns, P., Lavergne, J., Rappaport, F., and Junge, W. (1991). Stoichiometry of proton release during photosynthetic H_2O oxidation: A reinterpretation of the responses of Neutral red leads to a non-integer pattern. *Biochim. Biophys. Acta* **1057**, 313–319.

Joliot, P., and Kok, B. (1975). Oxygen evolution in photosynthesis. In Govindjee (Ed.), *Bioenergetics of Photosynthesis*. Academic Press, New York, pp. 387–411.

Joliot, P., Barbieri, G., and Chabaud, R. (1969). Un noveau modele des centres photochimique du systems II. *Photochem. Photobiol.* **10**, 309–329.

Kalosaka, K., Beck, W. R., Brudvig, G., and Cheniae, G. M. (1990). Coupling of the PS2 reaction center to the O_2-evolving center requires a very high affinity Ca^{2+} site. In M. Baltscheffsky (Ed.), *Current Research in Photosynthesis*. Kluwer Academic Publishers, Dordrecht. The Netherlands, pp. 721–724.

Kamen, M. (1963). *Primary Processes in Photosynthesis*. Academic Press, New York, p. 1.

Kavelaki, K., and Ghanotakis, D. F. (1991). Effect of the manganese complex on the binding of the extrinsic proteins (17, 23 and 33 kDa) of photosystem II. *Photosynth. Res.* **29**, 149–155.

Keegstra, K., Olsen, L. J., and Theg, S. M. (1989). Chloroplastic precursors and their transport across the envelope membranes. *Annu. Rev. Plant Physiol. Plant Mol. Biol.* **40**, 471–501.

Kelley, P. M., and Izawa, S. (1978). The role of chloride ion in photosystem II. I. Effects of chloride ion on photosystem II electron transport and on hydroxylamine inhibition. *Biochim. Biophys. Acta* **502**, 198–210.

Kim, D. H., Britt, R. D., Klein, M. P., and Sauer, K. (1990). The $g = 4.1$ EPR signal of the S_2 state of the photosynthetic oxygen evolving complex arises from a multinuclear Mn cluster. *J. Am. Chem. Soc.* **112**, 9389–9391.

Kim, D. H., Britt, R. D., Klein, M. P., and Sauer, K. (1992). The manganese site of the photosynthetic oxygen-evolving complex probed by EPR spectroscopy of oriented photosystem II membranes: The $g = 2$ and $g = 2$ multiline signals. *Biochemistry* **31**, 541–547.

Kobayashi, M., Maeda, H., Watanabe, T., Nakane, H., and Satoh, K. (1990). Chlorophyll α and β-carotene content in the D1/D2/cytochrome b559 reaction center complex from spinach. *FEBS Lett.* **260**, 138–140.

Kok, B., Forbush, B., and McGloin, M. (1970). Cooperation of charges in photosynthetic oxygen evolution. I. A linear four step mechanism. *Photochem. Photobiol.* **11**, 457–475.

Koppenol, W. H. (1978). Free energies of some interconversion reactions. *Photochem. Photobiol.* **28**, 431–433.

Kornberg, A. (1989). Never a dull enzyme. *Annu. Rev. Biochem.* **58**, 1–30.

Kretschmann, H., Pauly, S., and Witt, H. T. (1991). Evidence for a chemical reaction of hydroxylamine with the photosynthetic water splitting enzyme S in the dark—possible states of manganese and water in the S cycle. *Biochim. Biophys. Acta* **1059**, 208–214.

Kuwabara, T., and Murata, N. (1982). Inactivation of photosynthetic oxygen evolution and the concomitant release of three polypeptides in the photosystem II particles of spinach chloroplasts. *Plant Cell Physiol.* **23**, 533–539.

Lavergne, J. (1986). Stoichiometry of the redox changes of manganese during the photosynthetic water oxidation cycle. *Photochem. Photobiol.* **43**, 311–317.

Lavergne, J. (1987). Optical-difference spectra of the S-state transitions in the photosynthetic oxygen-evolving complex. *Biochim. Biophys. Acta* **894**, 91–107.

Levine, B. A., and Williams, R. J. P. (1982). Calcium binding to proteins and other large biological anion centers. In W. J. Cheung (Ed.), *Calcium and Cell Function*. Academic Press, London, Vol. 2, pp. 1–38.

Martin, R. B. (1984). Bioinorganic chemistry of calcium. *Met. Ions Biol. Syst.* **17**, 1–49.

Mayfield, S. P., Bennoun, P., and Rochaix, J. D. (1987a). Expression of the nuclear encoded OEE1 protein is required for oxygen evolution and stability of photosystem II particles in *Chlamydomonas reinhardtii*. *EMBO J.* **6**, 313–318.

Mayfield, S. P., Rahire, M., Frank, J. G. Zuber, H., and Rochaix, J. D. (1987b). Expression of the nuclear gene encoding the oxygen-evolving enhancer protein 2 is required for high levels of photosynthetic oxygen evolution in *Chlamydomonas reinhardtii*. *Proc. Natl. Acad. Sci. U.S.A.* **84**, 749–753.

Mei, R., and Yocum, C. F. (1990). Inorganic ions affect reductant mediated inhibition of the manganese cluster of PSII. In M. Baltscheffsky (Ed.), *Current Research in Photosynthesis*. Kluwer Academic Publishers, Dordrecht, The Netherlands, pp. 729–732.

Mei, R., and Yocum, C. F. (1991). Calcium retards NH_2OH inhibition of O_2 evolution activity by stabilization of Mn^{2+} binding to photosystem II. *Biochemistry* **30**, 7836–7842.

Messinger, J., Wacker, U., and Renger, G. (1991). Unusual low reactivity of the water oxidase in redox state S_3 toward exogenous reductants. Analysis of the NH_2OH- and NH_2NH_2-induced modification of flash-induced oxygen evolution in isolated spinach thylakoids. *Biochemistry* **30**, 7852–7862.

Michel, H., and Deisenhofer, J. (1986). X-ray diffraction studies on a crystalline bacterial photosynthetic reaction center: A progress report and conclusions on the structure of photosystem II reaction centers. In A. Staehelin and C. J. Arntzen (Eds.), *Photosynthesis III: Photosynthetic Membranes and Light Harvesting Systems*. Springer-Verlag, Berlin, pp. 373–381.

Miyao, M., and Murata, N. (1984a). Calcium can be substituted for the 24-kDa polypeptide in photosynthetic oxygen evolution. *FEBS Lett.* **168**, 118–120.

Miyao, M., and Murata, N. (1984b). Role of the 33 kDa polypeptide in preserving Mn in the photosynthetic oxygen-evolution system and its replacement by chloride ions. *FEBS Lett.* **168**, 281–286.

Miyao, M., and Murata, N. (1985). The Cl⁻ effect on photosynthetic oxygen evolution: Interaction of Cl⁻ with 18-kDa, 24-kDa and 33-kDa proteins. *FEBS Lett.* **180**, 303–308.

Miyao, M., and Murata, N. (1986). Light-dependent inactivation of photosynthetic oxygen evolution during NaCl treatment of photosystem II particles: The role of the 24-kDa protein. *Photosynth. Res.* **10**, 489–496.

Miyao, M., and Murata, N. (1989). The mode of binding of three extrinsic proteins of 33 kDa, 23 kDa, and 18 kDa in the photosystem II complex of spinach. *Biochim. Biophys. Acta* **977**, 315–321.

Miyao-Tokutomi, M., and Inoue, Y. (1991). Enhancement by chloride ions of photoactivation of oxygen evolution in manganese-depleted photosystem II membranes. *Biochemistry* **30**, 5379–5387.

Murata, N., Miyao, M., and Kuwabara, T. (1983). Organization of the photosynthetic oxygen evolution system. In Y. Inoue, A. R. Crofts, Govindjee, N. Murata, G. Renger and K. Satoh (Eds.), *The Oxygen Evolving System of Photosynthesis.* Academic Press, Tokyo, pp. 213–222.

Murata, N., Miyao, M., Omata, T., Matsunami, H., and Kuwabara, T. (1984). Stoichiometry of components in the photosynthetic oxygen evolution system of photosystem II particles prepared with Triton X-100 from spinach chloroplasts. *Biochim. Biophys. Acta* **765**, 363–369.

Nanba, O., and Satoh, K. (1987). Isolation of a photosystem II reaction center consisting of D-1 and D-2 polypeptides and cytochrome b559. *Proc. Natl. Acad. Sci. U.S.A.* **84**, 109–112.

Nixon, P., and Diner, B. (1992). Aspartate 170 of the photosystem II reaction center polypeptide D1 is involved in the assembly of the oxygen-evolving manganese center. *Biochemistry* **31**, 942–948.

Okamura, M. Y., Satoh, K., Isaacson, R. A., and Feher, G. (1987). Evidence of the primary charge separation in the D_1D_2 complex of photosystem II from spinach: EPR of the triplet state. In J. Biggins (Ed.), *Progress in Photosynthesis Research.* Nijhoff, Dordrecht, The Netherlands, pp. 379–381.

Ono, T. A., and Inoue, Y. (1983). Mn-preserving extraction of 33-, 24-, and 16-kDa proteins for O_2-evolving PSII particles by divalent salt-washing. *FEBS Lett.* **164**, 255–259.

Ono, T. A., and Inoue, Y. (1984). Reconstitution of photosynthetic oxygen evolving activity by rebinding of 33 kDa protein to CaCl₂-extracted PS II particles. *FEBS Lett.* **166**, 381–384.

Ono, T. A., and Inoue, Y. (1988). Discrete extraction of the Ca atom functional for O_2 evolution in higher plant photosystem II by a simple low pH treatment. *FEBS Lett.* **227**, 147–152.

Ono, T. A., and Inoue, Y. (1989). Roles of Ca^{2+} in O_2 evolution in higher plant photosystem II; Effects of replacement of Ca^{2+} by other cations. *Arch. Biochem. Biophys.* **275**, 440–448.

Ono, T. A., and Inoue, Y. (1990a). A marked upshift in threshold temperature for the S_1-to-S_2 transition induced by low pH treatment of PSII membranes. *Biochim. Biophys. Acta* **1015**, 373–377.

Ono, T. A., and Inoue, Y. (1990b). Abnormally stable S_2 formed in PSII during stringent depletion of Ca^{2+} by NaCl/EDTA wash under illumination. In M. Baltscheffsky (Ed.), *Current Research in Photosynthesis.* Kluwer Academic Publishers, Dordrecht, The Netherlands, pp. 741–744.

Ono, T. A., Zimmermann, J. L., Inoue, Y., and Rutherford, A. W. (1986). EPR evidence for a modified S-state transition in chloride-depleted photosystem II. *Biochim. Biophys. Acta* **851**, 193–201.

Pauly, S., Schlodder, E., and Witt, H. T. (1990). No evidence for specific function of Ca^{2+} and Cl⁻ in oxygen evolution of isolated PSII-complexes from thermophilic cyanobacteria. In M. Baltscheffsky (Ed.), *Current Research in Photosynthesis.* Kluwer Academic Publishers, Dordrecht, The Netherlands, pp. 745–748.

Pecoraro, V. L. (1988). Structural proposals for the manganese centers of the oxygen evolving complex: An inorganic chemist's perspective. *Photochem. Photobiol.* **48**, 249–264.

Penner-Hahn, J. E., Fronko, R. M., Pecoraro, V. L., Yocum, C. F., Betts, S. D., and Bowlby, N. R. (1990). Structural characterization of the manganese sites in the photosynthetic oxygen-evolving complex using X-ray absorption spectroscopy. *J. Am. Chem. Soc.* **112**, 2549–2557.

Petersen, J., Dekker, J. P., Bowlby, N. R., Ghanotakis, D. F., Yocum, C. F., and Babcock, G. T. (1990). EPR characterization of the CP47-D1-D2-cytochrome b559 complex of photosystem II. *Biochemistry* **29**, 3226–3231.

Radmer, R., and Ollinger, O. (1986). Do the higher oxidation states of the photosynthetic O_2-evolving complex contain bound H_2O? *FEBS Lett.* **195**, 285–289.

Rickert, K., Sears, J., Beck, W., and Brudvig, G. W. (1991). Mechanism of irreversible inhibition of O_2 evolution in photosystem II by tris (hydroxymethyl) aminomethane. *Biochemistry* **30**, 7888–7894.

Rogner, M., Chisholm, D. A., and Diner, B. A. (1991). Site-directed mutagenesis of the psbC gene of photosystem II: Isolation and functional characterization of CP43-less photosystem II core complexes. *Biochemistry* **30**, 5387–5395.

Rutherford, A. W. (1988). Photosystem II, the oxygen evolving photosystem. In S. E. Stevens and D. Bryant (Eds.), *Light Energy Transduction in Photosynthesis: Higher Plant and Bacterial Models*. American Society of Plant Physiologists, Rockville, MD, pp. 163–177.

Rutherford, A. W. (1989). Photosystem II, the oxygen-evolving enzyme. *Trends Biochem. Sci.* **14**, 227–232.

Rutherford, A. W., Zimmermann, J. L., and Boussac, A. (1992). Oxygen evolution. In J. Barber (Ed.), *The Photosystems: Structure, Function and Molecular Biology*. Elsevier, Amsterdam, pp. 179–229.

Sandusky, P. O., and Yocum, C. F. (1984). The chloride requirement for photosynthetic oxygen evolution. Analysis of the effect of chloride and other anions on the amine inhibition of the oxygen-evolving complex. *Biochim. Biophys. Acta* **766**, 603–611.

Sandusky, P. O., and Yocum, C. F. (1986). The chloride requirement for photosynthetic oxygen evolution: Factors affecting nucleophilic displacement of chloride from the oxygen evolving complex. *Biochim. Biophys. Acta* **849**, 85–93.

Saphon, S., and Crofts, A. R. (1977). Protolytic reactions in photosystem. II. A new model for the release of protons accompanying the photooxidation of H_2O. *Z. Naturforsch., C: Biosci.* **32C**, 617–626.

Satoh, K. (1983). Photosystem II reaction center complex purified from higher plants. In Y. Inoue, A. R. Crofts, Govindjee, N. Murata, G. Renger, and K. Satoh (Eds.), *The Oxygen Evolving System of Photosynthesis*. Academic Press, Tokyo, pp. 27–38.

Sharp, R. R., and Yocum, C. F. (1981). Factors affecting hydroxylamine inactivation of photosynthetic water oxidation. *Biochim. Biophys. Acta* **635**, 90–104.

Shen, J. R., Satoh, K., and Katoh, S. (1988a). Calcium contents of oxygen evolving photosystem II preparations from higher plants. Effects of NaCl-treatment. *Biochim. Biophys. Acta* **933**, 358–364.

Shen, J. R., Satoh, K., and Katoh, S. (1988b). Isolation of an oxygen-evolving photosystem II preparations containing only one tightly bound calcium atom from a chlorophyll b deficient mutant of rice. *Biochim. Biophys. Acta* **936**, 386–394.

Srinivasan, A. N., and Sharp, R. R. (1986a). Flash-induced enhancements in the proton NMR relaxation rate of photosystem II particles. *Biochim. Biophys. Acta* **850**, 211–217.

Srinivasan, A. N., and Sharp, R. R. (1986b). Flash-induced enhancements in the proton NMR relaxation rate of photosystem II particles: Response to flash trains of 1–5 flashes. *Biochim. Biophys. Acta* **851**, 369–376.

Styring, S., and Rutherford, A. W. (1987). In the oxygen-evolving complex of photosystem II the S_0 state is oxidized to the S_1 state by D^+ (Signal II_{slow}). *Biochemistry* **26**, 2401–2405.

Tamura, N., and Cheniae, G. M. (1988). Photoactivation of the water oxidizing complex: The mechanisms and general consequences to photosystem II. In S. E. Stevens and D. Bryant (Eds.), *Light Energy Transduction in Photosynthesis: Higher Plant and Bacterial Models*. American Society of Plant Physiologists, Rockville, MD, pp. 227–242.

Tamura, N., Radmer, R., Lantz, S., Cammarata, K., and Cheniae, G. (1986). Depletion of photosystem II—extrinsic proteins. II. Analysis of the PSII/water-oxidizing complex by measurements of N, N, N', N'-tetramethyl-p-phenylenediamine oxidation following an actinic flash. *Biochim. Biophys. Acta* **850**, 369–379.

Toyoshima, Y., Akabori, K., Imaoka, A., Nakayama, H., Ohkouchi, N., and Kawamori, A. (1984). Reconstitution of photosynthetic charge accumulation and oxygen evolution in $CaCl_2$-treated PSII particles. II: EPR evidence for reactivation of the $S_1 \to S_2$ transition in $CaCl_2$ treated PSII particles with the 17-, 23-, and 34-kDa proteins. *FEBS Lett.* **176**, 377–381.

Trebst, A. (1986). The topology of the plastoquinone and herbicide binding sites of photosystem II in the thylakoid membrane. *Z. Naturforsch., C: Biosci.* **41C**, 240–245.

Tso, J., Sivaraja, M., and Dismukes, G. C. (1991). Calcium limits substrate accessibility or reactivity at the manganese cluster in photosynthetic water oxidation. *Biochemistry* **30**, 4734–4739.

Van Gorkom, H. J. (1985). Electron transfer in photosystem II. *Photosynth. Res.* **6**, 97–112.

Vanngard, T., Hansson, O., and Haddy, A. (1992). EPR studies of manganese in photosystem II. In V. L. Pecoraro (Ed.), *Manganese Redox Enzymes*. VCH Publishers, New York, pp. 105–118.

Vass, I., Ono, T., and Inoue, Y. (1987). Stability and oxidation properties of thermoluminescent charge pairs in the O_2-evolving system depleted of Cl^- or the 33 kDa protein. *Biochim. Biophys. Acta* **892**, 224–235.

Velthuys, B. R. (1975). Binding of the inhibitor NH_3 to the oxygen-evolving apparatus of spinach chloroplasts. *Biochim. Biophys. Acta* **396**, 392–401.

Vermaas, W., Charite, J., and Shen, G. (1990). Glu-69 of the D2 protein in photosystem II is a potential ligand to Mn involved in photosynthetic oxygen evolution. *Biochemistry* **29**, 5325–5332.

Vermaas, W. F. J., Rutherford, A. W., and Hansson, O. (1988). Site directed mutagenesis in photosystem II of the cyanobacterium *Synochocystis* sp. PCC 6803: Donor D is a tyrosine residue in the D2 protein. *Proc. Natl. Acad. Sci. U.S.A.* **85**, 8477–8481.

Waggoner, C. M., and Yocum, C. F. (1987). Selective depletion of water-soluble polypeptides associated with photosystem II. In J. Biggins (Ed.), *Progress in Photosynthesis Research*. Nijhoff, Dordrecht, The Netherlands, pp. 685–688.

Waggoner, C. M., and Yocum, C. F. (1990). Calcium-activated oxygen evolution. In M. Baltscheffsky (Ed.), *Current Research in Photosynthesis*. Kluwer Academic Publishers, Dordrecht, The Netherlands, pp. 733–736.

Waggoner, C. M., Pecoraro, V., and Yocum, C. F. (1989). Monovalent cations (Na^+, K^+, Cs^+) inhibit calcium activation of photosynthetic oxygen evolution. *FEBS Lett.* **244**, 237–240.

Wales, R., Newman, B. J., Pappin, D., and Gray, J. C. (1989). The extrinsic 33 kDa polypeptide of the oxygen-evolving complex of photosystem II is a putative calcium-binding protein and is encoded by a multi-gene family in pea. *Plant Mol. Biol.* **12**, 439–451.

Yachandra, V. K., Guiles, R. D., Sauer, K., and Klein, M. P. (1986). The state of manganese in the photosynthetic apparatus. 5. The chloride effect in photosynthetic oxygen evolution. Is halide coordinated to the EPR-active manganese in the O_2-evolving complex? Studies of the substructure of the low-temperature multiline EPR signal. *Biochim. Biophys. Acta* **850**, 333–342.

Yachandra, V. K., DeRose, V. J., Latimer, M. J., Mukerji, I., Sauer, K., and Klein, M. P. (1991). A structural model of the oxygen evolving manganese cluster. *Photochem. Photobiol.* **53**, Supp., 98S.

Yeates, T. O., Komiya, H., Chirino, A., Rees, D. C., Allen, J. P., and Feher, G. (1988). Structure of the reaction center from *Rhodobacter sphaeroides* R-26 and 2.4.2: Protein-cofactor (bacteriochlorophyll, bacteriopheophytin, and carotenoid) interaction. *Proc. Natl. Acad. Sci. U.S.A.* **85**, 7993–7997.

Yocum, C. F. (1991). Calcium activation of photosynthetic oxygen evolution. *Biochim. Biophys. Acta* **1059**, 1–15.

Yocum, C. F., and Babcock, G. T. (1981). Amine-induced inhibition of photosynthetic oxygen evolution: A correlation between the microwave power saturation properties of signal IIf and photosystem II-associated manganese. *FEBS Lett.* **130**, 99–102.

Yocum, C. F., Yerkes, C. T., Blankenship, R. E., Sharp, R. R., and Babcock, G. T. (1981). Stoichiometry, inhibitor sensitivity and organization of manganese associated with photosynthetic oxygen evolution. *Proc. Natl. Acad. Sci. U.S.A.* **78**, 7507–7511.

Zimmermann, J. L., and Rutherford, A. W. (1984). EPR studies of the oxygen-evolving enzyme of photosystem II. *Biochim. Biophys. Acta* **767**, 160–167.

2

OXIDANTS IN THE UNPOLLUTED MARINE ATMOSPHERE

Anne M. Thompson

NASA/Goddard Space Flight Center, Greenbelt, Maryland 20771

Environmental Oxidants, Edited by Jerome O. Nriagu and Milagros S. Simmons.
ISBN 0–471–57928–9   © 1994 John Wiley & Sons, Inc.

1. OXIDANTS IN THE MARINE ATMOSPHERE

1.1. Marine Oxidants and Photochemistry

There are three major oxidants in the marine atmosphere: O_3, OH, and H_2O_2. They are photochemically related to one another and to many trace gases produced by natural and anthropogenic activities. The cycles relating these gases are illustrated in Figure 1.

Ozone is both the initiator and the end product of conversion among CH_4, nonmethane hydrocarbons (NMHC), H_2O, nitrogen oxides, and carbon monoxide. Photodissociation of O_3 begins when the small amount of UV ($\lambda < 320$ nm) transmitted through the stratosphere into the troposphere breaks O_3 into the oxygen molecule and the highly reactive $O(^1D)$ free radical:

$$O_3 + hv \longrightarrow O_2 + O(^1D)$$

The $O(^1D)$ radical interacts with water vapor to form two molecules of the hydroxyl radical (OH):

$$O(^1D) + H_2O \longrightarrow OH + OH$$

No measurements of OH have been reported in the marine atmosphere. However, because water is abundant in the lower troposphere, especially in marine environments, it is believed that most tropospheric OH is concentrated in the

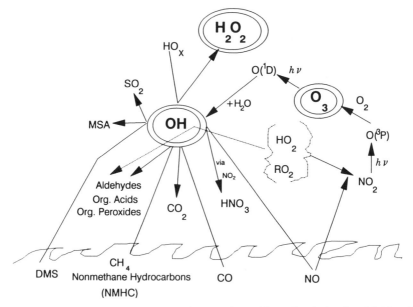

Figure 1. Oxidants in marine atmosphere and connection to biogeochemical cycles of C, N, and S.

tropics, where UV radiation is abundant and there are few pollutant sinks for OH.

The OH radical is the main component of atmospheric "oxidizing capacity," converting nitrogen oxides to nitric acid (HNO_3); nonmethane hydrocarbons to aldehydes, peroxides, other oxygenated compounds, and organic acids; CO to CO_2; and SO_2 to H_2SO_4. It also reacts with aldehydes, peroxides, and hundreds of other atmospheric transients. Reaction between OH and CH_4 starts off the methane oxidation chain, which leads to formation of CO, formaldehyde, and methylhydrogen peroxide (CH_3OOH).

Methane Oxidation Chain:

$$OH + CH_4(+O_2) \longrightarrow CH_3O_2 + H_2O$$
$$CH_3O_2 + NO \longrightarrow NO_2 + CH_3O$$

[Formation of formaldehyde: $CH_3O(+O_2) \longrightarrow HCHO$]
[Formation of CH_3OOH: $CH_3O_2 + HO_2(+M) \longrightarrow CH_3OOH + O_2 + M$]

Formation of CO or HO_2 from HCHO:

$$HCHO + hv \longrightarrow H_2 + CO$$
$$HCHO + hv(+O_2) \longrightarrow HO_2 + HCO$$

Conversion of NO to NO_2 can be accomplished by reaction of NO with HO_2 from CO oxidation or by another hydrocarbon free radical that originates from OH oxidation of a nonmethane hydrocarbon, RH:

$$CO + OH(+O_2) \longrightarrow CO_2 + HO_2$$
$$RH + OH(+O_2) \longrightarrow RO_2 + H_2O$$

Conversion of NO to NO_2:

$$HO_2 + NO \longrightarrow NO_2 + OH$$

or

$$RO_2 + NO \longrightarrow NO_2 + RO$$

Photodissociation of NO_2 leads to O_3 formation:

$$NO_2 + hv \longrightarrow O + NO$$
$$O + O_2(+M) \longrightarrow O_3 + M$$

The OH radical, and ultimately H_2O_2, is formed from O_3, the OH is formed as shown above, and the H_2O_2 is formed from combination of two HO_2 radicals:

$$HO_2 + HO_2 \longrightarrow H_2O_2 + O_2$$

Hydrogen peroxide is the oxidant responsible for converting SO_2 to sulfuric acid in clouds and raindrops at low pH. In the marine atmosphere, the major source of SO_2 is believed to be OH oxidation of dimethyl sulfide $[(CH_3)_2S]$, although the detailed mechanism is not fully understood (Fig. 1).

1.2. Oxidant Interactions with Global Chemical and Climate Change

Global levels of oxidants—the amounts of O_3, OH, and H_2O_2 in the background (nonurban) troposphere—are topics of current interest because of concern that they are perturbed by changing concentrations of CH_4 and other trace gases, as well as by stratospheric O_3 depletion and global warming.

For example, the reactions in Section 1.1. show that increases in atmospheric CO and CH_4 (the latter presently increasing 0.5–1% yr globally) would tend to add to the burden of tropospheric O_3. Estimates are also available for OH and H_2O_2 changes in unpolluted marine regions, for which CO and CH_4 increases are assumed to occur with little or no increase in NO levels (Thompson et al., 1989, 1990a). The latter assumption is made because increased emissions of NO are driven from continental regions (fossil-fuel combustion and biomass burning), with NO_x ($= NO + NO_2$) having too short a lifetime (a few hours) to be transported to remote marine regions. Perturbations in global O_3, OH, or H_2O_2 levels would signify changes in the atmosphere's oxidizing capacity. There is evidence of

Table 1   Ground-based   Observations   for   Ozone   and   Related   Trace-Gas Measurements

Project Name[a]	Dates of Operation	Sites
NOAA/CMDL	1960s–present (continuous monitoring)	Barrow, AK (76 N); Mauna Loa (20 N); Am. Samoa (14 S); South Pole (90 S)
ALE/GAGE	1977–	Oregon, Tasmania, Ireland, Barbados, Am. Samoa
WATOX	1986–1988	Bermuda with coordinated aircraft missions
AEROCE	1986–	Ireland; Bermuda; Barbados; Azores; Tenerife
MLOPEX	1988, 1991–1992	Mauna Loa Observatory (20 N)
NARE	1991–	Sable Island

[a] CMDL = Climate Monitoring and Diagnostics Laboratory (formerly GMCC, Geophysical Monitoring for Climate Change); ALE/GAGE = Atmospheric Lifetime Experiment/Global Aerosols and Gases Experiment; WATOX = Western Atlantic Oxidant Experiment; AEROCE = Atmospheric Ocean Chemistry Experiment; MLOPEX = Mauna Loa Observatory Photochemistry Experiment; NARE = North Atlantic Regional Experiment.

tropospheric O_3 and H_2O_2 increasing in some nonurban locations; possible trends in OH are less certain (Thompson, 1992).

2. OZONE IN THE MARINE ATMOSPHERE

Ozone is the most frequently measured oxidant in the marine atmosphere. It is monitored at the four baseline stations of the National Oceanic and Atmospheric Administration/Climate Monitoring and Diagnostics Laboratory (NOAA/CMDL) network (two high-latitude and two tropical stations), and it has been measured on numerous ship cruises since the mid-1970s. Since the late 1970s, O_3 in the marine boundary layer and free troposphere has also been measured in many aircraft experiments (Section 2.2.2.). Tables 1–3 summarize programs in which measurements of O_3 and other trace gases have been made.

2.1. Ozone at Monitoring Sites: Soundings and Surface Data

Surface O_3 has been monitored at CMDL [formerly Geophysical Monitoring for Climate Change (GMCC)] background sites since the early 1960s (NOAA/CMDL, 1990). Surface O_3 measurements are continuous and balloon-borne ozonesondes are launched at least several times a month at most locations. Three of the four CMDL stations are essentially remote marine locations, and the fourth, at the South Pole, is probably not very different from Antarctic oceanic regions. Since 1988, CMDL has expanded North Atlantic surface measurements through the AEROCE (Atmosphere/Ocean Chemistry Experiment) network (Table 1). Other stations at which CMDL is collecting data are also marine-influenced: Hilo, Hawaii; Lauder, New Zealand, and Syowa, Antarctica. Figure 2 shows annually averaged O_3 mixing ratio profiles from five of the marine stations. The two tropical sites (Samoa and Hilo) have the lowest total O_3.

Figure 3a shows annual cycles of O_3 at the CMDL sites. The two Northern Hemisphere CMDL sites may not be representative of Northern Hemisphere marine environments (NOAA/CMDL, 1990). The Barrow cycle is affected by boundary-layer O_3 depletion in early spring (see April mixing ratio in Fig. 3a), and Mauna Loa (3.4 km elevation) is well into the free troposphere much of the time (up to 50 ppbv, Fig. 3a). AEROCE sites (Fig. 3b) may be more representative of O_3 in NH marine conditions (Oltmans and Levy, 1993). A springtime maximum is evident in the Mace Head and Bermuda data. This is believed to be due to the 30 to 35 N jet stream mixing O_3-rich air downward from the stratosphere. Mace Head data show a secondary seasonal maximum in autumn. Examination of the O_3 and CO correlation for autumn 1990 shows contrasting patterns (Doddridge et al., 1994). Most of the data clusters in a "box" with CO at 100 to 150 ppbv and O_3 at 25 to 45 ppbv. However, at CO mixing ratios in excess of 200 ppbv, there are two clusters of values. Negative correlations (i.e., O_3 decreasing as the CO mixing ratio increases) are suggestive of polluted European air masses with initially high NO_x titrated out, possibly destroying O_3 in the

Table 2 Shipboard Observations for Ozone and Related Trace-Gas Measurements

Project Name[a]	Dates of Operation	Sites	Species Measured
R/V Knorr 73/7	1978	Kwajalein to American Samoa	O_3, NO, HNO_3
R/V Meteor 56/1	1980	Hamburg to Montevideo	O_3/CO/NO_x/HCHO/DMS/NMHC
RITS Series	1984–	Atlantic, $\pm 50°$ Pacific, $\pm 50°$	O_3/CO/CH_4/NO_x
SAGA-1,-2,-3	1984/1987/1990	Pacific	O_3/CO/CH_4; SAGA-3 (Table 5)
R/V Polarstern ANT87/5	1987	Atlantic	O_3/NO_x/H_2O_2/NMHC/CO/DMS
PSI	1989/1990/1991	Northeast Pacific	O_3/CO/SO_2/CO/DMS
Pac. MAGE/JGOFS	1992	Eastern Pacific	O_3/NO_x/CH_4/SO_2/CO/DMS
ASTEX/MAGE	1992	Eastern Atlantic	O_3/NO_x/SO_2/NO_y/PAN/CO/DMS/ NMHC/aerosols

[a] RITS = Radiatively Important Trace Species; SAGA = Soviet-American Gases and Aerosols; PSI = Pacific Sulfur/Stratus Investigation; JGOFS = Joint Global Ocean Flux Study; ASTEX = Atlantic Stratocumulus Transition Experiment; MAGE = Marine Aerosols and Gases Experiment.

Table 3   Aircraft Observations for Ozone and Related Trace-Gas Measurements

Project Name[a]	Dates	Location	Species Measured
GAMETAG	1978	Central Pacific	O_3, NO, HNO_3
Winter MONEX	1978–1979	Indian Ocean, West. Pacific	O_3, H_2O, NMHC, $J(O_3)$, $J(NO_2)$
GTE/ABLE 1	1983	Western Atlantic	O_3, CO
GTE/CITE 1	1983–1984	Eastern Pacific	O_3, CO, NO, CH_4, NMHC
DYCOMS	1985	Eastern Pacific	O_3, NO_x, HNO_3, H_2O, $DMS/CS_2/OCS/SO_2$
AGASP-1;-2,-3	1983/1985	Arctic sites —	O_3/Br-HC/aerosols/NMHC/NO_2
	1986/1989	U.S./Canada/Europe	$CO/HNO_3/PAN$
STRATOZ-III	1984	Atlantic, $\pm 60°$, Pacific, 0–60 S	NO, NO_y, CO, O_3, CO, CH_4, NMHC
TROPOZ	1987	Atlantic	O_3, NO, NO_y, CH_4
FIRE-1	1987	Eastern Pacific	O_3, H_2O, CO
GTE/CITE 2	1986	Eastern Pacific	$O_3/CO/NO_2/NO_y/HNO_3/PAN/NMHC$
GTE/CITE 3	1989	Western Atlantic (30–40 N) and 5 S	$O_3/CO/NO_2/NO_y/HNO_3/PAN/NMHC/DMS/CS_2/OCS/SO_2$, aerosols
PEM-WEST	1991	Western Pacific	$O_3/CO/NO_2/NO_y/HNO_3/PAN/NMHC/SO_2/H_2O_2/$aerosols
TRACE-A	1992	Atlantic	$O_3/CO/NO_2/NO_y/HNO_3/PAN/NMHC/H_2O_2/$aerosols
ASTEX/MAGE	1992	Eastern Atlantic	$O_3/CO/NO_2/NO_y/HNO_3/PAN/NMHC/DMS/SO_2$, aerosols

[a] GAMETAG = Global Atmospheric Measurements Experiment on Tropospheric Aerosols and Gases; MONEX =.Monsoon Experiment; GTE = Global Tropospheric Experiment; ABLE = Atlantic Boundary Layer Experiment; CITE = Chemical Intercomparsion and Testing Experiment; DYCOMS = Dynamics and Chemistry of Marine Stratocumulus Experiment; AGASP = Arctic Gases and Aerosols Sampling Project; FIRE = First ISSCP Regional Experiment; JGOFS = Joint Global Ocean Flux Study; ASTEX = Atlantic Stratocumulus Transition Experiment; MAGE = Marine Aerosols and Gases Experiment; PEM-WEST = Pacific Exploratory Mission; TRACE-A = Transport and Atmospheric Chemistry near the Equator Atlantic.

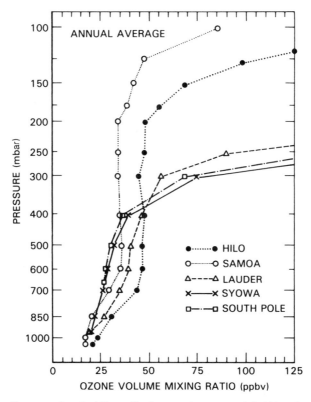

Figure 2. Annually averaged vertical O_3 profiles from marine or coastal sites (American Samoa, Hilo, Syowa) or locations remote from continental influences (South Pole, Lauder).

process. Positive correlations imply photochemical formation, presumably in fresher air masses.

It is noteworthy that, on average, the maximum surface O_3 in Bermuda does not coincide with high summertime O_3 in the eastern United States (Oltmans and Levy, 1992). Despite the potential for long-range transport from the United States (Moody et al., 1994), incursions of pollution are episodic exceptions to the persistent "Bermuda high," which tends to isolate Bermuda from the continent in the summer.

The springtime minimum in surface O_3 at Barrow appears to be related to a spike in boundary-layer bromine (Berg et al., 1983) associated with the melting of polar ice and seasonal algal emissions. For several consecutive days, surface-O_3 mixing ratios fall to zero (Oltmans et al., 1989 a, b). Several mechanisms involving bromine have been advanced, but none has been proven so far (Barrie et al., 1988; Sturges et al., 1992). In recent years, a seasonal depletion in surface O_3 has been observed at the South Pole (Schnell et al., 1991). Unlike the Barrow loss, which is apparently natural, the South Pole O_3 minima in spring and summertime is associated with the Antarctic ozone hole. With lower stratospheric O_3 greatly

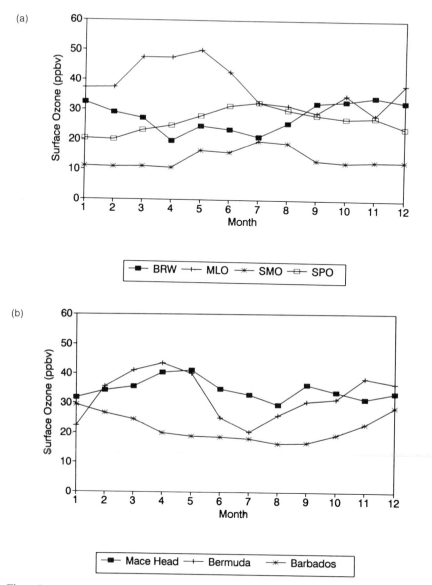

Figure 3. (a) Annually averaged surface O_3 for 1990 at the four baseline stations of NOAA/CMDL. BRW = Barrow, MLO = Mauna Loa, SMO = American Samoa, SPO = South Pole. (b) Annual surface O_3 cycles at AEROCE stations for 1990.

depleted, more UV penetrates to the troposphere, resulting in photochemical destruction of surface O_3 (see reactions in Section 1.1.) (Liu and Trainer, 1988).

2.2. Regional Ozone Observations from Satellite, Ship, and Aircraft Missions

A global view of marine O_3 comes from shipboard, aircraft, and satellite observations.

(a)   POLARSTERN ATLANTIC CRUISE (1987)

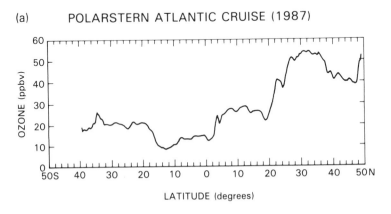

(b)   OZONE MIXING RATIO FROM SOUNDINGS ON R/V POLARSTERN (MARCH 1987)

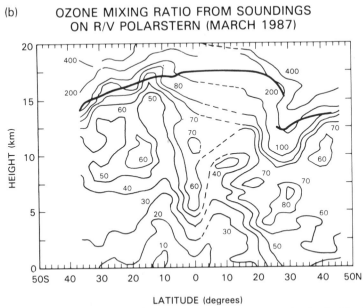

Figure 4. (a) Surface O_3 from an Atlantic transect oceanographic cruise in the Northern and Southern Hemispheres (March–April 1987, on German R/V *Polarstern*); note O_3 minimum at 15S. (b) O_3 latitude-altitude cross-section from ozonesondes launched on same *Polarstern* cruise as in (a). (Adapted from Smit et al., 1989. Copyright by A. Deepak Publishing.)

2.2.1. Oceanographic Cruises and Satellite Observations

Shipboard measurements show that the springtime O_3 maximum observed at monitoring sites is a general feature of the Northern Hemisphere. A greater degree of stratosphere–troposphere mixing in the Northern Hemisphere compared to the Southern Hemisphere leads to higher levels of O_3 in the Northern Hemisphere. An example is shown in Figure 4a, in which high concentrations of surface O_3 measured on a springtime south-to-north Atlantic oceanographic cruise are illustrated. On an annually averaged basis, surface O_3 is also greater in the Northern Hemisphere (Winkler, 1988). On the same cruise (*Polarstern ANT/V* in March–April 1987; Smit et al., 1989), ozonesondes were launched to obtain vertical profiles (Fig. 4b). Because stratosphere–troposphere mixing is greater in the Northern Hemisphere, the gradient between upper tropospheric and surface O_3 is less than in the Southern Hemisphere.

The O_3 minimum at 10 S is a prominent feature of near-equatorial O_3 on both Atlantic and Pacific cruises (Piotrowicz et al., 1986, 1991; Winkler, 1988; Johnson et al., 1990) Smit et al. (1989) ascribe the Atlantic minimum to photochemical destruction by UV photolysis in the presence of high relative humidity and low concentrations of O_3 precursors. This is certainly a factor contributing to low O_3 in general, but it probably cannot account for small regions of undetectably low O_3 that have been observed shipboard and from aircraft (Routhier et al., 1980). For example, Figure 5 shows a sharp 10- to 15-ppbv gradient at the ITCZ (Intertropical Convergence Zone), where the chemical environments on either side are probably not very different. Theories of halogen-mediated equatorial O_3 loss (Chameides and Davis, 1980) have been proposed but so far have not been supported by observations.

Transport probably plays a role in the equatorial O_3 minimum (Piotrowicz et al., 1991; Shiotani, 1992). Convection, for example, can displace O_3 in a matter of minutes. In an environment where the O_3 photochemical lifetime may be as short as four to five days, and there are not enough O_3 precursors to replenish O_3 quickly, convective updraft of boundary-layer O_3 can deplete surface O_3, causing it to "disappear" for several days. On the other hand, convection has also been linked to equatorial surface-O_3 maxima (Stallard et al., 1975; Winkler, 1988;

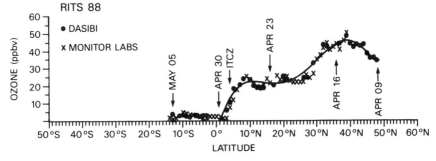

Figure 5. Latitude profile of surface O_3 from NOAA/RITS (Radiatively Important Trace Gases) 1988 Pacific cruise, showing 20 ppbv drop at ITCZ. (From Johnson et al., 1990.)

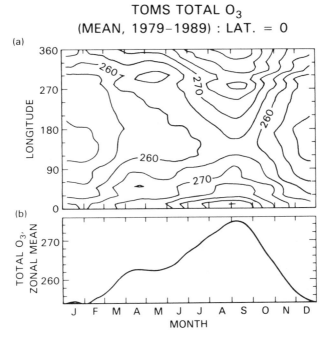

Figure 6. (*a*) Annual cycle of total O_3 column (in Dobson units, DU) at equator as seen from TOMS satellite. (*b*) Annual cycle of zonally averaged O_3. Cycle reflects influences of the QBO (quasibiennial oscillation), El Niño, and planetary-scale dynamics. (After Shiotani, 1992.)

Thompson et al., 1993). In a given set of conditions, tropospheric O_3 levels depend on whether the net effect of convection is to force O_3 out of the boundary layer or to bring down O_3-rich air. Fishman et al. (1987) reported profiles of low O_3 (and CO) extending from near the surface to 7 km in the vicinity of the ITCZ during a Pacific aircraft experiment. In addition to synoptic effects, Piotrowicz et al. (1991) have pointed out that the Walker circulation, with large-scale upward or downward motions driven by variations in the sea-surface temperature, may be responsible for whether equatorial total-column O_3 is a local maximum or minimum. Similarly, Shiotani (1992) has shown that equatorial O_3 responds to dynamical variations associated with El Niño and the quasibiennial oscillation, affecting both the magnitude of the mean and the location of the minimum O_3 column in the Pacific through changes in the tropopause height (Fig. 6).

2.2.2. *Ozone from Aircraft*

Table 3 lists aircraft experiments dedicated to marine O_3 transport and photochemistry. Some of the earlier experiments focused on O_3 transport, including deposition flux to the ocean. Using eddy correlation, Lenschow and co-workers

(1982) measured velocities for deposition to the ocean surface from 0.01 to 0.1 cm sec^{-1}, with most observations between 0.05 and 0.1 $cm\,sec^{-1}$.

At least four aircraft missions (DYCOMS, FIRE-1, ASTEX/MAGE, and TRACE-A) have been conducted in regions of marine stratocumulus. These areas, which occur off the western coast of continents from 20° to 40°, offer extensive horizontally homogeneous regions with persistent cloud cover, and a well-mixed boundary layer capped by a strong inversion (Lenschow et al., 1988; Cox et al., 1987). The typical O_3 mixing ratio profile (Fig. 7) is constant throughout the boundary layer, usually at unpolluted concentrations because the winds are normally not from the continent bordering the region. However, above the clouds there is rich structure in the O_3 profile, with air masses from several sources: moist, boundary-layer air; dry, high-O_3 air from the upper troposphere or lower stratosphere; and air from upstream convection (Paluch et al., 1992). Because O_3 is well conserved in marine stratocumulus, Kawa and Pearson (1989) used O_3 to derive an entrainment velocity of 1 to 3 cm sec^{-1} for mixing between the trade-wind inversion (just above cloud level) and the cloud-capped boundary layer.

The two marine stratocumulus regions west of Africa were sites of intensive aircraft measurements in 1992 (Table 3). ASTEX/MAGE (conducted in June 1992) used multiple airborne platforms in coordination with shipboard measurements to investigate the marine stratus region near the Azores. TRACE-A focused on the region off Angola, where satellite-derived tropospheric O_3 estimates are maximum (Fishman et al., 1991). This is a region of subsidence, upper-level convergence (Krishnamurti et al., 1993) and potentially enhanced photochemical O_3 formation (Pickering et al., 1994).

Correlations among profiles of O_3, H_2O, CO, and potential temperature have been used to study tropospheric mixing, stratosphere–troposphere exchange, and photochemistry. During the GTE/CITE-1 aircraft campaign flown out of California and Hawaii in 1983 (Table 3), both continental and marine air masses were sampled (*J. Geophys. Res.* [Spec. Issue] **92**, Feb. 1987). In the former, correlation between CO and O_3 was strong, and net photochemical formation of O_3 was frequently indicated (Fishman et al., 1987; Chameides et al., 1987a). Figure 8 shows strong CO and O_3 correlation in a marine air mass. In most of the air masses sampled, O_3 and NO were highly correlated with each other and both were anticorrelated with H_2O. Photochemical calculations (Section 2.3.) show that in the marine free troposphere, higher NO levels and net O_3 formation are associated with air parcels originating in the stratosphere, with NO sometimes enriched by lightning (Hipskind et al., 1987; Chameides et al., 1987b). The pristine marine boundary layer, where NO levels are low (~ 10 pptv), is a net sink for O_3.

On the STRATOZ-III campaign over the Atlantic, however, CO and O_3 were not always well correlated and NO_x and O_3 were frequently anticorrelated (Gerhardt et al., 1989). Some of the air masses included polluted air that masked local photochemical effects. Mixing ratios of NO_x were several hundred pptv, and

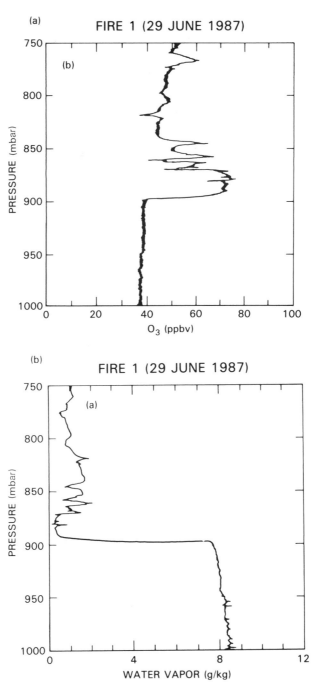

Figure 7. Vertical (*a*) O_3 and (*b*) water-vapor profiles in the vicinity of marine stratocumulus. Observations from the eastern Pacific during the FIRE-1 experiment, 1987. (From Paluch et al., 1992.)

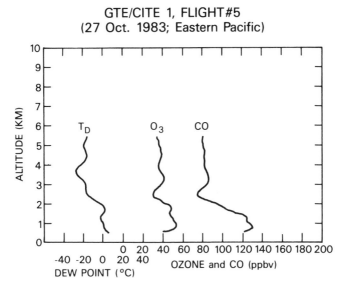

**GTE/CITE 1, FLIGHT#5
(27 Oct. 1983; Eastern Pacific)**

Figure 8. Profiles from GTE/CITE 1 aircraft mission taken in marine air mass > 250 km west of California, October 1983. Concentrations of O_3 and CO are well correlated. (After Fishman et al. 1987.)

NMHC concentrations were sufficient to contribute to net photochemical O_3 formation (Poppe et al., 1989).

Several types of marine air masses (maritime tropical, maritime polar, mixed polar-tropical, and mixed maritime-continental) were sampled in the eastern Pacific (California–Hawaii) during GTE/CITE-2 (*J. Geophys. Res.* [Spec. Issue] **95**, June 20, 1990). Of all the maritime air masses sampled, the negative correlation between O_3 and H_2O ($r = -0.62$) and the positive correlation between NO_x and O_3 ($r = 0.48$) are most significant (Carroll et al., 1990). For data with $NO_x \leqslant 75$ pptv (i.e., excluding the occasional pollution spike), the NO_x–O_3 correlation is 0.68. Correlations are highest in the lower middle troposphere; $r \sim 0.8$ for observations between 1 and 3 km in most maritime air masses. Positive correlation between CO and O_3 in mixed maritime and marine tropical air masses is pronounced (Fig. 9).

2.3. Ozone Measurements and Photochemical Theory

2.3.1. Shipboard Experiments

The first photochemical model descriptions of the nonpolluted marine environment were reported more than a decade ago (Graedel, 1979; Logan et al., 1981; Thompson and Cicerone, 1982). Model analysis of actual O_3 measurements in a nonpolluted marine environment was first performed by Liu et al. (1983), using data from the 1978 equatorial Pacific cruise (R/V *Knorr 73/7*). On that cruise, O_3

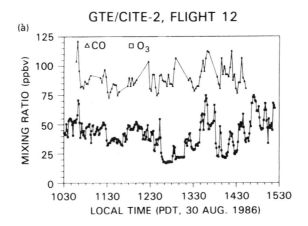

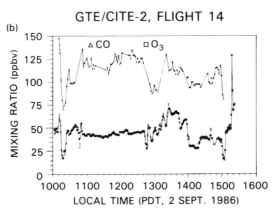

Figure 9. Aircraft observations of CO and O_3 in marine air masses from the GTE/CITE 2 mission in August–September 1986. Positive correlation of CO and O_3 is striking below 3 km. (After Carroll et al., 1990.)

averaged ~ 10 ppbv, and nitric oxide (reported for the first time in a pristine marine environment) was recorded at a 4 pptv noon maximum. Liu et al. (1983) estimated that 1 ppbv O_3/day was destroyed photochemically, giving a 10-day lifetime for surface O_3 in the study region (170 W in July–August).

Photochemical loss of O_3 in the marine boundary layer is largely due to UV photolysis causing OH formation, with catalytic destruction of O_3 by OH and HO_2:

$$O_3 + OH \longrightarrow HO_2 + O_2$$

$$O_3 + HO_2 \longrightarrow OH + 2O_2$$

Ozone formation occurs after NO_2 is photolyzed, with conversion of NO to NO_2 by HO_2, CH_3O_2, or R_iO_2 being the rate-determining step (reactions in Section 1.1.). Because NO levels are so low (< 5 pptv) in the equatorial Pacific (McFarland et al., 1979; Zafiriou et al., 1980), photochemical O_3 formation does not compensate for these losses of O_3. Therefore, loss of O_3 dominates. The typical diurnal cycle of O_3 is loss in the early morning followed by photochemical recovery (Liu et al., 1983; Thompson and Lenschow, 1984). This O_3 behavior was first reported by Stallard et al. (1975), but extensive observations were not described until Johnson et al. (1990) presented a summary of Pacific cruise data (Fig. 10). The Johnson et al. (1990) data show a lot of variability in the pattern of diurnal behavior, with the magnitude of the maximum-minimum spread varying at different latitudes and on different cruise tracks.

Shipboard experiments directed at trace-gas biogeochemical cycles (including C-, N-, and S-containing gases) and O_3 photochemistry have become increasingly complex (Table 2). Perhaps the most photochemically complete shipboard experiment to date was SAGA-3 (the third Soviet-American Gases and Aerosols), conducted in the equatorial Pacific in February to March 1990 (Johnson et al., 1993; Thompson et al., 1993). On leg 1, which took place between Hilo, Hawaii, and Pago Pago, American Samoa, shipboard measurements were made of O_3, CO, CH_4, nonmethane hydrocarbons (NMHC), NO, dimethyl sulfide (DMS), H_2S, H_2O_2, organic peroxides, and total-column O_3. A latitudinal gradient in O_3 was observed, with O_3 at 15 to 20 ppbv north of the ITCZ and 5 to 6 ppbv south of the ITCZ, but never $\leqslant 3$ ppbv as was observed on other equatorial Pacific cruises (Piotrowicz et al., 1986; Johnson et al., 1990). A diurnal pattern in O_3 was observed at all latitudes, with more O_3 at night than during the day (Fig. 10).

On SAGA-3 there was a UV filter instrument for measurement of total-column O_3. These measurements, which were in good agreement with total O_3 from TOMS, were further evidence that the equatorial Pacific is a region of low

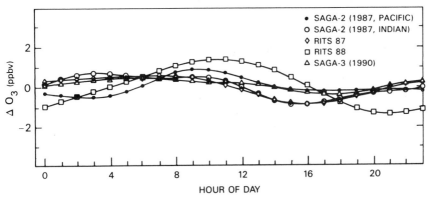

Figure 10. Diurnal behavior of surface O_3 from Pacific cruises measured by NOAA/PMEL. Deviation (in ppbv O_3) from daily mean is shown.

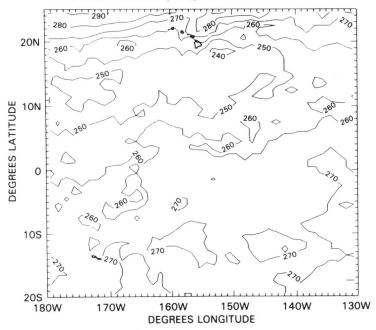

Figure 11. Low total ozone in the equatorial Pacific during SAGA-3 as seen by TOMS. Shipboard total O_3 column measurements on SAGA-3 cruise were within 10% of corresponding TOMS value. (Thompson et al., 1993.)

total O_3 (Fig. 11). Ozone in the SAGA-3 study region was in a low-O_3 phase associated with a slightly developed El Niño (Johnson et al., 1993). This is consistent with the correlation between total O_3 and the southern oscillation index (Piotrowicz et al., 1986; Shiotani, 1992). A very low southern oscillation index during the SAGA-3 cruise signifies equatorial warming in the eastern Pacific, with ascending air and reduced total-column O_3. The normal seasonal pattern of a high tropopause during SAGA-3 (Northern Hemisphere winter) may also have contributed to low O_3 (Fig. 6).

SAGA-3 gave a more comprehensive picture of the low-O_3, low-NO_x, low-hydrocarbon equatorial Pacific trace gases than had been available from previous cruises. A photochemical model with $O_3/NO/CO/CH_4/NMHC$ and gas-phase sulfur chemistry reproduced the observed diurnal O_3 pattern and observed mixing ratios of the reactive gases measured on the cruise (Thompson et al., 1993). Analysis of $O_3/CO/NO/NMHC/CH_4$ chemistry at typical equatorial concentrations shows a destruction rate of 1.2 ppbv O_3/day, when mean O_3 is ~ 10 ppbv (~ 8-day photochemical lifetime).

2.3.2. Aircraft Experiments

Studies of O_3 photochemistry throughout the marine boundary layer and free troposphere are conducted with aircraft and require highly sensitive and high-frequency instrumentation in order to resolve gradients over small distances. Since about 1980, when such devices were largely limited to O_3 detection, flight–worthy instruments for roughly a dozen photoreactive trace gases have been developed: CO, NO, NO_2, NO_y, DMS, CS_2, OCS, SO_2, PAN (peroxyacetyl-nitrate), HNO_3, and H_2O_2. The GTE/CITE (Global Tropospheric Experiment/Chemical Instrumentation Test and Evaluation) program has sponsored three standardization and intercomparison missions (McNeal et al., 1983; J. Geophys. Res. February 1987 and June 1990 issues).

For each GTE/CITE mission, O_3 correlations with other species and O_3 photochemistry have been analyzed (Fishman et al., 1987; Chameides et al., 1987a; Carroll et al., 1990; Davis et al., 1990). The following expression is used to calculate the net rate of O_3 formation:

$$P(O_3) = k_1[NO][HO_2] + \Sigma k_i[NO][R_iO_2] - [O_3]\{k_3[OH] + k_4[HO_2]\}$$
$$- k_5[O(^1D)][H_2O] \tag{1}$$

where R_iO_2, organic peroxy radicals, include primarily CH_3O_2, $C_2H_5O_2$, and CH_3CO_3 for the nonpolluted marine atmosphere. [See Donahue and Prinn (1990), Madronich and Calvert (1990), and Thompson et al. (1993) for discussions of other forms of R_iO_2 in the marine atmosphere.] The expression is evaluated with measurements of as many species as possible (typically NO, NO_2, O_3, CO, H_2O, CH_4, and some NMHC); a photochemical model supplies the transient species concentrations $[O(^1D), HO_2, R_iO_2]$. The key role of NO in O_3 formation is evident in the first two terms; likewise, the anticorrelation between O_3 and H_2O results from the reaction between O_3 and H_2O to form $O(^1D)$.

Figure 12 shows net O_3 production in maritime air masses sampled below 4.7 km on GTE/CITE 2. The amount of NO is one factor determining whether O_3 is formed or destroyed. In these data with mostly low NO concentrations, loss of O_3 dominates. However, Carroll et al. (1990) point out that even with NO concentrations at > 10 pptv (often considered the turning point from O_3 loss to production), the CITE 2 data usually show O_3 loss. Besides NO levels, the concentrations of H_2O, O_3, and NMHC play a role in whether or not there is net O_3 formation, and these factors vary substantially among the air masses sampled.

Observations in CITE 2 have also been used to evaluate the NO_2:NO ratio, the so-called photostationary ratio (Stedman and Jackson, 1975). The expression for NO_2:NO would be given by

$$\text{photostationary steady state (PSS)} = k[O_3]/J \tag{2}$$

if only reaction between NO and O_3 and the photodissociation of NO_2 determine the relationship between NO and NO_2. In Equation (2), k is the rate coefficient

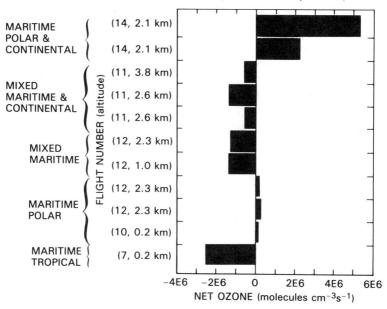

Figure 12. Net ozone production, $P(O_3)$, at different altitudes in different types of marine air masses from GTE/CITE-2. (From Carroll et al., 1990.)

for the reaction between NO and NO_2, and J is the rate coefficient for photodissociation of NO_2:

$$NO + O_3 \longrightarrow NO_2 \qquad \text{Rate} = k$$

$$NO_2 + h\nu \longrightarrow O + NO \qquad \text{Rate} = J$$

$$O + O_2(+ M) \longrightarrow O_3 + M$$

However, atmospheric measurements of NO_2, NO, and O_3 show that NO_2:NO rarely equals the expression in Equation (2). Because other processes convert NO to NO_2, a more accurate expression for the photostationary ratio is

$$NO_2/NO = (J)^{-1}\{k[O_3] + k_1[HO_2] + \Sigma k_i[R_iO_2]\} \tag{3}$$

where the free radicals R_iO_2 are the same as in Equation (1). On CITE 2, Chameides et al. (1990) measured NO_2:NO = 0.5–6, with most values in pristine marine air at ~ 1.5. Calculation of Equation (3) with a photochemical model based on measured O_3, NO_2, NO, H_2O, and NMHC gives values within 20 to 25% of the measured ratios (usually lower). The fact that model calculations of Equation (3) are close to the measured PSS is an important validation of model determination of $P(O_3)$ in Equation (1), because the same free radicals are part of the O_3 formation process.

2.3.3.   MLOPEX—Ozone at the Boundary Layer–Free Troposphere Interface

A unique marine environment was studied in 1988 at Mauna Loa on the MLOPEX [Mauna Loa Observatory Photochemical Experiment (*J. Geophys. Res.* **97** June 30, 1992)]. Because of the location of the observatory (3400 m) and peculiar meteorological conditions, both free tropospheric (from downslope winds, ~ 1900–1000 hr) and marine boundary-layer air (during upslope winds, ~ 1000–1900 hr) can be sampled on a daily basis.

A comprehensive set of reactive trace gases was measured during MLOPEX, including $O_3, CH_4, CO, NMHC$ and H_2O, all major components of the odd nitrogen family and NO_y (total reactive odd nitrogen). Ozone and NO_y levels during MLOPEX were similar to those from GTE/CITE 2 at the same altitude and were characteristic of free tropospheric air. Mean concentrations were 43 ppbv O_3 and ~ 250 pptv NO_y, the latter mostly supplied by the stratosphere (Hübler et al., 1992). Moderately low NO_x levels (mean value 30 pptv) with low NMHC concentrations (characteristic of an area remote from sources) lead to a small photochemical O_3 loss of 0.5 ppbv O_3/day (Liu et al., 1992; Ridley et al., 1992).

The measured photostationary ratio, $NO_2:NO$, for clear-sky conditions during MLOPEX was consistently ~ 2 and is explained photochemically by a high peroxy radical concentration (Ridley et al., 1992).

3.   HYDROXYL RADICAL IN THE MARINE TROPOSPHERE

No transient species is as essential to marine photochemistry as the OH radical. An instrument intercomparison campaign in the marine environment in 1983 showed that measurement of tropospheric OH is beyond the reach of shipboard or airborne technology (Beck et al., 1987), although long-path spectroscopic methods have been employed successfully at continental sites (Hofzumahaus et al., 1991; Mount and Eisele, 1992). Figure 1 shows the role that OH plays in the marine atmosphere. The OH radical reacts with hundreds of gases, for example, anything with an extractable hydrogen atom. In continental regions near emissions, OH controls the buildup of pollutants. In the marine atmosphere, hydrocarbons emitted by the ocean are oxidized to peroxides, aldehydes, and organic acids; NO_x is oxidized to nitric acid; and SO_2 is converted to sulfuric acid (and particulate sulfate) in a multistep process. Partially halogenated organics are also subject to OH attack.

The first modeling studies of the marine troposphere (Graedel, 1979; Logan et al., 1981; Chameides and Tan, 1981; Thompson and Cicerone, 1982) presented OH levels for idealized marine environments or were based on a few species measured on GAMETAG and the *Knorr 73/7* or 1980 *Meteor 56/1* cruises (Table 4). For example, Liu et al. (1983) estimated $2–3 \times 10^6$ molecules cm^{-3} for the *Knorr 73/7* cruise with only surface-O_3, NO, and HNO_3 measurements to guide the calculation. Since that time, more photochemically complex aircraft

Table 4 Model Calculated Tropospheric OH Using Measurements from Field Campaigns

Project Name	Date	Location	$[OH]^a$, (cm^{-3})	Ref.
R/V *Knorr 73/7*	July–Aug. 1978	Equatorial Pacific	2–3×10^6	Liu et al. (1983)
			1.5×10^6	Chameides and Tan (1981)
			1.5×10^6	Logan et al. (1981)
			1.5×10^6	Thomas and Cicerone (1982)
R/V *Meteor 56/1*	Oct.–Nov. 1980	Atlantic (50 N–35 S)	1.8×10^6	Thompson et al. (1990b)
GTE/CITE 1	1983–1984	Eastern Pacific	1–2×10^6	Chameides et al. (1987a)
MLOPEX	1988	Mauna Loa (20 N)	1.5–2.5×10^6	Liu et al. (1992)
SAGA-3	Feb.–Mar. 1990	Equatorial Pacific	6×10^5	Donahue and Prinn (1993)
			8×10^5	Yvon et al. (1993)
			9×10^5	Thompson et al. (1993)

^{a}Surface for *Knorr*, *Meteor*, and *SAGA-3*; MLOPEX at 3.4 km; lower free troposphere for GTE/CITE.

Table 5 Mixing Ratios for Photochemically Reactive Gases on SAGA-3 Cruise (1990)

Species	Mean Mixing Ratio		References
O_3	10 ppbv		Thompson et al. (1993)
NO	1.3 pptv		Torres and Thompson (1993)
$RONO_2$ (total)	5–15 pptv		Atlas et al. (1993)
CH_4	1.70 ppmv		Bates et al. (1993)
CO	75 ppbv		Bates et al. (1993)
H_2O_2	590 pptv		Thompson et al. (1993)
ROOH	610 pptv		
DMS	380 pptv		Huebert et al. (1993)
H_2S	3 pptv		Yvon et al. (1993)
Nonmethane Hydrocarbons			
	Low	*High*	
C_2H_4	50 pptv	420 pptv	
C_2H_6	790 pptv	790 pptv	
C_3H_6	55 pptv	150 pptv	
C_3H_8	64 pptv	85 pptv	Atlas et al. (1993); Donahue and Prinn (1993)

campaigns and cruises have been conducted, making a more constrained calculation of OH possible. For example, on GTE/CITE 1 in the eastern Pacific, Chameides et al. (1990) estimated OH to be $1-2 \times 10^6$ molecules cm^{-3} in the lower and middle troposphere from measurements of O_3, CO, NO, and temperature.

A primary objective of both the SAGA-3 (1990) Pacific cruise and MLOPEX (1988) was to measure sufficient species to constrain model calculation of OH. This requires, in addition to measurements of O_3, UV, and H_2O (for the rate of OH production), measurements of sink species for OH: CH_4, CO, peroxides, and NMHC. Table 5 shows the mean concentrations of these trace gases and of H_2S and DMS measured on SAGA-3 for equatorial conditions. There were two sets of NMHC concentrations on SAGA-3, and the OH calculation was parameterized with both of them. Surface-OH levels were computed to be $\sim 9 \times 10^5$ molecules cm^{-3} in both cases, but the relative importance of the OH sink terms differs. Methane, CO, and aldehyde and peroxide represent 60 to 70% of OH loss reactions; however, NMHC account for 17% of OH loss if higher alkene (C_2H_4 and C_3H_6) mixing ratios are used, and for only 4% of OH loss if lower alkene mixing ratios are assumed (Thompson et al., 1993).

Ozone and CO concentrations on SAGA-3 showed a latitudinal gradient with both trace gases decreasing from north to south. Maximum and minimum OH levels calculated with the model show that although higher CO levels in the Northern Hemisphere limit the OH lifetime, the higher O_3 levels give rise to

SAGA-3 MODEL ANALYSIS (1 March 1990 Conditions)

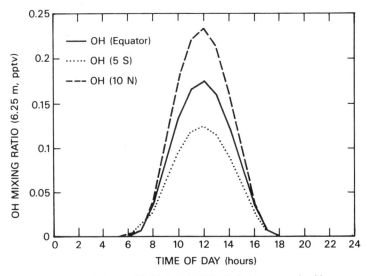

Figure 13. Diurnal OH variation on SAGA-3 with OH calculation constrained by measurements of O_3, CO, CH_4, NMHC, H_2O, UV, NO, and DMS. (From Thompson et al., 1993.)

higher OH. Figure 13 shows the diurnal variation in SAGA-3 surface-OH concentrations for conditions at 5 S, the equator, and 10 N.

Inasmuch as OH is a calculated rather than a measured quantity, it is important to estimate the uncertainty in the calculation. Besides the one-dimensional photochemical model (Thompson et al., 1993), two other photochemical models were used to compute OH, with results of 6×10^5 molecules cm^{-3} (Donahue and Prinn, 1993) and 8×10^5 molecules cm^{-3} (Yvon et al., 1993).

OH and HO_2 Radical Concentrations for SAGA-3

Latitude		Mixing Ratio—24-Hr Mean		
	O_3(ppbv)	CO(ppbv)	OH $(\times 10^5 \, cm^{-3})$	HO_2(pptv)
5 S	6.0	53	7.1	4.2
0	9.3	75	8.9	5.2
10 N	20	90	12	6.1

The range of values suggests a practical limit of 30 to 50% uncertainty in the OH estimate. More systematic uncertainty analyses put a 25 to 30% precision limit on model calculations of OH, due to imprecision in rate coefficients used in the typical model (Chameides and Tan, 1981; Thompson and Stewart, 1991).

Two models used to simulate MLOPEX measurements give OH values within 10% of one another (Liu et al., 1992; Ridley et al., 1992). Noontime concentrations are $4-5 \times 10^6$ molecules cm^{-3}. One reason for the close agreement might be the constraint of additional odd nitrogen species, including those incorporating OH chemistry (e.g., HNO$_3$ and PAN). Noontime HO$_2$ concentrations for MLOPEX were 30 pptv.

The dominance of O$_3$ in determining OH levels in low-NO$_x$, low-hydrocarbon environments is seen in MLOPEX as well as in SAGA-3. However, owing to the elevation of Mauna Loa, O$_3$ concentrations and computed OH levels during MLOPEX were twice as high as in the marine boundary layer at 10 to 20 N on SAGA-3.

4. HYDROGEN PEROXIDE: MEASUREMENTS AND MODELS

Hydrogen peroxide measurements in the marine atmosphere have been made on several ship cruises and three or four aircraft campaigns. The first report was made by Heikes et al. (1988) for measurements with the NOAA WP3-D off the eastern United States as part of winter WATOX in January 1986 (Table 1). Concentrations in the boundary layer were 0.05 ppbv, increasing with altitude to 0.2 ppbv at 3 km. The altitude gradient may be due to a shorter lifetime in the boundary layer, where aerosol scavenging (followed by oxidation of aqueous SO$_2$) cuts the H$_2$O$_2$ lifetime to less than one day. Photochemical model calculations show that there would be a maximum in OH, HO$_2$, and H$_2$O$_2$ at the 3- to 5-km level with O$_3$ and NO$_x$ as measured on WATOX-86. In contrast, on the summer WATOX campaign (eight flights in July 1988), Ray et al. (1990) measured mixing ratios of 1.0 ± 0.5 ppbv H$_2$O$_2$ with no vertical gradient up to 2.3 km, the upper limit of the flights. The lowest concentration profiles were made in air of marine (low O$_3$) or stratospheric (high O$_3$) origin. Of two flights with the highest H$_2$O$_2$ concentrations (2 ppbv), one had high concentrations of O$_3$, SO$_2$, NO$_y$, and H$_2$O believed to be associated with convection.

Continuous shipboard measurements of H$_2$O$_2$ and soluble organic peroxides (ROOH) show maxima in the late afternoon as the free radicals leading to peroxide formation buildup in daylight (Thompson et al., 1993). Surface deposition, aerosol, and cloud droplet scavenging probably contribute to some of the variability measured on the SAGA-3 cruise (see error bars in Fig. 14). Nonetheless, a photochemical model simulation with constant deposition and scavenging rates reproduces mean H$_2$O$_2$ and ROOH fairly well and captures their diurnal behavior.

There have been two recent reports of peroxide measurements at marine continental sites. Ayers et al. (1992) monitored peroxide (actually "total peroxide," presumed to consist of H$_2$O$_2$ and a small amount of CH$_3$OOH) along with O$_3$ at Cape Grim (Tasmania, 41 S, 145 E) from February 1991 to March 1992. The seasonal patterns of O$_3$ and H$_2$O$_2$ were opposite to one another, with a summertime O$_3$ minimum and H$_2$O$_2$ maximum. Ayers et al. (1992) attribute this

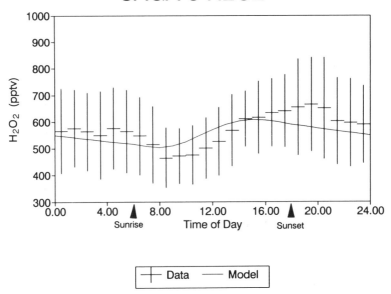

Figure 14. Surface peroxide measurements on SAGA-3 made with URI instrument. (From Thompson et al., 1993.)

to photochemical destruction of O_3 at 1.6 ppbv/day, giving minimum O_3 in late afternoon as H_2O_2 builds up. Total peroxide increases from an early morning minimum of 600 pptv to 1 ppbv by sunset, very similar to that observed on SAGA-3 (Fig. 14). The seasonality of O_3 at Cape Grim is largely due to reduced photochemical loss during the austral winter. Not only are O_3 concentrations higher, but the diurnal cycle of peroxide (as well as the mean mixing ratio) is much reduced.

Peroxide and organic peroxide measurements were also made as part of MLOPEX (Heikes, 1992). On average, the MLOPEX H_2O_2 mixing ratios are about twice as high as those for SAGA-3 and Cape Grim, although they are about half as great as those computed by a photochemical model (Liu et al., 1992; Ridley et al., 1992). Higher H_2O_2 concentrations were associated with upslope conditions (marine boundary-layer air, sometimes influenced by local pollution) and one episode of "photochemical haze," indicated by elevated hydrocarbon and NO_x levels. Lower H_2O_2 concentrations follow scavenging (and in-droplet reaction with SO_2) by clouds or precipitation. Photochemical models simulate observed mixing ratios of O_3, odd nitrogen species, and NMHC during MLOPEX quite well, but do a poor job of rationalizing peroxide observations, especially of ROOH (Heikes, 1992; Liu et al., 1992). Uncertainties in the formation and destruction rates assumed for CH_3OOH (the major component of ROOH) cannot account for the discrepancy. The H_2O_2:ROOH ratio for MLOPEX is $\sim 6{:}1$, although in 10 late afternoon periods it dropped to 2.5:1.

Different scavenging efficiencies of H_2O_2 and ROOH offer only a partial explanation for the ratio. Radical-radical permutation reactions (processes of the type $RO_2 + R'O_2$) have been suggested as competing with peroxide formation at low NO_x levels, reducing ROOH yields (Madronich and Calvert, 1990). On MLOPEX, mean NO_x was less than 40 pptv, but at the low NMHC concentrations measured at Mauna Loa, it is not clear that organoperoxy radical interactions would lead to a difference by a factor of 5 between computed and observed ROOH.

5. SUMMARY

The primary oxidants (O_3, OH, and H_2O_2) in the background (nonpolluted) marine troposphere have been described, including the photochemical transformations among them. Measurements of O_3 in the marine atmosphere from the surface, from aircraft, and from ships are comprehensive enough to give a good understanding of O_3 climatology. Seasonal, regional, and diurnal cycles of O_3 have been described and major patterns of dynamical and photochemical influences summarized. In a region of net photochemical destruction, for example, in the equatorial Pacific (where the O_3 mixing ratio has been measured at < 2 ppbv), extremely low O_3 may be related to convection or large-scale dynamical effects.

The hydroxyl radical has not been measured with any direct technique in the remote marine troposphere. On several recent campaigns in which sufficient measurements of other trace gases (notably O_3, NO, hydrocarbons, CO, H_2O) have been made to calculate OH from a model, there are greater differences among models than one would expect from model uncertainties alone.

Hydrogen peroxide measurements in the marine atmosphere have been made on a few campaigns. Vertical profiles show evidence of heterogeneous scavenging by aerosols and clouds; mixing ratios are frequently higher just above cloud level. The diurnal cycle of H_2O_2 and ROOH (soluble organic peroxides) in the remote marine boundary layer shows a buildup during the day that is to be expected from photochemistry and a loss at night, possibly due to surface deposition.

ACKNOWLEDGMENTS

Thanks to R. R. Dickerson, B. G. Doddridge, B. G. Heikes, and S. R. Piotrowicz for discussion of data and comments on the manuscript. This chapter was written as part of a NASA EOS Interdisciplinary Science Investigation on Air-Sea Fluxes of Biogeochemical Cycles (MBARI/ WHOI/GSFC).

REFERENCES

Atlas, E., Pollock, W., Greenberg, J., Heidt, L., and Thompson, A. (1993). Alkyl nitrates, nonmethane hydrocarbons and halocarbon gases over the Equatorial Pacific Ocean during SAGA-3. *J. Geophys. Res.* **98**, 16933–16948.

Ayers, G. P., Penkett, S. A., Gillett, R. W., Bandy, B., Galbally, I. E., Meyer, C. P., Ellsworth, C. M., Bentley, S. T., and Forgan, B. W. (1992). Evidence for photochemical control of ozone concentrations in unpolluted marine air. *Nature (London)* **360**, 446–448.

Barrie, L. A., Bottenheim, J. W., Schnell, R. C., Crutzen, P. J., and Rasmussen, R. A. (1988). Ozone destruction and photochemical reactions at polar sunrise in the lower Arctic troposphere. *Nature (London)* **334**, 138–141.

Bates, T. S., Kelly, K. C., and Johnson, J. E. (1993). Measurements of dissolved biogenic gases (DMS, CH_4, CO, CO_2) in the Equatorial Pacific during the SAGA-3 experiment. *J. Geophys. Res.* **98**, 16969–16978.

Beck, S. M. et al. (1987). Operational overview of NASA/GTE/CITE 1 airborne instrument intercomparisons: Carbon monoxide, nitric oxide and hydroxyl instrumentation. *J. Geophys. Res.* **92**, 1977–1986.

Berg, W. W., Sperry, P. O., Rahn, K. A., and Gladney, E. S. (1983). Atmospheric bromine in the Arctic. *J. Geophys. Res.* **88**, 6719–6736.

Carroll, M. A., et al. (1990). Aircraft measurements of NO_x over the eastern Pacific and continental United States and implications for ozone production. *J. Geophys. Res.* **95**, 10, 205–10, 233.

Chameides, W. L., and Davis, D. D. (1980). Iodine: Its possible role in tropospheric chemistry. *J. Geophys. Res.* **85**, 7383–7398.

Chameides, W. L., and Tan, A. (1981). The two-dimensional tropospheric model for OH: An uncertainty analysis. *J. Geophys. Res.* **86**, 5209–5223.

Chameides, W. L., Davis, D. D., Rodgers, M. O., Bradshaw, J., Sandholm, S., Sachse, G., Hill, G., Gregory, G., and Rasmussen, R. (1987a). Net ozone photochemical production over the eastern and central North Pacific as inferred from GTE/CITE 1 observations during Fall 1983. *J. Geophys. Res.* **92**, 2131—2152.

Chameides, W. L., Davis, D. D., Bradshaw, J., Rodgers, M., Sandholm, S., and Bai, D. B. (1987b). An estimate of the NO_x production rate in electrified clouds based on NO observations from the GTE/CITE 1 Fall 1983 field operation. *J. Geophys. Res.* **92**, 2153–2156.

Chameides, W. L., et al. (1990). Observed and model-calculated NO_2/NO ratios in tropospheric air sampled during the NASA GTE/CITE-2 field study. *J. Geophys. Res.* **95**, 10235–10247.

Cox, S. K., McDougal, D. S., Randall, D. A., and Schiffer, R. A. (1987). FIRE—The first ISCCP Regional Experiment. *Bull. Am. Meteorol. Soc.* **68**, 114–118.

Davis, D. D., Chameides, W. L., Bradshaw, J. D., Sandholm, S. T., Schendel, J., Sachse, G., Gregory, G. L., and Anderson, B. (1990). O_3 photochemical tendency in the tropical south Atlantic as determined from the NASA/GTE CITE-3 mission. *Eos, Trans. Am. AGU* **71**, 1258.

Doddridge, B. G., Dickerson, R. R., Spain, T. G., Oltmans, S. J., and Novelli, R. C. (1994). Carbon monoxide measurements at Mace Head, Ireland. In R. D. Hudson (Ed.), *Ozone in the Troposphere and Stratosphere*, NASA Conf. Publ. NASA, Washington, D.C. (in press).

Donahue, N. M., and Prinn, R. G. (1990). Nonmethane hydrocarbon chemistry in the remote marine boundary layer. *J. Geophys. Res.* **95**, 18387–18412.

Donahue, N. M., and Prinn, R. G. (1993). In-situ nonmethane hydrocarbon measurements on SAGA-3. *J. Geophys. Res.* **98**, 16915–16932.

Fishman, J., Gregory, G. L., Sachse, G. W., Beck, S. M., and Hill, G. F. (1987). Vertical profiles of ozone, carbon monoxide, and dew-point temperature obtained during GTE/CITE-1, October–November 1983. *J. Geophys. Res.* **92**, 2083–2094.

Fishman, J., Fakhruzzaman, K., Cros, B., and Nganga, D. (1991). Identification of widespread pollution in the Southern Hemisphere deduced from satellite analyses. *Science* **252**, 1693–1696.

Gerhardt, P., Poppe, D., and Marenco, A. (1989). Ozone, CO and NO_x distribution in the troposphere during STRATOZ-III. In R. D. Bojkov and P. Fabian (Eds.), *Ozone in the Atmosphere*. A. Deepak Publ., Hampton, VA, pp. 467–470.

Graedel, T. E. (1979). The kinetic photochemistry of the marine troposphere. *J. Geophys. Res.* **84**, 273–286.

Heikes, B. G. (1992). Formaldehyde and hydroperoxides at Mauna Loa Observatory. *J. Geophys. Res.* **97**, 18001–18013.

Heikes, B. G., Walega, J. G., Kok, G. L., Lind, J. A., and Lazrus, A. L. (1988). Measurements of H_2O_2 during WATOX-86. *Global Biogeochem. Cycles* **2**, 57–61.

Hipskind, R. S., Gregory, G. L., Sachse, G. W., Hill, G. F., and Danielson, E. F. (1987). Correlations between ozone and carbon monoxide in the lower stratosphere, folded tropopause and maritime troposphere. *J. Geophys. Res.* **92**, 2121–2130.

Hofzumahaus, A., Dorn, H.-P., Callies, J., Platt, U., and Ehhalt, D. H. (1991). Tropospheric OH concentration measurements by laser long-path absorption spectroscopy. *Atmos. Environ.* **25A**, 2017–2022.

Hübler, G. et al. (1992). Total reactive oxidized nitrogen (NO_y) in the remote Pacific troposphere and its correlation with O_3 and CO: Mauna Loa Observatory Photochemical Experiment 1988. *J. Geophys. Res.* **97**, 10427–10447.

Huebert, B. J., Howell, S., Laj, P., Johnson, J. E., Bates, T. S., Quinn, P. K., Yegorov, V. I., Clarke, A. D., and Porter, J. N. (1993). Observations of the atmospheric sulfur cycle on SAGA-3. *J. Geophys. Res.* **98**, 16985–16996.

Johnson, J. E., Gammon, R. H., Larsen, J., Bates, T. S., Oltmans, S. J., and Farmer, J. C. (1990). Ozone in the marine boundary layer over the Pacific and Indian Oceans: Latitudinal gradients and diurnal cycles. *J. Geophys. Res.* **95**, 11847–11856.

Johnson, J. E., Koropalov, V. M., Pickering, K. E., Thompson, A. M., and Bond, N. (1993). SAGA-3 experiment: Overview and meteorological and oceanographic conditions. *J. Geophys. Res.* **98**, 16893–16908.

Kawa, S. R., and Pearson, R., Jr. (1989). Budgets from the Dynamics and Chemistry of Marine Stratocumulus (DYCOMS) experiment. *J. Geophys. Res.* **94**, 9809–9817.

Krishnamurti, T. N., Fuelberg, H. E., Sinha, M. C., Oosterhof, D., Bensman, E. L., and Kumar, V. B. (1993). The meteorological environment of the tropospheric ozone maximum over the tropical south Atlantic Ocean. *J. Geophys. Res.* **98**, 10621–10641.

Lenschow, D. H., Pearson, R., Jr., and Stankov, B. B. (1982). Measurements of ozone vertical flux to ocean and forest. *J. Geophys. Res.* **87**, 8833–8837.

Lenschow, D. H., Paluch, I. R., Bandy, A. R., Pearson, R., Jr., Kawa, S. R., Weaver, C. J., Huebert, B. J., Kay, J. G., Thornton, D. C., and Driedger, A. R., III (1988). Dynamics and Chemistry of Marine Stratocumulus Experiment (DYCOMS). *Bull. Am. Meteorol. Soc.* **69**, 1058–1067.

Liu, S. C., and Trainer, M. (1988). Responses of the tropospheric ozone and odd hydrogen radicals to column ozone changes. *J. Atmos. Chem.* **6**, 221–234.

Liu, S. C., McFarland, M., Kley, D., Zafiriou, O., and Huebert, B. (1983). Tropospheric NO_x and O_3 budgets in the Equatorial Pacific. *J. Geophys. Res.* **88**, 1360–1368.

Liu, S. C. et al. (1992). A study of the photochemistry and ozone budget during the Mauna Loa Observatory Photochemistry Experiment. *J. Geophys. Res.* **97**, 10463–10471.

Logan, J. A., Prather, M. J., Wofsy, S. C., and McElroy, M. B. (1981). Tropospheric chemistry: A global perspective. *J. Geophys. Res.* **86**, 7210–7254.

Madronich, S., and Calvert, J. G. (1990). Permutation reactions of organic peroxy radicals. *J. Geophys. Res.* **95**, 5697–5715.

McFarland, M., Kley, D., Drummond, J. W., Schmeltekopf, A. L., and Winkler, R. H. (1979). Nitric oxide measurements in the equatorial Pacific region. *Geophys. Res. Lett.* **6**, 605–608.

McNeal, R. J., Mugler, J. P., Jr., Harriss, R. C., and Hoell, J. M., Jr. (1983). NASA Global Tropospheric Experiment. *Eos, Trans. Am. AGU* **64**, 561–562.

Moody, J. L., Oltman, S., Levy, H., II, and Merrill, J. (1994). A chemical climatology of tropospheric ozone: Tudor Hill, Bermuda: 1988–1991. In R. D. Hudson (Ed.), *Ozone in the Troposphere and Stratosphere*, NASA Conf. Publ. NASA, Washington, D.C. (in press).

Mount, G. H., and Eisele, F. L. (1992). An intercomparison of tropospheric OH measurements at Fritz Peak Observatory, Colorado. *Science* **256**, 1187–1189.

National Oceanic and Atmospheric Administration/Climate Monitoring and Diagnostics Laboratory (NOAA/CMDL) (1990). Summary Report No. 19. NOAA/CMDL, Boulder, CO.

Oltmans, S. J., and Levy, H., II (1992). Seasonal cycles of surface ozone over the western North Atlantic. *Nature (London)* **358**, 392–394.

Oltmans, S. J., and Levy II, H. (1993). Surface ozone measurements from a global network. *Atmos. Environ.* (in press).

Oltmans, S. J., Sheridan, P. J., Schnell, R. C., Peterson, R. E., Winchester, J. W., Barrie, L. A., Kahl, J. D., and Komhyr, W. D. (1989a). Springtime surface ozone fluctuations at high arctic latitudes and their possible relationship to atmospheric bromine. In R. D. Bojkov and P. Fabian (Eds.), *Ozone in the Atmosphere*. A. Deepak Publ., Hampton, VA, pp. 498–501.

Oltmans, S. J., Komhyr, W. D., Franchois, P. R., and Matthews, W. A. (1989b). Tropospheric ozone: Variations from surface and ECC ozonesondes. In R. D. Bojkov and P. Fabian, (Eds.), *Ozone in the Atmosphere*. A. Deepak Publ., Hampton, VA, pp. 539–543.

Paluch, I. R., Lenschow, D. H., Hudson, J. G., and Pearson, R., Jr. (1992). Transport and mixing processes in the lower troposphere over the ocean. *J. Geophys. Res.* **97**, 7527–7542.

Pickering, K. E., Thompson, A. M., McNamara, D. P., Schoeberl, M. R., Newman, P. A., Lait, L. R., Justice, C. O., and Kendall, J. D. (1994). A trajectory modeling investigation of the biomass burning–tropical ozone relationship. In R. D. Hudson (Ed.), *Ozone in the Troposphere and Stratosphere*, NASA Conf. Publ. NASA, Washington, D.C. (in press).

Piotrowicz, S. R., Boran, D. A., and Fischer, C. K. (1986). Ozone in the boundary layer of the equatorial Pacific Ocean. *J. Geophys. Res.* **91**, 13113–13119.

Piotrowicz, S. R., Bezdek, H. F., Harvey, G. R., Springer-Young, M., and Hanson, K. J. (1991). On the ozone minimum over the Equatorial Pacific Ocean. *J. Geophys. Res.* **96**, 18679–18687.

Poppe, D., Drummond, J., Ehhalt, D. H., and Marenco, A. (1989). Tropospheric ozone budget during STRATOZ-III: A model study. In R. D. Bojkov and P. Fabian (Eds.), *Ozone in the Atmosphere*. A. Deepak Publ., Hampton, VA, pp. 565–567.

Ray, J. D., Valin, C. C., Luria, M., and Boatman, J. F. (1990). Oxidants in the marine troposphere: H_2O_2 and O_3 over the western Atlantic Ocean, 1990. *Global Biogeochem. Cycles* **4**, 202–214.

Ridley, B. A., Madronich, S., Chatfield, R. B., Walega, J. G., Shetter, R. E., Carroll, M. A., and Montzka, D. D. (1992). Measurements and model simulations of the photostationary state during the Mauna Loa Observatory Experiment: Implications for radical concentrations and ozone production and loss rates. *J. Geophys. Res.* **97**, 10375–10388.

Routhier, F., Dennett, R., Davis, D. D., Wartburg, A., Haagenson, P., and Delany, A. C. (1980). Free tropospheric and boundary-layer airborne measurements of ozone over the latitude of range of 58°S and 70°N. *J. Geophys. Res.* **87**, 7307–7321.

Schnell, R. C., et al. (1991). Decrease of summer tropospheric ozone concentrations in Antarctica. *Nature (London)* **351**, 726–729.

Shiotani, M. (1992). Annual, quasi-biennial and El Niño–Southern Oscillation (ENSO) time-scale variations in Equatorial total ozone. *J. Geophys. Res.* **97**, 7625–7634.

Smit, H. G. J., Kley, D., McKeen, S., Volz, A., and Gilge, S. (1989). The latitudinal and vertical distribution of tropospheric ozone over the Atlantic Ocean in the southern and northern hemispheres. In R. D. Bojkov and P. Fabian (Eds.), *Ozone in the Atmosphere*. A. Deepak Publ., Hampton, VA, pp. 419–422.

Stallard, R. F., Edmond, J. M., and Newell, R. E. (1975). Surface ozone in the southeast Atlantic between Dakar and Walvis Bay. *Geophys. Res. Lett.* **2**, 289–292.

Stedman, D. H., and Jackson, J. O. (1975). The photostationary state in photochemical smog. *Int. J. Chem. Kinet., Symp.* **1**, 493–501.

Sturges, W. T., Cota, G. F., and Buckley, P. T. (1992). Bromoform emissions from Arctic ice algae. *Nature (London)* **358**, 660–662.

Thompson, A. M. (1992). The oxidizing capacity of the earth's atmosphere: Probable past and future changes. *Science* **256**, 1157–1165.

Thompson, A. M., and Cicerone, R. J. (1982). Clouds and wet removal as causes of variability in the trace-gas composition of the marine troposhere. *J. Geophys. Res.* **87**, 8811–8826.

Thompson, A. M., and Lenschow, D. H. (1984). Mean profiles of reactive trace gases in the unpolluted marine surface layer. *J. Geophys. Res.* **89**, 4788–4796.

Thompson, A. M., and Stewart, R. W. (1991). Effects of chemical kinetics uncertainties on calculated constituents in a tropospheric photochemical model. *J. Geophys. Res.* **96**, 13089–13108.

Thompson, A. M., Owens, M. A., and Stewart, R. W. (1989). Sensitivity of atmospheric hydrogen peroxide to global chemical and climate change. *Geophys. Res. Lett.* **16**, 53–56.

Thompson, A. M., Huntley, M. A., and Stewart, R. W. (1990a). Perturbations to tropospheric oxidants, 1985–2035: 1. Model calculations of ozone and OH in chemically coherent regions. *J. Geophys. Res.* **95**, 9829–9844.

Thompson, A. M., Esaias, W. E., and Iverson, R. L. (1990b). Two approaches to determining the sea-to-air flux of dimethyl sulfide: Satellite ocean color and a photochemical model with atmospheric measurements. *J. Geophys. Res.* **95**, 20551–20558.

Thompson, A. M., et al. (1993). Ozone observations and a model of marine boundary layer photochemistry during SAGA 3. *J. Geophys. Res.* **98**, 16955–16968.

Torres, A. L., and Thompson, A. M. (1993). Nitric oxide in the Equatorial Pacific boundary layer: SAGA-3 measurements. *J. Geophys. Res.* **98**, 16949–16954.

Winkler, P. (1988). Surface ozone over the Atlantic Ocean. *J. Atmos. Chem.* **7**, 73–91.

Yvon, S., Saltzman, E. S., Koropalov, V. M., and Cooper, D. J. (1993). Atmospheric hydrogen sulfide over the Equatorial Pacific (SAGA-3). *J. Geophys. Res.* **98**, 16985–16996.

Zafiriou, O. C., McFarland, M., and Bromund, R. H. (1980). Nitric oxide in seawater. *Science* **207**, 637–639.

3

OXIDATION PROCESSES IN THE ATMOSPHERE AND THE ROLE OF HUMAN ACTIVITIES: OBSERVATIONS AND MODEL RESULTS

Paul J. Crutzen, Jos Lelieveld,* and Christoph Brühl

Division of Atmospheric Chemistry, Max-Planck-Institute for Chemistry, 55020 Mainz, Germany

*Present address: Air Quality Department, Wageningen University, 6700 EV Wageningen, The Netherlands.

Environmental Oxidants, Edited by Jerome O. Nriagu and Milagros S. Simmons.
ISBN 0–471–57928–9 © 1994 John Wiley & Sons, Inc.

1. INTRODUCTION

The atmosphere, containing O_2 with a volume mixing ratio as high as 20.95%, may appear to be in a highly oxidative state. However, the fact is that O_2 itself hardly reacts with any of the gases that are emitted into the atmosphere by natural and anthropogenic processes. In addition, with only a few exceptions, the solubility of these gases in water is so low that they are not removed from the atmosphere to any significant degree by precipitation. One of the most surprising chemical characteristics of the atmosphere is thus that the breakdown of most of its compounds requires preprocessing by minute quantities of highly reactive species, in the first place the hydroxyl radical (OH). Hydroxyl, the "detergent of the atmosphere," is present in the troposphere at an average volume mixing ratio of only about 3×10^{-14} (Prinn et al., 1992). The OH radicals are in turn formed by the action of solar ultraviolet radiation of wavelengths shorter than about 315 nm on ozone (O_3), producing electronically excited $O(^1D)$ atoms, which react with water vapor (Levy, 1971):

$$O_3 + h\nu \longrightarrow O(^1D) + O_2 \qquad (\lambda < 315\,nm) \tag{1}$$

$$O(^1D) + H_2O \longrightarrow 2OH \tag{2}$$

Actually, OH transforms gases such as carbon monoxide (CO), methane (CH_4), and other higher hydrocarbons, as well as halogenated and sulfurated hydrocarbons, $NO_x (= NO + NO_2)$, and sulfur dioxide (SO_2), into soluble species that can be removed from the atmosphere by wet and dry deposition processes. Obviously, through the production of OH, tropospheric ozone in conjunction with solar UV radiation plays an essential role in cleansing the atmosphere of the huge amounts and great variety of gases that enter into it from the biosphere and "anthroposphere," despite the fact that both O_3 and UV are damaging to life on Earth. For example, UV radiation can cause skin cancer and reduce plant productivity [Scientific Committee on Problems of the Environment (SCOPE), 1992], and O_3 is phytotoxic and can damage lung tissue (Reich and Amundson, 1985; Lippmann, 1989). Note that O_3 in the troposphere, up to altitudes of 10 to 15 km, is most important in maintaining the atmospheric oxidation efficiency, whereas stratospheric O_3 (10–50 km) traps much of the biologically harmful UV-B radiation and reduces the oxidation efficiency of the atmosphere. As a result of a variety of human activities, there have been significant trends toward diminishing stratospheric ozone contents worldwide, and especially in the Southern Hemisphere (Stolarski et al., 1991). Table 1 shows some calculated changes in UV-B radiation at the earth's surface, weighted with a biological spectral response function, for springtime at about 30° N latitude, after prescribing reductions of stratospheric ozone in the model. These reductions, amounting to total-column ozone changes of -5% and -10%, may be conceived as crude approximations of stratospheric ozone decreases measured during the 1980s and those expected by the end of this century, respectively. The last two columns show

Table 1 Biologically Weighted[a] UV-B Irradiance and Photolysis Frequencies for the Formation of O(^{1}D) from O$_3$ at the Earth's Surface as a Function of Total Ozone[b]

Total ozone, DU[c]	UV-B (280–320 nm) 10^{-2} W/m^2		J[O$_3$ → O(^{1}D)] sec^{-1}	
	Noon	24 hr Average	Noon	24 hr Average
360	4.32	0.95	4.1×10^{-5}	1.3×10^{-5}
Percent change: − 5	10.6	10.8	8.0	8.4
Percent change: − 10	22.8	23.3	16.9	17.8

[a]DNA damage action spectrum; National Research Council (NRC), 1982.
[b]Calculations were done for about 30° N latitude at the beginning of May, assuming a surface albedo of 30% and a cloudless atmosphere. In both perturbation experiments, ozone was reduced in the lower and the upper stratosphere, approximating observed changes in the vertical profile. Solar zenith angle for noon is 15°.
[c]Dobson units.

the changes in the photolysis rates of O$_3$, leading to enhanced OH formation through the reaction of O(^{1}D) with water vapor [Reactions (1) and (2)]. For 5% and 10% total ozone reductions, our model predicts mean O(^{1}D) and OH production increases of about 8% and 18%, respectively. The corresponding increases in spectrally weighted UV radiation are about 11% and 23%. In both cases, about a doubling of the effects in relation to the initial ozone depletions is noted.

Tropospheric ozone concentrations in the Northern Hemisphere have been steadily increasing throughout the industrial era (e.g., Bojkov, 1988; Volz and Kley, 1988; Crutzen, 1988; Logan, 1989). This leads to enhanced OH *production*, because of enhanced solar ultraviolet photon fluxes as well as higher tropospheric ozone concentrations, but not necessarily to higher OH *concentrations*. The reason for this apparent discrepancy is that the main gases with which OH reacts in most of the troposphere, CO and CH$_4$, likewise show increasing concentrations in the atmosphere: CO in the Northern Hemisphere by 0.7 to 1.1% per year and CH$_4$ by 0.5 to 0.8% per year worldwide [Intergovernmental Panel on Climate Change (IPCC), 1992; Zander et al., 1989a, b]. No definite trends have been reported for CO at stations in the Southern Hemisphere (e.g., Dianov-Klokov et al., 1989). The upward trends in CH$_4$ have slowed from about 1.5% per year at the beginning of the 1960s to about 0.5 to 0.8% at the end of the 1980s [World Meteorological Organization (WMO), 1992, and references therein]. Thus, although the tropospheric production of OH radicals has increased, a considerable fraction (which can actually be more than 100%) is being used up to counteract increasing emissions of CO and CH$_4$, possibly even allowing for an overall decline in tropospheric OH concentrations and thus the oxidizing efficiency of the atmosphere. Other emissions, such as NO$_x$, enter into this balance as well. Ozone is thus a pivotal species in tropospheric chemistry and key questions are whether and where ozone is increasing or decreasing owing to human activities. These will be the central issues of this review paper.

2. OTHER OXIDANTS BESIDES OH

Although measurements and modeling studies point to gaseous hydroxyl as the dominant oxidizing agent initiating the degradation of most primary emittants in the atmosphere, the potential role of other highly reactive compounds, such as O_3, Cl, and NO_3 in the gas phase and OH, H_2O_2, and O_3 in the aqueous phase, has to be considered as well. During nighttime conditions, when significant concentrations of NO_3 may be formed in NO_x-polluted air masses by the reaction $NO_2 + O_3 \rightarrow NO_3 + O_2$, this radical may initiate oxidation of organic sulfur compounds, especially DMS (CH_3SCH_3), and of olefins and aldehydes (DeMore et al., 1992; Atkinson et al., 1992; Jensen et al., 1992). However, because of the short one- to two-day lifetime of NO_x in the atmosphere, the role of NO_3 as an oxidizing species will be limited to regions near pollution sources and maybe near active thunderstorms where NO is produced by lightning discharges (e.g., Kowalczyk and Bauer, 1982); from a global perspective, however, NO_3 is most likely not a significant oxidizer. The same probably applies to the role of chlorine. The only known significant tropospheric gas-phase source of chlorine atoms, the reaction $OH + HCl \rightarrow Cl + H_2O$, could produce sufficient amounts of chlorine atoms in the marine boundary layer if volume mixing ratios of HCl were of the order of several ppbv (Singh and Kasting, 1988). However, it is difficult to comprehend how such large HCl volume mixing ratios could be sustained in an environment loaded with sea-salt aerosol originating from seawater with a typical pH value around 8. Indeed, shipboard in situ measurements of HCl by Harris et al. (1992), using tunable diode laser technology, have shown the volume mixing ratios of HCl to be below 0.2 ppbv, the limit of detection of the experimental system. The only conceivable possibility for substantial chlorine formation would be associated with heterogeneous reactions of N_2O_5 with NaCl or HCl in or on seasalt-containing aerosol via $N_2O_5 + Cl^- \rightarrow NO_3^- + ClNO_2$, followed by rapid photodissociation of $ClNO_2$ into Cl and NO_2 (Finlayson-Pitts et al., 1989; Livingston and Finlayson-Pitts, 1991). However, such activation reactions would be dependent on the presence of significant amounts of NO_x and thus likewise be restricted to coastal regions near pollution sources.

The remaining gas-phase oxidant on our list, ozone itself, reacts at significant rates with some organic compounds that are emitted by natural vegetation, such as isoprene (C_5H_8), terpenes ($C_{10}H_{16}$), and other similar organic compounds. It is quite conceivable that O_3 plays an important role as a scavenger of organic compounds above forests. The same reactions may in turn provide a significant sink for ozone in the boundary layer above tropical and temperate forests (Crutzen et al., 1985). Ozone also plays a role in the removal of NO_x from the atmosphere due to the nighttime formation of N_2O_5 and HNO_3 via

$$NO_2 + O_3 \longrightarrow NO_3 + O_2 \tag{3}$$

$$NO_3 + NO_2 \rightleftharpoons N_2O_5 \quad \text{(strongly temperature dependent)} \tag{4}$$

$$N_2O_5 + H_2O \ (aq) \longrightarrow 2HNO_3 \tag{5}$$

(Platt et al., 1981), the latter reaction taking place on cloud droplets or H_2O-containing aerosol. Finally, in the aqueous phase and maybe on aerosol, important oxidation reactions by O_3 and H_2O_2 take place via

$$HSO_3^- + H_2O_2 \longrightarrow SO_4^{2-} + H^+ + H_2O \tag{6}$$

$$SO_3^{2-} + O_3 \longrightarrow SO_4^{2-} + O_2 \tag{7}$$

(Chandler et al., 1988). Also in this case, the availability of H_2O_2 depends primarily on the production of OH via Reactions (1) and (2) and subsequent formation of H_2O_2 either through gas-phase reactions

$$OH + CO(+ O_2) \longrightarrow CO_2 + HO_2 \qquad (2 \times) \tag{8}$$

$$HO_2 + HO_2 \longrightarrow H_2O_2 + O_2 \tag{9}$$

followed by the dissolution of H_2O_2 in water droplets, or through its formation within the droplets after diffusion of gas-phase HO_2 into these and the subsequent aqueous-phase reactions:

$$HO_2 \rightleftharpoons H^+ + O_2^- \tag{10}$$

$$HO_2 + O_2^- + H^+ \longrightarrow H_2O_2 + O_2 \tag{11}$$

Note that the O_2^- formed in Reaction (10) can react with aqueous-phase O_3, which is significant as a sink of ozone, but it also regenerates OH radicals that in turn oxidize other dissolved gases (Lelieveld and Crutzen, 1990).

In this article, we will focus on the role of OH, the main oxidizing compound of the atmosphere, and those gases that determine its concentrations in most of the troposphere, in the first place O_3, but also CO, CH_4, NO_x, and related compounds. We will discuss what changes in the concentration distributions of these compounds have or may have occurred as a result of human activities. Actually, O_3, CH_4, and CO concentration increases are obvious from measurements (IPCC, 1990); those of OH, however, have to be derived from model calculations, such as those presented later on. We will review current knowledge about the tropospheric budget of O_3 and present our model estimates of preindustrial and potential future concentration distributions of O_3, CO, CH_4, and OH. Finally, we consider the potential effect of the loss of stratospheric ozone and the resulting increase in UV radiation on the production and concentrations of OH radicals, tropospheric O_3, CO, and CH_4.

3. TROPOSPHERIC OZONE: ITS DISTRIBUTION AND CHEMISTRY

Since the beginning of this century it has been known that by far the most atmospheric ozone is located in the stratosphere (Strutt, 1918), where it is produced by the photodissociation of O_2 by solar radiation of wavelengths shorter than

240 nm (Chapman, 1930). As this radiation does not penetrate into the troposphere, it was believed until about 20 years ago that the relatively small amount of ozone in the troposphere, about 10% of the total, had to be entirely of stratospheric origin. Although already in the 1950s, Haagen-Smit and Fox (1956) had demonstrated that ozone could be produced in the heavily polluted air masses of Los Angeles and environs, chemical processes as sources or sinks for tropospheric ozone of global significance were only considered in the early part of the 1970s, when it was realized that the troposphere is the scene of intense photochemical activity, with ozone being one of the main actors (Levy, 1971; Crutzen, 1973, 1974).

Although there is now little doubt about the important role of tropospheric ozone in the overall chemistry of the atmosphere, quantitative understanding of its global tropospheric distribution is still insufficient, especially with respect to the tropics and the Southern Hemisphere. We will next give a brief overview of current knowledge of the sources and sinks of tropospheric ozone, the "classical" ones being downward flux from the stratosphere and destruction at the earth's surface and the "modern" ones being photochemical production and loss.

3.1. The "Classical" Sources and Sinks of Tropospheric Ozone

Following the discovery that water-vapor mixing ratios in the stratosphere are generally extremely low, Brewer (1949) proposed a general circulation model of stratosphere–troposphere exchange with rising air motions and dehydration at the cold tropical tropopause, continuing upward transport at low latitudes in the stratosphere, and a compensating downward return flow into the troposphere at middle and high latitudes that reaches a maximum during the late-winter and spring months in either hemisphere and brings with it ozone-rich air from the stratosphere into the extratropical troposphere.

Although the transport model of Brewer (1949) is correct for the average flow, research over the past decades has clearly shown that both the upward and especially the downward motions of air between the stratosphere and troposphere are highly irregular. In fact, the average flow is nothing more than the small residual of large fluctuating motions. As shown by observations of the tropospheric concentrations of specific species of stratospheric origin, such as ozone and natural or nuclear bomb test-generated radioactive material, it is clear that the extratropical transfer of stratospheric air in the troposphere is highly episodic, connected with so-called tropopause folding events (Danielsen, 1968; Dütsch, 1978; Fabian and Pruchnewicz, 1977). Likewise, studies in the tropics have shown that rapid upward transport of air, accompanied by dehydration of the lower tropical stratosphere, occurs through "overshooting" turrets of thunderstorm clouds in the most humid, convective regions. A detailed, but still somewhat qualitative, model for this mechanism was developed by Danielsen (1982, 1993), and explains the extreme dryness of the stratosphere. The preferred regions for such upward transport are Micronesia, Southeast Asia, and the Bay of Bengal and India, the latter during the summer monsoons (Newell and Gould-

Stewart, 1981). However, it is still an open question how this small-scale transport in the lower, tropical stratosphere is connected to the main circulation of the stratosphere. Likewise, estimates of air mass transport in these most convective "stratospheric fountain" regions are uncertain (Robinson, 1980; Newell and Gould-Stewart, 1981).

Although the mechanism of tropopause folding in extratropical cyclones is relatively well understood, the estimation of net vertical air mass fluxes is uncertain, particularly in the Southern Hemisphere (Danielsen and Mohnen, 1977; Holton, 1990). Estimates of stratosphere–troposphere mass exchanges have been based on meteorological as well as tracer observations. Downward transport of stratospheric ozone in the extratropics has been estimated by simultaneous measurements and transport modeling of conservative quantities, such as radioactive elements, potential vorticity, and ozone around a number of tropopause folds, yielding information about mixing across the boundaries of the folds (Danielsen, 1968; Danielsen and Mohnen, 1977; Shapiro, 1980; Vaughan and Tuck, 1985; Danielsen et al., 1987). Measurements of ground-level deposition of nuclear bomb test-produced ^{90}Sr, together with measurements of its concentration and that of ozone in the center of tropopause folds, likewise were used by Danielsen to estimate the flow of stratospheric air into the troposphere. The annual average downward flux of ozone in the Northern Hemisphere derived with this method was almost 8×10^{10} molecules/cm^2/sec ("mcs" hereafter). Recently, Murphy et al. (1993), using information on the production of odd nitrogen (NO$_y$) in the stratosphere by reaction between O(^{1}D) and N$_2$O (whose distributions were derived from satellite observations) and the observed nearly constant ratios of the concentrations of NO$_y$ and O$_3$ of $2.5-4 \times 10^{-3}$ in the lower stratosphere, have estimated a global average downward flux of ozone of $1.9-6.4 \times 10^{10}$ mcs. Using the same approach and our own estimates of stratospheric NO$_y$ production (Crutzen and Schmailzl, 1983), we obtain a range of $3-8 \times 10^{10}$ mcs. As there are indications that the stratospheric influx into the troposphere in the Northern Hemisphere is larger than in the Southern Hemisphere, a value in the upper part of this range corresponds quite well with the previously quoted estimate of 8×10^{10} mcs in the Northern Hemisphere by Danielsen et al. (1987). Holton (1990) used atmospheric circulation statistics for 1958 to 1973, derived from an extensive global rawinsonde network (Oort, 1983) and a model for the residual vertical circulation, to compute downward air mass fluxes at the 100-mb level at extratropical latitudes. Holton's results confirm that mean mass transports from the stratosphere into the troposphere are considerably larger in the Northern than in the Southern Hemisphere, particularly owing to tropopause folding events during wintertime cyclonic storms.

Earlier calculations with general circulation models (GCMs) yielded only about $4-7 \times 10^{10}$ mcs in the Northern Hemisphere and $2-4 \times 10^{10}$ mcs in the Southern Hemisphere (Mahlman et al., 1980; Gidel and Shapiro, 1980; Allam and Tuck, 1984). However, as a result of the coarse grid resolution of these models, they may have been deficient in simulating the small-scale tropopause folding events in sufficient detail. To overcome this problem, an estimate of the exchange of stratospheric air and ozone into the Northern Hemisphere's troposphere has also been derived from detailed, mesoscale meteorological modeling of a tropopause folding event (Ebel et al., 1992). The O_3 flux of $1.2-1.5 \times 10^{11}$ mcs thus derived for the Northern Hemisphere is substantially higher than the previous estimates, indicating the need for further exploration. Additional significant transfer of ozone into the troposphere may occur in "cut-off lows," that is, stationary (blocking) cyclonic systems with low temperatures and pressures (Bamber et al., 1984; Ebel et al., 1992). Further discussions on this important, but quantitatively far from well-established, phenomenon can be found in Vaughan (1988).

Until the beginning of the 1970s, it was generally believed that destruction at the earth's surface constitutes the only significant mechanism of removal of ozone from the troposphere. In fact, the highly oxidative O_3 molecules react rapidly on many surfaces and are efficiently taken up by vegetation. On the other hand, because of ozone's relatively low solubility, its removal from the atmosphere at water surfaces, notably the oceans, appears to be relatively small (Galbally and Roy, 1980). The most recent estimates indicate that the average surface-loss terms may amount to about 17×10^{10} and 5×10^{10} mcs in the Northern and Southern Hemispheres, respectively (Galbally and Roy, 1980). However, these figures need to be updated, taking into account the considerable body of new knowledge on ozone uptake rates by vegetation that has accumulated during the past 12 years.

3.2. Photochemical Sinks and Sources

As noted, through the production of OH by Reactions (1) and (2), ozone is the main species to initiate the photochemistry of the troposphere and determine the removal rates of most primary emitted gases in the atmosphere. The same reactions, however, also provide a sink for tropospheric ozone, which can be estimated relatively accurately given available information on the global distributions of ozone and water vapor. Current calculations indicate a global average loss of tropospheric ozone of 8×10^{10} mcs, and a primary production of OH that is correspondingly twice as large.

Additional destruction of ozone occurs whenever O_3 reacts with the HO_2 radical, which is formed as an intermediate in the oxidation cycles of carbon monoxide and hydrocarbons, for example, in the case of CO via

$$CO + OH \longrightarrow H + CO_2 \tag{8}$$

$$H + O_2 + M \longrightarrow HO_2 + M \tag{12}$$

$$HO_2 + O_3 \longrightarrow OH + 2O_2 \tag{13}$$

$$Net: \quad CO + O_3 \longrightarrow CO_2 + O_2$$

leading to net loss of O_3, with OH and HO_2 radicals serving as catalysts. As about 70% of the OH radicals react with CO in the "background" atmosphere, this reaction sequence, if it were the only one causing CO oxidation to CO_2, would lead to a globally averaged O_3 loss of about 3×10^{11} mcs, a figure derived from the measured distributions of CO and the calculated distributions of OH (to be discussed later in this paper). Adding only the additional loss of O_3 through its photodissociation in the presence of H_2O, producing OH [Reactions (1) and (2)], the potential ozone loss would even add up to more than 4×10^{11} mcs, much more than the estimated import of ozone from the stratosphere. Clearly, in order to balance the budget, substantial ozone production must also take place in the troposphere. Direct photolysis of O_2 is excluded, because of the absence of the necessary energetic photons. Indirect dissociation of O_2 can, however, take place in the various oxidation chains of CO (Fishman and Crutzen, 1978) and hydrocarbons, with NO and NO_2 alongside OH and HO_2 acting as catalysts. In the simplest case of CO oxidation, the reaction sequence is as follows:

$$CO + OH \longrightarrow H + CO_2 \tag{8}$$

$$H + O_2 + M \longrightarrow HO_2 + M \tag{12}$$

$$HO_2 + NO \longrightarrow OH + NO_2 \tag{14}$$

$$NO_2 + h\nu \longrightarrow NO + O \quad (\lambda < 405\,\text{nm}) \tag{15}$$

$$O + O_2 + M \longrightarrow O_3 + M \tag{16}$$

$$Net: \quad CO + 2O_2 \longrightarrow CO_2 + O_3$$

As the rate coefficient of Reaction (14) is about 4000 times faster than that of Reaction (13) the ozone production cycle is more important than the destruction cycle if the NO to O_3 concentration ratio is larger than 2.5×10^{-3}, or at NO volume mixing ratios above 5 to 10 pptv near the earth's surface (1 pptv $= 10^{-12}$ by volume). In the case of CH_4 oxidation, the possible sequences of reactions become much more lengthy, the simplest one being:

$$CH_4 + OH \longrightarrow CH_3 + H_2O \tag{17}$$

$$CH_3 + O_2 + M \longrightarrow CH_3O_2 + M \tag{18}$$

$$CH_3O_2 + NO \longrightarrow CH_3O + NO_2 \tag{19}$$

$$CH_3O + O_2 \longrightarrow CH_2O + HO_2 \tag{20}$$

$$HO_2 + NO \longrightarrow OH + NO_2 \tag{14}$$

$$NO_2 + h\nu \longrightarrow NO + O \quad (2\times) \tag{15}$$

$$O + O_2 + M \longrightarrow O_3 + M \quad (2\times) \tag{16}$$

$$CH_2O + h\nu \longrightarrow CO + H_2 \tag{21}$$

$$Net: \quad CH_4 + 4O_2 \longrightarrow CO + H_2 + H_2O + 2O_3$$

Additional production of ozone occurs if one takes into account the other branches of CH_2O oxidation to CO

$$CH_2O + h\nu \longrightarrow H + CHO \tag{22}$$

$$H + O_2 + M \longrightarrow HO_2 + M \tag{12}$$

$$CH_2O + OH \longrightarrow H_2O + CHO \tag{23}$$

$$CHO + O_2 \longrightarrow CO + HO_2 \tag{24}$$

$$HO_2 + NO \longrightarrow OH + NO_2 \tag{14}$$

followed by Reactions (15) and (16).

Altogether, in the presence of sufficient NO, for each CO molecule oxidized to CO_2 and each CH_4 molecule oxidized to CO, there could be a net production of 1 and 2.7 ozone molecules, respectively, leading to a maximum potential, globally averaged ozone production of 6×10^{11} mcs, which is too much to provide a balanced tropospheric ozone budget. It is clear from these arguments that ozone is being both produced and destroyed in the troposphere, depending on the prevailing concentrations of the NO_x ($NO + NO_2$) catalysts. As NO emissions into the atmosphere are to a large degree influenced by human activities, especially in the Northern Hemisphere, an increase in tropospheric ozone in the affected regions of the globe must be expected. Indeed, reviews of available ozone observations in Europe clearly indicate substantially lower concentrations prior to the World War II than at present (Volz and Kley, 1988; Bojkov, 1988; Crutzen, 1988). Despite some initial experimental uncertainties, such as those due to SO_2 and NO_2 interferences (Low et al., 1992), additional recent records indicate average increases of about 1% per year during the past decades at various "background" Northern Hemisphere ozone monitoring stations, extending into the free troposphere; no significant trends have, however, been noted in the Southern Hemisphere (Logan, 1985; Oltmans et al., 1989; WMO, 1992; Bojkov, 1988). The impact of increasing anthropogenic emissions of NO_x and hydrocarbons on tropospheric ozone is now very clearly observed in many industrial and surrounding rural regions, with ground-level ozone volume mixing ratios often reaching hazardous levels above 100 ppbv during meteorologically stable conditions, especially during summer (see, e.g., Logan, 1989; Guicherit and van Dop, 1977). Estimates by Liu (1988) indicate that the global contribution of regional ozone production in such industrially polluted areas, involving oxidation of rapidly reacting hydrocarbons, could be about 30% of that resulting from the oxidation of CO and CH_4 in background air.

In addition to the episodic "high society" ozone production, continental-scale ozone production has also been observed in the developing world during the dry season as a consequence of the annual burning of some $2-5 \times 10^{15}$ g of biomass carbon, causing the atmospheric release of essentially the same ozone-forming ingredients, NO_x, CO, and hydrocarbons, as in industrial regions (Crutzen et al., 1985; Crutzen and Andreae, 1990; Kirchhoff and Rasmussen, 1990; Kirchhoff et al., 1989; Rudolph et al., 1992; Andreae et al., 1992; Reichle et al., 1986, 1990; Hao et al., 1990).

The nitrogen oxide pair $NO + NO_2$ (NO_x) thus plays a pivotal role in the ozone formation chemistry of the troposphere. Mainly because of its short atmospheric residence time of only a few days due to the nighttime heterogeneous Reactions (3) to (5) and the daytime reaction

$$OH + NO_2 (+ M) \longrightarrow HNO_3 (+ M) \qquad (25)$$

its atmospheric concentrations are very variable in space and time, strongly depending on meteorological conditions and the location of the source regions, thereby making the interpretation of the measurements and modeling of the NO_x distributions extremely difficult. Some of the first 3-D modeling studies of global NO_x and NO_y distributions have been reported (e.g., Kasibhatla et al., 1991; Levy and Moxim, 1987, 1989a; Crutzen and Zimmermann, 1991; Penner et al., 1991). The main problem with these efforts is that they are based on 3-D chemistry-transport models that have low spatial resolution, with horizontal grids on the order of several hundred to a thousand kilometers and average NO_x injection rates adopted for the entire grid box, while in reality the NO_x input is highly irregular, coming from many point, line, and other limited-area sources. This model-inherent spreading out of the NO_x source over large areas favors its global dispersion and leads to over-estimation of OH and O_3 production rates and concentrations (Kanakidou and Crutzen, 1993). This problem will partly be overcome by the introduction of 3-D models with much finer resolution, at least in regions of strong NO_x emissions. In fact, regional air pollution models that have been developed at some research institutes are already able to zoom in on parts of the globe, such as North America, Europe, and northeast Asia, and have been used rather successfully for acid deposition and ozone episode studies (e.g., McKeen et al., 1991a, b; J. S. Chang et al., 1987; Y. S. Chang et al., 1990; Shin and Carmichael, 1992; Ebel et al., 1989).

4. MODEL SIMULATIONS OF REGIONAL AND GLOBAL-SCALE PHOTOCHEMISTRY

Regional-scale acid deposition models have been applied in particular to show the importance of long-range transport of pollutants (Carmichael et al., 1991; Shin and Carmichael, 1992). Brost et al. (1988) expanded a regional acid deposition model, covering North America, western Europe, and the Atlantic, to study the continental anthropogenic influences on the Atlantic Ocean atmosphere. It was shown that cloud venting of pollutants from the boundary layer into the free troposphere is a significant factor in enhancing pollutant lifetime, contributing to long-range transport. McKeen et al. (1991a, b) employed a comprehensive 3-D model of the eastern United States and Canada to study the relative importance of nonmethane hydrocarbons (NMHC) and NO_x for rural O_3 production. It was concluded that in rural North America, abatement of anthropogenic ozone formation can best be achieved by NO_x emission reductions because in rural environments, O_3 formation is not NMHC-limited by anthropogenic emissions, as these are supplied in copious amounts by vegetation (Chameides et al., 1988; Trainer et al., 1987). However, NMHC emissions appear to be more critical in urban environments, where NO_x is very abundant. Levy and Moxim (1987) were the first to use a global meteorology model to predict pollutant transport and deposition of anthropogenic nitrogen oxides. They particularly emphasized long-range transport, for example from the United States and Canada into the Atlantic Ocean, considering the dry deposition process as a major atmospheric pollutant-removal mechanism. Levy and Moxim (1989a, b) were able to reproduce with their model nitric acid concentrations at Mauna Loa in Hawaii, showing that HNO_3 and other odd nitrogen enhancements that occur during certain periods could be ascribed to NO_x emissions in the United States. Importantly, this indicates that even at remote locations in the "background" atmosphere significant influences of anthropogenic activity can occur. Levy and Moxim's model simulations also showed the possibility of considerable anthropogenic influences from the Asian continent over much of the northern Pacific Ocean. Kasibhatla et al. (1991), expanding upon the above global model, confirmed the importance of peroxyacetylnitrate (PAN) as a carrier of NO_x (Crutzen, 1979; Singh and Salas, 1983). The thermally unstable PAN, which is formed from NMHC and NO_2, can be carried rapidly into higher, colder air layers by convection. At lower temperatures its lifetime increase manyfold, so that it can be transported over large distances. In descending, adiabatically warming air parcels, remote from the regions, the NO_2 can be set free again from PAN. However, the lifetime of NO_2 in the lower atmosphere is short, about one day, so that the NO_2 is rapidly converted into HNO_3. Hence, the global-scale importance of PAN on the ozone and OH budgets still needs considerable analysis (Kanakidou et al., 1991; Kasibhatla et al., 1993). In the chemical GCM by Penner et al. (1991) PAN is neglected.

This description of the difficulties connected with 3-D modeling of NO_x distributions may have given the false impression that improved model performance is only a question of achieving finer spatial resolution. However, increased knowledge of photochemical processes, especially those involving multiple phases, is also of great importance. For instance, Dentener and Crutzen (1993) showed that surface reactions of N_2O_5 with H_2O on sulfate and other wetted aerosol via Reactions (3) to (5), especially those that occur during nighttime and at lower temperatures, may lead to a major loss of NO_x, especially in the free troposphere at middle and high latitudes during the "winter half year". This significantly affects ozone and hydroxyl concentrations and spreads through transport even into subtropical regions. Thus, in future it will be of considerable importance to greatly increase our knowledge of the aerosol distribution, including its size distribution and chemical composition, through substantially enhanced observations and modeling. Also, the very complex cloud chemical processes are usually not considered in large-scale models, first of all because these models often do not produce "realistic" clouds. For example, precipitation is often parameterized through removal of water after a preset threshold relative humidity has been exceeded. Moreover, the modeling of additional rapid reaction sequences in the aqueous phase consumes large resources of computer time. Nevertheless, it was shown by Lelieveld and Crutzen (1990, 1991) that in clouds any potential photochemical formation of ozone ceases. Dissolution of the soluble HO_2 radicals in cloud droplets effectively prevents this radical from reacting with NO, which is very insoluble, thus interrupting the gas-phase ozone production reactions [(14) to (16)]. Furthermore, in clouds, dissolved HO_2 strongly enhances the liquid-phase equivalent to the gas-phase ozone destruction reaction [(13)] following its aqueous-phase dissociation:

$$HO_2 \rightleftharpoons H^+ + O_2^- \tag{10}$$

$$O_2^- + O_3\,(aq) \longrightarrow O_3^- + O_2 \tag{26}$$

$$O_3^- + H_2O \longrightarrow OH + OH^- + O_2 \tag{27}$$

This reaction sequence is rapid, despite the relatively low solubility of O_3. The preferred reaction path of the OH radical that is generated in Reaction (27) is with dissolved, hydrolyzed formaldehyde, which yields HO_2, re-entering the cycle through equilibrium-reaction (10). An additional consequence is that formaldehyde is destroyed, which could otherwise have contributed to gas-phase formation of CO, HO_2, and O_3 through Reactions (21) to (24) and (14) to (16) under cloud-free conditions. Furthermore, as mentioned earlier, during the night NO_2 and O_3 can form N_2O_5, which is efficiently converted into HNO_3 in clouds via Reactions (3) to (5). Subsequent removal of HNO_3 by dry and wet deposition is thus a significant sink of NO_x. These heterogeneous and aqueous-phase reactions thus lead to removal of photochemically active NO_x, favoring both reduced production and enhanced destruction of O_3 and OH.

4.1. Model Simulations of Past and Future Tropospheric Ozone Changes

Substantial increases have occurred in the atmospheric concentrations of CO_2, CH_4, N_2O, industrial chlorofluorocarbon gases, and emissions of NO at the earth's surface due to fossil-fuel and biomass burning activities. This creates the potential for significant increases in tropospheric ozone concentrations in the Northern Hemisphere, as well as substantial ozone depletions in the stratosphere. In this review we will only discuss the latter in connection with its effects on $O(^1D)$ and OH production via Reactions (1) and (2), resulting from the increased penetration of photochemically active, ultraviolet solar radiation into the troposphere.

Given the information available from ice-core analyses of past atmospheric concentrations of CH_4 and estimates on the natural sources of NO, it is interesting to try to model the distributions of O_3, OH, and CO and to compare these with those derived for present conditions. Such an exercise has been performed with the aid of our global, low-resolution, three-dimensional chemistry-transport model Moguntia, which is run with a horizontal resolution of $10°$ by longitude and latitude and a vertical resolution of 100 mb. The model considers the photochemistry of the background troposphere, driven by solar ultraviolet radiation and involving the gases O_3, H_2O, CH_4, CO, and NO_x, as well as the reaction products derived from these, as discussed earlier in this paper Transport processes are described by using monthly average winds with the addition of an eddy diffusion parameterization, based on standard deviations and correlations of the wind components (Oort, 1983; Zimmermann, 1984; Zimmermann et al., 1989). As this parameterization in particular does not take care of very small-scale, rapid vertical transport by convective thunderstorms or near-frontal zones, a special scheme was derived, as described by Feichter and Crutzen (1990).

To simulate current and future concentration changes of several trace gases, we have adopted the IPCC (1992) emission scenarios for the years 1990 and 2050 in our 3-D model (Table 2), realizing, however, that these anticipated emissions are based on uncertain economic and demographic projections (World Bank, 1991; United Nations, 1992) that cannot account for unexpected political events

Table 2   Emission Scenarios Adopted from IPCC (1992)

Emission Source	CO (Tg yr^{-1})			CH$_4$ (Tg yr^{-1})			NO$_x$ (Tg N yr^{-1})		
	1850	1990	2050	1850	1990	2050	1850	1990	2050
Energy use		303	511		91	140		25	53
Biomass burning	170	693	847	7	28	34	2	9	11
Vegetation	100	100	100						
Natural HC[a]	300	300	300						
Anthropogenic HC[b]	25	234	515						
Oceans	40	40	40						
Wildfires	30	30	30						
Soils[c]							3	3	3
Lightning[d]							5	5	5
Domestic ruminants				15	78	167			
Wild ruminants[e]				6	6	6			
Rice paddies				20	60	87			
Animal wastes				7	26	54			
Landfills					38	93			
Domestic sewage					25	47			
Natural wetlands				155	155	155			
Total:	665	1700	2343	210	507	783	10	42	72

[a] Natural hydrocarbons accounted for in the model through CO emissions (Logan et al., 1981) are not included in IPCC (1992).

[b] Anthropogenic hydrocarbons (IPCC, 1992) are also accounted for through CO, assuming that oxidation of each hydrocarbon molecule yields four CO molecules.

[c] The soil NO_x emissions from IPCC (1992) have been reduced by 50% to account for NO_x removal in the vicinity of the emission sources, particularly in forest ecosystems (Sanhueza, personal communication).

[d] The lightning NO_x source has been adopted from Crutzen and Zimmermann (1991).

[e] The fraction by enteric fermentation from wild ruminants has been adopted from Crutzen et al. (1986).

or technological advances [International Energy Agency (IEA), 1991; World Resources Institute, 1990; Organization for Economic Cooperation and Development (OECD), 1991; Food and Agriculture Organization (FAO), 1991]. Table 2 shows that the most important emission sources of CO, CH_4, higher hydrocarbons, and NO_x, both in terms of extent and expected growth, are energy use and biomass burning. The latter is believed to be a dominant source of anthropogenic ozone precursors in tropical and subtropical regions (Crutzen and Andreae, 1990). Other strongly increasing CH_4 emission sources derive from ruminants, animals waste and domestic sewage disposal, landfills, and, to a lesser extent, rice cultivation.

For the preindustrial atmospheric simulations (reference year 1850), we have reduced agriculture-related emissions by a factor of 4, roughly proportional to population changes. Moreover, industrial sources (i.e., energy use, landfills, and domestic sewage) have been omitted (Table 2). The results are given in Table 3 and Figure 1. Calculated ozone concentrations in the lower troposphere appear to agree well with O_3 record reconstructions from measurements performed during the last two decades of the nineteenth century in the outskirts of Paris, indicating an average ozone level of about 10 to 15 ppbv (Volz and Kley, 1988). Including anthropogenic emissions in the simulations causes strong O_3 increases throughout the troposphere (Fig. 2), particularly in the Northern Hemisphere, which agrees well with the observational record (Feister and Warmbt, 1987; Attmannspacher et al., 1984; Crutzen, 1988; Logan, 1989). Table 3 shows that anthropogenic activities may be held responsible for global tropospheric total ozone increases of about 50% and for CO and CH_4 increases of more than a factor 2, the latter in agreement with ice-core measurements (Etheridge et al., 1988; Pearman and Fraser, 1988). Our results also indicate that the global tropospheric OH loading has decreased by 5 to 10%, particularly through reductions in the tropics and Southern Hemisphere, which are partially offset by increases in the Northern Hemisphere. The former are mainly due to the increasing concentrations of CO and CH_4, the latter to increases in NO and O_3. The model reproduces the observed CO peaks in northern midlatitudes and the tropics (Dianov-Klokov et al., 1989) but tends to be high at high southern latitudes.

The model-"predicted" atmospheric trace-gas burdens for the year 2050 are also given in Table 3. A major increase is calculated for NO_x (exceeding the current level by more than a factor of 2), which not only enhances the photochemical ozone formation potential of the troposphere (Fig. 3), but also contributes to acidification of precipitation. Our calculations also indicate large increases in CO abundance, about 0.6% per year in the Southern Hemisphere and a rate almost two times higher, of about 1% per year, in the Northern Hemisphere. The predicted rates of increase of CO as well as of CH_4 (the latter about 0.6–0.7% per year) are in fair agreement with currently measured trends (IPCC, 1992). This is of course encouraging, but no guarantee that our predictions are close to the truth. The atmospheric oxidation efficiency, as determined by tropospheric OH concentrations, decreases further throughout the next 60-year

Table 3 Calculated Total Hemispheric and Global Trace-Gas Burdens in the Troposphere (Tg)

Species	Southern Hemisphere			Northern Hemisphere			Global		
	1850	1990	2050	1850	1990	2050	1850	1990	2050
NO_x	0.1	0.2	0.2	0.3	1.8	4.2	0.3	2.0	4.4
HNO_3	0.5	0.6	0.7	0.5	1.7	3.0	1.0	2.3	4.2
O_3	110	153	179	136	222	285	246	375	554
H_2O_2	0.5	1.1	1.5	0.5	1.0	1.4	1.0	2.1	2.9
CH_2O	0.1	0.2	0.3	0.1	0.3	0.4	0.2	0.5	0.7
OH	2.5×10^{-4}	2.0×10^{-4}	1.8×10^{-4}	1.6×10^{-4}	1.5×10^{-4}	1.4×10^{-4}	4.1×10^{-4}	3.5×10^{-4}	3.2×10^{-4}
HO_2	1.7×10^{-2}	2.5×10^{-2}	2.8×10^{-2}	1.4×10^{-2}	1.9×10^{-2}	2.1×10^{-2}	3.1×10^{-2}	4.4×10^{-2}	4.9×10^{-2}
CO	58	46	199	127	255	418	185	401	617
CH_4	923	2010	2810	949	2120	2960	1872	4130	5770

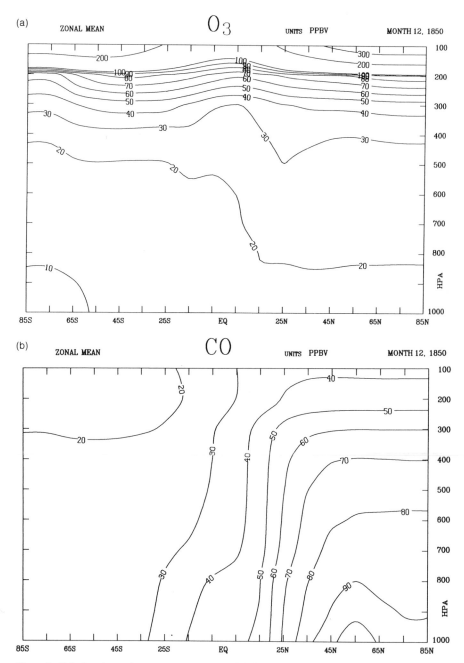

Figure 1. Calculated zonal average, pressure- and latitude-dependent O_3 (*a*) and CO (*b*) volume mixing ratios (ppbv) and daytime average OH (*c*) concentrations (10^6 molecules cm^{-3}) for December 1850, obtained by adopting the trace-gas emissions from the IPCC (1992) scenario for 1990 and reducing anthropogenic sources (see Table 2), in a global 3-D atmospheric chemistry-transport model.

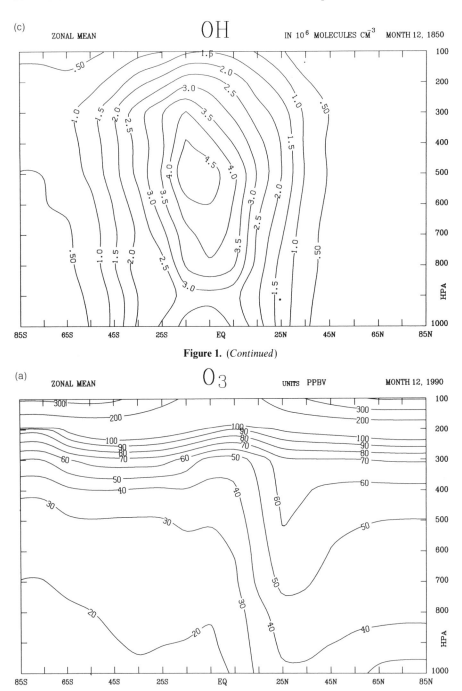

Figure 1. (*Continued*)

Figure 2. Calculated zonal average, pressure- and latitude-dependent O₃ (*a*) and CO (*b*) volume mixing ratios (ppbv) and daytime average OH (*c*) concentrations (10⁶ molecules cm⁻³) for December 1990, based on the IPCC (1992) emission scenario (Table 2).

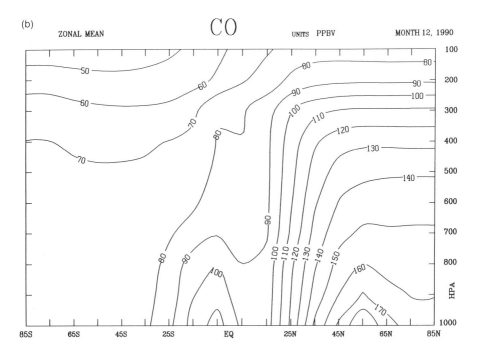

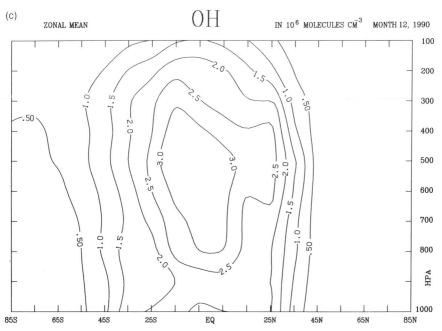

Figure 2. (*Continued*)

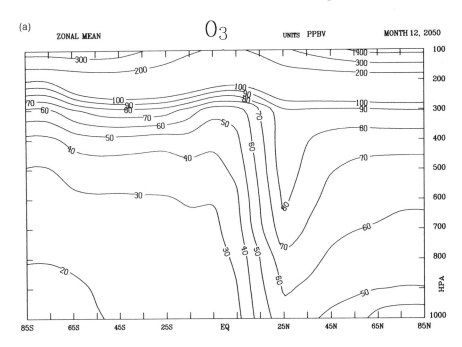

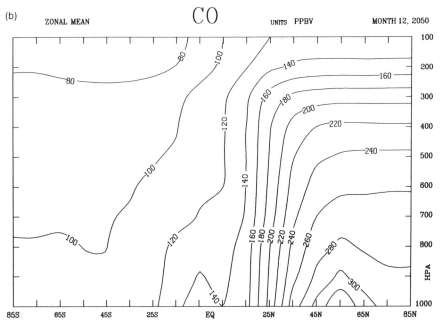

Figure 3. Model-predicted zonal average, pressure- and latitude-dependent O_3 (a) and CO (b) volume mixing ratios (ppbv) and daytime average OH (c) concentrations (10^6 molecules cm^{-3}) for December 2050, based on the IPCC (1992) emission scenario (Table 2).

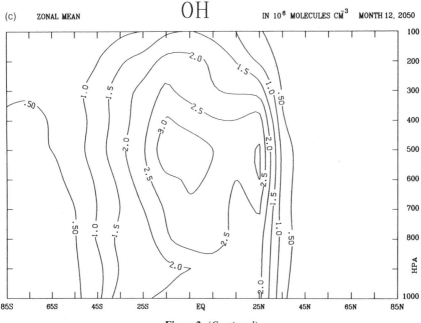

Figure 3. (*Continued*)

simulation, by about 8 to 9%. Actually, the calculated changes in tropospheric OH are rather small compared to the large anthropogenic perturbations of CH_4, CO, NO_x, and O_3. Clearly, the OH-reducing consequences of growing CO and CH_4 concentrations will be greatly lessened by the OH-increasing effect related to the upward trend of NO_x and O_3 concentrations (Thompson, 1992). Note, however, that these simulations did not include changes in UV radiation fluxes resulting from stratospheric ozone depletion, and water-vapor increases associated with climate warming. These will lead to enhanced OH formation. Hence, further model advancements, including stratospheric processes, are required to improve our assessments.

5. THE IMPORTANCE OF THE TROPICS IN ATMOSPHERIC CHEMISTRY

As the flux of solar ultraviolet radiation peaks in the tropics, because of minimum overhead ozone amounts and overall smaller solar zenith angles, the oxidizing power of the atmosphere, as determined by the reactivity of OH, reaches a maximum in the tropics and, during summertime, in the extratropics. The few direct observations of hydroxyl radical concentrations that are available (e.g., Dorn et al., 1988; Mount and Eisele, 1992) were not made in the background atmosphere and do not allow direct comparison with model results. No measurements have yet been made in background air. Because the modeling of short-lived gases like NO_x, which strongly affect OH concentrations, is extremely difficult, requiring the use of models with high spatial resolution, the correctness of the calculated OH distribution may be questioned. Fortunately, however, the calculated OH concentration distributions can be tested against global observations of methylchloroform (CH_3CCl_3), a gas that is mainly removed from the atmosphere by reactions with OH and that has only industrial emissions, which are known to within about 10%. Using this approach, the global, 24-hour average OH concentration is estimated at $8.1 \pm 0.9 \times 10^5$ molecules cm^{-3} (Prinn et al., 1987, 1992). The observed time series of methylchloroform mixing ratios in Ireland, Oregon, Barbados, Samoa, and Tasmania (Prinn et al., 1987) could generally be reproduced by our model within about 10%. However, recent experimental results by Talukdar et al. (1992) have yielded a 15% lower rate constant for the reaction of methylchloroform with OH, so that Prinn et al. (1992), as well as our model, may underestimate the global OH abundance. On the other hand, it may be that some CH_3CCl_3 is also lost from the atmosphere by uptake and hydrolysis in ocean waters (Butler et al., 1991). Altogether, however, we expect that our model approximately represents the main features of the current OH radical distribution. This is extremely useful, because it allows some rough estimation of the global distribution of the sinks of CH_4 and CO, which are removed from the atmosphere by reaction with OH. A particular advantage in this respect is that the temperature dependence of the reaction between CH_3CCl_3 and OH is very similar to that of the reaction between CH_4 and OH. Thus, with our global 3-D model we estimate that the reaction of CH_4 with OH (Vaghjiani and Ravishankara, 1991) removes about 410 Tg CH_4 per year. In addition, some 30 ± 25 Tg yr^{-1} CH_4 is destroyed in soils (Born et al., 1990), and about 30 Tg yr^{-1} of CH_4 is destroyed in the stratosphere, above the model domain, by reactions with OH and other radicals. Including the current annual increase of about 25 to 40 Tg CH_4 (Steele et al., 1987, 1992), the total CH_4 emissions to the atmosphere amount to 500 Tg yr^{-1}, similar to the estimate given by the IPCC (see Table 2). Although uncertain, Prinn et al. (1992) suggested the possibility that average OH concentrations may have increased during the past decade by as much as 1% per year, which would be more than needed to explain why the upward CH_4 trend during past decades of about 1% per year has decelerated to 0.5 to 0.8% per year since 1987 (Steele et al., 1992). However, our model simulations do not reproduce any such strong OH trend, which may partly be related to our neglect of stratospheric O_3 losses and the resulting tropospheric UV radiation increases.

Since the tropical regions contain the highest concentrations of OH (Fig. 3C), they contribute most strongly to the destruction of CO and CH_4 (see also Spivakovsky et al., 1990). Consequently, their production and release to the atmosphere also peak in the tropics. The most important tropical sources for CO are biomass burning and the oxidation of hydrocarbon gases, especially methane and isoprene (C_5H_8) emitted by vegetation (Zimmerman et al., 1978). Methane is mostly formed in the tropics by the decay of organic matter in the anaerobic sediments of natural wetlands and rice fields, in the rumen of cattle, and in landfills, and by biomass burning. At higher latitudes, CH_4 emissions from wetlands, cattle, landfills, fossil-fuel production sites, and leaks in natural gas distribution systems are most significant (Crutzen and Andreae, 1990; Crutzen et al., 1986; Cicerone and Oremland, 1988; Aselmann and Crutzen, 1989; Fung et al., 1991; IPCC, 1992). Considering that the tropics dominate the sinks and sources of CO and CH_4 and that the greatest industrial and agricultural developments will occur in these regions of the world, it is of the utmost importance that much study be devoted to tropical photochemistry.

6. RESPONSES OF TROPOSPHERIC PHOTOCHEMISTRY TO CHANGES IN UV AND HUMIDITY

Ground-based and satellite observations have provided ample evidence that in middle and high latitudes the total-column abundance of ozone is decreasing (Stolarski et al., 1991; WMO, 1992) despite ozone increases in many regions of the Northern Hemisphere's troposphere, which partially compensate for the stratospheric O_3 losses. The cause of the stratospheric ozone loss is emissions of industrial chlorofluorocarbons (CFC) and the photolytic breakdown of these gases in the stratosphere, leading to the formation of ozone-attacking ClO_x radicals (Molina and Rowland, 1974; WMO, 1992). As a consequence, it may be expected that the levels of UV-B and $O(^1D)$-producing solar radiation in the troposphere are increasing. However, measurements from an observational network in the United States did not indicate any significant trend in UV-B radiation (Scotto et al., 1988; Berger and Urbach, 1982). To explain this, Grant (1988) has suggested that in the vicinity of air pollution sources enhanced levels of NO_2, SO_2, and particulate material may cause additional UV-B absorption and scattering. An additional contributing factor to UV-B loss was identified by Brühl and Crutzen (1989), who argued that the observed increases in tropospheric ozone, particularly during summer in the Northern Hemisphere, may strongly compensate for the increase in UV-B radiation from stratospheric ozone loss. The mechanism for this is that short-wave radiation penetrating the troposphere is increasingly scattered by air molecules, cloud droplets, and aerosol particles. In high-sun conditions, this strongly enhances the path length of photons through the troposphere relative to the stratosphere, so that increases in tropospheric O_3 may cause a disproportionally strong additional UV-B absorption effect. In addition, Liu et al. (1991) have shown that anthropogenic increases in the concentration of aerosol particles can cause enhanced UV-B backscattering over polluted sections of the continents. Nevertheless, a few recent studies indicate that UV-B radiation fluxes have increased globally at the earth's surface, especially in the Southern Hemisphere (SCOPE, 1992; WMO, 1992; Seckmeyer and McKenzie, 1992).

Through the enhanced production of OH by Reactions (1) and (2), the overall photochemical activity of the troposphere is enhanced by increases in UV fluxes (see Table 1). It also depends on ambient levels of NO_x (Liu and Trainer, 1988). As explained earlier, at low-NO_x mixing ratios of, roughly 10 pptv or less, ozone destruction by Reaction (13) between HO_2 and O_3 is more important than O_3 production via the reaction [(14)] between NO and HO_2. In addition, Reactions (1) and (2) will lead to more O_3 destruction. Hence, increasing UV fluxes in low-NO_x environments enhance O_3 destruction, particularly in the lower part of the Southern Hemisphere's troposphere. In the Northern Hemisphere, growing NO_x emissions can enhance O_3 and OH formation. Thus, the influence of stratospheric ozone changes clearly needs to be incorporated into models describing the chemical composition of the atmosphere.

The same applies to the inclusion of tropospheric water-vapor increases due to anthropogenic climate warming. Climate modeling studies indicate that, regardless of possible temperature increases owing to anthropogenic greenhouse-gas emissions, the relative humidity in the troposphere remains on average approximately constant (Mitchell, 1989), implying larger water-vapor concentrations at higher temperatures. Because water vapor efficiently absorbs terrestrial infrared radiation, this internal climate system response significantly enhances the anthropogenically induced greenhouse effect, a positive feedback (Manabe and Wetherald, 1967). A 1-K temperature increase thus can lead to a 6% increase in the mean water-vapor concentration. Recent evaluation of satellite measurements have largely confirmed the water-vapor feedback (Raval and Ramanathan, 1989). Considering the likelihood of future anthropogenic temperature increases in the troposphere (IPCC, 1992), we anticipate that, through enhanced water-vapor concentrations, O_3 destruction and OH production may be enhanced via Reactions (1) and (2).

The H_2O and UV effects on OH tropospheric ozone and CH_4 trends have been demonstrated to be significant by time-dependent sensitivity runs with the coupled hemispherical 1D chemical climate model of Brühl and Crutzen (1988) for the IPCC scenario given above.

7. CONCLUSIONS

Observational and theoretical evidence for major anthropogenic perturbations of the chemical composition of the atmosphere on all scales is mounting. Analyses of air samples extracted from ice cores and some nineteenth century measurements suggest strong global increases of several radiatively and chemically active trace gases, such as CO_2, CH_4, and O_3, during industrial times. Most likely these trends will continue.

Applying a 3-D global chemistry-transport model and adopting the IPCC (1992) scenario for future trace-gas emissions, we calculated growth in the tropospheric abundance of CH_4 and O_3 by 40% and 25%, respectively, between 1990 and 2050. We also calculated significant reductions in OH concentrations near the earth's surface by 1 to 2% per decade. This rate of change may even accelerate after implementation of the London amendment to the Montreal protocol, precluding emissions of CFC gases after the year 2000, which would cause gradual recovery of stratospheric O_3 and decreased UV fluxes and OH formation in the troposphere. On the other hand, climate warming will most likely increase tropospheric water-vapor concentrations, resulting in enhanced OH formation. Actually, increasing tropospheric water-vapor concentrations, together with enhanced UV radiation owing to past stratospheric ozone losses, may have led to enhanced OH formation in the troposphere and may thus have contributed to the recently observed deceleration of the global CH_4 increase.

Some of the most important scientific problems to be solved have been identified in this paper. Perhaps the most serious problem is that we still lack adequate knowledge about the tropospheric distributions of such important gases as O_3 and CO, as well as the sources and sinks of these gases and of CH_4, NO, the hydrocarbons, etc. The situation is particularly severe in the tropics, that is in the region of the globe where the largest increases in trace-gas emissions will occur as a consequence of rapid population growth and requirements for a higher standard of living. Future research programs should thus be directed especially to improving our knowledge about the environmental chemistry in these regions of the world.

REFERENCES

Allam, R. J., and Tuck, A. F. (1984). Transport of water vapour in a stratosphere-troposphere general circulation model I. Fluxes. *Q. J. R. Meteorol. Soc.* **110**, 321–356.

Andreae, M. O., Chapuis, A., Cros, B., Fontan, J., Helas, G., Justice, C., Kaufman, Y.J., Minga, A., and Nganga, D. (1992). Ozone and Aitken nuclei over equatorial Africa: Air-borne observations during DECAFE 88. *J. Geophys. Res.* **97**, 6123-6136.

Aselmann, I., and Crutzen, P. J. (1989). Global distribution of natural freshwater wetlands and rice paddies, their net primary productivity, seasonality and possible methane emissions. *J. Atmos. Chem.* **8**, 307–358.

Atkinson, R., Baulch, D. L., Cox, R. A., Hampson, R. F., Kerr, J. A., and Troe, J. (1992). Evaluated kinetic and photochemical data for atmospheric chemistry: Supplement IV. *Atmos. Environ.* **26A**, 1187–1230.

Attmannspacher, W., Hartmannsgruber, R., and Lang, P. (1984). Langzeittendenzen des Ozons der Atmosphäre aufgrund der 1967 begonnenen Ozonmessreihen am Meteorologischen Observatorium Hohenpeissenberg. *Meteorol. Rundsch.* **37**, 193–199.

Bamber, D. J., Healey, P. G. W., Jones, B. M. R., Penkett, S. A., Tuck, A. F., and Vaughan, G. (1984). Vertical profiles of tropospheric gases: Chemical consequences of stratospheric infusions. *Atmos. Environ.* **18**, 1759–1768.

Berger, D. S., and Urbach, F. (1982). A climatology of sunburning ultraviolet radiation. *Photochemistry, Photobiology* **35**, 187–192.

Bojkov, R. D. (1988). Ozone changes at the surface and in the free troposphere. In I.S.A. Isaksen (Ed.), *Tropospheric Ozone.* Reidel Publ. Dordrecht, The Netherlands, pp. 83–96.

Born, M., Dörr, H., and Levine, I. (1990). Methane consumption in aerated soils of the temperate zone. *Tellus* **42B**, 2–8.

Brewer, A. W. (1949). Evidence for a worldwide circulation provided by the measurements of helium and water vapor in the stratosphere. *Q. J. R. Meteorol. Soc.* **75**, 351–363.

Brost, R. A., Chatfield, R. B., Greenberg, J. P., Haagenson, P. L., Heikes, B. G., Madronich, S., Ridley,

B. A., and Zimmermann, P. R. (1988). Three-dimensional modeling of transport of chemical species from continents to the Atlantic Ocean. *Tellus* **40B**, 358–379.

Brühl, C., and Crutzen, P. J. (1988). Scenarios of possible changes in atmospheric temperatures and ozone concentrations due to man's activities, estimated with a one-dimensional coupled photochemical climate model. *Climate Dynamics* **2**, 173–203.

Brühl, C., and Crutzen, P. J. (1989). On the disproportionate role of tropospheric ozone as a filter against solar UV-B radiation. *Geophys. Res. Lett.* **16**, 703–706.

Butler, J., Elkins, J., Thompson, T., Hall, B., Swanson, T., and Koropalov, V. (1991). Oceanic consumption of CH_3CCl_3: Implications for tropospheric OH. *J. Geophys. Res.* **96**, 22347–22355.

Carmichael, G. R., Peters, L. K., and Saylor, R. D. (1991). The STEM-II regional scale acid deposition and photochemical oxidant model. I. An overview of model development and applications. *Atmos. Environ.* **24A**, 2077–2090.

Chameides, W. L., Lindsay, R. W., Richardson, J., and Kiang, C. S. (1988). The role of biogenic hydrocarbons in urban photochemical smog: Atlanta as a case study. *Science* **241**, 1473–1474.

Chandler, A. S., Choularton, T. W., Dollard, G. J., Eggleton, A. E. J., Gay, M. J., Hill, T. A., Jones, B. M. R., Tyler, B. J., Bandy, B. J., and Penkett, S. A. (1988). Measurements of H_2O_2 and SO_2 in clouds and estimates of their reaction rate. *Nature (London)* **336**, 562–565.

Chang, J. S., Brost, R. A., Isaksen, I. S. A., Middleton, P., Stockwell, W. R., and Walcek C. J. (1987). RADM, a three-dimensional Eulerian acid deposition model. Part I: Physical concepts and model formulation. *J. Geophys. Res.* **92**, 14681–14700.

Chang, Y. S., Shin, W.-C., Carmichael, G. R., Kurita, H., and Ueda, H. (1990). Evaluation of the effect of emission reductions on pollutant levels in Central Japan. *Environ. Sci. Technol.* **24**, 1355–1366.

Chapman, S. (1930). A theory of upper-atmospheric ozone. *Mem. R. Meteorol. Soc.* **3**, 103–125.

Cicerone, R. J., and Oremland R. S. (1988). Biogeochemical aspects of atmospheric methane. *Global Biogeochem. Cycles* **2**, 299–328.

Crutzen, P. J. (1973). A dicussion of the chemistry of some minor constituents in the stratosphere and troposphere. *Pure Appl. Geophys.* **106–108**, 1385–1399.

Crutzen, P. J. (1974). Photochemical reactions initiated by and influencing ozone in unpolluted tropospheric air. *Tellus* **26**, 47–57.

Crutzen, P. J. (1979). The role of NO and NO_2 in the chemistry of the troposphere and stratosphere. *Annu. Rev. Earth Planet. Sci.* **7**, 443–472.

Crutzen, P. J. (1988). Tropospheric ozone: A review. In I.S.A. Isaksen (Ed.), *Tropospheric Ozone*. Reidel Publ., Dordrecht, The Netherlands, pp. 3–32.

Crutzen, P. J., and Andreae, M. O. (1990). Biomass burning in the tropics: Impact on atmospheric chemistry and biogeochemical cycles. *Science* **250**, 1669–1678.

Crutzen, P. J., and Schmailzl, U. (1983). Chemical budgets of the stratosphere. *Planet. Space Sci.* **31**, 1009–1032.

Crutzen, P. J., and Zimmermann, P. H. (1991). The changing photochemistry of the troposphere, *Tellus* **43A-B**, 136–151.

Crutzen, P. J., Delany, A. C., Greenberg, J., Haagenson, P., Heidt, L., Lueb, R., Pollock, W., Seiler, W., Wartburg A., and Zimmerman, P. (1985). Tropospheric chemical composition measurements in Brazil during the dry season. *J. Atmos. Chem.* **2**, 233–256.

Crutzen, P. J., Aselmann, I., and Seiler, W. (1986). Methane production by domestic animals, wild ruminants, other herbivorous fauna and humans. *Tellus* **38B**, 271–284.

Danielsen, E. F. (1968). Stratospheric-tropospheric exchange based on radioactivity, ozone and potential vorticity. *J. Atmos. Sci.* **25**, 502–518.

Danielsen, E. F. (1982). Statistics of cold cumulonimbus anvils based on enhanced infrared photographs. *Geophys. Res. Lett.* **9**, 601–604.

Danielsen, E. F. (1993). In situ evidence of rapid, vertical, irreversible transport of lower tropospheric air into the lower tropical stratosphere by convective cloud turrets and by larger scale upwelling in tropical cyclones. *J. Geophys. Res.* **98**, 8665–8682.

Danielsen, E. F., and Mohnen, V. A. (1977). Project Dustorm report: Ozone transport, in situ measurements and meteorological analyses of tropopause folding, *J. Geophys. Res.* **82**, 5867–5877.

Danielsen, E. F., Hipsking, R. S., Gaines, S. E., Sachse, G. W., Gregory, G. L., and Hill, G. F. (1987). Three-dimensional analysis of potential vorticity associated with tropopause folds and observed variations of ozone and carbon monoxide. *J. Geophys. Res.* **92**, 2103–2111.

DeMore, W. P., Sander, S. P., Golden, D. M., Hampson, R. F., Kurylo, M. J., Howard C. J., Ravishankara, A. R., Kolb, C. E., and Molina, M. J. (1992). *Chemical Kinetics and Photochemical Data for Use in Stratospheric Modeling*, Evaluation No. 10, JPL Publ. 92–20. NASA Jet Propulsion Lab., Pasadena, CA.

Dentener, F., and Crutzen, P. J. (1993). Reaction of N_2O_5 on tropospheric aerosols: Impacts on the global distributions of NO_x, O_3 and OH. *J. Geophys. Res.* **98**, 7149–7164.

Dianov-Klokov, V. I., Yurganov, L. N., Grechko, E. I., and Dzhola, A.V. (1989). Spectroscopic measurements of atmospheric carbon monoxide and methane. 1: Latitudinal distribution. *J. Atmos. Chem.* **8**, 139–151.

Dorn, H.-P., Callies, J., Platt, U., and Ehhalt, D. H. (1988). Measurements of tropospheric OH concentrations by long-path absorption spectroscopy. *Tellus* **40B**, 437–445.

Dütsch, H. U. (1978). Vertical ozone distribution on a global scale. *Pure Appl. Geophys.* **116**, 511–529.

Ebel, A., Neubauer, F. M., Raschke, E., and Speth, P. (1989). *Das EURAD-Modell: Aufbau und erste Ergebnisse*. No. 61. Mitteilungen aus dem Institut für Geophysik and Meteorologie der Universität zu Köln, Köln.

Ebel, A., Elbern, H., and Oberreuter, A. (1992). Stratosphere-troposphere air mass exchange and cross-tropopause fluxes of ozone. In E. V. Thrane, (Ed.), *Coupling Processes in the Lower and Middle Atmosphere*. Kluwer Academic Publishers, Dordrecht, The Netherlands.

Etheridge, D. M., Pearman, G. I., and de Silva, F. (1988). Atmospheric trace gas variations as revealed by air trapped in an ice core from Law Dome, Antarctica. *Ann. Glaciol.* **10**, 28–33.

Fabian, P., and Pruchniewicz, P. G. (1977). Meridional distribution of ozone in the troposphere and its seasonal variation. *J. Geophys. Res.* **82**, 2063–2073.

Feichter, J. and Crutzen, P. J. (1990). Parameterization of vertical tracer transport due to deep cumulus convection in a global transport model and its evaluation with [222]radon measurements. *Tellus* **42B**, 100–117.

Feister, U., and Warmbt, W. (1987). Long-term measurements of surface ozone in the German Democratic Republic. *J. Atmos. Chem.* **5**, 1–21.

Finlayson-Pitts, B. J., Ezell, M. J., and Pitts, J. M., Jr. (1989). Formation of chemically active chlorine compounds by reactions of atmospheric NaCl particles with gaseous N_2O_5 and $ClONO_2$. *Nature (London)* **337**, 241–244.

Fishman, J., and Crutzen, P. J. (1978). The origin of ozone in the troposphere. *Nature, (London)* **274**, 855–858.

Food and Agriculture Organization (FAO) (1991). *Forest Resources Assessment 1990 Project*, Forestry No. 7. FAO, Rome.

Fung, I., John, J., Lerner, J., Matthews, E., Prather, M., Steele, L. P., and Fraser, P. J. (1991). Three-dimensional model synthesis of the global methane cycle. *J. Geophys. Res.* **96**, 13033–13065.

Galbally, I. E., and Roy, C. R. (1980). Destruction of ozone at the earth's surface. *Q. J. R. Meteorol. Soc.* **106**, 599–620.

Gidel, L. T., and Shapiro, M. A. (1980). General circulation model estimates of the net vertical flux of ozone in the lower stratosphere and the implications for the tropospheric ozone budget. *J. Geophys. Res.* **85**, 4049–4058.

Grant, W. B. (1988). Global stratospheric ozone and UV-B radiation. *Science* **242**, 1111–1114.

Guicherit, R., and van Dop, H. (1977). Photochemical production of ozone in Western Europe (1971–1978) and its relation to meteorology. *Atmos. Environ.* **11**, 145–155.

Haagen-Smit, A. J., and Fox, M. M. (1956). Ozone formation in photochemical oxidation of organic substances. *Ind. Eng. Chem.* **48**, 1484–1501.

Hao W. M., Liu, M. H., and Crutzen, P. J. (1990). Estimates of annual and regional releases of CO_2 and other trace gases to the atmosphere from fires in the tropics, based on the FAO Statistics for the Period 1975–1980. In J. G. Goldammer (Ed.), *Fire in the Tropical Biota*. Springer-Verlag, Berlin, pp. 440–462.

Harris, G. W., Klemp, D., and Zenker, T. (1992). An upper limit on the HCl near-surface mixing-ratio over the Atlantic measured using TDLAS. *J. Atmos. Chem.* **15**, 327–332.

Holton, J. R. (1990). On the global exchange of mass between the stratosphere and troposphere. *J. Atmos. Sci.* **47**, 392–395.

Intergovernmental Panel on Climate Change (IPCC) (1990). *Climate Change, The IPCC Scientific Assessment*. WMO/UNEP, Cambridge University Press, Cambridge, UK.

Intergovernmental Panel on Climate Change (IPCC) (1992). *Climate Change. The Supplementary Report to the IPCC Scientific Assessment*. WMO/UNEP, Cambridge University Press, Cambridge, UK.

International Energy Agency (IEA) (1991). *Energy and Oil Outlook to 2005*. IEA, Paris.

Jensen, N. R., Hjorth, J., Lohse, C., Skov, H., and Restelli, G. (1992). Products and mechanisms of the gas phase reactions of NO_3 with CH_3SCH_3, CD_3SCD_3, CH_3SH and CH_3SSCH_3. *J. Atmos. Chem.* **14**, 95–108.

Kanakidou, M., and Crutzen, P. J. (1993). Scale problems in global tropospheric chemistry modeling: Comparison of results obtained with a three-dimensional model, adopting longitudinally uniform and varying emissions of NO_x and NMHC. *Chemosphere* **26**, 787–801.

Kanakidou, M., Singh, H. B., Valentin, K. M., and Crutzen, P. J. (1991). A two-dimensional study of ethane and propane oxidation in the troposphere. *J. Geophys. Res.* **96**, 15395–15413.

Kasibhatla, P. S., Levy, H., II, Moxim, W. J., and Chameides, W. L. (1991). The relative impact of stratospheric photochemical production on tropospheric NO_y levels: A model study. *J. Geophys. Res.* **96**(D10). 18631–18646.

Kasibhatla, P. S., Levy, H., II, and Moxim, W. J. (1993). Global NO_x, HNO_3, PAN, and NO_y distributions from fossil fuel combustion emissions: A model study. *J. Geophys. Res.* **98**, 7165–7180.

Kirchhoff, V. W. J. H., and Rasmussen, R. A. (1990). Time variations of CO and O_3 concentrations in a region subject to biomass burning. *J. Geophys. Res.* **95**, 7521–7532.

Kirchhoff, V. W. J. H., Setzer, A. W., and Pereira, M. C. (1989). Biomass burning in Amazonia. Seasonal effects on atmospheric O_3 and CO. *Geophys. Res. Lett.* **16**, 469–472.

Kowalczyk, M., and Bauer, E. (1982). *Lightning as a Source of NO in the Troposphere*, Rep. FAA-EE-82-4. U.S. Department of Transportation, Washington, DC.

Lelieveld, J., and Crutzen, P. J. (1990). Influences of cloud photochemical processes on tropospheric ozone. *Nature (London)* **343**, 227–233.

Lelieveld, J., and Crutzen, P. J. (1991). The role of clouds in tropospheric photochemistry. *J. Atmos. Chem.* **12**, 229–267.

Levy, H., II (1971). Normal atmosphere: Large radical and formaldehyde concentrations predicted. *Science* **173**, 141–143.

Levy, H., II, and Moxim, W. J. (1987). The fate of U.S. and Canadian combustion nitrogen emissions. *Nature (London)* **328**, 414–416.

Levy, H., II, and Moxim, W. J. (1989a). Simulated global distribution and deposition of reactive nitrogen emitted by fossil fuel combustion. *Tellus* **41B**, 256–271.

Levy, H., II, and Moxim, W. J. (1989b). Influence of Longrange transport of combustion emissions on the chemical variability of the background atmosphere. *Nature (London)* **338**, 326–328.

Lippmann, M. (1989). Health effects of ozone: A critical review. *JAPCA* **39**, 672–695.

Liu, S. C. (1988). Model studies of background ozone formation. In I.S.A. Isaksen (Ed.), In *Tropospheric Ozone*. Reidel Publ., Dordrecht, The Netherlands, pp. 303–318.

Liu, S.C., and Trainer, M. (1988). Responses of the tropospheric ozone and odd hydrogen radicals to column ozone change. *J. Atmos. Chem.* **6**, 221–234.

Liu, S. C., McKeen, S. A., and Madronich, S. (1991). Effect of anthropogenic aerosols on biologically active ultraviolet radiation. *Geophys. Res. Lett.* **8**, 2265–2268.

Livingston, F. E., and Finlayson-Pitts, B. J. (1991). The reaction of gaseous N_2O_5 with solid NaCl at 298K: Estimated lower limit to the reaction probability and its potential role in tropospheric and stratospheric chemistry. *Geophys. Res. Lett.* **18**, 17–20.

Logan, J. A. (1985). Tropospheric ozone: Seasonal behaviour, trends and anthropogenic influence, *J. Geophys. Res.* **90**(D6), 10463–10482.

Logan, J. A. (1989). Ozone in rural areas of the United States. *J. Geophys. Res.* **94**, 8511–8532.

Logan, J. A., Prather, M. J., Wofsy, S. C., and McElroy, M. B. (1981). Tropospheric chemistry: A global perspective. *J. Geophys. Res.* **86**, 7210–7254.

Low, P. S., Kelly, P. M., and Davies, T. D. (1992). Variations in surface ozone trends over Europe. *Geophys. Res. Lett.* **19**, 1117–1120.

Mahlman, J. O., Levy, H., II, and Moxim, W. J. (1980). Three-dimensional tracer structure and behaviour as simulated in two ozone precursor experiments. *J. Atmos. Sci.* **37**, 655–685.

Manabe, S., and Wetherald, R. T. (1967). Thermal equilibrium of the atmosphere with a given distribution of relative humidity. *J. Atmos. Sci.* **24**, 241–259.

McKeen, S.A., Hsie, E.-Y., Trainer, M., Tallamraju, R., and Liu, S. C. (1991a). A regional model study of the ozone budget in the eastern United States. *J. Geophys. Res.* **96**(D6), 10809–10845.

McKeen, S.A., Hsie, E.-Y., and Liu, S. C. (1991b). A study of the dependence of rural ozone on ozone precursors in the eastern United States. *J. Geophys. Res.* **96** (D8), 15377–15394.

Mitchell, J. F. B. (1989). The "greenhouse" effect and climate change. *Rev. Geophys.* **27**, 115–139.

Molina, M. J., and Rowland, F. S. (1974). Stratospheric sink for chlorofluoromethanes: Chlorine atom catalyzed destruction of ozone. *Nature (London)* **249**, 810–814.

Mount, G. H., and Eisele, F. L. (1992). An intercomparison of tropospheric OH measurements at Fritz Peak Observatory, Colorado. *Science* **256**, 1187–1190.

Murphy, D. M., Fahey, D. W., Proffitt, M. H., Liu, S. C., Chan, K. R., Eubank, C. S., Kawa, S. R., and Kelly, K. K. (1993). Reactive nitrogen and its correlation with ozone in the lower stratosphere and upper troposphere. *J. Geophys. Res.* **98**, 8751–8773.

National Research Council (NRC) (1982). *Causes and Effects of Stratospheric Ozone Reduction: An Update.* National Academy Press, Washington DC.

Newell, R. E., and Gould-Stewart, S. (1981). A stratospheric fountain? *J. Atmos. Sci.* **38**, 2789–2796.

Oltmans, S. J., Komhyr, W. D., Franchois, P. R., and Matthews, W. A. (1989). Tropospheric ozone: Variations from surface and ozonesonde observations. In R. Bojkov and P. Fabian (Eds.), *Ozone in the Atmosphere.* A. Deepak Publ., Hampton, VA, pp. 539–543.

Oort, A. H. (1983). *Global Atmospheric Statistics 1958–1973,* NOAA Prof. Pap., Vol. 14. National Oceanic and Atmospheric Administration, Washington, DC.

Organization for Economic Cooperation and Development (OECD) (1991). *Estimation of Greenhouse Gas Emissions and Sinks,* Final Report from the OECD Experts Meeting, Paris.

Pearman, G. I., and Fraser, P. J. (1988). Sources of increased methane. *Nature (London)* **332**, 489–490.

Penner, J. E., Atherton, C. S., Dignon, J., Ghan, S. J., and Walton, J. J. (1991). Tropospheric nitrogen: A three-dimensional study of sources, distributions, and deposition, *J. Geophys. Res.* **96**, 959–990.

Platt, U., Perner, D., Schröder, J., Kessler, C., and Toenissen, A. (1981). The diurnal variation of NO_3. *J. Geophys. Res.* **86**, 11965–11970.

Prinn, R., Cunnold, D., Rasmussen, R., Simmonds, P., Alyea, F., Crawford, A., Fraser, P., and Rosen, R. (1987). Atmospheric trends in methylchloroform and the global average for the hydroxyl radical. *Science* **238**, 945–950.

Prinn, R., Cunnold, D., Simmonds, P., Alyea, F., Boldi, R., Crawford, A., Fraser, P., Gutzler, D., Hartley, D., Rosen, R., and Rasmussen, R. (1992). Global average concentration and trend for hydroxyl radicals deduced from ALE/GAGE trichloroethane (methyl chloroform) data from 1978–1990. *J. Geophys. Res.* **97**, 2445–2462.

Raval, A., and Ramanathan, V. (1989). Observational determination of the greenhouse effect. *Nature (London)* **342**, 758–761.

Reich, P. B., and Amundson, R. G. (1985). Ambient level of ozone reduce net photosynthesis in tree and crop species. *Science* **230**, 566–570.

Reichle, H. G., Jr., Connors, V. S., Holland, J. A., Hypes, W. D., Wallio, H. A., Casas, J. C., Gormsen, B. B., Saylor, M. S., and Hesketh, W. D. (1986). Middle and upper tropospheric carbon monoxide mixing ratios as measured by a satellite-borne remote sensor during November 1981. *J. Geophys. Res.* **91**, 10865–10887.

Reichle, H. G., Jr., Connors, V. S., Holland, J. A., Sherrill, R. T., Wallio, H. A., Casas, J. C., Condon, E. P., Gormsen, B. B., and Seiler, W. (1990). The distribution of middle tropospheric carbon monoxide during early October 1984. *J. Geophys. Res.* **95**, 9845–9856.

Robinson, G. D. (1980). The transport of minor atmospheric constituents between troposphere and stratosphere. *Q. J. R. Meteorol. Soc.* **106**, 227–253.

Rudolph, J., Khedim A., and Bonsang, B. (1992). Light hydrocarbons in the tropospheric boundary layer over tropical Africa. *J. Geophys. Res.* **97**, 6181–6186.

Scientific Committee on Problems of the Environment (SCOPE) (1992). *Effects of Increased Ultraviolet Radiation on Biological Systems.* SCOPE/UNEP, Paris.

Scotto, J., Cotton, G., Urbach, F., Bergerand, D., and Fears, T. (1988). Biologically effective ultraviolet radiation surface measurements in the United States, 1974 to 1985. *Science* **239**, 762–764.

Seckmeyer, G., and McKenzie R. L. (1992). Increased ultraviolet radiation in New Zealand (45° S) relative to Germany (48° N). *Nature (London)* **359**, 135–137.

Shapiro, M. A. (1980). Turbulent mixing within tropopause folds as a mechanism for the exchange of chemical constituents between the stratosphere and troposphere. *J. Atmos. Sci.* **37**, 994–1004.

Shin, W.-C., and Carmichael, G. R. (1992). Sensitivity of acid production/deposition to emission reductions. *Environ. Sci. Technol.* **26**, 715–725.

Singh, H. B., and Kasting, J. F. (1988). Chlorine-hydrocarbon photochemistry in the marine troposphere and lower stratosphere. *J. Atmos. Chem.* **7**, 261–285.

Singh, H. B., and Salas, L. J. (1983). Peroxyacetylnitrate in the free troposphere. *Nature (London)* **302**, 326–328.

Spivakovsky, C. M., Yevich, R., Logan, J. A., Wofsy, S. C., and McElroy, M. B. (1990). Tropospheric OH in a three-dimensional chemical tracer model: An assessment based on observations of CH_3CCl_3. *J. Geophys. Res.* **95**, 18441–18471.

Steele, L. P., Fraser, P. J., Rasmussen, R. A., Khalil, M. A. K., Conway, T. J., Crawford, A. J., Gammon, R. H., Masarie, K. A., and Thoning, K. W. (1987). The global distribution of methane in the troposphere. *J. Atmos. Chem.* **5**, 125–171.

Steele, L. P., Dlugokencky, E. J., Lang, P. M., Tans, P. P., Martin, R. C., and Masarie, K. A. (1992). Slowing down of the global accumulation of atmospheric methane during the 1980s. *Nature (London)* **358**, 313–316.

Stolarski, R. S., Schoeberl, M. R., Newman, P. A., McPeters, R. D., and Krueger, A. J. (1991). Total ozone trends deduced from Nimbus-7 TOMS data. *Geophys. Res. Lett.* **18**, 1015–1018.

Strutt, R. J. (1918). Ultra-violet transparency of the lower atmosphere and its relative poverty in ozone. *Proc. R. Soc. London* **A94**, 260–268.

Talukdar, R. K., Abdelwakid, M., Schmoltner, A.-M., Watson, T., Montzka, S., and Ravishankara, A. R. (1992). Kinetics of the OH reaction with methyl chloroform and its atmospheric implications. *Science* **257**, 227–230.

Thompson, A. M. (1992). The oxidizing capacity of the earth's atmosphere: Probable past and future changes. *Science* **250**, 1157–1165.

Trainer, M., Williams, E. J., Parrish, D. D., Buhr, M. P., Allwine, E. J., Westberg, H. H., Fehsenfeld, F. C., and Liu, S. C. (1987). Models and observations of the impact of natural hydrocarbons on rural ozone. *Nature (London)* **329**, 705–707.

United Nations (1992). *Long-range World Population Projections*. United Nations, Population Division, New York.

Vaghjiani, G. L., and Ravishankara, A. R. (1991). New measurements of the rate coefficient for the reaction of OH with methane. *Nature (London)* **350**, 406–409.

Vaughan G. (1988). Stratosphere-troposphere exchange of ozone. In I.S.A. Isaksen (Ed.), *Tropospheric Ozone*. Reidel Publ., Dordrecht, The Netherlands, pp. 125–135.

Vaughan, G., and Tuck, A.F. (1985). Aircraft measurements near jet streams. In C.S. Zerefos and A. Ghazi (Eds.), *Atmospheric Ozone*. Reidel Publ., Dordrecht, The Netherlands, pp. 572–579.

Volz, A., and Kley, D. (1988). Evaluation of the Montsouris series of ozone measurements made in the nineteenth century. *Nature (London)* **332**, 240–242.

World Bank (1991). *World Development Report 1991*. Oxford University Press, New York.

World Meteorological Organization (WMO) (1992). *Scientific Assessment of Ozone Depletion: 1991*, Global Ozone Research and Monitoring Project, Report No, 25. WMO, Geneva.

World Resources Institute (1990). *World Resources 1990–91*. Oxford University Press, Oxford, UK.

Zander, R., Demoulin, P., Ehhalt, D. H., Schmidt, U., and Rinsland, C. P. (1989a). Secular increase of the total vertical column abundance of carbon monoxide above Central Europe since 1950. *J. Geophys. Res.* **94**, 11021–11028.

Zander, R., Demoulin, P., Ehhalt, D. H., and Schmidt, U. (1989b). Secular increase of the total vertical column abundance of methane derived from IR solar spectra recorded at the Jungfraujoch station. *J. Geophys. Res.* **94**, 11029–11039.

Zimmerman, P. R., Chatfield, R. B., Fishman, J., Crutzen, P. J., and Hanst, P. L. (1978). Estimates on the production of CO and H_2 from the oxidation of hydrocarbon emissions from vegetation. *Geophys. Res. Lett.* **5**, 679–682.

Zimmermann, P. H. (1984). Ein dreidimensionales numerisches Transportmodell für atmosphärische Spurenstoffe. Dissertation, University of Mainz, Germany.

Zimmermann, P. H., Feichter, J., Rath, H. K., Crutzen, P. J., and Weiss, W. (1989). A three-dimensional source-receptor model investigation using ^{85}Kr. *Atmos. Environ.* **23**, 25–35.

4

OZONE FORMATION IN URBAN PLUMES

I. Colbeck

Department of Chemistry and Biological Chemistry,
University of Essex, Colchester CO4 3SQ, United Kingdom

A.R. MacKenzie

Department of Chemistry, Centre for Atmospheric Science,
University of Cambridge, Cambridge CB2 1EW,
United Kingdom

Environmental Oxidants, Edited by Jerome O. Nriagu and Milagros S. Simmons.
ISBN 0–471–57928–9   © 1994 John Wiley & Sons, Inc.

1. INTRODUCTION

Ozone has long been recognized as a natural constituent of the atmosphere. The main sources for ozone in the troposphere, where about 10% of the atmospheric content is found, are influxes from the stratosphere and photochemical production involving nitrogen oxides and volatile organic substances, including carbon monoxide from natural and anthropogenic sources. Under particular meteorological conditions, ozone can be formed within the atmospheric boundary layer as a result of human activities. This phenomenon was first observed in Los Angeles in the 1940s (Haagen-Smit, 1952). It was initially assumed that such photochemical air pollution was unlikely to occur outside the United States, as the essential prerequisites, such as high temperatures, adequate sunlight, and ample emissions of precursors, were largely absent. Evidence of photochemical ozone in Europe was first reported in the early 1970s and has since been observed, on occasion, in many of the urban and industrial centers of the world [Photochemical Oxidants Review Group (PORG), 1987; Stevens, 1987; Chameides et al., 1992].

Photochemical air pollution is not merely a local problem. The spread of anthropogenic plumes away from urban centers results in regional-scale oxidant problems, such as these found in Europe and the northeastern United States (McKeen et al., 1991; Logan, 1985; Hov et al., 1986; Grennfelt et al., 1989). Ozone may be transported after its formation for distances up to 1000 km or more. Similarly, other oxidants (e.g., peroxyacetylnitrate, or PAN) may be transported over long distances. Ozone production is not confined to the plumes emanating from urban areas. It has also been connected with plumes from refineries, the petrochemical industry, and power plants, as well as from biomass burning in the tropics (Sexton and Westberg, 1983a).

The tracing of air masses back to a specific plume is difficult. The air mass is continuously modified due to mixing with pollutant emissions from ground sources and with other air parcels. Various techniques have been used, including constructing air mass trajectories (Lindsay and Chameides, 1988; Colbeck and Harrison, 1985b), airborne sampling (Clarke and Ching, 1983; Wakamatsu et al., 1983), and numerical modeling (MacKenzie et al., 1990; Hough and Derwent, 1987; Balko and Peters, 1983).

In this chapter we will concentrate primarily on urban plumes and describe the chemistry of ozone formation, summarize measurements of ozone and precursors in the plumes, and discuss control strategies.

2. TRANSPORT PROCESSES

Pollutants in the atmosphere are carried to other locations (advection), diluted to lower concentrations (diffusion), modified in their physicochemical form (conversion), and eventually removed at the earth's surface (deposition). The overall effect of these transport and transformation processes on a pollutant depends on

Table 1   Time and Space Scales for Air Pollution Transport

Category	Distances (km)	Time Scale
Local	0–10	1 hr–1 day
Urban	10–100	1 hr–1 day
Long-range		
Midrange	100–250	3 hr–1 day
Regional	200–1000	8 hr–4 days
Continental	1000–5000	1 day–2 weeks
Global	5000–400,000	2 weeks or more

the motions and other natural properties of the atmosphere, the presence of other pollutants, and the properties of the substance in question. All these processes affect the lifetime of the pollutant.

Pollution transport is affected not only by atmospheric motions and meteorological variables, which fluctuate over a wide spectrum of time and distance scales, but also by the emission patterns and rates of transformation and deposition of the pollutants. A number of different terms and values may be applied to the typical time and distance scales that characterize pollution transport, as shown in Table 1. Pollution transport may be divided into four types based on the distances involved: local, urban, long-range, and global. Of particular importance in this chapter is urban-scale transport.

3.   CHEMISTRY

There are two effects that urban plumes can have on ambient ozone levels: the near-instantaneous adjustment of the major constituents to a change in emission rates, and the dominating effect over time of far-field processes. Such far-field processes include the (nonlinear) degradative pathways of both anthropogenic and biogenic hydrocarbons, as well as the variation in chemical pathways over the diurnal cycle. We will deal first with the immediate effects of an air parcel entering a plume, and then discuss each of these far-field processes in turn.

3.1.   Near-Field Chemical Reequilibration and Ozone Production

The concentrations of some key species, as calculated by a reactive plume model (MacKenzie, 1989), are shown in Figure 1. On traversing an urban area, an air parcel is subject to a rapid change in pollutant inputs. This, in turn, changes

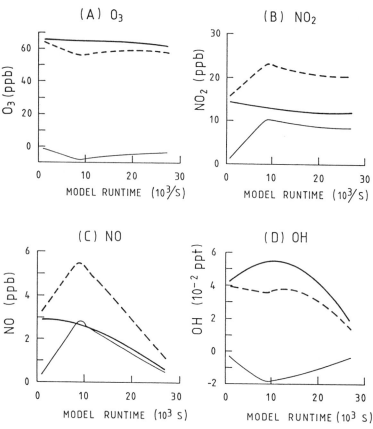

Figure 1. Base-case pollutant profiles from an expanding box model. Heavy solid line = rural backgrounds; broken line = plume concentration; light solid line = net concentration increment in plume. (A) Ozone; (B) nitrogen dioxide; (C) nitric oxide; (D) hydroxyl radical.

the rates of those reactions that govern the level of ozone present

$$NO_2 + h\nu \xrightarrow{\;O_2\;} NO + O_3 \qquad\qquad J_1 = 10^{-2} \qquad (R1)$$

$$NO + O_3 \longrightarrow NO_2 + O_2 \qquad\qquad k_2 = 10^{-14} \qquad (R2)$$

$$NO + RO_2 \longrightarrow NO_2 + HO_2 + R'CHO \quad k_3 = 10^{-11} \qquad (R3)$$

$$NO + HO_2 \longrightarrow NO_2 + OH \qquad\qquad k_4 = 10^{-11} \qquad (R4)$$

where RO_2 represents the generic organic peroxy radical, and order of magnitude reaction rate coefficients are given in cm^3 molecules sec units for midlatitude summer conditions at the earth's surface (Atkinson et al., 1992; Atkinson, 1990).

Peroxy radical self-reactions, cross-reactions, and reactions with ozone can be neglected in the high-NO environment of daytime urban plumes, although they can be important in the remote troposphere (Crutzen, 1988) and at night (see below).

Reactions (1) and (2) represent a very rapid null-cycle in ozone. Steady-state analysis yields the classical expression

$$[O_3] = \frac{J_1[NO_2]}{k_2[NO]} \tag{E1}$$

which dominates the determination of ozone levels in all environments. In Figure 1c the effect of increasing NO emissions is to decrease ozone concentrations (Fig. 1a) while the total amount of $NO_2 + O_3$ remains constant (i.e., stoichiometric titration). Ozone can only produced by the action of Reactions (3) and (4) changing the $[NO_2]:[NO]$ ratio without destroying ozone.

To see this effect we can expand the steady-state expression Equation (1) to take account of the fact that not all NO reacts with O_3. We then have:

$$[O_3]^2 - P_s[O_3] - P_s\left(\frac{k_4[HO_2] + k_3[RO_2]}{k_2}\right) = 0 \tag{E2}$$

where P_s is the right-hand side of Equation (1). The species HO_2 and RO_2 are not easily quantified, but we can replace them with reasonable steady-state expressions. First, HO_2 can be derived by balancing the source reactions (3) and

$$OH + O_3 \longrightarrow HO_2 + O_2 \qquad k_5 = 10^{-13} \tag{R5}$$

$$HCHO + h\nu \xrightarrow{\;2O_2\;} 2HO_2 + CO \qquad J_6 = 10^{-5} \tag{R6}$$

[Reaction (6) is the minor channel for formaldehyde photolysis: most is photolyzed to $CO + H_2$] with the sink Reaction (4). We are assuming that the reaction

$$HO_2 + O_3 \longrightarrow OH + 2O_2 \qquad k_7 = 10^{-15} \tag{R7}$$

is much slower than Reaction (4) in the urban atmosphere. This gives:

$$[HO_2] = \frac{k_3[RO_2]}{k_4} + \frac{J_6[HCHO]}{k_4[NO]} + \frac{k_5[OH][O_3]}{k_4[NO]} \tag{E3}$$

Hence :

$$[O_3]^2 - P_s\left(1 + \frac{k_5[OH]}{k_2[NO]}\right)[O_3] - P_s\left(\frac{2k_3[RO_2]}{k_2} + \frac{J_6[HCHO]}{k_2[NO]}\right) = 0 \tag{E4}$$

To develop this treatment further we must identify the sources and sinks of the recondite species RO_2. As will be discussed in following sections, the oxidation of emitted hydrocarbons is a multistep, nonlinear process. Here, for the sake of simplicity and noting that our present concern is with ozone production in the near-field, we can assume that the principal source of RO_2 is the reaction

$$HC_r + OH \xrightarrow{O_2} RO_2 + H_2O \qquad k_8 = 10^{-11} \qquad (R8)$$

where HC_r represents reactive hydrocarbons such as alkenes, aromatics, and oxygenates (Table 2). As written, Reaction (7) assumes that the major reaction pathway is hydrogen atom abstraction, though many hydrocarbons will in fact react by OH addition (see below). The net outcome of both pathways is the same, however, in terms of RO_2 production.

Table 2   Common Urban Hydrocarbons and Their Reactivities

Species	Conc., ppb C^a	$k_{OH} \times 10^{12b}$	$k_{NO_3} \times 10^{16b}$
Isopentane	45.3	3.9	1.6
n-Butane	40.3	2.54	0.65
Toluene	33.8	5.96	0.3
Propane	23.5	1.15	—
Ethane	23.3	0.268	0.01
n-Pentane	22.0	3.94	0.8
Ethylene	21.4	8.52	2.1
m, p-Xylene	18.1	23.6, 14.3	1, 3
2-Methylpentane	14.9	5.6	—
Isobutane	14.8	2.34	1
Ethyne	12.9	0.9	⩽ 0.2
Benzene	12.6	1.23	0.2
n-Hexane, 2 ethyl-1-butene	11.0	5.61	1.5
3-Methylpentane	10.7	5.7	—
1, 2, 4-Trimethylbenzene	10.6	32.5	15
Propylene	7.7	26.3	94
2-Methylhexane	7.3	—	—
o-Xylene	7.2	13.7	3
2, 2, 4-Trimethylpentane	6.8	3.68	—
Methylcyclopentane	6.4	—	—
3-Methylhexane	5.9	—	—
2-Methylpropene, 1-butene	5.9	51.4, 31.4	3100, 120
Ethylbenzene	5.9	7.1	6
m-Ethyltoluene	5.3	19.2	—
n-Heptane	4.7	7.15	1.4
α-Pinene	—	53.7	58000

^{a} From Seinfeld (1989). Concentrations of biogenic nonmethane hydrocarbons were not reported.
^{b} Rate constants are given in cm^3 molecules^{-1} sec^{-1} for 298 K and as the high-pressure limit where appropriate. Values are taken from Atkinson (1990) and Wayne et al. (1991).

The assumption that RO_2 is in steady state due to the balance of Reactions (3) and (8) allows us to substitute for it in Equation (4), giving:

$$[O_3]^2 - P_s[O_3]\left(1 + \frac{k_5[OH]}{k_2[NO]}\right) - P_s\left(\frac{2k_8[OH][HC_r] + J_6[HCHO]}{k_2[NO]}\right) = 0 \quad (E5)$$

This expression is informative because it demonstrates the dependence of ozone production on the $[HC_r]$:$[NO]$ ratio, a fact that is well documented (e.g., Seinfeld, 1989). An order of magnitude analysis of Equation (5) shows that, as we have already seen, the major factor controlling $[O_3]$ is P_s, but that, for ozone production to occur over and above the null cycling due to NO and NO_2, the ratio $[HC_r]$:$[NO]$ is crucial.

We can say more about this dependence by employing a simple steady-state approximation to the hydroxyl radical concentration (bearing in mind the caveats already noted for RO_2 and HO_2). The sources of OH in the urban plume will be Reactions (4) and (7), as well as

$$HONO + h\nu \longrightarrow OH + NO \qquad J_9 = 10^{-3} \quad (R9)$$

and

$$O_3 + h\nu \xrightarrow{H_2O} 2OH + O_2 \qquad J'_{10} = 10^{-6} \quad (R10)$$

where J'_{10} has been adjusted for that fraction of $O(^1D)$, produced from ozone photolysis, that goes on to react with H_2O (roughly 10%). The sinks for OH are Reactions (5) and (8), and

$$OH + NO_2 \xrightarrow{M} HNO_3 \qquad k_{11} = 10^{-11} \quad (R11)$$

so that our simple steady-state expression is:

$$[OH] = \frac{k_4[HO_2][NO] + k_7[HO_2][O_3] + J_9[HONO] + J'_{10}[O_3]}{k_5[O_3] + k_8[HC_r] + k_{11}[NO_2]} \quad (E6)$$

Such an expression is useful because it can give us further information on the behavior of the factor in Equation (5) that includes the $[HC_r]$: $[NO]$ ratio:

$$\frac{2k_8[OH][HC_r]}{k_2[NO]}$$

$$= \frac{2k_8}{k_2}\left(\frac{k_4[HO_2][NO] + J_9[HONO] + [O_3](k_7[HO_2] + J'_{10})}{k_5[O_3] + k_8[HC_r] + k_{11}[NO_2]}\right)\frac{[HC_r]}{[NO]} \quad (E7)$$

From this we can deduce the irreversible effect of hydrocarbons and NO_x ($NO + NO_2$) under different conditions. First, when $[HC_r] \gg [NO_2]$, $[O_3]$, the production of ozone will become independent of hydrocarbon concentrations; this is the *NO$_x$-limited scenario*. Conversely, when $[HC_r] \ll [NO_2]$, $[O_3]$, increasing NO$_x$ emissions will decrease the ozone production; this is the *HC-limited* or *NO-titration scenario*. Both these situations can be observed in urban plumes (see below).

The general behavior of urban ozone concentrations over a range of HC and NO$_x$ conditions is shown in Figure 2. This plot has been generated by using results from a numerical solution of a more complete chemistry than that discussed above, but the underlying behavior is the same. At the bottom right of the diagram, ozone concentrations become independent of hydrocarbon levels; at the top left, ozone concentrations vary inversely with NO$_x$ input. Maximum ozone concentrations result from a $[HC_r]$: $[NO_x]$ ratio of about 6.

3.2.   Anthropogenic Hydrocarbons

The general behavior of hydrocarbons in the urban atmosphere is shown schematically in Figure 3. This figure improves on the chemistry discussed above by including a feedback loop for the multistep hydrocarbon oxidation. An additional dimension would add an intersecting feedback loop for OH and HO$_2$, since they are consumed by reaction with organic moieties [Reaction(8)] but are reproduced by the reaction of RO$_2$ with NO [Reaction(3)] and by the photolysis of RCHO:

$$R_{c=n}CHO + h\nu \xrightarrow{\ O_2\ } R_{c=n-1}O_2 + HO_2 + CO \qquad J_{12} = 10^{-5} \quad (R12)$$

Table 2 shows a recent estimate of the 25 most common hydrocarbon compounds in the urban environment (Seinfeld, 1989). Most of the carbon emitted is in the form of straight-chain or branched alkanes; the rest is emitted as aromatics, alkenes, and ethyne. The first step in alkane degradation is hydrogen atom abstraction by OH [Reaction(8)]. For example:

$$H_3C\,CH(CH_3)CH_2CH_3 + OH \xrightarrow{\ O_2\ } H_3CC(O_2)(CH_3)CH_2CH_3 + H_2O$$

$$k_{13} = 10^{-12} \quad (R13)$$

The major route for isopentane degradation, shown above, is attack at the hydrogen atom attached to the tertiary carbon atom (Atkinson, 1990), in accordance with the radical-stabilizing effect of alkyl groups. Further reaction of the alkyl peroxy radical is via Reaction(3), although the formation of organic nitrates (RONO$_2$) can become a minor, and temporary, chain-termination step for long-chain hydrocarbons (Roberts, 1990).

A more important chain-terminating reaction, and NO$_x$ sink, is the formation of peroxycarboxylic nitric acid anhydrides. The action of OH on organic alde-

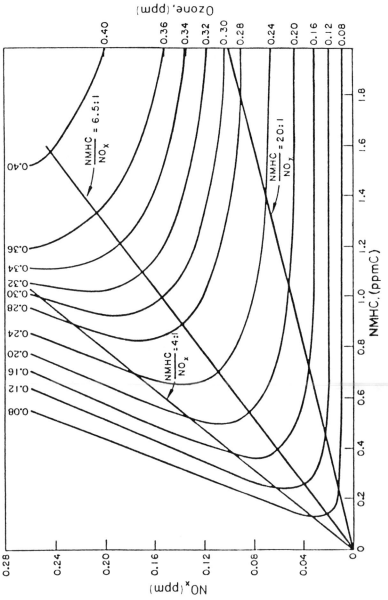

Figure 2. Ozone-isopleth diagram showing peak ozone concentration as a function of initial nonmethane hydrocarbon and NO_x concentrations.

103

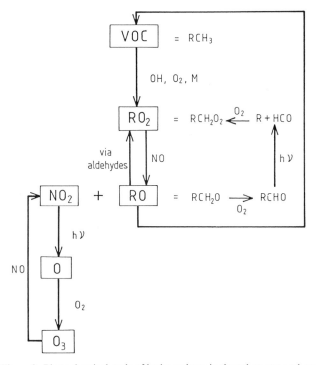

Figure 3. Photochemical cycle of hydrocarbons in the urban atmosphere.

hydes produces a carbonyl radical, which then goes on to form these mixed-acid anhydrides by reaction with O_2 and NO_2. For example, the most common compound of this type in the atmosphere, peroxyacetic nitric acid anhydride (PAN, commonly called peroxyacetyl nitrate), is formed from acetaldehyde:

$$CH_3CHO + OH \xrightarrow{O_2} CH_3C(O)O_2 + H_2O \quad k_{14} = 10^{-11} \quad (R14)$$

$$CH_3C(O)O_2 + NO_2 \xrightarrow{M} CH_3C(O)OONO_2 \quad k_{15} = 10^{-12} \quad (R15)$$

Substantial amounts of PAN can be generated in urban air (Roberts, 1990) and subsequently transported to the remote troposphere. There it makes up a sizeable proportion of the total NO_y ($= NO_x + N_2O_5 +$ nitrates) budget (Bottenheim and Gallant, 1987).

The main degradative pathway for unsaturated organic compounds is the addition of OH. Radical-stabilizing effects again influence the abundance of isomers. For example

$$CH_3CH = CH_2 + OH \longrightarrow 0.35\,CH_3CH(OH)CH_2 + 0.65\,CH_3CHCH_2OH$$

$$k_{16} = 10^{-11} \quad (R16)$$

(Atkinson, 1990). These hydroxy alkyl radicals rapidly add oxygen to form alkyl peroxy radicals, which react with NO and then decompose, releasing aldehydes

$$RCHR'(OH) + O_2 \xrightarrow{\quad M \quad} RCH(O_2)R'(OH) \qquad k_{17} = 10^{-11} \quad (R17)$$

$$RCH(O_2)R'(OH) + NO \longrightarrow RCH(O)R'(OH) + NO_2 \quad k_{18} = 10^{-11} \quad (R18)$$

$$RCH(O)R'(OH) \longrightarrow RCHO + R'OH \qquad k_{19} = 10^3 \quad (R19)$$

$$R'(OH) + O_2 \longrightarrow R''CHO + HO_2 \qquad k_{20} = 10^{-11} \quad (R20)$$

where R here can be hydrogen or an organic fragment, and R' is an organic fragment. The bifurcating oxidation of unsaturated hydrocarbons is thus more rapid than might be imagined from the step-wise process implied in Figure 3.

Reaction schemes like Reactions(16) through (20) may also be applied to the aromatic compounds, but it has not been shown that aromatic degradation involves the reaction of the aryl radical with O_2 (Atkinson, 1990). Reaction with NO_2 is probably competitive in the urban atmosphere. It is known that reaction of OH with the aromatic ring dominates over reaction with alkyl side-chains, and that the final products of the reaction include aldehydes, di-aldehydes, and α-dicarbonyls.

3.3. Biogenic Hydrocarbons

Unsaturated organic compounds from plants have been observed in the atmosphere since the mid 1950s (Went, 1955) and have been cited recently as one reason for the failure of certain U.S. cities to meet air quality standards for ozone (Chameides et al., 1988). This hydrocarbon source is composed primarily of isoprene and the products of its polymerization (Zimmerman, 1979). Emissions may consist of isoprene and a variety of higher terpenes/terpenoids, or the higher terpene substances alone; when isoprene is emitted it is always the major fraction (Table 3).

Isoprene (2-methyl butadiene) is very reactive toward OH

$$H_2CC(CH_3)CHCH_2 + OH \xrightarrow{\quad NO, 2O_2 \quad} 0.5\,CH_3C(O)CHCH_2\,(MVK)$$

$$+ 0.5\,CH_2C(CH_3)CHO\,(MACR) + HCHO + NO_2 + HO_2$$

$$k_{21} = 10^{-11} \quad (R21)$$

forming methyl vinyl ketone (MVK) and methacrolein (MACR), which are both reactive in their turn toward OH. Recently, Atkinson et al. (1989) identified 3-methylfuran as an additional product of Reaction(21), with a yield of about 4%. This aromatic compound has a different reactivity from that of MVK and MACR, so that its inclusion in isoprene oxidation mechanisms may have a small effect on our estimates of photochemical activity.

Table 3 Comparison of Emission Rate Estimates for Selected Plant Species at a Leaf Temperature of 28 °C and Light Intensity of 1000 μE m^{-1} s^{-1}

Plant Species	No.	Isoprene[a] (μgC/g·hr)	Monoterpenes (μgC/g·hr)									Total Emission Rates Detected (μgC/g·hr)	Percent Isoprene of Total Hydrocarbon Detected	Isoprene Emission Rate Ranking
			α-Pinene	Camphene	β-Pinene	Myrcene	Ocimene	β-Phellandrene	1,8-Cineole	Limonene	Sabinene			
Eucalyptus globulus	8	38.50 (1.10)	1.14 (1.11)	—	.58 (1.25)	—	.84 (1.24)	—	3.53 (1.24)	.69 (1.34)	—	45.28	85	High
Populus tremuloides	8	33.89 (1.08)	—	—	—	—	—	—	—	—	—	33.89	100	High
Populus deltoides	8	24.95 (1.09)	—	—	—	—	—	—	—	—	—	24.95	100	High
Rhamnus californica	5	19.80 (1.07)	—	—	—	—	—	—	—	—	—	19.80	100	(Medium)[b]
Platanus occidentalis	7	18.55 (1.29)	—	—	—	—	—	—	—	—	—	18.55	100	(Medium)
Salix nigra	12	17.00 (1.11)	—	—	—	—	—	—	—	—	—	17.00	100	(Medium)
Thelypteris decursive-pinnata	8	16.50 (1.05)	—	—	—	—	—	—	—	—	—	16.50	100	(Medium)

Species	n								Total	%	Ranking[b]
Quercus borealis	7	13.33 (1.16)	—	—	—	—	—	—	13.33	100	(Medium)
Liquidambar styraciflua	12	12.00 (1.10)	1.20 (1.17)	—	.35 (2.33)	—	—	.57 (1.98)	14.12	85	(Medium)
Picea engelmannii	12	11.00 (1.12)	1.40 (1.12)	.17 (2.14)	.65 (1.32)	.10 (1.97)	.19 (2.14)	—	13.51	81	Medium
Pueraria lobata	8	6.45 (1.36)	—	—	—	—	—	—	6.54	100	Medium
Picea sitchensis	12	2.70 (1.28)	.21 (1.31)	.13 (1.14)	.18 (1.31)	.32 (1.13)	.00[c] (2.44)	—	3.54	76	Medium
Glycine max	9	.018 (1.14)	—	—	—	—	—	—	0.018	100	Low
Cercis canadensis	4	.013 (1.19)	—	—	—	—	—	—	0.013	100	Low
Pinus elliottii	5	—	2.79 (1.19)	—	2.00 (1.17)	.19 (1.19)	.14 (1.21)	.01 (1.15)	5.13	0	Nondetectable
Acer saccharum	8	—	.45 (1.31)	.21 (1.14)	.38 (1.16)	—	—	.58 (1.44)	1.62	0	Nondetectable

[a] Dashes indicate that these compounds were not detected in the emission profiles by the methods used in this study. The values in parentheses are the standard geometric errors.

[b] Preliminary ranking resulting from screening encapsulation when different from final ranking.

[c] This compound was detected in the foliar emissions of only one of the 12 plants tested.

Source: After Evans et al. (1982).

Subsequent OH attack on MVK is by addition to an sp^2-hybridized carbon

$$OH + CH_3C(O)CHCH_2 \xrightarrow{NO,2O_2} CH_3C(O)CHO + HCHO + HO_2$$
$$+ NO_2 \quad k_{22} = 10^{-12} \tag{R22a}$$

or

$$\xrightarrow{NO,2O_2} HOCH_2CHO + CH_3CO_3 + NO_2 \tag{R22b}$$

where $k_a/k_b \sim 3/7$. A similar addition reaction of MACR has two routes for decomposition:

$$OH + CH_2C(CH_3)CHO \xrightarrow{NO,2O_2} CH_3COCHO + HCHO + HO_2$$
$$+ NO_2 \quad k_{23} = 10^{-12} \tag{R23a}$$

$$\xrightarrow{NO,2O_2} HOCH_2COCH_3 + HO_2 + CO$$
$$+ NO_2 \tag{R23b}$$

Reaction (23b) has now been established as the major pathway. Competing with this reaction, however, is the attack by OH on the aldehydic hydrogen of methacrolein, which can, in part, generate a PAN-like substance:

$$CH_2C(CH_3)CHO + OH \xrightarrow{O_2} CH_2C(CH_3)C(O)O_2$$
$$+ H_2O \quad k_{24} = 10^{-11} \tag{R24}$$

$$CH_2C(CH_3)C(O)O_2 + 3NO \xrightarrow{3O_2} CH_3C(O)CHO + 3NO_2 + HO_2$$
$$+ CO_2 \quad k_{25} = 10^{-12} \tag{R25}$$

$$CH_2C(CH_3)C(O)O_2 + NO_2 \xrightarrow{M} CH_2C(CH_3)C(O)O_2NO_2$$
$$k_{26} = 10^{-12} \tag{R26}$$

Further degradation is then by photolysis and attack by OH, as with the anthropogenic species.

A great many higher terpenes are emitted by plants, but one of typical reactivity, and often used as a surrogate in modeling and emission inventories, is α-pinene. Reaction of terpenes with ozone is significant, and about 25% of the emitted α-pinene reacts in this manner (Fig. 4a). The products are compounds containing four-membered rings and oxygenate moieties at opposite ends of the

(a)

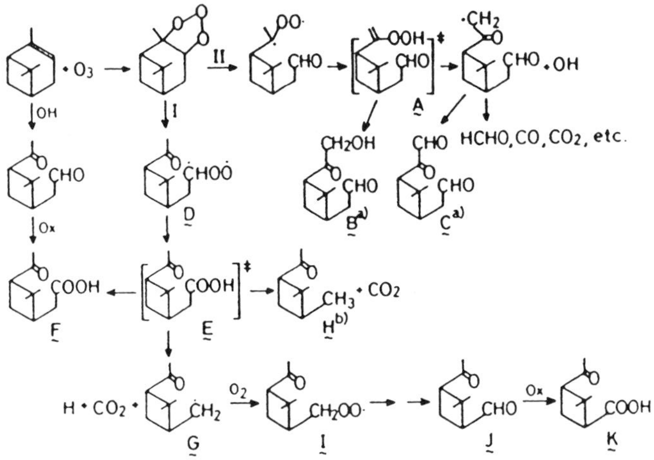

(b)

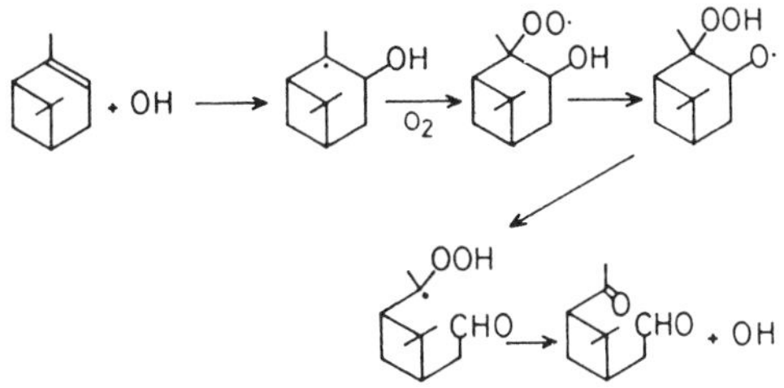

Figure 4. Reaction of α-pinene with (a) ozone and (b) OH. (Reprinted with permission from Hatakeyama et al. *J. Geophys. Res.* **94**, 13013, 1989.)

molecule. Our best guess is that these functional groups act independently so that the subsequent degradative pathway follows that of the most reactive moiety. Figure 4b shows the reaction of α-pinene with OH to form pinonaldehyde. This may then be photolyzed in two ways, or attacked by OH. Further degradation then occurs via reactions with NO and NO_2, as in the isoprene scheme. Note that both the O_3 and the OH reaction produce OH and so perpetuate the photochemistry that can produce ozone. The subsequent reactions of pinonaldehyde and its analogues generate ozone more directly by converting NO to NO_2 in the course of their degradation.

3.4. Diurnal Variations in Chemistry

As night falls, the photolytic production of OH and NO decreases rapidly, halting the oxidation of hydrocarbons via Reactions (3) and (8). All NO_x present is in the form of NO_2, which can further react with ozone to produce the nitrate radical:

$$NO_2 + O_3 \longrightarrow NO_3 + O_2 \qquad k_{27} = 10^{-17} \qquad \text{(R27)}$$

Although the rate coefficient for this reaction is small, both reactants are sufficiently abundant in the urban atmosphere to make the overall rate significant. Dinitrogen pentoxide, formed from the reaction of NO_3 with NO_2

$$NO_3 + NO_2 \underset{}{\overset{M}{\rightleftharpoons}} N_2O_5 \qquad k_{28} = 10^{-12} \qquad \text{(R28)}$$

is converted to nitric acid on liquid surfaces such as moist aerosol

$$N_2O_5(g) + H_2O(l) \longrightarrow 2HNO_3(l) \qquad k_{29} = 10^{-12} - 10^{-11} \quad \text{(R29)}$$

at a rate comparable to the daytime removal of NO_x [Reaction(11)]. This behavior, in itself, will influence subsequent photochemistry by changing the $[HC_r]:[NO_x]$ ratio, but NO_3 also has a more direct influence.

Table 2 lists the rate coefficients for the reaction of common atmospheric hydrocarbons with NO_3. Those of the alkenes are particularly large (this is also true for isoprene and α-pinene), suggesting that a significant amount of olefinic hydrocarbon can be degraded at night (Wayne et al., 1991; Platt et al., 1990). Reaction is by NO_3 addition and, again, follows the rules of carbon radical stabilization. Subsequent reaction is with O_2 to form nitrooxy-substituted alkyl peroxy radicals. These would react with NO in daylight but, at night, reaction with another molecule of NO_3 is likely:

$$RO_2 + NO_3 \longrightarrow RO + NO_2 + O_2 \qquad k_{30} = 10^{-12} \qquad \text{(R30)}$$

(The formation of peroxynitrates by reaction with NO_2 is also possible of course, but will be a temporary reservoir since peroxynitrates are thermally labile.) Similarly for HO_2 radicals formed from the oxidation of aldehydes and by Reaction (5):

$$HO_2 + NO_3 \longrightarrow OH + NO_2 + O_2 \qquad k_{31a} = 10^{-12} \qquad \text{(R31a)}$$
$$\longrightarrow HNO_3 + O_2 \qquad k_{31b} = \tfrac{1}{4}k_{31a} \qquad \text{(R31b)}$$

The low-NO environment at night will also permit peroxy radical self-reactions

$$2R_1R_2CHO_2 \longrightarrow 2R_1R_2CHO + O_2 \qquad \qquad \text{(R32a)}$$
$$\longrightarrow R_1R_2CHOH + R_1R_2CO + O_2 \qquad k_{32} = 10^{-13} \text{ (R32b)}$$
$$\longrightarrow R_1R_2CHOOCHR_1R_2 + O_2 \qquad \qquad \text{(R32c)}$$

(Atkinson, 1990) and

$$R_1R_2CHO_2 + HO_2 \longrightarrow R_1R_2CHOOH + O_2 \qquad k_{33} = 10^{-12} \quad \text{(R33)}$$

which will act as terminating steps in the oxidation if aqueous surfaces are present to wash out the peroxides. The organic nitrates formed by the action of NO_3 on hydrocarbons will probably have a longer lifetime against deposition, and so may remain in the atmosphere to be photolyzed in daylight

$$RONO_2 + hv \longrightarrow RO + NO_2 \qquad J_{34} = 10^{-6} \qquad (R34)$$

(Roberts, 1990).

The nighttime chemistry of urban plumes can be summarized, then, as comprising the slow oxidation of unsaturated hydrocarbons to nitrates, oxygenates, and peroxides by NO_2 and O_3. Intermediate in this process are peroxy radicals, which can build up to concentrations in excess of their daytime maximum (Platt et al., 1990), and the nitrate radical, which is the principal nocturnal agent of hydrocarbon oxidation. Hydrocarbon speciation will change during the course of a night, since unsaturated compounds are much more reactive toward NO_3 than alkanes. This change in composition may be reflected, in turn, in the potential for ozone production on subsequent days, because unsaturated compounds are also the most important hydrocarbons in the short-term production of ozone. This may be one reason why air parcels traversing an urban area at night are not observed to produce large ozone maxima even after many hours of illumination (see Fig. 7 below).

4. URBAN PLUMES

4.1. Measurements

The continuous emission of pollutants into a moving fluid will result in the formation of a pollutant plume that will dilute and disperse within a characteristic period indicative of the flow in the fluid. In the planetary boundary layer, and under stable conditions, such plumes are easily recognizable when emitted from point sources. However, diffuse-source regions, such as highly developed urban complexes, are of sufficient magnitude and structural diversity to make direct human observation impossible. Nevertheless, the analogy with point-source plumes has been made and has yielded meaningful results (Altshuller, 1986). Urban plumes can contribute significantly to the total ozone budget in rural environments (Altshuller, 1988). Concentrations across the plume are approximately constant, but they are a complex function of meteorological and kinetic parameters in the mean wind direction.

The literature on field studies of ozone in urban plumes is summarized in Table 4. To keep the table brief, those studies composed of many individual measurements are cited as giving a range of ozone increments. Thus, instead of in- and out-of-plume measurements, a single parameter, ΔO_3, is reported. Studies that report ozone concentrations downwind of urban centers without any information on the surrounding nonurban air quality are not reported. Up to

Table 4 Ozone Increments in Urban Plumes

Reference	Source Area	Date	Receptor Area	Ozone Concentration (ppb) Within Plume	Outside of Plume	Distance Downwind (km)	Time of Day
Sexton (1983)	Madison and Janesville, WI; Beloit and Rockford, IL	1977/78	Wisconsin and Illinois	$\Delta O_3 = 20\text{-}30^a$		50–65	Afternoon
Sexton and Westberg (1980a)	Chicago; NW Indiana	8/15/77	Wisconsin	150	70–80	170	1450–1720 CDT
Altshuller (1988)	St. Louis, MO	July 1975	St. Louis environs	$\Delta O_3 = 30\text{-}120$		Various	Various
White et al. (1977)	St. Louis, MO	7/18/75	New Springfield, IL	130	70	140	1631–1656 CDT
Karl (1978)	St. Louis, MO	8/19/75	N and W of St. Louis Illinois	180–250	125–150	25–40	1200–1600 CDT
Spicer et al. (1982a)	Springfield, IL	8/3/77		70–80	≤50	70	Afternoon
Lindsay and Chameides (1988)	Atlanta, GA	1983/84	NW or SE of Atlanta	$\Delta O_3 = 20\text{-}40$		10–20	Various
Chung (1977)	Toronto, Canada	Summer 1976	Ajax, near Toronto	$\Delta O_3 \leq 60$		20–50	Various
Harrison and Holman (1979)	Industrialized regions of England	1978	Heysham, Lancaster	$\Delta O_3 = 20\text{-}50$		80–400	Afternoon
Varey et al. (1988)	London	1982–86	London environs	$\Delta O_3 = (-20)\text{-}70$		30–90	Various

Reference	Location	Date	Measurement location	ΔO_3	O_3	Background O_3	Time
Colbeck and Harrison (1985a, b)	Midlands of England	July 1983	Stodday, Lancaster	$\Delta O_3 = 15{-}160$		30–140	Various
Van Valin and Luria (1988)	East Coast, U.S.A.	3/26/85 3/5/85	W Atlantic W Atlantic	5 65	50 35	100 Unknown	1255–1345 EST 1630–1720 EST
Spicer et al. (1982b)	Boston, MA	8/18/78	Atlantic Ocean	130	60	100	1540 EDT
Zeller et al. (1977)	Boston, MA	8/10/75	W of Boston	100–150	70–100	50–100	1130–1430 EDT
Spicer et al. (1977)	New York City	7/18/75	NW and W of Boston	150–200	60	400	2100 EDT
Spicer et al. (1979)	New York and adjacent areas	7/23/75	S Connecticut NE Connecticut NW of Boston	250–300 150–200 100–150	70–85 80–90 60–70	125 200 300	1500 EST 2100 EST 2400 EST
Cleveland et al. (1975, 1976, 1977)	New Jersey	July 1974	New Jersey	150–270	50–120	30–50	Afternoon
Chan et al. (1981)	Philadelphia, PA	7/16/79	Pennsylvania	150	60	40	1500–1500 LST
Wolff et al. (1977)	Philadelphia, PA	8/10/75	New Jersey	194	139[b]	45	Afternoon
Possiel et al. (1979)	Tulsa, AZ	1977	N of Tulsa	$\Delta O_3 = 60$	34	34	Afternoon

[a] ΔO_3 = Plume ozone − background ozone.
[b] Upwind of Philadelphia.

Source: After Altshuller (1986).

113

150 ppb is attributed to photochemistry in a plume crossing New Jersey (Cleveland et al., 1976); the maximum cited for a European city is 72 ppb downwind of London. Only Varey et al. (1988) cite negative values, and this is perhaps surprising given the high NO emissions of an urban center. As discussed in Section 3, NO can deplete ozone temporarily to form NO_2

$$NO + O_3 \longrightarrow NO_2 + O_2 \qquad (R2)$$

thus altering NO, NO_2, and ozone concentrations in the photostationary state. The presence of reactive hydrocarbons will shift these concentrations away from clean-air values, but this should not alter the fact that significant ozone depletion occurs in air impacted by large NO emissions (Ball, 1987). The reason for this lack of decremental plume ozone appears to be the method used to detect the plume. In many cases (e.g., Harrison and Holman, 1979), it is the presence of increased ozone levels that is used to identify the arrival of an urban plume at a receptor site. Only when additional meteorological information is available (Varey et al., 1988) or an unrelated chemical indicator, such as $CFCl_3$, is used (Sexton and Westberg, 1980a) can the plume be positively inferred without necessarily assuming that photochemistry has "matured" the air mass to reveal some or all of its potential oxidant loading. A further general point merits note. In the great majority of cases cited, ozone concentrations outside the plume are well above ambient background levels, and in some instances exceed the current U.S. National Ambient Air Quality Standard guideline of 120 ppb hourly mean, not to be exceeded more than once per year. Varey et al. (1988) provide evidence for a positive correlation between estimated urban ozone effects and 7-hr average ozone outside the plume. That is, photochemical ozone generation within an urban plume often appears to be embedded within a regional-scale episode. At present it is not clear whether this implies that maximum urban ozone effects require the injection of fresh precursors into an already polluted air parcel, or that the correlation is simply an artifact of other interrelating factors that affect both parameters.

It is well documented that ozone concentrations in urban areas tend to be less than at suburban sites. This is due to the scavenging effect of NO. Ball (1987) has reported that this scavenging effect is evident at downwind travel times of approximately 1.5 hr around London. Angle and Sandhu (1989) presented evidence that these effects do not extend beyond distances of 2 km from the edge of Edmonton's (Canada) built up area. Leahey and Hansen (1990) have found that the urban plume from Calgary (Canada) acts as an ozone sink for downwind distances of at least 40 km. This depletion results from the atmospheric dispersion processes around Calgary. At night the air is stable and downwind pollutant concentrations are greatest. During the day, the greater dispersion capacity of the atmosphere associated with unstable atmospheres quickly causes the urban plume to dissipate, resulting in low levels of NO_x. Occasional ozone episodes do occur in these two Canadian cities. Gladstone et al. (1991) used a box model to assess the extent of local photochemical ozone production. They concluded that a combination of a stagnant air mass with urban NMHC and NO_x inputs can lead to substantial ozone production near the urban source and elevated concentrations in the rural downwind air mass. The inclusion of biogenic hydrocarbon emissions resulted in significant enhancement of ozone production (an additional 10 ppb on day two) and a doubling of PAN concentrations. MacKenzie et al. (1991) studied the effect of BNMHC emissions around London and concluded that these emissions have a negligible effect on ozone production within that urban plume.

The treatment of the dispersion of urban pollution assuming the presence of a reasonably cohesive plume, analogous to those emitted by strong point sources, is now supported by a large body of fieldwork (Table 4). Maxium reported urban effects for ozone are 150 ppb in the United States (Cleveland et al., 1975) and 72 ppb in Europe (Varey et al., 1988), while decremental effects have largely been ignored. This omission springs from the common practice of inferring the presence of an urban plume from the detection of enhanced ozone levels. Given the NO_x-rich character of urban plumes, this is an unsatisfactory criterion. The negative effects should not be ignored, since the particular air quality problems encountered downwind of a city will depend on whether, and how often, the urban area acts as a sink for ozone.

4.2. Aircraft Measurements

Measuring sites at ground level do not give a representative picture of the horizontal and vertical distribution of ozone and its precursors in the atmosphere. Near the ground, ozone is rapidly destroyed by heterogeneous processes and by chemical quenching via reactions in the boundary layer, mainly by NO and olefinic hydrocarbons. Aircraft measurements have been used for some time to yield information on the vertical and horizontal profiles of ozone and its precursors (see Altshuller, 1986).

The results of aircraft measurements of plumes from cities in the United States are widely reported (see Altshuller, 1986). Generally, the urban plume tracking was only available through the afternoon or evening hours of the same day the air parcels left the city. Measurements demonstrate that ozone levels within the plume can exceed 120 ppb up to several hundred kilometers downwind of a city (Altshuller, 1986), and the combined influence of plumes from small cities is similar in magnitude to that associated with emissions from a large urban area (Sexton, 1983). Altshuller (1988) has made detailed studies of ozone formation in the urban plume of St. Louis. Plume widths of 40 to 60 km were observed 100 to 150 km downwind. From this figure he estimated that 5 to 10% of the land west of the Appalachian Mountains and east of 105° W was subject to excess ozone within urban plumes between April and October. The location of this 5 to 10% of land area will vary with shifts in wind direction.

Measurements downwind of Baltimore and Washington, D.C., yield hydrocarbon:NO_x ratios of 7 ± 2.9 and 9.5 ± 7.2, respectively. These fall within the range of 6 to 12 normally reported for large urban centers in the United States. Ozone concentrations in the combined Washington–Baltimore plume ranged from 120 to 220 ppb, while background levels generally ranged between 90 and 110 ppb. It is therefore reasonable to assume that ozone concentrations will increase by approximately 100 ppb in an air parcel that passes directly over Washington and Baltimore during the morning hours. On days when the plumes from these cities partially overlapped, Westberg (1985) was able to semiquan-titatively apportion the contribution to ozone formation of the precursors from each city. The maximum ozone concentration at the centerline point of the Washington plume was 112 ppb. Outside the plume, background levels were 76 ppb. In the section under the influence of both emissions, ozone levels were 123 ppb. Therefore, maximum ozone buildup in the combined plume was 47 ppb, an increase of 30% over the Washington plume alone. The result implies peak ozone levels in the two separate plumes were nearly identical.

In measurements of plumes downwind of Boston, NO_x concentrations varied between 27 to 131 ppb near Boston and between 5 to 10 ppb, 4 to 7 hr down-wind (Spicer, 1982). The average lifetime of NO_x under sunny conditions was calculated to be 5.5 hr. It can be concluded from this and other measurements (Sexton and Westberg, 1980a) that NO_x is rapidly depleted within urban plumes.

Measurements of the C_2 to C_6 hydrocarbons, NO_x, and ozone in Germany over the Rhine–Ruhr area have shown that the ratio of high-reactive to low-reactive hydrocarbons decreases form urban to rural areas (Neuber et al., 1982). Data from Cologne–Bonn indicate that this area can form excess ozone of up to 80 ppb. Airborne measurements have been carried out in the Netherlands since 1972 (van Duuren et al., 1982). Both urban and power plant plumes have been studied and the results agree well with investigations elsewhere. That is, close to sources, quenching by NO gives rise to plumes of diminished ozone concentra-tions; futher away, the contributions usually become positive. The travel time at which the contribution changes sign varies considerably.

In the United Kingdom there have been numerous airborne surveys of air pollutants (Bamber et al., 1984). These have been concerned primarily with the fate of NO_x emissions from power stations and large urban areas such as London (PORG, 1987, 1990). These flights have demonstrated quantitatively the depletion of ozone by nitric oxide-rich plumes at distances up to 100 km downwind of the source. The time scale over which this effects occurs is dependent on insolation and small-scale dispersion. The dispersion characteristics also influence the time scale over which in-plume ozone levels increase to ambient levels. Elevated levels of ozone have been observed both leaving and entering the United Kingdom. Figure 5 shows an ozone plume downwind of the Tyneside/Teeside area. The effect of high NO emissions in the Humberside area and high hydrocarbons from the Tyneside/Teeside area can be seen.

Aircraft measurements can be used to investigate the relationship between the three-dimensional distribution of ozone and meteorological conditions, in par-ticular the influence of sea breezes in pollution dispersion. Sea breezes tend to advect back to cities air masses whose origin, at least in the morning, was the same city from which land breezes advected pollutants seaward during the night and early morning. The phenomenon of photochemical ozone formation in sea

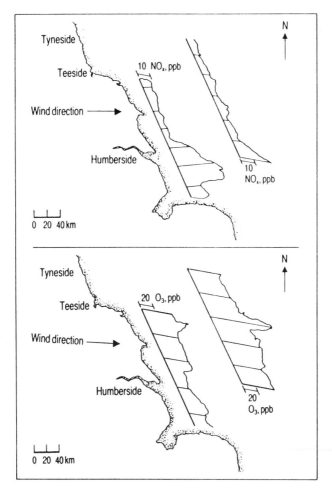

Figure 5. Measurements of NO_x and ozone downwind of central and northern England on June 18, 1980. (PORG, 1987.)

breezes has been observed in many coastal locations (Blumenthal et al., 1978; Wakamatsu et al., 1983; Gusten et al., 1988). In Los Angeles, Blumenthal et al. (1978) detected a highly polluted air mass above an inversion layer in the morning, and these pollutants increased oxidant concentrations by mixing in accordance with the elevation of the mixing layer. Wakamatsu et al. (1983) report similar results for Tokyo. They found high ozone levels in land breezes caused by transport of pollutants from the coastal industrial area by sea breezes the day before. Any expanse of water of sufficient size to influence the local meteorology can play a significant role in pollution dispersion and local air quality. Lyons and Cole (1976) have made detailed studies of summertime meteorology along the western shore of Lake Michigan and have concluded that the lake breeze plays an important role in controlling pollutant concentrations in this region. Airborne measurements of ozone showed a dramatic rise of over 50 ppb as the aircraft penetrated the lake breeze zone, while ground-level measurements have been reported to increase by up to 100 ppb on the passage of the land breeze (Westberg et al., 1981a). The effects of local climatology, such as sea breeze circulations, on pollutant concentrations make the formulation of control strategies for pollution abatement difficult.

4.3.   The London Plume

In 1982 a number of research institutions initiated the London Ozone Survey (LOS). Its primary objectives were to establish an intercalibrated network of ground-based monitoring sites in southeast England for ozone and other relevant pollutants, and from this data to investigate the behavior of ozone throughout the boundary layer with emphasis on both long-range transport and ozone formation in the urban plume (Varey et al., 1987). Varey et al. (1988) extracted

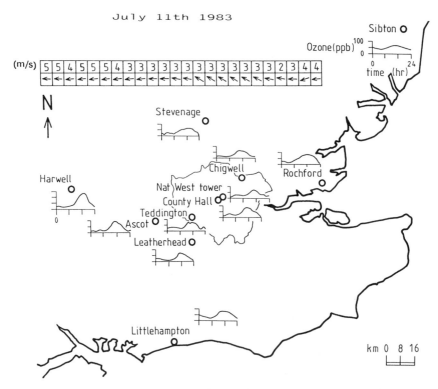

Figure 6. Hourly mean ozone concentrations for each site operating on July 11, 1983, forming input for the London Ozone Study. Wind speed and direction from the post office tower.

data for days between 1982 and 1985 in which ozone concentrations in some part of the region exceeded 80 ppb. Figure 6 shows the ground-level ozone data for one of these days together with wind data from the post office tower. To distinguish the generation of ozone due solely to the plume from that occurring on a regional scale, mean regional out-of-plume maxima were subtracted from the plume maximum to give an "urban effect" for ozone downwind of London. This urban effect correlates significantly with mean regional ozone, wind speed, maximum daytime temperature and time elapsing between passage of the air mass over the urban center and its measurement. The hourly mean maxima data can also be plotted on an emission/travel-time surface as illustrated in Figure 7. Here two areas of elevated maxima can be identified, both of which occur at approximately the same time of day, 14–1800 (BST). The circled area near the x-axis represents the optimum time for local ozone formation in or near the urban center: air parcels pass over the city at around solar noon and travel for only a few hours before reaching a maximum. The second circled area encompasses the optimum conditions for elevated ozone concentrations in the plume at longer travel times: an air parcel crossing the city earlier in the day requires more time to produce maximum ozone levels than in the first case. Presumably this is because downwind rural/suburban emissions make a significant contribution to the oxidant loading of a plume, or because the precursors emitted in the urban area require a greater integrated photon flux to generate ozone than in the former situation. Interestingly, neither area coincides with the morning or evening rush hour. Modeling studies suggest that this may be due to the strong ozone-depleting capacity of urban NO emissions in the short term (MacKenzie, 1989).

The diurnal cycle, at each site on each day, has been characterized by defining t_0, the start of the initial rise of gradient; m, the initial rate of rise; t_1, its finish time;

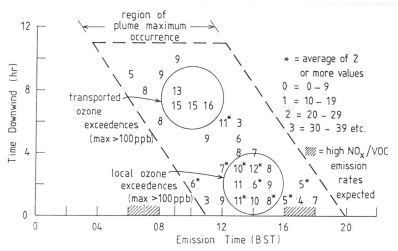

N.B. Actual time of plume maximum = emission time + travel time

Figure 7. The London urban ozone effect, ΔO_3, as a function of air parcel emission and travel times.

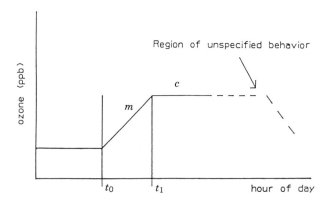

Figure 8. An idealized diurnal cycle for surface ozone concentrations.

and c, the final plateau concentration reached (Fig. 8). This process eliminates information on fine structure but it enables useful intersite comparison. As an illustration of the type of information made available through this analysis, a small sample period, July 4, 1983, to July 18, 1983, is presented. The results are shown in Table 5 and Figure 9.

The graphs suggest that sites within the London plume tend to begin their initial rise later in the day than corresponding sites not in the plume, but rise more steeply when the rise from low nighttime concentrations begins. Statistical testing on the data set yields significant differences between sites in and out of the plume, which supports these conclusions (In Table 5, the columns headed " + LON" include inner London sites in the calculations; those marked " − LON" do not.) The initial rise in ozone concentration in caused by the breakup of the stable nocturnal boundary layer, which allows ozone held aloft overnight to be mixed down to the surface. Like the emission time effect described above, this result is probably due to the titration of downward mixing ozone by the NO-rich urban plume.

The parameter marking the end of the initial rise in ozone concentrations, t_1, is similar for all sites regardless of plume position. This is to be expected since mixing to dissipate the nocturnal inversion is complete at approximately the same time independent of pollutant loading in the upper or lower layer. That the average plateau values do not differ greatly is also not surprising, since this analysis is primarily concerned with the early part of the day, when photochemistry has not had sufficient time to convert the London NO_x/hydrocarbon plume into an ozone plume.

The consistency of the results, both with and without the inner London sites, suggests that the analysis is not wholly dominated by urban/suburban site differences. Moreover, the similarity of standard deviations between sites in and out of the London plume discounts the possibility of a selective bias in the manual interpretation of the data. This should not be taken to imply that the approach is objective, but merely that the subjective interpretation of the LOS data has an

Table 5 London Ozone Study Fitted to an Idealized Diurnal Cycle

| | In-Plume | | | | Out-of-Plume | | Statistics[c] | | | |
| | +LON[b] | | −LON | | | | F-Test In-Plume vs Out-of-Plume | Pooled SD | t-Value | Significance |
Parameter[a]	Mean	SD	Mean	SD	Mean	SD				
n	57		34		48					
t_0	9.7	1.32	9.5	1.33	8.9	1.57	1.18	1.47	1.82	90%
t_1	14.2	1.31	14.2	1.44	13.9	1.68	1.17	1.59	0.842	—
m	15.5	5.28	15.2	5.03	11.9	4.49	1.12	4.72	3.12	95%
c	79.5	16.20	81.5	19.69	77.8	16.08	1.16	17.20	0.960	—

[a] n = number of samples; other parameters are defined in the text. Probability points for the double-sided t-distribution: $t_{0.05,\,60} = 2.00$; $t_{0.10,\,60} = 1.67$.

[b] +LON, data from urban monitoring sites, included; −LON, urban monitoring sites excluded.

[c] Statistics are for the comparison of out-of-plume sites with those within the plume boundaries but outside the city center (−LON).

121

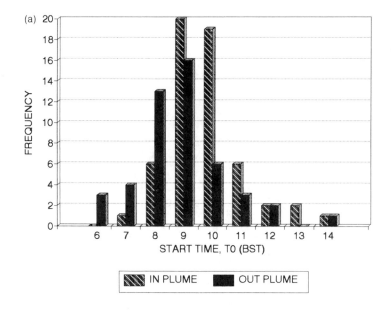

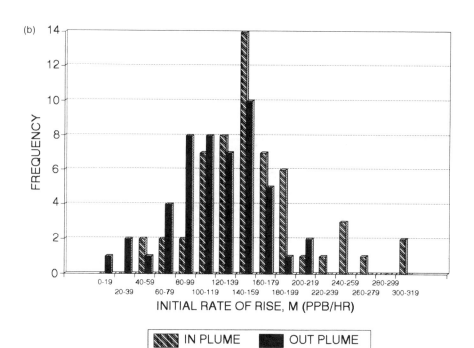

Figure 9. Frequency distribution of parameters in and out of the London plume. (a) Time of morning increase, t_0, BST; (b) initial rate of rise, m (ppb/hr); (c) plateau concentrations, c (ppb). For statistics of in- and out-of-plume differences, see Table 5.

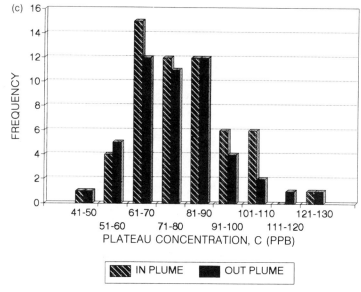

Figure 9. (*Continued*)

equivalent random error with respect to analysis of in-plume sites and those out of the plume.

Measurements made from aircraft enable the chemical history of urban emissions to be established as they are transported over hundreds of kilometers. Flights across London have measured ozone, sulfur dioxide, and other gases from a Jetstream aircraft. Figure 10 shows ozone results from a flight on July 12, 1983. On this day, winds were predominantly northeasterly and ozone levels in excess of 100 ppb were attained at ground-level sites downwind of London (Ball, 1987). The flight details show higher ozone concentrations on the downwind side of Greater London, providing further evidence of local ozone generation.

An expanding box model of the urban plume has been used to give more quantitative data on the factors influencing plume concentrations at the end of one day's travel (MacKenzie, 1989). A 2^6 factorial experiment was carried out using urban and rural emission rates typical of those in southeast England, and NO, CO, and hydrocarbons as variables. A sample of the model output is shown in Figure 11. It shows the dominance of urban emissions in the determination of plume maxima 100 km downwind of the upwind urban boundary. The effect of urban hydrocarbon emissions is seen by comparing the two graphs. The effect of urban NO emissions is seen in the two broad bands of model results within each graph. The breadth of each band is due to other, less important factors. Changing urban hydrocarbon emissions from low to high levels results in an increase in plume ozone of 10 to 25 ppb, depending on the level of NO emissions. That is, the NO and volatile organic compounds (VOC) emissions interact strongly and nonlinearly (MacKenzie et al., 1990).

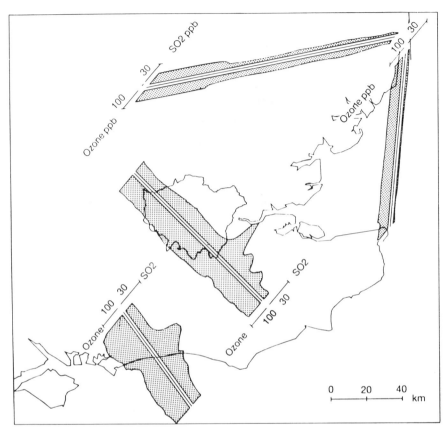

Figure 10. Ozone and sulfur dioxide concentrations during a period of northeasterly winds, indicating enhanced ozone levels downwind of London. (Reprinted with permission of *Sci. Total Environ.* **59**, 203, 1987.)

The Harwell photochemical trajectory model, which employs a detailed description of the atmospheric physics and chemistry of photochemical ozone formation in the United Kingdom but a highly simplified representation of the relevant transport processes, has been used to study, among other things, the role of different hydrocarbons in ozone formation, mechanisms for photochemical ozone production, and sensitivity and uncertainty analysis (Derwent and Hov, 1979, 1988; Derwent, 1990; Derwent and Hough, 1987; Hough and Derwent, 1987). Some of the results on the impact of motor-vehicle control technologies will be discussed later. The model gives good agreement between predicted and measured levels of ozone, PAN, and aerosol downwind of London. It shows, first, that it is possible to form significant ozone concentrations, up to 100 ppb, downwind of London, but that the formation of higher levels requires persistent and slow-moving anticyclonic systems. Second, aromatic hydrocarbons make the largest contribution to ozone and PAN formation during the first day's

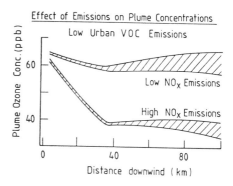

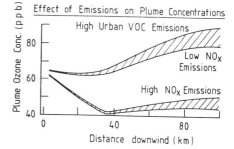

Figure 11. Effect of emissions on plume concentrations. (Reprinted with permission of *Sci. Total Environ.* **93**, 250, 1990.)

photochemistry. Finally, under most circumstances during the first day's photochemistry, control of hydrocarbon emissions would always decrease ozone formation, whereas control of NO emissions would in some cases increase and in other cases decrease ozone concentrations. Both the Harwell and our own model conclude that total hydrocarbon emissions, NO_x emissions, and photolysis rates are the most important factors in determining ozone concentrations downwind of London (MacKenzie, 1989; Hough and Derwent, 1987). Tracing ozone-rich air masses back to specific urban plumes is difficult. Backward air mass trajectories provide one method, aircraft measurements another. Models may be used in such a manner that the modeled air mass follows the atmospheric mean motion. Determination of the exact course of the trajectory is less of a constraint, however, if the plume is expanding into a relatively uniform rural environment, and if the plume ventilation rate can be established.

4.4. Fossil-Fuel Power Stations

The rapid formation of ozone from both large and small cities is favorable because of the presence of highly reactive hydrocarbons and optimum hydrocarbon to NO_x ratios. Typical urban areas have high emission rates for both hydrocarbons and NO_x, with hydrocarbon to NO_x emission ratios (carbon atoms:nitrogen atoms) of about 5:1 (Sillman et al., 1990b). Power station emissions, on the other hand, are high in NO_x and have a much smaller hydrocarbon to NO_x ratio. In fact, ozone formation from power station plumes will be dependent upon the availability of hydrocarbon emissions in areas traversed by the plume.

Plumes from power stations suppress ozone levels and reduce the NO_x:NO ratio compared with surrounding air. In a NO-rich plume, the dominant reaction is NO with ozone outside the plume to form NO_2. In this reaction, the sum of NO_2 and O_3, often termed total oxidant, remains constant. As the plume disperses downwind, it takes on more of the characteristics of the surrounding air. Modeling studies that treat the interaction of the plume and urban air explicitly indicate that even at distances at which ozone levels in plumes are restored to ambient levels, NO_x:hydrocarbon concentration ratios may be less favorable to oxidant production than those in the ambient atmosphere, and that this situation may persist to considerable distances from the sources. This should not be taken to mean that such plumes ameliorate the toxicological effects of secondary pollutant formation, since NO_2 is itself a toxic substance (PORG, 1990).

Although the major chemical processes in the plume occur on the 1–10 hr time scale, the chemical reactions coupling NO, NO_2, and O_3 are exceedingly rapid and may occur on time scales of a few seconds to minutes. Since the NO concentrations are in vast excess over the O_3 concentrations, the supply of O_3 is rapidly exhausted and the time scale of the $NO–O_3–NO_2$ systems extends to the time scale for entraining ozone into the plume. Hence, small-scale mixing processes can influence large-scale processes such as photochemical ozone formation. It is for this reason that large-scale Eulerian grid models often underestimate ozone concentrations near source areas and show an excessive production in rural areas. Recently, Sillman et al. (1990b) developed a model whereby rural chemistry and transport are simulated on a coarse grid ($400 \times 480\,km^2$), while chemistry and transport in urban and power station plumes are represented by detailed subgrid ($20 \times 20\,km^2$) models. A realistic fraction of pollutants react under high NO_x emissions, and NO_x is removed significantly before dispersal. Because of the nonlinear dependence of ozone production on NO_x, significant overestimates of ozone production are still likely when a single grid box contains subregions with both low ($< 1\,ppb$) and high ($> 3\,ppb$) NO_x (Sillman et al., 1990a).

The dependence on hydrocarbon emissions explains the variations in ozone concentrations between different plumes reported in the literature. In the eastern United States, the formation of excess ozone in the range 20 to 50 ppb above ambient background levels of 40 to 100 ppb has been observed several hours downwind of power stations (Davis et al., 1974; Gillani and Wilson, 1980; Gillani et al., 1978; Miller and Alkezweeny, 1980). In the southeastern United States, Meagher et al. (1987) have observed ozone concentrations above 100 ppb at several ground-level monitoring sites 3 to 8 hr downwind of power stations. However, over rural regions of the western United States (Hegg et al., 1977; Ogren et al., 1977) or over areas of ocean (Spicer et al., 1982b), excess ozone has not been observed in power station plumes.

The importance of the NMHC:NO$_x$ ratio is highlighted in the case of the Newport power station in Melbourne, Australia (Hess, 1989). The NMHC:NO$_x$ ratio is normally < 10:1 and, in this case, trajectories passing over Newport show a decrease in plume ozone of about 20 ppb. However, certain trajectories pass over Port Cook, an area with high concentrations of NMHC where the NMHC:NO$_x$ ratio can be > 52:1. In this case, ozone levels in the plume were enhanced above ambient.

All measurements generally indicate an initial ozone depletion within the plume due to its reaction with NO to produce NO$_2$. For distances > 5 km, consumption of ozone and production of NO$_2$ are generally comparable (Joos and Maffiolo, 1989). At much shorter distances, ozone consumption alone is unable to account for NO$_2$ production, and any difference probably results from the reaction between NO and O$_2$.

Airborne surveys around London allow the behavior of more dispersed sources within London to be compared with that of point-source emissions from the Thameside power station. Figure 12 shows a traverse 75 km downwind of London. The contrast between urban and plume air is clear. In the former, ozone has been generated, the NO$_2$:NO ratio was about 2, and NO$_2$ and O$_3$ peaked at 55 ppb; in the latter, ozone was suppressed, the NO$_2$:NO ratio was approximately 1, and NO$_2$ and O$_3$ about 27 ppb. Under conditions found in the United Kingdom, modeling studies indicate that power station plumes do not contribute significantly to ozone formation (Derwent, 1982).

4.5. Other Plumes

Ozone formation has been observed within several industrial plumes (Sexton and Westberg, 1983a). The plumes investigated have included those from petroleum refineries (Sexton and Westberg, 1983b) and automotive paint plants (Sexton and Westberg, 1980b). Although these sources are rich in NMHC, they tend to be composed of the less reactive alkanes mixed with limited amounts of NO$_x$. Industrial plumes are narrow and difficult to track. Additionally, the hydrocarbons and NO$_x$ are emitted from different sources at different heights. Hence, it is difficult to observe excess ozone from such plumes. Where it has been possible, ozone concentrations of 5 to 30 ppb above background have been reported (Sexton and Westberg, 1983a). Joos and Maffiolo (1989) reported ozone concentrations > 100 ppb 5 km downwind of an industrial area of Le Havre, France, despite low values of insolation. Analysis of air samples in the plume showed high concentrations of reactive hydrocarbons. Ozone concentrations outside the plume were not measured, but typically they would be expected to be < 45 ppb.

The burning of vegetation can supply substantial amounts of primary pollutants, including CO, NO$_x$, and hydrocarbons, to the atmosphere. In many parts of the tropics this is still the main cause of air pollution. Plumes from controlled fires in Washington State were found to form ozone levels 40 to 50 ppb above ambient background concentrations (Westberg et al., 1981b). Hydrocarbon analyses revealed the presence of many photochemically reactive olefins. Delany

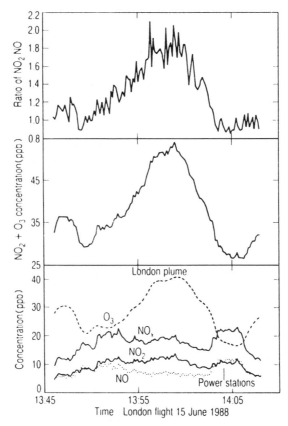

Figure 12. The degree of oxidation of NO to NO_2, NO_x, and O_3 concentrations measured during a traverse across the London and Thameside plumes on June 15, 1988. (PORG, 1990.)

et al. (1985) and Crutzen et al. (1985) found during the dry season in the Brazilian savannah an ozone excess of 20 ppb, with a mean background level of 50 ppb, resulting from bush fires. Ozone production is possible from the emission of substantial quantities of NO from biomass burning and the oxidation of reactive hydrocarbons and CO. Similarly, bush fires over Africa can result in elevated ozone concentrations. Cros et al. (1988) observed concentrations of 100 ppb in a layer residing just above the monsoon. In fact, Fishman et al. (1986), using satellite data, estimate that biomass burning results in an elevation of ozone throughout most of the tropical troposphere by 20 to 30 ppb.

The Kuwait oil field fires emitted vast quantities of sulfur dioxide, hydrogen sulfide, carbon monoxide, carbon dioxide, and NO_x, as well as particulate containing partially burned hydrocarbons. Airborne measurements in the near- and far-field plumes yielded information on their chemical composition (Johnson et al., 1991). In the near-field (120 km downwind), NO_x and O_3 concentrations were negatively correlated, indicative of the reaction between O_3 and NO, but

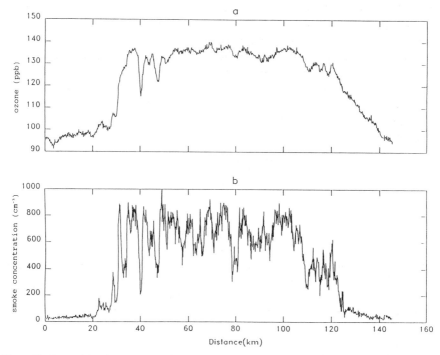

Figure 13. Ozone (*a*) and smoke (*b*) concentrations at an altitude of 4400 m through an area of smoke (about 1000 km from Kuwait) on March 29, 1991. [Reprinted by permission from *Nature* **353**, 618 (Copyright 1991, Macmillan Journals Limited).]

their sum showed little horizontal variation. Far-field measurements (1000 km downwind) found ozone concentrations increased by 40 to 60 ppb (Fig. 13). This increased ozone was formed from NO_x and hydrocarbons released from the oil fires. Analysis of hydrocarbons near the fires and in the plume showed relatively high levels of the longer lived hydrocarbons, such as ethane and propane, within the plume, but reduced concentrations of short-lived hydrocarbons, such as ethene and propene. These low levels of the shorter lived species suggest that significant degradation through OH attack has taken place, consistent with the increase in ozone within the plume.

5. CONTROL STRATEGIES

Photochemical ozone concentrations can, in principle, be reduced by controlling hydrocarbon and nitrogen oxide emissions. However, the development of an effective strategy for reducing ozone production by controlling anthropogenic emissions of these precursors has proven to be problematic (Lindsay et al., 1989). Despite this, control strategies for ozone have been given extensive consideration throughout the world, particularly in the United States, where they have been formulated and enforced over the last 25 years. A significant barrier to progress has been the complexity of the chemistry and transport processes involved.

The complex relationships between the precursor species and ozone are often represented by an isopleth diagram. Such isopleths are generally constructed from the results of numerical models (Fig. 2). Note that the effect on ozone of reducing hydrocarbons and NO_x changes considerably as the concentrations of hydrocarbons and NO_x change. In order to devise an effective strategy for ozone abatement in a given air shed, it is first necessary to determine where on the ozone isopleth diagram this air shed is located. This is particularly difficult because of the variability in time and space of source and sink terms and transport. Additionally, the shape of the isopleths generated from model calculations depends upon the formulation of the chemistry and the reactivity of the hydrocarbon mix.

Many different types of photochemical air quality models have been developed (see Seinfeld, 1988). Because of the difficulties of developing the complex chemical mechanisms and incorporating them into diffusion-advection models, models for reactive pollutants have lagged behind those for inert pollutants, which have been developed to a high level of sophistication.

All air quality models require the input of a number of parameters that are currently either poorly defined and/or have large uncertainty. It appears likely that present inventories significantly underestimate the hydrocarbon emissions from mobile sources (Pierson et al., 1990; Chameides et al., 1992). In the United Kingdom, hydrocarbon emission rates are quoted within a range of $\pm 30\%$. Additionally, areas of uncertainly of a similar range exist for the question of speciation. As well as man-made hydrocarbon sources, biogenic emissions have to be treated species by species.

The balance of present evidence is that natural hydrocarbons have no significant part to play in the regional-scale photochemical episodes observed over northwestern Europe. (MacKenzie et al., 1991; Simpson, 1992). However, in certain areas of the United States, emissions of biogenic hydrocarbons dominate anthropogenic emissions. In early studies, it was generally concluded that the impact of biogenic hydrocarbons on ozone production was small. Recently, this has been shown not to be the case in certain regions. Trainer et al. (1991) calculated that biogenic hydrocarbons played a dominant role in ozone production over a rural site in Pennsylvania. Chameides et al. (1988) studied the role of biogenic hydrocarbons in the Atlanta metropolitan area. They found that, even in an urban area, natural emissions can significantly alter ozone levels and their presence can exert a profound influence on the effectiveness of an ozone abatement strategy based on anthropogenic hydrocarbon reductions. Similarly, Lin et al. (1992) concluded that biogenic hydrocarbons significantly influence ozone production over rural sites in Ontario, Canada.

It is important to note that the biogenic contribution represents a background hydrocarbon concentration that cannot be removed from the atmosphere by emission control measures. Hence, if as a result of emission control efforts anthropogenic emissions are reduced, this background contribution will begin to represent a larger and more significant fraction of the total hydrocarbon reactivity. Chameides et al. (1992) predict that if anthropogenic emissions are totally eliminated, biogenic hydrocarbons will be able to generate ozone concentrations in excess of current guidelines, provided NO_x emissions are not controlled.

The conclusions concerning ozone control strategies obtained from numerical models depend on how well they reflect what would actually occur in the atmosphere. This is dependent upon the adequacy and completeness of the models themselves, as well as the processes represented in them and the simplifications made. The model results do not represent validated predictions of what will happen in the atmosphere in response to changes in emissions.

REFERENCES

Altshuller, A. P. (1986). Review paper: The role of nitrogen oxides in non-urban ozone formation in the planetary boundary layer over N. America, W. Europe and adjacent areas of ocean. *Atmos. Environ.* **20**, 245–268.

Altshuller, A. P. (1989). Some characteristics of ozone formation in the urban plume of St Louis, MO. *Atmos. Environ.* **22**, 499–510.

Angle, R. P., and Sandhu, H. S. (1989). Urban and rural ozone concentrations in Alberta, Canada. *Atmos. Environ.* **23**, 215–221.

Atkinson, R. (1990). Gas-phase tropospheric chemistry of organic compounds: A review. *Atmos. Environ.* **24A**, 1–41.

Atkinson, R., Aschmann, S. M., Tuazon, E. C., Arey, J., and Zielinska, B. (1989). Formation of 3-methylfuran from the gas-phase reaction of OH radicals with isoprene and the rate constant for its reaction with the OH radical. *Int. J. Chem. Kinet.* **21**, 593–604.

Atkinson, R., Baulch, D. L., Cox, R. A., Hampson, R. F., Kerr, J. A., and Troe, J. (1992). Evaluated kinetic and photochemical data for atmospheric chemistry: Supplement IV. *Atmos. Environ.* **26A**, 1187–1230.

Balko, J. A., and Peters, L. K. (1983). A modelling study of SO_x-NO_x-hydrocarbon plumes and their transport to the background troposphere. *Atmos. Environ.* **17**, 1965–1978.

Ball, D. J. (1987). A history of secondary pollutant measurements in the London region 1971–86. *Sci. Total Environ.* **59**, 181–206.

Bamber, D. J., Clark, P. A., Glover, G. M., Healey, P. G. W., Kallend, A. S., Marsh, A. R. W., Tuck, A. F., and Vaughan, G. (1984). Air sampling flights round the British Isles at low altitudes: SO_2 oxidation and removal rates. *Atmos. Environ.* **18**, 1777–1790.

Blumenthal, D. L., White, W. H., and Smith, T. B. (1978). Anotomy of a Los Angeles smog episode: Pollutant transport in the daytime sea breeze regime. *Atmos. Environ.* **12**, 893–907.

Bottenheim, J. W., and Gallant, A. J. (1987). The occurrence of peroxyacetyl nitrate over the Atlantic Ocean east of North America during WATOX-86. *Global Biogeochem. Cycles* **1**, 369–380.

Chameides, W. L., Lindsay, R. W., Richardson, J., and Kiang, C. S. (1988). The role of biogenic hydrocarbons in urban photochemical smog: Atlanta as a case study. *Science* **241**, 1473–1475.

Chameides, W. L., Fehsenfeld, F., Rodgers, M. O., Cardelino, C., Martinez, J., Parrish, D., Lonneman, W., Lawson, D. R., Rasmussen, R. A., Zimmerman, P., Greenberg, J., Middleton, P., and Wang, T. (1992). Ozone precursor relationships in the ambient atmosphere. *J. Geophys. Res.* **97**, 6057–6055.

Chan, M. W., Allard, D. W., and Possiel, N. C. (1981). Relative contributions of transported and locally produced ozone downwind of Philadelphia, Pennsylvania. *74th Annu. Meet. Air Pollut. Control Assoc.* Philadelphia, PA, Paper 81-21.1.

Chung, Y.-S. (1977). Ground-level ozone and regional transport of air pollutants. *J. Appl. Meteorol.* **16**, 1127–1136.

Clarke, J. F., and Ching, J. K. S. (1983). Aircraft observations of regional transport of ozone in the northeastern United States. *Atmos. Environ.* **17**, 1703–1712.

Cleveland, W. S., Kleiner, B., McRae, J. E., and Warner, J. L. (1975). The analysis of ground-level ozone data from New Jersey, New York, Connecticut and Massachusetts: Transport from the New York City metropolitan area. *Symp. Status. Environ., 4th*, pp. 70–90.

Cleveland, W. S., Kleiner, B., McRae, J. E., and Warner, J. L. (1976). Photochemical air pollution: Transport from the New York City area into Connecticut and Massachusetts. *Science* **191**, 179–181.

Cleveland, W. S., Kleiner, B., McRae, J. E., Warner, J. L., and Pasceri, R. E. (1977). Geographical properties of ozone concentrations in the northeastern United States. *J. Air Pollut. Control Assoc.* **27**, 325–328.

Colbeck, I., and Harrison, R. M. (1985a). The frequency and causes of elevated concentrations of ozone at ground-level at rural sites in north-west England. *Atmos. Environ.* **19**, 1577–1587.

Colbeck, I., and Harrison, R. M. (1985b). The photochemical pollution episode of 5–16 July in north-west England. *Atmos. Environ.* **19**, 1921–1929.

Cros, B., Delmas, R., Nganga, D., and Clairac, B. (1988). Seasonal trends of ozone in equatorial Africa: Experimental evidence of photochemical formation. *J. Geophys. Res.* **93**, 8355–8366.

Crutzen, P. J. (1988). Tropospheric ozone: An overview. In I. S. A. Isaksen (Ed.), *Tropospheric Ozone*. Reidel Publ., Dordrecht, The Netherlands.

Crutzen, P. J., Coffey, M. T., Delany, A. C., Greenberg, J., Haagenson, P., Heidt, L., Lueb, R., Mankin, W. G., Pollock, W., Seiller, W., Wartburg, A., and Zimmerman, P. (1985). Observations of air composition in Brazil between the equator and 20°S during the dry season. *Acta Amazonica* **15**, 77–119.

Davis, D. D., Smith, G., and Klauber, G. (1974). Trace gas analysis of power plant plumes via aircraft measurement: O_3, NO_x and SO_2 chemistry. *Science* **186**, 733–736.

Delany, A. C., Haagenson, P., Walter, S., Wartburg, A. F., and Crutzen, P. J. (1985). Photochemically produced ozone in the emission from large scale tropical vegetation fires. *J. Geophys. Res.* **90**, 2425–2429.

Derwent, R. G. (1982). *Computer Modelling Studies of Photochemical Air Pollution Formation in Power Station Plumes in the United Kingdom*, AERE-R-10631. Harwell Laboratory, Oxon, UK.

Derwent, R. G. (1990). The long-range transport of ozone within Europe and its control. *Environ. Pollut.* **63**, 299–318.

Derwent, R. G., and Hough, A. M. (1987). *The Impact of Emission Reduction Scenarios on Photochemical Ozone and Other Pollutants Downwind of London*, UK-AERE-R12B59. Harwell, Laboratory, Oxon, UK.

Derwent, R. G., and Hov, O. (1979). *Computer Modelling Studies of Photochemical Air Pollution in North-West Europe*, AERE-R- 12615. Harwell Laboratory, Oxon, UK.

Derwent, R. G., and Hov, O. (1988). Application of sensitivity and uncertainty analysis techniques to a photochemical ozone model. *J. Geophys. Res.* **93**, 5185–5199.

Evans, R. C., Tingey, D. T., Gumpertz, M. L., and Burns, W. F. (1982). Estimates of isoprene and monoterpene emission rates. *Bot. Gaz. (Chicago)* **143**, 304–310.

Fishman, J., Minnis, P., and Reichle, H. G. (1986). Use of satellite data to study tropospheric ozone in the tropics. *J. Geophys. Res.* **91**, 14451–14465.

Gillani, N. V., and Wilson, W. E. (1980). Formation of ozone and aerosols in power plant plumes. *Ann. N. Y. Acad. Sci.* **338**, 276–296.

Gillani, N. V., Husar, R. B., Husar, J. D., Patterson, D. E., and Wilson, W. E. (1978). Project MISTT:

Kinetics and particulate sulfur formation in a power plant plume out to 300 km. *Atmos. Environ.* **12**, 589–598.

Gladstone, K. P., Niki, H., Shepson, P. B., Bottenheim, J. W., Schiff, H. I., and Sandhu, H. S. (1991). Photochemical oxidant concentrations in two Canadian prairie cities: Model evaluation. *Atmos. Environ.* **25B**, 243–254.

Grennfelt, P., Hoem, K., Saltbones, J., and Schjoldager, J. (1989). *Oxidant Data Collection in OECD-Europe 1985–87 (OXIDATE)*, NILU 63/89. Norwegian Institute of Air Research.

Gusten, H., Heinrich, G., Cvitas, T., Klansic, L., Ruscic, B., Lalas, D. P., and Petrakis, M. (1988). Photochemical formation and transport of ozone in Athens, Greece. *Atmos. Environ.* **22**, 1855–1861.

Haagen-Smit, A. J. (1952). Chemistry and physiology of Los Angeles smog. *Ind. Eng. Chem.* **64**, 1342–1346.

Harrison, R. M., and Holman, C. D. (1979). The contribution of middle- and long-range transport of tropospheric photochemical ozone to pollution at a rural site in north-west England. *Atmos. Environ.* **13**, 1533–1545.

Hatakeyama, S., Izumi, K., Fukuyama, T., and Akimoto, H. (1989). Reactions of ozone with α-pinene and β-pinene in air: Yields of gaseous and particulate products. *J. Geophys Res.* **94**, 13013–13024.

Hegg, D. A., Hobbs, P. V., Radke, L. F., and Harrison, H. (1977). Ozone and nitrogen oxide in power plant plumes. In B. Dimitriades (Ed.), *Photochemical Oxidant Pollution and its Control*. EPA 6003–77–001a, Vol. 1, pp. 173–183.

Hess, G. D. (1989). Simulation of photochemical smog in the Melbourne airshed: Worst case studies. *Atmos. Environ.* **23**, 661–669.

Hough, A. M., and Derwent, R. G. (1987). Computer modelling studies of the distribution of photochemcial ozone production between different hydrocarbons. *Atmos. Environ.* **21**, 2015–2033.

Hov, O., Becker, K. H., Builtjes, P., Cox, R. A., and Kley, D. (1986). *Evaluation of the Photo-oxidants — Precursor Relationship in Europe*, Air Pollution Research Report 1. Commission of the European Communities, Directorate-General for Sci., Res. & Dev., Brussels.

Johnson, D. W., Kilsby, C. G., McKenna, D. S., Saunders, R. W., Jenkins, G. J., Smith, F. B., and Foot, J. S. (1991). Airborne observations of the physical and chemical characteristics of the Kuwait oil smoke plume. *Nature (London)* **353**, 617–621.

Joos, E., and Maffiolo, G. (1989). Airborne measurements of ozone concentrations in background air and power plant plumes. *Atmos. Environ.* **23**, 2249–2257.

Karl, T. R. (1978). Ozone transport in the St Louis area. *Atmos. Environ.* **12**, 1421–1431.

Leahey, D. M., and Hansen, M. C. (1990). Observational evidence of ozone depletion by nitric oxide at 40 km downwind of a medium size city. *Atmos. Eviron.* **24A**, 2533–2540.

Lin, D., Melo, O. T., Hastie, D. R., Shepson, P. B., Niki, H., and Bottenheim, J. W. (1992). A case study of ozone production in a rural area of central Ontario. *Atmos. Environ.* **26A**, 311–324.

Lindsay, R. W., and Chameides, W. L. (1988). High-ozone events in Atlanta, Georgia in 1983 and 1984. *Environ. Sci. Technol.* **22**, 426–431.

Lindsey, R. W., Richardson, J. L., and Chameides, W. L. (1989). Ozone trends in Altanta, Georgia: Have emission controls been effective? *J. Air Pollut. Control Assoc.* **39**, 40–43.

Logan, J. A. (1985). Tropospheric ozone: Seasonal behaviour, trends and anthropogenic infulence. *J. Geophys. Res.* **90**, 10463–10482.

Lyons, W. A., and Cole, H. S. (1976). Photochemical oxidant transport: Mesoscale lake breeze and synoptic-scale aspects. *J. Appl. Meteorol.* **15**, 733–743.

MacKenzie, A. R. (1989). Ph.D Thesis, University of Essex.

MacKenzie, A. R., Harrison, R. M., and Colbeck, I. (1990). The impact of local emissions on the formation of secondary pollutants in urban plumes. *Sci. Total Environ.* **93**, 245–254.

MacKenzie, A. R., Harrison, R. M., Colbeck, I., and Hewitt, C. N. (1991). The role of biogenic hydrocarbons in the production of ozone in urban plumes in southeast England. *Atmos. Environ.* **25A**, 351–359.

McKeen, S. A., Esie, E. Y., and Liu, S. C. (1991). A study of the dependence of rural ozone on ozone precursors in the eastern United States. *J. Geophys. Res.* **96**, 15377–15394.

Meagher, J. F., Lee, N. T., Valente, R. J., and Parkhurst, W. J. (1987). Rural ozone in the southeastern United States. *Atmos. Environ.* **21**, 605–615.

Miller, D. F., and Alkezweeny, A. J. (1980). Aerosol formation in urban plumes over Lake Michigan. *Ann. N. Y. Acad. Sci.* **338**, 219–232.

Neuber, E., Georgii, H., and Muller, J. (1982). Aircraft measurements of oxidants and precursors in the plumes of heavy industrialised areas. In B. Versino and H. Ott (Eds.), *Physico-chemical Behaviour of Atmospheric Pollutants.* Reidel Publ., Dordrecht, The Netherlands, pp. 469–481.

Ogren, J. A., Blumenthal, D. L., and Vanderpool, A. H. (1977). *Oxidant Measurements in Western Power Plant Plumes*, EPRI-421. Electric Power Research Institute, Palo Alto, CA.

Photochemical Oxidants Review Group (PORG) (1987). *Ozone in the United Kingdom*, Harwell Laboratory, Oxon, UK.

Photochemical Oxidants Review Group (PORG) (1990). *Oxides of Nitrogen in the United Kingdom.* Photochemical Oxidants Review Group Harwell Laboratory, Oxon, UK.

Pierson, W. R., Gertler, A. W., and Bradow, R. L. (1990). Comparison of the SCAQS tunnel study with on-road vehicle emission data. *J. Air Pollut. Control Assoc.* **40**, 1495–1504.

Platt, U., Le Bras, G., Poulet, G., Burrows, J. P., and Moortgat, G. K. (1990). Peroxy radicals from night-time reactions of NO_3 with organic compounds. *Nature (London)* **348**, 147–149.

Possiel, N. C., Eaton, W. C., Saeger, M. L., Sickles, J. E., Bach, W. D., and Decker, C. E. (1979). Ozone-precursor concentrations in the vicinity of a medium sized city. *72nd Annu. Meet. Air Pollut. Control Assoc.*, Paper 79–58.3.

Roberts, J. M. (1990). The atmospheric chemistry of organic nitrates. *Atmos. Environ.* **24A**, 243–287.

Seinfeld, J. H. (1988). Ozone air quality models: A critical review. *J. Air Pollut. Control Assoc.* **38**, 616–645.

Seinfeld, J. H. (1989). Urban Air Pollution: State of the Science. *Science* **243**, 745–752.

Sexton, K. (1983). Evidence of an additive effect for ozone plumes from small cities. *Environ. Sci. Technol.* **17**, 402–407.

Sexton, K., and Westberg, H. (1980a). Elevated ozone concentrations measured downwind of the Chicago-Gary urban complex. *J. Air Pollut. Control Assoc.* **30**, 911–914.

Sexton, K., and Westberg, H. (1980b). Ambient hydrocarbon and ozone measurements downwind of a large automotive painting plant. *Environ. Sci. Technol.* **14**, 329–332.

Sexton, K., and Westberg, H. (1983a). Photochemical ozone formation in urban and point-source plumes. *Environ. Sci. Technol.* **17**, 224–227.

Sexton, K., and Westberg, H. (1983b). Photochemical ozone formation from petroleum refinery emissions. *Atmos. Environ.* **17**, 467–475.

Sillman, S., Logan, J. A., and Wofsy, S. C. (1990a). The sensitivity of ozone to nitrogen oxides and hydrocarbons in regional ozone episodes. *J. Geophys. Res.* **95**, 1837–1851.

Sillman, S., Logan, J. A., and Wofsy, S. C. (1990b). A regional scale model for ozone in the United States with subgrid representation of urban and power plant plumes. *J. Geophys Res.* **95**, 5731–5748.

Simpson, D. (1992). Long-period modelling of photochemical oxidants in Europe. Model calculations for July 1985. *Atmos. Environ.* **26**, 1609–1634.

Spicer, C. W. (1982). Nitrogen oxide reactions in the urban plume of Boston. *Science* **215**, 1095–1097.

Spicer, C. W., Gemma, J. L., and Stickel, P. R. (1977). *The Transport of Oxidant Beyond Urban Areas. Data Analysis and Predictive Models for the Southern New England Study*, EPA-600/3-77-041.

Spicer, C. W., Joseph, D. W., and Sticker, P. R. (1979). Ozone sources and transport in the north-eastern United States. *Environ. Sci. Technol.* **13**, 975–984.

Spicer, C. W., Joseph, D. W., and Stickel, P. R. (1982a). An investigation of the ozone plume of a small city. *J. Air Pollut. Control Assoc.* **32**, 278–281.

Spicer, C. W., Koetz, J. R., Keigley, C. W., Sverdrup G. M., and Ward, G. F. (1982b). *Nitrogen Oxides Reactions Within Urban Plumes Transported Over the Ocean*, U.S. EPA Contract No. 68–02–2957.

Stevens, C. S. (1987). Ozone formation in the greater Johannesburgh region. *Atmos. Environ.* **21**, 523–550.

Trainer, M., Buhr, M. P., Curran, C. M., Fehsenfeld, F. C., Hsie, E. Y., Liu, S. C., Norton, R. B., Parrish, D. D., and Williams, E. J. (1991). Observations and modelling of reactive nitrogen photochemistry at a rural site. *J. Geophys. Res.* **96**, 3045–3063.

van Duuren H., Romer, F. G., Diederen, H. S. M. A., Guicherit, R., and van den Hout, K. D. (1982). Measurements by aeroplane of the distribution of ozone and primary pollutants. In B. Versino and H. Ott (Eds.), *Physico-Chemical Behaviour of Atmospheric Pollutants*. Reidel Publ., Dordrecht, The Netherlands, pp. 460–468.

Van Valin, C. C., and Luria, M. (1988). O_3, CO, hydrocarbons and dimethyl sulphide over the western Atlantic ocean. *Atmos. Environ.* **22**, 2401–2409.

Varey, R. H., Clark, P. A., Crane, A. J., Ellis, R. M., Sutton, S., Sprague, P. J., Ball, D. J., Laxton, D. P. H., Woods, P. T., Jolliffe, B. W., Michelson, E., Swan, N. R. W., Cox, R. A., and Sandalls, F. J. (1987). *The London Ozone Survey 1982–85*, Report TPRDL/3142/R87. Central Electricity Generating Board. (Now National Power Technology, Kelvin Avenue, Leatherhead, Surrey KT22 75E, UK.)

Varey, R. H., Ball, D. J., Crane, A. J., Laxen, D. P. H., and Sandalls, F. J. (1988). Ozone formation in the London plume. *Atmos. Environ.* **22**, 1335–1346.

Wakamatsu, S., Ogawa, Y., Murano, K., Goi, K., and Aburamoto, Y. (1983). Aircraft survey of the secondary photochemical pollutants covering the Tokyo metropolitan area. *Atmos. Environ.* **17**, 827–835.

Wayne, R. P., Barnes, I., Biggs, P., Burrows, J. P., Canosa-Mas, C. E., Hjorth, J., Le Bras, G., Moortgat, G. K., Perner, D., Poulet, G., Restelli, G., and Sidebottom, H. (1991). The nitrate radical: Physics, chemistry, and the atmosphere. *Atmos. Environ.* **25A**, 1–203.

Went, F. W. (1955). Air pollution. *Sci. Am.* **192**, 63–72.

Westberg, H. (1985). *Ozone Behaviour in the Combined Baltimore-Washington, D. C. Plume*, EPA/-600/3–85/070.

Westberg, H., Sexton, K., and Roberts, E. (1981a). Transport of pollutants along the western shore of Lake Michigan. *J. Air Pollut. Control Assoc.* **31**, 385–388.

Westberg, H., Sexton, K., and Flyckt, D. (1981b). Hydrocarbon production and photochemical ozone formation in forest burn plumes. *J. Air Pollut. Control Assoc.* **31**, 661–664.

White, W. H., Blumenthal, D. L., Anderson, J. A., Husar, R. B., and Wilson, W. E., Jr. (1977). Ozone formation in the St. Louis urban plume. In B. Dimitriades (Ed.), *Photochemical Oxidant Pollution and its Control*. EPA-600/3–77–001a, vol. 1, pp. 237–247.

Wolff, G. T., Lioy, P. J., Wight, G. D., and Pasceri R. E. (1977). Aerial investigation of the ozone plume phenomenon. *J. Air Pollut. Control Assoc.* **27**, 460–463.

Zeller, K. F., Evans, R. B., Fitzsimmons, C. K., and Siple, G. W. (1977). Mesoscale analysis of ozone measurements in the Boston environs. *J. Geophys. Res.* **82**, 5879–5887.

Zimmerman, P. R. (1979). *Testing of Hydrocarbon Emissions from Vegetation*, EPA/450/4–79/004. U. S. Environmental Protection Agency, Atlanta, Ga.

5

APPLICATION OF A REACTIVE PLUME MODEL TO A CASE STUDY OF POLLUTANT OXIDATION AND ACID DEPOSITION

Eric Joos

Electricité de France, Direction des Etudes et Recherches, 78400 Chatou, France

Christian Seigneur

ENSR Consulting and Engineering, Alameda, California 94501

Environmental Oxidants, Edited by Jerome O. Nriagu and Milagros S. Simmons.
ISBN 0–471–57928–9 © 1994 John Wiley & Sons, Inc.

1. INTRODUCTION

The understanding of the chemical and physical processes that lead to acid deposition has been the subject of considerable research over the past decade [e.g., National Acid Precipitation Assessment Program (NAPAP), 1991]. Although the past 10 years have produced significant advances in our ability to simulate the major atmospheric processes that govern the transport, transformation, and fate of the chemicals that produce acid deposition, some areas have been identified that need further work. To that end, several large research programs are still pursuing refinements of existing modeling tools and development of models that pertain to specific phenomena and that later can be incorporated into large-scale air quality models.

One of the areas of research that has been identified for further work is the development of reactive plume models that can be integrated into regional or mesoscale air quality models. Such air quality models provide a spatial resolution on the order of 100 km (Seigneur, 1990), that is, model grid cells have a volume on the order of more than 10^{11} m^3. Emissions are assumed to be released and mixed instantaneously in these grid cells. The dispersion and chemistry of pollutants emitted from large point sources can be significantly modified if there is instantaneous dispersion of the plume over such a large volume. Therefore, mathematical modeling of the plumes emitted from large point sources such as power plants appears to be an essential component in the improvement of regional air quality models. Such a subgrid-scale treatment of large reactive plumes has been implemented for photochemical air quality models. (e.g., Seigneur et al., 1983). However, the treatment of acid deposition requires taking into account, in addition to dispersion, dry deposition, and gas-phase chemistry, processes such as cloud microphysics, precipitation (wet deposition), and aqueous chemistry.

We present here a reactive plume model that incorporates all the desired processes and can, therefore, be used as the subgrid-scale component of a large-scale air quality model. The formulation of the model is presented in the following section. The application of the model to a case study is described in Section 3. Quantification of acid deposition over episodic and long-term average periods is presented in Sections 4 and 5, respectively.

2.   FORMULATION OF THE REACTIVE PLUME MODEL

The reactive plume model PARADE (Panache Réactif en Atmosphère avec Dépôts) was developed by Electricité de France (EDF) to quantify the short- to medium-range (i. e., up to 100–300 km) contribution of the emissions of a coal-fired power plant to acid deposition. A coal-fired power plant emits gases such as sulfur dioxide (SO_2), nitric oxide (NO), and nitrogen dioxide (NO_2), which undergo transport, dispersion, and transformation in the atmosphere and may lead to dry and wet deposition of acidic species such as sulfates and inorganic nitrates.

The present version of PARADE comprises a little over 10,000 instructions and requires a capacity, on a Cray XMP, of 1,250,000 words (10 megabytes). Execution time ranges from 100 to 1000 seconds, depending on the complexity of the simulation.

An earlier version of the model has been described by Joos et al. (1987). Further model development was performed by Joos and Seigneur (1988) and Joos (1989, 1990).

2.1.   Plume Transport and Dispersion

PARADE treats the transport of the plume by the mean wind and its horizontal and vertical dispersion in the mixing layer beyond the downwind distances corresponding to the final plume rise. It also simulates the various physicochemical interactions that can occur with ambient atmospheric compounds during plume dilution.

In the model, calculations are performed for a crosswind plume section advected by the mean wind. Horizontal dispersion and the dilution of the emitted atmospheric gases take place in an array of contiguous cells (with a maximum of 10 cells) that expand in width as the plume moves downwind. As the plume cross section is advected and the cells expand, the model takes into account mixing of the plume with ambient air as well as transport between cells.

Horizontal dispersion can be either prescribed by providing the evolution of plume width as a function of time, or calculated by means of dispersion coefficients that depend upon atmospheric stability. Horizontal dispersion is Gaussian for a chemically inert species.

Vertical resolution is required for simulating vertical dispersion and the interaction of the plume with clouds and rain. This vertical resolution is represented by means of contiguous horizontal layers (with a maximum of 10 layers, each layer consisting of the contiguous cells dealing with the horizontal dispersion), within which chemical species concentrations—including ambient air concentrations—and values of meteorological variables (i.e., temperature, relative humidity, pressure, liquid water content, updraft velocity in case of clouds, and preci- pitation characteristics) are allowed to vary with time. The vertical transport processes through these horizontal layers consist of vertical diffusion, convective transport due to the presence of clouds, and precipitation.

2.2. Physicochemical Interactions with a "Dry" Atmosphere

PARADE treats 90 gas-phase chemical reactions, including 10 photochemical reactions, involving about 60 ambient species, particularly organic compounds (Whitten and Gery, 1985). When the plume does not interact with clouds or precipitation (i.e., in the case of an unsaturated atmosphere), PARADE calculates the mass concentrations and sulfate/nitrate/ammonium/water chemical composition of aerosols along the plume cross section. This calculation is based on thermodynamic considerations. The thermodynamic equilibria considered involve the gas-phase ammonia and water-vapor concentrations present in the ambient atmosphere and the concentrations of sulfuric acid and nitric acid produced by the oxidation reaction of SO_2 and NO_2 (Saxena et al., 1986). When relative humidity is high ($> 70\%$), PARADE takes into account the major reaction of SO_2 that can occur in aqueous aerosols, particularly the oxidation of SO_2 by dissolved hydrogen peroxide (H_2O_2) (Saxena and Seigneur, 1987).

An option of PARADE allows the evolution of the aerosol size distribution to be calculated by taking into account diffusion, condensation, coagulation, and sedimentation processes and by using a sectional representation of the aerosol size distribution over the 0.01- to 10-μm size range (Seigneur et al., 1986).

2.3. Physicochemical Interactions with Clouds and Rain

When the plume interacts with precipitating or nonprecipitating clouds, a module of PARADE allows us to calculate the distribution of water among four phases—vapor, cloud droplets, raindrops, and ice—for each time step as a function of the vertically distributed characteristics of the cloud (updraft velocity, temperature, pressure, and total water content). In the case of rain, this module also provides the precipitation rate and the raindrop fall velocity as a function of altitude (Seigneur and Wegrecki, 1990). It should be noted that an option of PARADE allows the different cloud, and eventually, precipitation characteristics to be input directly as a function of altitude and time.

A module coupled to the cloud microphysics module treats aqueous-phase chemistry. This module takes into account mass transfer between the gas and the liquid phases. Mass transfer processes are based on Henry's law equilibrium constants for 14 species and, for others, on mass transfer limitation due to gas- or liquid-phase diffusion (Seigneur and Saxena, 1988). Sixteen ionic dissociation equilibria and about 30 chemical reactions between 40 species are treated within the aqueous phase. The oxidation of SO_2 takes place primarily through reactions involving dissolved hydrogen peroxide and ozone. The contribution of the latter reaction is significant, especially at high pH. Aqueous-phase nitrates result primarily during the day from the dissolution of nitric acid formed in the gas phase.

2.4. Determination of Dry and Wet Deposition

PARADE concludes the calculation cycle at each time step with the quantification of wet and/or dry deposition for the major species that contribute to acid deposition.

The calculation of wet deposition is performed by taking into account the raindrop fall velocities, the precipitation rate provided by the vertical transport module, and the raindrop chemical composition provided by the aqueous-phase chemistry module. For each cell in each layer (including the surface layer), PARADE also provides the chemical composition of precipitation. However, it is assumed that, in a given cell, the chemical composition of cloud droplets (if present) and raindrops is identical.

The calculation of dry deposition of gases and aerosols includes the effect of three resistances: aerodynamic resistance, surface-layer resistance, and transfer resistance. Therefore, the deposition velocity calculation integrates atmospheric stability, wind speed, season, and surface layer. Values of dry deposition of gases and aerosols may be provided directly to the code.

Two types of values are calculated by PARADE for dry and wet deposition: a local deposition rate under the plume cross section at a given time and, therefore, at a given downwind distance from the power plant; and a deposition rate averaged between initial time/location and local time/location.

3. DESCRIPTION OF THE CASE STUDY

The Blénod power plant was selected to provide the experimental case study for the application of PARADE. It is located near Pont-à-Mousson, between Metz and Nancy in eastern France. Due to the prevailing wind sector in this area (winds coming from the west), the emissions from this power plant may contribute to acid deposition on the Vosges forest.

The period considered for the calculation of deposition was six months, from October to March. This period was selected because, due to the extensive use of nuclear power in France and the smaller consumption of power during the spring and summer months, the coal-fired plants operate mostly in autumn and winter.

Such a case study reveals a great number of uncertainties and requires many simplifying assumptions concerning the values of the chemical and weather-related variables taken into account by the model. Accordingly, we deemed it reasonable to simulate only one base case over the six-month period, with typical characteristics of the dispersion of the plume and the physicochemistry of the ambient air. Likewise, the meteorological scenarios applied to this base case represent means over the period considered. The uncertainties and simplifications required for the modeling task are discussed in Sections 3 and 4.

3.1. Characterization of the Atmosphere: Meteorological Data

3.1.1. Mixing Layer/Temperatures

For cloudless atmospheres, we selected a neutral vertical atmospheric stability—Pasquill-Gifford class D—the temperature gradient with altitude being $-1\,°C/100\,m$. When stratus clouds are present, whether precipitating or not, we selected a more stable class of atmospheric stability—class E—the temperature gradient being in this case $-0.5\,°C/100\,m$.

The mixing layer, located between the ground and the first inversion, is 1000 m thick during both day and night (synoptic regime). It comprises three sublayers 300, 400, and 300 m thick. The temperatures in these three sublayers, which influence the gas- and liquid-phase chemistry as well as the formation of the aerosols, were calculated based on the monthly mean ground-level temperature at Metz and had a 3-hr resolution.

Averaging the values so obtained produced a mean temperature for the gas-phase chemical reactions of 1.2 °C for the "dry" atmospheres and 2.8 °C for the cloudy atmospheres. This mean temperature was held constant during the simulation.

3.1.2. Wind Directions and Speeds

The meteorological data available for the Metz station reveal, for the entire six-month period, a preferential sector of 180° to 280° (winds with origins ranging from the south to the west) corresponding to about 50% of the noncalm winds (> 1 m/sec) blowing during the year. Based on these data, we determined an average wind speed of 4 m/sec constant through all of the simulations. For the prevailing winds in autumn–winter, the mean frequencies of occurrence by direction are as follows:

Direction (°):	180	200	220	240	260	280
Frequency (%):	7.8	8.1	10.0	7.4	8.0	6.5

It must be noted that, since PARADE is a Lagrangian model, the wind speed is constant with altitude. Moreover, the wind direction does not have an explicit influence; it is taken into account only during the final integration of the depositions from the various scenarios.

3.1.3. *Clouds and Precipitation: Frequency of Occurrence and Characteristics*

The frequency of occurrence of clouds and rain in a given region over several months can only be approximately estimated, because of the small amount of relevant information available. Data from the Chooz site, which is located in the prevailing wind area, show that the proportion of nonovercast situations (i.e., clear or partly cloudy skies) was about 60%. In the case of partly cloudy skies, given the height of the clouds and their erratic presence, it appeared reasonable to assume that the clouds do not interact significantly with the plume. In the case of overcast skies (40% of the total time), only 75% are characterized by interaction of the plume with rain and/or clouds.

The frequency of occurrence of rain was estimated from data obtained at the Faulquemont meteorological station, located 40 km from Metz in the 240° direction. The mean annual precipitation, based on 39 years of observations (1951–1989), is 810 mm, of which 48% falls in the autumn and winter. Based on a precipitation rate of about 0.9 mm/hr—the mean rate observed in 1989 at Metz during the months considered—the frequency of occurrence of precipitating clouds is calculated to be 10%. Therefore, the model simulations involved the simplifying assumption that during the period considered and in the prevailing wind sector the plume interacts 10% of the time with precipitating clouds whose precipitation rate is 0.9 mm/hr.

The simulated meteorological scenarios and their frequency of occurrence are thus as follows:

Scenario 1—cloudless atmosphere or atmosphere with nonprecipitating clouds having no interactions with the plume: 70%

Scenario 2—atmosphere with nonprecipitating clouds and some interaction with the plume: 20%

Scenario 3—atmosphere with precipitating clouds and interactions with the plume: 10%.

The clouds interacting with the plume were considered to be only stratus, or low-convection, clouds because this cloud type occurs most frequently during the period considered. In the case of a nonprecipitating stratus cloud, the typical cloud is located in the top layer of the model and has a liquid water content of $0.5 \, g/m^3$. In the case of a precipitating stratus, the typical cloud is located in the top two layers of the model (including the one where the plume is located), and has a liquid water content of $0.5 \, g/m^3$ and a precipitation rate of 0.9 mm/hr.

3.2. Characterization of the Atmosphere: Physicochemical Data

3.2.1. *Ambient Concentrations*

Ambient concentration profiles were assumed to be homogeneous with altitude. The concentrations of chemicals that play an essential role in the plume/atmosphere physicochemical interactions were derived from the field campaigns (Joos et al., 1985a, 1986; Joos and Maffiolo, 1989) or from the literature (Nodop and Georgii, 1990) and are as follows [gaseous concentrations are expressed in ppbv (parts per billion—10^{-9}—by volume)]: ozone $(O_3) = 30$ ppbv; hydrogen peroxide $(H_2O_2) = 0.1$ ppbv; ammonia $(NH_3) = 2$ ppbv; nonmethane hydrocarbon compounds (NMHC) $= 47$ ppbvC. The ambient concentrations of sulfate aerosols and nitrate aerosols are 4 and 2 $\mu g/m^3$, respectively. The NO and NO_2 concentrations, respectively 2 and 5 ppbv, give a [NMHC] to $[NO_x]$ ratio of 6.7 to 1. The ambient concentration of SO_2 is 5 ppbv in the cloudless layers and 0.5 ppbv when stratus are present. The latter concentration was chosen to obtain, in the cloud aqueous phase, a concentration of dissolved hydrogen peroxide approximating the values found in the literature (e.g., Kelly et al., 1985) and measured in similar circumstances.

3.2.2. Dry Deposition Velocities

The calculation of the dry deposition flux is performed by multiplying, for each chemical species, its deposition velocity by its concentration in the first (surface) layer at a given distance from the power plant. The dry deposition velocity values were directly provided to PARADE and correspond to mean values estimated on the basis of a detailed study of the data reported in the literature (Sehmel, 1980; Dolske and Gatz, 1985; Voldner et al., 1986; Walcek et al., 1986; Nicholson and Davies, 1987; Nicholson, 1988). The daytime/nighttime velocities, in cm/sec, are as follows: SO_2: 0.5/0.25; NO_2: 0.25/0.15; NO: 0.15/0.10; HNO_3: 2/1.5; aerosols of sulfate and nitrate: 0.2/0.1.

3.2.3. Photochemical Reactions

Photochemical constants depend on solar insolation and hence on time of year. Preliminary simulations showed that the optimal date for calculating, in the base-case simulation, the photolytic constants was mid-February. The change in value of the zenith angle measuring the sun elevation in relation to the horizon line therefore constitutes a satisfactory average between the low values observed in December and January and the higher values observed in October and March.

   For overcast skies, as in scenarios 2 and 3, an attenuation coefficient of 0.6 was appropriately introduced for the photolytic constants. For scenario 1, the attenuation coefficient was set to 0.8 to take into account the attenuation due to aerosols or to clouds not interacting with the plume (Joos and Seigneur, 1988).

3.3. Plume Dispersion/Initial Concentrations

3.3.1. Plume Dispersion

The base-case simulation involves transport and dispersion of the plume up to a distance of 300 km, with an average wind of constant speed and direction. As described in Section 2, the lateral dispersion of the plume is either directly input to the model or calculated from Pasquill-Gifford-type dispersion coefficients that depend upon atmospheric stability. Preliminary simulations showed that the first alternative is preferable.

We used a relationship between the width L of the plume and the distance X from the source

$$L = a \cdot X^b$$

where a and b are the coefficients calculated from experimental data (Hanna et al., 1982). The widths of the cross sections taken during the plume-measuring flights at given distances from the plant were used only (1) if the weather conditions were sufficiently stationary, in particular with a relatively constant wind direction, to allow satisfactory characterization of the plume; (2) if the power plant load was deemed sufficient (between 700 and 1100 MW); and (3) if the atmosphere stability was neutral (Joos et al., 1985a, b, 1986, 1990).

A graphic representation of the plume width/distance pairs is given in Figure 1. The straight line $\log L = 0.921 \cdot \log X - 0.237$ is deduced from the graph with an excellent correlation factor of 0.92.

This relationship, established up to the maximum exploration distance of 100 km with a corresponding width of 23,500 m, is extrapolated up to 300 km with a width of 64,200 m. Note that the widths derived from formulas such as those of the Brookhaven National Laboratory or of Briggs (Hanna et al., 1982), at the same distances and with a neutral atmospheric stability, are clearly less than those calculated above. In fact, at 300 km, the values are smaller by a factor

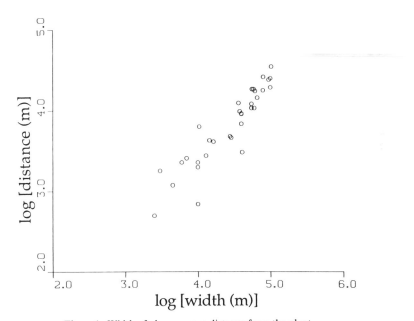

Figure 1. Width of plume versus distance from the plant.

of 2.5 and 3.5, respectively. This difference is certainly due to the conditions for which the formulas were derived, that is, emissions near ground level, plumes with no buoyancy, and measurements taken at small distances from the source. At 300 km, the width of 64,200 m corresponds to an angle of dispersion of 12° that is clearly smaller than the 20°-angle sector of wind directions. It is therefore unnecessary, at any given location, to integrate the wind directions when assessing the contribution to the deposition of plumes transported by the prevailing winds. Conversely, if the occurrence frequencies presented in Section 3.1 are used, deposition will be overestimated given the values of the angles mentioned above, since a single wind sector covers the dispersion of about two plumes.

The vertical dispersion of the plume between the various model layers is treated with coefficients of vertical diffusion that depend on atmospheric stability.

3.3.2. Concentrations in the Initial Layer

It is assumed that the power plant operates at the steady state throughout the period under consideration, with a load of 1000 MW. The initial plume cross section is taken at 2 km from the source and has an initial width of 650 m. The plume is then located in the second layer, between 300 and 700 m above ground level, with the following concentrations: $[SO_2] = 450$ ppbv; $[NO] = 360$ ppbv; $[NO_2] = 60$ ppbv. These values were calculated from the emission fluxes at the stack.

3.4. Uncertainties and Simplifying Assumptions

The simulations required for such a case study involve numerous uncertainties associated with either the meteorological variables or the physicochemical parameters. These uncertainties result from the great variety of possible situations and/or incomplete knowledge of the processes treated by the model. Limitations due to the model formulation must also be taken into account.

3.4.1. Main Uncertainties Associated with the Meteorological Variables

3.4.1.1. *Thermal Structure of the Atmosphere and Mixing Layer.* The temperature versus altitude profiles that govern the vertical dispersion of pollutants are highly variable in the planetary boundary layer (i.e., in the first 1000 m). They result from radiative exchanges between the ground and the atmosphere and present both daily and seasonal effects. At nightfall and during the night, a temperature inversion may occur at ground level and rise over time, thereby preventing the dry deposition of pollutants.

Since it is impossible to estimate the frequency of the various possible scenarios over the 6-month period under consideration, we consider only one mixing layer in the base case with a thickness of 1000 m during both daytime and nighttime. The model treats this mixing layer using three sublayers. The vertical temperature gradient is assumed to be adiabatic except when stratus clouds are present. This approximation is therefore likely to lead to an overestimate of dry deposition.

3.4.1.2. Wind Speeds and Directions. The wind speed and direction data used in this study are surface data available from the Metz meteorological station. Based upon these values, we determined a mean speed at plume height in the 180° to 280° wind sector. However, the wind's speed and direction can vary with altitude and time (wind shear). The lack of pertinent information, not to mention the difficulty of modeling nonsteady-state situations, prompted the use of the following simplifying assumption for the base case: in the mixing layer, the wind is constant in speed and direction from the initial cross section up to the end of the simulation (i.e., 300 km from the power plant, which corresponds to 21 hr of plume transport), a maximum acceptable duration for plume travel time.

3.4.1.3. Meteorological Scenarios and Cloud Characteristics. In two of the average meteorological scenarios applied to the base case, clouds that may or may not be precipitating interact with the plume. Although it was possible to estimate their frequency of occurrence, the choice of their characteristics—height, liquid water content, updraft velocity—and of the duration of their presence was a more difficult matter because of the very large number of possible situations and of the lack of data.

The choice of stratus and their characteristics is, a priori, a reasonable approximation given the time of year considered, but it is, nevertheless, an approximation whose validity is impossible to estimate at this point.

Regarding the time during which the stratus are present and the time during which they interact with the plume, we assumed in the scenario with non-precipitating clouds that they are present throughout the simulation (i.e., for 21 hr) with a continuous rainfall. Here again, the lack of pertinent data requires this simplifying assumption.

3.4.2. Main Uncertainties Associated with the Physicochemical Data

3.4.2.1. Ambient Concentrations. In the ambient air, a certain number of gaseous species give rise to physicochemical reactions with the nitrogen and sulfur oxides present in the power plant plume during its transport and dispersion. Some of these ambient species can be considered primary oxidants, such as:

Ozone, which oxidizes NO to NO_2 in the gas phase and oxidizes SO_2 in the liquid phase to sulfate

Hydrogen peroxide, which oxidizes SO_2 to sulfate in the liquid phase.

Others may be considered secondary oxidants, the presence of which depends on the photochemical activity:

The hydroxyl radical (OH), which oxidizes in the gas phase SO_2 to sulfates and NO_2 to nitric acid

The peroxyl radical (HO_2), which oxidizes in the gas phase NO to NO_2

The various hydrocarbon radicals produced by NMHC oxidation, which oxidize in the gas phase NO to NO_2.

The concentrations of the first set must be supplied to the model as input data along with the concentrations of all the other molecular species constituting primary reactants such as SO_2, NO, NO_2, CO, the NMHC, and NH_3, which may neutralize nitric acid produced by NO_2 oxidation to yield nitrate aerosols. The model computes the gas-phase concentrations of secondary oxidants in the plume from the concentrations of the molecular species. Incidentally, the ratio of the concentration of NMHC to NO_x characterizing the atmosphere is important because of its influence on the concentrations of ozone and oxidizing radicals OH and HO_2. This ratio is highly variable. The ratio selected here corresponds to an average reactivity. When precipitating or nonprecipitating clouds are present, PARADE computes the chemical composition of the water droplets of the stratus and that of the interstitial gaseous phase based on the concentration of primary gaseous reactants. The liquid-phase concentrations of H^+ (strong acidity), hydrogen peroxide, and ozone are particularly important since they govern the oxidation of dissolved SO_2 to sulfates.

The ambient concentrations of the primary reactants change considerably over time and space, especially as a function of altitude. They are a function of their emission rates, of their chemical or photochemical reactivities (and hence of the presence of the corresponding reactants and of the insolation), of the atmospheric stability governing their dispersion, and of their dry deposition velocity. The numerous data available in the literature generally relate to studies with specific conditions and/or specific times (of day and year) and locations. They are consequently difficult to use for conditions different from those for which they were obtained. Furthermore, many measurements are made at ground level where, because of the emission sources, the concentrations are often greater than at higher elevations.

In view of these substantial uncertainties, it appeared reasonable to use profiles of concentrations that are uniform with altitude and do not change during the simulation.

3.4.2.2. Dry Deposition Velocities. The dry deposition velocities depend on the pollutant being considered, the season, the amount of sunlight, the wind speed, the humidity, the temperature, the thermal structure of the atmosphere, and the characteristics of the surface receiving the deposited species (pasture, field, deciduous or evergreen forest, builtup areas, presence of water, snow, etc.). Thus, dry deposition rates are generally greater in spring and summer than in autumn and winter, and greater in the daytime than at night. They are also greater on farmland, for example, than on forests. In addition, dry deposition on vegetation is affected by some physiological mechanisms, especially stomatal processes (respiration, excretion).

The numerous values found in the literature are based on measurements that reflect a broad variety of situations combining the different parameters that govern deposition. However, all of the information about the relevant parameter is rarely supplied. Thus, a satisfactory synthesis is again difficult to achieve, especially since the different meteorological techniques used may lead, under similar conditions, to markedly different results. For some of the species, such as SO_2, the dry deposition velocities can vary by more than an order of magnitude, or even more than two orders of magnitude for particles, whose deposition velocity is also a function of their size.

The values selected for the base case thus constitute further approximations in the model. Moreover, PARADE, in its handling of plume dispersion, does not take into account possible variations in relief and, *a fortiori*, their effects on dry deposition or on the formation of orographic clouds that may give rise to precipitation.

In this section, we have recapitulated the principal simplifying assumptions and approximations needed to reasonably define the case study. Given the cumulative effect of all these uncertainties, and often the impossibility of making quantitative estimates of them, it does not appear warranted to refine any treatment of a specific parameter. Nevertheless, it should be noted that in general we have always favored those simplifying hypotheses that did not, a priori, induce an underestimation of deposition.

4. CALCULATION OF DRY AND WET DEPOSITION

For each of the three meteorological scenarios applied to the base case, we carried out four simulations to cover the daily cycle. The starting hours of these simulations were 12:00, 18:00, 0:00, and 6:00 (Greenwich mean time). Indeed, parameters such as temperature in the various layers exhibit daily variations that must be taken into account since temperature governs, for example, chemical processes in the liquid phase, the dry deposition rates, and the kinetic rates of the photochemical reactions.

In order to determine the net contribution of the plume to acid deposition, a second simulation was carried out under the same conditions, except for the substitution of ambient concentrations of SO_2 and NO_x for those in the initial cross section of the plume. The results of this background simulation were then substracted from the first simulation to provide the incremental contribution of the plume.

At a given distance from the power plant, for a given meteorological scenario, the dry or wet deposition of a species is the mean value taken over 24 hr of the instantaneous depositions (expressed in kilograms per hectare per minute) supplied by the four simulations associated with the four different emission times. The distances from the plant used for the integral calculation of the deposits are 10, 100, and 300 km.

The net contribution of the plume to the daily dry and wet deposition is presented in the following sections for the different weather scenarios.

4.1.   Daily Plume Contribution: Meteorological Scenario 1

For scenario 1, which corresponds to the case of a plume without any cloud inter-actions, the daily mean values of dry deposition, in grams of sulfur and nitrogen per square meter, are listed in Table 1 for the different distances considered.

The sulfate aerosol deposition remains clearly less than the SO_2 deposition, due to the latter's slow oxidation rate in the gas phase, especially in autumn and winter. At 300 km, the order of magnitude of the ratio of dry SO_2 deposition to sulfate aerosol deposition is about 250. The dry deposition of SO_2 is about seven times less at 300 km from the plant than at 10 km.

As for the nitrogen-bearing deposition, the deposition of nitrates (nitric acid plus nitrate aerosols) is always greater than the deposition of sulfate. It increases as one moves away from the power plant due to a nighttime chemistry different from that governed by sunlight. It is important to emphasize the combination of various uncertainties in several physicochemical data that affect the nitrate dry deposition rate. The formation of aerosols of nitrate from gaseous HNO_3 de-pends on the latter's concentration and on the concentration of the ambient NH_3 available after neutralization of the H_2SO_4 formed by the oxidation of SO_2. In addition, the thermodynamic equilibria taken into account in PARADE to compute the aerosols' chemical composition allow for the effect of relative humidity and temperature. In the present case, the contribution of aerosols to nitrate deposition varies from 25% to 65%. At night, the formation of HNO_3 results from the following reaction, which occurs heterogeneously at the surface of droplets or aqueous aerosols:

$$N_2O_5 + H_2O \longrightarrow 2HNO_3 \tag{1}$$

Therefore, nitrate formation depends on the relative humidity (70% on the average base case) and the N_2O_5 concentration. N_2O_5 is formed from the reac-tion between the radical NO_3, which is rapidly photolyzed during the daytime, and NO_2. The kinetic constant of this reaction has not yet been satisfactorily established; however, the value used in the reactional scheme is considered to be an upper limit (Sverdrup et al., 1987). Finally, certain dry deposition velocity values, especially for nitrate aerosols, are relatively imprecise.

Table 1   Daily Mean Dry Deposition from the Plume for Meteorological Scenario 1

Distances from Power Plant (km)	Daily Mean Dry Deposition				
	Sulfate Aerosols ($g\,S/m^2$)	Nitrate Aerosols ($g\,N/m^2$)	SO_2 ($g\,S/m^2$)	NO_2 ($g\,N/m^2$)	NO ($g\,N/m^2$)
10	2.5×10^{-7}	1.0×10^{-6}	3.2×10^{-3}	6.1×10^{-4}	7.8×10^{-5}
100	1.7×10^{-6}	4.0×10^{-6}	1.4×10^{-3}	2.9×10^{-4}	2.7×10^{-5}
300	2.0×10^{-6}	8.0×10^{-6}	4.5×10^{-4}	8.2×10^{-5}	8.1×10^{-6}

In summary, the main parameters that affect the calculation of dry nitrate deposition are as follows:

Relative humidity
Temperature
Ambient concentrations of NH_3 and HNO_3
Kinetic constant of Reaction (1)
Dry deposition velocities of nitric acid and especially of nitrate aerosols.

There is 7.5 times less dry deposition of NO_2 at 300 km than at 10 km from the plant; NO deposition is always much less (by about a factor of 10) than NO_2 deposition because the oxidation of NO by ozone is fast and, additionally, the velocity of dry deposition of nitric oxide is less than that of nitrogen dioxide. At 300 km from the plant, the order of magnitude of the dry deposition ratio of NO_2 and NO to nitrates is 10.

4.2. Daily Plume Contribution: Meteorological Scenario 2

In the case of plume interactions with nonprecipitating clouds in the upper layer, the mean daily values of dry deposition due to the plume are given in Table 2. In comparison with Table 1, it can be seen that first, the values of dry deposition of sulfates are three to six times greater. This difference is not due to the fallout of sulfates formed in the liquid phase in the cloud (they remain trapped in the droplets), but rather to the deposition of sulfates appearing beneath the cloud. In fact, the relative humidity was taken as greater than 80% and a nonnegligible liquid water content appears in the aerosols, which allows the SO_2 to dissolve and then to oxidize into sulfates. Second, the values of dry deposition of nitrates are two to four times less due to less photochemical activity and a more stable atmosphere that limits fallout to the ground. While the oxidation of SO_2 to sulfates is highly efficient in the liquid phase—via such oxidants as H_2O_2 or O_3—the nitrates present in the liquid phase result almost exclusively from the dissolution of the nitric acid formed in the gas phase or at the surface of droplets.

Table 2   Daily Mean Dry Deposition from the Plume for Meteorological Scenario 2

Distance from Power Plant (km)	Daily Mean Dry Deposition				
	Sulfate Aerosols ($g\,S/m^2$)	Nitrate Aerosols ($g\,N/m^2$)	SO_2 ($g\,S/m^2$)	NO_2 ($g\,N/m^2$)	NO ($g\,N/m^2$)
10	1.5×10^{-6}	5.0×10^{-7}	2.4×10^{-3}	2.6×10^{-4}	2.4×10^{-5}
100	5.0×10^{-6}	1.2×10^{-6}	1.1×10^{-3}	1.3×10^{-4}	1.0×10^{-5}
300	6.0×10^{-6}	6.5×10^{-6}	3.0×10^{-4}	4.7×10^{-5}	4.0×10^{-6}

Finally, the lesser dry deposition of SO_2, NO_2, and NO can be explained mainly by the differences in vertical stability.

4.3.   Daily Plume Contribution: Meteorological Scenario 3

The mean daily values of wet and dry deposition are presented in Table 3 in terms of a gram of sulfur and nitrogen per square meter, except for the H^+ ions. It appears that the wet deposition values of sulfates and nitrates are clearly greater, in a ratio of 1400 and 90 to 1, respectively, than their dry deposition at 100 km from the plant. The values of dry deposition of SO_2 and NO_x are smaller in scenario 3 than in scenario 1 for the same distances; as in the case of scenario 2, this is mainly due to a more stable atmosphere, which is less favorable to dry deposition in scenario 3 than in scenario 1, and to a partial washout of the SO_2. The nitrogen oxides are not much affected, because of their negligible solubility. The wet deposition of SO_2, which in fact consists of deposition of liquid SO_2 and of ionized forms [sulfite (SO_3^{2-}) and bisulfite (HSO_3^{2-})], is always greater than the dry deposition.

4.4.   Calculation of the Integrated Deposition Rates—Comparison with the Experimental Values Derived from the Literature

For a species k producing a plume contribution to dry deposition, the value of the plume deposition rate, integrated over 6 months for a given wind direction and a given distance from the plant, is given by the following equation:

$$DD_k = (0.7 \cdot DDS1_k + 0.2 \cdot DDS2_k + 0.1 \cdot DDS3_k) \cdot 365/2 \cdot f$$

where $DDS1_k$, $DDS2_k$, and $DDS3_k$ are the mean values of daily dry deposition of species k, corresponding respectively to meteorological scenarios 1, 2, and 3, at the distance considered (note that for some species—SO_2, NO_2, NO—the three terms must be taken into account), and f is the frequency of occurrence of the wind in the considered direction with respect to the power plant. Under the same conditions, the integrated value of wet deposition of the species k will be given by the following equation:

$$WD_k = 0.2 \cdot WD3_k \cdot 365/2 \cdot f$$

where $WD3_k$ is the mean value of daily wet deposition of species k, corresponding to scenario 3, at the distance considered.

Table 4 summarizes the calculated values of wet and dry deposition rates integrated from the daily values, as well as the equivalent annual mean experimental values based on measurements available in the literature. These literature data were divided by 2 for purposes of comparison with the values calculated in the 6-month simulations.

Table 3 Daily Mean Dry and Wet Deposition from the Plume for Meteorological Scenario 3

Distance from Power Plant (km)	Wet Deposition				Dry Deposition		
	Sulfates (g S/m²)	Nitrates (g N/m²)	H⁺ (g/m²)	SO₂ (g S/m²)	SO₂ (g S/m²)	NO₂ (g N/m²)	NO (g N/m²)
20	8.8×10^{-3}	1.1×10^{-4}	6.5×10^{-4}	2.6×10^{-3}	1.6×10^{-3}	2.6×10^{-4}	2.4×10^{-5}
100	2.4×10^{-3}	3.5×10^{-4}	2.9×10^{-4}	3.4×10^{-4}	9.2×10^{-4}	1.3×10^{-4}	1.0×10^{-5}
300	1.0×10^{-3}	2.9×10^{-4}	1.0×10^{-4}	8.7×10^{-4}	1.4×10^{-4}	4.7×10^{-5}	4.0×10^{-6}

Table 4 Comparison of Calculated and Measured Values of Wet and Dry Deposition

Distance from Power Plant in 220° Sector (km)	Wet Deposition			Dry Deposition				
	Sulfates (g S/m²)	Nitrates (g N/m²)	H⁺ (g/m²)	Sulfates (g S/m²)	Nitrates (g N/m²)	SO₂ (g S/m²)	NO₂ (g N/m²)	NO (g N/m²)
Calculated Values								
10	0.021 $(8.4\%)^{a}$	2.0×10^{-4}	1.2×10^{-3} (8.0%)	1.0×10^{-5} (0.1%)	1.5×10^{-5}	0.053 (17.7%)	9.2×10^{-3} (4.6%)	1.1×10^{-3} (7.3%)
100	0.011	6.4×10^{-4} (0.9%)	5.3×10^{-4}	4.0×10^{-5}	5.6×10^{-5}	0.024	4.4×10^{-3}	4.0×10^{-4}
300	3.4×10^{-3}	5.3×10^{-4}	1.8×10^{-4}	4.7×10^{-5} (0.1%)	1.3×10^{-4} (0.2%)	7.1×10^{-3}	1.3×10^{-3}	1.3×10^{-4}
Measured Values								
	0.25–0.85	0.07–0.45	0.015–0.060	0.05–0.07	0.07–0.15	0.30–0.45	0.20–0.25	0.015

ᵃThe plume's contribution to the aggregate deposition measured is indicated in parentheses.

Three types of integrated values were calculated in the 220° sector (south-westerly wind), where the frequency of occurrence is the highest (10%) at 10,100, and 300 km from the plant, the latter two points being in Germany, toward the Ruhr. For the highest calculated value of wet and dry deposition of each of the species considered, the percentage value of the plume's contribution to the aggregate deposition measured is indicated in parentheses in Table 4. The smallest experimental mean value was taken as reference.

The lowest experimental values of wet deposition of sulfates and nitrates were taken from measurements made in eastern France in 1985, by the EMEP network, in the context of the "Co-operative Programme for Monitoring and Evaluation of Long-Range Transmission of Air Pollutants in Europe" (Nodop and Georgii, 1990). The highest values in the deposition range are averaged values measured, in 1989 to 1990, at different locations in the Vosges Mountains (Dambrine and Nourrisson, 1991). As the data from the EMEP network also relate to the annual mean concentrations in the ambient air of SO_2, NO_2, and sulfate aerosols, we deduced dry deposition values using the same deposition velocities at those entered in the model. The dry deposition values for NO were deduced from mean annual concentrations measured in 1986 to 1987 in the Vosges Mountains (Drach and Target, 1987). The lowest mean value of H^+ was obtained from the mean annual concentration measured in precipitation in eastern France and the mean annual height of precipitation (Unsworth and Fowler, 1988). The highest mean value is an averaged value measured in the Vosges Mountains (Dambrine and Nourrisson, 1991). The wet deposition of SO_2 is in fact included in the experimental deposition values for sulfates, since oxidation of sulfites to sulfates, in the liquid phase, continues on the ground. The calculated value of wet deposition of sulfates thus includes both contributions. Lastly, we failed to find any values in the literature for dry nitrate deposition in eastern France. We therefore used mean values as measured in the Netherlands in 1986 (Erisman et al., 1989), duly corrected to allow for the different velocity of dry deposition of HNO_3. The values thus found agree with other measured values (Ro et al., 1988).

The comparison between calculations and measurements shown in Table 4 reveals that for the chosen distance/direction pairs, the plume's net contribution to dry sulfate deposition and to wet and dry nitrate deposition is always very small (< 1%). This result is mainly due to less photochemical activity and lower rates of dry deposition during the part of the year under consideration. Further calculations showed that these deposition rates would be greater by about one order of magnitude in the spring and summer. The plume's contribution to dry deposition of NO and NO_2 is 7.3 and 4.6%, respectively, at 10 km downwind from the plant (remember that we are using the smallest value of the interval as the experimental reference), and decreases to 0.9 and 0.7%, respectively, at 300 km downwind. The dry deposition of SO_2 from the plume represents the largest relative contribution: 17.7% at 10 km, 8% at 100 km, and still 2.4% at 300 km. Note that previous simulations showed that dry deposition of SO_2 and NO_x from the plume peaked at about 10 km from the plant. In terms of wet deposition of sulfates and H^+, the plume's contribution to the overall deposition is 8.4 and 8.0%, respectively, 10 km downwind. These figures both drop to less than 1.5% 100 km downwind.

All the contributions of the plume to the various deposition fluxes were obtained with a base case involving several simplifying assumptions. Three of these assumptions tend to overestimate deposition (especially dry deposition of SO_2, which is little affected by physicochemical processes). These assumptions were constant wind direction with altitude and over time; no temperature inversion near the ground, especially toward the end of the afternoon and at night; and continuous operation of the plant with a load of 1000 MW.

In reality, the first two assumptions are never true over periods of time as long as those considered here for dispersion of the plume and for integration of the deposition rates. Therefore, the uncertainty of the calculated values increases from the 10-km distance to the 300-km distance. Frequent wind shear and the more or less frequent occurrence of temperature inversions at low altitudes (< 300 m) will have the effect, respectively, of increasing the plume's lateral dispersion and preventing pollutants from reaching the ground for several hours of their transport time, during which they will continue to disperse. As a consequence, the deposition rates will be more spread out and therefore lower, although deposition will cover a larger area, especially since the wind speed may increase with altitude. On the other hand, it should be noted that the effects of complex terrain may increase deposition rates in some areas.

5. CONCLUSION

We used the PARADE model to study the contribution of the Blénod thermal power plant to wet and dry deposition of SO_2, NO_x, sulfates, nitrates, and H^+ ions at distances of 10, 100, and 300 km from the plant, over a period of 6 months corresponding to the autumn and winter seasons.

This study involved the simulation of three meteorological scenarios applied to a base case of plume dispersion and atmospheric physicochemistry. The values of the variables characterizing the scenarios were selected to be mean values for the period studied. Due to the complexity of the case to be modeled, which is characterized by a great variety of situations, and due to the lack of much meteorological and physicochemical data, such as approach requires a certain number of approximations and simplifying assumptions. The resulting uncertainty affecting the deposition values is nevertheless difficult to assess, especially for wet deposition.

The dry and wet deposition values calculated on the scale of a day were integrated over the six-month period, taking into account the frequency with which the weather scenarios occur in the prevailing wind sector with a mean direction of 220° (this direction of highest occurrence frequency corresponds to transport of the plume toward the Ruhr) and downwind distances of 10, 100, and 300 km.

The values thus obtained from the simulation of the base case were compared with the mean aggregate values for the corresponding deposition rates in eastern France derived from measurements available in the literature. The comparison made it possible to quantify the plume's contributions to aggregate deposition at the selected locations. The greatest contributions were obtained for dry deposition of SO_2 and, to a lesser extent, wet deposition of sulfates and H^+ and dry deposition of NO and NO_2.

We thus showed that PARADE can calculate, on a time scale of several months, the contribution of a plume to a region's acid deposition up to 100 km downwind and more. PARADE is also a useful tool for studying the plume's contribution to deposition in "episodic" conditions occurring over a few days and characterized by a specific weather situation and atmospheric chemistry. The uncertainties associated with the background meteorological and air quality conditions can be reduced, to a significant extent, by the use of regional meteorological models (Pielke, 1992) and air quality models (Seigneur and Wegrecki, 1990). PARADE can then, be used as the subgrid-scale component for the treatment of emissions of large point sources in a comprehensive three-dimensional Eulerian air quality model.

REFERENCES

Dambrine, E., and Nourrisson, G. (1991). *Réseau de mesure du dépôt atmosphérique d'éléments minéraux sur les écosystèmes forestiers vosgiens.* CRF Report, Nancy.

Dolske, D. A., and Gatz, D. F. (1985). A field intercomparison of methods for the measurement of particle and gas dry deposition. *J. Geophys. Res.* **90**, 2111–2118.

Drach, A., and Target, A. (1987). *Présentation des résultats de mesures obtenus aux stations du Donon et d'Aubure.* ASPA Report, Strasbourg.

Erisman, J. W., De Leeuw, F. A. A. M., and van Aalst, R. M. (1989). Deposition of the most acidifying components in the Netherlands during the period 1980–1986. *Atmos. Environ.* **23**, 1051–1062.

Hanna, S. R., Briggs, G. A., and Hosker, R. P., Jr. (1982). *Handbook on Atmospheric Diffusion.* U.S. Department of Energy, Washington, DC.

Joos, E. (1989). Modélisation des dépôts secs et humides d'un panache de centrale thermique classique. In L. J. Brasser and W. C. Mulder (Eds.), *Man and His Ecosystem.* Elsevier, Amsterdam, pp. 375–380.

Joos, E. (1990). "Contribution d'un panache de centrale thermique aux dépôts acides Calcul à l'aide du code PARADE de dépôts humides et de compositions chimiques de pluies. In G. Restelli and G. Angeletti (Eds.), *Proceedings of the 5th European Symposium on Physico-chemical Behaviour of Atmospheric Pollutants.* Kluwer Academic Publishers, Dordrecht, The Netherlands, pp. 490–496.

Joos, E., and Maffiolo, G. (1989). Airborne measurements of ozone concentrations in background air and power plant plumes. *Atmos. Environ.* **23**, 2249–2257.

Joos, E., and Seigneur, C. (1988). *Evaluation des possibilités du code de panache réactif PARADE dans sa version définitive,* EDF HE-35–88.21. Electricité de France, Chatou.

Joos, E., Charpentier, C., Leriquier, Y., Maffiolo, G., and Lesne, J. L. (1985a). *Campagne physicochimique de Blénod. Principaux résultats relatifs à l'évolution du panache dans l'atmosphère,* EDF HE/35–85.06. Electricité de France, Chatou.

Joos, E., Charpentier, C., Maffiolo, G., and Rouyer, G. (1985b). *Campagne physicochimique de Cordemais. Principaux résultats relatifs à l'évolution du panache dans l'atmosphère*, EDF HE/35-85.20. Electricité de France, Chatou.

Joos, E., Charpentier, C., Maffiolo, G., Millancourt, B., and Rouyer, G. (1986). *Campagne physicochimique du Havre. Principaux résultats relatifs à l'évolution du panache dans l'atmosphère*, EDF HE/35–86.14. Electricité de France, Chatou.

Joos, E., Mendonca, A., and Seigneur, C. (1987). Evaluation of a reactive plume model with power plant plume data—Application to the sensitivity analysis of sulfate and nitrate formation. *Atmos. Environ.* **21**, 1331–1344.

Joos, E., Millancourt, B., van Duuren, H., and Römer, F. G. (1990). Physico-chemical study by two aircraft of a plume from a coal-fired power plant. *Atmos. Environ.* **24A**, 703–710.

Kelly, T. J., Daum, P. H., and Schwartz, S. E. (1985). Measurements of peroxides in cloudwater and rain. *J. Geophys. Res.* **90**, 7861–7871.

National Acid Precipitation Assessment Program (NAPAP) (1991). *The U.S. National Acid Precipitation Assessment Program—1990 Integrated Assessment Report*. NAPAP Office of the Director, Washington, DC.

Nicholson, K. W. (1988). The dry deposition of small particles: A review of experimental measurements. *Atmos. Environ.* **22**, 2653–2666.

Nicholson, K. W., and Davies, T. D. (1987). Field measurements of the dry deposition of particulate sulphate. *Atmos. Environ.* **21**, 1561–1571.

Nodop, and K., Georgii, H. W. (1990). Air precipitation quality and trends in Europe. Results of the EMEP network. In G. Restelli and G. Angeletti (Eds.), *Proceedings of the 5th European Symposium on Physico-chemical Behaviour of Atmospheric Pollutants*. Kluwer Academic Publishers, Dordrecht, The Netherlands, pp. 44–50.

Pielke, R. A. (1992). *Status of Subregional and Mesoscale Models*, Vol. 2. Electric Power Research Institute, Palo Alto, CA.

Ro, C. U., Tang, A. J. S., Chan, W. H., Kirk, R. W., Reid, N. W., and Lusis M. A. (1988). Wet and dry deposition of sulfur and nitrogen compounds in Ontario. *Atmos. Environ.* **22**, 2763.

Saxena, P., and Seigneur, C. (1987). On the oxidation of SO_2 to sulfate in atmospheric aerosols. *Atmos. Environ.* **21**, 807–812.

Saxena, P., Hudischewskyj, A. B., Seigneur, C., and Seinfeld, J. H. (1986). A comparative study of equilibrium approaches to the chemical characterization of secondary aerosols. *Atmos. Environ.* **20**, 1471–1483.

Sehmel, G. A. (1980). Particle and gas dry deposition: A review. *Atmos. Environ.* **14**, 983–1011.

Seigneur, C. (1990). *Status of Subregional and Mesoscale Models*, EPRI EN-6649, Vol. 1. Electric Power Research Institute, Palo Alto, CA.

Seigneur, C., and Saxena, P. (1988). A theoretical investigation of sulfate formation in clouds. *Atmos. Environ.* **22**, 101–115.

Seigneur, C., and Wegrecki, A. M. (1990). Mathematical modelling of cloud chemistry in the Los Angeles Basin. *Atmos. Environ.* **24A**, 989–1006.

Seigneur, C., Tesche T. W., Roth, P. M., and Liu, M. K. (1983). On the treatment of point source emissions in urban air quality modeling. *Atmos. Environ.* **17**, 1655–1676.

Seigneur, C., Hudischewskyj, A. B., Seinfeld, J. H., Whitby, K. T., Whitby, E. R., Brock, J. R., and Barnes, H. M. (1986). Simulation of aerosols dynamics: A comparative review of mathematical models. *Aerosol Sci. Technol.* **5**, 205–222.

Sverdrup, G. M., Spicer, C. W., and Ward, G. F. (1987). Investigation of the gas phase reaction of dinitrogen pentoxide with water vapour. *J. Geophys. Res.* **19**, 191–205.

Unsworth, M. H., and Fowler, D. (1988). Deposition of pollutants on plants and soils; principles and pathways. In P. Mathy (Ed.), *Air Pollution and Ecosystems*. Reidel Publ., Dordrecht, The Netherlands, pp. 68–84.

Voldner, E. C., Barrie, L. A., and Sirois, A. (1986). A literature review of dry deposition of oxides of sulphur and nitrogen with emphasis on long-range transport modelling in North America. *Atmos. Environ.* **20**, 2101–2123.

Walcek, C. J., Brost, R. A., Chang, J. S., and Wesely, M. L. (1986). SO_2, sulfate and HNO_3 deposition velocities computed using regional landuse and meteorological data. *Atmos. Environ.* **20**, 949–964.

Whitten, G. Z., and Gery, M. W. (1985). *Development of CBM-X Mechanisms for Urban and Regional AQSMs.* U.S. Environmental Protection Agency, Research Triangle Park, NC.

6

THE IMPACT OF DYNAMICS AND TRANSPORT ON STRATOSPHERIC OZONE AND OTHER CONSTITUENTS

John Austin

Meteorological Office, Bracknell, Berkshire, RG12 2SZ, United Kingdom

1. **Introduction**
2. **Examples of the Influence of Dynamics and Transport**
    2.1. The Noxon Cliff
    2.2. Planetary Wave Breaking
    2.3. The Antarctic Ozone Hole
    2.4. Arctic Ozone Depletion
    2.5. Midlatitude Ozone Loss
    2.6. The Quasibiennial Oscillation and Semiannual Oscillation
3. **Future Developments**
    References

Environmental Oxidants, Edited by Jerome O. Nriagu and Milagros S. Simmons.
ISBN 0–471–57928–9   © 1994 John Wiley & Sons, Inc.

1. INTRODUCTION

Stratospheric ozone plays a vital role in protecting the biosphere from DNA damage by absorbing harmful solar ultraviolet (UV) radiation. The photochemistry of ozone is basically quite simple: UV radiation photolyzes oxygen molecules to form oxygen atoms, which react rapidly with more oxygen molecules to form ozone. In turn, ozone can be photolyzed to release back O and O_2. Since the pioneering work of Dobson and Harrison (1926), the absorption property of ozone at longer UV wavelengths has permitted the measurement of the total number of ozone molecules above the earth's surface. Based on laboratory measured rates for oxygen-only reactions and assuming balance between chemical production and destruction, Chapman (1930) calculated the amount of ozone that should be present. Improved laboratory measurements of the reaction rates subsequently revealed that this theory would predict the presence of considerably greater ozone amounts than are actually observed in the atmosphere, implying the presence of some other agent or agents that destroy ozone. Subsequently, ozone destruction cycles involving hydrogen, nitrogen, chlorine, and bromine catalysts were discovered.

Dobson and co-workers also observed that total ozone is generally smaller in tropical latitudes, where photolytic processes are most important, and larger in high latitudes, where photolytic processes are least important or even absent. Figure 1 shows total ozone amounts taken from recent observations [World

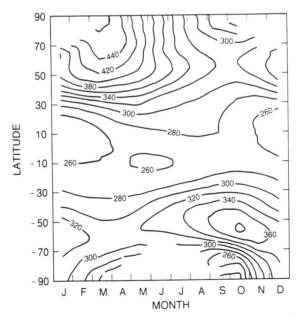

Figure 1. Variation in total ozone (Dobson units) with latitude and season derived from Total Ozone Mapping Spectrometer measurements between 1979 and 1987. (From WMO, 1989.)

Meteorological Organization (WMO), 1989]. The total ozone amount is maximum in the high-latitude Northern Hemisphere spring and is generally lower in the tropics by up to 40%. From these observations may be inferred the fact that atmospheric winds play a key role in transporting ozone away from the tropics to high latitudes. Since Northern Hemisphere transport is more vigorous, lower total ozone values appear in the Southern Hemisphere; therefore, the total ozone amounts do not show a symmetry between the hemispheres (Hou et al., 1991). In the Southern Hemisphere spring, transport is not sufficient to increase ozone in high latitudes, so there is a natural minimum there. More recently, the "Antarctic ozone hole" has appeared due to substantial chemical destruction. Atmospheric motions also have a more subtle effect in transporting the ozone-destroying catalysts, which can subsequently become more or less effective because of the change in local photochemical conditions.

The third way in which atmospheric dynamics can affect ozone chemistry is through local temperature changes, since many of the key chemical processes are strongly temperature dependent. As a result, photochemical equilibrium ozone amounts can change by 1% for each 1-K change in temperature. Thus, a typical temperature variation of 20 K around a latitude circle in the stratosphere can have a significant effect on the equilibrium ozone concentration. As a result of all these interactive processes, the calculation of atmospheric constituents requires the use of a "photochemical-transport model," which consists of a set of equations representing the most important processes and their solution by digital computer.

It is traditional to think of the various effects on ozone in terms of the photochemical timescale. When the photochemical timescale is very short, such as in the upper stratosphere, dynamics can influence ozone only through changes in temperature. When the photochemical timescale is comparable to the dynamical timescale of approximately one week, such as in the middle stratosphere, both transport effects and temperature effects are important. Finally, when the photochemical timescale is very long, as is generally the case in the lower stratosphere, only transport effects are important. Thus, some of the most challenging problems in ozone behavior occur in the middle stratosphere, or at about 30 km of altitude, where both dynamics and chemistry have similar timescales. More recently, however, the discovery of the Antarctic ozone hole (Farman et al., 1985) revealed problems in the lower stratosphere that had been completely unrecognized only a few years earlier. This has had the effect of extending the vertical domain over which dynamics and chemistry are known to interact appreciably. Ten years ago or more, there were relatively few measurements of chemical constituents on a global basis, and two-dimensional (latitude–height) zonally averaged models were generally considered capable of explaining the essential features of these observations. The more extensive measurements of the last decade have led to the realization that a fully three-dimensional description is required to take full account of the impact of atmospheric motions.

In Section 2, the most important situations in which transport and dynamics can influence stratospheric chemistry are discussed and are treated generally in the order in which the phenomena were recognized as important. These are the "Noxon cliff," a region of steep latitudinal gradients in NO_2; planetary wave breaking, the tendency of large wave disturbances to cascade to small scales; the Antarctic ozone hole, a severe depletion of ozone in the Antarctic spring; Arctic ozone depletion and midlatitude ozone loss; and the effects of tropical wind and temperature oscillations. In Section 3, future developments are discussed with particular emphasis on observational programs and the consequences and behavior of Arctic ozone.

2.   EXAMPLES OF THE INFLUENCE OF DYNAMICS AND TRANSPORT

2.1.   The Noxon Cliff

Perhaps the first phenomenon to expose weaknesses in our understanding of the influence of atmospheric dynamics on chemistry has come to be known as the "Noxon cliff." In a sequence of experiments, Noxon and collaborators (Noxon, 1979; Noxon et al., 1979a, b, 1983) found that under certain meteorological conditions, the total-column $[NO_2]$ above the ground was considerably reduced north of about 45° N over North America. A schematic representation of this is given in Figure 2a, taken from Solomon and Garcia (1983a) and based on a figure of Noxon's (1979).

A detailed appreciation of the nitrogen species is essential before we can understand ozone behavior, because these species have the largest impact of all the catalytic cycles that destroy ozone. One of the major discrepancies of two-dimensional models in the early 1980s was their failure to reproduce accurately the sharp latitudinal gradient in $[NO_2]$ observed in the Noxon cliff. Such gradients were usually calculated to be nearer the edge of the polar night, some 15° polewards of their observed latitude (see, e.g., Solomon and Garcia, 1983a). Two-dimensional (zonally averaged) models are unable to represent transport accurately and this proves to be a major problem close to the polar night.

The problem can be illustrated by the fact that at 75 ° N in mid-December, the period of darkness lasts many weeks, whereas air parcel trajectories do not maintain a constant latitude but often meander over more than 15° of latitude. Since air parcels in high latitudes are typically transported completely around the pole about once every 5 to 10 days, they experience several hours of daylight during each such period. Although this amount of daylight is modest, its effect is profoundly different from that of the purely zonal flow of two-dimensional models. Although the effect is most important near the solstice, Solomon and Garcia (1983b) and Callis et al. (1983) recognized that it was also important during January and February. Figure 2b shows the Solomon and Garcia model results for the cliff observed by Noxon (1979) in 1977. Model calculations along air parcel trajectories ending at the morning and afternoon twilight measurements are also shown in Figure 2b. Notice the steep gradient in NO_2-column amount between 43° N and 51° N. The model results are in broad agreement with

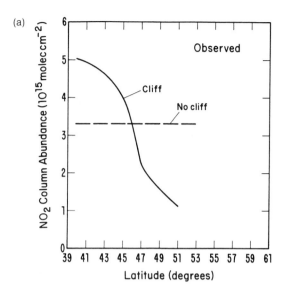

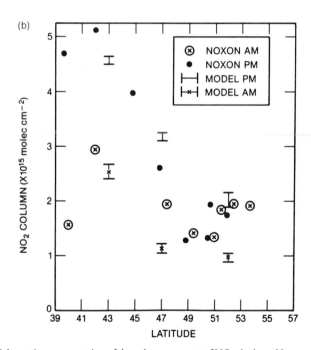

Figure 2. (*a*) Schematic representation of the column amount of NO_2 during a Noxon cliff and during noncliff conditions. (Adapted by Solomon and Garcia, 1983a, from a figure of Noxon's, 1979.) (*b*) Total-column abundance of NO_2 as a function of latitude observed by Noxon (1979) in February 1977, compared to the model calculations of Solomon and Garcia (1983b).

the measurements, although there are some important differences at the most northerly latitudes. As discussed by Noxon et al. (1979b), the steep gradients in $[NO_2]$ occurred when the circumpolar vortex (the region of low geopotential height fields containing a more or less isolated air mass) was displaced off the pole; this usually occurs with the vortex centered over Canada. Under these circumstances in late January or early February, air parcels just inside the edge of the vortex containing normal quantities of NO_2 will be transported to high latitudes, where they will spend a period of 24 hours or more in complete darkness while continuing their path around the vortex. The most important chemical reactions are

$$NO_2 + O_3 \longrightarrow NO_3 + O_2 \tag{1}$$

$$NO_2 + NO_3 + M \longrightarrow N_2O_5 + M \tag{2}$$

$$N_2O_5 + h\nu \longrightarrow NO_3 + NO_2 \tag{3}$$

where M is any molecule needed to satisfy energy and momentum conservation. After traversing the region of darkness, the air parcel has a low concentration of NO_2 but is relatively high in $[N_2O_5]$. In time, N_2O_5 photolyzes to release back NO_2, but for several days the air is deficient in NO_2, thus giving rise to the Noxon cliff, or a region in the atmosphere where the $[NO_2]$ horizontal gradients are significantly enhanced. Furthermore, the photolysis rate of N_2O_5 is temperature-dependent (Yao et al., 1982), so that the lower temperatures present in the vicinity of the Noxon cliff delay photolysis of N_2O_5. Hence, the displacement of the vortex from the pole causes the cliff to occur at lower latitudes than would otherwise be possible.

In addition to these gas-phase reactions it has recently been established that heterogeneous reactions could occur on the surfaces of atmospheric aerosol. In these conditions, the excess N_2O_5 could react with H_2O to form HNO_3 (Austin et al., 1986). Although this does not appear to be essential for the formation of the cliff itself, such a reaction could extend the longevity of the cliff by transferring the nitrogen into a species HNO_3 that photolyzes much more slowly than N_2O_5, thus maintaining low NO_2 amounts. Some evidence for the above heterogeneous reaction can be seen in Figure 2, which shows that the morning and afternoon measurements were very similar north of 49° N, in contrast to the model results, which showed a difference of about a factor of 2. The observed small diurnal variation at this time and place is consistent with the absence of N_2O_5, which would otherwise photolyze to increase $[NO_2]$ during the day.

The general aspects of cliff behavior have recently been investigated by Lary (1991), who used a numerical model to confirm that a cliff exists when the cross-polar flow is strong, which occurs when the circumpolar vortex is significantly displaced from the pole. If the cross-polar flow is weak, a cliff will not form in midlatitudes and ozone amounts will not be perturbed by these processes. When the vortex is displaced, an air parcel traversing through high latitudes does not have time to get into chemical equilibrium with the local environment. Since

NO_2 is the most important catalyst in controlling ozone in the stratosphere, after the formation of the cliff, the net production of ozone could increase for a period of several days. However, because of the generally sporadic nature of the Noxon cliff, each individual event would have a rather small effect on the total ozone in the atmosphere, but the overall effect for a whole season has yet to be quantified.

2.2. Planetary Wave Breaking

Significant further progress in the understanding of stratospheric chemistry came from the results of the Limb Infrared Monitor of the Stratosphere (LIMS) experiment (Gille and Russell, 1984). This instrument measured O_3, NO_2, HNO_3, and H_2O from the Nimbus 7 polar-orbiting satellite between October 1978 and May 1979. The advantage of observing from a polar-orbiting satellite is that almost complete global coverage can be obtained, although the horizontal resolution is relatively poor.

During January 1979, the high-latitude stratospheric temperatures increased very rapidly, resulting in a "stratospheric warming." The dynamics of this event were thoroughly analyzed in the papers by McIntyre and Palmer (1983, 1984) and subsequently by others. Their analysis suggested the presence of "planetary wave breaking" in which the atmospheric waves grew considerably in amplitude and eventually cascaded into much smaller scales. The situation is often described in terms of potential vorticity (PV), which is the fluid analog of angular momentum and is conserved along air parcel trajectories in the absence of dissipative processes. In practice, radiative terms contribute to PV on timescales of several weeks, while diffusion acts on the very smallest spatial scales. Because the relatively long-lived chemical species are also conserved along air parcel trajectories for periods up to several weeks, the analogy between PV and constituents such as ozone in the middle stratosphere is a natural and very powerful way of understanding tracer behavior. During planetary wave breaking, the polar vortex is eroded by the transport of high-PV air into middle latitudes. This takes the form of long, thin tongues of high-PV air surrounded by low-PV air, and these tongues become progressively smaller in scale and mix into the surrounding air. This essentially nonlinear process leads in time to a large region in midlatitudes where the stratosphere appears well mixed. In the middle stratosphere, where the photochemical timescale may be several weeks, these dynamical processes can produce well-mixed ozone fields equatorward of the wave-breaking region, with sharp gradients near the edge of the vortex.

Using LIMS data in a descriptive study of ozone variations, Leovy et al. (1985) showed that these processes occurred in the 1979 Northern Hemisphere winter. Figure 3a shows the ozone mixing ratio at 10 mbar on February 6, 1979, taken from Leovy et al. Indicated in the figure is the axis of high-ozone values, which extend from about 20° N almost into the polar regions and give rise to sharp gradients in ozone close to the vortex edge, given approximately by the 30.6-km geopotential height contour shown in Figure 3b. In contrast, there is a large

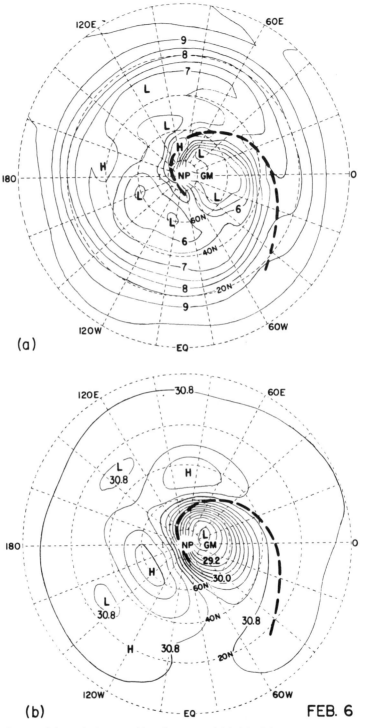

Figure 3. Ozone mixing ratio in ppmv (*a*) and geopotential height field in km (*b*) for the 10-mbar surface on February 6, 1979. The axis of a tongue of ozone-rich air is indicated by the heavy broken curves. (From Leovy et al., 1985.)

region centered near 40° N, 120° E with very low ozone gradients that McIntyre and Palmer (1984) term the "surf zone."

The study by Leovy et al. neglected the effects of photochemistry on the ozone amount but the later work by Austin and Butchart (1989) considered these processes. By integrating a comprehensive photochemical model along air parcel trajectories initially situated on the edge of the polar vortex in late January, Austin and Butchart attempted to quantify the effects of photochemistry and dynamical mixing. Comparisons between model results and colocated LIMS data for O_3, NO_2, HNO_3, and H_2O revealed significant discrepancies and it was concluded that mixing was the most likely explanation. Although on a 10-day timescale, mixing was generally greater than the effects of photochemistry in determining the ozone concentration in the surf zone, photochemistry could not be ignored in any quantitative assessment. The work of Butchart and Remsberg (1986) has shown that wave breaking is probably occurring during much of the Northern Hemisphere winter, and therefore any photochemical phenomenon occurring in the high-latitude region may be influenced in part by planetary wave breaking. In addition to the earlier studies described in this section, the concepts have recently been found to be useful in helping to interpret, for example, the effects of low polar ozone on midlatitudes, as described in Section 2.5.

2.3. The Antarctic Ozone Hole

Following the LIMS experiment, there were relatively few additional chemical measurements until the unexpected discovery of the Antarctic ozone hole (Farman et al., 1985). Figure 4 shows the total ozone over Halley Bay taken from Gardiner (1989). The upper panel shows the measurements of October ozone from 1957 to 1988, revealing a significant reduction of over 30% since the early 1970s. Shown in the lower panel are cumulative production and measurements of chlorofluorocarbon F11 ($CFCl_3$), which substantially increased during the same period. Figure 5 shows the vertical ozone profile also measured at Halley Bay during 1987, both before (August 15) and after (October 13) the formation of the ozone hole (Gardiner, 1988).

Soon after this discovery, observational campaigns were established. Observations of the column amounts, and in some cases the vertical distributions, of a number of species were made during the National Ozone Expedition (NOZE) based in McMurdo during the 1986 Antarctic spring (see, e.g., de Zafra et al., 1987; Solomon et al., 1987). These observations revealed that the concentrations of active chlorine species were considerably higher than predicted, while the nitrogen species were unexpectedly low in concentration. In the following spring, the Airborne Antarctic Ozone Experiment (AAOE) made a wide range of measurements from high-flying aircraft (see Tuck et al., 1989, and other papers in the special issues of *J. Geophys. Res.*, Vol. 94, Nos. D9 and D14) and, together with ground-based data from a second NOZE (see, e.g., Sanders et al., 1989), clearly established the reasons for the Antarctic ozone hole.

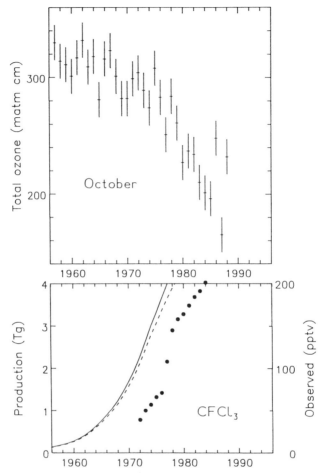

Figure 4. (*Top*) October mean total ozone at Halley Bay from 1957 onwards. (*Bottom*) Cumulative world production of $CFCl_3$ since the start of world production in 1931 (solid line) and the amount remaining, assuming a lifetime of 75 years (broken line). Solid circles show measurements of the same chemical in the Southern Hemisphere troposphere. (Both graphs from Gardiner 1989.)

In adequate sunlight such as in the middle or upper stratosphere, ozone can be destroyed by catalytic reactions involving chlorine:

$$Cl + O_3 \longrightarrow ClO + O_2 \tag{4}$$

$$ClO + O \longrightarrow Cl + O_2 \tag{5}$$

$$Net: \quad O_3 + O \longrightarrow 2O_2$$

However, these reactions are not effective in the lower stratosphere, because the

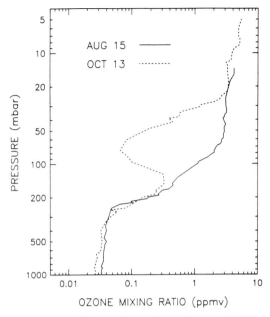

Figure 5. Ozone mixing ratio over Halley Bay, Antarctica, on August 15, 1987, and October 13, 1987. (From Gardiner, 1988.)

UV radiation needed to generate adequate concentrations of oxygen and chlorine is absorbed at higher altitudes. Instead, we find that at temperatures lower than about 195 K, such as frequently occur in the Antarctic lower stratosphere, the formation of polar stratospheric clouds (PSCs) is triggered. Chemical reactions can then take place on the surfaces of the cloud particles and liberate large amounts of active chlorine through reactions such as:

$$ClONO_2 + HCl \longrightarrow Cl_2 + HNO_3 \tag{6}$$

$$ClONO_2 + H_2O \longrightarrow HOCl + HNO_3 \tag{7}$$

Both Cl_2 and $HOCl$ photolyze rapidly, releasing chlorine atoms; however, if the temperature is low enough, the HNO_3 becomes embedded in the PSC crystal structure, leading to "denitrification" of the lower stratosphere. The presence of high chlorine and low nitrogen species allows the following reactions to proceed:

$$2 \times (Cl + O_3 \longrightarrow ClO + O_2) \tag{4}$$

$$ClO + ClO + M \longrightarrow Cl_2O_2 + M \tag{8}$$

$$Cl_2O_2 + h\nu \longrightarrow ClOO + Cl \tag{9}$$

$$ClOO + M \longrightarrow Cl + O_2 + M \tag{10}$$

$$Net: \quad 2O_3 \longrightarrow 3O_2$$

In conditions of low [O], ozone may also be destroyed via catalytic cycles involving Br and OH radicals in conjunction with chlorine species, although these cycles are less important. Thus, the original suggestion by Farman et al. that chlorine is responsible for the ozone depletion is essentially correct, although the details are not as originally envisaged by those authors. More details are given in the reviews by Gardiner (1989) and Solomon (1990).

The fact that the key chemical processes occur on PSCs, whose presence is determined by the dynamical situation, illustrates how the dynamics and chemistry interact. Enough is now known about these processes to carry out seasonal simulations with numerical models that include both dynamics and photochemistry. Cariolle et al. (1990) simulated the Antarctic ozone hole in a general circulation model assuming a simple sink to represent the special polar ozone chemistry, but Granier and Brasseur (1991) were the first to perform three-dimensional model simulations with reasonably detailed photochemistry. In their model, the formation of PSCs was internally determined from model temperature, water vapor, and nitric acid fields, and a simple representation of heterogeneous reactions on the surfaces of PSCs was included. They performed idealized simulations of both the Arctic and Antarctic spring, showing how ozone depletion was produced in the Antarctic but not in the Arctic by higher temperatures. The precise connection between the Arctic and Antarctic was revealed in a sequence of three-dimensional model calculations by Austin and Butchart (1992), using a more extensive model than that of Granier and Brasseur. Starting with realistic Southern Hemisphere conditions, they performed a range of experiments with differing planetary wave amplitudes. Figure 6 shows the minimum ozone column south of 57.5° S, calculated as a function of time. For small planetary wave amplitudes, model temperatures remained low and deep ozone holes were produced in the model. Increasing wave amplitude in the model had the effect of increasing model temperatures until eventually they became too high to support PSC formation. With the largest wave amplitude, comparable to wave amplitudes in the Northern Hemisphere, no ozone hole was produced despite initial conditions favorable to ozone destruction. Thus, it appears that low ozone in Antarctica can be attributed to the relatively small wave amplitudes that occur in the Southern Hemisphere combined with the high chlorine amounts.

Austin and Butchart carried out idealized simulations, but such sensitivity to meteorological conditions has also been demonstrated by Eckman et al. (1993) in three-dimensional photochemical model simulations of the 1987 and 1988 Antarctic ozone holes. Interannual variability in total ozone at Halley Bay (Figure 4) was very marked during the years 1986 to 1988. This was partly due to the single-point nature of the observation, which in 1986 did not represent typical ozone behavior over Antarctica. However, in 1988 planetary wave amplitudes were considerably larger than normal. Eckman et al. simulated a deep ozone hole for 1987 but only a relatively shallow ozone hole for 1988, in agreement with observations and broadly consistent with the idealized simulations of Austin and Butchart (Figure 6). Although in 1988 ozone depletion was

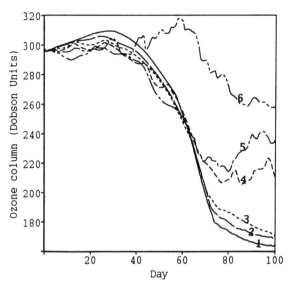

Figure 6. Minimum total ozone (Dobson units) south of 57.5° S, from 3-D photochemical model calculations of Austin and Butchart (1992) starting on August 1. The curves are labeled 1–6 corresponding to geopotential wave amplitudes of 0, 50, 100, 150, 200, and 250 m, respectively.

relatively modest, this certainly was not the beginning of a general recovery, since deep ozone holes were also experienced in the following three years (WMO, 1992). The fact that ozone has been very low during four of the past five years (1987–1991) is in contrast to the situation that prevailed in the mid-1980s, when the depth and area of the ozone hole exhibited a quasibiennial modulation, as discussed further in Section 2.6.

The Antarctic ozone hole appears now to be reasonably well understood; in a recent review, Schoeberl and Hartmann (1991) concluded that the maximum size of the ozone hole is fixed by the size of the region of very low temperatures and by the dynamical boundary of the polar vortex. Currently, the low-temperature region occupies about half the present area of the Antarctic vortex, but chemical perturbations can occur outside this region due to the intermittent appearance of PSCs associated with small-scale processes propagating from the lower atmosphere. As chlorine amounts increase over the next 10 years, Schoeberl and Hartmann (1991) predict that the Antarctic ozone hole might double in area. Such arguments have not yet been thoroughly tested by numerical modeling studies and, on the basis of the 3-D modeling studies of Granier and Brasseur (1991), may be unnecessarily pessimistic. Note, however, that the ozone hole is unlikely to get much deeper than the one that appeared, for example, in 1987, when between 60 and 90% of the ozone was destroyed within the 15- to 20-km altitude range where PSCs form (see, e.g., Gardiner, 1988, also Figure 5; Hofmann et al., 1989). Further significant destruction of column ozone would require PSCs to form at higher altitudes, which would require lower temperatures than are currently observed. Such lower temperatures, however, could occur in the future, as discussed in Section 3.

Although poorly understood five years ago, the Antarctic ozone hole has become an important test of the performance of any comprehensive dynamical-photochemical model, just as ten years ago, nature's dynamical experiment, the stratospheric warming, was the test of a dynamical model. Our understanding of the Antarctic ozone hole is also at a level similar to our understanding of the stratospheric warming ten years ago (see the review by McIntyre, 1982). For example, both are viewed as dramatic phenomena that considerably disrupt the climatological state of the atmosphere. At the time of McIntyre's review, the basic mechanisms of the stratospheric warming were established and there were prospects for understanding stratospheric warmings in some detail and forecasting them reasonably well. Ten years later, the same can be said for the Antarctic ozone hole.

2.4. Arctic Ozone Depletion

Although substantial ozone depletion has so far occurred only in the Antarctic, similar atmospheric situations might arise in the Arctic. The consequences for a large population are therefore of major concern and stimulated the establishment of the Airborne Arctic Stratospheric Expedition (AASE) based in Stavanger, Norway, in January 1989 (see the papers in the special issue of *Geophys. Res. Lett.*, Vol. 17, No. 4, 1990, with further papers in the special issue of *J. Geophys. Res.*, Vol. 97, No. D8, 1992). This project involved measurements of key stratospheric constituents from aircraft and in many ways was the Arctic equivalent of the highly successful AAOE. The main difference, however, was that since PSCs were already implicated in ozone destruction in Antarctica, it was necessary to make measurements earlier in the season, when PSCs were present. Also, the polar vortex is often substantially displaced from the pole, so that PSCs can occur at lower latitudes than in the Southern Hemisphere. Since ozone depletion also depends on sunlight, it can take place earlier in the Northern Hemisphere winter season. As in Antarctica, large abundances of ClO were observed in the lower stratosphere (Brune et al., 1990) together with low values of NO (Fahey et al., 1990), indicating that the Arctic polar vortex was extensively perturbed by heterogeneous chemistry. Model simulations of the chemical evolution of the vortex along air parcel trajectories (Jones et al., 1990; McKenna et al., 1990) were found to agree with the AASE observations of the critical species ClO when proper account was taken of the presence of PSCs. It is difficult to determine ozone-loss rates from observations over a short period of time because of the influence of small-scale atmospheric processes and measurement limitations. However, the model results indicated local ozone depletion rates in the lower stratosphere in excess of 1% per day (McKenna et al., 1990), approaching the values calculated for extended periods in the Antarctic (Austin et al., 1989).

Results conceptually similar to the AASE observations were also obtained for the large-amplitude wave case of the idealized model calculations of Austin and Butchart (1992), described in Section 2.3. The results showed that the model atmosphere was "primed for destruction," in that loss of nitrogen species had occurred and, for a period, the chlorine species were in active form. Ozone depletion was limited in the model, however, because of downward advection of nitrogen species and higher ozone values into the lower stratosphere. The three-dimensional model results have the advantage of being able to put local results such as trajectory results into an overall perspective. On the basis of the trajectory results, one is tempted to conclude that substantial ozone depletion would occur during the course of the winter and spring, but the more comprehensive results from the three-dimensional model showed that the chemically perturbed air is advected to lower levels where the temperatures are higher, and ozone-depletion mechanism is slowed down or no longer operates. Thus, ozone depletion, although it does occur locally, is significantly limited by current Northern Hemisphere dynamics, at least in the model calculations. In the Southern Hemisphere, downward transport is weaker than in the Northern Hemisphere because atmospheric dynamics are less vigorous, due ultimately to differences in the contrast between land and sea temperatures, among other things.

There is also observational evidence for the limiting effect of the meteorology of the Northern Hemisphere. For example, Figure 7 shows the ozone and temperature measurements of Koike et al. (1991) from balloonsondes launched from Kiruna, Sweden, during January and February 1990; these may be compared with similar results for the Southern Hemisphere shown in Figure 5 (note, however, the logarithmic scale for the ozone mixing ratio presented in Figure 5). The results show that the ozone mixing ratio between altitudes of 19 and 26 km were systematically lower within the polar vortex by up to 29%. Also included in Figure 7 are the simultaneous temperature measurements, which are lower than the condensation temperature for nitric acid trihydrate. This would allow for the formation of PSCs and the subsequent heterogeneous reactions. These results are highly suggestive of chemical ozone destruction, but because the full three-dimensional ozone field was not known, the results cannot be considered conclusive. Similar circumstantial evidence for ozone depletion in the Arctic has recently been supplied by Proffitt et al. (1992). Model results show a high correlation between $[O_3]$ and $[N_2O]$ in situations where no photochemistry is occurring. Similarly, the observations of Proffitt et al. (1990) show a very high correlation between these species in the high-latitude winter, but in the lower-stratosphere spring the correlation breaks down, implying that photochemical change has occurred. From this, Proffitt et al. deduced that in the lower stratosphere about 12 to 35% of the ozone was chemically depleted.

Although Proffitt et al. were unable to identify the cause of the ozone destruction, the fact that it occurred in the lower stratosphere, together with the results of the other studies made during the AASE, provide strong circumstantial evidence that heterogeneous chemistry is at least partly to blame. The estimated total depletions would be consistent with the Lagrangian studies of depletion

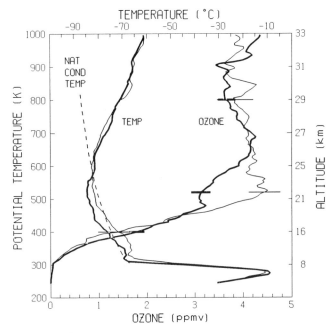

Figure 7. Average profiles of the ozone mixing ratio and temperature measured by Koike et al. (1991) inside (heavy line) and outside (light line) the vortex. Horizontal bars represent one standard deviation of the ozone mixing ratio. The broken line represents the nitric acid trihydrate condensation temperature for an H_2O mixing ratio of 4.5 ppmv and an HNO_3 mixing ratio of 10 ppbv.

rates, assuming that these rates were sustained for about 20 days. This is probably not an unreasonable length of time for the persistence of PSCs in the Arctic, since the highly disturbed nature of Northern Hemisphere dynamics in most years results in a stratospheric warming by late February, which substantially raises temperatures and evaporates any PSCs. Local ozone depletions on the order of 30% would account for total-column changes on the order of 10% when proper account is taken of the vertical distribution. This is currently considerably smaller than the depletion observed over Antarctica, but whether it will increase in the future is still of intense scientific interest.

2.5. Midlatitude Ozone Loss

Ozone has been observed remotely from space for over a decade and it is important to determine what trends have occurred during this period. Using observations from the Total Ozone Mapping Spectrometer, Stolarski et al. (1991) computed the rate of change as a function of latitude and month; the results for the 11 years of data from 1979 to 1990 are shown in Figure 8. The shaded area represents months and latitudes—mostly in the tropics—in which there is no

TOMS TOTAL OZONE TRENDS [%/YEAR]

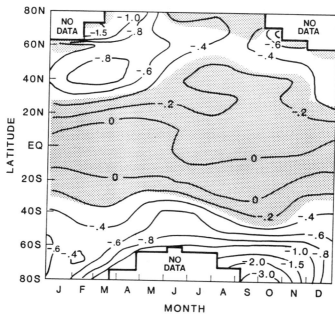

Figure 8. Trend in the TOMS total ozone measurements for the period 1979–1990. The unshaded area indicates where the trend is statistically different from zero at the 2σ level. (From Stolarski et al., 1991.)

statistically significant change. The remainder of the figure shows decreases in ozone that have contributed to a global average ozone loss of about 3% during the last decade. The largest decrease is over Antarctica during September and October; it exceeds 3% per year, or in excess of 33% for the entire measurement period, as discussed in detail in Section 2.3. The next largest decrease occurs in northern high latitudes in January and February; this change is obscured in part by the data gap caused by the inability of the satellite instrument to make observations in the polar night. These changes are broadly consistent with local ozone depletion rates of about 30%, as noted in Section 2.4.

Figure 8 also shows unexpected losses of about 8% in spring in the midlatitude Northern Hemisphere during the last decade. Pyle et al. (1991) suggest three possibilities for this occurrence: ozone depletion in polar regions followed by transport to midlatitudes, transport of chemically perturbed air from polar regions followed by ozone destruction in midlatitudes, or local perturbation of midlatitude chlorine chemistry by heterogeneous reactions on background sulfuric acid aerosol. All three may be occurring simultaneously in the Northern Hemisphere. In the Southern Hemisphere, ozone depletion occurs in May, before polar temperatures drop to levels at which PSC chemistry can commence, which suggests that heterogeneous reactions in midlatitudes are a strong possibility.

To aid our understanding of the observed ozone changes, it is instructive to investigate the dynamical processes of the polar vortices. McIntyre (1989) suggested that the Antarctic vortex behaves essentially as a "containment vessel" in which chemical reactions continue inside the vortex more or less isolated from the outside air throughout the winter. This view is supported by an extensive analysis of data (Schoeberl et al., 1992) from recent polar campaigns. Using air parcel trajectories, they calculated diffusion rates, which were found to decrease substantially at the inner edge of the vortices in both hemispheres. The smaller area covered by the Northern Hemisphere vortex compared with that of the Southern Hemisphere is a reflection of the greater level of planetary wave activity in the Northern Hemisphere, which erodes the vortex more rapidly. Nonetheless, Schoeberl et al. found the fundamental characteristics of the vortex in both hemispheres to be similar. A contrary view (Tuck, 1989; Proffitt et al., 1990) suggests that the Antarctic vortex is not actually a containment vessel but is more likely a "flow processor" in which transport in and out of the vortex is an important ingredient. Whatever the merits and demerits of these descriptions, it would appear that the Arctic vortex is more likely than the Antarctic to be a flow processor, which would allow the vortex to leak either depleted ozone air or chemically perturbed air and thus lead to an observed midlatitude ozone loss.

Using maps of lower stratospheric potential vorticity (PV), Tuck et al. (1992) found evidence of high-PV air from the Arctic transported to low latitudes. They concluded that tongues of PV were being peeled off the vortex and were accompanied by small-scale mixing, in a process reminiscent of that observed by McIntyre and Palmer (1983, 1984). Such a process would also act on chemically depleted ozone from the vortex, thereby transporting to middle latitudes air depleted in ozone. Unfortunately, Tuck et al. (1992) could provide no estimate of how much air is transported from the vortex. Nonetheless, they concluded that the Arctic vortex cannot be described as a containment vessel.

A slightly different picture is presented by Rood et al. (1992). Using observed wind fields to compute the transport terms, they performed simplified chemical calculations in which HCl is depleted when the temperatures are low, simulating heterogeneous effects. They then proceeded to determine the "leakiness" of the Arctic vortex. It is estimated that by late January, about 5 to 15% of air polewards of 20° has polar properties, with lower percentages possibly being more realistic for the atmosphere in view of numerical errors inherent in the modeling process. These numbers are at the lower end of the range of values that might be anticipated by Tuck et al. (1992), possibly because the Tuck et al. analysis was based on meteorological fields that may have contained higher contributions of the very small-scale processes than can occur in the atmosphere. On the other hand, it may be that the model resolution used by Rood et al. (2.5° longitude by 2° latitude) is insufficient to address the problem of small-scale mixing. Thus, at the present time, the quantitative aspects of a dynamically induced midlatitude ozone depletion are still an area of active research.

Work has also focused on the impact of the transport of ozone after the vortex breaks up in a stratospheric warming, either at the end of the winter or earlier.

Whether or not the atmosphere can be described as a containment vessel may have an important bearing on this question, since a containment vessel would presumably result in more dramatic ozone changes when the vortex breaks up. Prather et al. (1990) and Prather and Jaffe (1990) provided simplified modeling studies that showed that, in the case of the Antarctic, there would be measurable decreases in column ozone extending to 30° S in austral summer. The volume of Antarctic air is less than 10% of the Southern Hemisphere so that in general, the effects of the ozone hole on the mean ozone concentration in middle and low latitudes are likely to be only on the order of a few percent. These studies have looked at the problem of the dynamical propagation of a prescribed ozone hole. A more comprehensive chemical modeling study has been performed by Pitari et al. (1992b), who simulated the global distribution of ozone depletion explicitly. Their conclusion was that the depletion, largely but not wholly due to Antarctic ozone hole dilution, was about 1 to 1.5%. Indeed, local decreases of larger amounts have been observed at certain longitudes, for example, over Australia and New Zealand (Atkinson et al., 1989) and to a smaller extent over the middle and low latitudes of South America (Kane, 1991).

2.6. The Quasibiennial Oscillation and Semiannual Oscillation

As its name implies, the quasibiennial oscillation (QBO) is an irregular oscillation of the stratospheric tropical zonal winds; it was discovered independently by Reed et al. (1961) and Ebdon and Veryard (1961). Recent data (Naujokat, 1986) show that the equatorial zonal winds oscillate due to the QBO between about $25\,\mathrm{ms}^{-1}$ (westerly) and $35\,\mathrm{ms}^{-1}$ (easterly), with an average period of about 27 months. The semiannual oscillation (SAO) of the tropical zonal winds has a period of six months, with a peak amplitude of about $30\,\mathrm{ms}^{-1}$ in the upper stratosphere; it was discovered by Reed (1965) during a study of the amplitude of the QBO. Both oscillations are now understood to be caused by interactions between waves and the mean flow (see, e.g., Andrews et al., 1987). The effects of the QBO and SAO on constituents are twofold. First, the associated mean meridional motions can change the transport of long-lived species. Second, the oscillations in the mean zonal wind are accompanied by oscillations of up to 7 K in temperature, which can directly affect the photochemistry of constituents.

The QBO and SAO are weak or nonexistent outside of tropical latitudes, and therefore direct transport effects are confined almost entirely to the tropical regions. Observations of the long-lived tracers CH_4 and N_2O (Jones and Pyle, 1984) showed enhanced values at both 30° N and 30° S during April and November, a structure known subsequently as the "double peak." These observations were at first difficult to understand, since maximum concentrations were calculated by chemical models to occur only over the equator (Jones and Pyle, 1984). Inclusion of the SAO in numerical models (Gray and Pyle, 1986, 1987) subsequently led to a major improvement and to the conclusion that the SAO is responsible for double peak structures in the observed tracers. Although comparisons were made with observations of CH_4 and N_2O, any long-lived tracer

BB FILTERED TOMS TOTAL OZONE

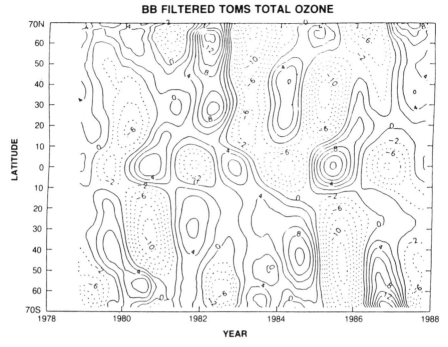

Figure 9. Broad-band filtered Total Ozone Mapping Spectrometer data (Dobson units) with the annual cycle removed. (From Lait et al., 1989.)

would be affected by the SAO, and this can have indirect photochemical effects on the equatorial concentrations of species such as O_3 and NO_2.

In a modeling study of the QBO, Gray and Ruth (1992) calculated NO_2 and HNO_3 local concentration changes exceeding 30% due to both transport and temperature effects, while for ozone the effects were calculated to exceed 20%. Figure 9 shows the total ozone measurements with the annual cycle removed, taken from Lait et al. (1989); it shows a very clear QBO signal in the tropics, with a larger signal in middle latitudes. The magnitude of the QBO signal in total ozone is somewhat smaller than the local effects calculated by Gray and Ruth (1992), because of the limited vertical extent of a significant QBO perturbation. The general behavior described by Lait et al. has been successfully modeled by Gray and Pyle (1989) using a two-dimensional photochemical-transport model. As illustrated by Gray and Pyle (1989), when the QBO is in the easterly phase in the lower stratosphere, the dominant motions are downward at the equator between about 17 and 23 km and upward at the same altitudes at about 30° N and 30° S. Above and below these levels, motions are in the opposite direction. In the westerly phase, the whole circulation is reversed. Since total ozone is dominated by lower stratospheric behavior, the values are slightly lower at the equator during westerly QBO years and higher during easterly QBO years.

The Gray and Pyle model was unable to reproduce the biennial oscillation in Antarctic ozone observed in the mid-1980s (Garcia and Solomon, 1987). How the QBO signal could be communicated to high latitudes is not at present understood. One possibility is that the tropical wind fields affect the propagation of midlatitude planetary waves, which in turn transport more or less ozone according to the QBO phase. Some evidence for this is available from observations (Labitzke, 1982) and modeling studies (Dameris and Ebel, 1990; Holton and Austin, 1991), which suggest that stratospheric warmings are more frequent in easterly QBO years. In turn, the frequency of stratospheric warmings depends on the planetary wave amplitudes, but a direct link between planetary wave changes resulting from the QBO and tracer transport has not yet been established.

Further observational evidence (Poole et al., 1989) shows that PSCs were more frequent over Antarctica during westerly QBO years, indicating that the temperatures were lower. Since PSCs are implicated in the formation of the Antarctic ozone hole (see Section 2.3), this is consistent with the approximately biennial signal in column ozone observed over Antarctica during the spring in the mid-1980s (Garcia and Solomon, 1987). So far, only correlations have been established, and the short period of time over which high correlations have been obtained is not entirely convincing. It is possible, for example, that some other physical phenomenon that happened to have a roughly biennial signal in the mid-1980s is responsible for the interannual variability seen in PSC frequency and the depth of the ozone hole. In order to be sure that the QBO is responsible for these variations, a modeling study is required imposing a QBO in the tropics and determining whether a temperature oscillation of the appropriate magnitude can result. Such a study would be complicated because radiative factors may be needed to amplify any high-latitude dynamical QBO signal, in the same way as might occur in the Arctic (see Section 3). More recently, however, the biennial signal in the depth of the ozone hole seems to have disappeared. This is probably because the effect of recent increases in chlorine has been to destroy almost all of the ozone in the lower stratosphere before the end of spring; therefore, if PSC amounts vary, there is still sufficient time to produce a deep ozone hole.

3. FUTURE DEVELOPMENTS

In recent years, a number of important measuring programs has stimulated concerns that manmade pollutants may be destroying the ozone layer. Stratospheric scientists are currently in the unprecedented position of having large quantities of data to analyze. Currently, only a fraction of this data has been fully investigated, and it may be anticipated that many more surprises are in store. It has become increasingly clear that in recent years progress in understanding stratospheric chemistry has been data-driven. This has stimulated the development of three-dimensional models capable of simulating dynamical-chemical interactions in detail, leading to advances in our understanding of these processes. Such models are expensive to run but modern supercomputers have developed substantially in the last 10 years, enabling our understanding of the science to develop in parallel.

Several issues are currently unresolved and these will clearly receive considerable attention during the next few years. The appearance of an ozone hole over Antarctica has raised legitimate fears of the possibility of a similar occurrence over the Arctic. Given current atmospheric conditions, it appears that the Northern Hemisphere circulation is sufficiently vigorous to drive chemically perturbed air away from the ozone destruction region, but whether this situation will persist is subject of intense current research. Chlorine levels in the atmosphere are expected to continue to rise during the next few decades, and this could enhance ozone destruction. However, as shown by Granier and Brasseur (1991), even increases in chlorine amounts of the magnitude anticipated from chlorofluorocarbon release estimates are unlikely on their own to decrease ozone substantially. Consequently, it would seem that the appearance of an Arctic ozone hole would require a significant change in stratospheric circulation. Such a change may be possible through a positive feedback from radiative processes. Increases of radiatively active gases such as CO_2 in the amounts anticipated over the next century (Houghton et al., 1990) could induce a cooling in the lower stratosphere, and this in turn could lead to an increased presence of the polar stratospheric clouds implicated in the destruction process. Finally, increased ozone destruction could lead to even less radiative heating of the lower stratosphere (Shine, 1986).

Such simple arguments, although worrying, do not take into account possible changes in the dynamical circulation (see, e.g., Rind et al., 1990) that could make up for any chemical ozone changes. It may well be that the atmosphere is stable to small perturbations, but if radiative perturbations increase beyond a certain amount, the Arctic may behave similarly to the Antarctic, where it appears that radiative perturbations enhance ozone destruction. How large does the radiative perturbation need to be before ozone destruction is significantly enhanced, and can we expect to reach this point based on current climate predictions? One of the difficulties in answering these sorts of questions is that we cannot be completely sure that our numerical models are adequate. First of all, it is essential to use three-dimensional models to represent dynamical processes properly.

Both Pitari et al. (1992a) and Austin et al. (1992) have attempted to address this problem. Using a climate model with simplified chemistry, Pitari et al. calculated only small increases in ozone depletion for a doubling of atmospheric CO_2. In contrast, Austin et al. used a comprehensive chemical model but idealized dynamics to demonstrate that an Arctic ozone hole, with almost complete ozone destruction in the lower stratosphere, is possible for doubled CO_2 amounts. Figure 10a shows the total ozone calculated by the model for current levels of CO_2 on April 14; Figure 10b shows the total ozone on the same day for doubled CO_2. In their model, Austin et al. specified tropospheric conditions, and since these can have a strong influence on stratospheric behavior, they could simulate specific winter conditions. With a simulation of quiescent conditions, which give rise to stratospheric temperatures that are lower than average but still within the range observed in the atmosphere, the probability of an Arctic

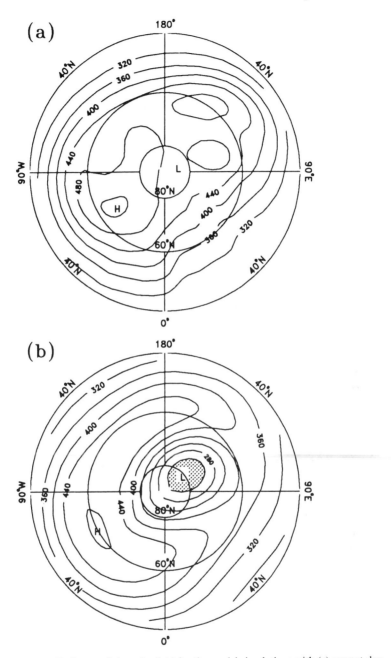

Figure 10. Total ozone (Dobson units) on April 14 for the model simulations with (*a*) present-day levels of CO_2 and (*b*) doubled CO_2. Shaded area denotes where the ozone column is under 240 DU. (Adapted from Butchart et al., 1993.)

ozone hole is increased. It is likely, therefore, that the different results obtained by the two studies are different aspects of the same problem. The study of Pitari et al. simulated what are probably the more normal effects of increased CO_2, while the study of Austin et al. is more applicable to those occasions, which occur perhaps 20% of the time, when the stratosphere is quiescent. In the latter simulation, the ozone perturbation was sufficient to delay the increase in temperatures that occurs in late winter, and the dynamics of the simulation appeared more like those of a typical Southern Hemisphere simulation (Butchart et al., 1993).

Even with modern supercomputers such comprehensive models consume large resources, and so relatively poor horizontal and vertical resolutions have to be used. This raises the question of whether small-scale processes can contribute to ozone depletion in significant amounts. The balance of opinion on the Antarctic would seem to suggest that 5° in latitude by 5° in longitude with 2 km vertical resolution is adequate to simulate the general features of the ozone hole. However, the Arctic may be influenced by smaller-scale processes, and a model resolution closer to 1° by 1° may be required, with the vertical resolution improved in proportion. This is well beyond our current computing capabilities.

Although many observation programs have been stimulated by the need to investigate polar ozone behavior, the launch of the Upper Atmosphere Research Satellite in September 1991 promises to reveal much about the dynamics and chemistry of the ozone layer in general. The satellite is measuring a wide range of chemical constituents, some of which were previously unmeasured but have theoretically been shown to play an important rôle in ozone photochemistry. Currently, only a fraction of the available data has been studied. Unfortunately, much of the data analysis has been complicated by the volcanic eruption in June 1991 in the Philippines of Mt. Pinatubo, which spewed out a large amount of aerosol that was transported to quite high latitudes. Previous work (Hofmann and Solomon, 1989) has shown the possible impact of volcanic aerosol on stratospheric chemistry, and an issue of current concern is whether such aerosol could enhance ozone destruction in the Arctic in a similar way that polar stratospheric clouds do in the Antarctic. However, it is important to recognize that such aerosol contains sulfur products, which, like anthropogenic sulfur products, absorb UV radiation (Liu et al., 1991). Therefore, the overall effect of these processes on UV levels at the earth's surface may be somewhat smaller than their effect on ozone concentrations alone.

In recent years, the phrase "dynamical chemical interactions" has become something of a cliché. I hope that in this chapter I have succeeded in elucidating many of the concepts that are frequently meant by this phrase, and provided insight into a stimulating area of current atmospheric research.

ACKNOWLEDGMENTS

I would like to thank Neal Butchart and Alda Oliveira for kindly reviewing this chapter. Additional invaluable comments were gratefully received from Brian Gardiner, Makoto Koike, Conway Leovy, and Susan Solomon.

REFERENCES

Andrews, D. G., Holton, J. R., and Leovy, C. B. (1987). *Middle Atmosphere Dynamics*. Academic Press, Orlando, FL.

Atkinson, R. J., Mathews, W. A., Newman, P. A., and Plumb, R. A. (1989). Evidence of the mid-latitude impact of Antarctic ozone depletion. *Nature (London)* **340**, 290–294.

Austin, J., and Butchart, N. (1989). A study of air particle motions during a stratospheric warming and their influence on photochemistry. *Q. J. R. Meteorol. Soc.* **115**, 841–866.

Austin, J., and Butchart, N. (1992). A 3-dimensional modeling study of the influence of planetary wave dynamics on polar ozone photochemistry. *J. Geophys. Res.* **97**, 10165–10186.

Austin, J., Garcia, R. R., Russell, J. M., III, Solomon, S., and Tuck, A. F. (1986). On the atmospheric photochemistry of nitric acid. *J. Geophys. Res.* **91**, 5477–5485.

Austin, J., Jones, R. L., McKenna, D. S., Buckland, A. T., Anderson, J. G., Fahey, D. W., Farmer, C. B., Heidt, L. E., Proffitt, M. H., Tuck, A. F., and Vedder, J. F. (1989). Lagrangian photochemical modeling studies of the 1987 Antarctic spring vortex. 2. Seasonal trends in ozone. *J. Geophys. Res.* **94**, 16717–16735.

Austin, J., Butchart, N., and Shine, K. P. (1992). Possible Arctic ozone hole in a doubled CO_2 climate. *Nature (London)* **360**, 221–225.

Brune, W. H., Toohey, D. W., Anderson, J. G., and Chan, K. R. (1990). In situ observations of ClO in the Arctic stratosphere: ER-2 aircraft results from 59°N to 80°N latitude. *Geophys. Res. Lett.* **17**, 505–508.

Butchart, N., and Remsberg, E. E. (1986). The area of the stratospheric polar vortex as a diagnostic for tracer transport on an isentropic surface. *J. Atmos. Sci.* **43**, 1319–1339.

Butchart, N., Austin, J., and Shine, K. P. (1994). Simulations of Arctic ozone depletion with current and doubled levels of CO_2. *Proc. Quadrenn. Ozone Symp.*, Charlottesville, VA, *1992* (in press).

Callis, L. B., Russell, J. M., III, Haggard, K. V., and Natarajan, M. (1983). Examination of wintertime latitudinal gradients in stratospheric NO_2 using theory and LIMS observations *Geophys. Res. Lett.* **10**, 945–948.

Cariolle, D., Lasserre-Bigorry, A., Royer, J.-F., and Geleyn, J.-F. (1990). A general circulation model simulation of the springtime Antarctic ozone decrease and its impact on mid-latitudes. *J. Geophys. Res.* **95**, 1883–1898.

Chapman, S. (1930). A theory of upper atmospheric ozone. *Mem. R. Meteorol. Soc.* **3**, 103–125.

Dameris, M., and Ebel, A. (1990). The quasi-biennial oscillation and major stratospheric warmings: A three dimensional model study. *Ann. Geophys.* **8**(2), 79–86.

de Zafra, R. L., Jaramillo, M., Parrish, A., Solomon, P., and Barrett, J. (1987). High concentrations of chlorine monoxide at low altitudes in the Antarctic spring stratosphere: Diurnal variation. *Nature (London)* **328**, 408–411.

Dobson, G. M. B., and Harrison, D. N. (1926). Measurements of the amount of ozone in the Earth's atmosphere and its relation to other geophysical conditions: Part I. *Proc. R. Soc. London* **A110**, 660–693.

Ebdon, R. A., and Veryard, R. G. (1961). Fluctuations in equatorial stratospheric winds. *Nature (London)* **189**, 791–793.

Eckman, R. S., Turner, R. E., Blackshear, W. T., Fairlie, T. D. A., and Grose, W. L. (1993). Some aspects of the interaction between chemical and dynamical processes relating to the Antarctic ozone hole. *Adv. Space Res.* **13**, 311–319.

Fahey, D. W., Kawa, S. R., and Chan, K. R. (1990). Nitric oxide measurements in the Arctic winter stratosphere. *Geophys. Res. Lett.* **17**, 489–492.

Farman, J. C., Gardiner, B. G., and Shanklin, J. D. (1985). Large losses of total ozone in Antarctica reveal seasonal ClO_x/NO_x interaction. *Nature (London)* **315**, 207–210.

Garcia, R. R., and Solomon, S. (1987). A possible relationship between interannual variability in Antarctic ozone and the quasi-biennial oscillation. *Geophys. Res. Lett.* **14**, 848–851.

Gardiner, B. G. (1988). Comparative morphology of the vertical ozone profile in the Antarctic spring. *Geophys. Res. Lett.* **15**, 901–904.

Gardiner, B. G. (1989). The Antarctic Ozone Hole. *Weather* **44**, 291–298.

Gille, J. C., and Russell, J. M., III (1984). The Limb Infrared Monitor of the Stratosphere: Experiment description, performance, and results. *J. Geophys. Res.* **89**, 5125–5140.

Granier, C., and Brasseur, G. (1991). Ozone and other trace gases in the Arctic and Antarctic regions: Three-dimensional model simulations. *J. Geophys. Res.* **96**, 2995–3011.

Gray, L. J., and Pyle, J. A. (1986). The semi-annual oscillation and equatorial tracer distributions. *Q. J. R. Meteorol. Soc.* **112**, 387–407.

Gray, L. J., and Pyle, J. A. (1987). Two-dimensional model studies of equatorial dynamics and tracer distributions. *Q. J. R. Meteorol. Soc.* **113**, 635–651.

Gray, L. J., and Pyle, J. A. (1989). A two-dimensional model of the quasi-biennial oscillation of ozone. *J. Atmos. Sci.* **46**, 203–220.

Gray, L. J., and Ruth, S. (1992). The interannual variability of trace gases in the stratosphere: A comparative study of the LIMS and UARS measurement periods. *Geophys. Res. Lett.* **19**, 673–676.

Hofmann, D. J., and Solomon, S. (1989). Ozone destruction through heterogeneous chemistry following the eruption of E1 Chichon. *J. Geophys. Res.* **94**, 5029–5041.

Hofmann, D. J., Rosen, J. M., Harder, J. W., and Hereford, J. V. (1989). Balloon-borne measurements of aerosol, condensation nuclei and cloud particles in the stratosphere at McMurdo station Antarctica during the spring of 1987. *J. Geophys. Res.* **94**, 11253–11269.

Holton, J. R., and Austin, J. (1991). The influence of the equatorial QBO on sudden stratospheric warmings. *J. Atmos. Sci.* **48**, 607–618.

Hou, A. Y., Schneider, H. R., and Ko, M. K. W. (1991). A dynamical explanation for the asymmetry in zonally averaged column abundances of ozone between northern and southern springs. *J. Atmos. Sci.* **48**, 547–556.

Houghton, J. T., Jenkins, G. J., and Ephraums, J. J. (Eds.) (1990). *Climate Change, the IPCC Scientific Assessment.* Cambridge University Press, Cambridge UK.

Jones, R. L., and Pyle, J. A. (1984). Observations of CH_4 and N_2O by the NIMBUS 7 SAMS: A comparison with in situ data and two-dimensional numerical model calculations. *J. Geophys. Res.* **89**, 5263–5279.

Jones, R. L., McKenna, D. S., Poole, L. R., and Solomon, S. (1990). Simulating the chemical evolution of the chemical composition of the 1988/1989 winter vortex. *Geophys. Res. Lett.* **17**, 549–552.

Kane, R. P. (1991). Extension of the Antarctic ozone hole to lower latitudes in the South-American region. *Pure Appl. Geophys.* **135**, 611–624.

Koike, M., Kondo, Y., Hayashi, M., Iwasaka, Y., Newman, P. A., Helten, M., and Aimedieu, P. (1991). Depletion of Arctic ozone in the winter of 1990. *Geophys. Res. Lett.* **18**, 791–794.

Labitzke, K. (1982). On the interannual variability of the middle stratosphere during the northern winters. *J. Meteorol. Soc. Jpn.* **60**, 124–139.

Lait, L. R., Schoeberl, M. R., and Newman, P. A. (1989). Quasi-biennial modulation of the Antarctic ozone depletion. *J. Geophys. Res.* **94**, 11559–11571.

Lary, D. J. (1991). Photochemical studies with a three-dimensional model of the atmosphere. Ph.D. Thesis, Cambridge University.

Leovy, C. B., Sun, C. R., Hitchman, M. H., Remsberg, E. E., Russell, J. M., III, Gordley, L. L., Gille, J. C., and Lyjak, L. V. (1985). Transport of ozone in the middle stratosphere: Evidence for planetary wave breaking. *J. Atmos. Sci.* **42**, 230–244.

Liu, S. C., McKeen, S. A., and Madronich, S. (1991). Effect of anthropogenic aerosols on biologically active ultraviolet radiation. *Geophys. Res. Lett.* **18**, 2265–2268.

McIntyre, M. E. (1982). How well do we understand the dynamics of stratospheric warmings? *J. Meteorol. Soc. Jpn.* **60**, 37–65.

McIntyre, M. E. (1989). On the Antarctic ozone hole. *J. Atmos. Terr. Phys.* **51**, 29–43.

McIntyre, M. E., and Palmer, T. N. (1983). Breaking planetary waves in the stratosphere. *Nature (London)* **305**, 593–600.

McIntyre, M. E., and Palmer, T. N. (1984). The "surf zone" in the stratosphere. *J. Atmos. Terr. Phys.* **46**, 825–849.

McKenna, D. S., Jones, R. L., Poole, L. R., Solomon, S., Fahey, D. W., Kelly, K. K., Proffitt, M. H., Brune, W. H., Loewenstein, M., and Chan, K. R. (1990). Calculations of ozone destruction during the 1988/1989 Arctic winter. *Geophys. Res. Lett.* **17**, 553–556.

Naujokat, B. (1986). An update of the observed quasi-biennial oscillation of the stratospheric winds over the tropics. *J. Atmos. Sci.* **43**, 1873–1877.

Noxon, J. F. (1979). Stratospheric NO_2. 2. Global behavior. *J. Geophys. Res.* **84**, 5067–5076.

Noxon, J. F., Whipple, E. C., Jr., and Hyde, R. S. (1979a). Stratospheric NO_2. 1. Observational method and behavior at mid-latitudes. *J. Geophys. Res.* **84**, 5047–5065.

Noxon, J. F., Marovich, E., and Norton, R. B. (1979b). Effects of a major stratospheric warming upon stratospheric NO_2. *J. Geophys. Res.* **84**, 7883–7888.

Noxon, J. F., Henderson, W. R., and Norton, R. B. (1983). Stratospheric NO_2. 3. The effects of large scale horizontal transport. *J. Geophys. Res.* **88**, 5240–5248.

Pitari, G., Palermi, S., Visconti, G., and Prinn, R. G. (1992a). Ozone response to a CO_2 doubling: Results from a stratospheric circulation model with heterogeneous chemistry. *J. Geophys. Res.* **97**, 5953–5962.

Pitari, G., Visconti, G., and Verdecchia, M. (1992b). Global ozone depletion and the Antarctic ozone hole. *J. Geophys. Res.* **97**, 8075–8082.

Poole, L. R., Solomon, S., McCormick, M. P., and Pitts, M. C. (1989). The interannual variability of polar stratospheric clouds and related parameters in Antarctica during September and October. *Geophys. Res. Lett.* **16**, 1157–1160.

Prather, M., and Jaffe, A. H. (1990). Global impact of the Antarctic ozone hole: Chemical propagation. *J. Geophys. Res.* **95**, 3473–3492.

Prather, M., Garcia, M. M., Suozzo, R., and Rind, D. (1990). Global impact of the Antarctic ozone hole: Dynamical dilution with a three dimensional chemical transport model. *J. Geophys. Res.* **95**, 3449–3471.

Proffitt, M. H., Margitan, J. J., Kelly, K. K., Loewenstein, M., Podolske, J. R., and Chan, K. R. (1990). Ozone loss in the arctic polar vortex inferred from high-altitude aircraft measurements. *Nature (London)* **347**, 31–36.

Proffitt, M. H., Solomon, S., and Loewenstein, M. (1992). Comparison of 2-D model simulation of ozone and nitrous oxide at high latitudes with stratospheric measurements. *J. Geophys. Res.* **97**, 939–944.

Pyle, J. A., Cox, R. A., Farman, J. C., Gray, L. J., Harris, N. R. P., Jones, R. L., O'Neill, A., Penkett, S. A., Roscoe, H. K., Shine, K. P., Warrilow, D. A., and Cayless, S. M. (1991). *Stratospheric Ozone 1991*. Report of the United Kingdom Stratospheric Ozone Review Group, Her Majesty's Stationery Office, London.

Reed, R. J. (1965). The quasi-biennial oscillation of the atmosphere between 30 and 50 km over Ascension Island. *J. Atmos. Sci.* **22**, 331–333.

Reed, R. J., Campbell, W. J., Rasmussen, L. A., and Rogers, D. G. (1961). Evidence of a downward propagating annual wind reversal in the equatorial stratosphere. *J. Geophys. Res.* **66**, 813–818.

Rind, D., Suozzo, R., Balachandran, N. K., and Prather, M. J. (1990). Climate change and the middle atmosphere. Part 1: The doubled CO_2 climate. *J. Atmos. Sci.* **44**, 475–494.

Rood, R. B., Nielsen, J. E., Stolarski, R. S., Douglass, A. R., Kaye, J. A., and Allen, D. J. (1992). Episodic total ozone minima and associated effects on heterogeneous chemistry and lower stratospheric transport. *J. Geophys. Res.* **97**, 7979–7996.

Sanders, R. W., Solomon, S., Carroll, M. A., and Schmeltekopf, A. L. (1989). Visible and near ultra-violet spectroscopy at McMurdo station, Antarctica. 4. Overview and daily measurements of NO_2, O_3 and OClO during 1987. *J. Geophys. Res.* **94**, 11381–11391.

Schoeberl, M. R., and Hartmann, D. L. (1991). The dynamics of the stratospheric polar vortex and its relation to springtime ozone depletions. *Science* **251**, 46–52.

Schoeberl, M. R., Lait, L. R., Newman, P. A., and Rosenfield, J. E. (1992). The structure of the polar vortex. *J. Geophys. Res.* **97**, 7859–7882.

Shine, K. (1986). On the modelled thermal response of the Antarctic stratosphere to a depletion of ozone. *Geophys. Res. Lett.* **13**, 1331–1334.

Solomon, S. (1990). Progress towards a quantitative understanding of Antarctic ozone depletion. *Nature (London)* **347**, 347–354.

Solomon, S., and Garcia, R. R. (1983a). On the distribution of nitrogen dioxide in the high latitude stratosphere. *J. Geophys. Res.* **88**, 5229–5239.

Solomon, S., and Garcia, R. R. (1983b). Simulation of NO_x partitioning along isobaric parcel trajectories. *J. Geophys. Res.* **88**, 5497–5501.

Solomon, S., Mount, G. H., Sanders, R. W., and Schmeltekopf, A. L. (1987). Visible spectroscopy at McMurdo station, Antarctica. 2. Observations of OClO. *J. Geophys. Res.* **92**, 8329–8338.

Stolarski, R. S., Bloomfield, P., and McPeters, R. D. (1991). Total ozone trends ·deduced from NIMBUS 7 TOMS data. *Geophys. Res. Lett.* **18**, 1015–1018.

Tuck, A. F. (1989). Synoptic and chemical evolution of the Antarctic vortex in late winter and early spring, 1987. *J. Geophys. Res.* **94**, 11687–11737.

Tuck, A. F., Watson, R. T., Condon, E. P., Margitan, J. J., and Toon, O. B. (1989). The planning and execution of ER-2 and DC-8 aircraft flights over Antarctica, August and September, 1987. *J. Geophys. Res.* **94**, 11181–11222.

Tuck, A. F., Davies, T., Hovde, S. J., Noguer-Alba, M., Fahey, D. W., Kawa, S. R., Kelly, K. K., Murphy, D. M., Proffitt, M. H., Margitan, J. J., Loewenstein, M., Podolske, J. R., Strahan, S. E., and Chan, K. R. (1992). PSC processed air and potential vorticity in the northern hemisphere lower stratosphere at mid-latitudes during winter. *J. Geophys. Res.* **97**, 7883–7904.

World Meteorological Organization (WMO) (1989). Report of the International Ozone Trends Panel, 1988. *World Meteorological Organization Global Ozone Research and Monitoring Project*, Report No. 18. WMO, Geneva.

World Meteorological Organization (WMO) (1992). Scientific assessment of ozone depletion: 1991. *World Meteorological Organization Global Ozone Research and Monitoring Project*, Report No. 25. WMO, Geneva.

Yao, F. Wilson, I., and Johnston, H. (1982). Temperature dependent ultraviolet absorption spectrum for dinitrogen pentoxide. *J. Phys. Chem.* **86**, 3611–3615.

7

THE PHYSICAL CHEMISTRY OF PHOTOCHEMICAL OXIDANT GENERATION IN NATURAL WATER SYSTEMS

Aldo Bruccoleri, Guiseppe Lepore, and Cooper H. Langford

Department of Chemistry, University of Calgary, Calgary, Alberta, Canada T2N 1N4

Environmental Oxidants, Edited by Jerome O. Nriagu and Milagros S. Simmons.
ISBN 0–471–57928–9 © 1994 John Wiley & Sons, Inc.

1. INTRODUCTION

The intention of this chapter is to provide a physical chemical view of the processes that underlie oxidative photochemistry in natural water systems. We will begin in the time domain of 20 psec and pursue the consequences of excitation out of the time domain of a second or two. This gives us a scope of 11 orders of magnitude. Were we to take a logarithmically similar view on the other side of the arbitrary divide of one second, that would implicate the nearly geologic time scale of more than 20,000 years, a long time in the domain of low temperature geochemistry. The results described will be almost entirely those derived in the laboratory by the techniques of "unnatural" science using picosecond lasers, high magnetic fields, and piezoelectric detectors. We do not urge any of our readers to try to implement such studies in the field. However, we do invite attention to the constraints they place on analysis of field data.

The first consideration in approaching the question of which laboratory data are useful and significant is the distribution of significant chromophores. The light entering the euphotic zone of natural waters includes the visible and near-ultraviolet regions. In consequence, the vast majority of dissolved small organic molecules and simple inorganic ions are not important light absorbers. Their spectra lie too deep inside the ultraviolet. The most important chromophores are, in fact, not unambiguously identifiable as dissolved species. They are colloidal and fall into two main classes. The first class is "dissolved" organic materials that belong predominately to that heterogeneous mixture of polydisperse polymers called humic substances. The second class is transition metal containing hydrous oxide colloids. The most important representatives of this second class are $Fe(III)$ and $Mn(IV)$ hydrous oxides. However, the most extensive laboratory data are available for TiO_2, which has been studied for its potential use as a photocatalyst. Since the more environmentally important hydrous oxides can be characterized by comparison and contrast with TiO_2, we will cite some of the laboratory data on TiO_2.

1.1. Humic Substances

Humic substances are a heterogeneous and polydisperse mixture of polymers in the molecular weight range from a few hundred to tens of thousands. For more than 100 years they have been subdivided into three main classes by a very simple scheme. The material that cannot be extracted into "solution" by aqueous NaOH is called humin. The fraction that can be extracted by base but is reprecipitated by acidification is called humic. The acid-"soluble" fraction is called fulvic. The studies we report are carried out mainly with a fulvic acid.

The spectrum of a typical (and "well characterized") fulvic acid is shown in Figure 1. It is quite striking that absorbance extends well into the red region and toward the near-infrared. This presents more of a puzzle than has usually been acknowledged. The functional group content, degradation, and fractionation studies suggest that the lowest energy transitions are expected to be those of the

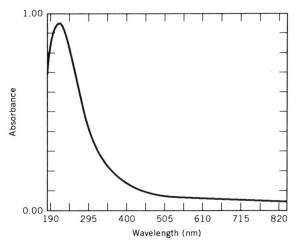

Figure 1. Absorption spectrum of Armadale fulvic acid (10 mg L^{-1}; pH 4). From Bruccoleri, unpublished data.

phenol carboxylate functional groups, which absorb in the near-UV region as monomers. Why does this small polymer display such profound red shifts? One factor is the degree of conjugation of the chromophores in the polymers. Another factor has been proposed by Joan Power that focuses on the opportunity for donor–acceptor interactions in the polymers. We give prominence to the accumulating evidence favoring this account of the origin of color because we think that it also aids in the understanding of the photochemistry where large quantum yields for primary events do not translate into high steady-state yields of expected reactive radicals. That is, the internal interactions in the polymers are probably central to the photochemistry.

1.2. Hydrous Oxides

Hydrous oxides of Fe(III) and Mn(IV) are colored and widely distributed. Colloidal particles are present in aerated waters exposed to light. Light absorbtion by the particles themselves is best described in solid-state terminology as "band-gap" excitation. A photon promotes an electron from the valence band to the conduction band (Fig. 5). In the hydrous oxides of transition metals in high oxidation states, the valence band is principally composed of oxide ion localized orbitals, and the conduction band is mainly metal d orbital in character. Electrons trapped at the surface may be expected to resemble reduced metal sites, for example, Fe(II). Surface trapped holes resemble adsorbed OH radicals. The interesting photochemistry is generated by interfacial electron transfer of the trapped carrier to species adsorbed from the solution. As noted above, the best-characterized example of such "semiconductor photochemistry" is TiO_2, which has received much attention for its potential application in waste treatment. We review some of this chemistry to point out the mechanistic features that may also be applicable to some of the hydrous oxide colloids important to natural water systems.

There is a second spectroscopic process that may also be important to photochemistry initiated on hydrous oxides. When an organic molecule that can function as a ligand (e.g., one with a carboxylate group) adsorbs, it may form a surface complex with the metal ion of the hydrous oxide lattice. Such a surface complex may be excited in a ligand-to-metal charge transfer band. The photochemical consequence may be reduction of the metal and oxidation of the ligand. There are significant indications that decarboxylation of carboxylic acids adsorbed on hydrous iron oxides occurs this way.

2.   THE HUMIC SUBSTANCES CHROMOPHORE CLASS

Humic substances occur in almost every aquatic environment and play a significant role in the regulation of the composition of natural water systems (1–9). They are a broad class of naturally occurring polymers that result from the modification of plant- and animal-derived organic matter by chemical and microbial action. The general character of humic substances in an extremely heterogeneous polydisperse mixture of substances of molecular weights ranging from several hundred to tens of thousands; they are hydrophilic, acidic, and contain phenol carboxylate, refractory carbohydrate, and aliphatic units joined together in an irregular manner by covalent linkages of varying strength. Beyond this, conformation and aggregational equilibria strongly influenced by hydrogen bonding and probably involving donor–acceptor interactions are important. Humics are characterized by variations in both physical and chemical properties in samples derived from different sources and to a certain extent even within a single sample. In particular, properties such as luminescence spectra, absorption spectra, particle size, and the reactivity of various functional groups vary with the source of the sample and, during titration, with a simple species such as H^+ or a metal cation. Humic materials, through their cation binding capacity, also contribute to the control of the pH of natural water systems and may play an important role in the uptake of metal ions by living organisms (3, 6, 7, 10–14).

Much work on the photochemistry of natural waters has demonstrated that under solar irradiation, waters containing humic substances generate a wide variety of reactive intermediates and radicals. The diversity of the photochemistry observed in natural waters cannot be fully understood without a more detailed understanding of the primary photophysical and photochemical pathways by which excited humic chromophores yield reactive states and intermediates (15–20). Such detailed analysis must depend on humic materials that are as well defined as possible. Among the best-characterized samples are the Armadale and Laurentian fulvic acids (11, 12) derived from Canadian podzols. These will receive emphasis here.

Table 1   Distribution of Functional Groups in Armadale Fulvic Acid

Functional Group	Abundance (mEq g^{-1})
Phenolics	3.3
Carboxylates	7.7
Aliphatic alcohols	3.6
Quinones	0.6
Ketones	0.3

2.1. Photophysical Characteristics

Recently, there has been extensive work dealing with the photophysical pathways open to well-characterized humic substances, including the Armadale and Laurentian fulvic acids, among others (2, 3, 11, 13). The fulvics have a broad absorption spectrum that extends with decreasing absorptivity into the far-red region. The spectrum (Fig. 1) resembles a light-scattering curve, but photothermal spectra make it clear that only a small to moderate part of the apparent absorptivity is attributable to light scattering.

Extensive degradative studies and functional group analysis of Armadale fulvic acid (1–9, 14) suggest a structure for the material with mainly aromatic low-excitation energy chromophores including phenolic, phenol carboxylate, or quinoid group (Table 1). The distribution of functional groups poses a difficulty for the explanation of humic substance spectra. In all cases, the dominant absorption of these fragments lies well inside the UV region even below 300 nm. One would therefore expect maximum fluorescence emission for excitation in the 250- to 300-nm wavelength range, since this corresponds to states of the organic chromophores, with an emission maximum in the near-UV. In fact, humic substance luminescence is most efficiently excited at 350 to 400 nm rather than at shorter wavelengths. The emission is broadband and extends from 400 to 500 nm (Fig. 2) (21). Luminescence lifetimes are distributed from values in the picosecond domain to those suggestive of phosphorescence.

The fluorescence properties of a number of phenol carboxylate model compounds have been reported and compared with those of fulvic acid. It has been pointed out by Buffle et al. (15) that the values for few of these compounds approach the excitation and emission wavelengths of fulvic acid. Also, the fluorescence intensity expressed proportional to dissolved organic carbon was much greater for the model compounds than for fulvic acid. This suggestion has directly confirmed low quantum yields for emission. Furthermore, at higher concentrations of fulvic acid (> 100 mg/L), the wavelengths of fluorescence excitation and emission shifted significantly to the red. This indicated that the fluorescence properties of the dissolved humic materials were strongly affected by aggregation phenomena.

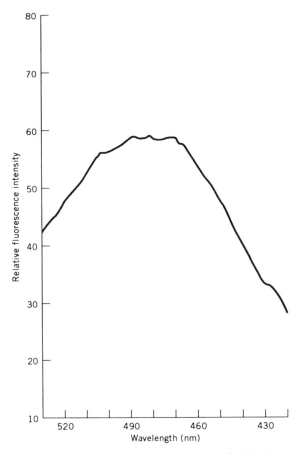

Figure 2. Emission spectrum of Armadale fulvic acid (25 mg L^{-1}; pH 4). From Lavigne (21).

Recent work by Wang et al. (11) involving spectroscopic studies of Laurentian fulvic acid provide direct evidence for the ground-state donor–acceptor interaction mechanism. In an attempt to evaluate the interactions between different fractions of fulvic acid, the UV-VIS absorption and fluorescence spectra for a mixture were monitored after different time intervals. The mixture was prepared by mixing the solution containing the highest molecular weight fraction with that containing the lowest molecular weight fraction in equal proportions. Comparison between the absorption spectra for the individual fractions and the mixture indicated an interaction that renders spectral summing nonlinear (Fig. 3) (11). Especially at the red end of the spectrum, the absorbance and fluorescence of the mixture depart from the mean of the separate absorbances of the high and low molecular weight fractions. In addition, the absorbance in the mixture increases over time after mixing. These features clearly indicate interactions of the smaller with the larger particles, and these aggregational interactions have the effect of enhancing the long-wavelength absorbance.

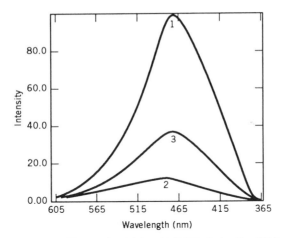

Figure 3. Fluorescence spectra of Laurentian fulvic acid (0.100 g L^{-1}, pH = 3.8) fractions and their mixture. (1) Lowest molecular weight fraction; (2) highest molecular weight fraction; (3) mixture (1:1). From Wang et al. (11).

As first suggested by Joan Power (3), known phenomena in polymer photophysics (16) suggest an explanation for the observed spectral properties of fulvic acid. A common process in the photophysics of synthetic polymers that contain aromatic functional units is the formation of donor–acceptor complexes in the ground state and excimers and exciplexes in the excited state. The formation of excimers determines the fluorescence properties of the materials. In aromatic systems, excimers form when chromophore ensembles at high density are irradiated (3, 13, 16). In these cases, the excimers arise due to face-to-face overlap of the aromatic rings in the excited state with a spacing on the order of 10 Å. Excimer fluorescence is very broad and red-shifted relative to the monomer fluorescence spectrum.

Similarly, formation of ground-state dimers (17,18) possessing the geometry of excimers arises from donor–acceptor interactions. These ground-state dimers may shift the absorption spectrum as much as 50 to 100 nm into the red. This is just the sort of shift needed to relate the fragment spectra to the absorbance across the visible seen in Figure 1.

If the chromophore pair forming an excited complex involves nonidentical chromophores, the result is an exciplex (an excited-state donor–acceptor complex). The formation of exciplexes is usually accompanied by a transfer of charge in the excited state. The exciplexes may then separate to yield radical ion pairs or cross between systems to form triplet exciplexes. An increased red shift in the ground-state absorption parallels the greater charge transfer character of the transitions in exciplexes.

The dimer–excimer–exciplex model provided by polymer photophysics is helpful in explaining the fluorescence and absorption properties of humic substances in general. It seems likely that the humic substances, because they are aggregates, offer many possibilities for achieving suitable conformations, since the chromophore density is very high within the aggregates. This point is confirmed by the observation that red shifts in the wavelengths of absorption and emission relative to independent chromophoric group values are observed in all humic substances (1, 13). The only other major mechanism for red shift generation, the conjugation of neighboring groups in the polymer, seems clearly insufficient to explain the whole effect given the degradation of these materials into predominately mononuclear aromatic fragments.

Donor–acceptor interactions can arise during aggregation but can also occur intramolecularly. This will be more probable in the larger polymers of the high molecular weight fractions. Further evidence of this so-called polymer model of fulvic acid photophysics may be seen in the results of studies involving the evaluation of primary photoproduct quantum yields of well-characterized fulvic acids (12, 13, 19).

2.2. Studies of Primary Quantum Yields in Fulvic Acid

Fulvic acid samples absorb strongly in the near-ultraviolet region of the spectrum. The primary photophysical processes that may occur after excitation are luminescence, intersystem crossing, and internal conversion from the singlet to the ground state. Time-resolved spectroscopy in the picosecond to microsecond time domain is needed to identify the primary processes. The primary photochemistry is photoionization, the formation of a solvated electron together with its associated cation radical, in less than 20 psec. The other initial photophysical

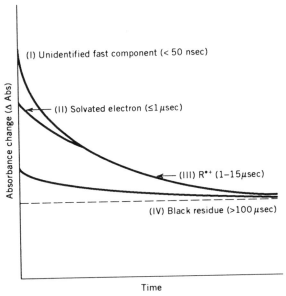

Figure 4. Schematic "map" of the transient species observed in the subnanosecond and nanosecond flash photolysis of Armadale fulvic acid. From Power et al. (13).

Table 2   Apparent Triplet Quantum Yields Calculated for
Various Solutions of Armadale Fulvic Acid

Solution	pH	Apparent Quantum Yield Φ^b	Ionic Strength
Fulvic acid	2.0	0.82	$< 10^{-3}$
	3.0	0.78	$< 10^{-3}$
	5.0	0.65	$< 10^{-3}$
	6.0	0.61	$< 10^{-3}$
	8.0	0.52	$< 10^{-3}$
	9.5	0.35	$< 10^{-3}$
Fulvic acid $(KCl)^a$	1.5	0.91	1.0
	8.5	0.91	1.0

[a] This solution was prepared with a concentration of 1 M potassium chloride as background electrolyte.
[b] The standard deviation in the apparent triplet quantum yields is 10%. Each quantum yield is an average calculated from 18 fulvic acid solutions prepared with concentrations $\leqslant 20\,mg\,L^{-1}$.

paths are triplet formation and radiationless decay to the ground state (Fig. 4) (13). Luminescence is minor at all pH values.

The work of Langford et al. (12, 19), using time-resolved photoacoustic spectroscopy and magnetic circular dichroism (MCD) spectroscopy, determined energies and quantum yields for the formation of triplet states in aqueous solutions of our two well-characterized (Laurentian and Armadale) fulvic acids. MCD spectra are useful for identification of spin-forbidden bands because transition intensities depend on the product of electric and magnetic dipole matrix elements, and spin-forbidden bands are relatively more intense than in ordinary absorption spectroscopy. Thus, it provided an estimate of the triplet energy. Pulsed photoacoustic spectroscopy with a time resolution of 2 μsec allowed estimation of the energy stored in triplets. Together, these two parameters afford triplet quantum yields. The major photophysical pathway is triplet formation, at least at low pH, as illustrated in Tables 2 and 3 (12, 19).

The fractionation of the Laurentian sample indicates the triplets are more associated with the lower molecular weight components of the mixture. This relates to an earlier observation that luminescence is concentrated in lower molecular weight fractions (11, 20). Photoionization seems less sensitive to molecular weight. Estimates of the quantum yield for the solvated electron and the corresponding radical have been made directly from time-resolved absorption spectra recorded at 20 psec; these are reported in Table 4 (19).

Together these pathways account for a total quantum yield of 0.10 ± 0.01. Adding the triplet yield of 0.82 ± 0.08 in acid solution gives 0.92 ± 0.09, which is within experimental error of accounting for all the photons. Thus, in acid solution, light absorption by fulvic acid gives, in decreasing order of importance, long-lived triplets, photoionization, and luminescence; radiationless decay to the

Table 3   Apparent Triplet Quantum Yields Calculated for Various Solutions of Laurentian Fulvic Acid

Solution[a]	pH	Apparent Quantum Yield Φ	Ionic Strength
Fulvic acid	2.0	0.79	$< 10^{-3}$
	5.0	0.60	$< 10^{-3}$
	9.5	0.28	$< 10^{-3}$
High fraction	2.0	0.26	$< 10^{-3}$
	5.0	0.21	$< 10^{-3}$
	9.5	0.15	$< 10^{-3}$
Low fraction	2.0	0.85	$< 10^{-3}$
	5.0	0.73	$< 10^{-3}$
	9.5	0.31	$< 10^{-3}$
High fraction (KCl)[b]	2.0	0.32	1.0
	5.0	0.27	1.0
	9.5	0.18	1.0
Low fraction (KCl)[b]	2.0	0.88	1.0
	5.0	0.88	1.0
	9.5	0.86	1.0
Fulvic acid (KCl)[b]	2.0	0.90	1.0
	5.0	0.91	1.0
	9.5	0.90	1.0

[a] High refers to the portion retained after Laurentian fulvic acid solutions were passed through a YM30 (nominal cutoff 30,000 daltons) filter membrane. Low refers to the freeze-dried filtrate of Laurentian fulvic acid solutions that passed through a YM2 (nominal cutoff 1000 daltons) filter membrane.
[b] These solutions were prepared with a concentration of 1 M KCl as background electrolyte.

ground state has a yield less than or equal to approximately 0.08 and may be near zero. The Laurentian fulvic acid behaves similarly.

The data at pH 2.0 provide an interesting reference base. They suggest that the behavior of fully protonated fulvic acid mixtures is dominated by intersystem crossing to triplets and photoionization. Picosecond absorption spectroscopy has revealed that both of these processes are quite rapid (< 20 psec) (1, 13). As pH is increased, the energy stored as triplets for periods longer than 2 μs (2 μsec is the minimum time resolution of the pulsed photoacoustic apparatus) decreases, that is, some processes of radiationless relaxation to the ground state become important. There are two possibilities. One is that internal conversion of singlets directly to the ground state becomes important. The second is that the radiationless decay of triplets to the ground state becomes more rapid, and an important fraction of triplets no longer lasts long enough to be identified as energy-storing species on the 2-μsec time scale. Since transient absorption spectra (1, 3, 13) suggest a relatively long triplet lifetime, the former process appears to be more important.

Table 4   Quantum Yields for the Formation of the Solvated Electron and the Corresponding Free Radical of Laurentian Fulvic Acid at 20 psec

Solution[a]	pH	$\Phi_{e, rad}$
Laurentian fulvic acid	4.0	0.12
	7.0	0.16
	9.0	0.18
High fraction	4.0	0.15
	9.0	0.20
Low fraction	4.0	0.18
	9.0	0.22

[a] High refers to the portion retained after Laurentian fulvic acid solutions were passed through a YM30 (nominal cutoff 30,000 daltons) filter membrane. Low refers to the freeze-dried filtrate of Laurentian fulvic acid solutions that were passed through a YM2 (nominal cutoff 1000 daltons) filter membrane. At a pH of 2.0, for the Armadale sample, the quantum yield for the formation of the solvated electron and radical cation (Table 5) (3, 13) is approximately 0.09 and the fluorescence yield is 0.014 (13).

The decrease of long-lived triplets with increasing pH has two possible molecular origins. One would focus on local chromophoric groups and emphasize the difference between the acid and base forms. Such an explanation would suggest a triplet yield–pH relationship with the functional form of an acid–base titration curve for the mixture. The titration curves are known and an approximate equivalence point for carboxylates is seen at mid-pH values of 5 to 6. A phenolate equivalence point is expected at high pH. The quantum yield–pH relationship does not reveal the "breaks" near equivalence points this model predicts. Rather, the decrease in quantum yield appears to be a smooth function of pH. The observation is not, however, definitive. A quantum yield–pH function composed of overlapping dependencies on carboxylate and phenolate titrations might be consistent with the present observations, especially given their rather large experimental uncertainty. Still, the present data do not favor this explanation.

The alternative molecular explanation of the quantum yield–pH function focuses on molecular conformation and aggregation, as does the theory of "color" outlined above. Deprotonation of fulvic acid causes polymers to adopt a more extended conformation and does not favor aggregation. Light-scattering studies (10) suggest a smooth increase in the effective particle size of Armadale fulvic acid with increases in pH. Since the behavior at high ionic strengths at all pH values mimics the low-pH, fully protonated sample, and it is known that high ionic strength favors aggregation and can mitigate the electrostatic factors that cause extended polymer conformations, the parallel between low-pH and ionic strength effects argues strongly for a conformation–aggregation explanation of the changing triplet quantum yields.

The likely origins of a conformation–aggregation effect on triplet yields are interesting and again support the hypothesis introduced by Power (1, 3, 13) that an important factor in both absorbance and luminescence spectra of humics is donor–acceptor interaction among constituent chromophores.

2.3. Photochemistry

The one primary photochemical event is photoionization. It is much less sensitive to environmental factors than triplet formation. This suggests, in agreement with picosecond spectra, a prompt process governed by local chromophore energetics. The polymer environment is not critical.

All remaining photochemistry, including the processes of trace contaminant oxidation initiated by humics, which have received much attention in the literature (22–27), must depend upon the subsequent fate of the three primary photoproducts: the triplets, the solvated electron, and the cation radicals. There is one unifying factor. In aerated waters (essentially all those substantially sunlight-illuminated), the reactive species present in the highest concentrations are dissolved O_2 (concentration at saturation is approximately 2.5×10^{-4} M) and the humic substances themselves. Thus, O_2 is the dominant scavenger of the photogenerated reactive species, which can compete with internal reactions of the humics themselves. An exceptional case is H^+ scavenging of the solvated electron at pH values less than approximately 4. We note that if oxygen reacts at the diffusion limit with a photogenerated species, the time constant for O_2 scavenging in air-saturated water at 25 °C is approximately 4 μsec. Thus, it is not critical whether triplet yields measured photoacoustically fall because intersystem crossing declines or because triplet lifetimes become short. The yield of triplets measured by time-resolved pulsed photoacoustic spectroscopy (12, 19) is the yield of oxygen-scavangable triplets to good approximation.

The primary photophysics and photochemistry thus suggest that overall photoprocesses will be mediated by reaction with the humic polymers themselves and three key reactions of O_2

$$^3FA + {}^3O_2(aq) \longrightarrow {}^1FA + {}^1O_2(aq) \tag{1}$$

$$e^- + {}^3O_2(aq) \longrightarrow O_2^-(aq) \tag{2}$$

$$R^{\cdot +}(aq) + O_2(aq) \longrightarrow RO_2^{\cdot +} \tag{3}$$

where ^3FA is triplet fulvic acid, e^- is the solvated electron, and $R^{\cdot +}$ is the primary radical cation photoproduct (1, 3, 4, 12, 13, 19). The first reaction generates a reagent for attack at unsaturated centers. Recently, Shao et al. (28) determined $^1O_2$ quantum yields in natural waters at different wavelengths from 280 to 700 nm and reported quantum yields between 0.001 and 0.026, which are very similar to those reported by Haag et al. (29) and Frimmel et al. (30). Interestingly,

Haag and Hoigne (31) studied the formation of $^1O_2$ from different fractions of humic substances and found that the lower molecular weight fractions of 100 to 500 Da were more efficient in forming $^1O_2$. This is consistent, assuming Equation (1) is correct, with the evidence from Langford et al. (19) that the lower molecular weight fractions have higher quantum yields for intersystem crossing than the higher fractions (Table 3).

The second reaction produces the superoxide ion, which disproportionates rapidly and efficiently to hydrogen peroxide (4, 32); approximately 24 to 41% of the superoxide was shown not to form hydrogen peroxide in studies by Petasne and Zika (4, 25). Recently, Zafiriou et al. (33) investigated total radical production with respect to superoxide production in marine surface waters by new electron paramagnetic resonance (EPR) detection methods. The total radical production involved the use of a nonselective free radical scavenger, nitric oxide; the superoxide radical production was measured by trapping it as a stable, isotopically labeled derivative, which avoids some uncertainties associated with indirect competition methods. The superoxide production was found to be approximately one-third of the total radical production. The total radical production rates ranged from undetectable, < 0.2 nmol/L/min with "1 simulated sun," to > 50 nmol/L/min/sun.

The organic peroxy radical is less characterized but quite reactive. Work by Power et al. (1, 3, 13) determined the quantum yield of formation for the solvated electron and associated radical cation upon irradiation of fulvic acid. Power suggests certain potential candidates for the radical co-products given the spectral characteristics and the phenol carboxylate character of fulvic acid (3, 34–36).

1. Phenolic compounds, when irradiated, efficiently yield solvated electrons and phenoxyl radicals:

$$C_6H_5OH \longrightarrow C_6H_5O \cdot + H^+ + e^-$$

2. Anisoles and related methoxy benzenes yield radicals of the phenoxymethyl type, with the loss of a proton:

$$C_6H_5OCH_3 \longrightarrow C_6H_5OCH_2 \cdot + H^+ + e^-$$

3. Phenoxy acids yield the phenoxymethyl radical via loss of CO_2 at the terminal carbon (3, 34):

$$C_6H_5OCH_2COO^- \longrightarrow C_6H_5OCH_2COO \cdot + e^-$$
$$C_6H_5OCH_2COO \cdot \longrightarrow C_6H_5OCH_2 \cdot + CO_2$$

All three types of radicals show broadband absorption from 350 to 500 nm (3). Power notes that phenoxyl radicals are insensitive to oxygen and persist for longer than 100 μsec. The phenoxymethyl radical is readily quenched by O_2 and possesses a lifetime of 25 μsec or less. The 480-nm spectral feature (Fig. 4) (13)

Table 5   Quantum Yields for Formation of the Solvated Electron for Armadale Fulvic Acid at 20 psec

Solution	pH	Φ_e
Armadale fulvic acid	2.0	0.09
	4.0	0.15
	7.0	0.20

observed at 1 μsec seems to be readily explainable in terms of species of this type. Power further suggests, on the basis of their long lifetimes, that phenoxyl radicals might contribute a component to the "black residue" absorbance (Fig. 4).

The elucidation of pathways for the degradation of specific compounds in water will depend on evaluation of the reactions of each of these species with the substrate and the competition provided by components of the humics themselves. Considering the primary quantum yields for the triplets (Tables 2 and 3) and the primary yields of the solvated electron and the corresponding radical (Tables 4 and 5), the literature values for steady-state yields of secondary photoproducts in water appear relatively low. For example, singlet oxygen yields are between 0.001 and 0.026 (28), and hydrogen peroxide yields are on the order of 10^{-4} (4) for irradiation of natural waters with wavelengths of about 340 nm and higher. The large difference between the primary yields of the initial photoproducts and the relatively low quantum yields of the secondary products would appear to indicate that competition provided by components of the humics is significant, given that the two major reactive species present in aerated natural waters are dissolved oxygen and the humic substances themselves. The immediate need is to enrich our understanding of the reactions of reactive intermediates with humics.

3. OXIDE PHOTOCATALYSTS

Heterogeneous photocatalysis involving excitation of oxide materials has been the subject of intense attention over the past 15 years, most recently in the waste treatment context. Photocatalytic processes are reactions that are driven in a spontaneous direction ($\Delta G < 0$), the radiant energy merely overcoming the energy of activation for the process. Oxygen oxidation of organic materials is in this class. The indirect phototransformation reactions observed on particulate oxide materials in the laboratory and in natural aquatic systems can be sub-divided into two dynamic mechanisms:

1. Surface reactions involving reactive intermediates generated by a semiconductor material upon band-gap illumination and surface trapping of photogenerated carriers, holes and electrons

2. Reactions of electronically excited surface species where light is absorbed in a transition of a surface complex.

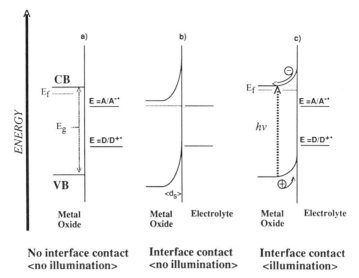

Figure 5. Energy diagram for a large n-type metal oxide particle illustrating the superposition of the energy bands of the metal oxide (valence band, VB; conduction band, CB; band-gap energy, E_g; Fermi level, E_f). (a) No metal oxide–electrolyte interface contact. (b) The solid particle in contact with the liquid at equilibrium in the dark, with the formation of a space charge of distance (d_s). (c) Band-gap illumination of the solid–liquid interface with the generation of electron–hole pairs in the band-bending region, resulting in charge separation followed by the minority carrier (hole) being driven to the interface.

3.1. Photophysics of Metal Oxides and Primary Photochemical Steps

The generally accepted mechanism for the semiconductor photoactivated process proceeds via the excitation of an electron from the valence band (with light energy $\geqslant$ optical band-gap energy of the semiconductor) to the conduction band, leaving a vacancy ("hole") in the valence band. This is illustrated in Figure 5 and in Scheme 1 (reactions S1–S16). In an intrinsic semiconductor (with few lattice defects), the hole can be regarded as a mobile entity because annihilation of a hole by nearby valence band electrons effectively translates the hole. Similarly, the electron can migrate in the conduction band. The photoproduced electron–hole pairs may migrate to and be trapped at the surface. Surface-trapped electrons are characteristically reduced metal sites and the surface-trapped hole of an oxide lattice resembles an adsorbed hydroxyl radical (HO·) denoted $\int \cdot OH$ (37).

Trapping the valence band holes and conduction band electrons opens channels for energetic interfacial oxidative and/or reductive processes with adsorbed species or surface groups, or with the bulk semiconductor material itself. The latter pathway may result in the photocorrosion (S4 and S5) (or dissolution) of a hydrous metal oxide such as Fe(III) and/or of Mn(IV) oxides found in natural systems (38–42). The quantum efficiency of redox reactions at the metal oxide–liquid interface will depend on how fast the carriers reach the surface of the

Substrate Adsorption

(S1) $-M-OH$ $+$ R_1 $\rightleftharpoons$ $-M-OH > -R_{1,ads}$

Band-Gap Excitation

(S2) $-M-OH$ $+$ $h\nu$ $\longrightarrow$ $h^+ + e^-$

Recombination

(S3) h^+ $+$ e^- $\longrightarrow$ Heat *or* $h\nu$

Photocorrosion

(S4) h^+ $+$ $-M-OH$ $\longrightarrow$ $-M-OH[M^{m+1}]$

(S5) e^- $+$ $-M-OH$ $\longrightarrow$ $-M-OH[M^{m-1}]$

Carrier Trapping

(S6) $-M-OH$ $+$ h^+ $\rightleftharpoons$ $-M\int\cdot OH$

(S7) $R_{1,ads}$ $+$ h^+ $\rightleftharpoons$ $R_{1,ads}^{\cdot+}$ (or) $R_{2,ads}$

(S8) $-M^m$ $+$ e^- $\rightleftharpoons$ $-M^{m-1}$

(S9) $-M^{m-1}$ $+$ O_2 $\rightleftharpoons$ $-M^m-O_2^{\cdot-}$

Radical Reactions

(S10) $-M\int\cdot OH$ $+$ $R_{1,ads}$ $\longrightarrow$ $R_{2,ads}$

(S11) $\cdot OH$ $+$ R_1 $\longrightarrow$ R_2

(S12) $-M^m-O_2^{\cdot-}$ $+$ $[H^+]$ $\rightleftharpoons$ $-M^m[HO_2\cdot]$

(S13) $-M^m[HO_2\cdot]$ $+$ $e^- + [H^+]$ $\longrightarrow$ $-M^m[H_2O_2]$

(S14) $-M^m[H_2O_2]$ $+$ e^- $\longrightarrow$ $-M[OH^-][\cdot OH]$

(S15) $-M^m-O_2^{\cdot-}$ $+$ $[H_2O_2]$ $\rightleftharpoons$ $-M^m[HO\cdot][HO^-] + O_2$

(S16) $[\cdot OH]$ $+$ $[H_2O_2]$ $\rightleftharpoons$ $(HO_2^{\cdot-}) + [H_2O]$

Scheme 1. Photocatalytic reaction scheme of a metal oxide. Species in brackets may be adsorbed or in the bulk aqueous phase.

solid and how rapidly they are captured through interfacial electron transfer by a thermodynamically appropriate electron donor or acceptor, in competition with the recombination rate within the particle.

In the absence of suitable hole or electron scavengers at the surface, carrier recombination processes will dominate (S3) (within a few nanoseconds), and the energy absorbed will dissipate as a photon or heat (43–45). The phototransformation of redox species is governed by the type of semiconductor and the energetic position of the bands. Photoprocesses are mainly determined by the minority carriers. Solution species are reduced at the surface of an illuminated *p*-type semiconductor, and oxidation occurs over illuminated *n*-type metal oxides where the photogenerated minority carriers are electrons and holes, respectively.

The energy distribution in the 315- to 400- nm range of the solar spectrum is approximately 2.7% of the total, and varies with the season (46). During peak sunlight in the summer season, surface water can receive up to approximately $1\,kW/m^2$ of sunlight, or about 2 einstein/m^2, within the wavelength region (300–500 nm) of interest for metal oxide photochemical reactions (31). Several metal oxides that can be activated by available solar radiation are listed in Table 6, with their respective optical band-gap energy.

Table 6   Optical Band-Gap Energy of Several Metal Oxide Materials

Metal Oxide	Optical Band-Gap Energy (eV)	Light Wavelength Equivalent (nm)
SiO_2	~ 8	~ 155
ZnO	3.35	370
TiO_2 (rutile)	3.0–3.3	376–413
TiO_2 (anatase)	3.2	388
WO_3	2.6	477
In_2O_3	2.5	496
PbO	2.76	499
α-Fe_2O_3 (hematite)	2.34	530
CdO	2.3	539
α-FeOOH (goethite)	2.10	591
γ-FeOOH (lepidocrocite)	2.06	602
β-PbO_2	1.7	729

Surface chemical reactions contribute to the movement across interfaces. The acquisition of charge through adsorption of anions and cations creates electrical potential gradients that can promote carrier migration. The depletion of reactants and accumulation of products by surface reactions create concentration gradients and, thereby, diffusion. The bulk structure of the oxide, whether it be amorphous or crystalline, is broken at the metal oxide–water interface. It is the chemistry and the three-dimensional structure of the surface groups exposed to the overlying solution that often differ in important respects from the mineral bulk.

When large metal oxide particles are immersed in an electrolyte solution, a strong electric field across the semiconductor is generated, creating an electric double layer described by the Gouy–Chapman–Stern model (Fig. 5b). The double layer acts much like a capacitor, where the metal oxide has an associated capacitance it ($0.001-1\,\mu F\,cm^{-2}$) that is much smaller than the double-layer capacitance ($10-100\,\mu F\,cm^{-2}$). Since these capacitors are in series, the smallest capacitor will govern the response of the system:

$$1/C_T = 1/C_{sc} + 1/C_{dl} \qquad (C_{sc} \ll C_{dl}) \qquad (4)$$

The bands in the bulk of the metal oxide will move with changes in applied potential (Fermi level), and the potential drop will appear in a region near the surface of the semiconductor (space charge layer) with very little potential drop on the solution side. The thickness of the space charge region is usually on the order of $0.1-200\,nm$, depending on the density of doping molecules and the dielectric constant of the solid material. At the metal oxide surface, electrons are depleted and band bending develops, with the Fermi level shifting to equilibrate with the redox potential of the electrolyte. The energetic position of the bands at the surface will depend upon the nature and type of interaction the surface has with the electrical double layer. As the amount of charged species in this layer changes, the position of the bands can change. This is especially true for metal oxides, which have surface oxide and hydroxide groups that respond to variations in the pH of the solution, typically following a 59 mV/pH change (47).

When the solid–liquid junction is illuminated, the photoproduced electron–hole pairs at the interface are separated in the space charge layer, with the majority carrier being driven into the bulk of the solid while the minority carrier is driven to the metal oxide–liquid interface (Fig. 5c). In Figure 5c, the holes reaching the surface are easily transferred because of the upward bending; however, electrons do not leave the particle, so it can become charged negatively. This leads to a flattening of the bands, so that finally the electrons can also cross the interface. Accordingly, the primary hole transfer leads to a compensation of the space charge.

In the illuminated zone of natural waters, most hydrous metal oxides exist as colloidal particles with diameters comparable to the thickness of space charge layers ($d_{particle} < d_{space\ charge}$). As a result, there is no real "interior" for carrier accumulation. Also, colloidal particles have a large surface available, and photogenerated electrons and holes can reach the surface very quickly before recombination.

The capture of reactive species (carriers) on the surface of colloidal oxides has been observed by nanosecond laser flash photolysis spectroscopy (48–57) with suitable organic scavengers (S7). Investigations with neat TiO_2 colloids have revealed that upon the illumination ($\lambda = 347.1$ nm) in the absence of organic scavengers, recombination was the dominant pathway. When the collodial solution contained polyvinyl alcohol (electron donor), a long-lived absorption signal (stable over several milliseconds) was present immediately after the flash (Fig. 6) (49, 50). The absorption spectrum was attributed to a reducing intermediate of the photolysis of TiO_2, an excess of conduction band electrons trapped on the surface as Ti(III) centers. Similar observations have been made with WO_3 colloids (51) and ZnO (48, 52) with suitable scavengers. Trapping of conduction band electrons is an extremely fast process that is accomplished in less than 20 psec; the lifetime of a single electron–hole pair on a particle is approximately 30 (± 15) nsec (44, 45). In the presence of molecular oxygen and/or other electron acceptors, the absorption observed in Figure 6 decays after the flash. The broad absorption band with a maximum at 650 nm produces a blue-colored suspension. The same color was also observed in γ-irradiation experiments on electron transfer from reducing organic radicals to TiO_2 colloids (58). The polyvinyl alcohol plays the main role of scavenger, where it is strongly adsorbed on the particle and able to impede carrier recombination by annihilating the holes. Thus, excess electrons remain on the colloidal particles for a long time. The primary scavenger reacting with the holes will determine the reaction yield of other additives with the electrons.

Several investigations of Fe_2O_3 colloids employing conductivity, spectroscopic, and γ-radiolysis studies revealed no free electrons for this metal oxide (39, 59, 60). Dimitrijević et al. (39) demonstrated, from studies on colloidal α-

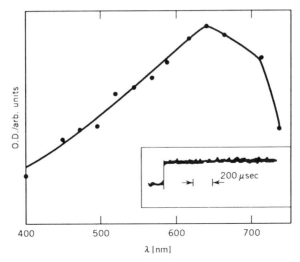

Figure 6. Transient absorption spectrum of a solution (pH 10) containing 6.3×10^{-3} M TiO$_2$ colloids and 5.0×10^{-3} M polyvinyl alcohol. The inset displays the time profile of the absorption signal. From Bahnemann et al. (49).

Fe$_2$O$_3$ (in neutral solution) with free radicals [(CH$_3$)$_2$COH] generated by ionizing radiation that accomplished electron transfer reactions with the colloidal particles of α-Fe$_2$O$_3$, that electrons donated by the free radicals penetrate deep into colloidal particles to reduce iron, producing a trapped Fe^{2+}. Furthermore, Moser and Grätzel (61) pointed out that single crystals and polycrystalline α-Fe$_2$O$_3$ electrodes show only small efficiencies as photoanodes for the oxidation of water and other substrates. This results from the low mobility of the charge carriers and short hole diffusion length.

Interfacial electron transfer processes can proceed extremely fast when a reactive molecule and photogenerated carrier are proximate to each other. Factors controlling interfacial electron transfer on metal oxide colloids have been probed with viologen compounds: methyl viologen, MV^{2+} (electron acceptor), and a semireduced methyl viologen, MV^{+}. Electron transfer from colloidal TiO$_2$ to MV^{2+} in solution was studied by flash photolysis, and it is known to undergo a reversible one-electron reduction with a well-defined and pH-dependent standard redox potential of the MV^{2+}/MV$^{\cdot+}$ ($E^{\circ} = 0.44$ V) couple. The reduced product, MV$^{\cdot+}$, was characterized by an optical absorption at 600 nm and its rate of formation was pH-dependent. Grätzel and co-workers (62) found that heterogenous electron transfer rates followed a Tafel relation at pH $\geqslant$ 5 and that the pH-dependent conduction band ($E_{CB} = 0.55$ V at pH 7) attained an electrochemical equilibrium with MV^{2+}/MV$^{\cdot+}$ at pH 5. It was also found that the position of the conduction band edge of the particle greatly influences the rate of interfacial electron transfer and that the rate constant for the reduction of MV^{2+} was linearly related to pH in a pH range of $4 \leqslant$ pH $\leqslant 10$. The shift of the conduction band position of colloidal particles with pH affects the overlap of the occupied donor levels in the metal oxide with those of the unoccupied acceptor levels in the electrolyte. Data from the two pH extremes imply that the electrostatic properties of the surface oxide material have been modified, which can hinder adsorption onto the oxide particle by MV^{2+}, and thus surface reactions become diffusion-controlled.

Several researchers have inferred (53, 54) that the rate of electron transfer to and from the metal oxide colloids can be significantly affected by the surface potential distribution of the oxide material combined with the charge of the solution species. Therefore, in order for efficient electron transfer to occur, a surface adsorption (encounter complex) must first arise between the metal oxide and the substrate present in the bulk solution. The rate of this process may be diffusion limited; this is determined by the viscosity of the medium, the radius of the reactants, and the electrostatic effect on the surface of the metal oxide, which may change with pH and ionic strength. However, Darwent et al. (53, 54) have pointed out that the surface area of colloids will have a significant effect on the rate of the interfacial electron transfer reactions, which are governed by the transfer rate itself and not by diffusion.

Similar investigations have been performed with other probe molecules (dyes), where the formation of the cation radical of these molecules within the laser pulse excitation demonstrated that the charge injection into a metal oxide is a fast process that occurs with a first-order rate constant of at least $5 \times 10^8 \sec^{-1}$ to $5 \times 10^{10} \sec^{-1}$ (45, 63–65). Since interfacial electron transfer is complete in less than 20 psec in favorable circumstances (45), pathways involving direct reaction of organic molecules with photogenerated surface oxidants are important. This would further suggest that with readily reactive molecules, suitable adsorption properties, and high surface coverage, quantum yields of unity can be reached. This was measured in the reactions of alcohols on several preparations of TiO_2 (66). It implies that interfacial electron transfer can compete efficiently with recombination when thermodynamics are favorable. These results have been obtained with both small colloids and large crystallites of TiO_2.

In the case of colloidal TiO_2, carrier trapping at the surface is fast and the lifetime of a single electron–hole pair on the particle is long enough for a substrate to encounter the surface after trapping.

Surface trapping of carriers on larger crystallites requires several nanoseconds (G.P. Lepore and C.H. Langford, unpublished results); accordingly, the sequence of reaction steps is more complex because of the space charge layer below the surface of the particle. The transit times for minority carriers to diffuse to the surface of big particles are longer and consequently may make recombination effects more important (67).

3.2.   Generation of Oxidants

In an aqueous medium, water is the dominant adsorbate. The redox potential of the valence band hole for several metal oxides (e.g., ZnO, TiO_2, α-Fe_2O_3) is sufficiently positive to oxidize water, yielding as the primary product of interfacial electron transfer an "*adsorbed*" hydroxyl radical (S6) (38, 68, 69). The immediate products of this adsorbed radical have been directly detected by spin-trapping electron paramagnetic resonance (70–72) and indirectly detected from the formation of hydroxylated organic intermediates (73–76) and H_2O_2 (73, 77, 78) in the photocatalytic degradation of numerous contaminants (Fig. 7). The formation of most photoproducts is explained by the successive reactions of hydroxyl radicals with the same molecule being progressively transformed in the adsorbed phase.

Lemaire and co-workers (79) have calculated initial quantum yields for the conversion of all three dichlorobenzene isomers to be quite low (0.046 $\pm$ 0.008 for *o*- and *p*-dichlorobenzene, 0.038 $\pm$ 0.008 for *m*-dichlorobenzene). The kinetics revealed that reactions occur mainly on the metal oxide surface. Several authors have reported that intermediates found in the liquid bulk of aqueous suspensions of metal oxide are consistent with those determined when similar aromatics reacted with a known source of hydroxyl radicals (80–82), which suggests that ·OH is the primary attacking species. However, other reported product distribution studies show that the species responsible for observable products in these experiments is not equivalent to the free hydroxyl radical in solution (72, 83, 84). However, the difference between a trapped hole on the surface of a wet oxide and an adsorbed hydroxyl radical may be mainly semantic (37).

Fox and co-workers (83) have illustrated the direct trapping of photo-generated holes by observing the phototransformation mechanism of diaryl and dibenzyl thioethers on the surface of suspended TiO_2 in dry aerated acetonitrile with time-resolved diffuse reflectance spectroscopy. The authors outlined a mechanistic pathway whereby oxidation of sulfur involved interfacial electron transfer to the photogenerated hole on the metal oxide (hole trapping), producing a cation radical intermediate (substrate). The subsequent step involves the cation radical trapping via an adsorbed oxygen or superoxide (produced from reactions S8–S9), and attack of the resulting species by the initial reactant (on the surface or liquid bulk). Moreover, cation radicals may also undergo nuclephilic attack, which can profoundly complicate mechanistic studies depending exclusively on product analysis, since similar products might resemble intermediates formed from hydroxyl radical attack. Fox has speculated that hydroxylation can involve the initial hydration of a primary singly oxidized cation radical, which can provide a hydronium ion that can be deprotonated to give the same adduct formed from direct ·OH attack. Thus, product analysis involving intermediates found in the liquid bulk may not invariably reflect primary photocatalytic events occurring in the illumination zone at the metal oxide interface.

The hydroxyl radical can also be formed via a secondary mechanistic route that must not be overlooked. Conduction band electrons can be scavenged by molecular oxygen to give rise to superoxide radical anions ($O_2^{\cdot-}$), which can disproportionate to give some hydroxyl radicals (·OH). Under acidic conditions, H^+ may protonate the superoxide ion to form the perhydroxyl radical $HO_2\cdot$ (reaction S12, $pK_a = 4.8$), which has also been detected in illuminated aqueous solutions of TiO_2 (71). Alternatively, the reaction of the hydroxyl radical with the

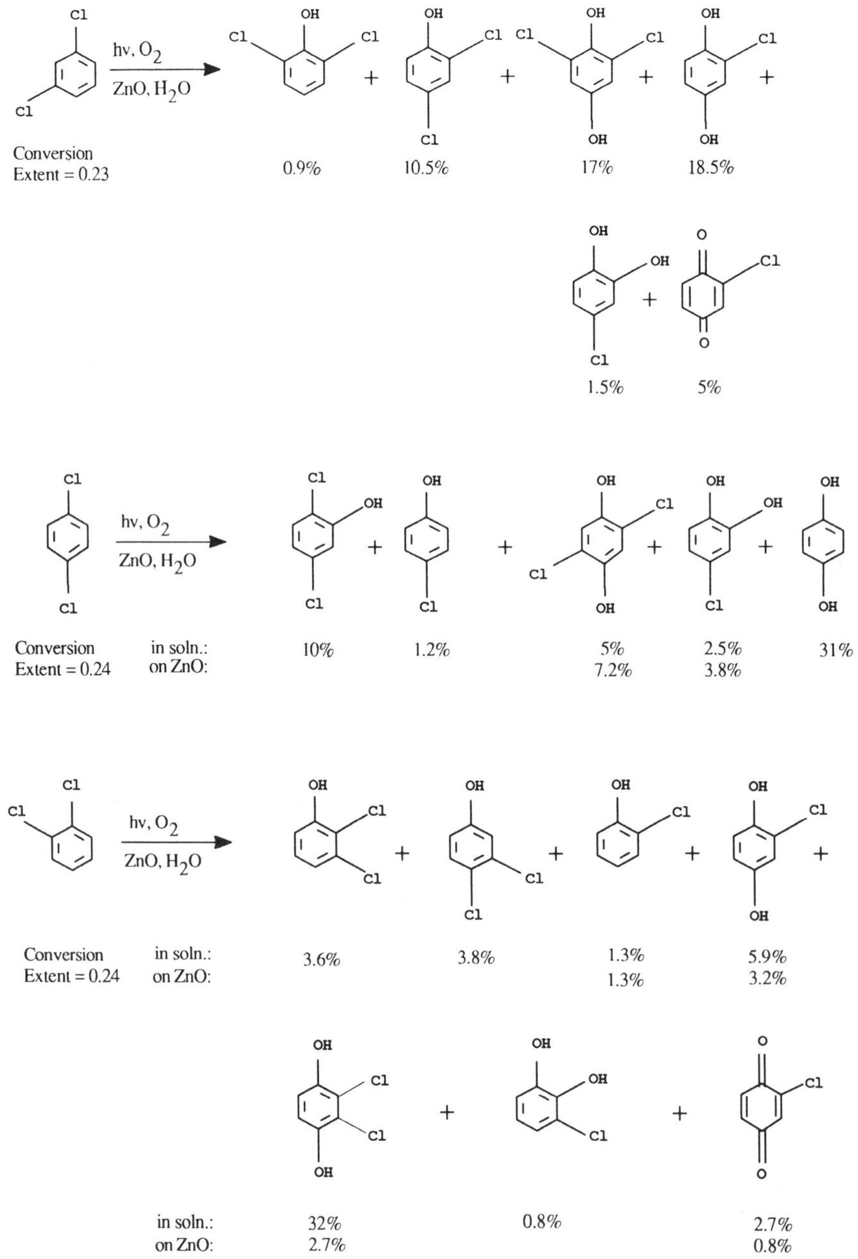

Figure 7. The photocatalytic transformation of dilute dichlorobenzene isomers in aqueous suspensions of ZnO. Product distribution of dichlorobenzene isomers. The isomers were individually illuminated for 10 minutes in aqueous suspensions of ZnO $(2\,g\,L^{-1})$. The substrate concentrations used were 1,3-dichlorobenzene $(3.0 \times 10^{-4}\,M)$; 1,2-dichlorobenzene $(4.6 \times 10^{-4}M)$; 1,4-dichlorobenzene $(1.5 \times 10^{-4}\,M)$. From Lemaire et al. (79).

photogenerated hydrogen peroxide (S15) can also yield the perhydroxyl intermediate. This intermediate leads to the eventual production of H_2O_2 (S13), which can also produce hydroxyl radicals (S14).

Whatever the detailed pathway, hydrogen peroxide is a ubiquitous substance in the sea and in freshwater that can be formed through homogeneous (4) and heterogeneous photochemical reactions (as mentioned above). Hydrogen peroxide is sufficiently long-lived that it can be transported away from the photic zone to dark regions of water bodies, where it can participate in various redox reactions (e.g., Fenton's reaction). Thus generation of hydroxyl radicals via H_2O_2 may well be an important route in natural waters. The importance of the routes of formation of ·OH via H_2O_2 compared to that of hole reactions remains ambiguous. Matthews has suggested that the superoxide routes are of negligible importance compared with the reactive holes (73). Okamoto et al. (57) measured the production of low concentrations of H_2O_2, which reached steady state in prolonged illuminations. These authors proposed that the formation of hydrogen peroxide follows the sequence (S12–S13), which requires two photons and is a sequence in balance with peroxide decomposition (S14–S15). Additional support for the formation of ·OH via H_2O_2 emerged from studies of the uptake of oxygen required to consume 1 mole of phenol in an illuminated aqueous suspension of TiO_2 (0.7 mol O_2 to 1 mol phenol). The conclusions may be appropriate for low concentrations and weakly adsorbed organic matter, yet be the exception as mentioned in Section 3.1.

The photoreduction of molecular oxygen over several oxides may not be so important, since the potential of the conduction band is more positive (i.e., α-Fe_2O_3 at pH 2, $E_{CB} = +0.3$ V) (47) than the reduction potential of O_2 in homogeneous solution (E° ≈ -0.15 V), especially since the O_2 value may be more positive when O_2 is chemically bound to the surface, its reduction by a conduction band electron is endothermic—in the case of α-Fe_2O_3, $\Delta G° = +44$ kJ/mole (85). It has been shown that the presence of molecular oxygen is necessary for photocatalytic oxidations to occur (77) or to prevent the reductive dissolution of hydrous oxides (86–88).

The hydroxyl radical is a strong oxidant that rapidly reacts with a wide variety of molecules. The nonspecific nature of HO· in solution reactions results in very low steady-state concentrations in the aqueous environment (89, 90), and HO·-initiated transformations may be slow relative to other pathways. Surface reactions of organic compounds can be initiated by these radicals; however, electron transfer rates are expected to correlate with over-potential following the Levich equation (91–93), so that an adsorbed organic molecule may capture holes from metal oxides faster than does water. Recent studies have clearly indicated that most reactions occur on the TiO_2 surface, as opposed to being solution-phase hydroxyl radical chemistry (83, 84, 94, 95). However, as noted above, the distinction between a surface-bound hydroxyl radical and an oxygen site-trapped hole may be little more than semantic (37).

Yamagata and co-workers (94) have demonstrated with EPR spin trapping and photocurrent measurements the photocatalytic oxidation of several different alcohols on TiO_2. Their findings indicate that photoproduced holes oxidize alcohols with an electron transfer to the valence band of the metal oxide, with the production of a carbon-centered radical as a primary step. The redox potential of that radical species determines whether the intermediate can inject another electron into the conduction band of the metal oxide. An alcohol with α-hydrogens was shown to produce a radical species that can inject an electron into the conduction band of TiO_2, while an alcohol without α-hydrogens cannot. The reaction process can become a radical chain process under oxygen. Another study found that "current doubling" (electron transfer and hole transfer) was not as large with TiO_2 as with ZnO (55).

The dissolution of hydrous oxides may present an alternative route to the generation of the reactive intermediates. Past studies have shown that Fe(II) (Fe^{2+} and complexes thereof) and H_2O_2 can generate $\cdot OH$ under acidic conditions, generally referred to as Fenton reactions (96–99). Several authors (100–102) have suggested that the reaction between H_2O_2 and a photochemically produced Fe(II) (photo-Fenton reaction) may be involved in the indirect photooxidation of several organic substances in freshwater and seawater (103). This sequence may introduce complications for the determination of quantum yield, since it involves reduction products in oxidant generation.

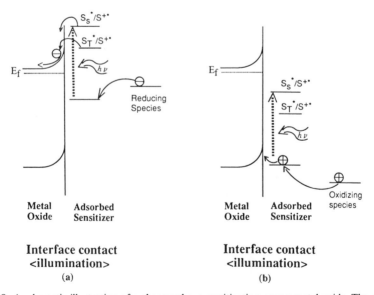

Figure 8. A schematic illustration of a chromophore sensitization over a metal oxide. The excited chromophore molecule can act as an electron donor (a) or electron acceptor (b). The excited state of a chromophore is depicted by S^*, where S_S^* and S_T^* represent the singlet and triplet excited state. In the presence of an electrolyte, the molecule can undergo redox reactions with the light-activated surface chromophore.

3.3. Photosensitizers

An important way to extend the response to energies below that for band-gap excitation is photosensitization. Photosensitization of a stable metal oxide is a phenomenon that extends the material's photoactive range, thus enabling photoelectrochemical reactions under illumination by visible light (45, 63, 104). The obvious chromophores found in natural waters are humic substances, discussed above because of their ability to initiate photochemical transformations of organic compounds in natural waters that lead to the eventual degradation of those compounds. Several processes can take place when an electron is raised to the excited state of the chromophore molecule. It may involve an electron transfer to the conduction band of the metal oxide, an electron transfer from the oxide particle to the empty level of the sensitizer, and the simple deexcitation of the chromophore molecule to its ground state. Moreover, the electron transfer from the excited state of the sensitizer to the conduction band of the metal oxide implies that the lowest vacant level of the sensitizer is situated above the conduction band (empty d-orbital) of the metal oxide (Fig. 8).

Most investigations to date have involved well-characterized chromophores to elucidate the photophysical properties of several metal oxide particles. However, the probing of photophysical phenomena with humics is the interesting exception that unites this discussion with Section 2. Vinodgopal et al. (105) have obtained evidence for direct charge transfer between colloidal ZnO and TiO_2 and humic substances (Suwannee River humic and fulvic acids) by using steady-state fluorescence quenching. This is illustrated in Figure 9, which shows a noticeable decline in fluorescence yield of the humics, which the authors attributed to the quenching of the excited state of the strongly adsorbed humic material by the semiconductor colloid. Stern–Volmer analysis suggests a high apparent binding constant, reflecting strong adsorption suggestive of chemisorption. However, it is unlikely that a single equilibrium constant describes adsorption of all components of the mixture, and there may be a residual fluorescence at high concentrations associated with the strongly fluorescent small-molecule fraction of the humic, which does not adsorb readily. If so, the large estimated binding constant is still too low.

As noted in Section 2, the transient absorption spectra from laser flash photolysis of neat fulvic acid reveal the presence of the solvated electron formed as a result of the photoionization of the fulvic acid at longer wavelengths (Fig. 10a). Zepp and co-workers have observed this short-lived transient in the laser flash photolysis of Suwannee River humics (26). With the addition of ZnO colloids in the fulvic acid solution (Fig. 10b), the solvated electron signal is quenched, supporting the assignment of the fluorescence quenching to electron transfer. The resulting species will be the humic cation radicals mentioned in Section 2 (48).

3.4. Surface Complex Photochemistry

A number of reports exist on the complexation reactions and photochemistry of metal hydrous oxides and organic molecules with functional groups that mimic

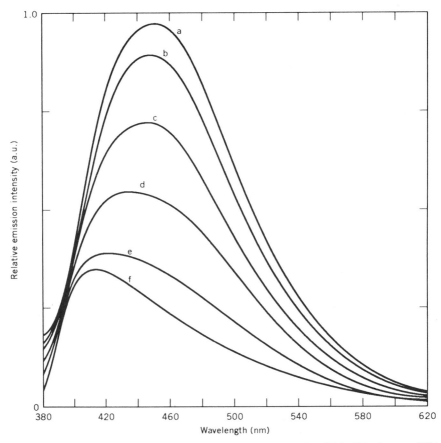

Figure 9. Fluorescence emission spectrum of Suwannee River fulvic acid in 1% v/v water-99% ethanol at various concentrations of colloidal ZnO. (*a*) neat fulvic acid; (*b*) 6.5×10^{-5} M; (*c*) 9.8×10^{-5} M; (*d*) 1.3×10^{-4} M; (*e*) 1.63×10^{-4} M; (*f*) 3.2×10^{-4} M. From Vinodgopal and Kamat (105).

humic substances (86, 102, 106–108). The coordination model for hydrous oxide surface sites outlines possible modes of interaction between dissolved species and surface sites. Since surface sites are hydrated and can undergo protonation–deprotonation reactions, they create a distribution of positive, negative, and neutral surface oxygen-donor ligands. The role of surface hydroxyl groups as principal reactive sites on metal oxides with amphoteric properties has been clearly established (109). Peri (110) has identified five different suface hydroxyl groups on $\alpha\text{-Al}_2\text{O}_3$; these surface configurations (sites) will play an important role in the surface complexation with ligand molecules prior to photolysis. Some feasible surface group configurations are illustrated in Figure 11 for hydrolyzed iron(III) oxide surfaces.

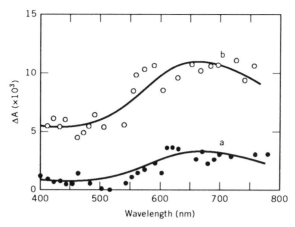

Figure 10. Transient absorption spectra observed upon laser pulse excitation of fulvic acid in ethanol (a) neat fulvic acid (b) after addition of 250 μL of 0.02 M ZnO. From Vinodgopal and Kamat (105).

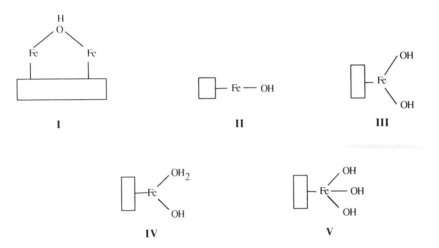

Figure 11. Some plausible group configurations for hydrolyzed Iron(III) oxide surfaces. From Faust et al. (112).

In the presence of dissolved organics, the solute molecules can displace surface-bound OH⁻ and H_2O species via ligand substitution and bind directly to the metal center to form an inner-sphere surface complex (Fig. 12a).

Spectroscopic investigations such as electron paramagnetic resonance (EPR), electron double-resonance (ENDOR) (113), Fourier transform infrared (FTIR) (114), and in situ X-ray absorption studies of surface complexes (EXAFS) (115) have confirmed the inner-sphere structure of numerous surface complexes. One of the best known photoreactions of inner-sphere complexes is photooxidation of

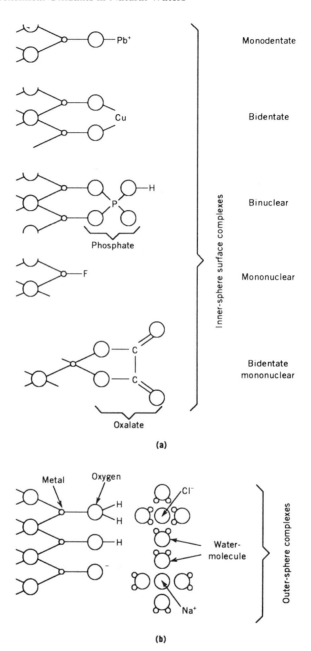

Figure 12. A schematic representation of two types of surface coordination complexes on a metal oxide surface. (a) Inner-sphere complex mechanisms comprise ionic and covalent bonding. (b) Outer-sphere surface complexation mechanisms of adsorption involve almost exclusively electrostatic interactions. From Stumm and Wieland (111).

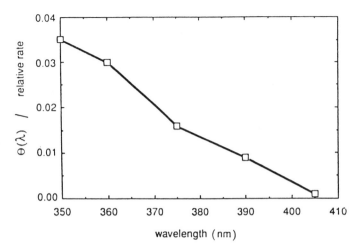

Figure 13. The relative quantum yield of light-induced reductive dissolution of α-Fe$_2$O$_3$ in the presence of oxalate as a function of wavelength. The experiment was conducted under N$_2$ atmosphere with a solution containing 0.5 g L^{-1} hematite, 3.3 mM at pH 3. The calculation of relative quantum yields is based on the assumption that Fe(III) oxalate surface complex is the chromophore, and the extinction coefficients estimated from absorption spectrum of dissolved ferrioxalate were used. From Sulzberger (119).

the ligand and photoreduction of the metal following excitation of the spectral bands assigned as ligand-to-metal charge transfer (LMCT). Such bands are characteristically allowed fairly intense transitions centered in the near-UV (300–400 nm) region. LMCT photochemistry has been observed on photolysis of surface complexes of iron oxide minerals with ligands such as carboxylic acids (86, 116) and mercapto acids (117).

Several authors have studied the wavelength dependence of the rate of photodissolution of particular iron hydrous oxides with different surface ligands [α-Fe$_2$O$_3$-bisulfite (87); α-Fe$_2$O$_3$-oxalate (108); γ-Fe$_2$O$_3$-EDTA (118)]. These investigations revealed that only light in the near-UV ($\lambda < 400$ nm), which corresponds to band-gap energy, leads to enhancement of the dissolution of the hydrous oxide in the presence of ligands (Fig. 13). Ideally, proof of the surface complex mechanism would come from a wavelength-dependence mapping of the LMCT band of the complex.

A study supporting a surface complex mechanism established the direct dependence of the dissolution rate of γ-FeOOH (lepidocrocite) on the surface-bound ferric citrate concentration. A first-order rate constant was obtained for the photodissolution process that was similar to that found for the decomposition of soluble ferric citrate (86). Of the two plausible routes for the redox photodissolution of hydrous metal oxide [i.e, iron(III) hydroxides]—ligand-to-metal charge transfer transition of a surface complex and/or direct band-gap excitation [lattice-oxygen to lattice-iron(III) charge transfer transition]—it is rarely clear which dominates. Nevertheless, most reports indicate that it is important for an electron donor to form an inner-sphere complex at the surface of the hydrous oxide for an efficient electron transfer to develop.

An outer-sphere surface complex may also form in which the inner coordination sphere of the metal center remains intact (Fig. 12b). The orbital overlap between the adsorbate and the surface metal center is lower in the outer-sphere complex than in the inner-sphere complex. However, outer-sphere electron transfer is a well-known process. The requirements for rapid reaction have been well established and systematized by the Marcus theory. However, outer-sphere metal-to-ligand charge transfer bands typically lie at wavelengths beyond 300 nm in the deep and largely unavailable UV. If outer-sphere reactions are to contribute, they will depend upon the bad-gap mechanism for generation of the primary oxidant.

4. CONCLUSIONS

This chapter has illustrated how oxidation of organic compounds and even water can be initiated photochemically in natural water systems. The main mediating species are $\cdot OH$, $\cdot OH_{(ads)}$, organic radical cations ($R^{\cdot+}$), singlet oxygen, and H_2O_2. Especially in the case of humic substances as light absorbers, the primary processes can be quite efficient. Elsewhere in this volume the steady-state levels of the oxidants and the pathways of a number of organic oxidations are discussed. In that context, processes are not found to be so efficient. The main gap that needs to be filled by future research on the basic mechanisms is the role played by the humic substances themselves in sweeping up and "deactivating" reactive species generated in the primary steps.

REFERENCES

1. Power, J. F., Sharma, D. K., Langford, C. H., Bonneau, R., and Joussot-Dubien, J. (1986). *Photochem. Photobiol.* **44**, 11–13.

2. Choudhry, G. G. (1981). *Toxicol. Environ. Chem.* **4**, 261–295.

3. Power, J. F. (1986). Ph.D. Thesis, Concordia University, Montreal.

4. Cooper, W. J., Petasne, R. G., Zika, R. G., and Fischer A. M. (1989). *ACS Symp. Ser.* **219**, 333–362.

5. Zika, R. G. (1987). *Rev. Geophys.* **25**, 1390–1394.

6. Gamble, D. S., Underdown, A. W., and Langford, C. H. (1980). *Anal. Chem.* **52**, 1901.

7. Langford, C. H., Gamble, D. S., Underdown, A. W., and Lee, S. (1983). In R. F. Christman and E. T. Gjessing (Eds.), *Aquatic and Terrestrial Humic Substances*. Ann Arbor Science Publishers, Ann Arbor, MI, Chapter 11.

8. Zafiriou, O. C., Joussot-Dubien, J., Zepp, R. G., and Zika, R. G. (1984). *Environ. Sci. Technol.* **18**, 358A–371A.

9. Waite, T. D., Sawyer, D. T., and Zafiriou, O. C. (1988). *Appl. Geochem.* **3**, 9–18.

10. Underdown, A. W. (1982). Ph.D. Thesis, Carlton University, Ottawa.

11. Wang, Z., Pant, B., and Langford, C. H. (1990). *Anal. Chim. Acta* **232**, 43–49.

12. Langford, C. H., Bruccoleri, A., and Arbour, C. (1990). *Environ. Technol.* **11**, 169–172.

13. Power, J. F., Sharma, D. K., Langford, C. H., Bonneau, R., and Joussot-Dubien, J. (1987). *ACS Symp. Ser.* **327**, 157–173.

14. Gamble, D., and Schnitzer, M. (1973). In P. S. Singer (Ed.), *Trace Metals and Metal Organic Interactions in Natural Waters*. Ann Arbor Science Publishers, Ann Arbor, MI.

15. Buffle, J., Deladoey, P., Zumstein, J., and Haerdi, W. (1982). *Schweiz. Z. Hydrol.* **44**, 326–361.

16. Mattes, S. L., and Farid, S. (1984). *Science* **226**, 917.

17. Haniman, A., and Rockett, B. W. (1974). *J. Photochem.* **2**, 405.

18. Allen, N. S., and McKellars, J. F. (1978). *Makromol. Chem.* **179**, 523.

19. Langford, C. H., Bruccoleri, A., and Sharma, D. K. (1992). *Prepr. Pap., 203rd Am. Chem. Soc. Natl. Meet.*, Vol. 32, No. 1, pp. 216–220.

20. Senesi, N. (1990). *Anal. Chim. Acta* **232**, 77–106.

21. Lavigne, J. A. (1987). Ph.D. Thesis, Concordia University, Montreal.

22. Zepp, R. G., and Baughman, G. L. (1978). In O. Hutzinger, I. H. Van Lelyveld, and B. C. J. Zoeteman (Eds.), *Acquatic Pollutants: Transformation and Biological Effects*. Pergamon, New York, pp. 237–263.

23. Zepp, R. G. (1980). In R. Haque (Ed.), *Dynamics, Exposure and Hazard Assessment of Toxic Chemicals*. Ann Arbor Science Publishers, Ann Arbor, MI, pp. 69–110

24. Choudhry, G. G. (1984). *Humic Substances: Structrual, Photophysical, Photochemical and Free Radical Aspects and Interactions with Environmental Chemicals*. Gordon & Breach, New York.

25. Petasne, R. G., and Zika, R. G. (1987). *Nature (London)* **325**, 516–518.

26. Zepp, R. G, Braun, A. M., Hoigne, J., and Leenheer, J. A. (1986). *Environ. Sci. Technol.* **21**, 485–490.

27. Faust, B. C., and Hoigne, J. (1987). *Environ. Sci. Technol.* **21**, 957–964.

28. Shao, C., Cooper, W. J., and Lean, D. R. S. (1992). *Prepr. Pap., 203rd Am. Chem. Soc. Natl. Meet.*, Vol. 32, No. 1, pp. 222–223.

29. Haag, W. R., Hoigne, J., Gassmann, E., and Braun, A. M. (1984) *Chemosphere* **13**, 641–650.

30. Frimmel, F., Bauer, H., Putzien, J., Murasecco, P., and Braun, A. M. (1987). *Environ. Sci. Technol.* **21**, 541–545

31. Haag, W. R., and Hoigne, J. (1986) *Environ. Sci. Technol.* **20**, 341–348.

32. Lean, D. R. S., Cooper, W. J., and Pick, F. R. (1992). *Prepr. Pap., 203rd Am. Chem. Soc. Natl. Meet.*, Vol. 32, No. 1., pp. 80–83.

33. Zafiriou, O. C., Ball, L. A., Dister, B., and Micinski, E. (1992). *Prepr. Pap., 203rd Am. Chem. Soc. Natl. Meet.*, Vol. 32, No. 1, pp. 76–79.

34. Grossweiner, L. I., and Joschek, H. I. (1966). *J. Am. Chem. Soc.* **88**, 3261–3281.

35. Grossweiner, L. I., and Joschek, H. I. (1965). *Adv. Chem. Ser.* **50**, 279–288.

36. Grabner, G., Rauscher, W., Zechner, J., and Getoff, N. (1980). *J. Chem. Soc., Chem. Commun.* **222**, 1.

37. Lawless, D., Serpone, N., and Meisel, D. (1991). *J. Phys. Chem.* **95**, 5166–5171.

38. Waite, T. D., and Morel, F. M. M. (1984). *Environ. Sci. Technol.* **18**, 860–868.

39. Dimitrijević, N. M., Savić, D., and Mićić, O. I. (1984). *J. Phys. Chem.* **88**, 4278–4283.

40. Blesa, M. A., and Màroto, A. J. G. (1986). *J. Chim. Phys.* **83**, 757–764.

41. Lume-Pereira, C., Baral, S., Henglein, A., and Janata, E. (1985). *J. Phys. Chem.* **89**, 5772–5778.

42. Baral, S., Lume-Pereira, C., Janata, E., and Henglein, A. (1985). *J. Phys. Chem.* **89**, 5779–5783.

43. Kasinski, J. J., Gomez-Jahn, L. A., Faran, K. J., Gracewski, S. M., and Dwayne Miller, R. J. (1989). *J. Chem. Phys.* **90**, 1253–1269.

44. Rothenburger, G., Moser, J., Grätzel, M., Serpone, N., and Sharma, D. K. (1985). *J. Am. Chem. Soc.* **107**, 8054–8058.

45. Arbour, C., Sharma, D. K., and Langford, C. H. (1990). *J. Phys. Chem.* **94**, 331–334.

46. Getoff, N. (1984). *Int. J. Hydrogen Energy* **9**, 997–1004.

47. Gerischer, H. (1979). *Top. Appl. Phys.* **31**.

48. Patrick, B., and Kamat, P. V. (1992). *J. Phys. Chem.* **96**, 1423–1428.

49. Bahnemann, D., Henglein, A., Lilie, J., and Spanhel, L. (1984). *J. Phys. Chem.* **88**, 709–711.

50. Bahnemann, D., Henglein, A., and Spanhel, L. (1984). *Faraday Discuss. Chem. Soc.* **78**, 151–163.

51. Nenadović, M. T., Rajh, T., Mićić, O. I., and Nozik, A. J. (1984). *J. Phys. Chem.* **88**, 5827–5830.

52. Bahnemann, D., Kormann, C., and Hoffmann, M. R. (1987). *J. Phys. Chem.* **91**, 3789–3798.

53. Darwent, J. (1986). *J. Chem. Soc., Faraday Trans. 2*, **82**, 2323–2335.

54. Brown, G. T., Darwent, J., and Fletcher, P. D. I. (1985). *J. Am. Chem. Soc.* **107**, 6446–6451.

55. Miyake, M., Yoneyama, H., and Tamura, H. (1977). *Chem. Lett.*, p. 635.

56. Heinz, F., Fitzmaurice, D. J., and Grätzel, M. (1990). *Langmuir* **6**, 198–206.

57. Okamoto, K.-I., Yamamoto, Y., Tanaka, H., Tanaka, M., and Itaya, A. (1985). *Bull. Chem. Soc. Jpn.* **58**, 2015–2022.

58. Henglein, A. (1982). *Ber. Bunsenges. Phys. Chem.* **86**, 241.

59. Gardner, R. F. G., Swett, F., and Tanner, D. W. (1963). *J. Phys. Chem. Solids* **24**, 1183–1196.

60. Kennedy, J. H., and Frese, K. W. (1978). *J. Electrochem. Soc.* **125**, 723.

61. Moser, J., and Grätzel, M. (1982). *Helv. Chim. Acta* **65**, 1436–1444

62. Duonghong, D., Ramsden, J., and Grätzel, M. (1982). *J. Am. Chem. Soc.* **104**, 2977–2985.

63. Kamat, P. V., Chauvet, J.-P., and Fessenden, R. W. (1986). *J. Phys. Chem.* **90**, 1389–1394.

64. Moser, J., and Grätzel, M. (1984). *J. Am. Chem. Soc.* **106**, 6557–6564.

65. Kamat, P. V. (1989). *J. Phys. Chem.* **93**, 859–864.

66. Lepore, G. P., and Langford, C. H. (1990). *Symp. Proc. Adv. Oxid. Processes, Treat. Contam. Water Air*, Toronto, Canada *1990*, Session 4, WasteWater.

67. Memming, R. (1988). *Top. Curr. Chem.* **143**, 79–112.

68. Tanaka, H., Tanaka, M., and Itaya, A. (1985). *Bull. Chem. Soc. Jpn.* **58**, 2015–2022.

69. Cunningham, J., and Srijaran, S. (1988). *J. Photochem. Photobiol. A* **42**, 329–335.

70. Ceresa, E. M., Burlamacchi, L., and Visca, M. (1983). *J. Mater. Sci.* **18**, 289–294.

71. Jaeger, C. D., Bard, A. J. (1979). *J. Phys. Chem.* **83**, 3146–3151.

72. Bolton, J. R. (1990). *Symp. Proce. Adv. Oxid. Processes, Treat. Contam. Water Air*, Toronto, Canada *1990*, Session 2, WasteWater Technology Centre of Environment Canada.

73. Matthews, R. W. (1984). *J. Chem. Soc., Faraday Trans. 1* **80**, 457–471.

74. Armstrong, W. A., and Grant, D. W. (1958). *Nature (London)* **182**, 747.

75. Armstrong, W. A., Black, B. A., and Grant, D. W. (1960). *J. Phys. Chem.* **64**, 1415–1419.

76. Sehili, T., Boule P., and Lemaire, J. (1988). *J. Photochem. Photobiol. A* **50**, 103–116.

77. Kormann, C., Bahnemann, D. W., and Hoffmann, M. R. (1988). *Environ. Sci. Technol.* **22**, 798–806.

78. Harbour, J. R., and Hair, M. L. (1979). *J. Phys. Chem.* **83**, 652–656.

79. Lemaire, J., Boule, P., Sehili, T., and Richard, C. (1991). In E. Pelizetti and M. Schiavello (Eds.), *Photochemical Conversion and Storage of Solar Energy*. Kluwer Academic Publishers, New York, pp. 477–495.

80. Fujihira, M., Satoh, Y., and Osa, T. (1982). *Bull. Chem. Soc. Jpn.* **55**, 666–671.

81. Cox, R. A., Derwent, R. G., and Williams, M. R. (1980). *Environ. Sci. Technol.* **14**, 57–61.

82. Barbeni, M., Minero, C., Pelizzetti, E., Borgarello, E., and Serpone, N. (1987). *Chemosphere* **16**, 2225–2237.

83. Fox, M. A., Draper, R. B., Dulay, M., and O' Shea, K. (1991). In E. Pelizetti and M. Schiavello (Eds.), *Photochemical Conversion and Storage of Solar Energy.* Kluwer Academic Publishers, New York, pp. 323–335.

84. Lepore, G., and Langford, C. H. (1989). *Water Pollu. Res. J. Can.* **24**, 537–551.

85. Kormann, C., Bahnemann, D. W., and Hoffmann, M. R. (1988). *J. Photochem. Photobiol. A* **48**, 161–169.

86. Waite, T. D., and Morel, F. M. M. (1984). *J. Colloid Sci.* **102**, 121–137.

87. Faust, B. C., and Hoffmann, M. R (1986). *Environ. Sci. Technol.* **20**, 943–948.

88. Leland, J. K., and Bard, A. J. (1987). *J. Phys. Chem.* **91**, 5076–5083.

89. Mill, T., Hendry, D, G., and Richardson, H. (1980). *Science* **207**, 886–887.

90. Haag, W., and Hoigné, J. (1985). *Chemosphere* **14**, 1659–1671.

91. Levich, V. G. (1966). *Adv. Electrochem. Electrochem. Eng.* **4**, 249.

92. Levich, V. G. (1970). In H. Eyring, D. Henderson, and W. Jost (Eds.), *Physical Chemistry: An Advanced Treatise.* Academic Press, New York, Vol. 9B, Chapter 12.

93. Dogonadze, R. R. (1971). In N. S. Hush (Ed.), *Reactions of Molecules at Electrodes.* Wiley (Interscience), New York, Chapter 3.

94. Yamagata, S., Nakabayashi, S., Spancier, K., and Fujishima, A. (1988). *Bull. Chem. Soc. Jpn.* **61**, 3429–3434.

95. Yamagata, S., Baba, R., and Fujishima, A. (1989). *Bull. Chem. Soc. Jpn.* **62**, 1004–1010.

96. Walling, C. (1975). *Acc. Chem. Res.* **8**, 125–131.

97. Walling, C., and Amarnath, K. (1982). *J. Am. Chem. Soc.* **104**, 1185–1189.

98. Walling, C., and Johnson, R. A. (1975). *J. Am. Chem. Soc.* **97**, 363–367.

99. Fenton, H. J. H. (1894). *J. Chem. Soc.* **65**, 899–910.

100. Skurlatov, Y. I., Zepp, R. G., and Baughman, G. L. (1983). *J. Agric. Food Chem.* **31**, 1065–1071.

101. Zepp, R. G., Faust, B. C., and Hoigné, J. (1992). *Environ. Sci. Technol.* **26**, 313–319.

102. Cunningham, K. M., Goldberg, M. C., and Weiner, E. R. (1988). *Environ. Sci. Technol.* **22**, 1090–1097.

103. Moffett, J. W., and Zika, R. G. (1987). *Environ. Sci. Technol.* **21**, 804–810.

104. Kamat, P. V., and Dimitrijević, N. M. (1990). *Sol. Energy* **44**, 83–98.

105. Vinodgopal, K., and Kamat, P. V. (1992). *203rd Am. Chem. Soc. Natl. Meet., Div. Environ. Chem., 1992,* Vol. 32, No. 1, pp. 199–203.

106. Buffle, J. (1988). In R. A. Chalmers and M. R. Massson (Eds.), *Complexation Reactions in Aquatic Systems: An Analytical Approach,* Ellis Horwood Ser. Anal. Chem. Wiley, New York.

107. Waite, T. D., Wrigley, I. C., and Szymczak, R. (1988). *Environ. Sci. Technol.* **22**, 778–785.

108. Siffert, C., and Sulzberger, B. (1991). *Langmuir* **7**, 1627–1634.

109. Schindler, P., and Stumm, W. (1987). *Aquatic Surface Chemistry.* Wiley (Interscience), New York, pp. 83–110.

110. Peri, J. B. (1965). *J. Phys. Chem.* **69**, 220–230.

111. Stumm, W., and Wieland, E. (1990). In W. Stumm (Ed.), *Aquatic Chemical Kinetics.* Wiley (Interscience), New York, pp. 367–400.

112. Faust, B. C., Hoffmann, M. R., and Bahnemann, D. W. (1989). *J. Phys. Chem.* **93**, 6371–6381.

113. Motschi, H. (1987). In W. Stumm (Ed.), *Aquatic Surface Chemistry.* Wiley (Interscience), New York, Chapter 5.

114. Hayes, K. F., Roe, A. L., Brown, G. E., Jr., Hodgson, K. O., Leckie, J. O., and Parks, G. A. (1987). *Science* **238**, 783–786.

115. Zeltner, W. A., Yost, E. C., Machesky, M. L., Tejedor-Tejedor, I., and Anderson, M. A. (1986). In J. A. Davis and K. F. Hayes (Eds.), *Geochemical Processes at Mineral Surfaces.* American Chemical Society, Washington, DC.
116. Sakata, T., Kawai, T., and Hashimoto, K. (1984). *J. Phys. Chem.* **88**, 2344–2350.
117. Waite, T. D., Torikov, A., and Smith, J. D. (1986). *J. Colloid Interface Sci.* **112**, 412–420.
118. Stramel, R. D., and Thomas, J. K. (1986). *J. Colloid Interface Sci.* **110**, 121–129.
119. Sulzberger, B. (1990). In W. Stumm (Ed.), *Aquatic Chemical Kinetics.* Wiley (Interscience), New York, pp. 401–432.

8

AQUEOUS SULFUR(IV) OXIDATION REVISITED

L. Robbin Martin

Mechanics and Materials Technology Center, The Aerospace Corporation, El Segundo, California 90245-4691

Environmental Oxidants, Edited by Jerome O. Nriagu and Milagros S. Simmons.
ISBN 0–471–57928–9 © 1994 John Wiley & Sons, Inc.

1. PURPOSE AND SCOPE

It has been a decade since the review "Kinetic Studies of Sulfite Oxidation in Aqueous Solution" was compiled (Martin, 1984). Since that time, a great deal of research has been done on the oxidation of sulfur(IV) and a much clearer and more detailed picture of this system has developed. The purpose of this chapter is to update the reader on research in this area and to summarize present knowledge about sulfur(IV) oxidation in solution, with an eye to atmospheric applications. In order to conserve space, frequent reference will be made to facts presented in the earlier review. We hope to do this without sacrificing clarity of presentation.

The work reviewed will be confined to oxidation of sulfur(IV) in aqueous solution by "stable," that is, nonradical species. Radical species are clearly involved in the oxidation of sulfur by molecular oxygen and in atmospheric aerosols, but since the experimental approach to radical reactions is very different, such studies will not be included here. We will, however, quote a number of rate constants for radical reactions. Also neglected will be processes presently considered too slow to be important in typical tropospheric clouds, such as oxidation by nitrogen species. This is not to say that such reactions will not be important in some special situations.

We will present an updated calculation of the relative importance of the various oxidation reactions at the end of the chapter.

2. THERMODYNAMICS AND EQUILIBRIA

As we will show later in this chapter, to understand the mechanistic details of sulfur oxidation, it is not sufficient to regard the equilibrium constants simply as fixed numbers, as we did in the earlier review. These constants are highly dependent on the ionic strength of the aqueous solution, and this behavior can provide clues about the mechanism and can explain some of the features of sulfur oxidation.

We present in Table 1 some important equilibrium constants and relationships for this system. The iron(III) equilibria are included here for convenience. The ionic strength dependence of some of the equilibria are shown in the Debye–Hückel plot in Figure 1. Note that the ionic strength corrections may be very large. The various sulfur complexes and their respective stability constants have been measured by Kraft and van Eldik (1989a, b) and by Conklin and Hoffmann (1988).

All constants in Table 1 are from Smith and Martell (1976), corrected to an ionic strength of 0.01, with the following exceptions: K_w is from Glasstone (1942). K_1 and K_4 are from Conklin and Hoffmann (1988) (CH). K_4, K_7, and K_8 are from Kraft and van Eldik (1989c) (KE). The observed constants in those papers are related to the tabulated constants by: $K_{obs} = K_5 f_1 a_2$. After correcting for pH, the remaining difference between the CH and KE values for K_{obs} is attributed to the difference in ionic strength in the two studies (0.4 and 0.1, respectively). The

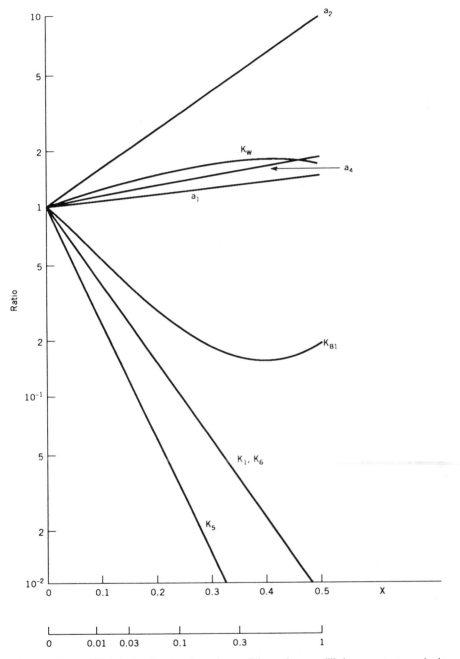

Figure 1. Debye–Hückel plot showing dependence of the various equilibrium constants on ionic strength, I. X is the variable $I^{1/2}/(1 + I^{1/2})$. (Data from Smith and Martell, 1976.) Data on K_5 based on the difference between the value for this constant in Conklin and Hoffmann (1988) at $I = 0.4$ and that in Kraft and van Eldik (1989a) at $I = 0.1$. Typical rate constants vary as $10^{-z_1 z_2 x}$. (Figure from Martin et al., 1991. Copyright American Geophysical Union.)

Table 1   Equilibria[a]

$$S(IV) = SO_2(aq) + HSO_3^- + SO_3^{2-} = SO_2(aq)\{1 + K_{A1}/(H^+) + K_{A1}K_{A2}/(H^+)^2\}$$
$$= SO_2(aq)\{Y\}$$
$$a_0 = SO_2(aq)/S(IV) = 1/\{Y\}$$
$$a_1 = HSO_3^-/S(IV) = K_{A1}/(H^+)\{Y\};$$
similarly,
$$a_2 = SO_3^{2-}/S(IV)$$

$$S(VI) = HSO_4^- + SO_4^{2-} = HSO_4^-\{1 + K_{A3}/(H^+)\}$$
$$a_3 = HSO_4^-/S(VI)$$
$$a_4 = SO_4^{2-}/S(VI)$$
$$Fe(III) = Fe^{3+} + FeOH^{2+} + Fe(OH)_2^+ + FeSO_3^+ + FeOHSO_3 + FeSO_4^+ + Fe(SO_3)_2^-$$
$$\qquad + Fe(SO_3)_3^{3-}$$
$$= Fe^{3+}\{1 + K_{B1}K_w/(H^+) + K_{B2}K_w^2/(H^+)^2 + K_1 a_2 S(IV) + K_5 a_2 S(IV)$$
$$\qquad + K_6 a_4 S(VI) + K_7 K_5 a_2 [S(IV)]^2 + K_8 K_7 K_5 a_2 [S(IV)]^3\}$$
$$= Fe^{3+}\{X\}$$
$$Fe(III)/Fe^{3+} = \{1 + \cdots\} = \{X\}$$
$$f_1 = Fe^{3+}/Fe(III) = 1/\{X\};$$
$$f_2 = FeOH^{2+}/Fe(III) = (K_{B1}K_w/H^+)(1/\{X\});$$
similarly,
$$f_3 = Fe(OH)_2^+/Fe(III);$$
$$f_4 = FeSO_3^+/Fe(III)$$
$$f_5 = FeOHSO_3/Fe(III);$$
$$f_6 = FeSO_4^+/Fe(III)$$
$$f_7 = Fe(SO_3)_2^-/Fe(III);$$
$$f_8 = Fe(SO_3)_3^{3-}/Fe(III)$$

$$K_{A1} = [H^+][HSO_3^-]/[SO_2(aq)] = 1.60 \times 10^{-2}\,M$$
$$K_{A2} = [H^+][SO_3^{2-}]/[HSO_3^-] = 9.25 \times 10^{-8}\,M$$
$$K_{A3} = [H^+][SO_4^{2-}]/[HSO_4^-] = 1.48 \times 10^{-2}\,M$$
$$K_w = [H^+][OH^-] = 1.2 \times 10^{-14}\,M^2$$
$$K_{B1} = [FeOH^{2+}]/[Fe^{3+}][OH^-] = 4.5 \times 10^{11}\,M^{-1}$$
$$K_{B2} = [Fe(OH)_2^+]/[Fe^{3+}][OH^-]^2 = 1.72 \times 10^{22}\,M^{-2}$$
$$K_1 = [FeSO_3^+]/[Fe^{3+}][SO_3^{2-}] = 6.6 \times 10^7\,M^{-1}$$
$$K_4 = [FeOHSO_3]/[FeOH^{2+}][SO_3^{2-}] = 7.3 \times 10^8\,M^{-1}$$
$$K_5 = [FeOHSO_3]/[Fe^{3+}][SO_3^{2-}] = K_4 K_{B1} K_w/[H^+]$$
$$K_6 = [FeSO_4^+]/[Fe^{3+}][SO_4^{2-}] = 4.7 \times 10^3\,M^{-1}$$
$$K_7 = [Fe(SO_3)_2^-]/[FeOHSO_3][S(IV)] = 245\,M^{-1}$$
$$K_8 = [Fe(SO_3)_3^{3-}]/[Fe(SO_3)_2^-][S(IV)] = 60\,M^{-1}$$

[a] The equilibrium constants are for an ionic strength of $I = 0.01$ M.

resulting dependence is consistent with Debye–Hückel theory ($z_1 z_2 = -6$). A plot of the sulfur(IV) equilibria is given in Martin (1984).

3.   UNCATALYZED OXIDATION BY MOLECULAR OXYGEN (AUTOXIDATION)

This problem was reviewed by Radojević (1984). That review concluded from a survey of many studies that the uncatalyzed oxidation of sulfite does indeed take

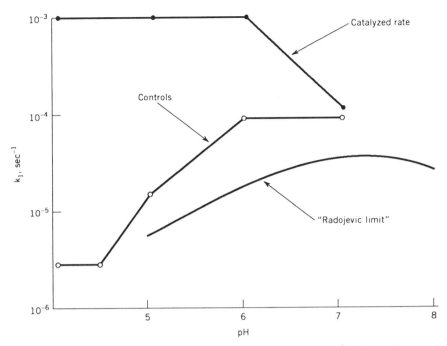

Figure 2. Rates from pH 4 to 7, $[S(IV)] = 1 \times 10^{-4}$ M, $[Fe(III)] = 1 \times 10^{-6}$ M. Controls are rates without added iron. The "Radojević limit" is uncatalyzed rate taken from Radojević (1984). (Figure from Martin et al., 1991. Copyright American Geophysical Union.)

place, and that the purest systems converged on the rate expression:

$$d[SO_4^{2-}]/dt = 0.32[SO_3^{2-}][H^+]^{1/2} \text{M sec}^{-1}$$

We refer to this as the "Radojević limit." It is plotted, along with some of our own unbuffered high-pH data (Martin et al., 1991), in Figure 2. Our data are about a factor of 5 above this limit, but our experiments are done with chloride ion present, which is also thought to be a catalyst (Clarke and Radojević, 1983). Therefore, we see no reason to question the validity of the Radojević limit at this time. Although very little iron or manganese is required to overtake the rate of this process in the environment, there may be situations where clouds exist with extremely low metal content, and thus this expression may have to be considered in the troposphere.

4. CATALYSIS BY IRON(III) IN SOLUTION

4.1. Low-pH Regime (0 to 3.6)

In 1982, the literature on the iron-catalyzed oxidation of sulfur(IV) was replete with apparently contradictory kinetic studies. At that time, the discrepancies

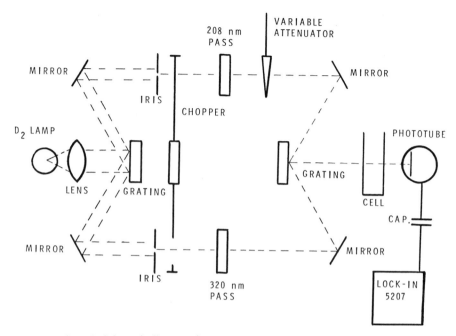

Figure 3. Schematic diagram of apparatus employed in the low-pH studies.

could not really be explained, except by contamination of the solutions or worse. By 1987, however, this reaction began to yield its secrets to intensive study (Martin and Hill, 1987a). We now believe that most of the apparent differences can be explained by four phenomena exhibited by this system:

1. Inhibition by ionic strength
2. Inhibition by sulfur(VI)
3. Self-inhibition [i.e., inhibition by sulfur(IV)]
4. Inhibition by organics at high pH but not at low pH

The first three effects may be seen in the "low"-pH studies, that is, for pH values from 0 to 3.6. The low-pH studies were performed on an apparatus shown schematically in Figure 3 and described in detail elsewhere (Martin and Hill, 1987b). Briefly, the apparatus works as follows. A beam of light from a deuterium UV lamp is split into two beams that then pass through interference filters, one centered at 280 nm on the SO_2 (aq) absorption and one centered at 320 nm, which is not absorbed. The beams are chopped out of phase by a two-window chopper and recombined to give a beam that is of uniform intensity but alternating in wavelength. If the alternating beam passes through a cell containing sulfur(IV), the alternation is converted to an amplitude modulation that is converted to a square wave by a phototube; this in turn is converted to a dc signal by a lock-in

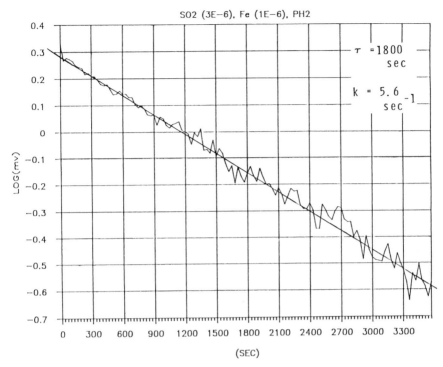

Figure 4. Semilogarithmic plot of the decay of sulfur(IV) at pH 2.0 set with HCl. Initial sulfur (IV) = 3×10^{-6} M, Fe(III) = 1×10^{-6} M. (From Martin and Hill, 1987a, by permission from Pergamon Press.)

amplifier. This arrangement tends to cancel out a number of drift and background effects, and the use of phase-sensitive detection permits a large degree of noise rejection. The final equivalent noise level is 3×10^{-8} mol L^{-1} of sulfur(IV) at pH 2, with a one-second integration time.

To minimize contamination effects, all solutions are prepared with "Milli-Q" ($\geqslant 20$ MΩ resistivity) water that is microfiltered and passed through a cartridge that removes organics. Sulfur(IV) solutions are prepared from gaseous SO_2 and the iron(III) solutions from ferric perchlorate. The pH is adjusted with "Suprapure"-grade hydrochloric acid from Merck, Darmstadt. No organics other than Teflon are used. No soap or detergent is used on glassware, only chromic acid.

Disposable glass pipets are prewashed in the purified water and discarded after use. Runs are conducted by mixing the reagents in a flask and pouring the mixture rapidly into the optical cell.

A typical run is shown in Figure 4. All low-pH runs except those with added complexing agents display this simple first-order behavior. Because we established in previous work (Martin, 1984) that the reaction is first-order in iron(III) and is inversely proportional to (H^+) from pH 0 to 3.6, runs may be characterized by the rate expression:

$$-d[S(IV)]/dt = k[Fe(III)][S(IV)][H^+]^{-1} \qquad (\text{mol } L^{-1} \sec^{-1})$$

Figure 5 shows the effect of sodium chloride and sodium sulfate on k at pH 2.0. The upper curve is a fit to the sodium chloride data. The abscissa for the upper curve is total chloride, so the points at 10^{-2} are for HCl with no added NaCl. Sodium perchlorate has an effect similar to sodium chloride, so the inhibition appears to be a simple ionic strength effect. The shape of the curve is given by the Debye–Hückel–Brønsted theory:

$$k = k_0 10^{-2.0[(I)^{1/2}/(1 + I^{1/2})]}$$

We are applying this theory to much higher ionic strengths than it is intended to describe, but if regarded as an empirical fit to the data, it works quite well.

The lower curve in Figure 5 is an empirical fit to the sulfate inhibition data, ignoring the inherent ionic strength effect of the sulfate:

$$k = k_0\{1/(1 + 150[S(IV)]^{2/3})\}$$

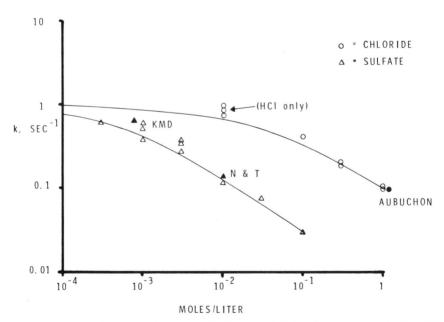

Figure 5. Effect of sulfate (open triangles) or chloride (open circles) on the rate constant k. pH = 2.0, initial sulfur(IV) = 1×10^{-3} M, Fe(III) = 5×10^{-5} M. Abscissa for the open circles is total chloride. Solid circle is from Aubuchon. Solid triangle N & T is from Neytzell-de-Wilde and Taverner. Solid triangle KMD is from Khorunzhii et al., corrected to pH 2. (From Martin and Hill, 1987a, by permission from Pergamon Press.)

At pH 2.0, the ionic strength of sodium sulfate is roughly three times its molality, so part of the sulfate inhibition is due to the primary salt effect. However, there is an additional inhibition due to sulfate specifically, and we believe this is a result of sulfate complexation of a catalytic species (see below).

The solid circle in Figure 5 is taken from the work of Aubuchon (1976), which was done in $I = 1.2$ sodium perchlorate acidified with perchloric acid. The solid triangle N & T is from Neytzell-de-Wilde and Taverner (1958), which was done in sulfuric acid. Those literature experiments were done at 10 °C and have been multiplied by 7.1 to bring the rates up to 25 °C. The solid triangle KMD is from Khorunzhii, Mot'ko, and Denisov (1983), done in sulfuric acid at pH 3 and 20 °C. It has been multiplied by 2.0 to correct for temperature. Thus, these three studies are consistent with ours, which were done at low ionic strength in HCl, when the two inhibiting effects are considered.

Even at the higher values of sulfate or chloride concentration, an inverse dependence on hydrogen ion is still seen, so the ionic strength and hydrogen ion effects are multiplicative. This is not true of the sulfate and ionic strength effects. When the ionic strength of a sulfate-inhibited run is increased, the net result is an increase in the rate. Thus the ionic strength and sulfate inhibitions are not mutually multiplicative, but rather have a slight tendency to cancel each other out, as shown in Figure 6. This effect may also be rationalized in terms of complex

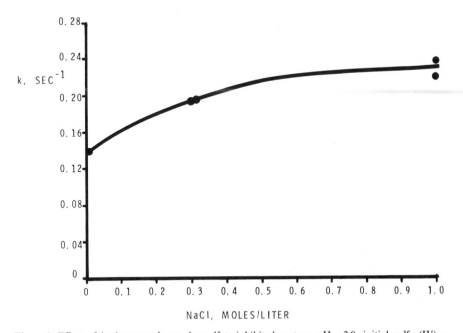

Figure 6. Effect of ionic strength on the sulfate-inhibited system. pH = 2.0, initial sulfur(IV) = 1×10^{-3} M, Fe(III) = 5×10^{-5} M, sulfate = 1×10^{-2} M. (From Martin and Hill, 1987a, by permission from Pergamon Press.)

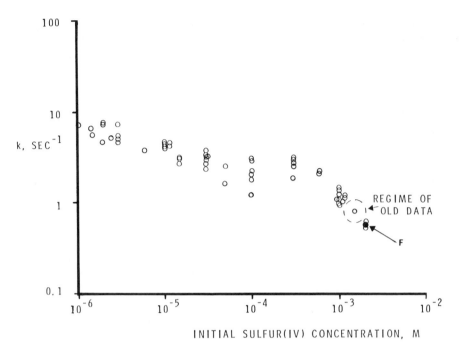

Figure 7. Effect of initial sulfur(IV) concentration on the rate constant. Broken circle is regime of stopped-flow data. Point marked F is our postulated location of the Fuzzi data point. pH = 2.0, Fe(III) ranges from 3×10^{-7} to 5×10^{-5} M. (From Martin and Hill, 1987a, by permission from Pergamon Press.)

formation, since ionic strength may destabilize a complex of oppositely charged ions.

There is also a dependence of the rate constant on the initial sulfur(IV) concentration. This behavior is illustrated in Figure 7, which shows k for a wide range of starting concentrations. The data in this figure are all at pH 2.0, with the iron concentration varying from 3×10^{-7} to 5×10^{-5} M. First-order dependence on iron was verified at a number of sulfur concentrations. This dependence of the rate constant on initial sulfur(IV) concentration is very surprising in view of the nearly perfect first-order decay seen in individual runs.

Our explanation for this effect, at least in the regime around 1×10^{-3} M sulfur(IV), is that *sulfite* is equally inhibiting as *sulfate*. Thus, the rate will be constant for a given set of starting conditions. We found support for this interpretation by doing some runs with a strong sulfate complexing agent—lanthanum ion—present (Fig 8). In such runs, the rate accelerates as the reaction proceeds.

The dashed circle in Figure 7 indicates the regime (1.5×10^{-3} M) of initial sulfur(IV) concentrations in the experiments described in our previous review (Martin, 1984) that had an average rate constant of $k = 0.82$. The point marked F in Figure 7 is an attempt to reconcile the data of Fuzzi (1978) with this scheme. In

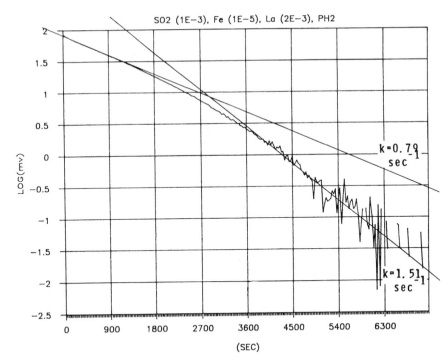

Figure 8. Kinetic data for pH 2.0, sulfur(IV) = 1×10^{-3} M, Fe(III) = 1×10^{-5} M, La(III) = 2×10^{-3} M. The rate accelerates because lanthanum scavenges accumulated sulfate as the reaction proceeds.

this case, we have positioned the point by correcting for pH and temperature, and since that experiment was done in sulfuric acid, by applying the inverse of the sulfate correction given above.

Thus, several apparent discrepancies in the literature may be explained by differences in the three inhibiting factors. We have not included literature data above pH 3 in this analysis, because the rate expression changes completely above pH 3.6. It is interesting to note that the rapid fall-off in the rate constant above 10^{-3} M sulfur(IV) has great potential for causing confusion, because that is the same concentration at which mass transfer of oxygen may become a limitation in some experiments. In the present experiments, solutions above 1×10^{-3} M sulfur(IV) were enriched with oxygen to avoid this problem.

It is illuminating to discuss the low-pH behavior of this system in terms of the species equilibria defined in the thermodynamics section. Figure 9 shows the iron(III) speciation as a function of sulfur(IV) concentration at pH 2.0 and an ionic strength of 0.01. The points in the figure are the superimposed data from Figure 7.

Examination of Figure 9 shows that, at pH 2.0, the iron–sulfite complexes are the major iron species above sulfur(IV) concentrations of 1×10^{-3} M. This fact

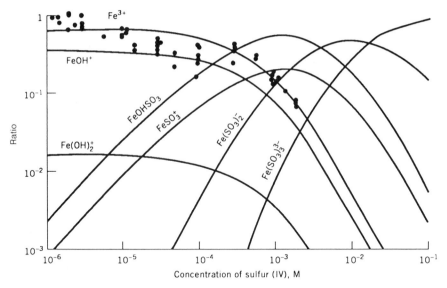

Figure 9. Iron(III) speciation at pH 2.0, $I = 0.01$ as a function of sulfur(IV) concentration. Based on the stability constants in Kraft and van Eldik (1989a) and Conklin and Hoffmann (1988). Superimposed data are rate of reaction at low sulfur(IV) concentrations, from Figure 7. (From Martin et al., 1991. Copyright American Geophysical Union.)

presents problems for any mechanistic description in this pH regime. If the reaction rate is proportional to the concentration of any iron species, we would expect a change in order with respect to sulfur(IV), occurring near 1×10^{-3} M. Our experiments show that the rate constant appears to drop near 10^{-3} M, but the reaction remains first order in sulfur. There is no reliable experimental information about the rate law at higher concentrations of sulfur(IV), because of the oxygen solubility problem.

We believe that the behavior near 10^{-3} M sulfur(IV) may be accounted for by formation of a 1:1 iron–sulfur(IV) complex with the complementary formation of a 1:1 iron–sulfur(VI) complex as the reaction proceeds. The reasoning is as follows.

The rate will in general be proportional to the concentration of some iron-containing species, given by one of the f_i formulas in Table 1. If we neglect the higher sulfur–iron complexes, these formulas have the approximate form:

$$f_i \sim (A)/\{X\} = (A)/\{(B) + (C \times S(IV)) + (D \times S(VI))\}$$

At pH 2.0, $I = 0.01$, the term C is

$$K_1 a_2 + K_5 a_2 = 5060 \, \text{M}^{-1}$$

and the term D is

$$K_6 a_4 = 2880 \, M^{-1}$$

These terms are sufficiently similar in magnitude that the formation of the iron–sulfate complex will tend to compensate for the loss of the iron–sulfite complexes as the reaction proceeds. During the course of the reaction:

$$[S(IV)] + [S(VI)] = constant = [S(IV)]_0$$

The coincidental similarity of the terms C and D near pH 2.0 means that the denominator $\{X\}$ common to all of the f_i terms will vary only slightly during the course of the reaction. This means that the order of the reaction with respect to sulfur will not change appreciably during the reaction, but the apparent rate constant will diminish at higher initial sulfur(IV) concentrations, in accordance with observations. This is illustrated by the superimposition of the kinetic data points (Martin and Hill, 1987a) on the curve for the equilibrium concentration of Fe^{3+} (f_1) in Figure 9.

At still higher sulfur(IV) concentrations, the 1:2 and 1:3 sulfur complexes should lead to changes in the order with respect to sulfur. There are insufficient data available at present to test this prediction.

4.1.1. Mechanisms at Low pH

A detailed discussion of the proposed mechanisms for this reaction at low pH appears in Martin et al. (1991), but here we will describe those mechanisms that, in our view, are most consistent with the observations to date.

We first discuss a mechanism proposed by Conklin and Hoffmann (CH) (1988) in their study of the stability constant of the iron(III)–sulfur(IV) complex:

$$HSO_3^- = SO_3^{2-} + H^+ \qquad K_{a2}$$
$$Fe^{3+} + H_2O = FeOH^{2+} + H^+ \qquad K_b$$
$$FeOH^{2+} + SO_3^{2-} = HOFeOSO_2 \qquad K_4$$
$$\rightarrow rearrangement \qquad k_1$$
$$\rightarrow addition \; of \; O_2$$
$$\rightarrow acid\text{-}catalyzed \; rearrangement$$
$$\rightarrow Fe(III) + S(VI)$$

Detailed examination of the CH mechanism shows that the rate is proportional to the concentration of the $FeOHSO_3$ complex, and therefore should be of the form:

$$d[SO_4^{2-}]/dt = k_1 f_5 [Fe(III)]$$

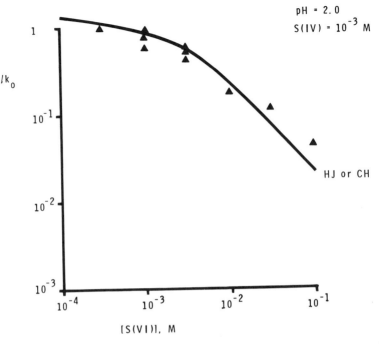

Figure 10. Effect of added sulfate on the rate of reaction at pH 2.0 and $I = 0.01$, $[S(IV)]_0 = 10^{-3}$ M. The solid curve is based on the modified HJ (Fe^{3+}/SO_3^{2-}) or the CH mechanisms. The ionic strength changes due to the sulfate are included in the calculation. (From Martin et al., 1991. Copyright American Geophysical Union.)

In this expression, the sulfur(IV) dependence is implicit in the term f_5, defined in Table 1. This expression predicts first-order in iron and sulfur at low concentrations. As the concentration of sulfur is raised, however, the denominator in the expression for f_5 will exhibit the effects described earlier; as $[S(IV)]$ is raised above 10^{-3} M, the apparent rate constant will fall, with the order remaining the same, at least over a limited region. Thus, this mechanism predicts the observed self-inhibition effect.

The predicted effect of the CH mechanism for sulfate inhibition is plotted in Figure 10 and the effect of ionic strength is plotted in Figure 11. Both of these effects are determined by the term f_5, and the effect of sulfate ion on the overall ionic strength is included in the sulfate calculation. Note that the agreement with experiment is good in both cases.

A different type of mechanism is one in which the rate is determined by the *rate of formation* of a complex. A mechanism of this type has been proposed by Hoffmann and Jacob (HJ) (1984). The key steps in the HJ mechanism are:

$$Fe^{3+} + HSO_3^- = FeSO_3^+ + H^+ \qquad k_4 \text{ (slow)}$$

$$FeSO_3^+ \rightarrow \text{internal redox}$$

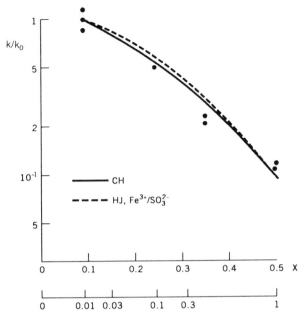

Figure 11. Effect of ionic strength, I, on the rate of reaction, Debye–Hückel-type plot. X is the variable $I^{1/2}/(1 + I^{1/2})$. pH = 2.0, [S(IV)] = 10^{-3} M. The solid curve is for the CH mechanism and the broken curve is for the modified HJ mechanism. (From Martin et al., 1991. Copyright American Geophysical Union.)

$$\rightarrow \text{addition of } HSO_3^-$$

$$\rightarrow \text{addition of } O_2$$

$$\rightarrow \text{formation of products}$$

$$\text{(all fast)}$$

The predicted rate expression will be:

$$R = k_4 f_1 a_1 [Fe(III)][S(IV)]$$

The rate of the corresponding complexation reaction for sulfur(IV) has been measured by Cavasino (1968) to be approximately $38 \, M^{-1} \, sec^{-1}$, which is the right order of magnitude to explain the measured rate.

If we generalize the HJ mechanism, there are four possible combinations of reactants: Fe^{3+} or $FeOH^{2+}$ with HSO_3^- or SO_3^{2-}. Examination of the pH dependence of these combinations shows that only the combination Fe^{3+}/SO_3^{2-} has a reasonable consistency with the data. (The dependence may be seen in Figure 18.) Using this version of the HJ mechanism, we have calculated predictions of the sulfate effect (shown in Figure 10) and the ionic strength effect (shown in Figure 11). In both cases, the predictions essentially overlap with the predictions for the CH mechanism, so these experimental tests do not distinguish between the two possible mechanisms.

The CH mechanism gives a somewhat steeper pH dependence than $(1/[H^+])$ at low pH, as may be seen in Figure 18 (This calculation includes the effect of increasing ionic strength at low pH.) The HJ mechanism gives better agreement at low pH, but falls off too rapidly with increasing pH. Therefore, neither mechanism is entirely satisfactory in this regard.

Additional mechanisms have been proposed by Huss et al. (1982a, b) and by Freiberg (1975) that give a rate proportional to the concentration of $FeSO_3^+$ times sulfur(IV). In our analysis, these would therefore predict second-order in sulfur(IV) at low concentrations, in contradiction experiment. Kraft and van Eldik (1989b, c) have proposed a mechanism involving the multiple sulfur complexes shown in Figure 9, thus predicting higher orders in sulfur(IV).

We believe that the correct mechanism must "shut down" at higher pH, since there is much evidence (shown below) for a free radical process above pH 4. The HJ mechanism does this automatically because of the dependence of the active species concentrations on pH. In the CH mechanism, the concentration of the iron–sulfur complex remains high as the pH is raised (see Figure 19), so a way must be found to shut down this mechanism at high pH. Examination of the CH mechanism shows that there is an acid-catalyzed process that takes place after the addition of oxygen, and possibly this could provide the necessary pH dependence.

At this time, a definitive decision about the mechanism for the low-pH regime is still difficult to make.

4.2. High-pH Regime (4 to 7)

Operation at "high" pH, that is, at pH 4 and above, requires the use of a recirculating, pH-controlled system for the aqueous medium. The reason for this is that oxidation of sulfur in this pH regime causes the pH to drop, and buffers inhibit the reaction. Use of a pH-stat based on sodium hydroxide addition eliminates the need for buffers, but requires that the solution be kept in motion. This is done with the system shown in Figure 12.

The low solubility of iron(III) above pH 3.6 presents special experimental problems. In order to obtain reproducible data, the following procedures had to be followed:

1. Fresh iron(III) perchlorate added to a pH 5 solution hydrolyzes over a period of 30 to 60 minutes. This can give a false optical signal that looks like sulfur decay. A correction can be made for this effect, but it is best dealt with by allowing the iron to hydrolyze for an hour before adding the sulfur. All of the data presented were taken in this latter mode.

2. Iron(III) added to the system at pH 5 is retained on the walls of the system, so that catalytic activity is "remembered" by the system, that is, plain water added

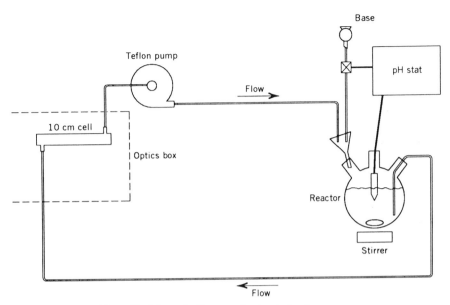

Figure 12. Schematic diagram of recirculating flow system.

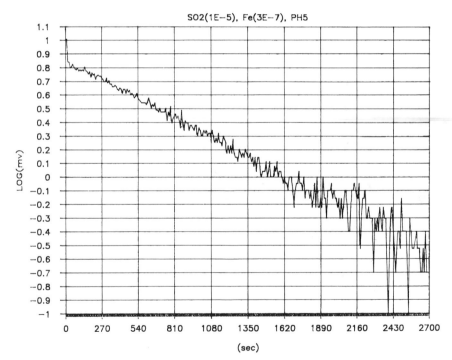

Figure 13. Typical data, pH = 5.0, Fe(III) = 3×10^{-7} M, initial sulfur(IV) = 1×10^{-5} M. (From Martin et al., 1991. Copyright American Geophysical Union.)

subsequently will appear to contain catalyst. The system has to be washed out completely with 1.5 M HCl after each run in order to remove the iron on the walls.

4.2.1.   High-pH Data

The effects of ionic strength, sulfate ion, and sulfur(IV) have not been studied at high pH. In our own experiments, the conditions have been chosen to be typical of cloud water: low ionic strength, low sulfate, and initial sulfur(IV) of 10^{-5} mol L^{-1}.

A typical run at pH 5 is shown in Figure 13. The apparent first-order dependence on sulfur is also supported by the plots of initial sulfur *versus* initial rate for both pH 5 and pH 6 in Figure 14. We occasionally see second-order time dependence for sulfur decay at high pH, but the majority of runs are first-order (see discussion below). Figure 15 provides evidence that only *dissolved* iron is effective as a catalyst. This is a plot of the rate as a function of added iron(III), and clearly shows that a plateau is reached at a concentration of about $1-3 \times 10^{-7}$ mol L^{-1} of iron(III). This plot does not rule out the possibility that colloidal iron

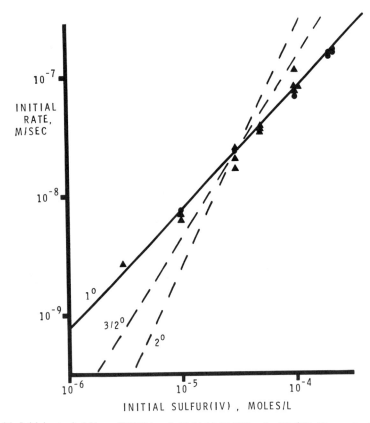

Figure 14. Initial rates in M/sec pH 5 (●) and pH 6 (▲). Fe(III) = 1×10^{-6} M. First-order in sulfur is indicated. (From Martin et al., 1991. Copyright American Geophysical Union.)

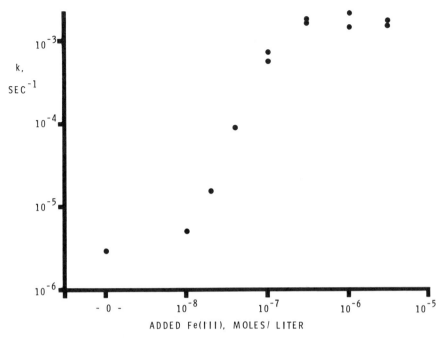

Figure 15. First-order rate at pH 5 as a function of added Fe(III). The data indicate second-order in dissolved iron until saturation is reached. Controls have no iron added. $[S(IV)]_0 = 10^{-5}$ M. (From Martin et al., 1991. Copyright American Geophysical Union.)

can be a catalyst, but it suggests that colloidal material is a much weaker catalyst than iron in solution.

Note also in Figure 15 the second-order in behavior in iron below saturation. This behavior was also seen in data taken at pH 4.0 and 4.5 (Martin et al., 1991). This is clearly different from the low-pH behavior, which is always first-order in iron.

At pH 6 and 7, we did not attempt to prepare solutions of iron below the saturation concentration. We were able to see a difference between controls without added iron and with added iron at pH 6, but essentially no difference at pH 7. These data were shown earlier in Figure 2, in which the rates are compared to the rate for the uncatalyzed reaction proposed by Radojević (1984).

A summary of the phenomenological rate expressions is given in Table 2.

4.2.2. Organic Inhibition

We present here a description of our own organic inhibition studies, taken from Martin et al. (1991). We took data on three organic acids—formic, acetic, and trichloroacetic—and three alcohols—ethyl, isopropyl, and allyl—to support the hypothesis that this system is far more sensitive to inhibition at high pH than it is at low pH.

Table 2   Summary of Empirical Rate Constants

When presented in "rate law" form, the expressions have a limited range of applicability; namely, they are valid only in the following concentration regimes:

$$[S(IV)] < 1 \times 10^{-5} \, mol \, L^{-1}$$

$$[Fe^{3+}]_{(total)} > 1 \times 10^{-7} \, mol \, L^{-1}$$

$$Ionic \, Strength_{(high \, pH)} < 1 \times 10^{-2} \, mol \, L^{-1}$$

$$[S(VI)] < 1 \times 10^{-4} \, mol \, L^{-1}$$

$$T = 25\,°C$$

For the different pH regimes (mol L^{-1} sec^{-1}):
pH 0 to 3.0:

$$-d[S(IV)]/dt = 6.0[Fe^{3+}][S(IV)][H^+]^{-1}$$

pH 4.0:

$$-d[S(IV)]/dt = 1 \times 10^9 [S(IV)][Fe^{3+}]^2 \qquad \text{(up to iron solubility limit)}$$

pH 5.0 to 6.0:

$$-d[S(IV)]/dt = 1.0 \times 10^{-3}[S(IV)]$$

pH 7.0:

$$-d[S(IV)]/dt = 1.0 \times 10^{-4}[S(IV)]$$

Iron does not appear in the pH 5 and 6 rates because we are assuming that a trace of iron will normally be present. At pH 7, no iron is required.

Inhibition plots are shown for the iron-catalyzed system in Figure 16. Clearly, in these systems the inhibition is substantial at high pH and negligible at low pH. While it is true that formic and acetic acids are ionized at the high pH but not at the low pH, this difference cannot explain the results for trichloroacetic acid, which is substantially ionized at both pH values, nor can it explain the results for the alcohols, which are not ionized at either pH. The form of the inhibition plots was chosen to fit typical mechanistic predictions for free radical schemes. The pH 5.0 results are summarized in Table 3 in terms of an inhibition parameter a, defined by

$$k^0/k = 1 + aC$$

where k^0 is the normal rate constant and k is the inhibited rate constant at the concentration C of organic molecule.

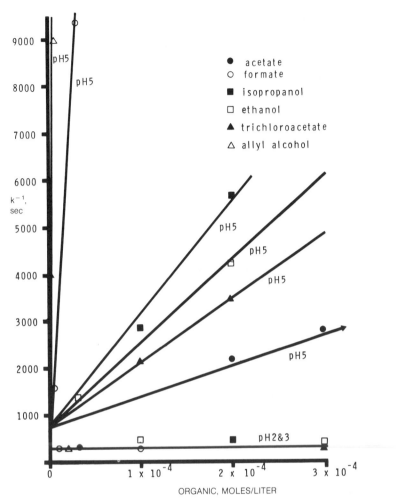

Figure 16. Inhibition data on various organic molecules at both high and low pH. High-pH data are all at pH 5, $[S(IV)] = 1 \times 10^{-5}$ M, $[Fe(III)] = 3 \times 10^{-7}$ M. Low-pH data are at pH 3, $[S(IV)] = 1 \times 10^{-4}$ M, $[Fe(III)] = 1 \times 10^{-6}$ or pH2, $[Fe(III)] = 1 \times 10^{-5}$ M. (From Martin et al., 1991. Copyright American Geophysical Union.)

It is interesting to note that the inhibition parameter correlates with the reactivity of the corresponding organic molecule with the sulfate radical ion (SO_4^-). In Figure 17, we show a log-log plot of the inhibition parameter a versus the reaction rate of the organic molecule with the sulfate radical ion, using rates from Neta et al. (1988) and Wine et al. (1989). The horizontal bars indicate the approximate range of reported rate constants. This kind of correlation does not prove that sulfate radical ion is part of the mechanism, because there could be similar correlations with other radicals, but the evidence suggests that the sulfate radical is involved.

Table 3   Inhibition Parameter a (pH 5.0)[a]

Molecule	a, M^{-1}
Acetate ion	8.7×10^3
Trichloroacetate ion	1.8×10^4
Ethyl alcohol	2.4×10^4
Isopropyl alcohol	3.2×10^4
Formate ion	4.0×10^5
Allyl alcohol	2.3×10^6

[a] $a = [(k^0/k) - 1]/[\text{Organic}]$.

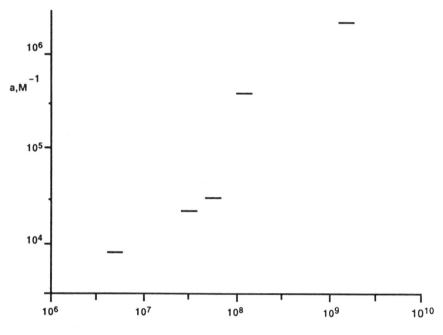

Figure 17. Log–log plot showing correlation of the organic inhibition parameter at pH 5 with the corresponding reaction rate constant of the organic molecule with the sulfate radical ion (SO$_4^-$). Horizontal bars, approximate range of reported rate constants. Reaction rate data from Wine et al. (1989) and Neta et al. (1988). (From Martin et al., 1991. Copyright American Geophysical Union.)

4.2.3. High-pH Summary

We draw the following conclusions from the data on the iron catalyzed oxidation at high pH:

1. The reaction rate depends on the amount of iron in solution at high pH, rather than on the total amount of iron present.

2. At high pH, the reaction is second-order in dissolved iron (zero-order in total iron above the saturation point) and first-order in sulfur, although some variability in the sulfur order is seen.

3. For two types of organic molecule, inhibition of the reaction occurs at high pH but not low pH, with a transition near pH 4.0.

The first conclusion is supported by the data shown in Figure 15, and by the fact that no dependence on added iron is seen at pH 6. In a high-pH study by Fuzzi (1978), a weak dependence on iron is reported at high pH, but it is clearly less than first-order. In any case, Fuzzi reached the same conclusion we make, namely, that only the iron in solution is effective as a catalyst. In a study by Brimblecombe and Spédding (1974a) at pH 4.9, a dependence on iron was found, but the highest iron concentration used in this study was 5×10^{-7} M, which is below the saturation value at this pH. Therefore, a plateau would not be expected.

The second conclusion is supported by the data in Figures 14 and 15, but is at variance with some previous work. There are only a few literature studies that address the order in iron and sulfur at pH 5 and above. In the study by Brimblecombe and Spédding (1974a) at pH 4.9, a first-order in iron and first-order in sulfur was found. This is in agreement with our sulfur order, but there is a discrepancy in iron order below saturation (theirs was first order, ours was second). That study was the first to recognize the important fact that some iron was already present in the water used for the reactions. Thus, the discrepancy might be related to the difficulty of establishing the iron concentration, but this interpretation is clearly tentative. In another study the same year, Brimblecombe and Spédding (1974b) concluded that the order in sulfur changes from first to second as the pH increases above 5.

A change in sulfur order was also observed by Fuzzi (1978) several years later. We have seen some indications of a higher order in sulfur, but by far the majority of runs is first-order in sulfur. One hypothesis to explain the discrepancies is that sulfur(IV) complexes the iron, especially at low ionic strength, and increases the total amount of iron in solution. This effect might be expected to be erratic, and we attribute some of the high-pH data scatter and order changes to this effect (see discussion below).

In any case, for atmospheric applications, a second-order process will tend to decline in significance relative to a first-order process since the concentrations are typically low.

The third conclusion is supported by the data in Figure 16. Inhibition by noncomplexing organic molecules certainly suggests that a free radical chain is taking place.

There are only a few quantitative literature studies on the inhibition of the iron-catalyzed oxidation by organics. Altwicker (1979, 1980) reported inhibition by organics. These results were obtained at pH 6.5 and 7.0, and are therefore consistent with our hypothesis, but the mechanism of inhibition is probably different for the molecules studied (i.e., complexation of the metal). Lim et al. (1982) in studies of the iron(II) catalysis reported inhibition by EDTA, 1,10-phenanthrolene, resorcinol, phloroglucinol, hydroquinone, and pyrocatechol at "low" pH (probably pH 2), but only slight inhibition by phenol. They point out that inhibition in these cases may be due to complexation of the iron by the organic and not by free radical scavenging. These authors proposed a combination of free radical and complex mechanisms for the iron(II)-catalyzed reaction (Huss et al., 1982a, b). [Note that all free radical sulfur oxidation schemes involve recycling between iron(II) and iron(III), and therefore, after an induction period, it should not matter which oxidation state is present at the start.]

A study by Lee and Rochelle (1987) shows inhibition of the reaction by carboxylic acids and demonstrates that these acids are themselves oxidized (decarboxylated) during the course of reaction, and that this oxidation is conjugated with the oxidation of the sulfur. This is also strong evidence for a free radical mechanism, and their paper presents quantitative expressions for the oxidation based on the mechanism. The study was done at pH 5.0, and is therefore also consistent with our hypothesis.

4.2.4.   Mechanisms at High pH

The changing order in iron(III) and its solubility limits make it somewhat difficult to compare reaction rates over a wide range of pH, but we have attempted to do so in Figure 18. The solid curve in this figure shows the first-order rate for disappearance of sulfur(IV) in a system containing 1×10^{-6} M Fe(III) and 1×10^{-5} M S(IV). Note that the rate is essentially independent of pH from pH 4 to 6 for the conditions we have chosen. This latter fact, combined with the evidence that only iron in solution is catalytic, severely limits the number of mechanisms that may be considered.

Figure 19 shows the speciation of iron(III) as a function of pH for 1×10^{-6} M Fe(III), 10^{-5} M S(IV), calculated from the formulas and data in Table 1. Note that the presence of sulfur(IV) has a large effect on the speciation and on the iron solubility at high pH. Since the concentrations of all iron(III) species fall when pH increases from pH 4 to 6, only a combination of base catalysis and the involvement of sulfite ion (SO_3^{2-}) can be consistent with a constant rate over this pH range. In particular, the simplest combination of factors with a "flat" pH dependence from pH 4 to 6 and the correct orders will be:

$$[Fe(OH)_2^+]^2 \, [OH^-] \, [SO_3^{2-}]$$

(Even this expression will be true only at low sulfur(IV) concentrations. At higher sulfur(IV) concentrations, the concentration of $Fe(OH)_2^+$ will be suppressed and will have a more complicated pH dependence. The noisy experimental behavior at high pH may be related to the degree to which iron is complexed with sulfur in this regime.)

Free radical mechanisms that have been proposed to date have the typical feature of a slow radical initiation step followed by a Bäckström–type chain involving sulfur radical ions (Bäckström, 1934). None of the schemes proposed to date are consistent with an initiation step involving the above combination of species. Furthermore, as pointed out by Hoffmann and Jacob (1984), most radical

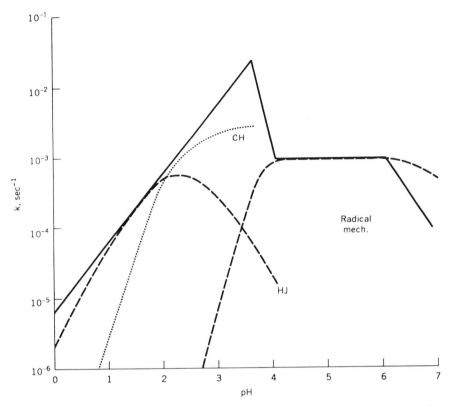

Figure 18. Solid curve is the composite pH dependence of reaction rate, based on the rate expressions in this chapter, for $[Fe(III)] = 1 \times 10^{-6}$ M, $[S(IV)]_0 = 1 \times 10^{-5}$ M. The partial contributions of the radical and the modified HJ mechanisms are indicated with broken lines, the CH mechanism with a dotted line. The high-pH behavior of the CH mechanism is not known. (From Martin et al., 1991. Copyright American Geophysical Union.)

schemes involve chain termination steps with radical-radical recombination. This latter feature leads to a half-order dependence on metal ion concentration, which has never been observed experimentally for iron catalysis.

In recent work, the half-order problem has been overcome in various ways. For example, a mechanism proposed by Lee and Rochelle (1987) has the initial (slow) step

$$Fe^{3+} + SO_3^{2-} = Fe^{2+} + SO_3^-$$

with the termination step:

$$Fe^{2+} + SO_4^- = Fe^{3+} + SO_4^{2-}$$

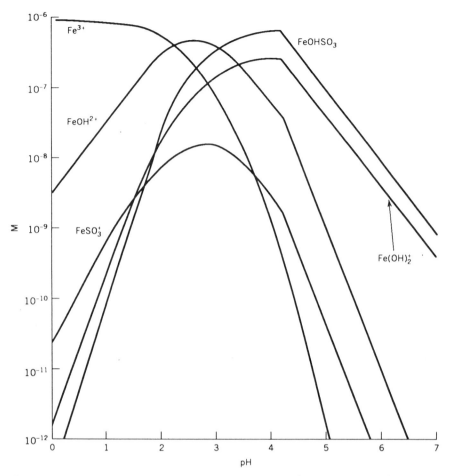

Figure 19. Iron(III) speciation at high pH, for $[Fe(III)] = 1 \times 10^{-6}$ M, $S(IV) = 1 \times 10^{-5}$ M, $I = 0.01$ (pH > 2). The high stability of the sulfur(IV) complexes will increase the equilibrium solubility of the iron above pH 3.6. (From Martin et al., 1991. Copyright American Geophysical Union.)

The initiation step would fall rapidly in rate between pH 4 and 6, and the scheme as a whole gives first-order in iron, so it is not consistent with our observations. Nevertheless, the termination step has the advantage that the iron is reoxidized and a half-order rate law is avoided. This reaction is very fast (Neta et al., 1988), and we will retain it in our own proposed reaction scheme.

A radical mechanism proposed by Huss et al. (1982a, b) avoids the half-order problem by assuming chain termination on trace organics.

We propose here a new initiation reaction that meets the pH dependence and reaction orders given by our experiments. This scheme has a chain propagation mechanism similar to that suggested by Huie and Neta (1987):

$$2Fe(OH)_2^+ = Fe_2(OH)_4^{2+} \qquad\qquad K_9$$

$$Fe_2(OH)_4^{2+} + OH^- = OH + Fe(OH)_2 + Fe(OH)_2^+ \qquad k_5$$

$$Fe(OH)_2 = Fe^{2+} + 2OH^- \qquad\qquad \text{(fast)}$$

$$OH + HSO_3^- = SO_3^- + H_2O \qquad\qquad k_6$$

$$SO_3^- + O_2 = SO_5^- \qquad\qquad k_7$$

$$SO_5^- + SO_3^{2-} = SO_4^{2-} + SO_4^- \qquad\qquad k_8$$

$$SO_4^- + SO_3^{2-} = SO_4^{2-} + SO_3^- \qquad\qquad k_9$$

$$SO_4^- + Fe^{2+} = Fe^{3+} + SO_4^{2-} \qquad\qquad k_{10}$$

$$SO_4^- + Organic = SO_4^{2-} + products \qquad\qquad k_{11}$$

$$SO_4^- = products \qquad\qquad k_{12}$$

$$k_6 = 4.5 \times 10^9 \text{ M}^{-1} \text{ sec}^{-1}$$
$$k_7 = 1.5 \times 10^9 \text{ M}^{-1} \text{ sec}^{-1}$$
$$k_8 = 1.3 \times 10^7 \text{ M}^{-1} \text{ sec}^{-1}$$
$$k_9 = 2 \times 10^9 \quad \text{ M}^{-1} \text{ sec}^{-1}$$
$$k_{10} = 9.9 \times 10^8 \text{ M}^{-1} \text{ sec}^{-1}$$
$$k_{11} = \text{(see Fig. 17)}$$
$$k_{12} = 410 \text{ sec}^{-1}$$

The free radical rates are taken from Neta et al. (1988), Huie and Neta (1987), Wine et al. (1989), and Tang et al. (1988).

This free radical chain is simplified, and other radical processes may be involved (Huie and Neta, 1987). Note that this scheme takes advantage of the fact that the peroxymonosulfate radical ion has a higher reactivity with sulfite than with bisulfite. The double iron complex we propose is not known but is plausible, and has been proposed as the first step in the precipitation of iron hydroxide (Flynn, 1984). $Fe_2(OH)_2^{4+}$ is known and $Fe(OH)_4^-$ is known (see Smith and Martell, 1976). Without a known stability constant, the thermodynamic variables for reaction k_5 are not certain, but we estimate that the Gibbs energy change is favorable for the production of the OH radical.

The rate expression for this mechanism may be partially solved by using the stationary-state approximation, the long-chain assumption, and the assumption that the rate of initiation equals the rate of termination at the stationary state. The predicted rate expression is:

$$d[SO_4^{2-}]/dt = \frac{2k_5k_9K_9[OH^-][Fe(OH)_2^+]^2[SO_3^{2-}]}{k_{12} + k_{10}[Fe^{2+}] + k_{11}[Organic]}$$

We give in Martin et al. (1991) detailed arguments to show that the unknown term $k_{10}[\text{Fe}^{2+}]$ is small compared to k_{12}. This gives an expression that is in qualitative agreement with our data. The pH dependence of this expression is fitted to the data and plotted in Figure 18. The sum of the radical mechanism and the CH or HJ mechanisms fairly well explains the overall pH dependence, although agreement below pH 4 is not very good. There is a possibility that still more reactions are contributing to the mechanism in this region.

We may use the experimental values for a and the literature values for k_{11} with five of the organic inhibitors to calculate the value of:

$$k_{12} = 600 \pm 300 \text{ sec}^{-1}$$

A value of approximately 400 sec^{-1} has been measured by Tang et al. (1988) for the first-order loss of sulfate radical ion. In that study, the reaction was believed to have hydroxyl radical as a product, which would not be a terminating step in our scheme. Our data can only be consistent with a first-order termination step for sulfate radical ion that has roughly the same rate as the process described by Tang et al., but with inert products.

Having the initiation step (k_5) create hydroxyl radicals is necessary to preserve the first-order dependence on sulfur(IV) in this scheme. If the initiation step involved sulfite ion, then a second-order dependence on sulfur(IV) would be seen. The experiments seem to suggest that this order may vary unpredictably, and therefore it is possible that more than one initiation path exists.

4.3. Summary of Inhibition Data

The empirical rate expressions are summarized in Table 2. The various inhibitions are as follows.

For higher ionic strength (verified at pH 2.0 only):

$$k = k_0 10^{-2.0[I^{1/2}/(1 + I^{1/2})]}$$

For higher sulfite or higher sulfate (at pH 2.0 only; for other conditions use full expression in Table 1):

$$k = k_0(1 + 3233[\text{S(IV)}] + 1841[\text{S(VI)}])^{-1}$$

For organic materials at pH greater than 4

$$k = k_0(1 + k_{11} k_{12}^{-1} [\text{Organic}])^{-1}$$

where k_{11} is the reaction rate of the organic molecule with the sulfate radical ion and k_{12} is 600 ± 300 sec^{-1}.

5.   CATALYSIS BY MANGANESE(II)

This reaction has been studied several times in the past with apparently conflict-ing results (Hoather and Goodeve, 1934; Neytzell-de-Wilde and Taverner, 1958; Coughanowr and Krause, 1965; Matteson et al., 1969; Huss et al., 1982a). There is evidence from studies at different sulfur concentrations that the reaction order changes with concentration (Martin, 1984; Ibusuki and Barnes, 1984). Since sulfur in atmospheric liquid water is extremely dilute (10^{-5} to 10^{-9} mol L^{-1} of liquid), a concentration dependence to the reaction order may profoundly change the calculated rate of oxidation in the atmosphere.

We have done some studies of this reaction (Martin and Hill, 1987c) with the differential optical system. As in the iron studies, for initial sulfur(IV) concentra-tions below 4×10^{-4} mol L^{-1}, there is sufficient molecular oxygen in solution in equilibrium with air to complete the oxidation. For runs up to 2×10^{-3} mol L^{-1} of sulfur(IV), the water was saturated with pure oxygen, eliminating mass transfer as a limiting factor to the kinetics. The reaction rate is zero order in oxygen, and it is possible to run right up to the stoichiometric limit for oxygen without observ-ing a slowing of the reaction rate of sulfur.

An example of a kinetic run is shown in Figure 20, which shows the decay of sulfur(IV) at pH 2 in the presence of 1×10^{-5} mol L^{-1} of Mn^{2+} ion. If these data

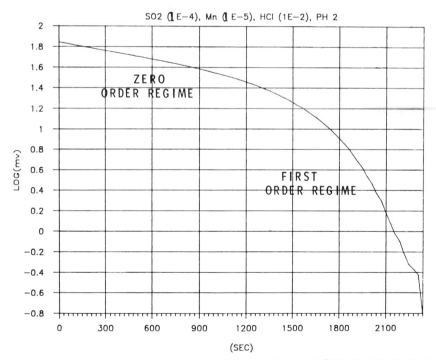

Figure 20. A kinetic run for sulfur(IV) $= 1 \times 10^{-4}$ M, Mn(II) $= 1 \times 10^{-5}$ M, pH $= 2.0$. Both the zero-order and the first-order regimes may be seen in this single run.

are plotted on a linear scale, the decay of sulfur at high concentrations (10^{-4} to 10^{-5} M) is a straight line, that is, zero-order time behavior. As the concentration falls below 10^{-5} M, the behavior becomes first-order in sulfur and appears linear on a semilogarithmic plot. This region may be seen at times after about 2000 sec in the figure.

At pH 2 and sulfur(IV) concentrations above 10^{-4} mol L^{-1}, the rate law for decay is:

$$d[S(IV)]/dt = -300[Mn^{2+}]^2 \quad (mol\ L^{-1}\ sec^{-1})(25\,°C)$$

At pH 2 and sulfur(IV) concentrations below 10^{-6} mol L^{-1}, the rate law for decay is:

$$d[S(IV)]/dt = -400[Mn^{2+}][S(IV)] \quad (mol\ L^{-1}\ sec^{-1})(25\,°C)$$

Note, however, that these rates are strongly dependent on ionic strength. We have measured the effect of ionic strength on the rate of this reaction in both the high (10^{-3} M) and low (10^{-6} M) sulfur concentration regimes. We find, in agreement with the study by Huss, Lim, and Eckert (1982a) (HLE), that the rate constant is very sensitive to the ionic strength of the reaction medium—100-fold slower at $I = 1.0$—and that this accounts entirely for the apparent pH dependence of this reaction rate. This effect explains a number of the conflicts in the literature and some of the unusual behavior.

Figure 21 shows values for the zero-order rate constant k_0 (high sulfur regime), in L mol^{-1} sec^{-1} at 25 °C, as a function of ionic strength for a number of added substances. Note that there is a strong inhibition of the reaction rate that depends on ionic strength but not on the specific ion added. The smooth curve in the figure

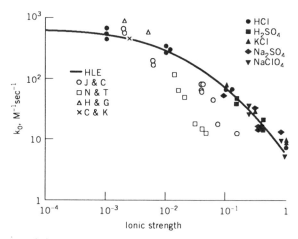

Figure 21. Ionic strength dependence of k_0 in zero-order (high sulfur) regime for manganese catalysis. $S(IV) \approx 10^{-3}$ M. Solid points are our data. Open points and the curve are from the literature. (From Martin and Hill, 1987c, by permission from Pergamon Press.)

is from HLE (1982a), converted from hours to seconds and neglecting a high sulfur term:

$$k = k_0^* 10^{-4.0[I^{1/2}/(1+I^{1/2})]}$$

$$k_0^* = 680 \pm 100 \qquad (\text{mol L}^{-1} \text{ sec}^{-1})(25\,^\circ\text{C})$$

(error estimate ours)

As in the iron studies, this equation is based on the Debye–Hückel–Brønsted theory of primary ionic strength effect and is being used at higher strengths than is justified by theory, but it works well as a fit both to the HLE data and to ours. Note that the acids fall on the same curve as the salts. This means that earlier reports of an apparent pH dependence of this reaction by us and by others are not correct; the behavior is better described as an ionic strength effect. This explains the phenomenon originally reported by Hoather and Goodeve (1934) that the reaction velocity does not depend on hydrogen ion produced during the course of reaction but does depend on the initial hydrogen ion concentration. This is because, in a typical reacting system, hydrogen ion may change a great deal but ionic strength changes very little.

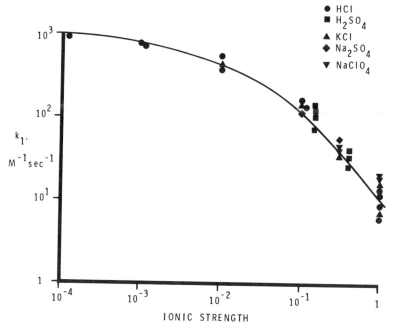

Figure 22. Ionic strength dependence of k_1 in first-order (low sulfur) regime for manganese catalysis. $S(IV) \approx 10^{-6}$ M. All points are our data. Curve has the same shape as in Figure 21, with the height adjusted to fit data. (From Martin and Hill, 1987c, by permission from Pergamon Press.)

Also shown in Figure 21 are points from Hoather and Goodeve (1934), divided by 4.6 to correct from 35 to 25 °C; from Johnstone and Coughanowr (1958) and Neytzell-de-Wilde and Taverner (1958), multiplied by 11.5 to correct from 10 to 25 °C; and from Coughanowr and Krause (1965). The ionic strength for these papers was estimated by assuming complete ionization of the first sulfuric acid proton, with the ionization of the second proton given by the ionization constant and the mean activity coefficients tabulated in Latimer (1952). This gives a minimum value for I in the sense that any ionic solutes not mentioned in those papers would increase the ionic strength above our estimate. The agreement is poor at high ionic strength, but all studies approach agreement at low ionic strength—the region typical of cloud water.

Figure 22 is a similar plot for the first-order decay of sulfur(IV) seen for sulfur concentrations below 10^{-6} M. In this regime:

$$d[S(IV)]/dt = -k_1[Mn^{2+}][S(IV)]$$
$$k_1 = k_1^* \, 10^{-4.07[I^{1/2}/(1+I^{1/2})]}$$
$$k_1^* = 1000 \pm 150 \qquad (mol\,L^{-1}\,sec^{-1})(25\,°C)$$

The smooth curve is the same ionic strength dependence as that in Figure 21, but arbitrarily adjusted in height to fit the data.

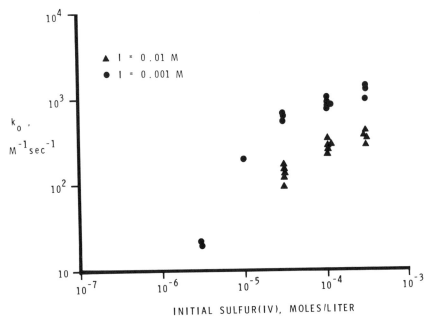

Figure 23. Manganese catalyst, intermediate regime. Dependence of zero-order rate on initial sulfur(IV) concentration for two fixed values of the ionic strength. (From Martin and Hill, 1987c, by permission from Pergamon Press.)

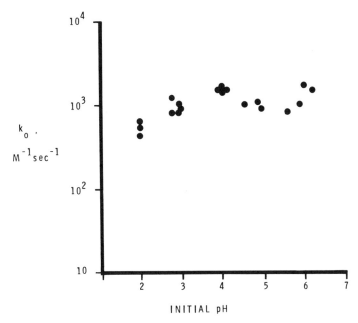

Figure 24. pH dependence of k_0 in zero-order regime for manganese catalysis. $S(IV) = 1 \times 10^{-4}$ M, $Mn(II) = 1 \times 10^{-5}$ M. Fall-off below pH 3 is due to increase in ionic strength. (From Martin and Hill, 1987c, by permission from Pergamon Press.)

If the reaction is initiated between these two regimes, the initial rate law is still zero order in sulfur, but the apparent rate constant is a function of the starting sulfur concentration, as shown in Figure 23. This remarkable behavior was first reported by Penkett et al. (1979) in rainwater, and presumably is due to the formation of a very stable intermediate but has not been fully explained.

Figure 24 shows the zero-order rate constant at fixed ionic strength for a wide range of pH. These points are for the initial rate in unbuffered systems, but no indication of an increase in rate is seen for either the zero- or first-order regimes at high pH. The rise above pH 4 reported by HLE (Huss et al. 1982a) is probably a high sulfur concentration phenomenon.

6. IRON–MANGANESE SYNERGISM

Since both iron and manganese are found in clouds in catalytically significant concentrations, there is a question about the additivity of the two catalytic reactions.

Barrie and Georgii (1976) suggested very early that there might be a synergism between iron(III) and manganese(II). Additional evidence came from Altwicker and Nass (1983), from Martin (1984), and from Ulrich et al. (1986). More recently, Ibusuki and Takeuchi (1987) did an extensive study of iron–manganese synergism at low concentrations of sulfur(IV). There are apparent contradictions between these different studies, and we did some work (Martin and Good, 1991) to examine the contradictions. In this case, the differences may be explained by the differences in sulfur(IV) concentration regimes. At very low concentrations of sulfur, we found behavior similar to that reported by Ibusuki and Takeuchi (1987).

The experimental procedure for the iron–manganese studies was similar to the high-pH iron-catalyzed studies. We had shown previously that for initial sulfur(IV) concentrations in the 10^{-3} to 10^{-4} M regime, the synergism between iron and manganese is very complex (Martin, 1984). This was not surprising, because the reaction order in sulfur(IV) is different for the two catalysts (first-order for iron and zero-order for manganese). However, as shown above, when the initial sulfur(IV) concentration is below the 10^{-5} regime, the manganese-catalyzed reaction becomes first-order in sulfur. In this regime, an entirely different synergism is seen. The reaction is first-order in sulfur for iron and for manganese,

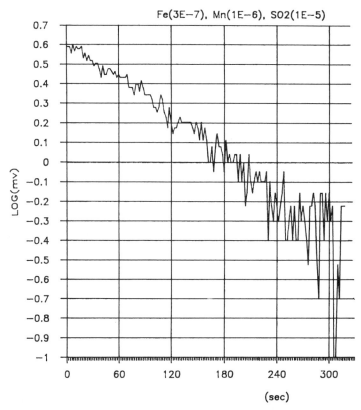

Figure 25. Typical synergism data showing decay of sulfur(IV) absorbance over time. Ordinate is log of the sulfur(IV) signal in millivolts. Initial sulfur(IV) concentration is $10\,\mu M$, iron is 0.3 μM, manganese is $1.0\,\mu M$, pH = 3.0. (From Martin and Good, 1991, by permission from Pergamon Press.)

and for the mixed system. This behavior may be seen in the decay plot shown in Figure 25.

If we anticipate that the overall rate law at low sulfur concentrations will have the form

$$-d[S(IV)]/[S(IV)]dt = k_1[Fe(III)] + k_2[Mn(II)] + k_3[Fe(III)][Mn(II)]$$

the data should fall in a straight line if we plot the quantity

$$-d[S(IV)]/[S(IV)]dt - k_1[Fe(III)] - k_2[Mn(II)]$$

as a function of the product:

$$[Fe(III)][Mn(II)]$$

The values for k_1 and k_2 are taken from above for an ionic strength of approximately 10^{-3}, pH 3.0, and $[S(IV)]_0 = 10^{-5}$ M:

$$k_1 = 2600 \text{ M}^{-1} \text{ sec}^{-1}$$

$$k_2 = 750 \text{ M}^{-1} \text{ sec}^{-1}$$

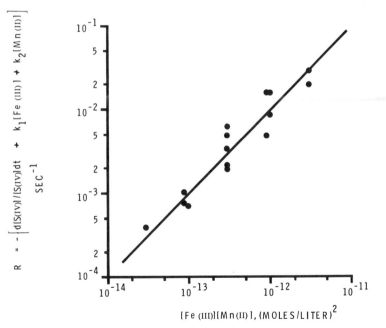

Figure 26. Summary of rate data for mixed catalyst system at pH 3.0, $[S(IV)]_0 = 1 \times 10^{-5}$ M. Ordinate is the first-order rate of decay with the partial contributions of iron and manganese subtracted. Abscissa is the product of the iron and manganese concentrations. Initial sulfur(IV) is 10 μM. (From Martin and Good, 1991, by permission from Pergamon Press.)

This is done in a log–log plot in Figure 26. The observed straight line supports this form of the rate law. The slope of the line yields a value for k_3 at pH 3.0:

$$k_3 = 1.0 \times 10^{+10} \, M^{-2} \, sec^{-1} \quad \text{(pH 3.0)}$$

This value is consistent with the value for the synergistic rate constant reported by Ibusuki and Takeuchi (1987). This agreement is possible only at high concentrations of the two catalysts because the form of the Ibusuki and Takeuchi rate law differs from the one above, that is, it does not explicitly remove the individual iron and manganese contributions to the rate of oxidation.

We also obtained data at pH 5.0. Since, at pH 5.0, the iron is saturated, we have subtracted a constant term for the iron

$$k_1 [Fe(III)] = 1 \times 10^{-3} \, sec^{-1}$$

and for the manganese, we use the same term as before:

$$k_2 = 750 \, M^{-1} \, sec^{-1}$$

In this case, the synergistic term will depend on the saturation concentration of iron(III). We believe this to be approximately $3 \times 10^{-7} \, mol \, L^{-1}$ for our conditions because the rate reaches a constant value at this amount of added iron (Fig. 15). With this assumption, the value of k_3 for pH 5.0 is:

$$k_3 = 2 \times 10^{+10} \, M^{-2} \, sec^{-1} \quad \text{(pH 5.0)}$$

Again, this is roughly comparable to the value obtained by Ibusuki and Takeuchi (1987), and indicates a substantial effect, even in the iron saturation region.

Having obtained a rate expression for the synergism, we were able to study the ionic strength and sulfate inhibition of the synergism. The experimental conditions were $1 \times 10^{-6} \, M \, Fe(III)$, $1 \times 10^{-6} \, M \, Mn(II)$, pH 3.0. The data are shown in Figure 27, which plots the logarithm of the rate as a function of the logarithm of the concentration of sodium perchlorate (ionic strength) and sodium sulfate (sulfate ion). Comparison with the inhibition data on iron shows that both the sulfate and ionic strength effects are similar to, but slightly weaker than, the corresponding effects with iron alone.

Figure 27 shows two broken Debye–Hückel curves for the ionic strength dependence (Glasstone, 1942). The upper curve is for an ion charge product of -1 and the lower for -2. Clearly, neither curve is satisfactory. The functional form of these curves is

$$k = k_0 10^{+n[(I^{1/2})/(1 + I^{1/2})]}$$

where n is the ion charge product (negative, in this case) of the two ions participating in the rate-limiting step of the reaction. This simple view of the

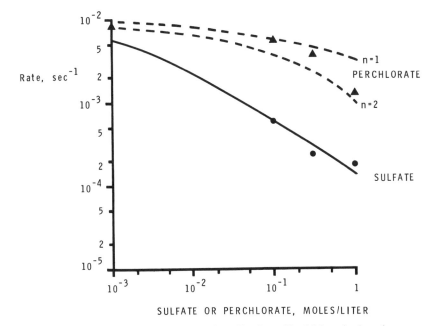

Figure 27. Inhibition data for ionic strength and for sulfate ion, pH = 3.0 for mixed catalyst system. Initial sulfur(IV) is 10 μM, iron is 1 μM, manganese is 1 μM. Dashed curves are from Debye–Hückel theory; solid curve is an empirical fit. (From Martin and Good, 1991, by permission from Pergamon Press.)

mechanism can be obscured by the effect of ionic strength on the stability constants of any complexes involved. Nevertheless, the ionic strength effect provides a means of testing mechanistic proposals.

The solid curve in Figure 27 is a purely empirical fit to the sulfate inhibition data:

$$k = k_0\{1/(1 + 75[S(VI)]^{2/3})\}$$

This fit is similar to that for the iron-catalyzed oxidation sulfate inhibition studies. This suggests that sulfate ion complexes iron and reduces the concentration of the catalytically active species, thus inhibiting the reaction.

Since we have shown that the iron-catalyzed oxidation of sulfur is very complicated, and that the mechanism probably changes from a molecular to a free radical process as the pH is raised, the synergism may similarly be complicated. There is insufficient information to propose a mechanism for this reaction at this time. Nevertheless, the existence of the synergism is now well established, and the data have been taken in a concentration regime that allows a plausible estimation of atmospheric consequences.

7. "FAST" OXIDANTS

7.1. Ozone

Hoffmann (1986) and Hoffmann and Calvert (1985) have reviewed the oxidation of sulfur(IV) by ozone, and we will briefly summarize that work here. This summary will allow us to mention an error in our earlier review (Martin, 1984, 1987). In the earlier review, the figure showing the ozone data (Figure 20 in that chapter) had a mislabeled abscissa; the actual experimental pH range was from 1 to 4 instead of 0 to 3. Because the pH dependence of the reaction goes roughly as $[H^+]^{-1/2}$ in this region, all of the rate constants are inflated by a factor of 3. This error did not extend to the numerical values for the rates, which were correct.

Probably the most important development in the kinetics of this reaction since the earlier review is the general acceptance of species-specific rate constants. Multiterm rate expressions for this reaction were used by Erickson et al. (1977), by Maahs (1983a, b), and by Hoigne et al. (1985). Hoffmann (1986) showed that all of these ozone results could reasonably and consistently be interpreted in terms of different reactivities of three sulfur(IV) species. This description of the pH dependence of the rate is more straightforward than empirical hydrogen ion terms, and we give this description precedence over our earlier one. Certainly, a species-specific rate constant lends itself more readily to computer modeling!

Therefore, we will quote from the Hoffmann review, using the notation from Table 1

$$- d[S(IV)]/dt = (k_0 a_0 + k_1 a_1 + k_2 a_2)[S(IV)][O_3]$$

where

$$a_0 = [SO_2 - Aq]/[S(IV)]$$
$$a_1 = [HSO_3^-]/[S(IV)]$$
$$a_2 = [SO_3^{2-}]/[S(IV)]$$

These terms include the complete pH dependence for this system.

The k_i terms correspond to the reaction rates of ozone with each species:

$$k_0 = (2.4 \pm 1.1) \times 10^4 \, mol \, L^{-1} \, sec^{-1}$$
$$k_1 = (3.7 \pm 0.7) \times 10^5 \, mol \, L^{-1} \, sec^{-1}$$
$$k_2 = (1.5 \pm 0.6) \times 10^9 \, mol \, L^{-1} \, sec^{-1}$$

Figure 28 is a plot of the overall pH dependence of the rate constant, that is, the sum of these three terms, for a closed system. (By specifying a closed system, we mean that the pH-dependent Henry's law solubilities of the gases are not

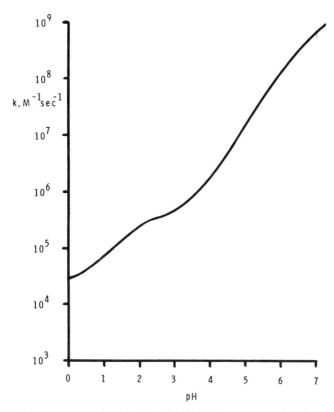

Figure 28. Effective rate constant for oxidation of sulfur(IV) by ozone as a function of pH [closed system, that is, sulfur(IV) is not a function of pH]. This is the sum of the three terms.

included; that is, no gases are allowed to enter or leave the aqueous medium as the pH is changed.)

These rate constants are for 25 °C. The Hoffmann review suggests the values of Erickson et al. (1977) for the activation energies: $E_a = 46.0 \text{ kJ mol}^{-1}$ for k_1 and 43.9 kJ mol^{-1} for k_2.

Proposed mechanisms for the three reaction paths are described in the Hoffmann paper (1986). A study done since the Hoffmann review (Nahir and Dawson, 1987) in a two-phase droplet system at very low concentrations reports somewhat lower rate constants from pH 4 to 6, but in such systems there is always the possibility of mass transport limitation to the overall rate.

7.2. Hydrogen Peroxide

Since our review of the hydrogen peroxide reaction, several studies have been done that give similar rate expressions for this reaction, including the work by Kunen et al. (1983), McArdle and Hoffmann (1983), and Jaeschke and Herrmann

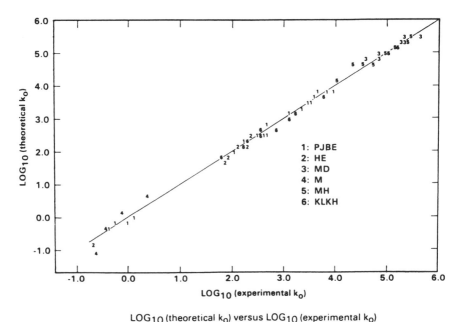

LOG$_{10}$ (theoretical k_0) versus LOG$_{10}$ (experimental k_0)

Figure 29. Overton's figure showing reconciliation of six different studies of the hydrogen peroxide oxidation of sulfur(IV). (From Overton, 1985, by permission of the author.)

(1986). An apparent discrepancy in the results for the mid-pH region (see Martin and Damschen, 1981) was resolved by Overton (1985), who showed that general acid catalysis of the reaction by the buffers could explain the differences between measurements. Figure 29 is quoted from the Overton paper, to show how all of the studies up to 1985 could be made to fit on a single line if general acid catalysis was taken into account. That paper also gives fits to the temperature dependence for several studies.

The Hoffmann and Edwards mechanism (1975) for this reaction therefore appears to be well established

$$HSO_3^- + H_2O_2 = A^- + H_2O \qquad k_f/k_r$$
$$A^- + H^+ = 2H^+ + SO_4^{2-} \qquad k_2$$
$$A^- + HB = 2H^+ + SO_4^{2-} + B^- \qquad k_B$$

where A^- is an intermediate (probably peroxymonosulfite ion).

The overall rate expression suggested by Overton is

$$d[SO_4^{2-}]/dt = k[H_2O_2][S(IV)]$$

where

$$k = \frac{a_1\{a[\text{H}^+] + b[\text{HB}]\}}{\{1 + c[\text{H}^+] + e[\text{HB}]\}}$$

and

$$a_1 = [\text{HSO}_3^-]/[\text{S(IV)}] \text{ as in Table 1}$$

$$a = k_f k_2/k_r = 1.56 \times 10^{+12} e^{-3032/T}$$

$$b = k_f k_B/k_r = 1.23 \times 10^{+12} e^{-5209/T}$$

$$\text{(acetate buffer)}$$

$$= 5.29 \times 10^{+9} e^{-4412/T}$$

$$\text{(phosphate buffer)}$$

$$= 1.61 \times 10^{+7} e^{-4034/T}$$

$$\text{(tris buffer)}$$

$$c = k_2/k_r = 13.8 \qquad \text{(no temperature dependence)}$$

$$e = k_b/k_r = cb/a$$

$$(a, b \text{ in L mol}^{-1} \text{sec}^{-1}; c, e \text{ in L mol}^{-1})$$

In a study done since the Overton paper by Lind et al. (1987), similar rate constants are observed in a very dilute bufferless system for pH 4.0 to 5.2.

In the Lind et al. study, rates are also reported for the oxidation of sulfur(IV) by methylhydroperoxide and by peroxyacetic acid. We quote their values for the mid-pH range:

$$-d[\text{peroxide}]/dt = k[\text{H}^+][\text{peroxide}][\text{S(IV)}]$$

$$k = 7.2 \times 10^7 \qquad \text{(hydrogen peroxide)}$$

$$k = 1.7 \times 10^7 \qquad \text{(methylhydroperoxide)}$$

$$k = 3.5 \times 10^7 \qquad \text{(peroxyacetic acid)}$$

In that paper (Lind et al., 1987), it is pointed out that the Henry's law coefficients for the latter two peroxides are much less than that for hydrogen peroxide, so the atmospheric contributions of these two peroxides to sulfur oxidation is not likely to compete.

Lastly, we mention a study by Betterton and Hoffmann (1988) on the oxidation of sulfur(IV) by peroxymonosulfate ion (HSO_5^-). This species has been predicted to be present in clouds on the basis of free radical models (Jacob, 1986). The rate law was

$$-d[\text{HSO}_3^-]/dt = \frac{k_1(k_2/k_{-1})[\text{H}^+]a_1[\text{S(IV)}]}{(1 + (k_2/k_{-1})[\text{H}^+])}$$

where a_1 is defined in Table 1, and:

$$k_1 = 1.21 \times 10^6 \qquad (\text{mol L}^{-1}\,\text{sec}^{-1})$$

$$k_2/k_{-1} = 5.9 \text{ L mol}^{-1}$$

This rate is somewhat lower than had been anticipated, so the atmospheric importance is not well established (Betterton and Hoffmann, 1988).

8. ESTIMATES OF ATMOSPHERIC SIGNIFICANCE

Our earlier review (Martin, 1984) gave formulas for calculating the atmospheric oxidation rate of SO_2 in a typical cloud. We will not repeat those formulas here, but we will give updated calculations of the rates. Since Henry's law coefficients are needed for such calculations, we will mention some studies on these.

For hydrogen peroxide, a measurement at 20 °C was made by Yoshizumi et al. (1984), and temperature-dependent measurements have been done by Hwang and Dasgupta (1985) and by Lind and Kok (1986). In our view, these measurements are all reasonably self-consistent, and we will quote the expression given by Lind and Kok:

$$K_H(\text{mol L}^{-1}\,\text{atm}^{-1}) = e^{[(6621/T) - 11.00]}$$

For comparison, the various papers give at 20 °C:

$$K_H = 1.45 \times 10^5 \qquad \text{(Yoshizumi et al., 1984)}$$
$$= 1.09 \times 10^5 \qquad \text{(Hwang and Dasgupta, 1985)}$$
$$= 1.09 \times 10^5 \qquad \text{(Lind and Kok, 1986)}$$
$$= 1.05 \times 10^5 \qquad \text{(Martin and Damschen, 1981)}$$

It should be mentioned that the Martin and Damschen number was based on data taken by Scatchard and Ticknor (1952) in concentrated solutions at temperatures above 45 °C! The close agreement of recent work with extrapolation of that data is a tribute to the care of the Scatchard work and to the accuracy of solution theory.

For ozone, some early work was reviewed by Hoffmann and Calvert (1985). A more recent temperature and ionic strength dependence has been done by Kosak-Channing and Helz (1983). Their expression is quoted here:

$$K_H = e^{[(+2297)/T - 2.659I + (688.0I)/T - 12.19]} \qquad (\text{mol L}^{-1}\,\text{atm}^{-1})$$

This expression has been inverted to correspond to our convention for the Henry's law constant. In this expression, I is the molar ionic strength and T is the absolute temperature.

For comparison, this expression gives at zero ionic strength and 25 °C

$$K_H = 1.13 \times 10^{-2} \quad (\text{mol L}^{-1} \text{atm}^{-1})$$

which is in good agreement with earlier values at this temperature. Note that ozone is roughly 10 times as soluble in water as molecular oxygen.

In reviewing the developments in the kinetics since 1984, it is clear that the only major changes are in the catalyzed autoxidation. The hydrogen peroxide rate we used in 1984 has been confirmed by several groups over a wide range of conditions. The ozone rate of Maahs (1983b) used previously is nearly identical to the presently accepted rate plotted in Figure 29.

Thus, the significant changes are in the manganese-catalyzed rate, the iron-catalyzed rate, and in the iron-manganese synergism. Compared to the previous calculations, the manganese rate will now be slightly more, and the iron rate and synergism will be more, than previously thought.

However, it is important to point out that the likely effect of organic inhibition on the predicted rates is very substantial. Since formic acid is commonly found in cloudwater and is a strong inhibitor of the iron-catalyzed oxidation, much of the anticipated increase in the iron-catalyzed rate could be erased by the inhibition. We are not aware of any data on the inhibition of the synergism by organics, but in view of the similar pH behavior of the synergism and the iron data, a good guess at this time is that the synergism will also be inhibited by formic acid.

Figure 30 is an updated version of the atmospheric rate calculation shown in the 1984 review. It presents a calculation of the oxidation rate of sulfur dioxide in a hypothetical cloud containing 1 mL of liquid water per cubic meter of air at 25 °C. The pH is treated as an independent variable. The conditions are:

$[SO_2] = 5$ ppb

$[Fe(III)] = 1 \times 10^{-6}$ mol L^{-1} of liquid or saturated, whichever is lower

$[Mn(II)] = 1 \times 10^{-7}$ mol L^{-1} of liquid

The iron and manganese concentrations are three times higher than before, and represent amounts typically reported in cloudwater (Jaeschke, 1982; Munger and Eisenreich, 1983). The solid lines are the calculations without organic inhibition. The broken line shows the effect of formic acid on the rate. Environmental measurements of formic acid in clouds and fog range from about 10^{-5} to 10^{-3} mol L^{-1} of liquid (Keene et al., 1983; Munger et al., 1989; Keene and Galloway, 1986). Since formic acid is a strong inhibitor of the iron-catalyzed reaction at high pH, the rate can range from roughly 1/5 to 1/400 of the rate without formic acid. This effect may explain some of the difficulty in demonstrating the participation of the iron-catalyzed process in the troposphere.

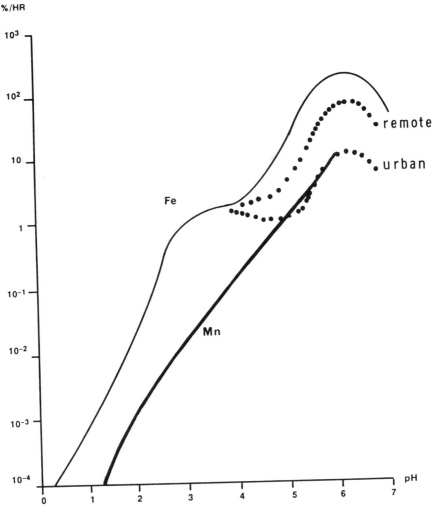

Figure 30. Estimate of the oxidation rate of sulfur dioxide to sulfate in a hypothetical cloud containing 1 mL of liquid water per cubic meter of air. Based on 5 ppb gaseous SO_2, 1×10^{-6} M Fe(III) available in the droplets. Manganese rate based on 1×10^{-7} M Mn(II) available in the droplets. Dashed curve shows the effect of ambient formic acid on the calculated iron-catalyzed rate. This calculation assumes that the reactions are taking place in isolation from other systems, which may not be the correct assumption for free radical reactions.

ACKNOWLEDGMENTS

Support from the U.S. Environmental Protection Agency is gratefully acknowledged (grant number R-816211–01–0, Project Officer L. G. Swaby). Some of the research described has been supported by U.S. EPA cooperative agreement (CR812301–01–0, Project Officer M. C. Dodge) and by contracts from the Electric Power Research Institute (RP 2023–07, Project Manager D. A. Hansen). This chapter has not been subject to agency review; therefore, it does not necessarily reflect agency views and no official endorsement should be inferred.

REFERENCES

Altwicker, E. R. (1979). Oxidation/inhibition of sulfite ion in aqueous solution, *AIChE J.* **75**, 145–149.

Altwicker, E. R. (1980). Oxidation and oxidation-inhibition of sulfur dioxide, *Adv. Environ. Sci. Eng.* **3**, 80–91.

Altwicker, E. R., and Nass, K. K. (1983). Evidence for enhanced mass transfer and synergistic catalysis of aqueous phase sulfur dioxide oxidation by mixtures of manganese and iron. *Atmos. Environ.* **17**, 187–190.

Aubuchon, C. (1976). The rate of iron-catalyzed oxidation of sulfur dioxide by oxygen in water, Ph.D. Thesis, Johns Hopkins University, Baltimore, MD.

Bäckström, H. (1934). Der Kettenmechanisms bei der Autoxydation von Natriumsulfitlosungen, *Z. Phys. Chem.* **25B**, 122–138.

Barrie, L. A., and Georgii, H. W. (1976). An experimental investigation of the absorption of sulphur dioxide by water drops containing heavy metal ions. *Atmos. Environ.* **16**, 743–749.

Betterton, E. A., and Hoffmann, M. R. (1988). Oxidation of aqueous SO_2 by peroxymonosulfate. *J. Phys. Chem.* **92**, 5962–5965.

Brimblecombe P., and Spédding, D. J. (1974a). The catalytic oxidation of micromolar aqueous sulphur dioxide. I. *Atmos. Environ.* **8**, 937–945.

Brimblecombe, P., and Spédding, D. J. (1974b). The reaction order of the metal ion catalyzed oxidation of sulphur dioxide in aqueous solution, *Chemosphere* **1**, 29–32.

Cavasino, F. P. (1968). A temperature jump study of the kinetics of the formation of the monosulfato complex of iron(III), *J. Phys. Chem.* **72**, 1378–1384.

Clarke, A. G., and Radojević, M. (1983). Chloride ion effects on the aqueous oxidation of SO_2, *Atmos. Environ.* **17**, 617–624.

Conklin, M. H., and Hoffmann, M. R. (1988). Metal ion-sulfur(IV) chemistry. 3. Thermodynamics and kinetics of transient iron(III)-sulfur(IV) complexes, *Environ. Sci. Technol.* **22**, 899–907.

Coughanowr, D. R., and Krause, F. E. (1965). The reaction of SO_2 and O_2 in aqueous solutions of MnO_4. *Ind. Eng. Chem. Fundam.* **4**, 61–66.

Erickson, R. E., Yates, L. M., Clark, R. L., and McEwen, D. (1977). The reaction of sulfur dioxide with ozone in water and its possible atmospheric significance. *Atmos. Environ.* **11**, 813–817.

Flynn, C. M. (1984). Hydrolysis of inorganic iron(III) state, *Chem. Rev.* **84**, 31–41.

Freiberg, J. (1975). The mechanism of iron catalyzed oxidation of SO_2 in oxygenated solutions, *Atmos. Environ.* **9**, 661–672.

Fuzzi, S. (1978). Study of iron(III) catalyzed sulphur dioxide oxide oxidation in aqueous solution over a wide range of pH. *Atmos. Environ.* **12**, 1439–1442.

Glasstone, S. (1942). *An Introduction to Electrochemistry*, Van Nostrand, Princeton, NJ.

Höather, R. C., and Goodeve, C. F. (1934). The oxidation of sulphurous acid. III. Catalysis by manganous sulphate. *Trans. Faraday Soc.* **30**, 1149–1156.

Hoffmann, M. R. (1986). On the kinetics and mechanism of oxidation of aquated sulfur dioxide by ozone. *Atmos. Environ.* **20**, 1145–1154.

Hoffmann, M. R., and Calvert, J. G. (1985). *Chemical Transformation Modules for Eulerian Acid Deposition Models. Vol. II: The Aqueous Phase Chemistry* (Report). Atmospheric Sciences Research Laboratory, Office of Research and Development, U.S. Environmental Protection Agency, Research Triangle Park, NC.

Hoffmann, M. R., and Edwards, J. O. (1975). Kinetics of the oxidation of sulfite by hydrogen peroxide in acidic solution. *J. Phys. Chem.* **79**, 2096–2098.

Hoffmann, M. R., and Jacob, D. J. (1984). Kinetics and mechanisms of the catalytic oxidation of dissolved sulfur dioxide in aqueous solution: An application to nighttime fog water chemistry. In J. G. Calvert (Ed.), *SO_2, NO and NO_2 Oxidation Mechanisms: Atmospheric Considerations*, Butterworth, Boston, pp. 101–172.

Hoigne, J., Bader, H., Haag, W. R., and Staehelin, J. (1985). Rate constants of reactions of ozone with organic and inorganic compounds in water. III. Inorganic compounds and radicals. *Water Res.* **19**, 993–1004.

Huie, R. E., and Neta, P. (1987). Rate constants for some oxidations of S(IV) by radicals in aqueous solutions, *Atmos. Environ.* **21**, 1743–1747.

Huss, A., Jr., Lim, P. K., and Eckert, C. A. (1982a). Oxidation of aqueous sulfur dioxide. 1. Homogeneous manganese(II) and iron(II) catalysis at low pH. *J. Phys. Chem.* **86**, 4224–4228.

Huss, A., Jr., Lim, P. K., and Eckert, C. A. (1982b). Oxidation of aqueous sulfur dioxide. 2. High pressure studies and proposed reaction mechanisms. *J. Phys. Chem.* **86**, 4229–4233.

Hwang, H., and Dasgupta, P. K. (1985). Thermodynamics of the hydrogen peroxide-water system. *Environ. Sci. Technol.* **19**, 255–258.

Ibusuki, T., and Barnes, H. M. (1984). Manganese(II) catalyzed oxidation in aqueous solution at environmental concentrations, *Atmos. Environ.* **18**, 145–151.

Ibusiki, T. and Takeuchi, K. (1987). Sulfur dioxide oxidation by oxygen catalyzed by mixtures of manganese(II) and iron(III) in aqueous solutions at environmental reaction conditions. *Atmos. Environ.* **21**, 1555–1560.

Jacob, D. J. (1986). Chemistry of OH in remote clouds and its role in the production of formic acid and peroxymonosulfate. *J. Geophys. Res.* **91**, 9807–9826.

Jaeschke, W. A. (1982). Measurements relating to the heterogeneous SO_2 oxidation in the atmosphere. *Atmos. Trace Const. Proc. Two-Annu. Colloq., 5th, 1981*, pp. 41–52.

Jaeschke, W. A. and Herrmann, G. J. (1986). SO_2-oxidation by hydrogen peroxide in suspended droplets. *Proc. Symp. Acid Rain, Am. Chem. Soc. New York Meet, 1986*, pp. 583–590.

Johnstone, H. F. and Coughanowr, D. R. (1958). Absorption of sulfur dioxide from air. *Ind. Eng. Chem.* **50**, 1169–1172.

Keene, W. C., and Galloway, J. N. (1986). Considerations regarding sources for formic and acetic acids in the troposphere. *J. Geophys. Res.* **91**, 14466–14474.

Keene, W. C., Galloway, J. N., and Holden, J. D., Jr. (1983). Measurement of weak organic acidity in precipitation from remote areas of the world. *J. Geophys. Res.* **88**, 5122–5130.

Khorunzhii, B. I., Mot'ko, S. M., and Denisov, V. V. (1983). Catalytic activity of iron salts in liquid phase oxidation of sulfur dioxide. In I. P. Mukhlenov (Ed.), *Kataliz. i Katalizatory, Leningrad Technol. Inst.*, Leningrad, pp. 12–18.

Kosac-Channing, L. F., and Helz, G. R. (1983). Solubility of ozone in aqueous solutions of 0–0.6 M ionic strength at 5–30 °C *Environ. Sci. Technol.* **17**, 145–149.

Kraft, J., and van Eldik, R. (1989a). Kinetics and mechanism of the iron(III) catalyzed autoxidation of sulfur(IV) oxides in aqueous solution. 1. Formation of transient iron(III)–sulfur(IV) complexes, *Inorg. Chem.* **28**, 2297–2305.

Kraft, J., and van Eldik, R. (1989b). Kinetics and mechanism of the iron(III)-catalyzed autoxidation of sulfur(IV) oxides in aqueous solution. 2. Decomposition of transient iron(III)–sulfur(IV) complexes, *Inorg. Chem.* **28**, 2306–2312.

Kraft, J., and van Eldik R. (1989c). The possible role of iron(III)–sulfur(IV) complexes in the catalyzed autoxidation of sulfur(IV) oxides. A mechanistic study, *Atmos. Environ.* **23**, 2709–2713.

Kunen, S. M., Lazrus, A. L., Kok, G. L., and Heikes, B. G. (1983). Aqueous oxidation of SO_2 by hydrogen peroxide. *J. Geophys. Res.* **88**, 3671–3674.

Latimer, W. M. (1952). *The Oxidation States of the Elements and Their Potentials in Aqueous Solutions.* Prentice-Hall, Englewood Cliffs, NJ, 2nd ed., pp. 354–356.

Lee, Y. J., and Rochelle, G. T. (1987). Oxidative degradation of organic acid conjugated with sulfite oxidation in flue gas desulfurization: Products, kinetics, and mechanism, *Environ. Sci. Technol.* **21**, 266–272.

Lim, P. K., Huss, A., Jr., and Eckert, C. A. (1982). Oxidation of aqueous sulfur dioxide, 3. The effects of chelating agents and phenolic antioxidants, *J. Phys. Chem.* **86**, 4233–4237.

Lind, J. A., and Kok, G. L. (1986). Henry's law determinations for aqueous solutions of hydrogen peroxide, methylhydroperoxide, and peroxyacetic acid. *J. Geophys. Res.* **91**, 7889–7895.

Lind, J. A., Lazrus, A. L., and Kok, G. L. (1987). Aqueous phase oxidation of sulfur(IV) by hydrogen peroxide, methylhydroperoxide, and peroxyacetic acid. *J. Geophys. Res.* **92**, 4171–4177.

Maahs, H. G. (1983a) Measurement of the oxidation rate of sulfur(IV) dioxide by ozone in aqueous solution and their relevance to sulfur dioxide conversion in nonurban tropospheric clouds. *Atmos. Environ.* **17**, 341–345.

Maahs, H. G. (1983b). Kinetics and mechanics of the oxidation of S(IV) by ozone in aqueous solution with particular reference to SO_2 conversion in nonurban tropospheric clouds. *J. Geophys. Res.* **88C**, 10721–10732.

Martin, L. R. (1984). Kinetic studies of sulfite oxidation in aqueous solution. In J. G. Calvert (Ed.), *SO_2, NO and NO_2 Oxidation Mechanisms: Atmospheric Considerations*, Butterworth, Boston, pp. 63–100.

Martin, L. R. (1987). Comment on "On the kinetics and mechanism of oxidation of aquated sulfur dioxide by ozone," *Atmos. Environ.* **21**, 1485.

Martin, L. R., and Damschen, D. E. (1981). Aqueous oxidation of sulfur dioxide by hydrogen peroxide at low pH. *Atmos. Environ.* **15**, 1615–1621.

Martin, L. R., and Good, T. W. (1991). Catalyzed oxidation of sulfur dioxide in solution: The iron-manganese synersism. *Atmos. Environ.* **25A**, 2395–2399.

Martin, L. R., and Hill, M. W. (1987a). The iron catalyzed oxidation of sulfur: Reconciliation of the literature rates, *Atmos. Environ.* **21**, 1487–1490.

Martin, L. R., and Hill, M. W. (1987b). Optical measurement of aqueous kinetics at micromolar concentrations, *J. Phys. E* **20**, 1383–1387.

Martin, L. R., and Hill, M. W. (1987c). The effect of ionic strength on the manganese catalyzed oxidation of sulfur(IV), *Atmos. Environ.* **21**, 2267–2270.

Martin, L. R., and Hill, M. W., Tai, A. F., and Good, T. W. (1991). The iron catalyzed oxidation of sulfur(IV) in aqueous solution: Differing effects of organics at high and low pH, *J. Geophys. Res.* **96**, 3085–3097.

Matteson, M. J., Stoeber, W., and Luther, H. (1969). Kinetics of the oxidation of sulfur dioxide by aerosols of manganese sulfate, *Ind. Eng. Chem. Fundam.* **8**, 677–687.

McArdle, J. V., and Hoffmann, M. R. (1983). Kinetics and mechanism of the oxidation of aquated sulfur dioxide by hydrogen peroxide at low pH. *J. Phys. Chem.* **87**, 5425–5429.

Munger, J. W., and Einsenreich, S. J. (1983). Continental-scale variations in precipitation chemistry, *Environ. Sci. Technol.* **17**, 32A–42A.

Munger, J. W., Collet, J., Jr., Daube, B. C., Jr., and Hoffmann, M. R. (1989). Carboxylic acids and carbonyl compounds in Southern California clouds and fogs. *Tellus* **41B**, 230–242.

Nahir, T. M., and Dawson, G. A. (1987). Oxidation of sulfur dioxide by ozone in highly dispersed water droplets. *J. Atmos. Chem.* **5**, 373–383.

Neta, P., Huie, R. E., and Ross, A. B. (1988). Rate constants for reactions of inorganic radicals in aqueous solution. *J. Phys. Chem, Ref. Data* **17**, 1027–1284.

Neytzell-de-Wilde, F. G., and Taverner, L. (1958). Experiments relating to the possible production of an oxidising acid leach liquor by auto-oxidation for the extraction of uranium. *Proc. U. N. Int. Conf. Peaceful Uses of At. Energy, 2nd, 1958,* Vol. 2, pp. 303–317.

Overton, J. H., Jr. (1985). Validation of the Hoffmann and Edward's S(IV)–H_2O_2 mechanism. *Atmos. Environ.* **19**, 687–690.

Penkett, S. A., Jones, B. M. R., and Eggleton, A. E. J. (1979). A study of SO_2 oxidation in stored rainwater samples, *Atmos. Environ.* **13**, 139–147.

Radojević, M. (1984). On the discrepancy between reported studies of the uncatalysed aqueous oxidation of SO_2 by O_2, *Environ. Technol. Lett.* **5**, 549–566.

Scatchard, G., and Ticknor, L. B. (1952), Vapor–liquid equilibrium. VIII. Hydrogen peroxide–water mixtures. *J. Am. Chem. Soc.* **63**, 3715–3720.

Smith, R. M., and Martell, A. E. (1976). *Critical Stability Constants, Vol. 4: Inorganic Complexes,* Plenum, New York.

Tang, Y., Thorn, R. P., Mauldin, R. L., III, and Wine, P. H. (1988). Kinetics and spectroscopy of the SO_4^- radical in aqueous solution, *J. Photochem. Photobiol.* **44**, 243–258.

Ulrich, R. K., Rochelle, G. T., and Prada, R. E. (1986). Enhanced oxygen absorption into bisulfite solutions containing transitions metal ion catalysts. *Chem. Eng. Sci.* **41**, 2183–2192.

Wine, P. H., Tang, Y., Thorn, R. P., Wells, J. R., and Davis, D. D. (1989). Kinetics of aqueous phase reactions of the SO_4^- radical with potential importance in cloud chemistry, *J. Geophys. Res.* **94**(D1), 1085–1094.

Yoshizumi, K., Aoki, K., Nouchi, I., Okita, T., Kobayashi, T., Kamakura, S., and Tajima, M. (1984). Measurements of the concentration in rainwater and of the Henry's law constant of hydrogen peroxide. *Atmos. Environ.* **18**, 395–401.

9

OXIDATIVE MECHANISMS OF PHOTOTOXICITY

Richard A. Larson and Karen A. Marley

University of Illinois, Institute for Environmental Studies, Urbana, Illinois 61801

Environmental Oxidants, Edited by Jerome O. Nriagu and Milagros S. Simmons.
ISBN 0–471–57928–9 © 1994 John Wiley & Sons, Inc.

1.   BACKGROUND

1.1.   Introduction

The profound effects of light on cells have been known for millennia. Photobiologists and other scientists, however, have always directed the majority of their attention to the elucidation of the roles of light in "normal" metabolic and physiological processes—photosynthesis, phototaxis, vision, bioluminescence, etc. Although it has been recognized for many years that very short-wave UV radiation is lethal, the effects of longer wavelengths on the mechanisms of action of toxic compounds have not attracted a great deal of attention. Historically, the phenomenon of "photodynamic toxicity" dates back to the early years of the twentieth century, when it was demonstrated that certain dyes become cytotoxic in the presence of light and oxygen. However, in general, the effects of light as a potentially important variable in environmental and biochemical toxicology are seldom taken into account.

Life originated in the presence of high-energy UV radiation, and even the earliest organisms must have developed methods for absorbing, reflecting, or otherwise minimizing its hazardous consequences. The earliest life forms on Earth probably also lived in oxygen-free environments. However, the development of photosynthetic, oxygen-discharging metabolism forced organisms to evolve even more elaborate means of protection against reactive forms of oxygen. Some of these reactive oxygen species are now known to be produced, at least in part, by light-dependent mechanisms. Therefore, the combination of light and oxygen is a potent vehicle for potentially hazardous effects to living things.

1.2.   Sunlight

In addition to visible light and infrared (IR) radiation (heat), the sun emits radiation in the 290- to 400-nm (ultraviolet; UV) region that reaches the earth's surface. The characteristics of sunlight have been reviewed by Finlayson-Pitts and Pitts (1986). Approximately 4% of the total energy contained in sunlight occurs in the UV region. This radiation has a high potential for inducing chemical reactions due to its elevated energy content relative to visible and IR radiation. UV radiation has the potential to cause direct damage to biochemically important molecules that absorb it. Of course, as the first law of photochemistry states, light must be absorbed by a system in order for chemical reactions to have a chance to occur. As a first approximation, then, molecules must have absorption above 290 nm (the shortest wavelength found in sunlight at the earth's surface) in order to be affected by solar UV. Among the common biochemical constituents of cells, most lipids have little or no absorption at these wavelengths; DNA also absorbs very weakly; proteins show some absorption due to their content of tyrosine and (especially) tryptophan; and flavins, metalloporphyrins, and other colored molecules absorb significantly. Consequently, in discussions of phototoxicity, the cellular constituents are the "targets" and are generally considered unreactive toward light, but in the presence of light-absorbing molecules ("photosensitizers"), they can be damaged or destroyed.

Distinctions are sometimes made between the long-wavelength "UV-A" (400–320 nm) and the more energetic, short-wavelength "UV-B" (320–290 nm) radiation that is more strongly absorbed by many pollutants and biomolecules. For example, UV-B wavelengths are more important in sunburn and vitamin D biosynthesis, whereas UV-A is more significant in melanin formation (tanning). In both spectral regions, however, the fundamental photochemical processes are similar, although different receptor molecules (chromophores) may be involved. Current calculations predict that short-wave solar ultraviolet (UV-B) light intensities at the earth's surface will increase as stratospheric ozone concentrations diminish due to current and future inputs of reactive chemical species such as chlorofluorocarbons and oxides of nitrogen. Many laboratory experiments using selective light sources such as mercury arcs suggest biological damage due to increases in UV-B is likely to occur in many types of organisms.

1.3. Cellular Targets

A single definition for phototoxicity has been elusive. "Toxic" implies a biological mechanism of action that presumably requires a measurable impairment of some cellular fitness, function, or metabolism, but different specialists have different criteria for measuring these injurious responses. Dermatologists, for example, concerned with human tissue responses, narrowly define a phototoxic agent as a chemical compound or mixture that produces accelerated sunburn-like symptoms in a single dose, with hyperpigmentation upon healing. By this definition, furocoumarins are the archetypal phototoxic agents. Other compounds that cause less acute symptoms of irritation, or require repeated doses to be effective, are referred to as contact photoallergens. Chemical ecologists take a different perspective, and usually assume that phototoxic compounds have a beneficial, defensive role for the organisms, usually plants, that biosynthesize them, but a detrimental effect on the antagonistic organisms that are exposed to them.

Biological procedures for identifying phototoxic compounds have inherent disadvantages in that the observed end point, for example, the death of a bacterial cell, may not have a well-defined relationship to a particular photochemical event. The use of chemical models has the advantage of permitting more carefully controlled environments in which photochemical mechanisms can be sorted out in a less ambiguous manner, although it is recognized that the use of overly simplistic models may lead to erroneous hypotheses. In this paper, therefore, our perspective will be broad, and basically biochemical; we will refer to phototoxic compounds as those that react with biologically important target molecules in the presence of light to induce changes that are potentially harmful to the cell. The underlying assumption of this definition is that phototoxic molecules have intracellular mechanisms of action that can be discovered in in vivo experiments. In other words, a phototoxic compound could in principle be recognized in a preliminary chemical screening test, without first probing the response of a living organism (although such a test would, of course, be necessary at some stage). This sense of the definition is analogous to using short-term tests with microorganisms to identify suspected animal carcinogens.

Three mechanisms for phototoxicity need to be considered. These possibilities are modified from a similar treatment by Ito (1978).

(i) The first possibility is that the phototoxic agent remains outside the cell, generating reactive species in solution (or even in the gas phase?) that diffuse into the cells of the target organism and react to induce some toxic lesion. An example of this mechanism might be a molecule such as curcumin (see Section 2.3.1) that produces H_2O_2 on illumination.

(ii) The second possibility is that the agent binds or becomes localized (by hydrophobic or coulombic interactions) at the cell surface or membrane, absorbs light there, and transfers either energy or an electron, hydrogen atom, etc., from or to a target molecule, which is damaged. Because of these proximity effects, the likelihood of reactive transient species actually being able to damage biomolecules, rather than decomposing in solution, would be greatly increased. Anionic porphyrins (Section 2.5.1) exemplify the sort of phototoxic compound that might be strongly attracted to membrane lipids bearing positive charge by virtue of their lecithin or sphingomyelin content. The interiors of membranes are, of course, much more hydrophobic than their surfaces, so phototoxic molecules of different polarities could still exert their effects at the membrane level.

(iii) The third possibility is that the agent penetrates into the interior of the cell and becomes associated with an intracellular target, possibly a protein (inducing enzymatic damage) or the nucleus (inducing genetic damage). Cationic porphyrins (Section 2.5.1), which bind strongly to the polyanionic macromolecule DNA, would be good examples of this type of phototoxic agent.

For a given phototoxic molecule, one or more of these possibilities may be important. When an external compound enters into the dynamic biological milieu of a living cell, pharmacokinetic processes will operate to govern the rates of transfer of the compound into and through the cell, and to fix the location and concentration of its molecules in the cellular environment. If, in the course of these processes, the phototoxic compound is able to absorb light, the possibility of a damaging cellular interaction will exist. In addition, concentrations of oxygen vary greatly among different cellular regions, making photoreactions involving oxygen species more or less probable. It should also be kept in mind that the photochemical behavior of a compound can be profoundly affected by its surroundings. Not only its absorption spectrum but the fates of its excited states may be drastically altered as it passes from one environment, such as an aqueous solution, into another, such as the hydrophobic interior of a DNA helix. A difficult extrapolation process needs to be made when one's perspective changes from observing simple photochemical measurements in pure solvents to considering the profound inhomogeneities of the cell. The challenge for the photochemist and photobiologist is to arrive at sensible and comprehensible mechanisms that are capable of being tested experimentally and that permit explanatory and predictive theories to be produced.

1.4. The Formation of Free Radicals by Light Absorption

Photochemically damaging events are initiated by the uptake of the electronic energy of a photon by a light-absorbing molecule. In the UV region of the electromagnetic spectrum, the energy of such photons is sufficient to break covalent bonds, although it is unusual for their energy to be absorbed that efficiently. Typical UV-absorbing groups in organic molecules commonly found in the environment are combinations of $C=C$, $C=O$, $C=N$, and $N=N$ multiple bonds. For example, an isolated $C=O$ group weakly absorbs light around 280 nm, just outside the solar region; but the absorbance maximum of a similar group conjugated with other double bonds is shifted to longer wavelengths. Thus, acetophenone ($PhCOCH_3$) absorbs at 320 nm. The absorption of a photon by a molecule results in its excitation to a species with a new electronic configuration and often a quite different geometry. The chemical reactivity of excited species is almost always quite different from that of the molecule in the ground state.

A photoexcited carbonyl group has some of the characteristics of a diradical with a structure that can be approximated as $R_2C^{\cdot}—O^{\cdot}$; actually this representation is an oversimplification of the true electron configuration, which is technically a pair of adjacent antibonding orbitals. However, the simplified structure is helpful in illustrating the kinds of reactions that excited carbonyl compounds can undergo.

Photoionization is another common fate of electron-rich excited states (electron donors). This consists of the emission of an electron from the excited state, either to the medium (especially water) or to an acceptor molecule, with the resultant formation of a radical cation and a solvated electron or radical anion, respectively (Joschek and Grossweiner, 1966). These reactions can occur either unimolecularly (that is, the donor absorbs light and emits an electron) or bimolecularly, where the donor and acceptor form a charge-transfer complex that is the light-absorbing species, and electron transfer occurs within the complex (Mattay, 1987). These reactions are very often strongly solvent-dependent and do not occur at all in nonpolar solvents; water is the medium in which most photoionization processes have been described. To a large extent, this is because water is an efficient medium for the separation of ionic particles such as radical ions, and thus inhibits the back-transfer process that can take place within an ion pair.

Many dyes and other compounds, in water, photoionize (emit an electron) efficiently. These electrons become solvated by water to form a relatively stable species, the hydrated electron (e_{aq}^-), which has been extensively studied by pulse radiolysis techniques. The species is a very potent reducing agent, in fact the strongest known reductant; it can be intercepted by a variety of dissolved substrates, including molecular oxygen. In the presence of oxygen concentrations typical of surface waters and cells (10^{-5}–10^{-4} M), superoxide is formed at a very high rate (see Section 1.5). In order for organic molecules to compete for e_{aq}^-, their concentrations must be considerably higher than that of oxygen; most common bimolecules have rate constants of about three orders of magnitude slower, although those for some quinones, unsaturated carboxylic acids, and nucleic acid derivatives such as adenine monophosphate are kinetically within an order of magnitude of oxygen's. Electron transfer to substrates other than oxygen, however, can be greatly enhanced by preliminary complex formation (Slifkin, 1971). Certain photochemically reactive dyes that have been shown to cause strand breakage in DNA probably act by this mechanism. In addition, many photochemically active substances such as riboflavin, a relatively electron-poor molecule, effectively form charge-transfer intermediates (in the ground state) with π-acids (electron-rich species) such as indoles, anilines, and phenols (Heelis, 1983). When the complex absorbs light, an electron can be transferred from the donor molecule to the acceptor, and if the partner species separate, two new free radicals (a radical anion and cation) are released into solution (Eriksen et al., 1977). Either or both of these radicals may react with molecular oxygen to form peroxy radicals.

Electron transfer can profoundly change the reactivity of a molecule. Consider the case of an electron-rich species such as a phenol absorbing light and emitting an electron that is taken up by an electron-poor acceptor. In the process, the phenol is converted to a radical cation that is, by definition, electron-deficient; conversely, the acceptor becomes electron-rich. Therefore, the electron configuration of the ground-state species is almost entirely transformed after excitation, leading to greatly dissimilar chemical characteristics. Radical cations are often attacked by nucleophiles such as superoxide, water, alcohols, or carboxylates (Mattes and Farid, 1984); radical anions may transfer an electron to oxygen to produce superoxide or accept a proton from the radical cation.

Electronically excited states are usually better electron donors or acceptors than their corresponding ground states. Accordingly, even when there is little or no tendency for ground-state molecules to interact, complexes may also form efficiently in the excited state (Mattes and Farid, 1984). Such complexes may be between either two identical species (excimers) or two different molecules (exciplexes). The fate of these complexes may be conversion to other excited states (energy transfer) or radical ions (electron transfer), or covalent adduct formation (Mattay, 1987).

1.5. Hydrogen Peroxide, Superoxide, and Hydroxyl Radical

From the free radicals that are produced by light absorption, further reactions can be observed, especially in the presence of oxygen. Superoxide, hydrogen peroxide, and hydroxyl radical form by reaction sequences beginning with photoejected electrons and oxygen. These transient forms are capable of damaging biomolecules critical to physiological function.

Molecular oxygen is an efficient scavenger of electrons (the rate constant in water is 2×10^{10} M^{-1} sec^{-1}, practically diffusion-controlled)(Buxton et al., 1988). The product of the reaction, superoxide anion radical, is usually depicted as O_2^- or $\cdot O_2^-$. Its conjugate acid, the hydroperoxyl radical (HOO$\cdot$), forms in weakly acidic solution with a pK_a of 4.8. In pure water at pH 9 or below, the principal fate of $\cdot O_2^-$ is a self-redox disproportionation, or "dismutation":

$$HOO\cdot + \cdot O_2^- \xrightarrow{H^+} H_2O_2 + O_2$$

The metalloenzyme SOD, found in virtually all aerobic organisms, catalyzes the dismutation reaction very efficiently. This has led many biologists to believe that $\cdot O_2^-$ or its reaction products are potentially dangerous to the cell (see, e.g., Fridovich, 1981). Neither $\cdot O_2^-$ nor HOO$\cdot$ is a particularly reactive species, however, except toward a few cellular constituents like quinones and ascorbate (von Sonntag, 1987), and it is usually assumed that the principal cytotoxic substances derived from them are hydrogen peroxide and the hydroxyl radical, HO$\cdot$ (or $\cdot$OH). The nonphotochemical ("dark") redox chemistry of transition metals has been shown to include the so-called Fenton reaction, a catalyzed decomposition of H_2O_2:

$$M^{n+1} + H_2O_2 \longrightarrow M^{n+1} + HO^- + HO\cdot$$

The hydroxyl radicals generated in this reaction with iron(II), copper(I), and other metal ions are extremely reactive and unselective oxidants toward almost all organic (and some inorganic) compounds. With electron-rich compounds like phenols and anilines, HO$\cdot$ reacts at virtually diffusion-controlled rates, but even with organic compounds not usually thought of as susceptible to oxidation, such as n-alkanes, it reacts rapidly at rates only about two orders of magnitude slower. (Hydroxyl radicals, unlike most other radical species, are unreactive with molecular oxygen, but they do react at rapid rates with many reduced metal cations and with anions such as bicarbonate.) Accordingly, it is expected that HO$\cdot$, if formed within a cell or at a cell surface, would react within a very few molecular diameters of the site of its generation, that is, it would have a small "reaction volume" (a property of reactive transients that depends on their diffusivity and lifetime).

Hydroxyl radicals can also be formed in a fast reaction ($k = 9.5 \times 10^9 \, M^{-1} \, sec^{-1}$) between hydrated electrons and hydrogen peroxide (Greenstock and Wiebe, 1981):

$$e_{aq}^- + HOOH \longrightarrow HO^- + \cdot OH$$

This reaction might take place in anoxic conditions if [HOOH] was greater than [O_2]. If an organic peroxide or hydroperoxide were substituted for H_2O_2 as the electron acceptor, an alkoxy radical, $RO\cdot$, would be produced. These radicals, although powerful oxidants, are not as reactive or unselective as $\cdot OH$.

Hydroxyl radicals have two principal modes of reaction with organic compounds, hydrogen atom (or electron) abstraction and (somewhat faster) addition to double bonds. The two processes may be illustrated by considering the products of reaction of $HO\cdot$ with toluene, which include o-, m-, and p-cresols, dimethylbiphenyls, and bibenzyl, as well as products of further oxidation (Hoshino et al., 1978).

1.6. Reactions of Free Radicals with Molecular Oxygen

Carbon-centered free radicals, $R_3C\cdot$, typically react with molecular oxygen at extremely fast, almost diffusion-controlled, rates. Because of the diradical character of oxygen molecules, there is little or no activation energy (spin barrier) for such an odd-odd electron pairing reaction. The reaction products, peroxy radicals, are often abbreviated $ROO\cdot$, and are important intermediates in many biological and abiotic processes such as lipid peroxidation and the weathering of synthetic polymers and petroleum products. Peroxy radicals have many possible routes of reaction; when present in high concentration, such as would be found in a thin film of lipids or hydrocarbon molecules, the step that is often most important is a hydrogen abstraction:

$$ROO\cdot + R'H \longrightarrow ROOH + R'\cdot$$

The products ROOH are called hydroperoxides. In this (usually slow) reaction, R'H, the hydrogen atom donor, may represent a molecule of the solvent or some other compound with a relatively active hydrogen atom. The formation of new radicals in this step provides other species able to continue to react with O_2.

Hydroperoxides are normally unstable to thermal or photochemical decomposition and often decompose to provide additional radical species. One such example is UV-induced homolysis of the O—O bond:

$$ROOH \xrightarrow{hv} RO\cdot + \cdot OH$$

Although the absorption of solar UV by hydroperoxides is usually quite small, the quantum yield of homolysis for the photons that are absorbed is usually quite high. Reactions of this type are unusually favorable for the occurrence of chain processes, since they increase the number of reactive free radicals that are capable of initiating chains of their own.

Addition reactions of peroxy radicals with olefins have been reported (Mayo, 1958; Hamberg and Gotthammar, 1973). This sort of reaction is probably important in peroxidation of unsaturated lipids. Among the products are epoxides, possibly formed by elimination of alkoxy radicals:

$$\text{ROO} \cdot + \text{\Large$\diagdown$}\text{C}=\text{C}\text{\Large$\diagdown$} \longrightarrow \text{ROO}-\text{C}-\text{C} \cdot \longrightarrow \text{RO} \cdot + \overset{\text{O}}{\overset{\triangle}{\text{\Large$\diagdown$}\text{C}-\text{C}\text{\Large$\diagdown$}}}$$

Unimolecular or bimolecular elimination reactions of peroxy radicals can be extremely fast processes. In one commonly observed unimolecular case, the radical possesses structural features that permit efficient loss of $\text{HOO} \cdot$ and the concomitant oxidation of the radical's precursor. For example, the peroxy radical derived from ethanol eliminates $\text{HOO} \cdot$ and is converted to acetaldehyde (Bothe et al., 1983):

$$\overset{\displaystyle \overset{\text{OO} \cdot}{|}}{\text{H}_3\text{C}\dot{\text{C}}\text{OH} \longrightarrow \text{H}_3\text{CCHOH}} \longrightarrow \text{HOO} \cdot + \text{H}_3\text{CCHO}$$

The elimination need not take place in an aliphatic structure; peroxy radicals derived from subsequent $\text{HO} \cdot$ and O_2 addition to aromatic rings undergo analogous eliminations, whereby, for example, benzene can be converted to phenol.

Bimolecular decay of peroxy radicals is also commonly observed. The intermediate in this reaction is generally accepted to be a tetroxide, that is, a species with four sequential oxygen atoms (Russell, 1957). These intermediates break down by several routes to afford oxidized products and O_2 or H_2O_2, for example:

$$2\text{R}_2\text{CHOO} \cdot \longrightarrow [\text{R}_2\text{CHOO}-\text{OOCHR}_2] \longrightarrow 2\text{R}_2\text{C}=\text{O} + \text{H}_2\text{O}_2$$

or $$\longrightarrow 2\text{R}_2\text{CHOH} + \text{O}_2$$

1.7. Energy Transfer—Photosensitizers

Another mechanism for biological interaction of light with molecules is an indirect effect that occurs when sunlight is absorbed by molecules and converts them to excited states capable of transferring some of their energy to ground-state acceptors. Traditionally, this mechanism has been discussed in terms of energy transfer from a "sensitizing" donor to ground-state oxygen, resulting in the ground state of the donor and an excited state of oxygen referred to as singlet oxygen ($^1O_2$), although, in principle, the energy could also be transferred to another acceptor such as a biomolecule. Photoactivated molecules of many types can interact with ground-state oxygen to form excited $^1O_2$ (Foote and Wexler, 1964; Larson and Berenbaum, 1988). These substances are referred to as photosensitizers. Among the naturally occurring substances that have been reported to act as photosensitizing agents under certain conditions are porphyrin derivatives, bile pigments, chlorophyll degradation products, flavins, some quinones, the humic materials from soils and colored waters, and tetracycline antibiotics. Environmental pollutants or synthetic substances such as polycyclic aromatic hydrocarbons (PAHs), synthetic dyes and drugs, and some phenols also act as sensitizers. Berenbaum and Larson (1988) demonstrated that leaves of *Zanthoxylum americanum* (Rutaceae) generated singlet oxygen at their surface in the presence of simulated sunlight, perhaps at levels sufficiently high to induce damage in cells of organisms on or near the leaf surface; nonrutaceous plants lacking photosensitizers did not have this ability.

Singlet oxygen, with its fully spin-paired electrons, has a different electron configuration than ground-state oxygen (a diradical), and as a consequence can take part more effectively in reactions with some classes of biomolecules (Foote, 1981; Straight and Spikes, 1985). It oxidizes protein amino acids, membrane lipids, and DNA bases at widely varying rates (Table 1). Only those protein amino acids containing heterocyclic rings or sulfur atoms appear to be attacked by $^1O_2$ at significant rates. In DNA and RNA, guanine has the lowest oxidation potential of the bases and is most reactive toward $^1O_2$, as it is with many other oxidizing agents (Cadet and Teoule, 1978). Unsaturated lipids react rather slowly

Table 1   Second-Order Rate Constants for the Reaction of $^1O_2$ with Selected Biomolecules

	Compound	Rate constant($M^{-1} sec^{-1}$)
Amino acids	Histidine	1.3×10^8
	Tyrosine	2.7×10^7
	Tryptophan	2.5×10^8
	Methionine	2.2×10^7
	Alanine	2.0×10^6
Nucleic acid bases	Guanine	$<1 \times 10^6$
	Thymine	$<1 \times 10^6$
Lipids	Oleic acid	1.3×10^5
	Linoleic acid	2.2×10^5
	Linolenic acid	2.9×10^5
	Cholesterol	6.6×10^4
	β-Carotene	1.3×10^{10}

Source: Wilkinson and Brummer (1981).

with $^1O_2$, although this mechanism is often invoked to explain photochemically initiated lipid peroxidation.

In aqueous medium, $^1O_2$ is rapidly quenched back to the ground state with a lifetime of only about 4 μsec (Rodgers and Snowden, 1982), meaning that only highly reactive acceptors would be kinetically able to compete. The mean path length for $^1O_2$ molecules in aqueous environments would be on the order of 100 nm (Lindig and Rodgers, 1979). In organic media, however, the lifetime of $^1O_2$ is greater by one to two orders of magnitude. Dahl (1993) has argued that the low concentration and limited lifetime of $^1O_2$, together with the different kinetics observed in reactions initiated by "pure" $^1O_2$ generated by illuminating dry sensitizer versus dissolved sensitizer, imply alternative or multiple mechanisms of cytotoxicity.

2. PHOTOTOXICITY OF PARTICULAR MOLECULES

2.1. Coumarins and Derivatives

2.1.1. Furocoumarins

Furocoumarins (also known as psoralens or furanocoumarins) are the best known and most studied class of compounds associated with phototoxicity. Psoralen (7H-furo[3,2-g][1]benzopyran-7-one, **1**), a constituent of *Psoralea corylifolia* and many other plants, is the parent compound; at least 119 known naturally occurring furocoumarins and 104 dihydrofurocoumarins have been reported (Murray et al., 1982). The compounds may assume a "linear" form in which the three rings are arranged in parallel, as in psoralen, or take the "angular" isomeric form typified by angelicin (**2**), where two ring junctions form an angle. In general, these compounds have low water solubility (usually less than 10^{-4} M) and probably tend to partition quite strongly into hydrophobic regions of the cell.

1   **2**

The absorption spectra of both linear and angular furocoumarins normally show three absorption maxima of decreasing intensity at about 220 ($\varepsilon \simeq 20,000$), 245 ($\varepsilon \simeq 12,000$), and 305 ($\varepsilon \simeq 6000$) nm. In addition, a small broadening or shoulder near 330 nm ($\varepsilon \simeq 4000$) is often observed. In alkaline solution, opening of the lactone ring of the compound results in considerable reduction in the intensity of the 305-nm band and appearance of a new maximum around 355 nm. The intensity of absorption in the UV-A region is of some importance because of the use of the long-wave mercury line at 365 nm in photochemotherapy (especially for skin diseases such as psoriasis) using these compounds. The precise nature of the photochemical process that occurs at this wavelength is not entirely certain, although most authors refer to "photobinding" to DNA, generally assuming a [2 + 2] cycloaddition to bases such as thymine, a reaction that occurs in vitro. At any rate, linear furocoumarins absorb quite weakly at 365 nm relative to shorter wavelengths; molar extinction coefficients from only about 300 (Lai et al., 1982) to 1000 (Potapenko *et al.,* 1984) have been reported. Substituted angular furocoumarins have extinction coefficients ranging from 20 to about 700 at this wavelength (Craw *et al.,* 1984a). It is, of course, possible that chemotherapeutically useful complexes between the psoralen derivative and some target molecule might be formed that would absorb more strongly in the 365-nm region, but no evidence exists to support this hypothesis.

The lowest excited singlet state of furocoumarins is generally accepted to be a $\pi \rightarrow \pi^*$ transition, probably including the olefinic π-electrons from the lactone ring, since this double bond often takes part in [2 + 2] cycloaddition reactions with acceptor molecules (Mantulin and Song, 1973; Song and Tapley, 1979; Lai et al., 1982). Like many heterocyclic compounds, however, they also possess a low-energy $n \rightarrow \pi^*$ state in close proximity to the $\pi \rightarrow \pi^*$ transition, which theoretically should tend to enhance fluorescence in nonpolar solvents relative to intersystem $S_1 \rightarrow T_1$ crossing (population of the chemically reactive triplet state). Experimentally, it is found that the photochemical and photophysical properties of these compounds are greatly affected by molecular environmental parameters such as changes in solvent or proximity of potential micelle-forming or complexing agents. For example, the quantum yield of psoralen fluorescence is about 10 to 20 times greater in polar solvents such as trifluoroethanol or methanol than in 3-methylpentane or CCl_4 (Lai et al., 1982; Krasnovsky et al., 1986). It is not certain what biological consequences, if any, may result from changes in the polarity of the cellular environment furocoumarins happen to occupy.

In addition to psoralen, the primary furocoumarins associated with phototoxicity are 5-methoxypsoralen (5-MOP or bergapten) and 8-methoxypsoralen (8-MOP or xanthotoxin), both found in numerous plants. A large number of structurally related synthetic analogues have also been described as probes in elucidating the mechanism of action and/or as alternatives to the medical use of these compounds [such as pyrrolocoumarins (Rodighiero et al., 1987), pyridopsoralens (Dall'Acqua, 1988), and benzofurocoumarins (Palumbo et al., 1986)].

Furocoumarins are most commonly found in the plant families Rutaceae, Apiaceae, and Umbelliferae (Epstein, 1991). Skin responses to plants containing high amounts of these compounds (such as the wild parsnip, *Pastinaca sativa*) are redness, swelling, and slight itching, sometimes described as similar to an acute sunburn. Onset of symptoms begins within 12–36 hr after contact only in areas of contact (not systemic), and healing occurs with hyperpigmentation (tanning). Although not always easily ascertained from casual contact with plant sources, the magnitude of the response is dependent on the amount of furocoumarin that is present. Controlled studies have shown that activation by UV light is also required (Pathak and Fellman, 1960).

Most phototoxicity studies of furocoumarins to date have emphasized their nonoxidative, non-oxygen-requiring photochemical reactions. Almost all of these reactions appear to occur with formation of a cyclobutane ring, that is, a [2 + 2] cycloaddition between the double bond of the lactone ring or the furan ring with a double bond in the target molecule. In a few cases, though, coupling products between the furocoumarin and the target molecule involving the formation of a new single bond have been described, such as the photoaddition products between 8-MOP and 2′-deoxyadenosine (3, Cadet and Vigny, 1990). Covalent cyclobutane-type adducts have been identified in the photoreactions of furocoumarins with olefins (Shim and Kim, 1983; Takano et al., 1984; Otsuki, 1987), pyrimidines (Musajo and Rodighiero, 1972), and DNA (Straub et al., 1981). Studies with monochromatic radiation have demonstrated that DNA base–furocoumarin reactions occur at 365 nm, but use of shorter wavelengths, such as 313 nm, has been reported to result in photoreversal of the cycloaddition, with the formation of the starting materials (Cadet and Vigny, 1990). Many of the cycloaddition reactions are inhibited by oxygen; however, fatty acids (oleic, linoleic, linolenic, and arachidonic acids) also form cycloaddition products with 8-MOP and 4, 5′, 8-trimethylpsoralen (TMP) in moderately dilute solutions with or without oxygen present; the quantum yield was as high as 0.003 for TMP plus oleic acid (Specht et al., 1988).

3

Cycloaddition reactions usually have been observed in organic solvents, in the solid state or in frozen aqueous solution. Quantum yields as high as 0.08 have been reported for the photoaddition of psoralen to native DNA under these conditions (Hearst, 1981). Although some photocycloadditions between DNA bases and psoralens have been observed in dilute aqueous solution, the quantum yield was extremely low, 0.00018 for 8-MOP and 0.000047 for 5-MOP (Arnaud et al., 1981). Preliminary binding or "intercalation" of the furocoumarin with DNA before photoaddition (Musajo and Rodighiero, 1972), exciplex formation

(Song, 1984), or cross-linking reactions of cellular targets by furocoumarins, especially with DNA, have been suggested to occur; however, there is evidence that other mechanisms may also be involved. Recent work has shown that furocoumarins can cleave DNA, presumably by free-radical mechanisms (Kagan et al., 1992; Shim and Yun, 1992; Oroskar et al., 1993).

It is also clear that psoralens are capable of generating activated oxygen species such as superoxide and singlet oxygen, making oxidative modes of phototoxicity possible (Poppe and Grossweiner, 1975; Cannistraro and van de Vorst, 1977; Joshi and Pathak, 1983). For example, the enzyme lysozyme was inactivated by 8-MOP in the presence of oxygen (Poppe and Grossweiner, 1975). Furocoumarin-protein photochemistry has been well summarized in recent review articles (Midden, 1988; Potapenko, 1991). Oxidative modifications that have been described include photopolymerization, formation of psoralen–protein and protein–protein linkages, and photodestruction of particular amino acids. Hydrophobic proteins and lipoprotein complexes are apparently extremely susceptible to psoralen-initiated photochemical damage, suggesting that preliminary binding or complex formation between the psoralen and relatively nonpolar regions of the protein may be an important initial step in the process (Sa E Melo et al., 1982). Damage to membrane constituents by psoralens in the presence of oxygen has also been attributed, at least in part, to oxidative reactions such as lipid peroxidation (Blan and Grossweiner, 1987; Dall'Acqua and Martelli, 1991), which may result in disturbance of membrane morphology or function, increases in permeability, or outright rupture of organelles or cells.

The oxygen dependence of furocoumarin-sensitized reactions is not always clear, and some of these reactions have at times been described as "non-photodynamic" (Musajo and Rodighiero, 1972). Various strains of bacteria sensitive to irradiation by 254-nm light (in the absence of furocoumarins) also seem to be particularly sensitive to inactivation by furocoumarins and UV-A in an anoxic environment (Oginsky et al., 1959; Tuveson et al., 1986). However, many studies (see Potapenko, 1991, for review) have been able to identify oxygen-enhanced phototoxicity, and usually conclude that singlet oxygen is the primary reactive oxygen species (Grossweiner, 1984). The accelerating effect of D_2O, protection under nitrogen or in presence of $^1O_2$ acceptors, such as DABCO, has usually been cited as evidence for the involvement of $^1O_2$. However, these additives are not always specific for $^1O_2$, and other inhibitors normally thought of as radical traps (SOD and BHT, for example) have also been claimed to protect against psoralen photodamage. These issues have been discussed by Potapenko (1991) and others.

Free-radical pathways for phototoxicity are favored by some authors, who have postulated that psoralens photoionize or otherwise form radicals, derived from either the psoralen or another molecule, that supposedly react with membrane constituents or other biomolecules to form covalently bound products (Meffert et al., 1982; Midden, 1988). Furocoumarin triplets have been shown to be efficient acceptors of electrons from donors such as thymine, tryptophan, 3,4-dihydroxyphenylalanine (DOPA), and various solvents (Midden, 1988;

Potapenko, 1991). On the other hand, cation radicals of the compounds have also been postulated as intermediates under some conditions, especially in the presence of oxygen as an electron acceptor (Cannistraro and van de Vorst, 1977; Decuyper et al., 1983). The formation of superoxide by furocoumarin photolysis has been reported by several authors, usually on the basis of a supposedly specific colored product from nitroblue tetrazolium (Joshi and Pathak, 1983) or because furocoumarin toxicity was inhibited by SOD. Angular furocoumarins were said to produce more $\cdot O_2^-$ than linear types (Pathak and Joshi, 1984). Both the radical cation and anion derived from 8-MOP and related compounds have been directly observed in deoxygenated water by laser flash photolysis (Sloper et al., 1979) and pulse radiolysis studies (Bensasson et al., 1984; Solar and Quint, 1992).

Finally, both 8-MOP and the angular furocoumarin angelicin (2) have been reported to form oxygenated intermediates, perhaps peroxides, upon illumination that are capable of dark reactions with membrane lipids to form cleavage products (Potapenko and Sukhorukov, 1984; Potapenko, 1991). The product mixtures that form when furocoumarins are photolyzed by themselves in solution are complex and solvent-dependent, and include products of ring cleavage, dimers, and unstable intermediates with peroxidic character.

Among other cellular constituents that have been reported to be damaged under photooxidative conditions with psoralens are amino acids, including cystine, histidine, methionine (Veronese et al., 1979), tryptophan (Megaw et al., 1980), and DOPA (de Mol and Beijersbergen van Henegouwen, 1979), flavins (Musajo and Rodighiero, 1962), and thiols (d'Ischia et al., 1989). Few products have been isolated from these reactions, although methionine was converted to the corresponding sulfoxide in the presence of various furocoumarins (Veronese et al., 1982). This product would be consistent with any of a number of oxidants, including, but certainly not limited to, $^1O_2$. Little work on the mechanisms of these reactions, as such, has been reported; however, DOPA was shown to act as a donor of an electron to the excited triplet state of 8-MOP, producing the radical cation of DOPA and the radical anion of the psoralen (Craw et al., 1984b). Potapenko (1991) calculated on the basis of photophysical and kinetic arguments that $^1O_2$ must have been an insignificant ($< 5\%$) contributor to the photolysis of DOPA by 8-MOP, although it may have been more important (30–40%) in tryptophan photodecomposition. Further evidence for mechanisms other than $^1O_2$-mediated sensitization in amino acid photolysis by furocoumarins was obtained by Veronese et al. (1979), who demonstrated that the relative rates of disappearance of several amino acids in the presence of illuminated 8-MOP were different from those observed when methylene blue or rose bengal, dyes known to be efficient producers of $^1O_2$, were used.

It has been proposed that furocoumarins preferentially react with cell membranes and enzymes leading to erythemal reactions under oxygen-dependent conditions, while under conditions of low or no oxygen, cross-linking reactions to DNA would occur (Pathak and Joshi, 1984). However, the few studies with whole cells that have been performed show that reactions with DNA, RNA, proteins, and lipids all occur without a great deal of selectivity (Midden, 1988; Beijersbergen van Henegouwen et al., 1989; Dall' Acqua and Martelli, 1991). It is apparent that many competing oxygen-requiring and oxygen-independent pathways for reactions of psoralen derivatives with cellular constituents are possible. Obviously, the full story of furocoumarin phototoxicity has by no means been clarified and much further research is necessary.

2.1.2. Coumarins and Chromones

Coumarins, widespread natural products found in more than 70 plant families, have some photochemical similarities to furocoumarins. Coumarin itself (**4**) shows absorption maxima at 278 nm ($\varepsilon = 10{,}500$) and 310 nm ($\varepsilon = 6000$). The photophysical properties of coumarin and some of its derivatives, such as 5,7-dimethoxycoumarin (5,7-DMC) have been reviewed (Song, 1984). The hydroxy and methoxy coumarins are highly fluorescent compounds, especially 7-hydroxycoumarin (umbelliferone); coumarin itself is not highly fluorescent unless irradiated in dilute NaOH solutions by UV light (the hydrolysis product, o-coumaric acid, is the actual fluorophore).

4

Although structurally related to furocoumarins, coumarins have generally been found not to be phototoxic in the strict sense used by dermatologists, but many have been characterized as being photoallergic. For example, coumarins were found to be inactive (topically or orally) when tested for their ability to produce erythema on guinea pig or human skin (Pathak and Fitzpatrick, 1959); however, 7-methoxycoumarin was found to be phototoxic when tested in an assay developed to predict phototoxicity (mouse ear swelling) with both UV-B and UV-A (290–400 nm). 8-MOP was phototoxic only with UV-A (320–400 nm) (Gerberick and Ryan, 1989). Some coumarins [6,7-dihydroxycoumarin (esculetin) and 5,6-dihydroxycoumarin (daphnetin)] have been found to be "photocontact sensitizers" (Hausen and Kallweit, 1986). Compounds with substituents at the 6- or 7-position of the coumarin ring were active, but when hydrogenated were inactive (Kaidbey and Kligman, 1981).

Free radicals were detected by electron paramagnetic resonance (EPR) in frozen aqueous ethanol, glycine, or alanine solutions containing coumarin or 4- or 6-hydroxycoumarins, following irradiation with UV-A light (> 320 nm) (Kirkiacharian et al., 1969). Photoaddition products of coumarins (coumarin, 5,7-DMC) have been isolated: cycloadducts with alkenes (Shim and Chi, 1978), coumarin cyclodimers and cycloadducts with pyrimidines (Shim and Im, 1976), and a coupling product between dimethoxycoumarin and adenosine (Cho et al.,

1987). However, very little work has been done on the photooxidative reactions of these compounds.

Khellin (**5**), a natural furochromone isolated from the plant *Ammi visnaga* (Shonberg and Sina, 1950), closely resembles furocoumarins in structure but appears to have very different photoreactivity. The absorption and fluorescence spectra of khellin are quite solvent-dependent, showing blue shifts as the polarity of the solvent decreases. In ethanol, the absorption maximum is at 330 nm ($\varepsilon = 4400$); in hexane, the maximum is at 320 nm. The fluorescence quantum yields are very low in nonpolar solvents and decrease even more as the solvent polarity increases (Shim and Kang, 1987). Further studies (Kang et al., 1991) on the photophysical (laser flash photolysis, triplet quantum yield) and photochemical (adduct formation) properties showed that khellin did not photoionize and had very low reactivity toward cycloaddition reactions. From observations that the fluorescence quantum yield increased and the triplet quantum yield decreased as the solvent polarity increased, it was hypothesized that the triplet state of khellin would not form in the polar environment outside of the DNA helix.

5

Khellin has been used successfully to induce repigmentation in vitiligo patients with very few side effects, and was especially noted for not inducing erythemal reactions in patients following their exposure to sunlight (Abdel-Fattah et al., 1982; Hönigsmann and Ortel, 1985). Even in patients with normal skin, erythema did not occur, and even more significantly, delayed hyperpigmentation did not occur (Morliere et al., 1988).

There are several reports of photo-cross-linking of khellin to DNA, but the efficiencies of these reactions appear to be quite low (Cassuto et al., 1977; Morliere et al., 1988). Studies with viruses (Hudson and Towers, 1988) have shown khellin to have slight photoactivity. A [2 + 2] photoadduct of khellin to thymine was isolated from a frozen aqueous solution irradiated with UV-A (Abeysekera et al., 1983); however, another study could not demonstrate any photo-cross-linking to DNA (Altamirano-Dimas et al., 1986).

In agreement with the low yields of photo-cross-linking, khellin also had very low quantum yields of photodegradation or photosensitized reaction with histidine or tryptophan, tyrosine, and DOPA (Morliere et al., 1988). The authors suggested that this was due to inefficient formation of $^1O_2$ or $\cdot O_2^-$.

2.2.   Alkaloids

2.2.1.   Furoquinoline Alkaloids

Together with linear furocoumarins, many members of the Rutaceae family also contain furoquinoline alkaloids with the general structure **6**. The close structural relationship between the two classes of heterocycles has prompted several groups to investigate the phototoxicity and photomutagenicity of dictamnine (**7**). The compound has been demonstrated to become associated with DNA in the absence of light and then to covalently bind, presumably by forming cycloadducts of the [2 + 2] type, upon exposure to 320- to 400-nm UV (Pfyffer et al., 1982). The compounds also show photogenotoxicity to *Escherichia coli* (Ashwood-Smith et al., 1982) and the alga *Chlamydomonas reinhardtii* (Schimmer and Kühne, 1991). In addition, dictamnine has antiviral activity in the presence of UV-A light (Hudson and Towers, 1988). The role of oxygen in these reactions has not been systematically investigated.

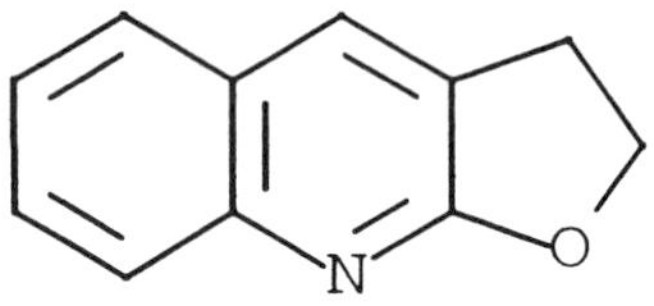

6

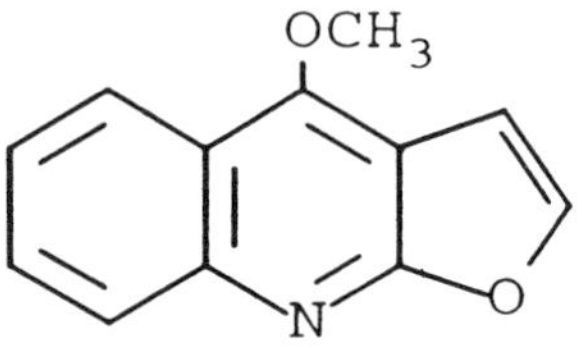

7

2.2.2.   Berberine

Berberine (**8**), an isoquinoline alkaloid found in many species of barberry plants, is a yellow compound with absorption both in the visible region at 420 nm ($\varepsilon = 4100$) and in the UV-A region at 342 nm ($\varepsilon = 21,400$). Therapeutic applications of berberine include medicinal use as an antibacterial and antimalarial compound and treatment of ulcerative skin infections (Varma, 1927). Complexes of berberine with DNA have been reported, and it was suggested that the intercalation of the berberine molecule into the DNA strand stabilized the DNA, preventing replication and thus inhibiting the growth of the protozoa responsible for malarial infections (Hahn, 1969).

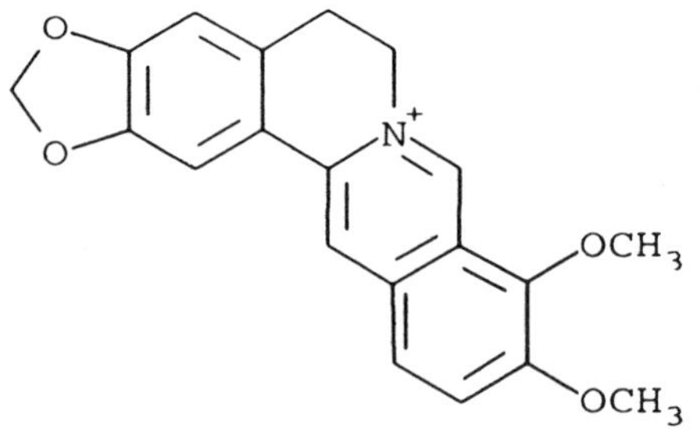

8

Investigations of light-induced activity were suggested by the intense fluorescence of this compound. It was found to be phototoxic to mosquito larvae and to increase chromosome aberrations in Chinese hamster ovary cells in the presence of UV light (Philogène et al., 1984); however, toxicity to bacteria was not enhanced in the presence of light (Tuveson et al., 1986). Photoreactivity by the production of $^1O_2$ was suggested based on the degradation of 2,5-dimethylfuran in aqueous solutions (Philogène et al., 1984) and phosphorescence emission at 1270 nm in dichloromethane solution (Arnason et al., 1992).

2.2.3. Quinine

Quinine (**9**), a quinoline alkaloid isolated from the bark of *Cinchona* trees, has been used widely as an antimalarial drug for many years; however, its mode of activity is still under investigation. Quinine [$\lambda_{max} = 334, \varepsilon = 4400$ (methanol)] is a highly fluorescent compound, emitting in a broad band centered at 450 nm and often used as a fluorescene calibration standard. Intercalated complexes of quinine with DNA (Estensen et al., 1969), with subsequent base-pair mutation (Obaseiki-Ebor and Obasi, 1987), have been observed. Another hypothesis suggests that a complex of quinine-ferriprotoporphyrin IX is formed that then damages the membrane of the malarial parasite (Fitch, 1983). Additionally, it is thought that the weakly basic compound can accumulate to high concentrations in the acidic food vacuole of the parasite, with subsequent damage to the parasite and rupture of the red blood cell (de Duve et al., 1974).

9

A major concern in the use of quinine is the side effect of phototoxic reactions, which include cutaneous flushing, skin rashes, and visual impairment. In an assessment of drugs for photochemical activity, quinine was reported as having significant photosensitizing ability, sensitizing the photooxidation of 2,5-dimethylfuran, tryptophan, and several purine derivatives in aqueous solution (Moore, 1980). On the basis of azide quenching, it was suggested that the reaction involved $^1O_2$. Like the fluorescence, the photochemical reactivity in these reactions varied in a pH-dependent manner, with the highest photooxidative activity occurring at pH 2 and then declining; fluorescence quantum yields followed the same trend. It was noted that the absorbance of the buffered solutions also changed, with the UV maximum shifting from 334 nm at pH 6 to 348 nm at pH 2, with a consequent broadening of the absorption band possibly resulting in more absorption of light and hence increased reactivity.

Photopolymerization of acrylamide did not occur under deaerated conditions in the presence of quinine, indicating that free radicals were not being formed (Moore, 1977). It was later shown that small but measurable rates of polymerization did occur, but only as the pH of the solution increased, reaching a maximum at pH 11.5 (Moore and Hemmens, 1982). Thus, the ionization state of quinine is particularly important relative to its photochemical reactivity.

Interestingly, there have been reports (see Stenberg and Travecedo, 1970) of increased antimalarial activity from solutions of quinine that were preirradiated. For example, a solution of quinine was irradiated for 38 days and then given to ducks infected with malaria; the irradiated solution was still two-thirds as active as the original solution, although only one-third of the quinine was still present (Kyker et al., 1947). Based on these reports, there have been several attempts to determine the photoproducts of quinine irradiation.

Deoxyquinine (**10**), a unique photoreduction product of quinine, was isolated from aqueous solution, although the solution was strongly acidic (2 N HCl); in dilute acidic solutions, products could not be isolated, due to a rapid reaction leading to polymeric products (Stenberg et al., 1969). Further work showed that the parent compounds, 2- and 4-hydroxymethylquinoline were also photoreduced, as were other cinchona alkaloids (quinidine, cinchonine, and cinchonidine) (Stenberg and Travecedo, 1970). Product yields incrased in propanol, lending support to a proposed mechanism of hydrogen atom abstraction. In the presence of aqueous (presumably oxygenated) citric acid, quinine afforded the photoalkylated adduct **11**, whose structure suggests a hydrogen-atom transfer from the central carbon of citric acid to the excited state of quinine, followed by radical–radical coupling (Laurie et al., 1986).

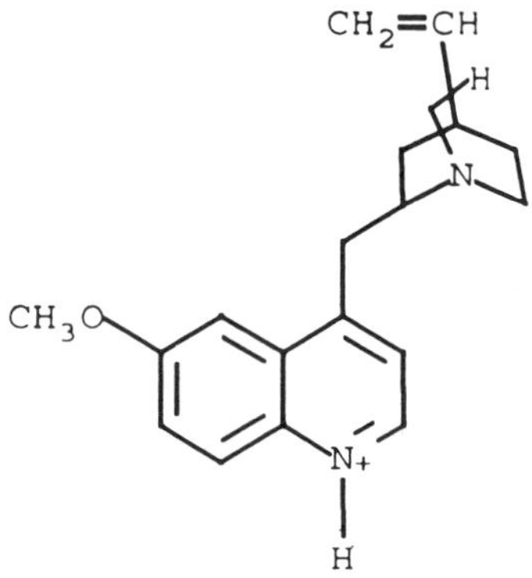

10

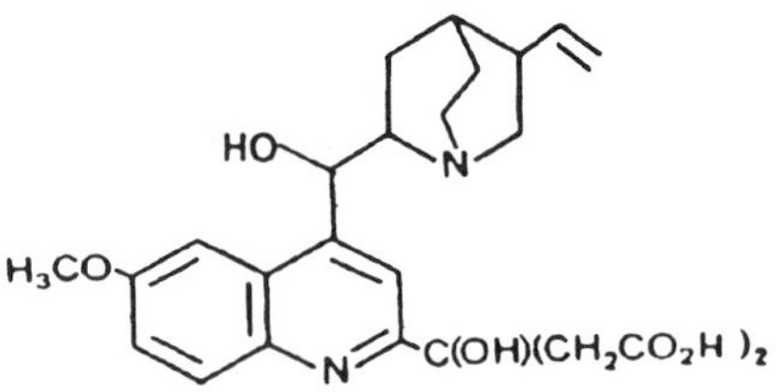

11

In polar solvents, quinine rearranged to form a 2′-oxo derivative (**12**), which was also detected in rats exposed to UV light after oral administration of quinine (Poehlmann et al., 1986). Under anoxic conditions in methanol, a fragmentation reaction occurred giving 6-methoxyquinoline and an aldehyde derived from the quiniclidine portion of the molecule (Yoon and Epling, 1980), which the authors ascribed to a radical mechanism.

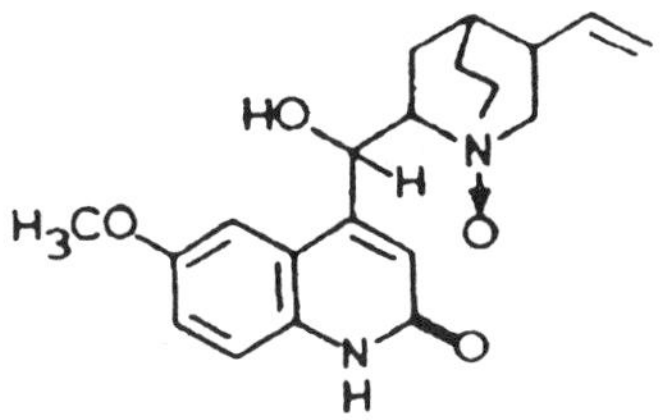

12

In tests of a series of quinolinemethanol compounds for photooxidation of tryptophan and cysteine, quinoline itself was slightly photoactive, suggesting the involvement of the heterocyclic nitrogen (Fels, 1968). Quinoline has been shown to photoionize (Potapov and Sorokin, 1970) and when irradiated on a TiO_2 surface, a radical cation and superoxide were formed (Sancier and Morrison, 1979). Pyrimidines and quinolines are photochemically reactive from the $n \rightarrow \pi^*$ state of the nitrogen, similar to carbonyl compounds (Prathapan et al., 1990).

In addition to quinine, a variety of substituted quinoline and quinolinemethanol compounds have been synthesized for use as antimalarial agents; in many cases, these compounds are even more phototoxic than quinine (Epling and Sibley, 1987).

2.2.4   β-Carboline Alkaloids

β-Carbolines are potent psychoactive drugs found in various plants in the Rutaceae and Simaroubaceae families (Waterman and Grundon, 1983). These compounds have been studied extensively for their medicinal properties and have not been reported as being phototoxic to humans or animals (Airaksinen and Kari, 1981); however, toxicity to livestock ingesting the plant *Peganum harmala* has been noted (Patterson, 1964).

β-Carboline alkaloids are structurally related to indoles and to tryptophan. The absorption and fluorescence spectra of these compounds depend strongly

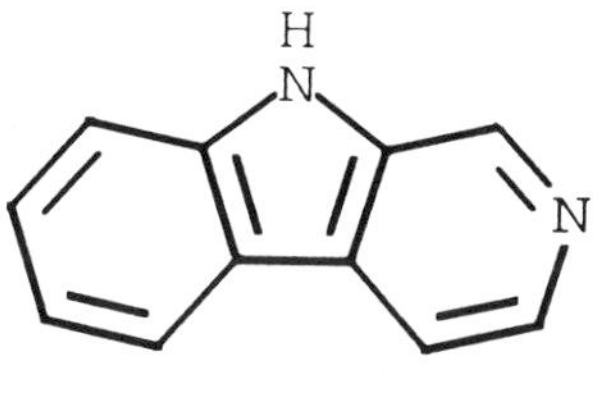

13

upon the nature of the solvent and also upon the protonation state of the compound (Vert et al., 1983). Typical absorption spectra of the neutral species in aqueous solution [for norharmane (13), the parent compound] show a maximum at 350 nm ($\varepsilon = 5300$, with shoulder at 340 nm) and 236 nm ($\varepsilon = 45,600$, with shoulder at 302 nm).

Because β-carbolines have some similarity to other well-known planar tricyclic photosensitizers (furocoumarins and acridines), and because it was reported that they "bind" to DNA, McKenna and Towers (1981) speculated that these compounds were phototoxic and were able to demonstrate this with several species of bacteria and fungi. Chromosome changes and inhibition of mitosis in Chinese hamster ovary cells also occurred (Towers and Abramowski, 1983); therefore, these compounds were termed "photogenotoxic." Further evidence of DNA alteration was obtained in a series of tests in which the compounds were tested as antiviral agents. It was found that the cell membrane was not altered, but that expression of viral activity was modified (Hudson et al., 1986a); further work showed that protein or RNA synthesis could not be detected in these strains after treatment (Hudson et al., 1986b), leading the authors to conclude that DNA was the most likely target, but that proteins could also be affected. None of these studies tested photoactivity in the absence of oxygen.

Free-radical intermediates have been postulated in several of the reported photoreactions of β-carbolines. When norharmane, harmine, or harmine was illuminated in a nonpolar solvent such as CH_2Cl_2, dimers (14) were isolated, but in a polar solvent (ethanol), dimers were not formed (Balsells and Frasca, 1983). The reaction was also quenched in acidic solutions and in the presence of cyclohexene. Oxygen did not inhibit or enhance the dimer formation. A free-radical mechanism was proposed. Further evidence of radical formation was reported when it was demonstrated that photopolymerization of vinyl monomers was initiated by various β-carbolines (Chae, 1986).

14

It was claimed that, in comparison to 8-methoxypsoralen, this class of compounds were efficient producers of $^1O_2$ and superoxide (Chae and Ham, 1986). In an attempt to correlate in vivo phototoxicity with in vitro photoreactivity, five compounds were assayed with a series of *E. coli* strains and against the noctuid moth *Trichoplusia ni* (Larson et al., 1988). In both organisms, the order of activity was similar: harmane–norharmane > harmine–harmaline > harmalol.

However, although the phototoxicity to *E. coli* was clearly shown to be dependent upon the presence of oxygen, the degree of effect did not correspond to the relative efficiency of the production of either singlet oxygen or superoxide. The activity was correlated with the hydrophobicity of the compounds, suggesting that association with a nonpolar intracellular target might have been a requirement for activity. It was also noted that the compounds that were the least phototoxic, harmaline and harmalol, were photodegraded the fastest, and thus might have deteriorated before reaching the target.

2.2.5. *Sanguinarine*

Sanguinarine (**15**), a benzophenanthrene alkaloid isolated from the roots of *Sanguinaria canadensis* (bloodroot), is also present in the stems and leaves of the plant (Slavík, 1967) and is a major component of the leaves and stems of *Macleaya microcarpa* and *M. cordata* (Reinhart et al., 1991). Like other quaternary benzophenanthridine alkloids, it shows absorption throughout the 300- to 380-nm region with several overlapping maxima.

15

Sanguinarine does have antibacterial and anti-inflammatory properties and is used commercially in toothpastes and dental rinses (Boulware et al., 1985). There are no reported phototoxic effects concerning this application; however, there are historical reports (Millspaugh, 1892) of its use as an escharotic (caustic) on skin surfaces, which might suggest some mechanisms of attack on cell surface constituents. Preparations tested for skin use as an antifungal (Vichkanova and Adgina, 1971) and antitumor (Elliott, 1985) agent were effective, with no reports of associated phototoxicity. Very little work has been reported on the photochemical and photophysical properties of sanguinarine.

Various studies have suggested that sanguinarine intercalates with DNA; evidence for this has been based on changes in the absorbance and fluorescence measurements of sanguinarine in solution with DNA and on changes in viscosity (Maiti et al., 1982). It has also been shown that sanguinarine inhibits ATPase (Straub and Carver, 1975) and uncouples respiration and oxidative phosphorylation in mitochondria (Faddeeva et al., 1987).

In a study to screen for phototoxins, sanguinarine chloride was shown to be phototoxic to various strains of *E. coli,* especially catalase-deficient mutants (Tuveson et al., 1989). In aqueous solutions and in the presence of SOD, H_2O_2 was rapidly produced, which also corresponded to a rapid loss of sanguinarine.

The proposed mechanism was photoionization of sanguinarine with the release of e_{aq}

$$\text{Sang} \underset{hv}{\overset{hv}{\rightleftharpoons}} \text{Sang*} + e_{aq}^{-} \underset{}{\overset{O_2}{\rightleftharpoons}} \cdot O_2^{-} \overset{\text{SOD}}{\longrightarrow} H_2O_2$$

with subsequent back electron transfer that would regenerate sanguinarine in solutions without the SOD present (when catalase was added, H_2O_2 was not detected).

Evidence of triplet-state formation was obtained by laser flash photolysis of a solution in methanol, which gave a moderately intense absorption signal that could be quenched by an added scavenger, 1,3-cyclohexadiene ($k_q = 3.4 \times 10^8\ M^{-1}sec^{-1}$) and also by oxygen ($k_q = 6 \times 10^9\ M^{-1}sec^{-1}$) (Arnason et al., 1992). These studies were also able to demonstrate that from the triplet state, sanguinarine was a good electron donor ($k_r = 1.14 \times 10^{10} M^{-1}sec^{-1}$ with methyl viologen) and a poor electron acceptor (no reaction with 0.3 M triethylamine).

The efficiency of $^1O_2$ production was very poor using furfuryl alcohol as a singlet oxygen acceptor in aqueous solution (Tuveson et al., 1989), but when irradiated in dichloromethane solution a quantum yield of 0.16 was reported, as determined by phosphorescence emission at 1270 nm (Arnason et al., 1992).

2.3. Carbonyl Compounds

The photochemistry of ketones has been thoroughly investigated (see comments in Section 1.3), but only in recent years has much attention been paid to naturally occurring compounds. Since virtually all aldehydes and ketones are capable of $n \to \pi^*$ excitation by virtue of their absorption bands at 280 to 320 nm, the possibility of free-radical reactions involving the diradical-like triplet state should be kept in mind. It has been demonstrated that lipophilic ketones related to benzophenone, for example, initiated the peroxidation of lipids by a hydrogen atom abstraction mechanism (Marković et al., 1990).

2.3.1. Curcumin

Curcumin (16) is an orange pigment found in turmeric (*Curcuma longa*), where it constitutes 1 to 5% of the dry weight of the rhizomes. It absorbs strongly in the

16

violet region in the visible spectrum, with a maximum at about 420 nm ($\varepsilon = 16,000$). Curcumin undergoes rather rapid direct photolysis, with a half-life of about 50 min in polar solvents (isopropanol, methanol) and 2–5 min in nonpolar solvents (acetonitrile, chloroform, and ethyl acetate) (Tønnesen et al., 1986a).

Illuminated curcumin solutions show antimicrobial activity against several species, including *Salmonella typhimurium* (Ames TA 100 strain) and *E. coli* (strains 753 and 765). Attack was observed on DNA, as indicated by experiments with repair-deficient and lactose fermentation-deficient *E. coli* (strain K12) (Tønnesen et al., 1986b). Gram-positive bacteria were more sensitive to curcumin phototoxicity, which required oxygen, than gram-negative bacteria (Srinivasan, 1991).

The quantum yield of $^1O_2$ production from curcumin in aprotic solvents is 0.5, but it drops to about one-tenth of that in oxygenated water, in which electron transfer to form $\cdot O_2^-$ (and hydrogen peroxide) becomes predominant (Chignell et al., 1990). This result would tend to support earlier evidence (Dahl et al., 1989) that curcumin produced a long-lived toxic substance, namely H_2O_2, that could diffuse into target cells.

2.3.2. *Citral*

The first aldehyde that was demonstrated to be phototoxic is citral (**17**), a major constituent of the leaves and fruits of citrus plants. Like most carbonyl compounds, citral exhibits a weak maximum ($\varepsilon = 62$) at 320 nm due to the typical $n \rightarrow \pi^*$ transition. It was shown to produce H_2O_2 when illuminated in oxygenated alcoholic solvents, probably by the mechanism indicated in Figure 1. Catalase-deficient strains of *E. coli* were shown to be susceptible to citral photokilling in the presence of air, supporting the H_2O_2 pathway for phototoxicity, but the hydrogen atom donors in cells (that would take the place of the alcohols used in the in vitro experiments) are not yet known. Citral was also phototoxic to several strains of fungi of the genus *Fusarium,* known root pathogens of citrus plants, but two fruit pathogens of the genus *Penicillium* were resistant (Asthana et al., 1992).

17

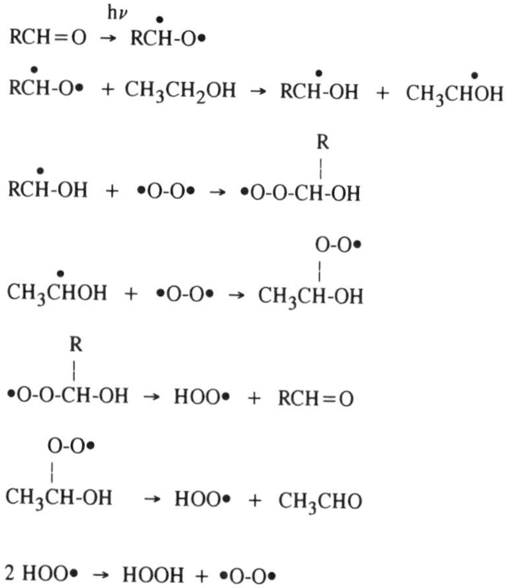

$$\text{RCH}{=}\text{O} \xrightarrow{\ h\nu\ } \text{RCH-O}\bullet$$

$$\overset{\bullet}{\text{RCH}}\text{-O}\bullet \ + \ \text{CH}_3\text{CH}_2\text{OH} \ \rightarrow \ \overset{\bullet}{\text{RCH}}\text{-OH} \ + \ \text{CH}_3\overset{\bullet}{\text{CHOH}}$$

$$\overset{\bullet}{\text{RCH}}\text{-OH} \ + \ \bullet\text{O-O}\bullet \ \rightarrow \ \bullet\text{O-O-}\overset{\overset{\textstyle R}{|}}{\text{CH}}\text{-OH}$$

$$\text{CH}_3\overset{\bullet}{\text{CHOH}} \ + \ \bullet\text{O-O}\bullet \ \rightarrow \ \text{CH}_3\overset{\overset{\textstyle \text{O-O}\bullet}{|}}{\text{CH}}\text{-OH}$$

$$\bullet\text{O-O-}\overset{\overset{\textstyle R}{|}}{\text{CH}}\text{-OH} \ \rightarrow \ \text{HOO}\bullet \ + \ \text{RCH}{=}\text{O}$$

$$\overset{\overset{\textstyle \text{O-O}\bullet}{|}}{\text{CH}_3\text{CH}}\text{-OH} \ \rightarrow \ \text{HOO}\bullet \ + \ \text{CH}_3\text{CHO}$$

$$2 \ \text{HOO}\bullet \ \rightarrow \ \text{HOOH} \ + \ \bullet\text{O-O}\bullet$$

Figure 1. Probable mechanism of formation of H_2O_2 by photolysis of citral in ethanol. (From Asthana et al. (1992). Reprinted by permission of *Photochemisty and Photobiology*.)

2.3.3.   α-Diketones

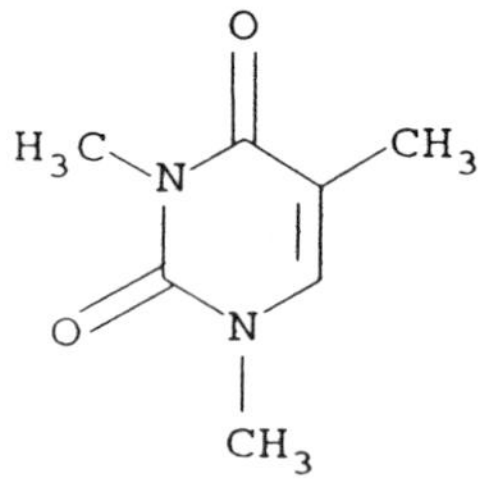

Biacetyl, $CH_3CO-COCH_3$, occurs naturally as a constituent of vegetable oils, in butter, and as a microbial fermentation product. The absorption spectra of α-diketones are notable for a broad but weak absorption band ($\varepsilon \simeq 20$) in the visible region around 400 to 440 nm, which may be due to excimers. Cleavage of the intercarbonyl linkage results in the formation of acetyl radical, which reacts rapidly with oxygen, if present, to form the acetylperoxy radical $CH_3CO-OO\cdot$, an active oxidant. Several workers have shown that biacetyl photoinactivates enzymes by reactions with many amino acids such as histidine, arginine, cystine, serine, glycine, tyrosine, methionine, and tryptophan (Mäkinen and Mäkinen, 1982). Although $^1O_2$ quenchers were reported to provide some protection against inactivation, the spectrum of amino acids undergoing damage is not suggestive of a pure $^1O_2$ mechanism, and radical processes would seem likely to be important contributors.

Biacetyl (PhCO) or benzil (COPh) sensitized the selective photooxygenation of the double bond of the nucleic acid base analog N,N-dimethylthymine (18) to a mixture of products whose formation was rationalized by the intermediacy of an unstable epoxide (Wang et al., 1981).

2.3.4. Quinones

Hypericin (19) is an abundant constituent of a few species in the genus *Hypericum* (St. John's wort) that are important toxicants to livestock. The compound, which absorbs strongly in the visible region (λ_{max} 592 nm, $\varepsilon = 42,000$), is localized in glands that appear as reddish spots on the leaves, sepals, and stems of the plant. The action spectrum for phototoxicity (which may be manifested at less than micromolar concentrations) corresponds closely to the absorption spectrum (Giese, 1971). The phototoxicity of hypericin appears to be completely oxygen-dependent. Hypericin promotes the photooxidation of methyl linoleate and tyramine by an apparent $^1O_2$ process (Seely, 1977; Knox and Dodge, 1985).

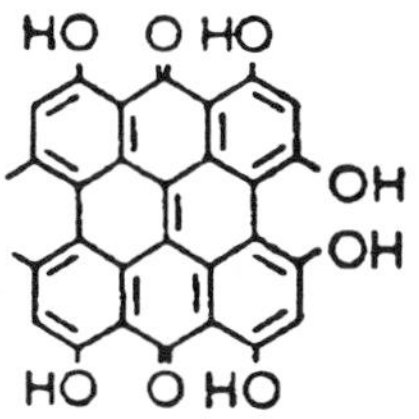

19

Cercosporin (20) is a product of the plant pathogenic fungus *Cercospora kikuchii*, which causes "purple speck disease" in soybeans, and other species of the same genus. Its UV-visible spectrum shows broad absorption throughout the visible region, with a maximum at around 480 nm and some absorbance as far out as 620 nm. This quinone is phototoxic to several strains of bacteria as well as to mice. The toxicity is oxygen-dependent (Yamazaki et al., 1975) and appears to involve membrane peroxidation reactions, since rupture of the plasma membrane of plant cells has been reported (Daub, 1982). Both $^1O_2$ and $\cdot O_2^-$ have been

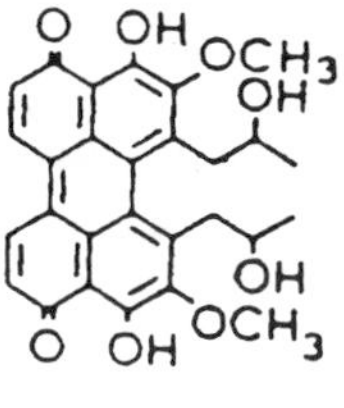

20

demonstrated to be produced when cercosporin is illuminated; in the presence of reductants such as ergothioneine or uric acid, $\cdot O_2^-$ production was greatly enhanced at the expense of $^1O_2$ (Hartman et al., 1988).

2.3.5. *Aryl Ketones from Encelia Species*

Several methyl ketones conjugated with benzofuran or chromene ring systems, isolated from several species in the genus *Encelia*, were shown to be phototoxic in tests with microorganisms. These compounds were said to show absorption maxima in the 320- to 350-nm range. The benzofuran 6-methoxyeuparin (**21**) showed activity against two yeasts, the gram-negative bacterium *Pseudomonas fluorescens* and the gram-positive organism *Bacillus subtilis*. The chromenes encecalin (**22**) and 7-demethylencecalin (**23**), in contrast, were inactive when tested with gram-negative bacteria, but were also active with *Bacillus* and the yeasts (Proksch et al., 1983). Oxygen was presumably present, but data were not reported that showed whether it was a requirement, nor were mechanistic speculations presented. It seems likely, however, that these ketones, like other phototoxic carbonyl compounds, could act by radical mechanisms including hydrogen atom abstraction.

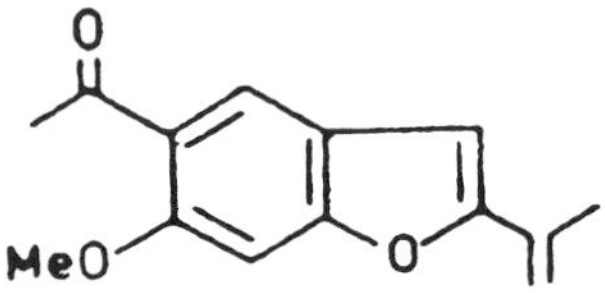

21 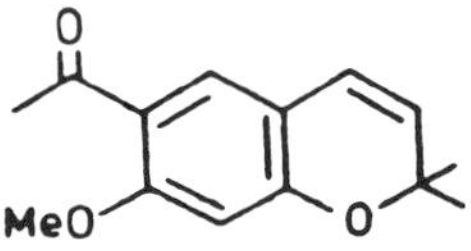   22 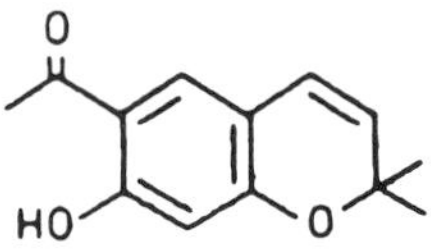   23

2.4. Thiophenes and Polyacetylenes

The roots of marigolds (*Tagetes* spp.), as well as other species in the family Compositae (Bohlmann and Zdero, 1985), contain a variety of conjugated sulfur heterocycles, including α-terthienyl (α-T, **24**) and its probable precursor, **25**. They are potent absorbers of sunlight UV, with the molar extinction coefficient of α-T being 23,000 at its maximum of 351 nm, and also weakly (but perhaps significantly) absorb long-wavelength visible light (ε_{720} for α-T in CCl_4 was 0.2). These compounds were shown to have nematocidal activity that was strongly intensified by light (Gommers and Geerligs, 1973). Oxygen was required for α-T phototoxicity to a nematode, and singlet oxygen was proposed as the active toxic agent (Gommers et al., 1980), although later workers have disputed this claim (Cooper and Nitsche, 1985). Its triplet state is efficiently quenched by oxygen, producing

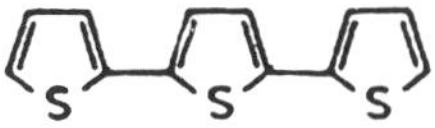

24    25

singlet oxygen in vitro, at least in organic solvents (McLachlan et al., 1984; Reyftmann et al., 1985; Evans et al., 1986; Scaiano et al., 1990), with a quantum yield as high as 0.8. Although the radical cation of α-T has been observed, it supposedly does not transfer an electron efficiently to oxygen (in organic solvents) to form $\cdot O_2^-$ (Reyftmann et al., 1985; Scaiano et al., 1989). Membrane and enzyme impairment, but little or no DNA damage, has been observed by most workers in a variety of investigations with α-T, consistent with the hydrophobic compound becoming associated with lipids or lipophilic membrane constituents before exerting phototoxicity (Cooper and Nitsche, 1985); however, it was shown to sensitize single-strand breaks in the supercoiled pBR322 DNA by an oxygen-independent process (Wang et al., 1991). α-T evidently attacked membrane consituents, however, only in the presence of oxygen (Downum et al., 1982; Tuveson et al., 1986).

Polyacetylenes, which occur in marigolds but also in many other plant species and families, are biosynthetically related to the thiophenes. Both long-chain (fatty acid-related) polyacetylenes and aromatically conjugated structures have been described (Bohlmann, 1971). These compounds may display phototoxic activity. For example, phenylheptatriyne (PHT, 26), a constituent of *Bidens pilosa*, is phototoxic to many organisms including bacteria, fungi, algae, protozoa, nematodes, insects, and fish (Towers, 1982; Kagan et al., 1984). The absorption spectrum of 26 shows three intense maxima at 290 ($\varepsilon = 17,000$), 310 ($\varepsilon = 25,000$), and 330 ($\varepsilon = 15,000$) nm, which closely coincide with the action spectrum for phototoxicity to mosquito larvae Arnason et al., 1981).

$$\langle\overline{}\rangle\text{-C≡C-C≡C-C≡C-CH}_3$$

26

To date, the only linear polyacetylene that has been screened for phototoxicity is falcarindiol (27) (Wat et al., 1980). It is known to have antibiotic activity in the dark, and since it does not have an absorption maximum at longer wavelengths than 300 nm, it is perhaps not surprising that its toxicity to erythrocytes was not increased in the light.

$$\underset{\displaystyle 27}{\text{CH}_3\text{-(CH}_2)_6\text{-CH=CH-}\overset{\displaystyle\text{OH}}{\overset{\displaystyle|}{\text{CH}}}\text{-(C ≡ C)}_2\text{-}\overset{\displaystyle\text{OH}}{\overset{\displaystyle|}{\text{CH}}}\text{-CH=CH}_2}$$

The photochemical reactions of polyacetylenes have not been very thoroughly studied. PHT was claimed to react in the absence of oxygen with 2 mol tetramethylethylene, forming an unusual adduct containing cyclopropane rings (Shim and Lee, 1986). More electron-deficient olefins, however, did not react.

The role of oxygen or oxidative reactions in the mechanisms of phototoxicity of polyacetylenes is still uncertain. PHT was reported to exert oxygen-independent toxicity toward *E. coli* by one research group (Towers, 1982) but not another (Kagan et al., 1984; Gong et al., 1988), which claimed that oxygen was required and repeated the finding using human erythrocytes. The latter workers also suggested, on the basis of the photochemical conversion of adamantylidene-adamantane to a dioxetane, that singlet oxygen was produced by illuminated PHT (Kagan et al., 1984), and other evidence supports this suggestion (Weir et al., 1985). One report suggests that PHT, as well as α-T, was a good electron photodonor to methyl viologen (paraquat) in methanol. Evidence has been provided that PHT (or its photodecomposition products is capable of transferring a hydrated electron to oxygen, forming superoxide (Evans et al., 1986; Gong et al., 1988). Therefore, the possible roles of singlet oxygen and free radicals in the phototoxicity of these compounds remain to be established.

2.5. Other Naturally Occurring Compounds

2.5.1. Tetrapyrroles

Porphyrins are derivatives of the tetrapyrrole ring system **28**, where X may be either two hydrogen atoms or a metal cation. The porphyrins make up a group of colored compounds that are essential constituents of respiratory enzymes and are accordingly found in all living cells. Although they may exist in the metal-free state "hematoporphyrins" or "protoporphyrins," depending on the nature of the side-chains), the commonest forms within living tissues incorporate metals. Many metal ions can form stable complexes with the four pyrrolic nitrogen atoms; in blood, the principal metal is iron (II), leading to ferroprotoporphyrin, or heme (**29**), whereas in plants, chlorophyll (**30**) incorporates a magnesium atom.

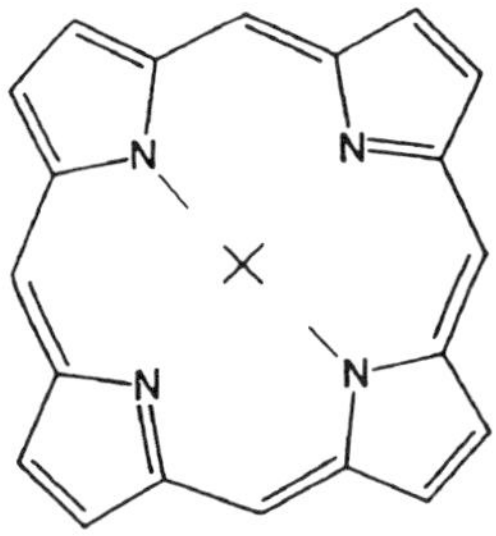

28

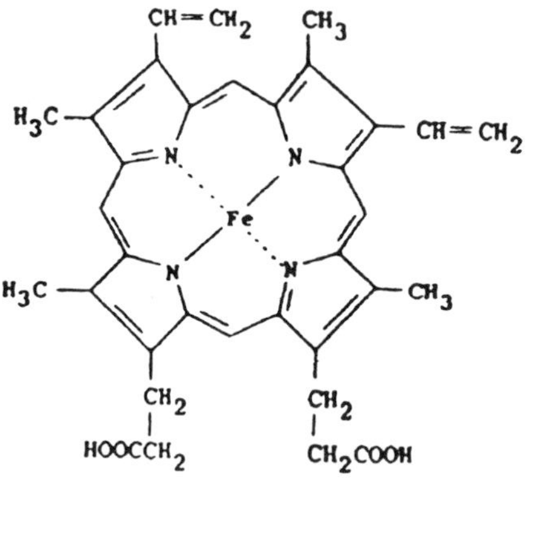

29

30

The visible spectrum of heme shows two fairly strong maxima at 550 and 575 nm ($\varepsilon = 5500$), and if the central iron atom becomes oxidized to the ferric ($+3$) state, the maximum shifts to 580 nm ($\varepsilon = 10,500$). Chlorophyll a, the principal form in higher plants, has maxima near 660 and 430 nm ($\varepsilon = 70,000$ and 90,000, respectively.) Magnesium-free chlorophyll derivatives (pheophytins) are produced when plant cells lyse, and are commonly found in the environment. The spectrum of pheophytin A shows maxima at around 670 and 410 nm ($\varepsilon = 40,000$ and 90,000).

As a constituent of the protein hemoglobin, heme is essential in the transport of oxygen through the bloodstream, and it is also a catalytically important cofactor for catalase and a multitude of other oxygen-handling enzymes. The intense colors of porphyrins and their ubiquity in nature makes them important light-gathering compounds, and their photochemical reactions are diverse and important, both in health and disease. In general, they have been shown to be good sensitizers for $^1O_2$ formation (Girotti, 1983; Moan, 1986) unless they contain paramagnetic metal ions such as Cu(II), Mn(II), or Ni(II). In addition, they are a source of oxygen radicals such as $HO\cdot$ and $\cdot O_2^-$ by electron-transfer mechanisms (Mauzerall and Feher, 1964; Buettner and Oberley, 1979) and (in the absence of O_2) are themselves electron acceptors, forming radical anions in the presence of reducing agents such as polyphenols, ascorbate, and thiols in illuminated aqueous solution (Felix et al., 1983).

Chlorophyll functions in the plant by donating an electron from its excited state to acceptors in the photosynthetic reaction center, or chloroplast. Under normal conditions the electron-transfer reaction is highly productive and results in a high quantum efficiency for the formation of reduced carbon compounds. Side reactions, however, or impaired chloroplast configurations, may result in the "escape" of excited-state energy or an electron from chlorophyll to undesired target molecules, and have potentially destructive effects. Both chlorophyll a and its magnesium-free derivative, pheophytin A, have been shown to produce $^1O_2$, with pheophytin being about twice as effective (Chauvet et al., 1981). In addition, formation of $\cdot O_2^-$ from chlorophyll photolysis has been demonstrated by You and Fong (1986) in spin-trapping experiments. It has been suggested that β-carotene and related carotenoids, universal constituents of chloroplasts, have a protective function because of their ability to quench singlet oxygen and free radicals (Burton and Ingold, 1984).

Phototoxicity of porphyrins (which mostly but not entirely depends on the presence of oxygen) has been attributed to lysis of membranes, inactivation of enyzmes, and damage to DNA; to a large extent, the target of choice may be a function of the charge on the porphyrin, with cationic porphyrins known to form tightly bound complexes with DNA and negatively charged porphyrins being principally localized at membrane sites (Kochevar and Dunn, 1990). Protoporphyrin, a negatively charged compound, was shown to induce the formation of a cholesterol 5-hydroperoxide in illuminated red blood cell ghosts; the compound is considered to be a specific $^1O_2$-derived product (Lamola et al., 1973). In the case of DNA, the ionic attraction mechanism was tested by adding the anionic surfactant SDS to the test system; when quantities of the surfactant were high enough to displace a significant fraction of the cationic porphyrin being tested, cleavage was greatly diminished (Praseuth et al., 1986).

Chlorophyll precursors (magnesium tetrapyrroles), under some conditions, have been shown to accumulate in the leaves of certain plants treated with δ-aminolevulinic acid (31), an early intermediate in the biosynthesis of chlorophyll (Rebeiz et al., 1988). Exposure of these compounds to light produces singlet oxygen, which was suggested to be responsible for the observed herbicidal effect.

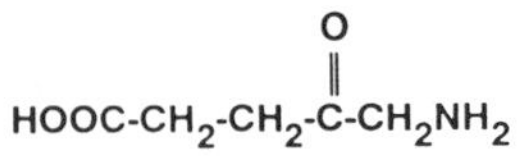

$$\underset{\text{HOOC-CH}_2\text{-CH}_2\text{-C-CH}_2\text{NH}_2}{\overset{\overset{\displaystyle O}{\|}}{}}$$

31

Bilirubin (32), a linear tetrapyrrole found in bile, is the principal end product of heme catabolism. The compound's spectrum displays a broad and intense absorption band in the visible light region from 400 to 500 nm, with a maximum ($\varepsilon = 60,700$) at 453 nm. As might be expected from its structure, bilirubin has very low water solubility and tends to partition into hydrophobic environments. Illumination of bilirubin in the presence of oxygen has been demonstrated to

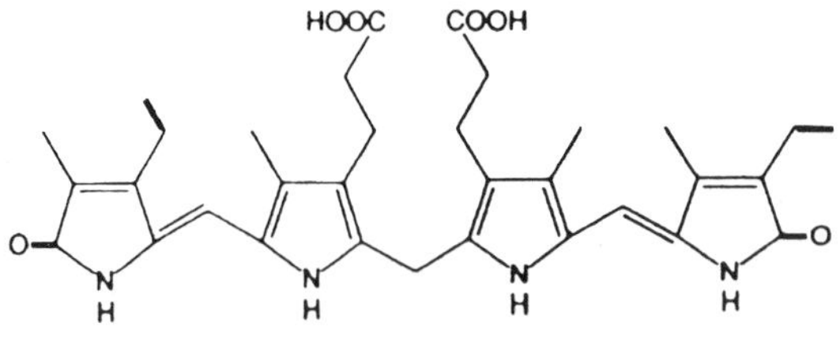

32

produce both $^1O_2$ and $\cdot O_2^-$ (McDonaugh, 1971; Rosenstein et al., 1983). In addition, it becomes photooxidized, with the formation of peroxides of uncertain structure that were shown to induce DNA strand breaks in the dark (Rosenstein et al., 1983). Bilirubin also binds to and reacts with proteins, photooxidizing tryptophan residues in human serum albumin, for example, by an apparent $^1O_2$ mechanism (Jori and Spikes, 1981).

2.5.2. Acrylic and Cinnamic Acid Derivatives

α, β-Unsaturated carboxylic acid derivatives, especially when conjugated with aromatic or heterocylic rings, may absorb in the solar UV and take part in photochemical reactions that have potential for cellular damage. A tremendous variety of cinnamic acid derivatives (**33**), for example, are produced by higher plants (Robinson, 1991); although they have repeatedly been shown to exert antimicrobial (particularly antifungal) activity (Friend, 1979), the role of light in the observed toxicity has seldom been considered. The compounds normally exist in the thermally stable E, or trans, configuration but uptake of a photon (and formation of the diradical-like $\pi \rightarrow \pi^*$ excited state) normally leads to the establishment of a cis-trans equilibrium that may lie significantly toward the side of the Z (cis) form. Little work on the photochemical reactions of these compounds has been reported beyond the isomerization step; however, recent data suggest that they may interact with biomolecules. Illumination of p-methoxycinnamic acid (**34**), a constituent of *Veronica virginica*, in the presence of DNA led to covalent incorporation of the compound; the quantum yield for addition was much lower at 308 nm ($\sim 1 \times 10^{-5}$) than at 226 nm ($\sim 1 \times 10^{-4}$). The presence of air did not affect the rate of extent of uptake. At least some of the adducts appeared to be [2 + 2] cyclodimers (Mohammad et al., 1991).

33   34

Urocanic acid (**35**), a histidine metabolite, became covalently bound to DNA or to some base analogues (e.g., N,N-dimethylthymine, **18**) when illuminated at 289 nm (Morrison and Deibel, 1986). The products appeared to be largely [2 + 2] photodimers. Indoleacrylic acid (**36**), a derivative of trytophan with a wide distribution in plants, showed similar and even more efficient [2 + 2] photobinding, probably mostly to thymidine residues, in deaerated DNA. However, large levels of incorporation into polyadenosine and polyguanine were also observed if air was present. The authors suggested that in this case, electron transfer

35   36

reactions rather than photocycloadditions were taking place (Farrow et al., 1990).

2.6. Polycyclic Aromatic Hydrocarbons and Petroleum

Crude petroleum is a mixture of thousands of compounds; aliphatic and aromatic hydrocarbons are the most abundant constituents, but many heterocyclic, phenolic, and acidic constituents have been identified. Both crude and refined petroleum products absorb sunlight strongly due to their content of substituted naphthalenes, naphthols, and higher aromatic hydrocarbons and heterocycles. Low-boiling petroleum distillates (e.g., kerosene) were demonstrated some time ago (Crafts and Reiber, 1948) to contain biocidal components after exposure to sunlight and air. Number 2 fuel oil, a higher-boiling distillate product that is usually enriched in aromatic compounds, also was converted by sunlight to peroxides, phenols, and other water-soluble compounds that are highly toxic to marine invertebrates, fish, algae, and yeasts (Larson et al., 1979).

The photochemical oxidation of petroleum, a complicated and inadequately understood subject, has been reviewed by Payne and Phillips (1985). Both singlet oxygen and free radical pathways appear to contribute to the mechanism of photooxidation (Larson et al., 1979; Lichtenthaler et al., 1989). The general classes of compounds produced when complex petroleum products are exposed to solar UV include carboxylic acids, carbonyl compounds, alcohols (including phenols), and hydroperoxides. All of these substances have a potential for toxicity, but the hydroperoxide fraction of a photooxidized number 2 fuel oil was by far the most toxic to microorganisms; furthermore, treatment of the photo-oxidized fuel oil with a mild reducing agent, which would convert hydroperoxides to alcohols, resulted in significant reduction or elimination of the toxic response (Larson et al., 1977). Mass spectroscopic investigation of the hydroperoxide fraction suggested the presence of alkylated tetrahydronaphthalene-1-hydroperoxides (**37**). The hydroperoxide fraction was also found to contain mutagenic constituents when tested in a *Saccharomyces* gene conversion test (Callen and Larson, 1978).

Some aromatic hydrocarbons found in petroleum and/or coal tar also are phototoxic, or may form toxic photoproducts, under certain conditions. Although the primary photochemistry of many such hydrocarbons has been studied, few have been tested under biologically relevant conditions. Anthracene

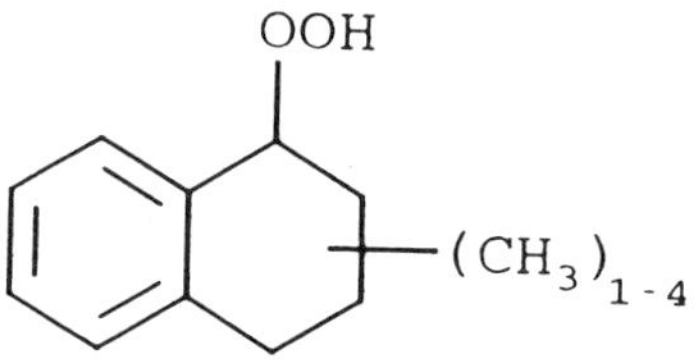

37

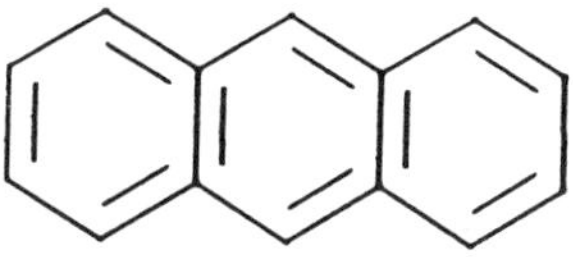

38

(**38**) and a few other PAHs have been shown to be phototoxic to many aquatic organisms including algae, zooplankton, and fish (Oris and Giesy, 1987; Kagan et al., 1987). After illumination with UV-A in the presence of primate epithelial cells in tissue culture, anthracene became covalently bound to their DNA. The authors postulated attack of the cation radical of the hydrocarbon on the DNA bases (Blackburn and Taussig, 1975). Anthracene also induced hemolysis of human erythrocytes by an oxygen-dependent reaction, and (possibly by nonoxidative mechanisms) induced cleavage of supercoiled DNA and inactivated *Haemophilus influenzae*, transforming DNA (Tuveson et al., 1990). These experiments suggest that anthracene may have several cellular targets. Anthracene may generate singlet oxygen upon irradiation and is also quite susceptible to photodecomposition, especially in the presence of oxygen, when it forms a moderately stable endoperoxide that rearranges to anthraquinone (Sigman et al., 1991). It is not known whether either of these photoproducts could contribute to the observed toxicity.

The polycyclic hydrocarbon fluoranthene (**39**), when sprayed onto the leaves of plants, induced severe foliar injury in the presence of UV. The authors suggested that a $^1O_2$ mechanism for toxicity was most consistent with their results. Fluoranthene was also one of the most phototoxic PAHs toward a variety of aquatic organisms (Kagan et al., 1987).

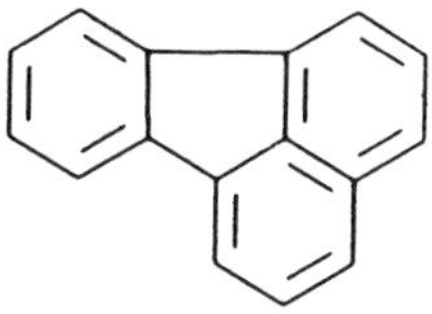

39

2.7. Dyes

Xanthene dyes such as rose bengal (**40**) have been used for many years as sources of singlet oxygen (Neckers, 1989). Rose bengal is indeed an effective producer of $^1O_2$, with a quantum yield of about 0.7–0.8. In oxygenated aqueous solution, however, photoionization is also significant, as about 20% of the rose bengal triplets transfer electrons to molecular oxygen to produce $\cdot O_2^-$ (Lee and Rodgers,

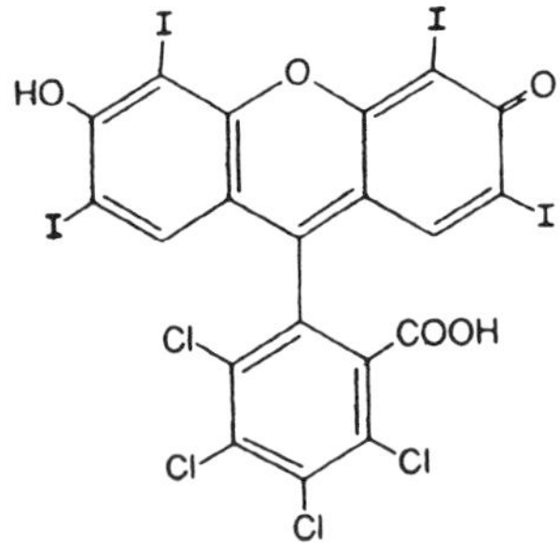

40

1987). A study of the oxidation of N, N-diethylhydroxylamine by rose bengal in acetonitrile and in water clearly demonstrated that, although the principal end products were the same, in the organic solvent a $^1O_2$ mechanism was operative, whereas in water an electron transfer mechanism was dominant (Bilski et al., 1991). Rose bengal, a hydrophobic anion, tends to be largely associated with sites in the membrane, causing lysis and similar injury due to attack on reactive lipids and protein amino acids (Pimprikar and Coign, 1987); however, DNA damage has also been reported. Several investigations of rose bengal-induced DNA cleavage have indicated that a non-$^1O_2$ mechanism predominates, since radical quenchers had a greater inhibitory effect than $^1O_2$ quenchers. In fact, oxygen did not even seem to be required for a portion of the cleavage to occur. The results are consistent with direct interactions, probably involving electron transfer, between the DNA and the dye triplet (Kochevar and Dunn, 1990). These data are supported by the studies of Dahl (1993), who showed that illumination of dry rose bengal separated by a short distance from target bacteria was ineffective at initiating killing of the cells, whereas rose bengal in solution was much more toxic. He explained the results by proposing that a gas-phase toxicant other than $^1O_2$, perhaps $\cdot OOH$ or H_2O_2, was being formed in the solution experiments.

Xanthene dyes have been demonstrated to have several important mechanisms of toxicity to various organisms (Pimprikar and Coign, 1987). For example, damage at the membrane occurs in some yeast cells, whereas intracellular toxicity, perhaps involving inactivation of essential enzyme systems such as acetylcholinesterase, seems to predominate in other cases.

Tricyclic, cationic dyes such as methylene blue (**41**), acridine orange (**42**), and proflavine (**43**) have been known to the phtotoxic for many years. The principal

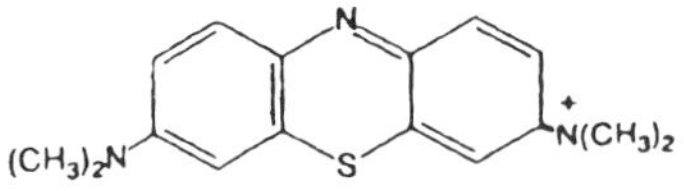

41

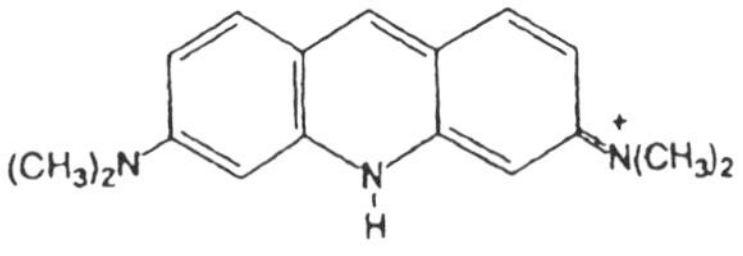

42

$$H_2N \quad \overset{+}{N}H_2$$

43

target for molecules of this structural type seems to be DNA; the flat, coplanar structure of the dyes lends itself to intercalation, and the positively charged amino group may assist in the close approach to the negatively charged phosphate backbone. The observed damage to DNA, which includes strand breakage and base destruction, seems largely due to oxidative reactions occurring at or near guanine sites. Mechanisms invoking singlet oxygen, electron transfer, and hydrogen atom abstraction have all been suggested (Kochevar and Dunn, 1990).

2.8. Drugs and Cosmetics

A wide variety of topically applied or ingested pharmaceutical preparations have been demonstrated to exhibit light-induced effects in humans, ranging from skin rashes to genotoxicity. Antibiotics such as the tetracyclines, chloramphenicol, and promazines, antidepressants such as the dibenzocycloheptadienes, estrogenic steroids (norethistrone), calcium antagonists (nifedipine), anti-inflammatory agents, and ingredients of perfumes and sunscreens are among the many classes that are photochemically active. In general, only a very limited number of research studies have addressed the biochemical mechanisms associated with the observed reactions, and there have been even fewer studies that have addressed the role of oxygen. Some of these studies, for example those regarding the psoralens and of quinine, have been addressed in earlier sections of this chapter. The reader is referred to recent reviews (Kochevar and Dunn, 1990; Cadet and Vigny, 1990; Beijersbergen van Henegouwen, 1991) for surveys of the limited information that is available.

2.9. Metal Complexes

Transition metal ions in solution are often susceptible to a variety of photoredox reactions that have the potential to produce free radicals and other active species. Most of these reactions occur within the coordination complexes of the metal ions, which may involve water molecules or other ligands. The electronic spectra of these (often colored) complexes are usually quite intricate, featuring electronic transitions involving only the metal or ligand as well as ligand-to-metal electron transfers. Although many of these metal complexes are potential phototoxins, only limited data on their interactions with biomolecules are available. By way of introduction, we will present a few such results with emphasis on iron(III) phototoxicity.

In aqueous solution, iron(III) is hydrated to a variety of different complexes whose stability depends on pH. The complex $Fe(OH)^{2+}$, which predominates in pure, weakly acidic water solutions, for example, has an absorption maximum at 297 nm and the dimer $Fe_2(OH)_2^{4+}$ absorbs at 335 nm (Knight and Sylva, 1975). The former complex has been demonstrated (Faust and Hoigné, 1990) to be a significant source of ·OH in atmospheric water droplets:

$$Fe(OH)^{2+} \xrightarrow{\ h\nu\ } Fe^{2+} + \cdot OH$$

Ions with complexing ability compete for water molecules surrounding iron cations. These complexes, which may also be thought of as ion pairs, typically absorb light at much greater wavelengths than the aequo complexes. Ferric chloride, for example, has an absorbance maximum at 320 nm. Redox reactions of these complexes are also well known; for example, $FeCl_3$ phtoinitiates vinyl polymerization because of the formation of ·Cl radicals (Evans et al., 1951).

Ferric oxalate complexes have been used for many years as actinometers. An electron transfer reaction appears to occur within the complex, resulting in the reduction of iron and the oxidation of the carboxylate ligand to CO_2. Similar processes occur in the sunlight photolysis of many other iron–carboxylate systems such as iron(III)–EDTA and iron(III)–NTA chelates (Langford et al., 1973). Like iron complexes, Cu(II) complexes with carboxylic acids and even with more reduced ligands such as ethylenediamine produce CO_2 upon irradiation (Balzani and Carassiti, 1970).

Metal ions complexed to DNA and other ligands have been shown to undergo photoinduced electron transfer reactions that could have potentially injurious consequences. For example, copper(II) chloride, when irradiated with 254-nm UV light in the presence of DNA fragments, resulted in the formation of alkali-labile strand breaks (Rossman, 1989). Iron(III) perchlorate and UV-A (320–400 nm) light were shown to strongly diminish the transforming activity of *H. influenzae* DNA in the presence of oxygen; hydroxyl radical scavengers inhibited only a small fraction of the inactivation (Larson et al., 1992). Treatment of *E. coli* plasmid DNA with iron(III) and light resulted in the conversion of the covalently closed circular form of the plasmid to open circles and ultimately to the linear form. In model systems, iron(III) was shown to photoreact with the DNA backbone, causing nicking and double-strand breakage. The results are consistent with a mechanism involving a preliminary complexation of iron(III) by DNA, followed by generation of reactive free radicals other than ·OH.

The interactions of metallic cations with nucleic acids and nucleotides appear to be largely governed by the complexing characteristics of their phosphate and base moieties; the hydroxyl groups of ribose and deoxyribose are poor ligands for metal coordination. The bases have a number of potential sites for coordination, and it has been demonstrated that they strongly bind "soft," polarizable heavy metals such as mercury and lead, but many transition metal ions are also somewhat electrostatically attracted to the strongly anionic phosphate region.

"Hard" alkaline earth metal ions such as calcium and magnesium are associated almost exclusively with the phosphates.

ACKNOWLEDGMENTS

We thank May Berenbaum and Jennifer Nevius for helpful comments on the manuscript, and the U.S. Department of Agriculture for financial support (Grant AG89-37280-4897). This paper is dedicated to R. W. Tuveson (1931–1992), whose work on phototoxicity was an inspiration to us and many other scientists.

REFERENCES

Abdel-Fattah, A., Aboul-Enein, M. N., Wassel, G., and El-Menshawi, B. (1982). An approach to the treatment of vitiligo by khellin. *Dermatologica* **165**, 136–140.

Abeysekera, B. F., Abramowski, Z., and Towers, G. H. N. (1983). Genotoxicity of the natural furochromones, khellin and visnagin, and the identification of a khellin-thymine photoadduct. *Photochem. Photobiol.* **38**, 311–315.

Airaksinen, M. M., and Kari, I. (1981). β-Carbolines, psychoactive compounds in the mammalian body. Part 1: Occurrence, origin and metabolism. *Med. Biol.* **59**, 21–34.

Altamirano-Dimas, M., Hudson, J. B., and Towers, G. H. N. (1986). Induction of cross-links in viral DNA by naturally-occurring photosensitizers. *Photochem. Photobiol.* **44**, 187–192.

Arnason, J. T., Swain, T., Wat, C. K., Graham, E. A., Partington, S., and Towers, G. H. N. (1981). Mosquito larvicidal activity of polyacetylenes from species in the Asteraceae. *Biochem. Syst. Ecol.* **9**, 63–68.

Arnason, J. T., Guérin, B., Kraml, M. M., Mehta, B., Redmond, R. W., and Scaiano, J. (1992). Phototoxic and photochemical properties of sanguinarine. *Photochem. Photobiol.* **55**, 35–38.

Arnaud, R., Deflandre, A., Lang, G., and Lemaire, J. (1981). Photosensitizing properties of furocoumarins. IV. Photochemistry of methoxypsoralens. *J. Chim. Phys. Phys.-Chim. Biol.* **78**, 597–605.

Ashwood-Smith, M. J., Towers, G. H. N., Abramowski, Z., Poulton, G. A., and Liu, M. (1982). Photobiological studies with dictamnine, a furoquinoline alkaloid. *Mutat. Res.* **102**, 401–412.

Asthana, A., Larson, R. A., Marley, K. A., and Tuveson, R. W. (1992). Mechanisms of citral phototoxicity. *Photochem. Photobiol.* **56**, 211–222.

Balsells, R. E., and Frasca, A. R. (1983). Photochemical dimerization of β-carboline alkaloids. *Tetrahedron* **39**, 33–39.

Balzani, V., and Carassiti, V. (1970). *Photochemistry of Coordination Compounds.* Academic Press, London.

Beijersbergen van Henegouwen, G. M. J. (1991). Systemic phototoxicity of drugs and other xenobiotics. *J. Photochem. Photobiol.* **B10**, 183–210.

Beijersbergen van Henegouwen, G. M. J., Wijn, E. T., Schoonderwoerd, S. A., and Dall'Acqua, F. (1989). Method for the determination of *in vivo* irreversible binding of 8-methoxypsoralen (8-MOP) to epidermal lipids, proteins and DNA/RNA of rats after PUVA treatment. *J. Photochem. Photobiol.* **B3**, 631–635.

Bensasson, R. V., Chalvet, O., Land, E. J., and Ronfard-Haret, J. C. (1984). Triplet, radical anion and radical cation spectra of furocoumarins, *Photochem. Photobiol.* **39**, 287–291.

Berenbaum, M. R., and Larson, R. A. (1988). Flux of singlet oxygen from leaves of phototoxic plants. *Experientia* **44**, 1030–1032.

Bilski, P., Li, A. S. W., and Chignell, C. F. (1991). The photo-oxidation of N,N-diethylhydroxylamine by rose bengal in acetonitrile and water. *Photochem. Photobiol.* **54**, 345–352.

Blackburn, G. M., and Taussig, P. E. (1975). Photocarcinogenicity of anthracene. Photochemical binding to deoxyribonucleic acid in tissue culture. *Biochem. J.* **149**, 289–291.

Blan, Q. A., and Grossweiner, L. I. (1987). Singlet oxygen generation by furocoumarins: Effect of DNA and liposomes. *Photochem. Photobiol.* **45**, 177–183.

Bohlmann, F. (1971). Acetylenic compounds in the Umbelliferae. In V. H. Heywood (Ed.), *Biology and Chemistry of the Umbelliferae.* Academic Press, New York, pp. 279–291.

Bohlmann, F., and Zdero, C. (1985). Naturally occurring thiophenes. In S. Gronowitz (Ed.), *The Chemistry of Heterocyclic Compounds.* Wiley (Interscience), New York, Vol. 44, pp. 261–323.

Bothe, E., Schuchmann, M. N., Schulte-Frohlinde, D., and von Sonntag, C. (1983). Hydroxyl radical-induced oxidation of ethanol in oxygenated aqueous solutions. A pulse radiolysis and product study. *Z. Naturforsch, B: Anorg. Chem., Org. Chem.* **38B**, 212–219.

Boulware, R. T., Southard, G. L., and Yankell, S. L. (1985). Sanguinaria extract, a new agent for the control of volatile sulfur compounds in the oral cavity. *J. Soc. Cosmet. Chem.* **63**, 297–302.

Buettner, G. R., and Oberley, L. W. (1979). Superoxide formation by protoporphyrin as seen by spin trapping. *FEBS Lett.* **98**, 18–20.

Burton, G. W., and Ingold, K. U. (1984). β-Carotene: An unusual type of lipid antioxidant. *Science* **224**, 569–573.

Buxton, G. V., Greenstock, C. L., Helman, W. P., and Ross, A. B. (1988). Critical review of rate constants for reactions of hydrated electrons, hydrogen atoms, and hydroxyl radicals in aqueous solution. *J. Phys. Chem. Ref. Data* **17**, 514–886.

Cadet, J., and Teoule, R. (1978). Comparative study of oxidation of nucleic acid components by hydroxyl radicals, singlet oxygen and superoxide anion radicals. *Photochem. Photobiol.* **28**, 661–667.

Cadet J., and Vigny, P. (1990). The photochemistry of nucleic acids. In H. Morrison (Ed.), *Bioorganic Photochemistry: Photochemistry and the Nucleic Acids.* Wiley, New York, pp. 1–272.

Callen, D. F., and Larson, R. A. (1978). Toxic and genetic effects of fuel oil hydroperoxides in *Saccharomyces cerevisiae. J. Toxicol. Environ. Health.* **4**, 913–917.

Cannistraro, S., and van de Vorst A. (1977). ESR and optical absorption evidence for free radical involvement in the photosensitizing action of furocoumarin derivatives and for their singlet oxygen production. *Biochim. Biophys. Acta* **476**, 166–177.

Cassuto, E., Gross, N., Bardwell, E., and Howard-Flanders, P. (1977). Genetic effects of photoadducts and photocrosslinks in the DNA of phage exposed to 360 nm light and trimethylpsoralen or khellin. *Biochim. Biophys. Acta* **475**, 589–600.

Chae, K. H. (1986). Evidence for photochemical radical formation by β-carbolines. *Bull. Korean Chem. Soc.* **7**, 253–254.

Chae, K. H., and Ham, H. S. (1986). Production of singlet oxygen and superoxide anion radicals by β-carbolines. *Bull. Korean Chem. Soc.* **7**, 478–479.

Chauvet, J. -P., Villain, F., and Viovy, R. (1981). Photooxidation of chlorophyll and pheophytin. Quenching of singlet oxygen and influence of the micellar structure. *Photochem. Photobiol.* **34**, 557–565.

Chignell, C. F., Reszka, K., Bilski, P., Motten, A., Sik, R., and Dahl, T. A. (1990). Singlet oxygen generation and radical reactions photosensitized by curcumin. *Photochem. Photobiol.* **51S**, 6S.

Cho, T. H., Shim, S. C., and Shim, H. K. (1987). Isolation and characterization of the photoadducts of 5,7-dimethoxycoumarin and adenosine. *Photochem. Photobiol.* **46**, 305–309.

Cooper, G. K., and Nitsche, C. I. (1985). α-Terthienyl, phototoxic allelochemical. *Bioorg. Chem.* **13**, 362–374.

Crafts, A. S., and Reiber, H. G. (1948). Herbicidal properties of oils. *Hilgardia* **18**, 77–156.

Craw, M., Truscott, T. G., Dall'Acqua, F., Guiotto, A., Vedaldi, D., and Land, E. J. (1984a). Methylangelicins: Possible correlation between photophysical properties of the triplet excited state and photobiological activity. *Photobiochem. Photobiophys.* **7**, 359–365.

Craw, M., Chedekel, M., Truscott, T. G., and Land, E. J. (1984b). The photochemical interaction between the triplet state of 8-methoxypsoralen and the melanin precursor L-DOPA. *Photochem. Photobiol.* **39**, 155–159.

Dahl, T. A. (1993). Examining the role of singlet oxygen in photosensitized cytotoxicity. In G. R. Helz (Ed.). *Aquatic and Surface Photochemistry.* Lewis Publisher, Boca Raton, FL.

Dahl, T. A., McGowan, W. M., Shand, M. A., and Srinivasan, V. S. (1989). Photokilling of bacteria by the natural dye curcumin. *Arch. Microbiol.* **151**, 183–185.

Dall'Acqua, F. (1988). Dark and photochemical interactions between monofunctional furocoumarins and DNA. *Biochem. Pharmacol.* **37**, 1793–1794.

Dall'Acqua, F., and Martelli, P. (1991). Photosensitizing action of furocoumarins on membrane components and consequent intracellular events. *J. Photochem. Photobiol.* **B8**, 235–254.

Daub, M. E. (1982). Peroxidation of tobacco membrane lipids by the photosensitizing toxin, cercosporin. *Plant Physiol.* **60**, 1361–1364.

Decuyper, J., Piette, J., and van de Vorst, A. (1983). Activated oxygen species produced by photoexcited furocoumarin derivatives. *Arch. Int. Physiol. Biochem.* **91**, 471–476.

de Duve, C., deBarsy, T., Poole, B., Trouet, A., Tulkens, P., and vanHoof, F. (1974). Lysosomotrophic agents. *Biochem. Pharmacol.* **23**, 2495–2531.

de Mol, N. J., Beijersbergen van Henegouwen, G. M. J. (1979). Formation of singlet molecular oxygen by 8-methoxypsoralen. *Photochem. Photobiol.* **30**, 331–335.

d'Ischia, M., Napolitano, A., and Prota, G. (1989). Psoralens sensitize glutathione photooxidation *in vitro. Biochim. Biophys. Acta* **993**, 143–147.

Downum, K. R., and Nemec, S. (1987). Light-activated antimicrobial chemicals from plants: Their potential role in resistance to disease-causing organisms. *A. C. S. Symp. Ser.* **339**, 281–294.

Downum, K. R., Hancock, R. W. E., and Towers, G. H. N. (1982). Mode of action of α-terthienyl on *Escherichia coli*: Evidence for a photodynamic effect on membranes. *Photochem. Photobiol.* **36**, 517–523.

Elliott, J. Q. (1985). Skin tumor removal and healing compositions. U.S. Pat. 4,515,779 A (cited in *Chem. Abstr.* **103** (4; 27317r).

Epling, G. A., and Sibley, M. T. (1987). Photosensitized lysis of red blood cells by phototoxic antimalarial compounds. *Photochem. Photobiol.* **46**, 39–43.

Epstein, W. L. (1991). Cutaneous responses to plant toxins. Ch. 6 In, R. F. Keeler and A. T. Tu (Eds.), *Handbook of Natural Toxins.* Dekker, New York, Vol. 6.

Eriksen, J., Foote, C. S., and Parker, T. L. (1977). Photosensitized oxygenation of alkenes and sulfides via a non-singlet-oxygen mechanism. *J. Am. Chem. Soc.* **99**, 6455–6456.

Estensen, R. D., Krey, A. K., and Hahn, F. E. (1969). Studies on a deoxyribonucleic acid-quinine complex. *Mol. Pharmacol.* **5**, 532–541.

Evans, C., Weir, D., Scaiano, J. C., MacEachern, A., Arnason, J. T., Morand, P., Hollebone, B., Leitch, L. C., and Philogéne, B. J. R. (1986). Photochemistry of the botanical phototoxin α-terthienyl and some related compounds. *Photochem. Photobiol.* **44**, 441–451.

Evans, M. G., Santappa, M., and Uri, N. (1951). Photoinitiated free radical polymerization of vinyl compounds in aqueous solution. *J. Polym. Sci.* **7**, 243–246.

Faddeeva, M. D., Belyaeva, T. N., and Sokolovskaya, E. L. (1987). Influence of some low-molecular-weight compounds (including DNA ligands) on membrane-bound sodium-potassium ATPase of bovine cerebral cortex. *Tsitologiya* **29**, 576–581.

Farrow, S. J., Mohammad, T., Baird, W., and Morrison, H. (1990). Photolytic covalent binding of indoleacrylic acid to DNA. *Photochem. Photobiol.* **51**, 263–271.

Faust, B. C., and Hoigné J. (1990). Photolysis of Fe(III)-hydroxy complexes as sources of OH radicals in clouds, fog, and rain. *Atmos. Environ.* **23**, 235–240.

Felix, C. C., Reszka, K., and Sealy, R. C. (1983). Free radicals from photoreduction of hematoporphyrin in aqueous solution. *Photochem. Photobiol.* **37**, 141–147.

Fels, I. G. (1968). The photoactivity of quinolinemethanols. *J. Med. Chem.* **11**, 887–888.

Finlayson-Pitts, B. J., and Pitts, J. N. Jr. (1986). *Atmospheric Chemistry.* Wiley (Interscience), New York.

Fitch, C. D. (1983). Mode of action of antimalarial drugs. *Ciba Found. Symp.* **94**, 222–232.

Foote, C. S. (1981). Photooxidation of biological model compounds. In M. A. J. Rodgers and E. I. Powers (Eds.), *Oxygen and Oxy-radicals in Chemistry and Biology.* Academic Press, New York, pp. 429–440.

Foote, C. S., and Wexler, S. (1964). Olefin oxidations with excited singlet molecular oxygen. *J. Am. Chem. Soc.* **86**, 3879–3880.

Fridovich, I. (1981). Role and toxicity of superoxide in cellular systems. In M. A. J. Rodgers and E. I. Powers (Eds.), *Oxygen and Oxy-radicals in Chemistry and Biology.* Academic Press, New York, pp. 197–204.

Friend, J. (1979). Phenolic substances and plant disease. *Recent Adv. Phytochem.* **12**, 557–588.

Gerberick, G. F., and Ryan, C. A. (1989). A predictive mouse ear-swelling model for investigating topical phototoxicity. *Food Chem. Toxicol.* **27**, 813–819.

Giese, A. C. (1971). Photosensitization by natural pigments. In A. C. Giese (Ed.), *Photophysiology.* Academic Press, New York, Vol. 6, pp. 77–129.

Girotti, A. J. (1983). Mechanisms of photosensitization. *Photochem. Photobiol.* **38**, 745–751.

Gommers, F. J., and Geerligs, J. W. G. (1973). Lethal effect of near ultraviolet light on *Pratylenchus penetrans* from roots of *Tagetes. Nematologica* **19**, 389–393.

Gommers, F., Bakker, J., and Smit, L. (1980). Effects of singlet oxygen generated by the nematicidal compound α-terthienyl from *Tagetes* on the nematode *Aphelenchus avenae. Nematologica* **26**, 369–375.

Gong, H.-H., Kagan, J., Seiyz, R., Stokes, A. B., Meyer, F. A., and Tuveson, R. W. (1988). The phototoxicity of phenylheptatriyne: Oxygen-dependent hemolysis of human erythrocytes and inactivation of *Escherichia coli. Photochem. Photobiol.* **47**, 55–63.

Greenstock, C. L., and Wiebe, R. H. (1981). Kinetic studies of peroxides and peroxy radicals and their reactions with biological molecules. In M. A. J. Rodgers and E. L. Powers (Eds.), *Oxygen and Oxy-radicals in Chemistry and Biology.* Academic Press, New York, pp. 119–131.

Grossweiner, L. (1984). Mechanisms of photosensitization by furocoumarins. *Natl. Cancer Ins. Monogr.* **66**, 47–54.

Hahn, F. E. (1969). Complexes of quinine and berberine with DNA. *Prog. Antimicrob. Anticancer Chemother.* **6**, 416–422.

Hamberg, M., and Gotthammar, B. (1973). A new reaction of unsaturated fatty acid hydroperoxides: Formation of 11-hydroxy-12,13-epoxy-9-octadecenoic acid from 13-hydroperoxy-9,11-octadecadienoic acid. *Lipids* **8**, 737–744.

Hartman, P. E., Dixon, W. J., Dahl, T. A., and Daub, M. E. (1988). Multiple mechanisms of photodynamic action by cercosporin. *Photochem. Photobiol.* **47**, 699–703.

Hausen, B. M., and Kallweit, M. (1986). The sensitizing capacity of coumarins. II. *Contact Dermatitis* **15**, 289–294.

Hearst, J. E. (1981). Psoralen photochemistry and nucleic and structure. *J. Invest. Dermatol.* **771**, 39–44.

Heelis, P. F. (1983). The photophysical and photochemical properties of flavins (isoalloxazines). *Chem. Soc. Rev.* **11**, 15–39.

Hönigsmann, H., and Ortel, B. (1985). Khellin photochemotherapy of vitiligo. *Photodermatology* **2**, 193–194.

Hoshino, M., Akimoto, H., and Okuda, M. (1978). Photochemical oxidation of benzene, toluene, and ethylbenzene initiated by OH radical in the gas phase. *Bull. Chem. Soc. Jpn.* **51**, 718–724.

Hudson, J. B., and Towers, G. H. N. (1988). Antiviral properties of photosensitizers. *Photochem. Photobiol.* **48**, 289–296.

Hudson, J. B., Graham, E. A., and Towers, G. H. N. (1986a). Antiviral effect of harmine, a photoactive β-carboline alkaloid *Photochem. Photobiol.* **43**, 21–26.

Hudson, J. B., Graham, E. A., Fong, R., Hudson, L. L., and Towers, G. H. N. (1986b). Further studies on the antiviral activity of harmine, a photoactive β-carboline alkaloid. *Photochem. Photobiol.* **44**, 483–487.

Ito, T. (1978). Cellular and subcellular mechanisms of photodynamic action: The $^1O_2$ hypothesis as a driving force in recent research. *Photochem. Photobiol.* **28**, 493–508.

Jori, G., and Spikes, J. D. (1981). Photosensitized oxidations in complex biological structures. In M. A. J. Rodgers and E. L. Powers (Eds.), *Oxygen and Oxy-radicals in Chemistry and Biology.* Academic Press, New York, pp. 119–131.

Joschek, H.-I. and Grossweiner, L. I. (1966). Optical generation of hydrated electrons from aromatic compounds. II. *J. Am. Chem. Soc.* **88**, 3261–3268.

Joshi, P. C., and Pathak, M. A. (1983). Production of singlet oxygen and superoxide radicals by psoralens and their biological significance. *Biochem. Biophys. Res. Commun.* **112**, 638–646.

Kagan, J. (1987). Phenylheptatriyne: Occurrence, synthesis, biological properties, and environmental concerns. *Chemosphere* **16**, 2405–2416.

Kagan, J., Gabriel, R., and Reed, S. (1980). *Alpha*-terthienyl, a non-photodynamic phototoxic compound. *Photochem. Photobiol.* **31**, 465–469.

Kagan, J., Tadema-Wielandt, K., Chan, G., Dhawan, S. N., Jaworski, J., Prakash, I., and Arora, S. K. (1984). Oxygen requirement for near-UV mediated cytotoxicity of phenylheptatriyne to *Escherichia coli, Photochem. Photobiol.* **39**, 465–467.

Kagan, J., Kagan, E. D., Kagan, I. A., and Kagan, P. A. (1987). Do polycyclic aromatic hydrocarbons, acting as photosensitizers, participate in the toxic effects of acid rain? *ACS Symp. Ser.* **327**, 191–204.

Kagan, J., Chen, X., Wang, T. P., and Forlot, P. (1992). Psoralens cleave pBR322 DNA under ultraviolet radiation. *Photochem. Photobiol.* **56**, 185–194.

Kaidbey, K. H., and Kligman, A. M. (1981). Photosensitization by coumarin derivatives. Structure-activity relationships. *Arch. Dermatol.* **117**, 258–163.

Kang, H. J., Shin, E. J., and Shim, S. C. (1991). Comparison of photophysical and photochemical properties of khellin and 8-methoxypsoralen. *Bull. Korean Chem. Soc.* **12**, 554–559.

Kirkiacharian, B., Santus, R., and Helene, C. (1969). Free radical formation by photosensitization with coumarin derivatives. *Ann. Pharm. Fr.* **27**, 129–134.

Knight, R. J., and Sylva, R. N. (1975). Spectrophotometric investigation of iron(III) hydrolysis in light and heavy water at 25 °C. *J. Inorg. Nucl. Chem.* **37**, 779–783.

Knox, J. P., and Dodge, A. D. (1985). Isolation and activity of the photodynamic pigment hypericin. *Plant, Cell Environ.* **8**, 19–25.

Kochevar, I. E., and Dunn, D. A. (1990). Photosensitized reactions of DNA: Cleavage and addition. In H. Morrison (Ed.), *Bioorganic Photochemistry: Photochemistry and the Nucleic Acids.* Wiley, New York, pp. 273–315.

Krasnovsky, A. A., Sukhurov, V. L., Egorov, S. Y., and Potapenko, A. Y. (1986). Generation and quenching of singlet molecular oxygen by furocoumarins. Direct measurements. *Stud. Biophys.* **114**, 149–158.

Kyker, G. C., McEwen, M. M., and Cornatzer, W. E. (1947). The formation of antimalarial agents by ultraviolet decomposition of quinine. *Arch. Biochem.* **12**, 191–199.

Lai, T.-I., Lim, B. T., and Lim, E. C. (1982). Photophysical properties of biologically important molecules related to proximity effects: Psoralens. *J. Am. Chem. Soc.* **104**, 7631–7635.

Lamola, A. A., Yamane, T., and Trozzolo, A. M. (1973). Cholesterol hydroperoxide formation in red cell membranes and photohemolysis in erythropoietic protoporphyria. *Science* **179**, 1131–1133.

Langford, C. H., Wingham, M., and Sastri, V. S. (1973). Ligand photooxidation in copper(II) complexes of nitrilotriacetic acid. Implications for natural waters. *Environ. Sci. Technol.* **7**, 820–822.

Larson, R. A. (1986). Insect defenses against phototoxic plant chemicals. *J. Chem. Ecol.* **12**, 859–870.

Larson, R. A., and Berenbaum, M. R. (1988). Environmental phototoxicity. *Environ. Sci. Technol.* **22**, 354–360.

Larson, R. A., Hunt, L. L., and Blankenship, D. W. (1977). Formation of toxic products from a #2 fuel oil by photooxidation. *Environ. Sci. Technol.* **11**, 492–496.

Larson, R. A., Bott, T. L., Hunt, L. L., and Rogenmuser, K. (1979). Photooxidation products of a fuel oil and their antimicrobial activity. *Environ. Sci. Technol.* **13**, 965–969.

Larson, R. A., Marley, K. A., Tuveson, R. W., and Berenbaum, M. R. (1988). β-Carboline alkaloids: Mechanisms of phototoxicity to bacteria and insects. *Photochem. Photobiol.* **48**, 665–674.

Larson, R. A., Lloyd, R. E., Marley, K. A., and Tuveson, R. W. (1992). Ferric ion-photosensitized damage to DNA by hydroxyl and non-hydroxyl radical mechanisms. *J. Photochem. Photobiol.* **B14**, 345–357.

Laurie, W. A., McHale, D., and Saag, K. (1986). Photoreactions of quinine in aqueous citric acid solution. *Tetrahedron* **42**, 3711–3714.

Lee, C. C., and Rodgers, M. A. J. (1987). Laser flash photokinetic studies of rose bengal sensitized photodynamic interactions of nucleotides and DNA. *Photochem. Photobiol.* **45**, 79–86.

Lichtenthaler, R. G., Haag, W. R., and Mill, T. (1989). Photooxidation of probe compounds sensitized by crude oils in toluene and as an oil film on water. *Environ. Sci. Technol.* **23**, 39–45.

Lindig, B. A., and Rodgers, M. A. J. (1979). Laser photolysis studies of singlet molecular oxygen in aqueous micellar dispersions. *J. Phys. Chem.* **83**, 1683–1688.

Maiti, M., Nandi, R., and Chaudhuri, K. (1982). Sanguinarine: A monofunctional intercalating alkaloid. *FEBS Lett.* **142**, 280–284.

Mäkinen, K. K., and Mäkinen, P.-L. (1982). Diketones as photosensitizing agents: Application to α-amino acids and enzymes. *Photochem. Photobiol.* **35**, 761–765.

Mantulin, W. W., and Song, P. S. (1973). Excited states of skin-sensitizing coumarins and psoralens. Spectrographic studies. *J. Am. Chem. Soc.* **95**, 5122–5129.

Marković, D. Z., Durand, T., and Patterson, L. K. (1990). Hydrogen abstraction from lipids by triplet states of derivatized benzophenone photosensitizers. *Photochem. Photobiol.* **51**, 389–394.

Mattay, J. (1987). Charge transfer and radical ions in photochemistry. *Angew. Chem., Int. Ed. Engl.* **26**, 825–845.

Mattes, S. L., and Farid, S. (1984). Exciplexes and electron transfer reactions. *Science* **226**, 917–921.

Mauzerall, D., and Feher, G. (1964). A study of the photoinduced porphyrin free radical by electron spin resonance. *Biochim. Biophys. Acta* **79**, 430–432.

Mayo, F. R. (1958). The oxidation of unsaturated compounds. 7. The oxidation of aliphatic unsaturated compounds. *J. Am. Chem. Soc.* **80**, 2497–2500.

McDonaugh, A. F. (1971). The role of singlet oxygen in bilirubin photo-oxidation. *Biochem. Biophys. Res. Commun.* **44**, 1306–1311.

McKenna, D. J., and Towers, G. H. N. (1981). Ultra-violet mediated cytotoxic activity of β-carboline alkaloids. *Phytochemistry* **20**, 1001–1004.

McLachlan, D., Arnason, J. T., and Lam, J. (1984). Role of oxygen in photosensitizations with polyacetylenes and thiophene derivatives. *Photochem. Photobiol.* **39**, 177–182.

Meffert, H., Bohm, F., Roder, B., and Sonnichsen, N. (1982). Is excited singlet oxygen involved in PUVA therapy effect? *Dermatol. Monatsschr.* **168**, 387–393.

Megaw, J., Lee, J., and Lerman, S. (1980). NMR analyses of tryptophan-8-methoxypsoralen photoreaction products formed in the presence of oxygen. *Photochem. Photobiol.* **32**, 265–269.

Midden, W. R. (1988). Chemical mechanisms of the bioeffects of furocoumarins: The role of reactions with proteins, lipids, and other cellular constituents. In F. P. Gasparro (Ed.), *Psoralen DNA Photobiology*. CRC Press, Boca Raton, FL. Vol. 2, pp. 1–49.

Millspaugh, D. V. (1892). *American Medicinal Plants*. J. C. Yorston and Co., Philadelphia (reprinted by Dover Publ., New York, 1974).

Mitchell, J. C., and Rook, A. (1979). *Botanical Dermatology*. Greenglass Publ., Vancouver, Canada.

Moan, J. (1986). Porphyrin photosensitization and phototherapy. *Photochem. Photobiol.* **43**, 681–690.

Mohammad, T., Baird, W. M., and Morrison, H. (1991). Photochemical covalent binding of *p*-methoxycinnamic acid to calf thymus DNA. *Bioorg. Chem.* **19**, 88–100.

Moore, D. E. (1977). Photosensitization by drugs. *J. Pharm. Sci.* **66**, 1282–1284.

Moore, D. E. (1980). Photosensitization by drugs: Quinine as a photosensitizer. *J. Pharm. Pharmacol.* **32**, 216–218.

Moore, D. E., and Hemmens, V. J. (1982). Photosensitization by antimalarial drugs. *Photochem. Photobiol.* **36**, 71–77.

Morliere, P., Höningsmann, H., Averbeck, D., Dardalhon, M., Hüppe, G., Ortel, B., Santus, R., and Dubertret, L. (1988). Phototherapeutic, photobiologic, and photosensitizing properties of khellin. *J. Invest. Dermatol.* **90**, 720–724.

Morrison, H., and Deibel, R. M. (1986). Photochemistry and photobiology of urocanic acid. *Photochem. Photobiol.* **43**, 663–665.

Murray, R. D. H., Mendez, J., and Brown, S. A. (1982). *The Natural Coumarins. Occurrence, Chemistry and Biochemistry*. Wiley (Interscience), New York.

Musajo, L., and Rodighiero, G. (1962). The skin-sensitizing furocoumarins. *Experientia* **18**, 153–161.

Musajo, L., and Rodighiero, G. (1972). Mode of photosensitizing action of furocoumarins. In A. C. Giese (Ed.), *Photophysiology*. Academic Press, New York, Vol. 7, Chapter 5.

Neckers, D. (1989). Rose bengal. *J. Photochem. Photobiol.* **A47**, 1–29.

Obaseiki-Ebor, E. E., and Obasi, E. E. (1987). Quinine induced *Escherichia coli* DNA base-pair substitution mutation. *Bull. Environ. Contam. Toxicol.* **38**, 422–425.

Oginsky, E. L., Green, G. S., Griffith, D. G., and Fowlks, W. L. (1959). Lethal photosensitization of bacteria with 8-methoxypsoralen to long wavelength ultraviolet radiation. *J. Bacteriol.* **78**, 821–833.

Oris, J. T., and Giesy, J. P., Jr. (1987). The photo-induced toxicity of polycyclic aromatic hydrocarbons to larvae of the fathead minnow (*Pimephales promelas*). *Chemosphere* **16**, 1395–1404.

Oroskar, A. A., Gasparro, F. P., and Peak, M. J. (1993). Relaxation of supercoiled DNA by aminomethyl trimethylpsoralen and UV photons: Action spectrum. *Photochem. Photobiol.* **57**, 648–654.

Otsuki, T. (1987). Photochemical reaction of furocoumarin (psoralen). The reactive site in its photoexcited state. *Chem. Lett.*, pp. 453–454.

Palumbo, M., Rodighiero, P., Gia, O., Guiotto, A., Magno, S., and Marciani Magno, S. (1986). Benzofurocoumarins: New monofunctional DNA-photobinding agents. *Photochem. Photobiol.* **44**, 1–4.

Pathak, M. A., and Fellman, J. H. (1960). Activating and fluorescent wavelengths of furocoumarins: Psoralens. *Nature (London)* **185**, 383–388.

Pathak, M. A., and Fitzpatrick, T. B. (1959). Relationship of molecular configuration to the activity of furocoumarins which increase the cutaneous responses following long wave ultraviolet radiation. *J. Invest. Dermatol.* **32**, 255–267.

Pathak, M. A., and Joshi, P. C. (1984). Production of active oxygen species (singlet oxygen and superoxide anion radical) by psoralens and ultraviolet radiation (320–400 nm). *Biochim. Biophys. Acta* **798**, 115–126.

Patterson, R. E. (1964). *Texas Plants Poisonous to Livestock*. Texas Agricultural Experiment Station, College Station.

Payne, J. R., and Phillips, C. R. (1985). Photochemistry of petroleum in water. *Environ. Sci. Technol.* **19**, 569–579.

Pfyffer, G. E., Panfil, I., and Towers, G. H. N. (1982). Monofunctional covalent photobinding of dictamnine, a furoquinoline alkaloid, to DNA as target *in vivo*. *Photochem. Photobiol.* **35**, 63–68.

Philogène, B. J. R., Arnason, J. T., Towers, G. H. N., Abramowski, Z., Campos, F., Champagne, D., and McLachlan, D. (1984). Berberine: A naturally occurring phototoxic alkaloid. *J. Chem. Ecol.* **10**, 115–123.

Pimprikar, G. D., and Coign, M. J. (1987). Multiple mechanisms of dye-induced toxicity to insects. In J. R. Heitz and K. R. Downum, eds., Light-activated pesticides. *ACS Symp. Ser.* **339**, 134–155.

Poehlmann, H., Theil, F. P., and Pfeifer, S. (1986). Photochemical reactivity of quinine- and quinidine-*N*-oxides *in vitro* and *in vivo*. *Pharmazie*. **41**, 859–862.

Poppe, W., and Grossweiner, L. I. (1975). Photodynamic sensitization by 8-methoxypsoralen via singlet oxygen mechanism. *Photochem. Photobiol.* **22**, 217–219.

Potapenko, A. Y. (1991). Mechanisms of photodynamic effects of furocoumarins. *J. Photochem. Photobiol.* **B9**, 1–34.

Potapenko, A. Y., and Sukhorukov, V. L. (1984). Photooxidative reactions of psoralens. *Stud. Biophys.* **101**, 89–98.

Potapenko, A. Y., Sukhorukov, V. L., and Davidov, B. V. (1984). A comparison between skin-sensitizing (334 nm) activities of 8-methoxypsoralen and angelicin. *Experientia* **40**, 264–265.

Potapov, V. K., and Sorokin, V. V. (1970). Ionic-molecular reactions taking place during the photoionization of aromatic compounds and alcohols. *Dokl. Akad. Nauk SSSR* **195**, 616–619.

Praseuth, D., Gaudemer, A., Verlhac, J.-B., Kraljic, I., Sissoëff, I., and Guillé, E. (1986). Photocleavage of DNA in the presence of synthetic water-soluble porphyrins. *Photochem. Photobiol.* **44**, 717–724.

Prathapan, S., Loft, S., and Agosta, W. C. (1990). Photochemical hydrogen abstraction by singlet and triplet $n\pi^*$ states of aromatic nitrogen: Fragmentation of 4-alkylpyrimidines and 2-alkylquinolines. *J. Am. Chem. Soc.* **112**, 3940–3944.

Proksch, P., Proksch, M., Towers, G. H. N., and Rodriguez, E. (1983). Phototoxic and insecticidal activities of chromenes and benzofurans from *Encelia*. *J. Nat. Prod.* **46**, 331–334.

Rebeiz, C. A., Juvik, J. A., and Rebeiz, C. C. (1988). Porphyric insecticides. 1. Concept and phenomenology. *Pestic Biochem. Physiol.* **30**, 11–27.

Reinhart, P., Harkdrader, R., Wylie, R., Yewey, G., and van Horne, K. C. (1991). Sanguinarine levels in biological samples by high-performance liquid chromatography. *J. Chromatogr.* **570**, 425–434.

Reyftmann, J. P., Kagan, J., Santus, R., and Morliere, P. (1985). Excited state properties of α-terthienyl and related molecules. *Photochem. Photobiol.* **41**, 1–7.

Robinson, T. (1991). *The Organic Constituents of Higher Plants*, 6th ed. Cordus Press, North Amherst, MA.

Rodgers, M. A., and Snowden, P. T. (1982). Lifetime of $O_2(^1\Delta_g)$ in liquid water as determined by time-resolved infrared luminescence measurements. *J. Am. Chem. Soc.* **104**, 5541–5543.

Rodighiero, P., Chilin, A., Pastorini, G., and Guiotto, A. (1987). Pyrrolocoumarin derivatives as potential photoreagents toward DNA. *J. Heterocycl. Chem.* **24**, 1041–1043.

Rosenstein, B. S., Ducore, J. M., and Cummings, S. W. (1983). The mechanism of bilirubin-photo-sensitized DNA strand breakage in human cells exposed to phototherapy light. *Mutat. Res.* **112**, 397–406.

Rossman, T. G. (1989). On the mechanism of the comutagenic effect of Cu(II) with ultraviolet light. *Biol. Trace Elem. Res.* **21**, 383–388.

Russell, G. A. (1957). Deuterium-isotope effects in the autoxidation of aralkyl hydrocarbons. Mechanism of the interaction of peroxy radicals. *J. Am. Chem. Soc.* **79**, 3871–3877.

Sa E Melo, T., Morliere, P., Goldstein, S., Santus, R., Dubertret, L., and Lagrange, D. (1984). Binding of 5-methoxypsoralen to human serum low density lipoproteins. *Biochem. Biophys. Res. Commun.* **120**, 670–676.

Sancier, K. M., and Morrison, S. R. (1979). ESR study of photooxidation of quinoline on rutile and effects of surface states. *Surf. Sci.* **83**, 29–44.

Scaiano, J. C., Evans, C., and Arnason, J. T. (1989). Characterization of the α-terthienyl radical cation: Evidence against electron transfer to oxygen *in vitro*. *J. Photochem. Photobiol.* **B3**, 411–418.

Scaiano, J. C., Redmond, R. W., Mehta, B., and Arnason, J. T. (1990). Efficiency of the photoprocesses leading to singlet oxygen generation by α-terthienyl: Optical absorption, optoacoustic calorimetry and infrared luminescence studies. *Photochem. Photobiol.* **52**, 655–659.

Schimmer, O., and Kühne, I. (1991). Furoquinoline alkaloids as photosensitizers in *Chlamydomonas reinhardtii*. *Mutat. Res.* **249**, 105–110.

Seely, G. R. (1977). Mechanisms of the photosensitized oxidation of tyramine. *Photochem. Photobiol.* **26**, 115–123.

Shim, S. C., and Chi, D. Y. (1978). Photocycloaddition of 5,7-dimethoxycoumarin to tetramethylethylene. *Chem. Lett.*, pp. 1229–30.

Shim, S. C., and Im, K. R. (1976). Studies on the photoreactions of coumarins and furocoumarins. *J. Korean Chem. Soc.* **20**, 236–239.

Shim, S. C., and Kang, H. K. (1987). Photophysical properties of khellin. *Bull. Korean Chem. Soc.* **8**, 341–344.

Shim, S. C., and Kim, Y. Z. (1983). The photocycloaddition reaction of 8-methoxypsoralen with olefins. *J. Photochem.* **23**, 83–92.

Shim, S. C., and Lee, T. S. (1986). Photoreaction of 1,6-disubstituted 1,3,5,-hexatriynes with some olefins. *Bull. Korean Chem. Soc.* **7**, 304–308.

Shim, S. C., and Yun, M. H. (1992). Adenosine mediated photoreaction of 4,5',8-trimethylpsoralen with alcohols. *Abstr. 20th Annu. Meet. Am. Soc. Photobiol.*, p. 51S.

Shonberg, A., and Sina, A. (1950). Khellin and allied compounds. *J. Am. Chem. Soc.* **72**, 1611–1616.

Sigman, M. E., Zingg, S. P., Pagni, R. M., and Burns, J. H. (1991). Photochemistry of anthracene in water. *Tetrahedron Lett.* **41**, 5737–5740.

Slavík, J. (1967). Alkaloids of Papaveraceae. XXXVII. Alkaloids of *Stylomecon heterophylla, Hylomecon vernalis* and *Sanguinaria canadensis. Collect. Czech. Chem. Commun.* **32**, 4431–4438.

Slifkin, M. A. (1971). *Charge Transfer Interactions of Biomolecules.* Academic Press, New York.

Sloper, R. W., Truscott, T. G., and Land, E. J. (1979). The triplet state of 8-methoxypsoralen. *Photochem. Photobiol.* **29**, 1025–1029.

Solar, S., and Quint, R. (1992). Radiation induced primary processes of aqueous 8-methoxypsoralen. A pulse radiolysis study. *Radiat. Phys. Chem.* **39**, 171–175.

Song, P.-S. (1984). Photoreactive states of furocoumarins. *Natl. Cancer Inst. Monogr.* **66**, 15–19.

Song, P.-S., and Tapley, K. J., Jr. (1979). Photochemistry and photobiology of psoralens. *Photochem. Photobiol.* **29**, 1177–1197.

Specht, K. G., Kittler, L., and Midden, W. R. (1988). A new biological target of furanocoumarins: Photochemical formation of covalent adducts with unsaturated fatty acids. *Photochem. Photobiol.* **47**, 537–541.

Srinivasan, V. S. (1991). Oxygen derivatives in electrochemical and photochemical studies. *Spectrum* 4 (Winter), 15–16.

Stenberg, V. I., and Travecedo, E. F. (1970). Nitrogen photochemistry. V. A new photochemical reduction of the cinchona alkaloids, quinine, quinidine, cinchonidine, and cinchonine. *J. Org. Chem.* **35**, 4131–4136.

Stenberg, V. I., Travecedo, E. F., and Musa, W. E. (1969). *Tetrahedron Lett.* **25**, 2031–2033.

Straight, R. C., and Spikes, J. D. (1985). Photosensitized oxidation of biomolecules. In A. A. Frimer (Ed.), *Singlet Oxygen*. CRC Press, Boca Raton, FL pp. 91–143.

Straub, K. D., and Carver, P. (1975). Sanguinarine, inhibitor of (sodium-potassium ion)-dependent ATPase. *Biochem. Biophys. Res. Commun.* **62**, 913–922.

Straub, K. D., Kanne, D., Hearst, J. E., and Rapoport, H. (1981). Isolation and characterization of pyrimidine-psoralen photoadducts from DNA. *J. Am. Chem. Soc.* **103**, 2347–2355.

Takano, J., Kitahara, T., Shirai, K., and Mitsuazawa, S. (1984). The photoreaction of psoralen with cyclohexene. *Proc. Fac. Sci. Tokai Univ.* **19**, 95–98.

Tønnesen, H. H., Karlsen, J., and Beijersbergen van Henegouwen, G. (1986a). Studies on curcumin and cucurminoids. VIII. Photochemical stability of curcumin. *Z. Lebensm.-Unters.-Forch.* **183**, 116–122.

Tønnesen, H. H., de Vries, H., Karlsen, J., and Beijersbergen van Henegouwen, G. (1986b). Studies on curcumin and cucurminoids. IX. Investigation of the photobiological activity of curcumin using bacterial indicator systems. *J. Pharm. Sci.* **76**, 371–373.

Towers, G. H. N. (1982). Photosensitizers from plants and their photodynamic action. *Prog. Phytochem.* **6**, 183–202.

Towers, G. H. N. (1984). Interactions of light with photochemicals in some natural and novel systems. *Can. J. Bot.* **62**, 2900–2911.

Towers, G. H. N., and Abramowski, Z. (1983). UV-mediated genotoxicity of furanoquinolines and of certain tryptophan-derived alkaloids. *J. Nat. Prod.* **46**, 576–581.

Tuveson, R. W., Berenbaum, M. R., and Heininger, E. (1986). Inactivation and mutagenesis by phototoxins using *Escherichia coli* strains differing in sensitivity to near- and far-ultraviolet light. *J. Chem. Ecol.* **12**, 933–948.

Tuveson, R. W., Larson, R. A., Marley, K. A., Wong, G. -R., and Berenbaum, M. R. (1989). Sanguinarine, a phototoxic H_2O_2-producing alkaloid. *Photochem. Photobiol.* **50**, 733–738.

Tuveson, R. W., Wang, G.-R., Wang, T. P., and Kagan, J. (1990). Light-dependent cytotoxic reactions of anthracene. *Photochem. Photobiol.* **52**, 993–1002.

Varma, R. L. (1927). Berberine phosphate in Oriental sore. *Indian Med. Gaz.* **62**, 84–85.

Veronese, F. M., Schiavon, O., Bevilacqua, R., and Rodighiero, G. (1979). Drug-protein interaction: 8-Methoxypsoralen as photosensitizer of enzymes and amino acids. *Z. Naturforsch., C: Biosci.* **34C**, 392–396.

Veronese, F. M., Schiavon, O., Bevilacqua, R., Bordin, F., and Rodighiero, G. (1982). Photoinactivation of enzymes by linear and angular furocoumarins. *Photochem. Photobiol.* **36**, 25–32.

Vert, F. T., Sanchez, I. Z., and Torrent, A. O. (1983). Acidity constants of β-carbolines in the ground and excited singlet state. *J. Photochem.* **23**, 355–368.

Vichkanova, S. A., and Adgina, V. V. (1971). Antifungal properties of sanguinarine. *Antibiotiki (Moscow)* **16**, 609–612.

von Sonntag, C. (1987). *The Chemical Basis of Radiation Biology*. Taylor & Francis, London.

Wang, S. Y., Ryang, H. S., Burrows, E. P., and Nagamatsu, T. (1981). α-Diketone sensitized photo-epoxidation of pyrimidines. In M. A. J. Rodgers and E. L. Powers (Eds.), *Oxygen and Oxy-radicals in Chemistry and Biology*. Academic Press, New York, pp. 475–478.

Wang, T. P., Kagan, J., Tuveson, R. W., and Wang, G. R. (1991). α-Terthienyl photosensitizes damage to pBR322 DNA. *Photochem. Photobiol.* **53**, 463–467.

Wat, C. -K., MacRae, W. D., Yamamoto, E., Towers, G. H. N., and Lam, J. (1980). Phototoxic effects of naturally occurring polyacetylenes and α-terthienyl on human erythrocytes. *Photochem. Photobiol.* **32**, 167–172.

Waterman, P. G., and Grundon, M. F. (Eds.) (1983). *Chemistry and Chemical Taxonomy of the Rutales.* Academic Press, New York.

Weir, D., Scaiano, J. C., Arnason, J. T., and Evans, C. (1985). Photochemistry of the phototoxic polyacetylene phenylheptatriyne. *Photochem. Photobiol.* **42**, 223–230.

Wilkinson, F., and Brummer, J. G. (1981). Rate constants for the decay and reactions of the lowest excited singlet state of molecular oxygen in solution. *J. Phys. Chem. Ref. Data* **10**, 809–999.

Yamazaki, S., Okubo, A., Akiyama, Y., and Fuwa, K. (1975). Cercosporin, a novel photodynamic pigment isolated from *Cercospora kikuchii. Agric. Biol. Chem.* **39**, 287–291.

Yoon, U. C., and Epling, G. A. (1980). The investigation of photochemical reactions of phototoxic antimalarial compounds. *Arch. Pharmacol. Res.* **3**, 87–88.

You, J.-L., and Fong, F. K. (1986). Superoxide photogeneration by chlorophyll A in water-acetone. Electron spin resonance studies of radical intermediates in chlorophyll A photoreaction *in vitro. Biochem. Biophys. Res. Commun.* **139**, 1124–1129.

10

EFFECTS OF OZONE IN HUMAN STUDIES

Neil Alexis, Susan M. Tarlo, and Frances Silverman

The Gage Research Institute, Toronto, Ontario, Canada M5T 1R4

Environmental Oxidants, Edited by Jerome O. Nriagu and Milagros S. Simmons.
ISBN 0–471–57928–9   © 1994 John Wiley & Sons, Inc.

1. INTRODUCTION

Much is already known about the physical properties, the chemistry, and some of the health effects of ozone (O_3) in both humans and animals. However, a lot of what we presently know concerning human health relates to transient, short-term, reversible effects that follow exposures lasting from 5 min to 6.6 hr. These health effects include changes in vital capacity, flow resistance, reactivity to bronchoactive challenge, epithelial permeability, and respiratory symptoms. These effects can all be observed within the first few hours after exposure begins and some effects persist for hours or days after exposure ceases. In addition, repetitive daily exposures will initially exacerbate and prolong these observed transient changes, causing a cumulative effect; following this, an adaptation phenomenon occurs, and effects are then reduced over time.

Interpreting the health significance of transient functional responses to O_3 is difficult. It is complicated by many factors such as the large interindividual variablity in pulmonary baseline function and responsiveness to ozone, as well as the need to control for both the transient effects of recent nonexperimental acute exposures and the cumulative effects of chronic outdoor exposures at any given measurement point. What cannot be ignored, however, is the ample evidence that some healthy adults who exercise heavily for 1 to 2 hr during periods of elevated O_3 concentrations experience pronounced symptoms and decreases in lung function. These effects have been demonstrated in adults and children even at O_3 concentrations comparable to peak levels found in many Canadian and American cities. Discerning populations that are at risk or sensitive to the effects of elevated O_3 levels, compared to elevated levels of other airborne irritants, appears to be a difficult task. However, a subgroup of the general population that regularly exercises or works outdoors is at increased risk because of the larger dose received from increased ventilation.

Many of the human health effects of O_3 discussed in this chapter are short-term, transient, and appear to be reversible. This has led some investigators to dismiss the relative importance of these effects and instead focus their attention on other irritants with known chronic effects. For other research groups, the challenge lies in revealing the significance of the transient pulmonary function effects and determining their role and contribution to the development of chronic lung disease.

What follows is a discussion of the human health effects of O_3, beginning with a short background discussion of source and tissue dose, continuing with a brief discussion of potentially at-risk populations and typical health effect study designs, and ending with a discussion of the specific health effects of O_3.

2. BACKGROUND SOURCES OF OZONE

Ambient O_3 is generated from several different sources. One source is stratospheric O_3, especially in springtime when the stratosphere–tropospheric air exchange is greatest. The other sources of O_3 result from photochemical reaction sequences requiring the input of organic vapors, nitrogen oxides (NO_xs), and actinic radiation (Lippmann, 1989). Olefinic hydrocarbons, formaldehyde, and *m*-xylene are very efficient contributors to O_3 formation; however, the role of NO_x is especially critical. Altshuller (1987) notes that unless NO_x concentrations exceed about 0.02 to 0.03 ppb, photochemical O_3 loss exceeds photochemical O_3 production. NO_x concentrations of 5 to 10 ppb are typical of rural areas within more heavily populated areas in the United States, Canada, and Europe. It has been estimated that at rural sites in the United States where NO_x concentrations were in the range of 1 to 10 ppb, 5 to 7 ppb of O_3 was formed per ppb of NO_x reacted (Altshuller, 1987).

Ozone is reactive with sunlight, so its levels tend to decrease in the evening but can remain at elevated concentrations in the ambient air above the mixing layer. This O_3 can then contribute to the following day's O_3 ground levels when air mixing increases. It can be seen then how multiday summer episode exposures occur. Rao (1988) reported a high likelihood of O_3 greater than 80 ppb continuing for three or more days once it has been around for at least one day.

A number of variables will govern the peak concentration of O_3 in a day. They are the baseline level of O_3 in the air above the mixing layer, the photochemical production rate during the day, and the concentration of O_3-scavenging chemicals, such as nitric oxide (NO) and ethylene (Lippmann, 1989).

The typical O_3 exposure pattern found in heavily populated areas in the United States and Canada has a daily plateau after 10:00 a.m. in which the maximum 8-hr average exposure level is approximately 90% of the maximum U.S. 1-hr exposure standard (120 ppb) (Rombout et al., 1986). Thus the local generation of O_3 represents only part of the broad daily curve arising from a series of upwind sources and photochemistry. The size of any particular O_3 peak depends on the concentration of precursor reactants in the incoming air and the concentration of reactants found in the local air. Atmospheric conditions then dictate either single or multiple periods outdoors during which an individual will be exposed to O_3 at or above the 1-hr maximum Canadian and U.S. standard of 80 and 120 ppb, respectively.

Since O_3 peaks can occur at any time throughout the day, individuals participating in outdoor activities and outdoor employment, such as construction and public works, may potentially be exposed to high O_3.

3. TISSUE DOSES

The only significant route of exposure to O_3 is inhalation. Exposure dose is determined by the concentration at the nose and mouth as well as by the volume of air inhaled and the pattern of uptake of O_3 molecules along the respiratory tract. People who exercise or work outdoors will increase their ventilation rate and thus their total dose of O_3. The removal efficiency of O_3 from inspired air at concentrations of 100, 200, and 400 ppb in the nose only, mouth only, and through oronasal breathing was studied by Gerrity et al. (1988). The mean extrathoracic (all airways superior and anterior to the posterior pharynx) removal efficiency was 39.6%; for the intrathoracic route (all other portions of the tract, including the larynx), it was 91% in healthy, nonsmoking males. Thus his study found that the extrathoracic airways of humans removed approximately 40% of inhaled O_3, while 95% of the remaining O_3 presented to the posterior pharynx was removed via the intrathoracic airway. From this work and based on the models developed by Miller and colleagues (1978), they estimated that humans have about twice the deposition rate at the respiratory acinus as rats with the same exposure. This is consistent with the larger pulmonary function responses seen in humans compared to rats. This important work suggests that chronic animal studies are likely to yield conservative estimates of the effects in humans in areas with high ambient exposure, such as southern California.

4. TRANSIENT REVERSIBLE PULMONARY FUNCTION RESPONSES

Research focused on O_3 and human health effects has extensively assessed transient responses to acute exposure. In addition to lung volume and flow changes, other responses to acute and subacute exposures have been studied, such as mucociliary and early alveolar zone particle clearance (Foster et al., 1987), functional responses in macrophages, and changes in lung cell secretions (Frager et al., 1979; Kenoyer et al., 1981; Lippmann et al., 1987; Schlessinger and Driscoll, 1987), but this body of work has largely been carried out in animals. The data from animal studies showing structural changes and transient functional and cellular responses suggest possible links between O_3 exposure and respiratory diseases such as chronic obstructive pulmonary disease in humans, but at present there is not enough clear evidence for such a tie.

Inhalation of O_3 causes a concentration-dependent mean decrement in vital capacity and flow rate during a forced expiratory maneuver, and the decrement increases with increasing minute ventilation (Hazucha, 1987). Several studies on both adults and children have consistently shown decreases in forced vital capacity (FVC), forced expiratory volume in 1 sec (FEV_1), and forced expiratory flow, mid–expiratory phase ($FEF_{25\%-75\%}$) (Folinsbee et al., 1988; Gibbons and Adams, 1984; Horstman et al., 1990; Kreit et al., 1989; McDonnell et al., 1983; Kulle et al., 1985; Spektor et al., 1988a). The extremely large data base on transient effects of O_3 comes from a combination of study designs, most notably controlled human exposures and natural human exposures (Tables 1A and 1B). The former are studies in which selected human volunteers are placed inside a controlled environmental chamber and exposed to a pollutant at a specific concentration for a defined period of time. Many of the protocols in this design involve an exercise regime to best replicate the dose received by an active person outdoors. Typical flow/volume measurements are made [FVC, FEV_1, $FEF_{25\%-75\%}$, PEFR (peak expiratory flow rate)] before and after exposure.

Table 1A Chamber Studies of Mean Pulmonary Function Changes per ppb O_3

Study	Exercise Level (L/min)	Exposure Duration (min) [exercise]	O_3 (ppb)	FVC (mL/ppb)	FEV_1 (mL/ppb)	$FEF_{25\%-75\%}$ (mL/sec/ppb)
Avol et al. (1985)	32	60	150[b]	-0.7	-0.8	-0.7
McDonnell et al. (1983)	39	150[60]	120[b]	-0.3	-0.5	-0.6
Linn et al. (1988)	68	120[60]	160[a]	-0.7	-0.6	-1.1
Kulle et al. (1985)	68	120[60]	150[a]	-0.5	-0.2	-2.1
McDonnell et al. (1983)	65	120[60]	180[a]	-1.8	-1.6	-2.9
	65	120[60]	120[a]	-1.4	-1.3	-3.0
Avol et al. (1984)	57	60[60]	160[a]	-1.5	-1.5	—
Gibbons and Adams (1984)	55	60[60]	150[a]	-1.1	-1.0	-0.6
Folinsbee et al. (1988)	40	395[300]	120[a]	-3.8	-4.5	-5.0

[a] Adults.
[b] Children.

Source: Adapted from Lippmann (1989).

Table 1B Field Studies of Mean Pulmonary Function Changes per ppb O$_3$

Study	Exercise Level (L/min)	Exposure Duration (min) [exercise]	O$_3$ (ppb)	FVC (mL/ppb)	FEV$_1$ (mL/ppb)	FEF$_{25\%-75\%}$ (mL/sec/ppb)
Avol (1987)	22	60[60]	113[b]	−0.3	−0.3	—
Spektor et al. (1988a)	22(est.)	110–150	19–113[b]	−1.0	−1.4	−2.5
Lippmann et al. (1983)	22(est.)	150–550	46–110[b]	−1.1	−0.8	—
Kinney et al. (1988)	15(est.)	1440	7–78[b]	−0.9	−1.0	−1.9
Lioy et al. (1985)	15(est.)	150–550	20–145[b]	−0.1	−0.3	−0.6
Spektor et al. (1988b)	64	26	21–124[a]	−2.9	−3.0	−9.7
	78	29	21–124[a]	−2.1	−1.4	−6.0
Avol et al. (1985)	57	60[60]	153[a]	−1.2	−1.3	—

[a] Adults.
[b] Children.

Source: Adapted from Lippmann (1989).

Other functional tests can be utilized in chamber studies and these include (1) nitrogen washout test to look at small airway function; (2) single breath of 0.3% carbon monoxide to determine diffusing capacity at the alveolar-capillary membrane; (3) helium (He) and sulfur hexafluoride (SF_6) test to measure inhomogeneities in ventilation using radioactive xenon (Xe) and external gamma emission imaging equipment; (4) administration of bronchodilators or bronchoconstrictors to detect airway hyperresponsiveness; and (5) measurement of the rate of clearance from the lung of gamma emitting ^{99m}Tc-diethylenetriaminepentaacetate, inhaled as a droplet aerosol, to determine the permeability of the respiratory epithelium.

Hazucha (1987) synthesized the results from over 75 chamber studies looking at the acute effects of 2-hr O_3 exposure on normal subjects. For each pulmonary function variable and ventilation level, a quadratic function was fitted to the data using regression procedures. The results showed that the slope (quadratic coefficient) for each variable within a group and almost all variables between groups were significantly different from zero and from each other ($p < 0.0001$). The results also showed that for the very high minute ventilation level, the O_3 concentration at which slopes became significantly different from zero for all pulmonary function variables was 100 ppb. Controlled human exposure studies, then, are most useful for examining the nature and extent of transient functional changes resulting from one or a few acute exposures. Table 1A lists chamber exposure studies with O_3 levels less than 180 ppb showing mean functional changes per ppb O_3 in FVC, FEV_1, and $FEF_{25\%-75\%}$.

The other experimental design assesses the natural human exposure, studying populations for evidence of health effects associated with ambient levels of pollutants. Among the difficulties in this design are identifying an accessible population at risk whose relevant exposures can be defined and adequately characterized, and specifying measurable indices of responses that may be expected to occur as a result of the exposures of interest. Several investigators have carried out such designs in adults and children showing significant pulmonary function changes (Table 1B). In a comparison of adult studies, the maximum mean functional change for FVC and FEV_1 was reported by Spektor et al. (1988b) at -2.1 and -1.4 mL/ppb, respectively, in subjects exposed to 21 to 124 ppb for an average of 29 minutes during exercise [Minute ventilation (expired) (VE_{min}) $= 20-153$ L/min]. The maximum reported mean FVC decrement (mL/ppb) in studies of children was reported by Lippmann et al. (1983) at -1.1 (mL/ppb). Lippmann's subjects were exposed for 150 to 550 min to a concentration range of 46 to 110 ppb O_3 with moderate exercise ($VE_{min} = 22$ L/min). For the maximum mean FEV_1 and $FEF_{25\%-75\%}$ decrement in field studies with children, Spektor et al. (1988a) reported -1.4 mL/ppb and -2.5 mL/sec/ppb, respectively, after his subjects were exposed for 150 to 550 min at a concentration range of 19 to 113 ppb O_3 with moderate exercise ($VE_{min} = 22$ L/min).

Natural human exposure studies are most useful for identifying the magnitude and extent of the acute responses to naturally occuring pollutants among a population of people engaged in normal outdoor activities. They do not, however, provide any information on the possible influence of prior chronic exposures on acute responses to the exposure of interest on a particular day. It is also difficult to separate responses to the pollutant of interest from the response to other pollutants, as well as effects of uncontrollable factors like temperature, humidity, and individual activity level.

Hazucha's (1987) regression data indicated earlier that the highest minute ventilation level in chamber studies ($VE_{min} = > 64$ L/min) was associated with the largest pulmonary functional decrement (FEV_1) at any given O_3 concentration. It can be established then that functional responses to O_3 increase with level of exercise and ventilation. McDonnell et al. (1983) and Kulle et al. (1985), using chamber exposure protocols, were the first to observe that O_3's effects on respiratory function accumulate over more than 1 hr. Later, Spektor et al. (1988a) determined that the effects of O_3 on respiratory function accumulate over more than 1 hr under ambient conditions. He noted that children at an outdoor camp for four weaks had a greater functional response with less exercise and comparable O_3 concentrations than during chamber studies performed for one or two hours. The greater response in the camp children had to be generated by other factors such as greater cumulative exposure or potentiation of response by other pollutants. From the work of Folinsbee et al. (1988), it is now clear that functional decrements become progressively greater after each hour of O_3 exposure, but these effects are transient since no residual functional decrements on days following exposure are seen. Horstman et al. (1989) conducted a follow-up study to Folinsbee et al. and found a very similar response at 120 ppb, that is, an FEV_1 decline of 12.3%, while exposures at 80 and 100 ppb showed lesser changes that also became progressively greater after each hour of exposure. It is also clear that toward the end of the accumulation effect, recovery of pulmonary function occurs (Folinsbee and Hazucha, 1989). Multiple linear regression analysis on the mean hourly responses of Horstman's subjects (Horstman et al., 1989) was performed by R. I. Larsen (personal communication in Lippmann, 1989). He demonstrated that exposure time is almost equally important as exposure concentration when the concentrations are in the range of normal peak ambient levels. The study by Horstman et al. also suggests that an appropriate averaging time for exposure assessments of O_3-induced transient functional decrements is greater than or equal to 6 hr. It can also be said that, based on increased pulmonary responses in natrual human exposure studies (where subjects have lower activity levels than subjects in chamber studies), results from chamber studies using O_3 in purified air underestimate the responses that occur among populations outdoors.

In summary, the minimal exposure conditions required to elicit a mean maximal decrement in pulmonary function (FVC, FEV_1) response to O_3 are as follows: in purified air in controlled chambers, 180 ppb O_3 with intermittent heavy exercise ($VE_{min} = 65$ L/min) for 2 hr; 120 ppb with moderate exercise ($VE_{min} = 40$ L/min) for 6.6 hr; in ambient air outside, an average of 100 ppb in adults with very heavy exercise ($VE_{min} = 64$ L/min) for 0.5 hr; an average of 100 ppb in healthy children engaging in normal summer camp activities with moderate exercise ($VE_{min} = 22$ L/min) for four weeks.

5. EXERCISE PERFORMANCE

Studies of exercise performance following O_3 exposure are summarized in Table 2. The effects of oxidants on exercise performance were suggested in an epidemiological study by Wayne et al. (1967), who showed that a percentage of high school track team members who failed to improve their performance had increased oxidant level exposures 1 hr before their races. Ozone, in particular, was looked at by Folinsbee et al. (1977) in determining oxidant effects on maximum exercise performance in healthy subjects. They found a 10% decrease in VO_2max and maximum attained workload, and a 16% decrease in maximum ventilation following a 2-hr exposure to 750 ppb O_3 with intermittent exercise. Earlier, this same group (Folinsbee et al., 1975) noted an increase in respiratory frequency and a decrease in tidal volume during intermittent exercise ($2.5 \times$ resting VE_{min}) following exposure to 350, 500, or 750 ppb O_3.

Schelegle and Adams (1986) exposed 10 young male endurance athletes to filtered air and O_3 during 30 min of exercise. All 10 subjects completed the exercise protocol with filtered air exposure, but one, five, and seven of them did not complete the exercise regime for 120, 180, and 240 ppb O_3, respectively. Similarly, Linder et al. (1988) found that maximum performance time was reduced (11%) for the 16–20 min progressive maximum exercise program in young female adults exposed to 130 ppb O_3 (Table 2).

Gong et al. (1986) demonstrated that exposure to 200 ppb O_3 for 1 hr significantly impaired maximal performance in endurance athletes continuously exercising at a heavy workload for 60 min in a hot environment (Table 2). In a comparative study, Gong et al. (1986) stated that exposure to 120 ppb O_3 for 1 hr did not limit exercise under the same conditions. Therefore it appears that a threshold may exist for significant impairment of exercise performance, and it probably lies between 120 and 200 ppb.

In attempting to define other extraneous factors that may contribute to athletic performance decline with O_3, Gibbons and Adams (1984) examined the effect of heat and determined that heat accentuates subjective limitations more than it enhances pulmonary function impairment, such that heat coinciding with photochemical oxidant episodes (300 ppb O_3 for 1 hr, $VE_{min} = 55$ L/min) is likely to result in more severe impairment of exercise performance. Gong et al. (1988), in further trying to elucidate the mechanisms underlying exercise performance decrement and O_3, pretreated nonasthmatic athletes with a β-adrenergic agonist compound (albuterol) and evaluated exercise performance and pulmonary function responses. The contribution of β-adrenergic mechanisms in the acute airway responses to O_3 in nonasthmatics appeared to be minimal, since albuterol failed to block impairment in pulmonary function and exercise.

Since exercise is a common part of exposure protocols in chamber studies, it is important to note that 85% of healthy adults are oronasal breathers when exercise-induced ventilation quadruples resting values (Niinimaa et al., 1980). It therefore can be questioned whether partially bypassing the nose as the primary form of breathing can increase toxic pulmonary effects, since the nose is an

Table 2 Exercise Performance Following Ozone Exposure

Reference	O_3 (ppb)	Exercise Level (L/min)	Population	Effect[b]
Wayne et al. (1967)	NR^a	NR	Normal	↓ Athletic performance
Folinsbee et al. (1977)	750	20 (est.)	Normal	↑ VO₂max, ↓ MAW, ↓ VENTmax, ↓ HR
Schelegle and Adams (1986)	120, 180, 240	54, 120	Normal (endurance athletes)	↓ Athletic performance
Linder et al. (1988)	120–130	30–120	Normal	↓ MPT
Gong et al. (1986)	200	89, 150	Normal (cyclists)	↓ VO₂, ↓ Vt, ↓ Athletic performance, ↑ SYM, ↓ FEV₁, ↑ BR
Gong et al. (1988)	210	780	Normal (cyclists)[c]	↓ FVC, ↓ FEV₁, ↓ FEF₂₅%₋₇₅%, ↑ SYM, ↑ BR

[a] NR, not recorded,

[b] ↑, increase; ↓, decrease; VO₂max, maximum oxygen consumption; MAW, maximum attained workload; VENTmax, maximum ventilation; HR, heart rate; MPT, maximum performance time; Vt, tidal volume; SYM, symptoms; BR, bronchial reactivity.

[c] Pretreated with albuterol (180 μg).

effective scrubber of some inhaled pollutants. However, Hynes et al. (1988) failed to find significant differences in toxic pulmonary response or symptomology between modes of breathing (nasal versus oral) following acute 30-min O_3 exposures of 400 ppb with continuous exercise ($VE_{min} = 30$ L/min).

6. SYMPTOM RESPONSES

Symptoms such as cough, substernal pain, wheezing, shortness of breath, pain on deep inspiration, and laryngitis have all been closely associated with mean pulmonary function decrements in adults exposed to acute O_3 doses in controlled chamber studies (Linn et al., 1980; Folinsbee et al., 1988; Horstman et al., 1990) (Table 3). The symptoms graded as mild are a mild cough; moderate symptoms are cough, pain on deep inspiration, and shortness of breath; severe symptoms are repeated cough, severe pain on deep inspiration, and breathing distress; incapacitating symptoms are severe cough, severe pain on deep inspiration, and obvious distress (Lippmann, 1989). The question of whether these symptoms are related to O_3-induced functional responses was addressed by Schelegle et al. (1987), who showed significant effects of pretreatment with indomethacin in reducing overall symptom severity while inhibiting O_3-induced pulmonary function decrements at 350 ppb for a 1-hr exposure ($VE_{min} = 60$ L/min).

However, not all studies point to a correlation between symptoms and pulmonary function indices, especially in children (Koenig et al., 1985; Kreit et al., 1989) (Table 3). Hayes et al. (1987) found only a weak to moderate correlation between FEV_1 changes and symptom severity when his individual subject data were analyzed.

McDonnell et al. (1985) reported no changes in frequency or severity of cough in children (aged 8 to 11) exposed for 2.5 hr at 120 ppb O_3 while intermittently exercising ($VE_{min} = 39$ L/min) and having a statistically significant decrease in FEV_1. Similarly, Avol et al. (1985) induced no changes in symptoms in continuously exercising adolescents in ambient air at 144 ppb O_3, although they showed a statistically significant decrease (4%) in group mean FEV_1. In addition, a human ambient study conducted by Spektor et al. (1988a) in children did not find any symptomatic responses despite the occurrence of relatively large decrements in pulmonary function.

Many epidemiological studies have demonstrated increased respiratory symptoms with O_3 exposure similar to the controlled chamber studies, but the relationships defined are qualitative associations with ambient oxidant levels greater than 100 ppb (Hammer et al., 1974). By and large, the symptoms reported in individuals exposed to O_3 in purified air are similar to those found for ambient air exposures, with the exception of eye irritation, which has not been consistently reported for O_3 exposures alone in laboratory studies. Other oxidants in ambient air, such as aldehydes and peroxyacetylnitrate, are mainly responsible for eye irritation.

Table 3 Ozone Concentration/Exercise Level/Airway Responses in Asthmatics and Nonasthmatics

Reference	O_3 (ppb)	Exposure (hr)	Exercise Level (L/min)	Population	Effect[a]
Koenig et al. (1985)	120	1	10 (est.)	Normal	$-FEV_1$, $-$PEF, $-FEF_{50\%}$, $-FEF_{75\%}$
				Asthmatic	$-FEV_1$, $-$PEF, $-FEF_{50\%}$, $-FEF_{75\%}$, ↑SYM
Koenig et al. (1987)	120, 180	0.67	32.5	Normal	$-FEV_1$, $-$PEF, $-FEF_{50\%}$, $-FEF_{75\%}$
				Asthmatic	$-FEV_1$, $-$PEF, $-FEF_{50\%}$, $-FEF_{75\%}$
Linn et al. (1980)	220	2	20 (est.)	Normal	↓FVC, $↓FEV_1$, ↓PEF, $↓FEF_{50}$, ↑SYM
				Asthmatic	↓FVC, $↓FEV_1$, ↓PEF, $↓FEF_{50\%}$, ↑SYM
Kreit et al. (1989)	400	2	30	Normal	$↓FEV_1$, $↓FEV_{1\%}$, $↓FEF_{25\%-75\%}$, $-$SYM
				Asthmatic	↑BR, $▼FEV_1$, $▼FEV_{1\%}$, $▼FEF_{25\%-75\%}$, $-$SYM, ↑BR
Eschenbacher et al. (1989)	400	2	30	Normal	$↓FEV_1$, $→ FEV_{1\%}$
				Asthmatic	$▼FEV_1$, $▼FEV_{1\%}$
Holtzman et al. (1979)	600	2	30	Normal	↑BR
				Asthmatic	↑BR
Hackney et al. (1975a, b)	500	4	20 (est.)	Normal	↑SYM
				Asthmatic	▲SYM
Hazucha (1987)	100	2	<23, 24–43, 44–63, > 64	Normal	↓FVC, $↓FEV_1$, $↓FEF_{25\%-75\%}$
Folinsbee et al. (1988)	120	6.6	42.6	Normal	$↓FEV_1$, ↑SYM, ↑BR
Horstman et al. (1990)	80, 100, 120	6.6	39	Normal	$↓FEV_1$, ↓FVC, $↓FEF_{25\%-75\%}$, ↑SYM, ↑BR

[a] ↑, increase; ↓, decrease; $-$, no change from baseline; ▲ greater increase; ▼ greater decrease; BR, bronchial reactivity; SYM, symptoms, est., estimated; FEV_1 forced expiratory volume in 1 second; PEF, peak expiratory flow; $FEF_{50\%}$, forced expiratory flow at 50% vital capacity; $FEF_{75\%}$, forced expiratory flow at 75% vital capacity; FVC, forced vital capacity; $FEF_{25\%-75\%}$, forced expiratory flow between 25 and 75% of vital capacity; $FEV_{1\%}$, FEV_1/FVC flow at 75% of vital capacity.

7. SENSITIVE/TARGET POPULATIONS

Due to O_3's ubiquitous nature and the difficulty of controlling its outdoor level, it is important to identify any potentially sensitive populations with or without preexisting disease, and to understand the effects O_3 has on them.

Many studies have documented a wide range of reproducible responses among healthy subjects (McDonnell et al., 1985), most notably a concentration-dependent mean decrement in volumes and flow rates during forced expiratory maneuvers.

In comparisons of functional responsiveness in potentially sensitive populations and healthy populations, it has been shown that smokers (Kagawa, 1984; Shephard et al., 1983), people with chronic obstructive pulmonary disease (COPD) (Linn et al., 1983; Solic et al., 1982), people with allergic rhinitis (McDonnell et al., 1987), and the elderly (Drechsler-Parks et al., 1987; Reisenauer et al., 1988) all have no greater functional responsiveness to O_3 and, in some cases, they have less. With respect to asthma, several studies have compared the airway response to acute O_3 exposure in asthmatics and nonasthmatics (Table 3). The majority of the studies show that asthmatics do not have increased pulmonary function responsiveness to O_3 (Koenig et al., 1985, 1987; Linn et al., 1980). As shown in Table 3, the 1985 and 1987 Koenig studies found no consistent significant pulmonary function changes in healthy versus asthmatic subjects, while the Linn et al. (1980) study found that both nonasthmatics and asthmatics showed similar forced expiratory responses and symptoms. As shown in Table 3, two studies reported that asthmatics have an increased pulmonary function response to O_3 compared to normal subjects. Kreit et al. (1989) and Eschenbacher et al. (1989) both demonstrated a greater decrease in flow-volume measurements in asthmatics versus nonasthamatics.

Parameters such as bronchial responsiveness, symptoms, and biochemical markers for inflammation can be seen as a potential basis for discerning populations of concern. Table 3 shows that Holtzman et al. (1979) and Kreit et al. (1989) attempted unsucessfully to discern asthmatics and nonasthmatics in their response to O_3 based on changes in bronchial reactivity. Conversely, Koenig et al. (1985) reported higher ratings on symptom scales in asthmatic versus healthy adolescents. Similarly, Hackney et al. (1975a) earlier demonstrated an increased symptom rating in subjects with preexisting hyperreactive airways; more recently, however, Kreit and colleagues (1989) were unable to show any significant difference in symptom scores between asthmatics and nonasthmatics (Table 3).

Biochemical markers for inflammation have recently been compared in asthmatics and nonasthmatics after O_3 exposure. McBride (1991) reported increased markers for inflammation in nasal lavage of asthmatics. Koren et al. (1989) reported increased levels of inflammatory cells and soluble factors in the bronchoalveolar lavage (BAL) of healthy young men after a 2-hr exposure to 400 ppb O_3 with intermittent exercise ($VE_{min} = 35.2$ L/min) compared to the levels in subjects exposed to air only.

Therefore, in the case of O_3, no special functional responsiveness has been definitively established among the potentially more sensitive groups with preexisting disease. Moreover, such parameters as bronchial reactivity and biochemical markers for inflammation are also unable to define distinctly sensitive populations.

At present, the primary focus in research studies is the healthy population of people who exercise regularly outdoors, because of their potentially higher O_3 exposures and doses. Because the highest levels normally occur in the more highly populated regions of the country, a significant percentage of urban populations is exposed during the summer months to O_3 levels that have known health effects.

8.   EFFECTS ON BRONCHIAL REACTIVITY AND INFLAMMATION

Although bronchial hyperreactivity to inhaled irritants is a characteristic feature of asthma, it was Golden et al. (1978) who showed nonspecific bronchial hyperreactivity in healthy subjects after a 2-hr exposure to 600 ppb O_3 at rest. Shortly thereafter, Holtzman et al. (1979) concluded that a 2-hr exposure to 600 ppb O_3 with intermittent exercise increases reactivity via cholinergic, postganglionic pathways. This occurred similarly in both atopic and nonatopic subjects.

Seltzer et al. (1986) sought to examine the possible cause of O_3-induced airway reactivity by assessing whether there was an association between O_3-induced hyperreactivity and the inflammatory reaction known to be triggered by common stimuli such as allergen or viral infections, as well as by O_3. They found that in subjects exposed to 400 or 600 ppb O_3 for 2 hr while intermittently exercising, hyperresponsiveness to methacholine was associated with an influx of polymorphonuclear neutrophils (PMN) into the airways and with changes in the levels of some cyclooxygenase metabolites of arachidonic acid. The minimal doses of O_3 capable of eliciting bronchial reactivity in healthy subjects was evaluated by Horstman et al. (1989), who showed increases in bronchial responsiveness at concentrations of 80, 100, and 120 ppb with exposure times of 6.5 hr and exercise levels of 40 L/min each hour for 50 min.

Since no association between pulmonary function decrements and airway reactivity had yet been demonstrated, the temporal pattern of airway hyperreactivity due to O_3 exposure was investigated by Folinsbee and Hazucha (1989). A temporal association was a logical first step. They revealed that airway responsiveness to methacholine increased immediately after 70-min exposures to 350 ppb O_3 while exercising for two 30-min periods ($VE_{min} = 40$ L/min). It is unclear whether this effect persisted for 18 hr like the other pulmonary function impairments.

Earlier work by Folinsbee et al. (1988) failed to show a relationship between O_3-induced changes (at 120 ppb for 5 hr, $VE_{min} = 40$ L/min) in methacholine reactivity and changes in FVC or FEV_1. The mechanism underlying induced airway hyperreactivity to O_3 exposure in humans remains unclear. Eschenbacher et al. (1989) and Ying et al. (1990) failed to prevent O_3-induced increases in airway responsiveness with indomethacin pretreatment in healthy subjects, thereby excluding the metabolites of the arachidonic acid chain as important mediators in this effect. In both studies, subjects were exposed to 400 ppb O_3 for 2 hr with intermittent exercise ($VE_{min} = 30$ L/min).

Other investigators have implicated cholinesterase activity as an important contributor to O_3-induced hyperreactivity, but this work has been in animal models (Gorden et al., 1981). As far as location of effect is concerned, Beckett et al. (1988) have implicated the small airways of the lung as the site of hyperreactivity.

In summary, many investigators have demonstrated increased airway reactivity from O_3 exposure at varying concentrations: 350 ppb for a 1-hr exposure (Folinsbee and Hazucha, 1989), 120 and 200 ppb for a 1-hr exposure (Gong et al., 1988), and 180 ppb for a 2-hr exposure (McDonnell et al., 1987). The mechanism of and relationship to pulmonary function decrement induced by O_3 remain speculative at present.

Nonspecific airway hyperreactivity is a characteristic feature of asthma and, as discussed in a previous section, O_3 is one of the agents known to initiate it. Several other known stimuli of inflammation, such as viral infection and allergens, also cause airway hyperreactivity. Seltzer et al. (1986) examined the hyperreactivity–inflammation relationship in healthy humans and were the first to show that O_3-induced airway reactivity to methacholine is associated with neutrophil influx into the airways and with changes in cyclooxygenase metabolites of arachidonic acid. In particular, they found significant increases in the concentration of prostaglandins E_2, $F_{2\alpha}$, and thromboxane B_2 in lavage fluid of O_3-exposed subjects. Seltzer exposed her subjects for 2 hr with intermittent exercise at 400 and 600 ppb O_3. Kirby et al. (1987) investigated atopic asthmatics and nonasthmatics without deliberate O_3 exposure to determine if there was evidence of cellular inflammation in the BAL of atopic asthmatics and to correlate the cell populations with measurements of airway hyperresponsiveness. Kirby's group concluded that there was evidence of cellular inflammation in the airways of stable asthmatics and that inflammation was significantly correlated with measurements of airway hyperresponsiveness.

Other investigators have sought alternative assay techniques, which are less invasive than BAL, to study O_3-induced inflammation in humans. Graham et al. (1988) and Bascomb et al. (1990) employed nasal lavage assays used to find indications of upper respiratory tract (URT) acute inflammatory response in hopes of providing useful information regarding the local cellular effect of O_3 on the respiratory tract of humans. Graham and colleagues found that a 4-hr exposure to 500 ppb O_3 at rest induced an inflammatory response in the URT of humans, as measured by nasal lavage PMN counts. Koren et al. (1989) also described inflammatory and biochemical changes in the airways following 400 ppb O_3 exposure for 2 hr at an exercise level of 35 L/min. Specifically, they found in BAL fluid an 8.2-fold increase in the percentage of PMN leukocytes in the total cell population; a 2-fold increase in the levels of proteins, albumins, and IgG, indicative of increased vascular permeability [as evidenced earlier in Kehrl's studies (Kehrl et al., 1987)]; as well as a 6.4-fold increase in fibronectin, a 2.1-fold increase in tissue factor, and a 1.8-fold increase in factor VII. The last two are initiating proteins of the extrinsic coagulation pathway. There was a 2-fold increase in the level of prostaglandin E_2 and a similar elevation of the complement component C3a. Levels of leukotrienes C4 and B4 were not affected even though recent evidence has shown leukotrienes C4, D4, and E4 to be capable of participating in the induction of three related processes observed during the immediate reaction in bronchial asthma: edema formation, mucous secretion, and muscle contraction (Piacentini and Kaliner, 1991).

Further work by Koren et al. (1989) using ambient O_3 levels (100 ppb) for 6.6 hr revealed a close observed-to-expected ratio of increased PMN counts in BAL, suggesting that lung inflammation from inhaled O_3 has no threshold down to ambient levels. In fact, Devlin et al. (1991) later confirmed that exposure of humans to as little as 80 ppb for 6.6 hr is sufficient to initiate an inflammatory reaction in the lung. Devlin et al. found increases in the BAL fluid of PMNs, prostaglandin E_2, interleukin-6 and $\alpha 1$ antitrypsin and decreased phagocytosis via the complement receptor after a 2-hr exposure to 400 ppb O_3 with moderate exercise ($VE_{min} = 40$ L/min). Protein and fibronectin were not significantly elevated.

In summary, the weight of evidence shows clear functional and biochemical responses in humans that can accumulate over multiple hours and last for hours or days after exposure has terminated. In humans, both functional changes and inflammatory processes have been shown to occur following 6.6 hr of O_3 exposure at levels as low as 80 ppb.

9.   EFFECTS ON AIRWAY PERMEABILITY

As previously mentioned, the permeability of the respiratory epithelium in humans can be determined from the externally measured rate of clearance from the lung of gamma-emitting ^{99m}Tc-diethylenetriaminepentaacetate (^{99m}Tc-DTPA) inhaled as a droplet aerosol. Kehrl et al. (1987) conducted a randomized, crossover, double-blind study with eight healthy, nonsmoking young men exposed for 2 hr to 400 ppb O_3 while exercising intermittently at 66.8 L/min. They measured the pulmonary clearance of ^{99m}Tc-DTPA by sequential posterior lung imaging and found seven out of eight had increased clearance along with respiratory symptoms and a 14% mean FVC decrement.

Thus, high O_3 exposure, as well as causing decrements in respiratory function, also causes an increase in permeability. Such an increase could enhance the uptake of other inhaled toxic pollutants or allow the release of inflammatory cells (neutrophils) onto the airway surfaces, with subsequent arachidonic acid metabolite reactions.

10.   EFFECTS ON PARTICLE CLEARANCE

The effects of O_3 on particle clearance have been more extensively studied in animals than humans. However, Foster et al. (1987) studied the effect of 2-hr exposures to 200 or 400 ppb O_3 with intermittent, light exercise on the rates of tracheobronchial mucociliary particle clearance in healthy adult males. They found an accelerated rate of particle clearance from both central and peripheral airways at 400 ppb O_3, as well as a 12% drop in FVC. In addition, 200 ppb O_3 produced an accelerated rate of particle clearance without any pulmonary function decline. This suggests that significant changes in the lungs' ability to clear particles occur prior to any other pulmonary function change in the respiratory airways.

Further work delineating a dose/response relationship in humans with respect to particle clearance is needed to further assess the effects of different concentrations in light of animal evidence suggesting a concentration-dependent trend of reduced clearance rate with increasing concentration (Schlesinger and Driscoll, 1987). Lippmann et al. (1987) found with other irritants (H_2SO_4 and cigarette smoke) accelerated clearance in humans at lower concentrations and slower clearance at higher concentrations. The underlying mechanism responsible for O_3's effects on this defense mechanism has not been fully elucidated, but what is known is as follows. Inhaled bacteria and viruses are removed from the lung by two mechanisms. First, ciliated cells of the bronchi and trachea propel mucus carrying entrapped particles and infectious agents upward from the distal regions of the lung to be swallowed and excreted. Second, alveolar macrophages engulf, kill, and remove infectious agents from the respiratory regions of the lung by migrating with the particle to either the mucous-thinning layer or the lymphatic system to be drained from the lung.

Short-term clearance is mostly due to removal by mucus of particles from the lung; the decreased short-term clearance demonstrated in animal studies as a result of increasing O_3 exposure is an anticipated yet unproven response in humans based on animal morphological observations. Ciliated cells are damaged by O_3 inhalation through necrosis and sloughing and shortening of the cilia, which results in decreased transport of particles (Menzel, 1984).

Long-term clearance of particles from the lung is a function carried out by alveolar macrophages (AM). AMs are damaged by O_3 inhalation. Animal studies show reduced phagocytosis, disrupted AM membranes, reduced AM lysosomal hydrolase activity, and reduced AM numbers (Menzel, 1984).

11.   ADAPTATION

The adaptation phenomenon associated with O_3 is, simply put, that repetitive daily exposures at a level that produces a functional response upon a single exposure (350 ppb) will result in an enhanced response on the second day (30–50%) (Folinsbee and Hazucha, 1989), with diminishing responses on the third and fourth days and virtually no response on the fifth day (Farrell et al., 1979; Folinsbee et al., 1980; Hackney et al., 1977). The adaptation to exposure stops about a week after exposure ceases (Horvath et al., 1981; Kulle et al., 1982). Interestingly, recent research in animals has revealed that persistent damage to lung cells accumulates even as functional adaptation takes place. This suggests the need to seriously consider the importance of transient functional decrements in humans.

The adaptation response to acute O_3 has been shown only in chamber studies. Controlled human studies by Lioy et al. (1985) on camp children who experienced a consecutive four-day high-pollution episode (max. O_3 concentration = 145 ppb) revealed a week-long baseline shift in pulmonary function (FVC, FEV_1, PEFR, $FEF_{25\%-75\%}$) the following week. Since higher concentrations have been used in chamber adaptation studies and Lioy's "persistence effect" was not found, it is possible that other potentiating factors are at work in outdoor ambient conditions.

Kulle et al. (1984) also looked at the adaptation response to 410 ppb O_3 exposure, but in the distinctive population of subjects with chronic bronchitis. He found that individuals with chronic bronchitis adapt rapidly to 410 ppb O_3 (by the second day) and lose this adaptive phenomenon within four days. Horvath et al. (1981) also concluded that more sensitive subjects (i.e, those not recently exposed to O_3 therefore not desensitized) require more daily sequential exposures to 500 ppb O_3 in order to maintain functional adaptation. One study has reported a seasonal adaptation effect to O_3 exposure. Linn et al. (1988) described a group of O_3-sensitive subjects who had a greater functional decrement following 2 hr of exposure at 180 ppb in a chamber in the spring than under the same conditions in autumn or winter. The subjects showed equivalent responses again the following spring. Thus, long-term adaptation to O_3 may be associated with a seasonal variation effect.

12. MECHANISMS OF AIRWAY CHANGE

It has already been established that irritants such as SO_2, NO_2, and H_2SO_4 produce greater functional responses among asthmatics than among healthy subjects. This is not true for O_3. Both Folinsbee et al. (1988) and Horstman et al. (1989) have found no apparent relationship between O_3 exposure, pulmonary function responses, and bronchial responsiveness in individual subjects, both healthy and asthmatic. Thus, there is a great deal known about O_3 exposure and respiratory function but very little known about the mechanisms responsible for the measured effects.

One investigative path taken by a few researchers has been pretreatment with drugs with specific and known modes of action in order to measure any inhibition of O_3's effects. For example, Beckett et al. (1985) attempted to implicate the parasympathetic nervous system in the bronchial reactive response to O_3 by examining the effect of atropine, a muscarinic receptor blocker, on responses to 400 ppb O_3. Atropine pretreatment prevented the significant increase in airway resistance with O_3 exposure and partially blocked the decrease in forced expiratory flow rates, but did not prevent the fall in FVC, changes in respiratory frequency and tidal volume, or the frequency of reported symptoms.

This suggests that the increase in pulmonary resistance from O_3 exposure is mediated by a parasympathetic mechanism, and that changes in other measured variables (lung volumes) are not dependent on muscarinic cholinergic receptors of the parasympathetic nervous system. Gong et al. (1988) studied the contribution of β-adrenergic mechanisms to the respiratory responses of O_3 in a study that looked at pulmonary function, histamine bronchoprovocation challenge, athletic performance, and symptoms in nonasthmatic athletes exposed to 210 ppb O_3 for 60 min. The results are reported in Table 2 and suggest that the contribution of the β-adrenergic mechanisms in the acute airway response to O_3 is minimal.

As mentioned previously, Seltzer et al. (1986) found significantly increased levels of prostaglandins E_2 and F2-α and thromboxane B2 in the BAL of subjects acutely exposed to 400 ppb O_3. In addition, Coleridge et al. (1976) and Roberts et al. (1985) have shown that prostaglandins E_2 and $F_{2\alpha}$ can stimulate pulmonary neural afferents that initiate responses very characteristic of acute O_3 exposure [i.e., decreased lung volume (FVC) and flow rates (FEV_1, $FEF_{25\%-75\%}$)]. In addition, inhaled $PGF_{2\alpha}$ has been shown to increase bronchial responsiveness in normal human subjects (Walters et al., 1981). These findings suggest the involvement of the prostaglandins (E_2, $F_{2\alpha}$) in pulmonary function responses subsequent to acute O_3 exposure.

Schelegle et al. (1987) attempted to determine whether a 1-hr exposure to 350 ppb O_3, at an exercise level of 60 L/min, could induce pulmonary function effects in healthy young males and be inhibited by pretreating the subjects with a known prostaglandin synthetase inhibitor, indomethacin. He found significant differences in FEV_1 and FVC in the no-drug group versus the indomethacin group, and in the placebo versus the indomethacin group. Therefore, it appears that the cyclooxygenase products of arachidonic acid (sensitive to indomethacin inhibition) play a significant role in the pulmonary function response to acute O_3 exposure.

Eschenbacher et al. (1989) also showed that indomethacin partially prevented O_3-induced changes in lung function in normal subjects. Ying et al. (1990) concluded that O_3-induced changes in lung function (FVC, FEV_1) are mediated only in part by cyclooxygenase products, but the O_3-induced increase in bronchial responsiveness occurs by some mechanism other than the arachidonic acid metabolite chain. Ying exposed subjects pretreated with indomethacin or placebo for 2 hr to 400 ppb O_3 with intermittent exercise ($VE_{min} = 30$ L/min). He found that indomethacin did not significantly alter the O_3-induced increase in bronchial responsiveness to methacholine [i.e., there was no significant difference in PC100* sRaw (specific airway resistance) between the indomethacin day and the placebo and no-drug days].

It is therefore fairly clear that cyclooxygenase products of arachidonic acid are

---

* PC100 = Provocation concentration that causes a 100% increase in specific airway resistance.

involved to some degree in the acute pulmonary function response to O_3. Current research is specifically focusing on lipid mediators such as leukotrienes, prostaglandins, thromboxane, and platelet activating factor, all produced by cells involved in the mediation of pulmonary inflammation. The mechanism by which the release of these lipid mediators in the lung leads to pulmonary function decrement from acute O_3 exposure is still not entirely clear.

Evidence of the involvement of a neural link has also gained some attention. Lee et al. (1979) and Hazucha et al. (1986) suggest that O_3-induced pulmonary function decrements and ventilatory pattern changes are neurally mediated. As previously cited, Coleridge et al. (1976) and Roberts et al. (1985) both conclude that cyclooxygenase products stimulate neural afferents in the lung. This, combined with the observation that indomethacin pretreatment reduces O_3-induced pulmonary function decrement, suggests that cyclooxygenase products released consequent to O_3-induced tissue damage may stimulate neural afferents in the lung that result in pulmonary function decrements.

Lippmann (1989) summarizes Bates's current hypothesis of the mechanisms underlying pulmonary responses to inhaled O_3 in the following way. After O_3 exposure, the inspiratory capacity is first reduced as a consequence of a lower maximal negative intrapleural pressure on full inspiration. Maximal inspiratory and expiratory mouth pressures are not affected. He emphasizes that the FEV_1:FVC ratio is not initially affected after O_3 exposure, which is to say that the FEV_1 and FVC fall together. Bates suggests that stimulation of the C-fiber system in the airways must lead to a breaking effect on the inspiratory muscles as a first consequence of O_3 exposure, and that this probably occurs as a result of induced inflammation. Increased respiratory rate after O_3 exposure, increased airway reactivity, increased airway permeability, and the inability of β-adrenergic stimulants to prevent changes induced by O_3 are all consistent with Bates's hypothesis. The mechanism of reduced responsiveness after repeated exposure may be due to a thicker lining of mucus over the surface of the airway or to actual cell replacement after exposure.

13.   BIOCHEMICAL AND STRUCTURAL CHANGES

One possible mechanism of O_3 toxicity appears to be related to an attack on the epithelial cellular membrane, resulting in lipid peroxidation and free radical production. Thus, O_3 damages the cells and fluid components lining the lung. Inflammatory reactions and permeability changes in the epithelial cells have been described in humans exposed to O_3 as much as 18 hr after exposure ceases (Koren et al., 1989), and at concentrations as low as 80 ppb for 2 hr (Devlin et al., 1991).

14.   EFFECTS OF TEMPERATURE AND RELATIVE HUMIDITY

Among the covariates encountered when studying outdoor ambient pollution levels and their health effects are temperature and relative humidity. Although it is very difficult, if not impossible, to control for these parameters in natural human study designs, chamber studies do offer the possibility of selectively controlling or changing them.

Some investigators have speculated about the effect that temperature and relative humidity may have on the synergistic effect of mixture gases, that is, O_3 and SO_2 (Hazucha and Bates, 1975). Bedi et al. (1982) examined the effects of high temperature (35 °C) and high relative humidity (85%) on the synergistic effect of 400 ppb O_3 and 400 ppb SO_2. The results were unable to confirm any synergistic effects between O_3 and SO_2. Thus, high temperature and high relative humidity did not contribute to synergism and might even have inhibited it.

Gong et al. (1986) concluded that exposure to 200 ppb O_3 for 1 hr with continuous exercise ($VE_{min} = 89$ L/min) significantly impaired maximum performance in endurance athletes continuously exercising in a hot environment (31 °C), but temperature was not specifically controlled for in this study. Therefore, conclusions cannot be drawn without caution. Gibbons and Adams (1984) investigated the effect of ambient heat (24 and 35 °C) on O_3 effects during 1 hr of prolonged moderate exercise training ($VE_{min} = 55$ L/min; $VO_2max = 2.79$ L/min). Results showed a significant interaction of 150 and 300 ppb O_3 and ambient heat, resulting in increased respiratory frequency and decreased alveolar volume [measured from the end-tidal CO_2 fraction ($FETCO_2$) at the mouth for 1 min at 10-min intervals and corrected to alveolar values], whereas flow-volume measurements showed only a trend toward on O_3–temperature effect. The authors concluded that ambient heat stress enhances subjective limitations and produces certain physiological changes in combination with O_3, and that this is likely to affect exercise performance.

In addition, large-scale epidemiological studies performed in British Columbia and Ontario, Canada, have examined some environmental covariates such as temperature as they relate to increased asthma attacks and emergency room visits. Bates et al. (1990) found O_3 and temperature to be strongly correlated with O_3 levels higher in the summer months; however, visits to the emergency room for respiratory complaints were not correlated with temperature or O_3 level. Bates and Sizto (1989), in the Ontario Air Pollution Study, performed a multiple regression analysis on all environmental variables and determined that they accounted for 5.6% of the variability in admissions to the hospital for respiratory complaints. Further, they determined that when temperature was forced into the regression equation first, it accounted for only 0.89% of the variability.

Therefore, the roles of temperature and relative humidity in O_3 effects on pulmonary function have not been completely defined, but at present they do not seem to be major contributing factors. The fact that high O_3 is strongly correlated with periods of high temperature remains undisputed.

15. OTHER POLLUTANT EFFECTS

In examining the health effects of O_3, it is important to note that O_3 is only one component of photochemical smog and cannot be separated out in natural surroundings. It is therefore of interest and importance to look at possible toxicological interactions among mixtures of O_3 and other pollutants in urban air environments. Not all urban air environments are uniform with respect to the dominant photochemical compounds occupying them. Chamber studies can be designed and performed to investigate potentiation effects of other pollutants. Table 4 summarizes studies of interactions of O_3 and other pollutants.

Hazucha and Bates (1975) were among the first to study the interaction between O_3 and other pollutants. They showed a 20% decline in $FEF_{25\%-75\%}$ with exposure to a mixed gas ($O_3 + SO_2$) compared to O_3 alone, and a 32% decline in $FEF_{25\%-75\%}$ compared to SO_2 alone. At around the same time, Hackney et al. (1975a, b) investigated gas mixtures of O_3, NO_2, and CO. They found, in contrast to Hazucha and Bates, that the addition of a second oxidant pollutant (NO_2 at 300 ppb for 2 hr) and then a third oxidant pollutant (CO at 30,000 ppb for 2 hr) produced no additional detectable effects (Table 4). Bell et al. (1977), attempting to replicate the findings of Hazucha and Bates, did not find such severe acute toxicological effects with exposure to the same mixture and found only slightly more severe effects than with O_3 alone. Kagawa (1983) also demonstrated slightly enhanced effects with O_3 and SO_2, but at a lower concentration (Table 4); moreover, the enhanced effect was demonstrated in only half the subjects compared to the effects seen with O_3 alone. The combination of O_3 and NO_2 also resulted in slightly greater decreases in airway conductance [Gaw/Vtg (thoracic gas volume)] in half the subjects than did O_3 alone. These results therefore do not show a definite additive effect from NO_2 or SO_2. Consistent reports from Bedi et al. (1979, 1982) failed to show any synergistic effects of O_3 and SO_2 in chamber studies, and also ruled out factors such as temperature and relative humidity as synergistic potentiators (Table 4). More recently, however, Koenig et al. (1990) showed significant changes in pulmonary function in a group of asthmatics exposed to 120 ppb O_3 followed by 100 ppb SO_2, compared to O_3 and SO_2 alone (Table 4).

Other chamber studies by Stacey et al. (1983) have shown that the mean responses to 100 ppb O_3 and 100 $\mu g/m^3$ H_2SO_4 after 2 hr of exposure at rest in 231 subjects were -9.0% and -11.5% for FVC and FEV_1, respectively; 5.7% and -7.7% for O_3 alone; and -1.4% and -1.2% for sham exposures. These results, however, did not reach statistical significance. Horvath et al. (1986) looked at 480 ppb O_3 and 220 ppb peroxyacetyl nitrate [(PAN) O_3 + PAN is the most common mixture of total ambient oxidants] for 2 hr and reported a 10% greater decrement in lung function measurements with PAN than with O_3 alone. Avol et al. (1984), however, did not find a potentiated response in subjects exposed for 1 hr to ambient southern Californian air containing an average of 153 ppb O_3 while exercising heavily, versus the same subjects exposed to purified air at 160 ppb O_3. Southern Californian air has higher NO_2 and lower acid aerosol than the northeastern United States.

Some natural human experiments have attempted to show potentiation of O_3-induced functional decrements. The study by Spektor et al. (1988b) in rural New York involved healthy adult nonsmokers engaged in daily outdoor exercise programs who were exposed to ambient mixtures containing low concentrations of acidic aerosols, NO_2, and O_3. Overall, the functional responses were subjec-

Table 4 Studies of Interactions of Ozone and Other Pollutants

Reference	O_3 (ppb)	Other Pollutants (ppb)	Exercise Level (L/min)[a]	Population	Effects[b]
Hazucha and Bates (1975)	370	$SO_2/370$	NR[a]	Normal	$\downarrow FEF_{25\%-75\%}$
Hackney et al. (1975 a,b)	250,370,500	$NO_2/300$, $CO/30{,}000$	20(est.)	Normal	$-FVC, -FEV_1, -FEF_{50\%}-FEF_{75\%}$
				Asthmatic	$-FVC, -FEV_1, -FEF_{50\%}-FEF_{75\%}$
Bell et al. (1977)	370–400	$SO_2/370$	NR	Normal	$\downarrow FEF_{25\%-75\%}$
Kagawa (1983)	150	$SO_2/150$, $NO_2/150$	NR	Normal	$\downarrow Gaw/Vtg$
Stacey et al. (1983)	100	$H_2SO_4/100$ $\mu g/m^3$	10(est.)	Normal	$\downarrow FVC, \downarrow FEV_1$
Bedi et al. (1982)	400	$SO_2/400$	30	Normal	$-FVC, -FEV_1, -FEF_{25\%-75\%}, -FEF_{75\%}$
Koenig et al. (1990)	120	$SO_2/100$	30	Asthmatic	$\downarrow FEV_1, \uparrow RT, \downarrow FEF_{50\%}$

[a] NR, not recorded.
[b] $\uparrow$ increase; $\downarrow$ decrease; Gaw/Vtg, airway conductance; RT, total respiratory resistance.

341

tively observed to be as large (for FEV_1) or larger (for FVC, $FEF_{25\%-75\%}$, and PEFR) compared to controlled 1- and 2-hr chamber studies with O_3 alone in similar concentrations.

Detels et al. (1987) and Lioy et al. (1985) showed a week-long shift in PEFR in children the week after a four-day pollution episode with elevated O_3 and H_2SO_4. A study by Raizenne et al. (1989) revealed a similar shift in PEFR among girls at a summer camp following a brief pollution episode. Further studies are needed to evaluate the role of O_3 as compared with that of other pollutants, such as SO_2 and sulfates, in these changes.

In summary, controlled chamber studies have not reproducibly demonstrated significant synergism in functional responses between O_3 and SO_2, NO_2, or H_2SO_4. Natural human studies have implicated potentiation among other pollutants such as NO_2 or H_2SO_4, but due to inherent limitations in design, these cannot be used to conclusively demonstrate potentiation.

One area gaining increasing attention is the potentiation of allergic response by O_3. It is well established that O_3 can increase bronchial reactivity in both healthy (Golden et al., 1978) and asthmatic (Kreit et al., 1989) subjects, but it is still unclear whether low concentrations of O_3 can trigger an increase in responsiveness of the airways to other agents such as exercise and viruses. Recent evidence by Molfino et al. (1991) suggests that exposure to low O_3 concentrations (120 ppb), similar to those occurring in urban areas, for only 1 hr at rest can increase the bronchial responsiveness to allergen in atopic asthmatic subjects. Bascombe et al. (1990), however, could not increase the acute response to nasal challenge with antigen in allergic-rhinitis subjects preexposed to 500 ppb O_3 at rest for 4 hr.

16. EPIDEMIOLOGICAL STUDIES—CHRONIC EFFECTS

As already discussed, several epidemiological studies have assessed acute effects of O_3 exposure. The majority of these studies were on children. Kinney et al. (1988) performed in effect a metanalysis on these epidemiological studies and found consistency among results from studies performed under varying environmental conditions by different investigators.

Epidemiological studies on chronic O_3 effects are fewer in number. Detels et al. (1987) performed a study that compared respiratory function at two points five years apart in a high-O_3 California community (Glendora), and in a low-O_3 California community (Lancaster). He found a lower baseline pulmonary function [FEV_1, FVC, $FEF_{25\%-75\%}$, $FEF_{50\%}$ (forced expiratory flow at 50% vital capacity), $FEF_{75\%}$ (forced expiratory flow at 75% vital capacity)] in Glendora, with a greater rate of decline over the five-year period. In comparing a study by Detels et al. (1987) with that of another epidemiological cohort study in Tucson, Arizona, Knudson et al. (1983) observed three different rates of pulmonary function decline (FEV_1, FVC, $FEF_{25\%-75\%}$). This suggests an exposure–response relationship over an extended period of time.

Two population groups were also compared by Kilburn et al. (1985), one in Long Beach, California, and the other in Michigan. The O_3 levels were not known for either location during the study, but that of Long Beach is similar to that of Lancaster, California, and that of Michigan is generally much lower than that of California. Results showed the Long Beach population to have significantly lower pulmonary function values (FEV_1, $FEF_{25\%-75\%}$) than the matched cohort in Michigan.

In summary, epidemiological studies of chronic oxidant exposure indicate a progressive decline in pulmonary function (FEV_1, FVC, $FEF_{25\%-75\%}$) over time. This is consistent with animal data showing progressive and cumulative structural and biochemical damage under conditions of chronic exposure.

17. OTHER EFFECTS

17.1. Effects on the Immune System

Humans are protected from inhaled infectious agents by an array of defense mechanisms. Two of these (ciliated cells and alveolar macrophages) were previously discussed with respect to particle clearance from the lung. Unfortunately, the large body of data regarding the influence of inhaled O_3 on lung immunity in animals is difficult to extrapolate to humans, since human studies are minimal.

According to a recent report however, Zwick et al. (1991) investigated changes in lymphocyte subpopulations in children who, under similar environmental conditions, were exposed to high and low concentrations of O_3 for five weeks. They found a significant difference in lymphocyte subpopulations between the children exposed to high concentrations and the children exposed to low concentrations. The high-concentration group (60–188 ppb O_3) had a decreased number of T-helper cells ($OKT4^+$), an increased number of T-suppressor cells ($OKT8^+$), and a decreased number of natural killer cells (OKNK). These results are consistent with some animal data, but much more work is needed in human studies before conclusions can be drawn.

17.2. Cognitive and Neurobehavioral Effects

This area of O_3 and neurobehavioral health effects has not generated much interest, particularly in the last 20 years. However, some reports do exist in the literature documenting subjective descriptions of mental dysfunction as a result of O_3 exposure. One such study by Griswold et al. (1956) reports subject's description after exposure to 2000 ± 1000 ppb O_3 for 1.5 hr, as "lessening of mental ability to concentrate and absorb thought ... a very marked effect in coordination and articulation and expression of oral thoughts during the last half hour of exposure until retiring that evening." Other subjective descriptions came from Lagerwerff (1963), who looked at O_3 effects (200–5000 ppb) on visual parameters. Although he found no significant changes, several subjects reported feeling tired, and run down and having difficulty concentrating. It was also noted that these subjects did not read or perform any work requiring mental concentration. The feeling of lethargy persisted for one to two days.

Hore and Gibson (1968) specifically examined intelligence testing and O_3 exposure. Using only verbal and nonverbal reasoning tests, they found no noticeable effect on mental functioning during 70-min O_3 exposures at rest at concentrations of 200 and 300 ppb.

It would seem reasonable to suggest that with current sophisticated testing procedures for neurobehavioral performance effects, a more detailed, complete, and controlled investigation can be carried out to clarify this area further.

18. SUMMARY AND CONCLUSION

Background O_3 is associated with stratospheric O_3 intrusion into the troposphere, natural reactive organic gases (including methane), and NO_x emissions, and anthropogenic NO_x and reactive organic gas emissions. The critical factor controlling average O_3 concentrations is the concentration of NO_xs, while O_3 peaks depend on the concentration of precursor reactants in the incoming and local air. An individual's total dose of O_3 will largely be determined from outdoor exposure; thus, people who work and play outdoors will receive the highest doses. Consequently, this subpopulation has been identified as being at increased risk from elevated ground-level O_3 concentrations.

Measurable effects following acute O_3 exposure are short-term and transient, and appear to be reversible. These effects include changes in lung volume and flow resistance, increased epithelial permeability, increased reactivity to bronchoconstricting agents such as histamine or methacholine, and increased respiratory symptoms. These effects may accumulate and persist for several hours or days after exposure ceases. An adaptation phenomenon usually appears on the third day of high doses of repetitive daily exposures and lasts for approximately one week, whereupon pulmonary function recovers. Most of the data gathered on transient pulmonary functional effects come from controlled human exposure studies using an environmental chamber. This design is useful in providing information but is limited because prior chronic exposures affect transient responses to a single exposure, and at present our ability to define the magnitude and nature of prior chronic exposures is inadequate. Natural human study designs have inherent limitations, most notably the inability to quantify exposure and separate out pollutants of interest. However, they do provide suggestive evidence of potentiating factors with respect to other pollutants on O_3 effects, whereas chamber studies have failed to demonstrate any synergism in functional response to O_3 and SO_2, NO_2, and H_2SO_4. One of the consistent effects mentioned from acute O_3 exposure is the increase in respiratory symptoms. These include cough, substernal pain, shortness of breath, and pain on deep inspiration. Their relationship to pulmonary function decrement, however, is not consistent and cannot be demonstrated in children.

Ozone's effects on exercise performance have been documented. A threshold concentration between 120 and 200 ppb appears to impair maximal performance in endurance athletes. Increased bronchial reactivity from O_3 exposure has been demonstrated at varying concentrations, yet the mechanisms remain unclear. Its relationship to pulmonary function decrement also remains undefined at present. What has been demonstrated, though, is an increase in epithelial permeability due to O_3, as well as an acute inflammatory response. Further, it has been shown that the increased airway reactivity is associated with the inflammatory reaction, as evidenced by neutrophil influx into the airways. Specific lipid mediators have been found in elevated concentrations in BAL and nasal lavage fluid of subjects exposed to 400 and 500 ppb O_3, respectively. These include prostaglandins, thromboxane, leukotrienes, and platelet-activating factor. Several studies have attempted to elucidate the mechanisms underlying O_3-induced responses. Results show that the cyclooxygenase products of arachidonic acid are involved in pulmonary function responses subsequent to acute O_3 exposure, but the contribution of β-adrenergic mechanisms to airway reactivity is minimal. Other authors have speculated on the involvement of a neural link whereby cyclooxygenase products stimulate neural afferents in the lung, resulting in pulmonary function decrement. The definitive mechanism underlying O_3-induced changes in pulmonary function, flow resistance, increased bronchial reactivity to bronchoactive challenge, increased epithelial permeability, and increased respiratory symptoms remains incompletely defined.

Low concentrations of O_3 increase particle clearance from the human lung. Higher concentrations, yet to be investigated in humans but shown in animals, decrease both short-term and long-term clearance by damaging ciliated cells and reducing phagocytosis, respectively. Other effects are cognitive and neurobehavioral performance, but this area requires further research using current testing procedures. To date, only subjective descriptions of neurobehavioral performance effects have been documented. Lastly, epidemiological studies have pointed to possible long-term pulmonary function decrement (FEV_1, FVC, $FEF_{25\%-75\%}$) from chronic O_3 exposure. This seems to be consistent with structural and biochemical animal studies, but additional epidemiological research is required before this important conclusion can be drawn in humans.

Elevated O_3 levels have been shown to cause adverse health effects in humans. It is therefore essential that a thorough and complete understanding of the human health effects of O_3 be developed so that appropriate policy initiatives may be promulgated and maximum protection to at-risk populations be provided.

REFERENCES

Altshuller, A. P. (1987). Estimation of the natural background of ozone present at surface rural locations. *JAPCA* **37**, 1409.

Avol, E. L., Linn, W. S., Venet, T. G., Shamoo, D. A., and Hackney, J. D. (1984). Comparative respiratory effects of ozone and ambient oxidant pollution exposure during heavy exercise. *JAPCA* **34**, 804.

Avol, E. L., Limm, W. S., Shamoo, D. A., Valencia, L. M., Anzar, U. T., and Hackney, J. D. (1985). Respiratory effects of photochemical oxidant air pollution in exercising adolescents. *Am. Rev. Respir. Dis.* **132**, 619.

Avol, E. L., Linn, W. S., Shamoo, D. A., Spier, C. E., Valencia, L. M., Venet, T. G., Trim, S. C., and Hackney, J. D. (1987). Short-term respiratory effects of photochemical oxidant exposure in exercising children. *JAPCA* **37**, 158.

Bascomb, R., Naclerio, R. M., Fitzgerald, T. K., Kagey-Sobotka, A., and Proud, D. (1990). Effect of ozone inhalation on the response to nasal challenge with antigen of allergic subjects. *Am. Rev. Respir. Dis.* **142**, 594.

Bates, D. V., and Sizto, R. (1989). The Ontario air pollution study: Identification of the causative agent. *Environ. Health Perspect.* **79**, 69.

Bates, D. V., Baker-Anderson, M., and Sizto, R. (1990). Asthma attack periodicity: A study of hospital emergency visits in Vancouver. *Environ. Res.* **51**, 51.

Beckett, W. S., McDonnell, W. F., Horstman, D. H., and House, D. E. (1985). Role of the parasympathetic nervous system in acute lung responses to ozone. *J. Appl. Physiol.* **59**, 1879.

Beckett, W. S., Freed, A. N., Turner, C., and Menkes, H. A. (1988). Prolonged increased responsiveness of canine peripheral airways after exposure to ozone. *J. Appl. Physiol.* **64**(2), 605.

Bedi, J. F., Folinsbee, L. J., Horvath, S. M., and Ebenste, R. S. (1979). Human exposure to sulfur dioxide and ozone: Absence of a synergistic effect. *Arch. Environ. Health* **34**, 233.

Bedi, J. F., Horvath, S. M., and Folinsbee, L. J. (1982). Human exposure to sulfur dioxide and ozone in a high temperature-humidity environment. *Am. Ind. Hyg. Assoc. J.* **43**, 26.

Bell, K. A., Linn, W. S., Hazucha, M., Hackney, J. D., and Bates, D. (1977). Respiratory effects of exposure to ozone plus sulfur dioxide in southern Californians and eastern Canadians. *Am. Ind. Hyg. Assoc. J.* **38**, 696.

Coleridge, H. M., Coleridge, J. C. G., Ginzel, K. H., Baker, D. G., Banzett, R. B., and Morrison, M. A. (1976). Stimulation of "irritant" receptors and afferent C-fibers in the lungs of prostaglandins. *Nature (London)* **264**, 451.

Detels, R., Tashkin, D. P., Sayre, J. W., Rokaw, S. N., Coulson, A. H., Massey, F. J., and Wegman, D. H. (1987). The UCLA population studies of chronic obstructive respiratory disease. 9. Lung function changes associated with chronic exposure to photochemical oxidant: A cohort study among neversmokers. *Chest* **92**, 594.

Devlin, R. B., McDonnell, W. F., Mann, R., Becker, S., House, D. E., Schreinemackers, D., and Koren, H. S. (1991). Exposure of humans to ambient levels of ozone for 6.6 hours causes cellular and biochemical changes in the lung. *Am. J. Respir. Cell Mol. Biol.* **4**, 72.

Drechsler-Parks, D. M., Bedi, J. F., and Horvath, S. M. (1987). Pulmonary function responses of older men and women to ozone exposure. *Exp. Gerontol.* **22**, 91.

Eschenbacher, W. L., Ying, R. L., Kreit, J. W., and Gross, K. B. (1989). Ozone induced lung function changes in normal and asthmatic subjects and the effect of indomethacin. In T. Schneider, S. D. Lee, G. J. R. Wolters, and L. D. Grant (Eds.), *Atmospheric Ozone Research and Its Policy Implications.* Elsevier, Amsterdam, p. 493.

Farrell, B. P., Kerr, H. D., Kulle, T. J., Sauder, L. R., and Young, J. L. (1979). Adaptation in human subjects to the effects of inhaled ozone after repeated exposure. *Am. Rev. Respir. Dis.* **119**, 725.

Folinsbee, L. J., and Hazucha, M. J. (1989). Persistence of ozone-induced changes in lung function and airway responsiveness. In T. Schneider, S. D. Lee, G. J. R. Wolters, and L. D. Grant (Eds.), *Atmospheric Ozone Research and Its Policy Implications.* Elsevier, Amsterdam, p. 483.

Folinsbee, L. J., Silverman, F., and Shephard, R. J. (1975). Exercise responses following ozone exposure. *J. Appl. Physiol.* **38**, 996.

Folinsbee, L. J., Horvath, S., Raven, P., Bedi, J., Morton, A., Drinkwater, B., Bolduan, N., and Gliner, J. (1977). Influence of exercise and heat stress on pulmonary function during ozone exposure. *J. Appl. Physiol.* **43**, 409.

Folinsbee, L. J., Bedi, J. F., and Horvath, S. M. (1980). Respiratory responses in humans repeatedly exposed to low concentrations of ozone. *Am. Rev. Respir. Dis.* **121**, 431.

Folinsbee, L. J., McDonnell, W. F., and Horstman, D. H. (1988). Pulmonary function and symptom response after 6.6 hour exposure to 0.12 ppm ozone with moderate exercise. *JAPCA* **38**, 28.

Foster, W. M., Costa, D. L., and Langenback, E. G. (1987). Ozone exposure alters tracheobronchial mucociliary function in humans. *J. Appl. Physiol.* **63**, 996.

Frager, N. B., Phalen, R. F., and Kenoyer, J. L. (1979). Adaptation to ozone in reference to mucociliary clearance. *Arch. Environ. Health* **34**, 51.

Gerrity, T. R., Weaver, R. A., Bernsten, J., House, D. E., and O'Neil, J. J. (1988). Extrathoracic and intrathoracic removal of ozone in tidal-breathing humans. *J. Appl. Physiol.* **65**, 393.

Gibbons, S. I., and Adams, W. C. (1984). Combined effects of ozone exposure and ambient heat on exercising females. *J. Appl. Physiol.* **57**, 450.

Golden, J. A., Nadel, J. A., and Boushley, H. A. (1978). Bronchial hyperirritability in healthy subjects after exposure to ozone. *Am. Rev. Respir. Dis.* **118**, 287.

Gong, H., Bradley, P. W., Simmons, M. S., and Tashkin, D. P. (1986). Impaired exercise performance and pulmonary function in elite cyclists during low level exposure in a hot environment. *Am. Rev. Respir. Dis.* **134**, 726.

Gong, H., Bedi, J. F., and Horvath, S. M. (1988). Inhaled albuterol does not protect against ozone toxicity in nonasthmatic athletes. *Arch. Environ. Health* **43**, 46.

Gorden, T., Taylor, B.F., and Amdar, M. O. (1981). Ozone inhibition of tissue cholinesterase in guinea pigs. *Arch. Environ. Health* **36**, 284.

Graham, D., Henderson, F., and House, D. (1988). Neutorophil influx measured in nasal lavages of humans exposed to ozone. *Arch. Environ. Health* **43**, 228.

Griswold, S. S., Chambers, L. A., and Motley, H. L. (1956). Report of a case of exposure to high ozone concentrations for 2 hours. *AMA Arch. Ind. Health* **108**.

Hackney, J. D., Linn., W. S., Mohler, J. G., Perderson, E. E., Breisacher, P., and Russo, A. (1975a). Experimental studies on human health effects of air pollutants. *Arch. Environ. Health* **30**, 379.

Hackney, J. D., Linn W. S., Law, D. C., Karuza, S. K., Greenway, H., Buckley, R. D., and Pederson, E. (1975b). Experimental studies on human health effects of air pollutants. *Arch. Environ. Health* **30**, 385.

Hackney, J. D., Linn, W. S., Mohler, J. G., and Collier, C. R. (1977). Adaptation to short term respiratory effects of ozone in men exposed repeatedly. *J. Appl. Physiol.* **43**, 82.

Hammer, D. I., Hasselblad, V., Portnay, B., and Wehler, P. F. (1974). Los Angeles student nurse study: Daily symptoms reporting and photochemical oxidants. *Arch. Environ. Health* **28**, 255.

Hayes, R. S., Moezzi, M., Wallsten, T. S., and Winkler, R. L. (1987). *An Analysis of Symptom and Lung Function Data from Several Human Controlled Ozone Exposure Studies*, draft final report. Systems Applications, San Rafael, CA.

Hazucha, M. J. (1987). Relationship between ozone exposure and pulmonary function changes. *J. Appl. Physiol.* **62**, 1671.

Hazucha, M. J., and Bates, D. V. (1975). Combined effects of ozone and sulfur dioxide on human pulmonary function. *Nature (London)* **257**, 26.

Hazucha, M. J., Bates, D. V., Bromberg, P. A. (1986). Mechanism of action of ozone on the human lung. (abstr.) *Am. Rev. Respir. Dis.* **133**, A214.

Holtzman, M. J., Cunningham, J. H., Sheller, J. R., Irsigler, G. B., Nadel, J. A., and Boushey, H. (1979). Effect of ozone on bronchial reactivity in atopic and nonatopic subjects. *Am. Rev. Respir. Dis.* **120**, 1059.

Hore, T., and Gibson, D. E. (1968). Ozone exposure and intelligence tests. *Arch. Environ. Health* **17**, 77.

Horstman, D. H., McDonnell, W. F., Folinsbee, L. J., Salaam, A., and Ives, P. (1989). Changes in pulmonary function and airway reactivity due to prolonged exposure to typical ambient O_3 levels. In T: Schneider, S. D. Lee, G. J. R. Wolters, and L. D. Grant (Eds.), *Atmospheric Ozone Research and Its Policy Implications.* Elsevier, Amsterdam, p. 755.

Horstman, D. H., Folinsbee, L. J., Ives, P. J., Abdul, S., and McDonnell, W. F. (1990). Ozone concentration and pulmonary response relationships for 6.6 hours with 5 hours of moderate exercise to 0.08, 0.10, and 0.12 ppm. *Am. Rev. Respir. Dis.* **142**, 1158.

Horvath, S. M., Gliner, J. A., and Folinsbee, L. J. (1981). Adaptation to ozone: Duration of effect. *Am. Rev. Respir. Dis.* **123**, 496.

Horvath, S. M., Bedi, J. F., and Drechsler-Parks, D. M. (1986). Effects of peroxyacetyl nitrate alone and in combination with ozone in healthy young women. *JAPCA* **36**, 265.

Hynes, B., Silverman, F., Cole, P., and Corey, P. (1988). Effects of ozone exposure: A comparison between oral and nasal breathing. *Arch. Environ. Health* **43**, 357.

Kagawa, J. (1983). Respiratory effects of 2 hours exposure with intermittent exercise to ozone, sulfur dioxide and nitrogen dioxide alone and in combination in normal subjects. *Am. Ind. Hyg. Assoc. J.* **44**, 14.

Kagawa, J. (1984). Exposure-effect relationship of selected pulmonary function measurements in subjects exposed to ozone. *Int. Arch. Occup. Environ. Health* **53**, 345.

Kehrl, H. R., Vincent, L. M., Kowalsky, R. J., Horstman, D. H., O'Neil, J., McCartney, W. H., and Bromberg, P. A. (1987). Ozone exposure increases respiratory epithelial permeability in humans. *Am. Rev. Respir. Dis.* **135**, 1174.

Kenoyer, J. L., Phalen, R. F., and Davis, J. R. (1981). Particle clearance form the respiratory tract as a test of toxicity: Effect of ozone on short and long term clearance. *Exp. Lung Res.* **2**, 111.

Kilburn, K. H., Warshaw, R., and Thronton, J. C. (1985). Pulmonary functional impairment and symptoms in women in the Los Angeles harbor area. *Am. J. Med.* **79**, 23.

Kinney, P. L. (1986). Short term lung function associations with air pollution in Kingston and Hurriman, Tennessee Doctoral Dissertation, Harvard School of Public Health, Cambridge, MA.

Kinney, P. L., Ware, J. H., and Spengler, J. D. (1988). A critical evaluation of acute ozone epidemiology results. *Arch. Environ. Health* **43**, 168.

Kirby, J. G., Hargreave, F. E., Gleich, G. J., and O'Byrne, P. M. (1987). Bronchoalveolar cell profiles of asthmatic and nonasthmatic subjects. *Am. Rev. Respir. Dis.* **136**, 379.

Knudson, R. J., Lebowitz, M. D., Holberg, C. J., and Burrows, B. (1983). Changes in the normal maximal expiratory flow-volume curve with growth and aging. *Am. Rev. Respir. Dis.* **127**, 725.

Koenig, J. Q., Covert, D. S., Morgan, M. S., Horike, M., Horike, N., Marshall, S. G., and Pierson, W. E. (1985). Acute effects of 0.12 ppm ozone or 0.12 ppm nitrogen dioxide on pulmonary function in healthy and asthmatic adolescents. *Am. Rev. Respir. Dis.* **132**, 648.

Koenig, J. Q., Covert, D. S., Marshall, S. G., Van Belle, G., and Pierson, W. E. (1987). The effects of ozone and nitrogen dioxide on pulmonary function in healthy and in asthmatic adolescents. *Am. Rev. Respir. Dis.* **136**, 1152.

Koenig, J. Q., Covert, D. S., Hanley, Q. S., Van Belle, G., and Pierson, W. E. (1990). Prior exposure to ozone potentiates subsequent response to sulfur dioxide in adolescent asthmatic subjects. *Am. Rev. Respir. Dis.* **141**, 377.

Koren, H., Devlin, R. B., Graham, D., Mann, R., McGee, M., Horstman, D. H., Kozumbo, W. J., Becher, S., House, D. E., McDonnell, W. F., and Bromberg, P. A. (1989). Ozone-induced inflammation in the lower airways to human subjects. *Am. Rev. Respir. Dis.* **139**, 407.

Kreit, J. W., Gross, K. B., Moore, T. B., Lorenzen, T. J., D'Arcy, J., and Eschenbacher, W. L. (1989). Ozone induced changes in pulmonary function and bronchial responsiveness in asthmatics. *J. Appl. Physiol.* **60**, 217.

Kulle, T. J., Sauder, L. R., Kerr, H. D., Farrell, B. P., Bermel, M. S., and Smith, D. M. (1982). Duration of pulmonary function adaptation to ozone in humans. *Am. Ind. Hyg. Assoc. J.*, **43**, 832.

Kulle, T. J., Milman, J. H., Sauder, L. R., Kerr, H. D., Farrell, B. P., and Miller, W. R. (1984). Pulmonary function adaptation to ozone in subjects with chronic bronchitis. *Environ. Res.* **34**, 55.

Kulle, T. J., Sauder, L. R., Hebel, J. R., and Chatham, M. D. (1985). Ozone response relationships in healthy nonsmokers. *Am. Rev. Respir. Dis.* **132**, 36.

Lagerwerff, J. M. (1963). Prolonged ozone inhalation and its effects on visual parameters. *Aerosp. Med.* **34**, 479.

Lee, L. Y., Dumont, C., Djokic, T. D., Menzel, T. E., and Nadel, J. A. (1979). Mechanism of rapid shallow breathing after ozone exposure in conscience dogs. *J. Appl. Physiol.* **46**, 1108.

Leibowitz, M. D., Holberg, C. J., Boyer, B., and Hayes, C. (1985). Respiratory symptoms and peak flow associated with indoor and outdoor pollutants in the southwest. *JAPCA* **35**, 1154.

Linder, J., Herren, D., Monn, C., and Wanner, H. U. (1988). The effect of ozone on physical activity. *Sportmed* **36**, 5.

Linn, W. S., Jones, M. P., Bachmyer, E. A, Spier, C. E., Mazur, S. F., Avol, E. L., and Hackney, J. D. (1980). Short-term respiratory effects of polluted air: A laboratory study of volunteers in a high oxidant community. *Am. Rev. Respir. Dis.* **121**, 243.

Linn, W. S., Shamoo, D. A., Venet, T. G., Spier, C. E., Valencia, L. M., Anzar, U. T., and Hackney, J. D. (1983). Response to ozone in volunteers with chronic obstructive pulmonary disease. *Arch. Environ. Health* **38**, 278.

Linn, W. S., Avol, E. L., Shamoo, D. A., Peng, R. C., Valencia, L. M., Little, D. E., and Hackney, J. D. (1988). Repeated laboratory ozone exposures of volunteer Los Angeles residents: An apparent seasonal variation in response. *Toxicol. Ind. Health* **4**, 504.

Lioy, P. J., Wollmuth, T. A., and Lippmann, M. (1985). Persistence of peak flow decrement in children following ozone exposures exceeding the national ambient air quality standard. *JAPCA* **35**, 1068.

Lippmann, M. (1989). Health effects of ozone. A critical review. *JAPCA* **39**, 672.

Lippmann, M., Lioy, P. J., Leikauf, G., Green, K. B., Baxter, D., Morandi, M., Pasternak, B. S., Fife, D., and Speizer, F. E. (1983). Effects of ozone on the pulmonary function of children. *Adv. Environ. Toxicol* **5**, 423.

Lippmann, M., Gearhart, J. M., and Schlessinger, R. B. (1987). Basis for a particle size-selective TLV for sulfuric acid aerosols. *Appl. Ind. Hyg.* **2**, 188.

McBride, D. E. (1991). Upper and lower respiratory effects of ozone in asthmatics and nonasthmatic individuals. Master's Thesis, Department of Environmental Health, University of Washington, Seattle.

McDonnell, W. F., Horstman, D. H., Hazucha, M. J., Seal, E., Haak, D., Salaam, A., and House, D. E. (1983). Pulmonary effects of ozone exposure during exercise: Dose-response characteristics. *J. Appl. Physiol.* **54**, 1345.

McDonnell, W. F., Horstman, D. H., Abdul, S., and House, D. E. (1985). Reproducibility of individual responses to ozone exposure. *Am. Rev. Respir. Dis.* **131**, 36.

McDonnell, W. F., Horstman, D. H., Abdul, S., Ragio, L. J., and Green, J. A. (1987). The respiratory responses of subjects with allergic rhinitis to ozone exposure and their relationship to nonspecific airway reactivity. *Toxicol. Ind. Health* **3**, 507.

Menzel, D. B. (1984). *Ozone: An Overview of Its Toxicity in Man and Animals.* Hemisphere Publishing Corp., New York.

Miller, F. J., Menzel, D. B., and Coffin, D. L. (1978) Similarity between man and laboratory animals in regional pulmonary deposition of ozone. *Environ. Res.* **17**, 84.

Molfino, N. A., Wright, S. C., Katz, I., Tarlo, S., Silverman, F., McMlean, P. A., Szali, J. P., Raizenne, M., Slutsky, A. S., and Zamel, N. (1991). Effects of low concentrations of ozone on inhaled allergen responses in asthmatic subjects. *Lancet* **338**, 199.

Niinimaa, V., Cole, P., Mintz, S., and Shephard, R. J. (1980). The switching point from nasal to oronasal breathing. *Respir. Physiol.* **42**, 61.

Piacentini, G. L., and Kaliner, M. A. (1991). The potential roles of leukotrienes in bronchial asthma. *Am. Rev. Respir. Dis.* **143**, S96.

Raizenne, M. E., Burnett, R. T., Stern, B., Franklin, C. A., and Spengler, J. D. (1989). Acute lung function responses to ambient acid aerosol exposures in children. *Environ. Health Perspect.* **79**, 179.

Rao, S. T. (1988). Prepared discussion: Ozone air quality models. *JAPCA* **38**, 1129.

Reisenauer, C. S., Koenig, J. Q., McManus, M. S., Smith, M. S., Kusic, G., and Pierson, W. E. (1988). Pulmonary response to ozone exposure in healthy individuals aged 55 years or greater. *JAPCA* **38**, 51.

Roberts, A. M., Schultz, H. D., Green, J. F., Armstrong, D. J., Kaufman, M. P., Coleridge, H. M., and Coleridge, J. C. G. (1985). Reflex tracheal contraction evoked in dogs by bronchodilator prostaglandins E2 and I2. *J. Appl. Physiol.* **58**, 1823.

Rombout, P. J. A., Lioy, P. J., and Goldstein, B. D. (1986). Rational for an 8 hour ozone standard. *JAPCA* **36**, 913.

Schelegle, E. S., and Adams, W. C. (1986) Reduced exercise time in competitive simulations consequent to low level ozone exposure. *Med. Sci. Sports Exercise* **18**, 408.

Schelegle, E. S., Adams, W. C., and Seifkin, A. D. (1987). Indomethacin pretreatment reduces ozone-induced pulmonary function decrements in human subjects. *Am. Rev. Respir. Dis.* **136**, 1350.

Schlessinger, R. B., and Driscoll, K. E. (1987). Mucociliary clearance from the lungs of rabbits following single and intermittent exposures to ozone. *J. Toxicol. Environ. Health* **20**, 125.

Seinfeld, J. H. (1988). Ozone air quality models. A critical review. *JAPCA* **38**, 616.

Seltzer, J., Bigby, B. G., Stulburg, M., Holtzman, J., and Wadel, J. A. (1986). Ozone-induced change in bronchial reactivity to methacholine and airway inflammation in humans. *J. Appl. Physiol.* **60**, 1321.

Shephard, R. J., Urch, B., Silverman, F., and Corey, P. N. (1983). Interaction of ozone and cigarette smoke exposure. *Environ. Res.* **31**, 125.

Solic, J. J., Hazucha, M. J., and Bromberg, P. A. (1982). The acute effects of 0.2 ppm ozone in patients with chronic obstructive pulmonary disease. *Am. Rev. Respir. Dis.* **125**, 664.

Spektor, D. M., Lippmann, M., Lioy, P. J., Thurston, G. D., Citak, D., James, D. J., Vock, N., Speizer, F. E., and Hayes, C. (1988a). Effects of ambient ozone on respiratory function in active normal children. *Am. Rev. Respir. Dis.* **137**, 313.

Spektor, D. M., Lippmann, M., Thurston, G. D., Lioy, P. J., Stecko, J., O'Connor, G., Garshick, E., Speizer, F. E., and Hayes, C. (1988b). Effects of ambient ozone on respiratory function in healthy adults exercising outdoors. *Am. Rev. Respir. Dis.* **138**, 821.

Spengler, J. D., Garsd, A., and Ozkaynak, J. (1985). *Statistical Analysis of the Lake Couchiching Health and Aerometric Data*, Phase 1 Report. Report produced for the Health Protection Branch Department of National Health and Welfare, Ottawa, Canada.

Stacey, R. W., Seal, E., House, D. E., Green, J., Roger, L. J., and Raggio, L. (1983). A survey of effects of gaseous and aerosol pollutants on pulmonary function of normal males. *Arch. Environ. Health* **38**, 104.

Walters, E., Purish, R., Bevan, C., Davies, B. H., and Smith, A. P. (1981). Induction of bronchial hypersensitivity: Evidence for a role for prostaglandins. *Thorax* **36**, 5721.

Wayne, W. S., Wehrle, P. F., and Carroll, R. E. (1967). Oxidant air pollution and athletic performance. *JAMA, J. Am. Med. Assoc.* **199**, 901.

Ying, R. L., Gross, K. B., Terzo, T. S., and Eschenbacher, W. L. (1990). Indomethacin does not inhibit the ozone-induced increase in bronchial responsiveness in human subjects. *Am. Rev. Respir. Dis.* **142**, 817.

Zwick, H., Popp, W., Wagner, C., Reiser, K., Schmoger, J., Bock, A., Herkner, K., and Radunsky, K. (1991). Effects of ozone on the respiratory health, allergic sensitization, and cellular immune system in children. *Am. Rev. Respir. Dis.* **144**, 1075.

11

HEALTH EFFECTS AND TOXICOLOGY OF OZONE AND NITROGEN DIOXIDE

Mohammad G. Mustafa

Department of Environmental Health Sciences, School of Public Health, University of California, Los Angeles, Los Angeles, California 90024-1772

Environmental Oxidants, Edited by Jerome O. Nriagu and Milagros S. Simmons.
ISBN 0–471–57928–9 © 1994 John Wiley & Sons, Inc.

1. INTRODUCTION

Oxidants of photochemical smog generally include ozone (O_3), nitrogen dioxide (NO_2), peroxyacyl nitrates (PAN), hydrogen peroxide, alkyl peroxides, nitrous and nitric acids, formaldehyde and formic acid, and traces of other compounds [National Research Council (NRC), 1977; Environmental Protection Agency (EPA), 1978, 1986; World Health Organization (WHO), 1979]. These chemicals, which can coexist, are responsible for the oxidizing properties of photochemical smog, and are collectively referred to as oxidants because they are capable of removing electrons from other molecules. Of these oxidants, the toxicity of O_3 and NO_2 has been studied in relatively greater detail because of their abundance and earlier identification as constituents of photochemical smog and as having enormous potential effects on humans, plants, and ecosystems.

Ozone and NO_2 are toxic gases, and their inhalation can cause respiratory irritation and injury. Because the urban atmosphere may contain elevated concentrations of these gases, the pulmonary effects of inhaled oxidants have received wide attention. Numerous animal studies and limited human studies have revealed that a variety of short-term health effects and chronic lung diseases, perhaps including increased tumor incidence, can result from exposure to these oxidants. The toxic effects of these gases are attributed to their ability to induce free radical reactions in the biological system. The purpose of this presentation is to review the toxicity and health effects of O_3 and NO_2, including antioxidant protection, as observed in experimental animals. To the extent feasible, the effects of these two oxidants will be discussed separately.

2. OCCURRENCE AND EXPOSURE SOURCES

Although the two oxidants O_3 and NO_2 coexist in photochemical smog, their origins are diverse. Both can occur naturally, but their contribution to air pollution is related to human activities. In addition to exposure to photochemical smog, there are other circumstances in which environmental and/or occupational exposure to O_3 and NO_2 can occur. The oxidants of photochemical smog, which irritate the eyes and throat, evoke asthma attacks in sensitive persons, and reduce visibility, were first noted in the Los Angeles Basin. This type of smog has now been recognized in most major urban areas of the world.

2.1. Ozone

Ozone occurs as a natural component of the atmosphere. However, its formation in the upper atmosphere (stratosphere) and in the lower atmosphere (lower troposphere) are distinctly different, as are the biological consequences. Ozone in the stratosphere is known to protect humans and other life forms from harmful cosmic radiation [National Academy of Sciences (NAS), 1984; EPA, 1987]. In contrast, ozone in the lower troposphere is a common cause of urban air pollution, posing a threat to human health and property (NRC, 1977; EPA, 1978, 1986; WHO, 1979).

2.1.1. Formation in the Stratosphere

Atmospheric O_3 concentrations are highest in the stratosphere, where it is thought to be generated in a cyclic reaction involving a photochemical dissociation of molecular oxygen (O_2) to atomic oxygen (O) by ultraviolet radiation [Reaction (1)] and a chemical combination of atomic oxygen with molecular oxygen [Reaction (2)]. Ozone thus formed is in dynamic equilibrium with its natural destruction, due to a photochemical dissociation by ultraviolet radiation [Reaction (3)]. Some of the stratospheric O_3 evidently reaches the lower troposphere, resulting in the global O_3 background of up to 0.04 ppm (78 $\mu g/m^3$) (NRC, 1977; EPA, 1978).

$$O_2 \xrightarrow{h\nu(\lambda < 242 \text{ nm})} O + O \tag{1}$$

$$O_2 + O \xrightarrow{M} O_3 \tag{2}$$

where M = a third-body molecule absorbing the excess energy of the reaction.

$$O_3 \xrightarrow{h\nu(\lambda < 310 \text{ nm})} O_2 + O \tag{3}$$

The stratospheric O_3 layer acts as a shield for the earth, protecting its surface and preventing adverse effects on human health (skin cancer and suppression of immune response) and on terrestrial and aquatic ecosystems from ultraviolet radiation (EPA, 1987). There is international concern that human-made chemicals (e.g., chlorofluorocarbons and oxides of nitrogen) might react with stratospheric O_3, causing a depletion of the O_3 layer and allowing harmful ultraviolet radiation to reach the earth's surface (NAS, 1984; EPA, 1987).

2.1.2. Formation in Smog

Ozone in urban air (lower troposphere) is a major cause of air pollution and concern for health effects. The occurrence of photochemical smog dates back to the early 1940s. The presence of O_3 in the air of Los Angeles was first recognized by its damaging oxidative effects on rubber products (NRC, 1977; EPA, 1978). How O_3 could possibly occur in ambient air remained a mystery until the work of Haagen-Smit (1952, 1963), which demonstrated that irradiation of a mixture of NO_2 and hydrocarbons produced O_3. Many of the basic reactions involved in the photochemical production of O_3 and other oxidants have by now been worked out (NRC, 1977; EPA, 1978; Logan et al., 1981; Whitten, 1983; Atkinson and Lloyd, 1984), but understanding the vastly complex of photochemistry that gives rise to a variety of secondary pollutants remains a challenge.

The precursors of photochemical oxidants are oxides of nitrogen (NO_x), volatile organic compounds (VOC) (e.g., vapor-phase hydrocarbons and halogenated organics), hydroxyl (HȮ), and other radicals, O_2, and sunlight. Human-made emissions of NO_x and VOC are the major contributors to elevated ambient O_3 levels, as well as a series of other photochemical oxidants. Photoactivation, photodecomposition, and free radical chain reactions are among the important reaction mechanisms by which they are formed.

Briefly, O_3 in smog is produced in a cyclic process involving NO_2 decomposition and regeneration. First, driven by the energy of sunlight, NO_2 undergoes photolysis, producing an atomic oxygen [Reaction (4)] that then reacts with a molecular oxygen to form O_3 [similar to Reaction (2)]. In the absence of any competing reactions, a rapid reaction between O_3 and NO completes the cycle, regenerating NO_2 and O_2 [Reaction (5)]. As a result of these reactions, a steady-state condition is established among NO_2, NO, and O_3 concentrations. In this steady state, any buildup of O_3 will depend upon the ratio of $[NO_2]$ to $[NO]$, that is, the higher the ratio the greater the O_3 concentration. Because Reaction (4) is rapid and O_3 is stoichiometrically destroyed by the reaction with NO [Reaction (5)], O_3 concentration in smog cannot rise until most of the NO has been converted to NO_2 by other reactions.

$$NO_2 \xrightarrow{h\nu(\lambda 295-430\,nm)} NO + O \qquad (4)$$

$$O_2 + O \xrightarrow{M} O_3 \qquad (2)$$

$$O_3 + NO \longrightarrow NO_2 + O_2 \qquad (5)$$

In smog, there are many chemical species that undergo photolysis, giving rise to a variety of radicals. These radicals participate in the secondary formation of NO_2 and/or in alternate pathways for the conversion of NO to NO_2 without destroying O_3. The result is a higher ratio of $[NO_2]$ to $[NO]$ and a buildup of O_3 concentration in smog. An important mechanism contributing to O_3 buildup is the formation and participation of HȮ and hydroperoxyl (HȮ$_2$) radicals in photochemical reactions. These radicals are formed by photolysis of various compounds, for example, formaldehyde (HCHO), nitrous acid (HONO), and others. In a chain reaction, HȮ gives rise to HO_2, which then oxidizes NO to NO_2 and is itself reduced to HȮ [Reactions (6)]. This chain reaction is a major pathway for the oxidation of NO in smog. In addition to HCHO, higher aldehydes (RCHO) and hydrocarbons (RH) give rise to alkoxyl (RȮ) and alkoperoxyl (RȮ$_2$) radicals, which then oxidize NO to NO_2 [Reaction (7)].

$$HȮ + NO \longrightarrow NO_2 + HȮ \qquad (6)$$

$$RȮ_2 + NO \longrightarrow NO_2 + RȮ \qquad (7)$$

2.1.3. *Exposure Potential and Regulation*

Photochemical smog is the major cause of O_3 exposure in the general population. Large urban areas of the world, characterized by automobile-dependent transportation and/or petroleum-dependent energy production, generally experience photochemical smog and hence elevated levels of O_3. Because the precursors (NO_x and VOC) much react in the sun to produce O_3, daily peak O_3 concentrations usually occur from late morning through the afternoon. Ozone concentrations are higher in summer than in winter because solar radiation is more intense and of longer duration in summer. In the South Coast Air Basin (SCAB) of California, O_3 levels remain between 0.1 and 0.2 ppm (196–392 $\mu g/m^3$) almost year-round, and in summer the levels are much higher.

Other sources of ozone exposure include gamma-radiation (food preservation) plants, commercial UV lamps for sterilization, dermatologic phototherapy equipment, and high-voltage electric equipment, in which O_3 is generated by radiation or electric arc as in Reactions (1) and (2). Ozone also occurs in air and water purification plants and in the oil, wax, textile, and inorganic synthesis industries, where O_3 is used on a large scale. Usually, the workers in such plants and industries are the likely subjects of O_3 exposure. Another source is stratospheric O_3, which can pose a health risk to both crew and passengers in high-altitude flights because the air intake for airplane cabins includes O_3 (Reed et al., 1980).

Because of toxicity and potential health hazards to humans, O_3 levels in the environment and workplace are regulated by governments (Table 1). The U.S. Occupational Safety and Health Administration (OSHA) standard for personnel exposure is 0.1 ppm (200 $\mu g/m^3$) O_3 as a time-weighted average during an 8-hr work day 40-hr work week, with a ceiling of 0.25 ppm (490 $\mu g/m^3$). The U.S. Environmental Protection Agency (EPA) standard for O_3 is 0.12 ppm (240 $\mu g/m^3$) average for 1 hr, not to be exceeded more than once per year; the California state standard is 0.1 ppm (200 $\mu g/m^3$) average for 1 hr. These standards, of course, are frequently exceeded in the Los Angeles Basin. The World Health Organization

Table 1   Air Quality Standards for Ozone (Photochemical Oxidants)

Country (Organization)	Standard and Criteria [$\mu g/m^3$ (ppm)[a]]
United States (EPA)[b,c]	240 (0.12), 1-hr avg
United States (OSHA)[d]	200 (0.10), 8-hr avg
Japan	160 (0.08), maximum permitted level
World Health Organization (WHO)	100–200 (0.05–0.10), 1-hr limit

[a] 1 ppm = 1960 $\mu g/cm^3$ (rounded to 2 mg/cm^3).

[b] U.S. EPA declares O_3 episodes as follows: alert at 0.10 ppm (200 $\mu g/cm^3$), 1-hr average; warning at 0.4 ppm (780 $\mu g/cm^3$), 1-hr average; emergency action at 0.5 ppm (980 $\mu g/cm^3$), 1-hr average; and significant harm to health at 0.6 ppm (1180 $\mu g/cm^3$), 1-hr average [South Coast Air Quality Management District (SCAQMD), 1980, 1983].

[c] State of California has a standard of 0.10 ppm (200 $\mu g/cm^3$), 1-hr average; California declares O_3 episodes as follows: stage 1 (health advisory) at 0.2 ppm (390 $\mu g/cm^3$), 1-hr average; stage 2 (warning) at 0.35 ppm (690 $\mu g/cm^3$), 1-hr average; and stage 3 (emergency) at 0.5 ppm (980 $\mu g/cm^3$), 1-hr average (SCAQMD, 1980, 1983).

[d] Occupational Safety and Health Administration standard for occupational safety.

(WHO) Task Group recommends a 1-hr exposure limit of 0.05–0.10 ppm (100–200 µg/m^3) O_3 for the protection of public health.

2.2. Nitrogen Dioxide

Oxides of nitrogen (NO_x), generally a mixture of nitric oxide (NO) and nitrogen dioxide (NO_2), are produced by both natural and human-made processes. However, the human-made sources are the cause of air pollution. Oxides of nitrogen are produced during combustion of fossil fuels such as coal, oil, and gasoline. Nitrogenous compounds in fossil fuels contribute, but the heat of fuel combustion causes the chemical combination of atmospheric N_2 and O_2, and any combustion process that generates sufficient heat can produce NO_x. In a combustion process, the formation of NO from N_2 and O_2 of air [Reaction (18)] is an endothermic process; the higher the temperature, the faster the NO formation. NO_2 is formed by a combination of NO and O_2 [Reaction (9)], which is an exothermic (spontaneous) process. NO and NO_2 are therefore interdependent gases. They coexist, but NO is the major component of NO_x, and its oxidation during combustion or subsequently in the air rapidly increases the NO_2 concentration. In photochemical smog, NO oxidation to NO_2 may involve more complex reactions. However, the rate of NO conversion to NO_2 and the various factors that influence the conversion reaction are not fully clear.

$$N_2 + O_2 \longrightarrow 2NO \tag{8}$$

$$2NO + O_2 \longrightarrow 2NO_2 \tag{9}$$

Emissions from mobile sources (automobiles and other vehicles) and stationary sources (power plants and chemical and manufacturing industries) are by far the largest contributors of No_x in the urban atmosphere. In addition, NO and NO_2 can arise from other sources, including industries that manufacture and use HNO_3, silent electric arc-welding devices, detonating explosives, diesel equipment, and stacked hay in silos. Tobacco smoke and cooking and heating by gas burning in homes are other significant sources of NO_x, causing indoor air pollution.

Earlier concern about NO_x emissions was primarily linked to the formation of other photochemical oxidants, particularly O_3 and peroxyacyl nitrates (PAN). In photochemical smog, NO_x evidently undergo a series of oxidation–reduction reactions, which are of serious air pollution significance. NO_x, particularly after interaction with O_3, may generate the following gases, radicals, acids, and salts: NO, NO_2, NO_3, N_2O_4, N_2O_5, HNO_2, HNO_3, NO_2^-, and NO_3^-. The presence of high levels of NO_x, particularly NO_2 in urban air, and the expansion of NO-producing activities (e.g., power generation by synthetic fuels and coal combustion and increased deployment of diesel engines in automobiles) have the potential for direct health effects from oxidizing air pollutants. Any health effects due to NO_x are generally attributed to NO_2, and there is very little known about whether NO produces any toxicity. Some of the higher oxides of nitrogen (e.g., NO_3, N_2O_5, and PAN) can have health effects.

NO_2 is a ubiquitous air pollutant. People living in rural (or low pollution) areas may be exposed to a background level of 0.05–0.08 ppm (100–150 $\mu g/m^3$) NO_2. Urban concentrations can range from 0.08 to 0.2 ppm (150–376 $\mu g/m^3$), with spikes reaching 1 ppm (1880 $\mu g/m^3$). In urban areas, daily peak NO_2 concentrations usually occur in early morning and early evening, coinciding with rush hour traffic. Because the intensity of solar radiation (i.e., the photochemical utilization of NO_2) is less, NO_2 concentrations tend to be higher in fall and winter than in other seasons. In homes burning gas for cooking and heating, indoor NO_2 levels can be several orders of magnitude greater than the background levels. Both indoor and outdoor NO_2 exposures are of direct health concern. In the United States, Canada, Europe, and other industrialized areas, NO_2 contributes to acid rain, thereby raising concerns about health and other biological effects.

High-level NO_2 exposure can occur in the workplace through, for example, handling fresh silage, welding in a confined space, combustion of nitrogen-

Table 2   Air Quality Standards for Nitrogen Dioxide

Country (Organization)	Standard and Criteria [$\mu g/m^3$ (ppm)[a]]
United States (EPA)[b,c]	100 (0.05), annual avg of 24-hr values
United States (OSHA)[d]	9400 (5.0), 15-min avg
	5640 (3.0), 8-hr avg
Germany	100 (0.05), annual avg of half-hour values
	300 (0.16), half-hour, 98% limit value
Switzerland	50 (0.03), annual avg
European Commission	50 (0.03), annual, 50% guide value
	135 (0.07), 1-hr, 98% guide value
	200 (0.10), 1-hr, 98% limit value
Japan	80–120 (0.04–0.06), 24 hr, interval for maximum level
World Health Organization (WHO)	190–320 (0.10–0.17), 1-hr, not to be exceeded more than once per month.

[a] 1 ppm = 1880 $\mu g/cm^3$ (rounded to 2 mg/cm^3).

[b] U.S. EPA declares NO_2 episodes as follows: alert at 0.60 ppm (1130 $\mu g/cm^3$), 1-hr average or 0.15 ppm (280 $\mu g/cm^3$), 24-hr average; warning at 1.2 ppm (2260 $\mu g/cm^3$), 1-hr average or 0.30 ppm (560 $\mu g/cm^3$), 24-hr average; emergency action at 1.6 ppm (3000 $\mu g/cm^3$), 1-hr average or 0.40 ppm (750 $\mu g/cm^3$), 24-hr average; and significant harm to health at 2.0 ppm (37680 $\mu g/cm^3$), 1-hr average or 0.50 ppm (940 $\mu g/cm^3$), 24-hr average [South Coast Air Quality Management District (SCAQMD), 1980, 1983].

[c] State of California has a NO_2 standard of 0.25 ppm (470 $\mu g/cm^3$), 1-hr average; (SCAQMD, 1980, 1983).

[d] Occupational Safety and Health Administration standard for occupational safety.

containing materials, and handling HNO_3 in the chemical industry. Fodder stored in silos for anaerobic fermentation undergoes a side-reaction in which thermophilic bacteria convert nitrate and nitrite into NO_2. Farm workers can unknowingly be exposed to a dangerous concentration of NO_2. The high temperatures of electric arc welding can produce a high concentration of NO_2 from N_2 and O_2 in air [Reactions (8) and (9)]. Combustion of nitrogen-containing materials (e.g., nitrocellulose, dynamite, and explosives) can give rise to a dangerous level of NO_2 (Lowry and Schuman, 1956). Chemical industries (e.g., those that produce HNO_3 or use HNO_3 for cleaning and carrying out nitration of organic compounds) can give out high concentrations of NO_2 through spillage and handling.

NO_2 as an air pollutant is regulated (Table 2). The U.S. EPA and California State standards for NO_2 are 0.05 ppm ($100 \, \mu g/m^3$) annual average and 0.25 ppm ($470 \, \mu g/m^3$) 1-hr average, respectively. In the Los Angeles Basin, the state NO_2 standard is often exceeded in late fall and winter, and the levels can attain 0.34–0.41 ppm ($640–770 \, \mu g/m^3$) [South Coast Air Quality Management District (SCAQMD), 1980, 1983]. In Europe, the environmental concentration of NO_x is on the rise, and so is the need for governmental regulation [Council of the European Communities (CEC), 1985].

3. TOXICITY AND MECHANISMS OF TOXICITY

Both O_3 and NO_2, being toxic gases, can cause mortality in animals if breathing air contains O_3 at 4–10 ppm ($8–20 \, mg/m^3$) and NO_2 at 50–150 ppm ($94–282 \, mg/m^3$). However, the sensitivities are different for different animal species. Massive lung injury involving pulmonary edema and hemorrhage and pleural effusion is the apparent cause of animal mortality (Coffin and Stokinger, 1977; NRC, 1977; EPA, 1978, 1986, and references cited therein).

Earlier studies with O_3 and NO_2, which are of basic toxicological importance, have involved oxidant concentrations several orders of magnitude higher than those generally encountered in photochemical smog. These studies established that an exposure to O_3 at > 1 ppm ($> 2 \, mg/m^3$) can be sublethal or lethal to animals, depending upon length of exposure (a few hours to a few days), the animal species (including strain, age, sex, preexisting diseases, and nutritional status), and other conditions (temperature, humidity, etc.) (Coffin and Stokinger, 1977; Mustafa and Tierney, 1978; EPA, 1978, 1986). NO_2, which may be 10- to 15-fold less toxic than O_3, will prove sublethal or lethal to animals at concentrations > 15 ppm ($> 28 \, mg/m^3$) (Mustafa et al., 1984). From the standpoint of environmental and occupational exposure, studies involving < 1 ppm ($< 1.96 \, mg/m^3$) O_3 and < 5 ppm ($< 9.4 \, mg/m^3$) NO_2 concentrations are relevant and will be discussed in this presentation. To date, inhalation exposure studies have been carried out extensively in a variety of animal species, and to a limited extent in human volunteers.

3.1. Pulmonary Absorption

It is important to consider the respiratory uptake of inhaled oxidants. The absorbed dose determines the tissue distribution, that is, the amount reaching the target tissues, and this in turn determines the degree of injury in pulmonary and extrapulmonary sites.

3.1.1. Ozone Uptake

Ozone is a relatively insoluble gas (aqueous solubility 0.494 mL/100 mL at 0 °C, 1 atm pressure). Inspired O_3 is dissolved partly in the mucous layer and partly in surface tissue of the upper airways, but due to its limited solubility, a large portion of O_3 reaches the lower airways (i.e., deep into the lung). Nonetheless, the nasopharyngeal absorption of O_3 diminishes in concentration by the time it reaches the tracheal lumen. Approximately 50% of the inspired O_3 is taken up by the nasopharynx, and 50% reaches the tracheal lumen, lower respiratory tract, and alveolar ducts. Overall, approximately 90% of inhaled O_3 (at a concentration <1 ppm or 2 mg/m^3) is absorbed by the lung during normal breathing (Yokoyama and Frank, 1972; Miller, 1979). Because O_3 is very reactive, cellular damage can potentially occur in the nasopharynx, conducting airways, and alveoli adjacent to the terminal bronchioles, although focal lesions are observed at the bronchioloalveolar junctions. Once impacted on the mucous layer and tissue surface of the respiratory tract, O_3 will break down to O_2 in the process of reaction with biomolecules. Since O_2 is universally present and likely to participate in the reaction, it is difficult to identify or quantify any products of O_3 reaction in vivo. Free radicals and reactive intermediates are apparently produced and transported, which explains many of the oxidant effects in the lung and extrapulmonary tissues.

3.1.2. Nitrogen Dioxide Uptake

Because NO_2 is more water-soluble than O_3, it is partly dissolved in the mucus and absorbed along the conducting airways as it reaches the terminal bronchioles and alveolar ducts. The absorption of NO_2 in tissues (or tissue fluid) may be related to its chemistry in an aqueous medium, in which it reacts with water, forming a mixture of nitrous acid (HNO_2) and nitric acid (HNO_3) [Reaction (10)]. Because HNO_2 is relatively unstable, it decomposes to form HNO_3 and NO [Reaction (11)]. Nitric oxide in air becomes NO_2 and dissolves again as in Reaction (10). Thus, HNO_3 is the ultimate reaction product of NO_2 or NO_x in an aqueous medium.

$$2NO_2 + H_2O \longrightarrow HNO_3 + HNO_2 \qquad (10)$$

$$3HNO_2 \longrightarrow HNO_3 + 2NO + H_2O \qquad (11)$$

Absorption of inhaled NO_2 has been studied in both animals and humans. The degree of absorption depends upon NO_2 concentrations in the inspired air. In animals, NO_2 exposures at 130–220 ppm (224–414 mg/m^3) have shown 31–78%

absorption of inhaled NO_2. At lower concentrations (0.1–1.0 ppm or 0.19–1.88 mg/m^3), the absorption has been 90%. In humans, up to 90% absorption of inhaled NO_2 at 0.29–7.2 ppm (0.55–13.54 mg/m^3) exposure levels has been observed (Goldstein et al., 1977, 1980, and references cited therein). Inhaled NO_2 is distributed in the lung fairly evenly, and is further distributed in the extrapulmonary sites. Goldstein et al. (1977, 1980) have explained that inhaled NO_2 produces HNO_2 and HNO_3 in the lung tissue fluid [as in Reaction (10)], which then diffuse to the circulating blood. The NO_2 absorbed in the lung eventually appears in the urine in the form of nitrites (NO_2^-) and nitrates (NO_3^-) (Lindvall, 1985, and references cited therein).

In subsequent studies using isolated perfused lung (IPL), Postlethwait and Mustafa (1981) analyzed the products of inhaled NO_2 absorption. The IPL system has been particularly important in separating the possible roles of lung tissue and blood in metabolizing absorbed NO_2. When the lung was ventilated with air containing NO_2, while at the same time being perfused with a physiologic medium free of red blood cells (RBC), the product of absorbed NO_2 was detected as NO_2^-, most of which appeared in the perfusate and a small amount in lung tissue. There was hardly any detectable NO_3^- in either perfusate or lung tissue. With the progress of ventilation, the NO_2^- concentration in perfusate increased linearly with time and in proportion to NO_2 concentration in ventilation air. However, when the perfusion medium contained RBC, the NO_2 absorption product in perfusate was all NO_3^-; only lung tissue contained some NO_2^-. It was also found that, based on the yield of products, solutions containing proteins or reducing substances (e.g., glutathione, amines) absorbed NO_2 more effectively than a plain aqueous medium or solutions containing simple salts, and the products were mostly NO_2^- with only a minor proportion of NO_3^-. Only when the solutions contained RBC was the product entirely NO_3^-. The reactions can be represented as follows:

$$NO_2 + DH \longrightarrow HNO_2 + D^+ \tag{12}$$

$$HNO_2 \longrightarrow NO_2^- + H^+ \tag{13}$$

$$2NO_2^- + 2O_2Hb \longrightarrow 2NO_3^- + 2Hb^+ + O_2 \tag{14}$$

where D is an election donor (i.e., a reducing substance).

Oxidation of NO_2^- occurs almost instantly in the presence of O_2 and hemoglobin (Hb) or oxyhemoglobin (O_2Hb) of RBC. Thus, inhaled NO_2, upon being absorbed in the lung, is converted to NO_2^- (which is a reduction process). Reducing substances in lung tissue evidently act as election donors. The NO_2^- then diffuses into the vascular space, where in the presence of RBC it is oxidized to NO_3^-. The oxidation of NO_2^- to NO_3^- in lung tissue would be possible by either enzymatic or nonenzymatic means, but such a system was evidently lacking in the rat lung tissue in which this study was carried out.

Because NO and NO_2 co-occur in the ambient air, a mention should be made of NO absorption. Nitric oxide is much less soluble (4.7 mL/100 mL of water at

20 °C, 1 atm pressure) than NO_2. Therefore, its pulmonary absorption when inhaled with NO_2 (or as NO_x) will be less than NO_2 (Goldstein et al., 1980, and references cited therein).

3.2. Mechanisms of O_3 Toxicity

Ozone is one of the most powerful oxidants known. The molecule contains two unpaired electrons, but it is not a free radical or biradical. However, it displays some of the characteristics of a biradical, such as the ability to abstract a hydrogen atom from a biomolecule. It is generally believed that O_3 can cause oxidative destruction of biomolecules both by a direct reaction and through the formation of free radicals and reactive intermediates. However, the mechanisms of O_3 toxicity in biological systems are complex. To date, various mechanisms, which generally explain the acute O_3 toxicities, have been proposed and reviewed (Menzel 1970, 1976, 1984; Cross et al., 1976; Mudd and Freeman, 1977; Coffin and Stokinger, 1977; NRC, 1977; Mustafa and Tierney, 1978; EPA, 1978, 1986; Goldstein, 1979a; WHO, 1979; Pryor et al., 1983; Mehlman and Borek, 1987; Lippmann, 1989; Mustafa, 1990). These mechanisms include free radical reaction, lipid peroxidation, oxidation of functional groups, alteration of membrane permeability, respiratory inflammation, and secondary processes.

3.2.1. Free Radical Reaction

Many of the biological effects of O_3 can be explained on the basis of free radical reactions. Ozone has been considered radiomimetic because some of its effects are similar to those of ionizing radiation, such as the formation of reactive intermediates and free radicals, induction of lipid peroxidation and chromosomal aberration, and behavior toward antioxidants (NRC, 1977; EPA, 1978; 1986, and references cited therein).

Ozone is able to abstract a hydrogen atom from a biomolecule, leading to chain reactions in the presence of air (O_2). It can react with alkanes, alkenes, amines, sulfhydryls, and other biomolecules to produce reactive intermediates and radicals (Koppenol, 1982; Pryor et al., 1983). The O_3 reaction with alkanes produces a charge-transfer complex and then a hydrotrioxide (ROOOH) intermediate, which decomposes to form free radical intermediates. The O_3 reactions with alkenes (olefins) produces nonradical products via the Criegee mechanism and free radical intermediates via the charge-transfer mechanism. These free radicals are thought to be responsible for most of the damage in tissue lipids and membranes caused by O_3.

In aqueous solution, O_3 decomposes to give H_2O_2 and O_2^- (superoxide) and the $H\dot{O}$ and $H\dot{O}_2$ radicals (Joigne and Bader, 1979; Koppenol, 1982; Carmichael et al., 1982; Glaze, 1986). Free radicals formed during O_3 reaction with linoleic acid and polyunsaturated fatty acids (PUFA) have been detected (Goldstein et al., 1968; Pryor et al., 1981). The O_3-induced free radicals in biological systems are thought to result from the oxidative decomposition of PUFA. From a nonradical O_3 attack on PUFA, lipid peroxides (ROO) and hydroperoxides (ROOH) are generated, which on further reactions with O_3 or O_2 give rise to free radicals such as the $H\dot{O}$, alkoxyl $(R\dot{O})$, and alkoperoxyl $(RO\dot{O})$ radicals (Menzel, 1970, 1976, 1984; Pryor, 1981; Kurz and Pryor, 1978; Pryor et al., 1981, 1983). Lipid peroxides and hydroperoxides are toxic and capable of causing cellular damage. Although O_3 may be relatively selective in its attack on biomolecules, the variety of free radicals it produces ($H\dot{O}$, $H\dot{O}_2$, $R\dot{O}$ and $RO\dot{O}$) are both nonselective and fast-reacting. The direct (nonradical) O_3 reaction and the propagating free radical reactions result in peroxidation of membrane and tissue lipids, oxidation of functional groups and loss of activity, alteration of membrane permeability, and ultimately cell death.

3.2.2. *Lipid Peroxidation*

Peroxidation of membrane and tissue lipids is thought to be an important mechanism of O_3 injury. In a (nonradical) attack on unsaturated fatty acids or PUFA, O_3 adds to the carbon–carbon double bond, producing a trioxygen intermediate (called 1,2,3-trioxolane), which rearranges to form an ozonide by the Criegee mechanism (Menzel, 1976; Pryor, 1981; Pryor et al., 1983, Chow, 1983); this decomposes to yield lipid peroxides or hydroperoxides. Fatty acids with one set of double bonds and cholesterol produce epoxides; those with two or more sets of double bonds give rise to dialdehydes (e.g., malonaldehyde) and hydroperoxides. In vitro O_3 exposure studies with fatty acids and red blood cells have provided some of the evidence for lipid peroxidation (Goldstein and Balchum, 1967; Goldstein et al., 1968; Goldstein, 1973). In vivo O_3 exposure causing lipid peroxidation has been detected in animal lung tissue and lavage fluid, conjugated dienes (Goldstein et al., 1969), epoxides of fatty acids and cholesterol (Sevanian et al., 1979), thiobarbituric acid reactants (mostly malonaldehyde) (Chow and Tappel, 1972, 1973; Sagai et al., 1987), and exhalation of ethane and pentane (Dumelin et al., 1978a, b). Even though lipid peroxidation is an important mechanism for O_3 injury in the lung, a quantitative determination of in vivo lipid peroxidation products remains difficult and controversial (Gelmont et al., 1981; Mustafa, 1985).

3.2.3. *Oxidation of Functional Groups*

Ozone can oxidize various functional groups (sulfhydryl, amine, alcohol, and aldehyde) that are present in proteins, enzymes, nucleic acids, membranes, and other small and large biomolecules. Both direct and free radical-mediated oxidations may occur, resulting in the disruption of cellular and metabolic functions.

A cell contains low-molecular-weight (or soluble) sulfhydryl (SH) compounds, including reduced glutathione (GSH) [referred to as nonprotein SH (NPSH)], and macromolecular (or insoluble) SH compounds, including SH groups of proteins [referred to as protein SH (PSH)] (Mustafa, 1985). Sulfhydryls are susceptible to O_3 exposure, and in an aqueous solution, GSH is oxidized to GSSG (oxidized glutathione) [Reaction (15)]. Although sulfonate (RSO_3H) as an in vitro oxidation product has been reported (Menzel, 1971), such an irreversible oxidation of GSH was not observed from an in vivo exposure (DeLucia et al., 1972, 1975). In lung tissue, following an in vivo O_3 exposure, there was a loss of

NPSH or GSH but no increase in GSSG. Instead, a mixed disulfide (PSSG) between PSH and NPSH or GSH may result [Reaction (16)]. Other studies have shown that acute O_3 exposure results in an oxidation of GSH or NPSH and inhibition of SH-dependent enzyme activities (Coffin and Stokinger, 1977; Freeman and Mudd, 1981, and references cited therein). Since the functions of proteins, enzymes, coenzymes, and other biomolecules that contain SH groups at the active sites can be lost or altered, SH oxidation represents a mechanism of O_3 toxicity.

$$2GSH \xrightarrow{\text{oxidation}} GSSG + H_2O \tag{15}$$

$$PSH + GSH \xrightarrow{\text{oxidation}} PSSG + H_2O \tag{16}$$

Ozone reacts with proteins, particularly destroying some of the amino acids (cysteine, methionine, tryptophan, and tyrosine), which may explain the loss of enzyme activity after O_3 exposure. After an acute O_3 exposure, a loss of enzyme activities has been observed in lung homogenate, mitochondria, microsomes, lysosomes, and cytosol (Hurst and Coffin, 1971; Palmer et al., 1971; DeLucia et al., 1972; Mustafa et al., 1973; Mustafa and Cross, 1974; Goldstein et al., 1975; Mudd and Freeman, 1977; Freeman and Mudd, 1981, and references cited therein). Other enzyme systems inhibited include lung prostaglandin synthetase, cholinesterase, and α-1-antiproteinase (Menzel, 1976; Johnson, 1980; Gordon et al., 1981; Madden et al., 1987). Loss of enzyme activity is therefore an important mechanism of O_3 toxicity.

3.2.4. Alteration of Membrane Permeability

Lipids and proteins are two major components of membranes. Oxidation of unsaturated fatty acids in lipids and susceptible amino acids in proteins may alter membrane properties, particularly the ability to regulate permeability. Ozone readily attacks the membranes of cells and intracellular organelles, which contain large amounts of PUFA in the phospholipid structure. Acute O_3 exposure has been found to cause increased permeability of lung mitochondria to external NADH and cytochrome c (Mustafa and Cross, 1974); decreased membrane fluidity of alveolar macrophages (Witz et al., 1983); and increased permeability of the airway and alveolar mucosa [i.e., increased transport of external dye (phenol red) into the blood circulation and increased transport of serum albumin and immunoglobulin G (IgG) into the airways and alveoli] (Alpert et al., 1971; Reasor et al., 1979; Williams et al., 1980; Hu et al., 1982; Bhalla and Crocker, 1986; Bhalla et al., 1986; Kehrl et al., 1987). Thus, damage to membrane structure, causing alterations in permeability, is an important means of O_3 toxicity.

3.2.5. Inflammation

Inflammation involves a complex series of events, including dilation of the blood vessels with increased permeability and exudation of plasma fluids into the site of injury. The plasma fluids include proteins, soluble factors, and migratory cells [e.g., polymorphonuclear leukocytes (PMNs), macrophages]. Ozone inhalation has been shown to induce inflammation of the lung, associated with a reversible state of airway hyperresponsiveness to bronchoconstrictor drugs, in several animal species and humans (Seltzer et al., 1986; Koren et al., 1989; Lippmann, 1989, and references cited therein). In O_3-induced inflammation, there is an increase in plasma constituents (proteins, proteolytic enzymes, and soluble factors) and in the number of PMNs and other migratory cells in both the bronchial epithelium and bronchoalveolar lavage (BAL). The plasma proteins can play a role in fibrotic and fibrinolytic processes in the lung, and the proteolytic enzymes can cause tissue damage in the respiratory epithelium. The soluble factors, which include various arachidonic acid metabolites (prostaglandins, thromboxane, hydroxyeicosatetraenoic acids, etc.), are potent mediators of PMN chemotaxis, but they can also cause increased vascular permeability and airway hyperresponsiveness (Ford-Hutchinson et al., 1980; Seltzer et al., 1986; Koren et al., 1989). PMNs can release mediators that contribute to further inflammation and airway reactivity. The mediators include arachidonic acid metabolites that act as chemotactic factors for PMNs and eosinophils, and reactive oxygen intermediates and destructive enzymes that can damage the respiratory epithelium. Ozone-induced inflammation of the lung has received particular attention because it helps in understanding O_3 effects in humans. The histological and biochemical changes associated with inflammation are possibly responsible for the development of more complex pathologic conditions in the lung resulting from recurrent or chronic O_3 exposure.

3.2.6. Secondary Processes

Ozone exposure can induce various other nonspecific processes that propagate O_3 toxicity. These include pharmacologic action or neurohormonal mechanisms, which are apparently secondary to O_3 reactions. They produce nonspecific effects associated with biochemicals, such as histamine and thyroxine. A release of histamine in the lung has been observed immediately after O_3 exposure and might be responsible for part of the acute toxicity produced (Dixon and Mountain, 1965). The effects of a lethal (4–10 ppm) or sublethal (1–3 ppm) O_3 exposure were considerably lessened in animals whose thyroid gland was removed (Fairchild, 1963; Fairchild and Graham, 1963). While the thyroidectomy decreased acute O_3 toxicity, injection of thyroid hormone (thyroxine or triiodothyronine) increased the animals' lung injury or mortality from O_3 exposure (Fairchild et al., 1964).

3.3. Mechanisms of NO_2 Toxicity

Like O_3, most of the biological effects of NO_2 can be explained by its ability to induce free radical reactions, including lipid peroxidation, sulfhydryl oxidation, membrane permeability alteration, respiratory inflammation, and secondary stresses. However, there are some important differences in the mechanisms of action of NO_2 and O_3. Nitrogen dioxide is itself a free radical, and its mode of initiating a free radical chain reaction is different. It is more soluble than O_3, its absorption along the respiratory tract is greater, and, unlike, O_3, it forms products that are easily quantifiable.

3.3.1. Free Radical Reaction

Both NO_2 and its precursor, NO, are themselves free radicals because each possesses an odd electron. Despite being a free radical, NO does not dimerize and is relatively unreactive to most organic molecules. Its oxidation product, NO_2, is much more reactive, and NO_2 dimerizes to nirogen tetroxide (N_2O_4), in which the odd electrons are paired.

Nitrogen dioxide reacts with both alkenes and alkanes by free radical mechanisms (Pryor and Lightsey, 1981; Pryor et al., 1982, and references cited therein). The alkenes include unsaturated fatty acids and model organic compounds. The alkanes include polycyclic aromatic hydrocarbons (PAH), which upon reaction with NO_2 produce nitroaromatics. The NO_2 reaction with alkenes is of particular significance in lipid peroxidation in the lung.

It is known that NO_2 reacts with alkenes by addition to the double bond [Reaction (17)]. Reactions of this sort have generally been carried out with high (20–30%) concentrations of NO_2, which are of no relevance to the NO_2 concentrations encountered in the environment or the workplace. Nonetheless, it has been assumed that this addition reaction is the mechanism by which NO_2 causes autoxidation of unsaturated fatty acids and other alkene compounds.

$$\begin{array}{c} & & & & NO_2 \\ & & & & | \\ -C = C- \; + \; NO_2 \longrightarrow \; -\dot{C} - C- \\ \;\;\; | \;\;\;\; | & & \;\;\;\; | \;\;\;\; | \\ \;\;\; H \;\;\; H & & \;\;\; H \;\;\; H \end{array} \qquad (17)$$

$$\begin{array}{c} -C = C- \; + \; NO_2 \longrightarrow \; -C - \dot{C}- + HNO_2 \\ \;\;\; | \;\;\;\; | & & \;\;\;\; | \\ \;\;\; H \;\;\; H & & \;\;\; H \end{array} \qquad (18)$$

More recently, the studies of Pryor et al. (1982) have shown that NO_2 at low concentrations can abstract a hydrogen atom (H) from an alkene [Reaction (18)]. Using a series of NO_2 concentrations (as high as 70% to as low as 70 ppm), they have shown that at high concentrations the NO_2–alkene reaction occurs predominantly by addition to the double bond, and at low concentrations the reaction occurs predominantly by H abstraction. The two pathways of this initiation reaction apparently compete with each other. A switch from one pathway to the other may not alter the kinetics of NO_2-induced oxidation, but the radical products formed are different. In the NO_2 addition pathway [Reaction (17)], the radical produced can react with another NO_2 to form a dinitro or nitro-nitrite compound, or it can react with O_2 to form a nitro-hydroperoxide or other oxygenated compounds. In the H abstraction pathway [Reaction (18)], a radical and HNO_2 are produced; the radical can combine with NO_2 to form an unsaturated nitro or nitrite compound, or it can react with O_2 to form an unsaturated hydroperoxide or nitrate ester. Since environmental and occupational exposures to NO_2 occur at low concentrations, it is plausible that H abstraction is the pathway for the initiation reaction for lipid peroxidation in the lung. In this pathway, the HNO_2 produced may cause nitrosation of amines.

3.3.2.   *Lipid Peroxidation*

Toxicity of NO_2 leading to cell injury and death is thought to result from NO_2-induced lipid peroxidation. Although the NO_2 reaction with unsaturated fatty acids (alkenes) is relatively slow compared to O_3, it can initiate autoxidation of some alkenes at concentrations as low as 0.1 ppm ($0.188 \, mg/m^3$) (Chao and Jaffe, 1972; Mead et al., 1980). Reaction of NO_2 with unsaturated fatty acids (linoleic, linolenic, or arachidonic) can cause cis to trans isomerization, which alters the geometry of fatty acids in the membrane (Thomas et al., 1968; Menzel, 1976). Various in vitro studies using monolayers of unsaturated fatty acids on silica gels, artificial membranes (phospholipid vesicles), or tissue membranes (erythrocytes, microsomes, or mitochondria) have shown the peroxidation process by free radical mechanisms. The products identified are nitrohydroperoxides and fatty acid epoxides and/or hydroperoxides (Menzel, 1976; Sevanian et al., 1979; Mead et al., 1980). Most of these products have been identified in lipids from lung tissue or lavage fluid after in vivo animal exposures to NO_2. Other evidence of in vivo lipid peroxidation includes formation of conjugated dienes and fatty acid epoxides in lung lipids (Thomas et al., 1968; Sevanian et al., 1982 a, b), presence of thiobarbituric acid (TBA) reactants in lung tissue homogenate (Sagai et al., 1982, 1984; Ichinose and Sagai, 1982), presence of ethane or pentane in exhaled air (Dillard et al., 1980; Sagai et al., 1981; Sagai and Ichinose, 1987), and decrease in unsaturated fatty acids or alteration in fatty acid composition of lipids from lung tissue and lavage fluid (Menzel et al., 1972).

3.3.3.   *Oxidation of Functional Groups*

Like O_3, the toxicity of NO_2 is attributed to its ability to oxidize SH groups of low-molecular weight substances (e.g., GSH and cysteine) and other reducing compounds (e.g., NADH and NADPH). In vitro studies have shown that NO_2 can inhibit enzyme activities (Ramazzotto et al., 1971). Although there are no data from in vivo exposure studies, it is plausible that an acute NO_2 exposure can oxidize functional groups, causing inhibition of enzyme activities and alteration of intermediary metabolism.

3.3.4.   *Alteration of Membrane Permeability and Inflammation*

These two phenomena associated with NO_2 exposure have not been studied as extensively as those associated with O_3 exposure, but they are important in explaining the health effects of NO_2. Exposure to NO_2 causes an increase in epithelial and capillary membrane permeability. This is evidenced by leakage of plasma proteins and labeled albumin across the airway and/or alveolar membrane (Sherwin and Richters, 1971; Sherwin and Carlson, 1973; Selgrade et al., 1981; Guth and Mavis, 1985). Respiratory inflammation, which accompanies epithelial cell injury by NO_2 exposure, is characterized by increased congregation of alveolar macrophages in the alveolar ducts, recruitment of PMNs, and histopathologic changes (Sherwin et al., 1968; Blair et al., 1969; Gardner et al., 1969; DeNicola et al., 1981).

4. PULMONARY EFFECTS OF SHORT-TERM EXPOSURE

Ozone and NO_2 in vitro can attack cells or tissues from any organ, but in an in vivo exposure they exert toxic effects primarily on the lung; effects on other organs result largely via secondary mechanisms. Most exposure studies have been carried out for relatively short periods of time (i.e., hours, days, weeks, or a few months) using O_3 or NO_2 concentrations approximating or up to several orders of magnitude higher than those encountered in the urban or workplace environment. At ambient or near-ambient concentrations (O_3 at 0.08–0.8 ppm or 156–1568 $\mu g/m^3$; NO_2 at 0.3–1.5 ppm or 564–2820 $\mu g/m^3$), pulmonary effects characterized by biochemical, morphological, and functional changes can occur, hence the concern for the health effects of oxidant exposure in urban smog or in the workplace.

As O_3 or NO_2 is inhaled and the oxidant is to a degree absorbed along the respiratory tract (including the nasopharynx), its concentration decreases distally. The extent of lung injury depends upon the amount of oxidant reaching the lower airways. Thus, lesions can be found in the trachea, bronchi, bronchioles, alveolar ducts, and proximal alveoli, with maximal injury at the junction of terminal bronchioles and alveolar ducts.

Short-term exposure to O_3 or NO_2 (e.g., days or weeks) results in biochemical and morphological changes that are essentially biphasic. Initially, there is an injury phase involving damage to or loss of cells in conjunction with depression of enzyme activity. After one day of exposure a repair phase is evident, characterized by extensive multiplication of stem cells and increase of enzyme activity. The overall changes (initial decreases and subsequent increases in enzyme activity) are dependent upon oxidant concentrations, length of exposure, and animal species and strain. The effects of O_3 and NO_2 exposure will be discussed as alterations in lung biochemistry, morphology, histology, and host defense.

4.1. Biochemical Changes

Biochemical studies with O_3 and NO_2 exposure include the effects on various parameters ranging from simple enzyme activities to complex matabolic pathways and products of biochemical importance. These studies have been carried out to delineate lung injury from oxidant exposure, including mechanisms of injury. An extensive body of data on this subject has been reviewed (Cross et al.,

1976; Coffin and Stokinger, 1977; NRC, 1977; Mudd and Freeman, 1977; Mustafa and Tierney, 1978; EPA, 1978, 1986; WHO, 1979; Goldstein, 1979a; Melton, 1982; Lee et al., 1983; Lee, 1985; Menzel, 1976, 1984; Mehlman and Borek, 1987; Wright et al., 1990; Mustafa, 1990). Likewise, several reviews on biochemical effects on NO_2 are available (Morrow, 1975, 1984; Menzel, 1976; NAS, 1977; WHO, 1977; EPA, 1982; Guidotti, 1978; Mustafa and Tierney, 1978; Dawson and Schenker, 1979; Lee, 1980; Schneider and Grant, 1982; Lindvall, 1985). In the following sections, biochemical changes will be presented as O_3 and NO_2 effects on enzyme activity and intermediary metabolism, including variations in oxidant effects with changing exposure conditions.

4.1.1. Ozone

Effects of Varying Exposure Conditions. There are several conditions that influence the effects of O_3 exposure. These include oxidant concentration, length of exposure, and continuous versus intermittent exposure, and they have been studied using various enzymatic and other metabolic parameters.

Ozone effects are concentration-dependent. In several studies, O_3 concentrations ranged from 0.1 to 0.8 ppm (196–1568 $\mu g/m^3$), and the biochemical changes were proportional to O_3 concentrations and statistically significant at all levels except 0.1 ppm (Chow and Tappel, 1972, 1973; Mustafa et al., 1973, 1977; Mustafa, 1975; Mustafa and Lee, 1976; Plopper et al., 1979). The observations were similar for morphological studies (Dungworth et al., 1975; Schwartz et al., 1976).

The time course of biochemical changes was studied for exposure periods ranging from 1 to 30 days (Mustafa and Lee, 1976; Mustafa et al., 1977). A small ($< 10\%$, nonsignificant) depression of enzyme activity was observed after one day of exposure, followed by an increase reaching a plateau after three to four days. The time course of morphological changes was similar, that is, a cell injury phase was followed by a cell proliferation phase (Stephens et al., 1974a; Dungworth et al., 1975; Schwartz et al., 1976). These observations were explained as the characteristic response of lung tissue to oxidant injury. The biochemical and morphological changes induced during the first few days of O_3 exposure possibly bring about a state of adaptation, whereby continued exposure results in a balance of cell injury and repair.

The biochemical changes due to < 1 ppm ($< 1960 \mu g/m^3$) O_3 exposures were generally reversible. Enzyme activity, which increased within a few days of O_3 exposure, returned to control levels after a few days of the cessation of injury with a relative completion of the repair phase. This was corroborated morphologically (Plopper et al., 1978). However, there were situations in which the biochemical changes could be irreversible (e.g., the collagen formed in the lung was not degraded rapidly) (Hussain et al., 1976)

Oxidation and Energy Metabolism. Oxygen consumption, coupled phosphorylation, and related enzyme activities in mitochondria of lung cells were found susceptible to O_3 exposure. High-level O_3 exposure (2–4 ppm for < 12 hr) caused a decrement of mitochondrial O_2 consumption and loss of coupled phosphorylation (Mustafa et al., 1973; Mustafa and Cross, 1974). Loss of mitochondrial function might be related to O_3-mediated sulfhydryl oxidation and lipid peroxidation. With lower-level O_3 exposure (0.2–0.8 ppm for a few days) mitochondrial O_2 consumption and marker enzyme activities increased significantly (Mustafa et al., 1973, 1977; Mustafa and Lee, 1976). However, such an increase appeared to be due to more mitochondria per unit of lung tissue, that is, an increase in mitochondrial population and/or mitochondia-rich cell population due to O_3 exposure.

Glucose Metabolism. Both the glycolytic and the hexose monophosphate (HMP) shunt pathways were elevated with exposure to O_3. In the glycolytic pathway, increased glucose consumption resulted in increased production of both pyruvate and lactate (Mustafa et al., 1977). In the HMP shunt pathway, two marker enzymes (glucose-6-phosphate and 6-phosphogluconate dehydrogenases) showed remarkable increases with oxidant exposure (Chow and Tappel, 1972, 1973; Chow et al., 1976; Mustafa and Lee, 1976; Mustafa et al., 1977; Lunan et al., 1977; Tyson et al., 1982; Elsayed et al., 1983). The increased activity of the HMP pathway may be relevant to its role in antioxidant defense.

Lipid Metabolism. Because unsaturated lipids are most susceptible to oxidant attack, O_3 effects on lung lipids have been studied with considerable interest. Typically, fatty acid composition and the oxidation (or peroxidation) products of lung lipids have been examined in both lung tissue and lavage fluid. Fatty acid composition after O_3 exposure showed variable changes, for example, a decrease in oleic and linoleic acids but an increase in arachidonic acid, or little or no change in some of these acids (Roehm et al., 1972; Menzel et al., 1972; Shimasaki et al., 1976).

Protein Metabolism. Ozone effects on lung protein synthesis have been studied based on amino acid (labeled leucine or proline) incorporation into proteins. Protein synthesis was found to increase with O_3 exposure and the time course was similar to that of various enzyme activities (Chow et al., 1976; Hussain et al., 1976; Mustafa et al., 1977; Myers et al., 1984). Two types of lung proteins, collagen and glycoproteins, were examined. Collagen synthesis (based on labeled proline incorporation and increased hydroxyproline content and/or prolyl hydroxylase activity) was found to increase with O_3 exposure (Hussain et al., 1976; Last et al., 1979, 1984; Last and Greenberg, 1980). After termination of O_3 exposure (i.e., during postexposure recovery), the markers for synthesis (e.g., prolyl hydroxylase activity) returned to control levels, but the hydroxyproline content remained unchanged (Hussain et al., 1976). The production of mucous glycoproteins and their secretions by tracheal explants increased with O_3 exposure (Last et al., 1977; Last and Kaizu, 1980).

Other Enzyme Activities. In addition to the enzyme activities and metabolic pathways discussed above, many other enzyme activities are used as markers of lung injury. Monoamine oxidase (MAO) activity increased (Mustafa et al., 1977), microsomal cytochrome P-450-dependent enzyme activities decreased (Palmer et al., 1971; Goldstein et al., 1975), and lysomal enzyme activity was found to increase in the lung following O_3 exposure (Dillard et al., 1972; Chow et al., 1974). An increase in lysomal enzyme activity has the potential for a secondary damage to lung tissue. Lactate dehydrogenase activity, a marker for cell injury, increased in the lung with O_3 exposure (Chow and Tappel, 1972).

4.1.2. Nitrogen Dioxide

Oxidative Metabolism. In an in vitro setup, tissue preparations (homogenate and mitochondria) directly exposed to NO_2 showed a decrease in O_2 consumption or substrate utilization (Ramazzotto et al., 1971). Alterations in mitochondrial membrane structure and inhibition of enzyme activity might be responsible for the decreased O_2 consumption or substrate utilization. However, in vivo NO_2 exposure even at relatively high concentrations (30–50 ppm or 56–94 mg/m^3) did not alter O_2 consumption in lung homogenate or mitochondria (Mustafa et al., 1979). In vivo exposure to NO_2 at lower concentrations (< 15 ppm or 28 mg/m^3) and for longer periods (up to seven days) resulted in an increase of O_2 consumption in lung tissue (Mustafa et al., 1979, 1980a, b). Increased oxidative metabolism reflects an increase in mitochondrial population, which in turn may be a consequence of renewal and/or hyperplasia of metabolically active cells induced by NO_2 injury.

Glucose Metabolism. Injury to lung tissue stimulates glucose metabolism, which is conveniently studied using isolated lung perfusion and/or tissue slices. An acute lung injury due to high-level NO_2 exposure caused as much as 50% of glucose consumed to be converted to lactate. Glucose consumption was also stimulated after relatively low-level NO_2 exposure (< 15 ppm or 28 mg/m^3), with an increased production of lactate and pyruvate (Mustafa et al., 1979, 1980a, b; Ospital et al., 1981). Typically, approximately 50% of glucose consumed was converted to lactate and pyruvate, the rest was oxidized to CO_2 and/or incorporated into lipids and proteins of lung tissue (Ospital et al., 1981). Increased glucose consumption after NO_2 exposure correlated with increased cell population in the alveolar epithelium [i.e., increased staining of alveolar type 2 cells for lactate dehydrogenase (LDH)] (Sherwin et al., 1972).

Glucose consumption in the lung was also stimulated via the hexose monophosphate (HMP) shunt pathway following NO_2 exposure. This stimulation was determined by increased activity of two marker enzymes—glucose-6-phosphate dehydrogenase (G6PD) and 6-phosphogluconate dehydrogenase (6PGD) (Mustafa et al., 1979, 1980a, b)—and by the ratio of labeled CO_2 from glucose-1-^{14}C to that from glucose-6-^{14}C (Tierney, 1974). The increased glucose metabolism by the HMP shunt pathway might be related to production of increased amounts of NADPH and pentose sugar for biosynthesis of cellular components in response to lung tissue injury.

Lipid Metabolism. Lung tissue lipids and surfactant phospholipids are targets of oxidant exposure. Alterations in lung tissue lipids and surfactant phospholipids (fatty acid composition and rate of synthesis) after NO_2 exposure were reported by various investigators (Thomas et al., 1968; Thomas and Rhoades, 1970; Menzel et al., 1972; Trzeciak et al., 1977; Blank et al., 1978; Kobayashi et al., 1980; Leung, 1983; Guth and Mavis, 1985).

Nucleic Acid and Protein Metabolism. Studies of the synthesis and metabolism of nucleic acids and proteins in the lung are limited. An increase in protein and DNA synthesis (as determined by incorporation of labeled precursors) was observed after NO_2 exposure (Kleinerman, 1970; Hacker et al., 1976). This increase might be related to repair of injured cells and/or multiplication of stem cells in response to oxidant lung injury.

Sulfhydryl Metabolism. Exposure to NO_2 at relatively low concentrations for a few days resulted in an increase in NPSH and GSH levels in lung tissue. Enzyme activities related to sulfhydryl metabolism, glutathione peroxidase, glutathione reductase, and disulfide reductase also increased in response to NO_2 exposure (Chow et al., 1974).

4.2. Morphological Changes

Earlier studies with O_3 or NO_2 involved animal exposure at relatively high concentrations, which resulted in massive lung injury with extensive tissue damage and severe pulmonary edema. Subsequently, as more attention was given to realistic exposure conditions and O_3 or NO_2 concentrations were scaled to approach ambient concentrations or exceed them by several orders of magnitude, the morphological changes were discernible. Exposure of animals to O_3 at 1 ppm ($2\,mg/m^3$) or less and to NO_2 at 20 ppm ($3.8\,mg/m^3$) or less has shown that oxidant injury to the respiratory organ is not linear or uniform. The injury is focal, and the portion of the respiratory organ that is particularly damaged is the junction of the conducting airways and the gas-exchange area, that is, the terminal bronchioles and alveolar ducts (also referred to as central acinii).

4.2.1. Ozone

In the normal lung, the ciliated cells are distributed throughout the airway epithelium and type 1 cells are distributed throughout the alveoli, covering the epithelium. Exposure to O_3 causes selective damage to the ciliated cells and the alveolar type 1 cells, particularly those located at the central acinii. In the conducting airways, cells that escape injury are the nonciliated cells, that is, the basal, serous, and mucous (goblet) cells in the upper airways and the Clara cells in the bronchiolar regions. In the gas-exchange area, alveolar type 2 cells survive oxidant injury (Stephens et al., 1974a, b; Dungworth et al., 1975; Evans et al., 1976a, b; Schwartz et al., 1976; Mellick et al., 1977; Barry et al., 1983). It is not known exactly why certain types of cells are more susceptible than others, but it is believed that the surviving cells are metabolically very active and thereby withstand oxidant injury.

Morphological and biochemical changes correlate well. Ozone-induced cellular damage is followed by an extensive division (hyperplasia) of stem cells. In the tracheobronchial region, the basal cells divide and differentiate to give rise to the ciliated cells. In the bronchiolar region, the Clara cells divide and differentiate to give rise to the ciliated cells. In the gas-exchange area, the alveolar type 2 cells divide and transform to the type 1 cells. The time course of these changes is similar to that of the biochemical alterations discussed above. Because young cells take the place of injured and/or dead cells, they may be at least partially responsible for the metabolic changes observed after O_3 exposure.

4.2.2. Nitrogen Dioxide

Exposure to NO_2 produces focal lung injury involving bronchioles, alveolar ducts, and adjacent alveoli. At the initial phase (within one day), ciliated cells in the bronchioles and type 1 epithelial cells in the alveoli are injured, and the dying or dead cells slough off the epithelium. In the upper airways (trachea and bronchi), cilia break up and become shortened and mucous secretion increases. As NO_2 exposure continues, a repair phase begins with the multiplication of stem cells (Clara cells in the bronchioles and type 2 epithelial cells in the alveoli), which survive the exposure. As the stem cells multiply, a hyperplasia becomes evident within three to four days of continued exposure (Stephens et al., 1971, 1972; Yuen and Sherwin, 1971; Evans et al., 1972, 1973, 1975, 1977; Cabral-Anderson et al., 1977). Using labeled thymidine (a DNA precursor) and autoradiography, Evans et al. (1973, 1975, 1977) have demonstrated that upon multiplication of stem cells the bronchiolar and alveolar epithelia return toward the normal stage, with no further cell multiplication even with continued exposure. These cellular changes are reversible in that the respiratory epithelium returns to near normal after cessation of NO_2 exposure.

In addition to cellular changes, alterations in connective tissues (collagen and elastin) have been observed in several animal species and under various NO_2 exposure conditions (Freeman et al., 1968, 1969, 1972; Stephens et al., 1971; Drozdz et al., 1977; Kleinerman, 1977). Emphysema-like pathogenic changes have been observed in some animal species (Freeman and Haydon, 1964; Freeman et al., 1968, 1969; Drozdz et al., 1977). Most of these structural changes appear to be reversible, that is, they return to near normal upon termination of NO_2 exposure.

4.3. Histological Changes

Accumulation of the inflammatory cells (e.g., alveolar macrophages, leukocytes, and polymorphonuclear cells) near the terminal bronchioles or respiratory bronchioles and in the alveolar ducts and proximal alveoli is an important characteristic of O_3 and NO_2 effects in the lung. The inflammatory cell response generally accompanies or follows alveolar epithelial cell damage. The inflammatory cells are thought to be recruited to the site of lung injury in response to a toxic irritation, but they also have the potential to cause further lung tissue damage by the destructive enzymes and the reactive oxygen species they release. In several studies, the number of inflammatory cells per centriacinar alveolus was found proportional to O_3 concentration at levels between 0.2 and 0.8 ppm (394–788 μg/m^3) (Schwartz et al., 1976; Brummer et al., 1977). Likewise, an accumulation of alveolar macrophages and an increase in PMNs after NO_2 exposure were noted.

4.4. Host Defense Changes

Inhaled particles, including infectious agents (bacteria, viruses, etc.), are removed from the lung by two different mechanisms: mucociliary transport in the airways and alveolar macrophage phagocytosis in the parenchyma. In the conducting airways (trachea, bronchi, and bronchioles), inhaled particles and infectious agents, upon being trapped by the mucous blanket and propelled upward by the constant cilia movement, are brought to the throat where they are either swallowed or excreted as sputum. In the parenchyma (alveolar gas-exchange area), the alveolar macrophages engulf particles and infectious agents (i.e., kill and/or digest live organisms) and then migrate either to the lymphatic system to be drained out of the lung or to the airways to be removed by the mucociliary system. The two lung defense mechanisms together keep the lung sterile (free of infection), but an exposure to oxidants or other toxic agents can adversely affect these defense mechanisms, increasing susceptibility to respiratory infection.

In a respiratory infectivity test, which is considered very sensitive, two groups of animals, one exposed to oxidants under defined conditions and the other similarly exposed to clean air, are challenged with an infectious agent (aerosols of bacteria or viruses). Sometimes, the animal groups are first challenged with an infectious agent and one group is then exposed to oxidant and the other to clean air. In a typical infectivity test, 15–20% of the animals in the clean-air group should die, ensuring virulence of the infectious agent and susceptibility of the host, and more (up to 100%) in the oxidant-exposed group should die, demonstrating any adverse effect on lung clearance.

4.4.1. Ozone

Ozone exposure causes a damage to ciliated cells, although mucous secretion is generally increased. Any damage to ciliated cells (which can range from breaking or shortening of cilia to death and sloughing of ciliated cells) results in decreased transport of the mucous blanket, and hence decreased clearance of particles and infectious agents. Both in vitro and in vivo O_3 exposure can adversely affect particle clearance by alveolar macrophages, and impaired ability of these phagocytes to kill or inactivate bacteria is the likely cause. Increased susceptibility to respiratory infection associated with O_3 exposure has been amply documented in animal models, and exposure to as low as 0.08–0.10 ppm (157–196 μg/m^3) O_3 has been found to have an inhibitory effect (Ehrlich et al., 1979; Ehrlich, 1980; Kenoyer et al., 1981; EPA, 1978, 1986; Foster et al., 1987).

4.4.2. Nitrogen Dioxide

Exposure to NO_2 causes damage to cilia and ciliated cells, and thereby depresses the airway mucociliary clearance of particles and infectious agents. Susceptibility to respiratory infection is one of the most sensitive tests for NO_2 effects in animals, and increased respiratory infections have been observed even after exposure to ambient and near-ambient concentrations of NO_2. Extensive studies examining the effect of NO_2 on a host's susceptibility to inhaled infectious agents have been reviewed by several authors (Ehrlich, 1966, 1980; Goldstein et al., 1973; Coffin et al., 1976; Gardner et al., 1979; EPA, 1982). The infectivity tests have been carried out successfully with a variety of animal species (mice, hamsters, and monkeys) and infectious agents (bacteria and viruses), and increased susceptibility to respiratory infection has been observed with NO_2 concentrations as low as 0.3–0.5 ppm (570–940 µg/m^3).

5. PULMONARY EFFECTS OF COMBINED EXPOSURE

In photochemical smog, O_3 and NO_2 occur interdependently, and in a typical smog exposure both gases are likely to be encountered. Attention has been given to the combined effects of O_3 and NO_2, but earlier studies have shown rather variable results. In human exposure studies, some of the physiological effects observed were of the same magnitude as those of O_3 alone (Hackney et al., 1975a, b; von Nieding et al., 1979; Folinsbee et al., 1981), or they were synergistic compared to those of O_3 alone (Kagawa and Tsuru, 1979). An exposure to relatively low concentrations of NO_2 alone did not seem to have any detectable effects. In animal exposure studies, the morphological changes in the lung appeared mainly due to O_3 alone (Freeman et al., 1974), but the biochemical changes were additive (Yokoyama et al., 1980). In pulmonary host defense studies with animals, the combined effects were found to be the same as those of O_3 or NO_2 alone (Goldstein et al., 1974), or they were additive or possibly synergistic (Ehrlich et al., 1977, 1979; Ehrlich, 1980; Graham et al., 1987). In an in vitro study with erythrocytes, several biochemical parameters were found to change additively, but lipid peroxidation showed synergism (Goldstein, 1976, 1979b).

In later studies, a series of biochemical parameters and several physical and physiological indicators were examined after exposure of animals to O_3, NO_2, or the two combined (Mustafa et al., 1984; Mautz et al., 1988; Lee et al., 1989, 1990). It was observed that O_3 exposure alone caused a relatively larger change in these parameters than did NO_2 exposure, and that most of the changes with either O_3 or NO_2 were not statistically significant. Exposure to a combination of O_3 and NO_2 produced changes that were greater and statistically significant. A further analysis of the data indicated that most of the changes produced by the combined exposure were synergistic.

5.1. Mechanism of O_3–NO_2 Synergism

The interaction between O_3 and NO_2 possibly occurs in variable degrees, producing biological effects that may be additive for some parameters and synergistic for others. Where synergism occurs, it appears that one oxidant potentiates the effects of the other when the two oxidants are inhaled as a mixture.

The phenomenon is not fully understood, but an explanation can be made on the basis of chemical changes in the oxidant mixture outside as well as inside the lung.

As O_3 or NO_2 enters the lung, it causes injury to the respiratory epithelial cells, particularly at the terminal bronchioles and the alveolar ducts. When exposure to a mixture of O_3 and NO_2 occurs, a variety of chemical species and free radicals are produced that will react with the respiratory epithelium. Formation of N_2O_5 and NO_3 as important intermediate species has been recognized in laboratory and environmental chamber studies (Diggle and Gage, 1955; Demerjian et al., 1974; Pitts, 1983; Platt et al., 1984). The gas-phase reaction products of O_3 and NO_2 are shown in Reactions (19) through (22). It is likely that some N_2O_5 and NO_3 are produced at the bronchioloalveolar junctions and that they then react immediately with the epithelial cells, greatly enhancing the focal injury.

$$O_3 + NO_2 \longrightarrow NO_3 + O_2 \tag{19}$$

$$O_3 + 2NO_2 \longrightarrow N_2O_5 + O_2 \tag{20}$$

$$NO_3 + NO_2 \xrightarrow{\text{M}} N_2O_5 \tag{21}$$

$$N_2O_5 + O_3 \xrightarrow{\text{M}} 2NO_2 + 2O_2 \tag{22}$$

In addition, the reactions of O_3, NO_2, N_2O_5, and NO_3 in the moist environment at the bronchioloalveolar junctions give rise to other chemical species, including HNO_2, HNO_3, NO_2^-, NO_3^-, and $H\dot{O}$ radical [Reactions (12) through (14) and (23) through (27)]. In lung tissue, NO_2 and NO_3 can be reduced to NO_2^- and NO_3^-, respectively, in the presence of an electron donor (DH), that is, a reducing substance, which then becomes a radical cation, D^+ [Reactions (12) and (26)]. Isolated lung perfusion studies have shown that the product of NO_2 absorption (or reaction) in lung tissue is NO_2^-, and it is not oxidized to NO_3^- [Reaction (14)] unless red blood cells (RBC) are also present (Postlethwait and Mustafa, 1981, 1983, 1989). Oxidation of NO_2^- to NO_3^- and conversion of ferrohemoglobin (Hb) or oxyhemoglobin (O_2Hb) to ferrihemoglobin (Hb^+ or metHb) represent a rather complex reaction system, but the overall process can be shown as Reaction (14) (Oda et al., 1975, 1980a, b; Chiodi et al., 1983; Kosaka and Tyuma, 1987; Klimmek et al., 1988). However, NO_3 radical can directly oxidize a biomolecule (DH) and become reduced [Reaction (26)] without the presence of RBC.

$$N_2O_5 + H_2O \longrightarrow 2NO_3^- + 2H^+ \tag{23}$$

$$NO_3 + H_2O \longrightarrow NO_3^- + H\dot{O} + H^+ \tag{24}$$

$$NO_2 + H\dot{O} \longrightarrow NO_3^- + H^+ \tag{25}$$

$$NO_3 + DH \longrightarrow HNO_3 + D^+ \tag{26}$$

$$HNO_3 \longrightarrow NO_3^- + H^+ \tag{27}$$

Formation of intermediate chemical species discussed above may explain the synergistic effects of the O_3 plus NO_2 system compared to O_3 or NO_2 alone. The toxicity caused by O_3, NO_2, and their radical and nonradical reaction products must be understood separately. Once the toxic potential of at least the major components are known, the phenomenon of additive or synergistic effects can be better explained.

6. PULMONARY EFFECTS OF LONG-TERM EXPOSURE

Most experimental studies involving animals and controlled human subjects provide data on O_3 and NO_2 toxicity that are derived from relatively short-term exposures, often using oxidant concentrations higher than those encountered in the ambient atmosphere. Nonetheless, they are essential for the formulation of regulations and measures against both short-term and long-term health effects.

6.1. Chronic Effects

6.1.1. Ozone

Relatively long-term or chronic effects of O_3 are not well documented, and experimental studies involving long-term exposure are limited. Epidemiological studies suggest that chronic oxidant exposure affects the baseline respiratory functions (Lippman, 1989), although studies linking increased incidence of chronic disease with oxidants of photochemical smog are not very convincing. Based on relatively short-term animal exposure studies, it is predicted that acceleration of the deterioration of lung function over time (i.e., as individuals age) might result from chronic oxidant exposure. There is a depletion of lung cells and loss of lung reserves, which may not be obvious clinically but might produce a cumulative effect over a period of some years (Oomichi and Kita, 1974; Sherwin et al., 1983). Destructive and/or chronic obstructive lung diseases (e.g., emphysema, chronic bronchitis, and development of airway hypersensitivity or asthma) might be the ultimate result of chronic oxidant exposure (Coffin and Stokinger, 1977; NRC, 1977; EPA, 1978, 1986; WHO, 1979; Goldstein 1979a; Eustis et al., 1981; Melton, 1982; Filipowicz and McCauley, 1986; Gross and White, 1987; Mehlman and Borek, 1987; Lippmann, 1989). A continued (or cumulative) effect of oxidative and free radical reactions might be the underlying cause of lung cell depletion and functional deterioration.

6.1.2. Nitrogen Dioxide

Pulmonary effects of chronic NO_2 exposure at relatively low concentrations (0.05–5 ppm or 100–10,000 $\mu g/m^3$) include increase in airway resistance, epithelium thickness, and permeability of epithelial membrane; depression of mucociliary transport (i.e., reduced clearance of inhaled particles); increase in lung weight; alteration of various classes of phospholipids and collagen content; and development of emphysema (Giordano and Morrow, 1972; Arner and Rhoades, 1973; Drozdz et al., 1977; Trzeciak et al., 1977; Ayaz and Csallany, 1978; Morrow, 1984; Lindvall, 1985; EPA, 1982, and references cited therein). Some of the extrapulmonary effects include decreased growth (in animals), enlargement of spleen (reflecting red blood cell destruction), nonspecific immunological changes (e.g., decreased formation of serum antibodies), and behavioral and other central nervous system changes (Csallany and Ayaz, 1978a, b; Kuraitis et al., 1981, Morrow, 1984; Lindvall, 1985; EPA, 1982, and references cited therein). In addition to chronic pulmonary effects produced by NO_2 (or O_3), the oxidants exacerbate preexisting lung diseases such as asthma, chronic bronchitis, and emphysema (von Nieding and Wanger, 1979).

6.2. Tumorigenic Effects

6.2.1. Ozone

A carcinogen (or tumorigen) can characteristically cause several types of cellular reactions, including induction of electron transfer or free radical mechanism, chromosomal alteration or mutagenesis, hyperplasia, inflammation, and ornithine decarboxylase (ODC) activity. Ozone exposure causes hyperplasia and inflammation at the sites of lung injury (Stephens et al., 1974a, b; Dungworth et al., 1975; Chow et al., 1976; Evans et al., 1976a, b; Mustafa and Lee, 1976; Schwartz et al., 1976; Mustafa et al., 1977; Plopper et al., 1978; Seltzer et al., 1986; Koren et al., 1989). Hyperplasia can be induced repeatedly in the lung by recurrent or intermittent exposure. Inflammatory cells release free radicals (e.g., O_2^- and $H\dot{O}$), which may cause further tissue injury, including DNA breaks. Ozone causes chromosomal alterations, which may be due to an attack on chromosomes either directly or via free radicals generated by O_3 (Zelac et al., 1971a, b; Marz et al., 1975; Guerrero et al., 1979; Borek et al., 1988). A stimulation of lung ODC activity has been observed with O_3 exposure (Elsayed, 1987; Elsayed et al., 1990). Ozone, therefore, is a potential carcinogen or promoter of the carcinogenic process.

Several experimental studies have been conducted to explore the carcinogenic potential of O_3. Mice exposed to synthetic smog (ozonized gasoline) for a year had a greater incidence of lung adenomas compared to control mice breathing clean air (Kotin and Falk, 1956; Kotin et al., 1958). The synthetic smog was intended to be qualitatively similar to Los Angeles smog, except O_3 concentration was high (4 ppm or $7840 \mu g/m^3$). Mice exposed to ambient smog of Los Angeles for 7 to 11 months showed a trend toward increased incidence of lung adenomas compared to those kept in clean air (M. B. Gardner, 1966). The oxidant concentration in the smog ranged between 0.04 and 0.25 ppm (78 and $490 \mu g/m^3$). Two other studies in which O_3-induced lung tumor incidence was significant were carried out by Werthamer et al. (1970) and Nettesheim et al. (1970). In the latter study, mice were exposed to an artificial smog (ozonized gasoline) containing 1 ppm ($1960 \mu g/m^3$) O_3 for up to 18 months. The incidence of lung tumors (adenomas and adenocarcinomas) was significantly greater in the smog-exposed mice than in the clean-air controls.

Subsequently, Hassett et al. (1985a) examined the carcinogenic potential of O_3 using A/J strain mice. A group of 45 mice was exposed to 0.50 ppm (980 µg/m³) O_3 intermittently for six months; another group of 45 mice breathed clean air for the same period. The lung tumor incidence was significantly higher in the O_3-exposed mice. In another study, Last et al. (1986) exposed A/J strain and Swiss–Webster (SW) mice to 0.8 ppm O_3 intermittently for approximately four months, and observed a significant increase in lung adenomas in A/J strain mice but not in SW mice compared to the corresponding controls. Since SW mice are relatively resistant to spontaneous or induced tumorigenesis, a four month exposure may not have been chronic enough. The findings of the two laboratories agree that O_3 increases tumorigenesis in a susceptible mouse strain. The subject of O_3 tumorigenicity/carcinogenicity has recently been reviewed (Maugh, 1982; Mustafa et al., 1988; Witschi, 1988; Lippman, 1989).

6.2.2 Nitrogen Dioxide

Exposure to NO_2 results in a variety of biochemical and cellular changes in the lung, including induction of free radical reactions (Menzel, 1976; Pryor and Lightsey, 1981; Pryor et al., 1982), stimulation of aryl hydrocarbon hydroxylase activity (Husain and Dehnen, 1978), formation of nitrosamines (Chaudhari and Dutta, 1981; Iqbal et al., 1980, 1981; van Stee et al., 1983), induction of mutagenicity (von Nieding, 1978; Tsuda et al., 1981), stimulation of DNA synthesis (Creasia et al., 1977), and induction of hyperplasia (Evans et al., 1972, 1973, 1975, 1976a, b, 1977; Sherwin et al., 1972; Stephens et al., 1972; Cabral Anderson et al., 1977). Such observed similarities between NO_2-induced cellular effects and cancer-causing reactions lead to the presumption that NO_2 may play an important role in pulmonary and/or systemic carcinogenesis. Recommendations for research on the carcinogenic potential of NO_2 were made by the World Health Organization (WHO, 1977). Richters and Kuraitis (1981) presented evidence that exposure of animals to near-ambient concentrations of NO_2 facilitated cancer cell colonization (i.e., metastasis) in the lung. The finding was confirmed with O_3 exposure at near-ambient concentrations (Kobayashi et al., 1987). Thus, there appeared to be a new type of relationship between air pollution and lung cancer (Richters and Richters, 1983). In a chronic NO_2 exposure study, lung tumors (adenomas) in A strain mice were significantly increased in the exposed group compared to controls (Adkins et al., 1986). In this study, NO_2 exposure at 10 ppm (18.8 mg/m³), but not at 5 ppm (9.4 mg/m³) or 1 ppm (1.88 mg/m³), was effective in enhancing tumor production.

7. EXTRAPULMONARY EFFECTS

Although the lung is the primary target for inhaled O_3 and NO_2 toxicity, the effects of these oxidants have been documented in other organs. Effects of NO_2 in extrapulmonary sites can be explained to some extent based on the distribution of absorbed NO_2 beyond the lung, but how O_3 toxicity reaches beyond lung tissue is not clear. It is possible that O_3 reaction products (as opposed to O_3 itself) cross the alveolar air–blood barrier, causing damage to various extrapulmonary tissues and organs.

7.1. Hematopoietic System

The effect of O_3 on RBC and various other hematologic parameters has been studied after both in vitro and in vivo exposure. These parameters include RBC morphology, survival, osmotic fragility, and oxyhemoglobin affinity; level of GSH, ascorbate, hemoglobin, Heinz bodies, and methemoglobin; level of serum lipids, cholesterol, albumin, and globulin; and RBC or plasma enzyme activities (acetylcholine esterase, glucose-6-phosphate dehydrogenase, creatine phosphokinase, glutamate-pyruvate transaminase, pyruvate kinase, lactate dehydrogenase, and glutathione peroxidase). Ozone had either depressing or stimulating effects on these parameters, depending upon the O_3 exposure conditions and the nutritional status of the animals or controlled human subjects. For example, RBC morphology was altered, survival was decreased, and osmotic fragility was increased; GSH level was decreased or not changed; serum cholesterol was increased; acetylcholine esterase activity was decreased and the activities of glutathione peroxidase, pyruvate kinase, and lactate dehydrogenase were increased (EPA, 1978, 1986, and references cited therein).

The effect of NO_2 on a variety of hematologic parameters has been examined after exposure of both experimental animals (mouse, rat, guinea pig, rabbit, dog, and monkey) and human subjects to NO_2 at ambient, near-ambient, or relatively higher concentrations. The findings include alterations in blood cell population (change or a decrease in erythrocytes, lymphocytes, eosinophils, and platelets and change or an increases in leukocytes) and alterations in blood proteins, enzymes, and other components (change or an increase in hemoglobin and hematocrit values, an increase in glutathione peroxidase, and an increase in D-2,3-diphosphoglycerate) (EPA, 1982, and references cited therein). Reactions of NO_x with RBC have received particular attention. In vivo exposure of animals to NO resulted in a detectable amount of nitrosylhemoglobin (NOHb) and methemoglobin (metHb) in the blood. In vitro exposure of blood to NO showed that the Hb affinity for NO was much greater than for O_2 or CO, and that the product was largely NOHb in the absence of O_2 and largely metHb in the presence of O_2 (Rowland and Gause, 1971; Azoulay et al., 1977; Oda et al., 1975, 1980a, b; Nakajima et al., 1980). In vitro exposure of blood to NO_2 produces metHb (Chiodi et al., 1983), but in vivo exposure of animals to NO_2 has produced metHb in some studies (Toothill, 1967; Oda et al., 1980a, b; Nakajima et al., 1980) but not in others (Wagner et al., 1965). Although variable, the effects of inhaled NO_2 on the blood and the interaction of NO_2-generated intermediates with the blood are important determinants of health effects.

7.2. Central Nervous System

Although the Central Nervous System (CNS) is adversely affected by O_3 exposure and various physical and behavioral effects (e.g., dizziness and visual impairment) have been noted, the biochemical basis for these effects is not clear. In animal studies, motor activity was found to be decreased by O_3 exposure (Konisberg and Bachman, 1970). A change in behavior associated with decreased motor activity was reported by several laboratories (Weiss et al., 1981; Tepper et al., 1982, 1983). Variability in enzyme activity (i.e., decrease in catechol methyltransferase and increase in monoamine oxidase) was also noted (Trams et al., 1972).

Studies of the effects of NO_2 exposure on the CNS involving changes in behavior and other neurological markers are limited. Various short-term and chronic exposures of animals have shown minor deteriorations in conditioned behavior and alterations in a series of brain tissue enzyme activites (EPA, 1982, and references cited therein).

7.3. Other Organs and Tissues

Other extrapulmonary organs and tissues that have been examined for O_3 effects include the cardiovascular system, spleen, liver, endocrine system, and reproductive system. In animal studies, heart rate and blood pressure were variably affected by O_3 exposure (Costa et al., 1983; Friedman et al., 1983). Spleen was found to be enlarged after a subchronic O_3 exposure (Hassett et al., 1985b), and this might be related to O_3 effects on the blood. The O_3 effects in the liver were variable. Evidence for alterations in xenobiotic metabolism (such as an increase in pentobarbital-induced sleeping time) was reported (Gardner et al., 1974; Graham et al., 1981, 1982a), but measurement of cytochrome P-450 concentration or benzopyrene hydroxylase activity showed no change (Graham et al., 1982b).

Ozone exposure was found to cause morphological changes in the parathyroid and a decrease in thyrotropin and thyroid hormones (Clemons and Garcia, 1980; Clemons and Wei, 1984). Studies of O_3 effects on the reproductive system are limited. Some of the important effects noted were decreased maternal weight gain and increased fetal resorption, increased neonatal mortality, and slower development of offspring (Kavlock et al., 1979, 1980). In addition, chromosomal alterations were observed in peripheral blood lymphocytes and other mammalian cells (Zelac et al., 1971a, b; Marz et al., 1975; Gooch et al., 1976; Tice et al., 1978; Guerrero et al., 1979). The molecular mechanism for O_3 effects on chromosomes is not clear, but single-strand breaks in DNA may contribute.

Other extrapulmonary effects of NO_2 include an increase in pentobarbital-induced sleeping time in female mice (reflecting a possible effect on liver enzyme function) (Miller et al., 1980), splenomegaly (reflecting possible RBC destruction and/or renewal) (Kuraitis et al., 1981), proteinuria (urinary excretion of albumin and globulin) (Sherwin and Layfield, 1974), and urinary excretion of collagen products (hydroxyproline and hydroxylysine) (Hatton et al., 1972), nitrites, nitrates, and electrolytes (EPA, 1982, and references cited therein).

8. ANTIOXIDANT PROTECTION

Since oxidative, peroxidative, and free radical reactions are important mechanisms of O_3 and NO_2 toxicity, investigations have been carried out to discern antioxidant defense against oxidative damage. The initial attack of O_3 or NO_2 on a biomolecule generates a free radical ($R^{\cdot}$), which can start a chain reaction. In a chain reaction, the radical will pick up an O_2 molecule, producing a peroxy radical ($ROO^{\cdot}$) [Reaction (28)]. The peroxy radical will pick up a hydrogen atom to form an unstable hydroperoxide (ROOH), producing a new free radical ($X^{\cdot}$) [Reaction (29)]; the hydroperoxide breakdown will produce two more free radicals [Reaction (30)].

$$R^{\cdot} + O_2 \longrightarrow ROO^{\cdot} \tag{28}$$

$$ROO^{\cdot} + XH \longrightarrow ROOH + X^{\cdot} \tag{29}$$

$$ROOH \longrightarrow RO^{\cdot} + HO^{\cdot} \tag{30}$$

Both free radicals and unstable intermediates are toxic, with a devastating effect on cell structure and function. An antioxidant defense mechanism, therefore, should have the ability to prevent the formation of toxic agents via the propagative chain reaction and to destroy the toxic agents once they are formed. Living cells contain a metabolic denfense mechanism (consisting of a variety of enzyme activities and metabolites) that is intrinsic to cells but appears to be a secondary defense against oxidant damage. The metabolic defense mechanism generally eliminates most of the free radicals and peroxides formed in the cell, although some free radicals (e.g., $H\dot{O}$ and $H\dot{O}_2$) may not be destroyed by this means. Supplementation of cells or organisms with antioxidant substances has been attempted as a primary defense mechanism to prevent the formation of free radicals and/or limit free radical damage. Externally supplied antioxidants often act in conjunction with the metabolic defense mechanisms and provide a strong antioxidant protection.

8.1. Antioxidant Supplements

8.1.1. Ozone

Various natural and synthetic antioxidants have been used as supplements (dietary or otherwise) to prevent and/or decrease oxidant toxicity. These are vitamin E (tocopherols), selenium, vitamin C (ascorbate), vitamin A and β-carotene, p-aminobenzoic acid, phenolic antioxidants (butylated hydroxytoluene, butylated hydroxyanisole), and glutathione and other sulfhydryl compounds (Tappel, 1968; Goldstein et al., 1970; Menzel et al., 1972; Menzel, 1976, 1984; Roem et al.,

1972; Chow and Tappel, 1972, 1973; Fletcher and Tappel, 1973; Chow et al., 1979, 1981; Lunan et al., 1977; Willis and Kratzing, 1974, 1976; Fukase et al., 1978; Mustafa, 1975, 1990; Mustafa and Lee, 1976; Mustafa et al., 1977; Donovan et al., 1977; Dumelin et al., 1978a, b; Chow, 1979, 1983; Goldstein, 1979a; Plopper et al., 1979; Kratzing and Willis, 1980; Elsayed et al., 1982, 1983, 1984, 1988; Tyson et al., 1982; Ballew et al., 1983; Dubick et al., 1985; Machlin and Bendich, 1987; Heffner and Repine, 1989; Pryor, 1991). These substances, usually administered as nutritional supplements or as therapeutic agents, are known to act as free radical scavengers and/or antioxidants against nonradical oxidation. It is plausible that these compounds, particularly tocopherols (ArOH), denote a hydrogen to a free radical and stop or contain the chain reaction [Reaction (31) and (32)] (Burton et al., 1983; Wilson, 1983; Heffner and Repine, 1989). In that process, they may become radicals (ArȮ) themselves, that is, the antioxidants are oxidized and are then either regenerated or, upon dimerization [as in Reaction (33)], lost from the system (Tappel, 1968; Machlin and Bendich, 1987; Heffner and Repine, 1989). For tocopherols, an in vivo regeneration pathway has been proposed by Tappel (1968) and experimentally supported by others (Packer et al., 1979; Chen et al., 1980; Leung et al., 1981; Slater, 1984; Lambelet et al., 1985; McCay, 1985; Heffner and Repine, 1989).

$$\dot{HO} + ArOH \longrightarrow HOH + Ar\dot{O} \tag{31}$$

$$\dot{RO_2} + ArOH \longrightarrow RO_2H + Ar\dot{O} \tag{32}$$

$$Ar\dot{O} + Ar\dot{O} \longrightarrow ArOOAr \tag{33}$$

Most typically, vitamin E is administered to experimental animals through dietary regimen. The effects of O_3 can be elegantly demonstrated in vitamin E-deficient animals compared with vitamin E-supplemented animals. Various degrees of protection (e.g., prevention of massive lung injury or death due to high-level exposure and decrease in biochemical and morphological signs of injury due to low-level exposure) have been observed. However, the protective role of vitamin E against oxidant (O_3) effects is not clearly demonstrated in humans, thus making any extrapolation from animals to humans imprecise. There may be several explanations for this discrepancy. Experimental animals can be strictly controlled with dietary regimens leading to vitamin E deficiency. It is difficult to restrict the diet of humans and induce a vitamin E deficiency by that means. Thus, a supplemented vitamin E level in humans may not make any obvious difference in the protection already afforded by the basal level. A lack of dietary vitamin E may not produce any obvious or acute signs of disease, probably because of the presence of other mechanisms of protection against lipid peroxidation and/or oxidant damage. Furthermore, the physiological and/or biochemical parameters examined in humans may not be sensitive enough to pick up the differences in effect between the basal (or average) and the supplemented level.

In humans, vitamin E has found a clinical or pharmacological use against a variety of diseases, including premature aging, all of which may have a relationship with free radical reactions (Walton and Packer, 1980; Witting, 1980; Bieri et al., 1983; Pryor, 1991). Vitamin E is thought to have no known human toxicity, although some clinical disorders are claimed to be associated with high doses (Ferrell, 1980). It is not unreasonable to assume that a vitamin E supplementation has potential benefits against oxidant air polution (O_3), but a judicial dose that can provide optimal protection has not been established. Based on the premise that some degree of vitamin E deficiency may be widespread in humans with symptoms of chronic lung disease, heart disease, or premature aging and that exposure to oxidizing air pollutants (O_3 and NO_2) may initiate and/or exacerbate these symptoms, a dietary supplement of vitamin E may be a prudent means of prophylaxis against chronic free radical damage.

In recent years, there has been concern that populations exposed to photochemical smog or other environmental oxidants may require increased levels of dietary antioxidants (e.g., vitamin E). The daily dosage received by the general population may appear marginal considering, that the daily recommended dietary allowance of this vitamin ranges from 4 to 15 IU for infants, children, and adults (NAS, 1980). Therefore, some investigators have recommended large doses of vitamin E (as much as 1 g, roughly equivalent to 1000 IU daily) and other antioxidants for protection against oxidant effects and other diseases (Roberts, 1981; Lambelet et al., 1985; McCay, 1985). However, the protection offered by vitamin E or any of the other antioxidants is not absolute. Even with a high degree of supplementation, lung injury from oxidant exposure occurs (Mustafa and Tierney, 1978; Elsayed et al., 1988).

Similarly, selenium, which is an integral part of glutathione peroxidase, has been found to influence O_3 effects. Animals deficient in dietary selenium were more susceptible to oxidants than those supplemented with this nutritional element (Elsayed et al., 1982, 1983, 1984), and this susceptibility appeared to be due to a decreased function of glutathione peroxidase (Elsayed et al., 1982). Selenium as selenocysteine and selenomethionine acts as a potent water-soluble free radical scavenger (Menzel, 1970, and references cited therein). The antioxidant role of selenium is also evident from its ability to prevent the symptoms of vitamin E deficiency, such as are caused by oxidant (O_3) stress (Menzel, 1970; Elsayed et al., 1983, 1984). This is referred to as a sparing action between selenium and vitamin E. However, selenium and vitamin E do not fully replace each other as antioxidants. Also, selenium is toxic in excess doses, and therefore cannot be widely used for protection against oxidant air pollution (Goyer and Mehlman, 1977).

Vitamin C plays an antioxidant role against oxidant toxicity (Donovan et al., 1977; Chow, 1979; McCay, 1985; Machlin and Bendich, 1987; Heffner and Repine, 1989; Mustafa, 1990). It is present in the normal lung (Willis and Kratzing, 1974, 1976), and its level is found to decrease with O_3 exposure (Kratzing and Willis, 1980; Dubick et al., 1985). Supplementation of vitamin C in animals offers protection against acute O_3 injury (Coffin and Stokinger, 1977; Mustafa and Tierney, 1978, and references cited therein). Although vitamin C has the ability to break free radical chain reactions (Walton and Packer, 1980; Slater,

1984; McCay, 1985; Heffner and Repine, 1989), it is not considered a major antioxidant, because it can also act as a pro-oxidant (Forman and Fisher, 1981; Halliwell et al., 1987; Heffner and Repine, 1989). Ascorbate can reduce ferric iron while itself becoming an ascorbyl radical, which forms superoxide and $\overset{\cdot}{H}O$ radicals in an aerobic environment (Tappel, 1968; Heys and Dormandy, 1981; Rawley and Halliwell, 1983).

Glutathione and other sulfhydryl compounds capable of furnishing —SH or —SS— groups have been found to offer a variable degree of protection against oxidant toxicity (Coffin and Stokinger, 1977; Heffner and Repine, 1989, and references cited therein). The exact mode of action is not known, but thiyl radicals that may act as free radical scavengers are thought to occur.

8.1.2. Nitrogen Dioxide

Studies involving NO_2 exposure and dietary modulation of antioxidants are somewhat limited. Appreciable degrees of protection against both acute and chronic NO_2 toxicity have been noted in animals fed an adequate or high-dose vitamin E diet compared to animals receiving a diet marginal or deficient in vitamin E (Menzel et al., 1972; Fletcher and Tappel, 1973; Chow et al., 1974; Ayaz and Csallany, 1978; Csallany and Ayaz, 1978a, b; Dillard et al., 1980; Elsayed and Mustafa, 1982; Sevanian et al., 1982a, b; Sagai et al., 1981, 1982, 1984; Sagai and Ichinose, 1987; Guth and Mavis, 1986). Lung injury from NO_2 exposure, measured in terms of biochemical and morphological changes, was potentiated by dietary vitamin E deficiency, and the degree of protection increased as vitamin E levels reached an adequate or higher than adequate range. A higher than adequate or therapeutic level of vitamin E apparently did not offer a proportionately high degree of protection.

Findings that vitamin E deficiency potentiates and that supplementation lessens lung injury are interpreted as support for the role of lipid peroxidation in NO_2 toxicity. The direct evidence for lipid peroxidation as a mechanism of NO_2 toxicity includes a demonstration of NO_2-induced lipid peroxidation in model systems in vitro and its prevention or retardation by vitamin E, and a demonstration of lipid peroxidation products after NO_2 exposure in vivo, particularly their increase in vitamin E deficiency. Some examples of various protective measures are as follows. NO_2 exposure caused growth reduction and lipofuscin (pigment) accumulation in various organs of mice, but dietary vitamin E supplementation improved growth and prevented lipofuscin accumulation (Csallany and Ayaz, 1978b); formation of lipid peroxidation products (fatty acid epoxides in lung tissue or pentane in exhaled breath) was decreased in vitamin E-supplemented rats (Dillard et al., 1980; Sevanian et al., 1982a, b); alteration of lung enzyme activities (reflecting tissue injury) were minimal in vitamin E-supplemented rats and mice (Ayaz and Csallany, 1978; Elsayed et al., 1982; Sagai et al., 1981, 1982, 1984; Sagai and Ichinose, 1987).

8.2. Metabolic Defense

8.2.1. Ozone

Various enzyme activities and metabolites (generated in situ by metabolic activities) contribute to antioxidant defense mechanisms. One such mechanism, referred to as the GSH redox cycle, is thought to play a pivotal role in antioxidant defense against lipid peroxides and other toxic and related oxygenated intermediates. In this cycle, GSH is oxidized to GSSG while reducing a peroxide in a GP-catalyzed reaction [Reaction (34)]. GSH is then regenerated through a GR-catalyzed reaction [Reaction (35)] using the reducing equivalents of NADPH furnished by glucose oxidation via the pentose phosphate shunt pathway [Reactions (36) and (37)]. Activities of various enzymes [glutathione peroxidase and associated thiol metabolizing enzymes (glutathione reductase, disulfide reductase, thiol-disulfide transhydrogenase), peroxidase, catalase, superoxide dimutase, and key enzymes of the pentose shunt pathway (glucose-6-phosphate and 6-phosphogluconate dehydrogenases)] have been studied as a part of the metabolic defense mechanism in the lung (Chow and Tappel, 1972, 1973; DeLucia et al., 1972; Chow et al., 1974, 1976, 1981; Mustafa and Lee, 1976; Mustafa et al., 1977; Lunan et al., 1977; Chow, 1979; Plopper et al., 1979; Elsayed et al., 1982, 1983; Tyson et al., 1982; Machlin and Bendich, 1987; Heffner and Repine, 1989). Typical antioxidant metabolites studied are GSH and NADPH. In addition to antioxidant protection, metabolic defense mechanisms can promote the repair of injured cells and tissues and enable the lung to withstand continuing injury.

$$2GSH + ROOH \xrightarrow{\text{GP}} GSSG + ROH + H_2O \tag{34}$$

$$GSSG + NADPH + H^+ \xrightarrow{\text{GR}} 2GSH + NADP^+ \tag{35}$$

$$NADP^+ + G6P \xrightarrow{\text{G6PD}} NADPH + H^+ 6PG \tag{36}$$

$$6GP + NADP^+ \xrightarrow{\text{6GPD}} R5P + NADPH + H^+ \tag{37}$$

where ROOH is a hydroperoxide; GP is glutathione peroxidase; GR is glutathione reductase; G6P is glucose-6-phosphate; 6PG is 6-phosphogluconate; G6PD and 6PGD are dehydrogenases pertaining to G6P and 6PG, respectively; and R5P is ribulose-5-phosphate.

The enzyme activities and metabolites associated with metabolic defense are increased with O_3 exposure (Mustafa and Lee, 1976; Mustafa et al., 1977). Whether this elevation is a protective mechanism against oxidant stress (i.e., the enzymes are synthesized de novo or metabolic pathways are activated for the purpose of defense) is not certain. It appears more likely that it is a response of lung tissue to oxidant damage in which the metabolically active cells (e.g., the alveolar type 2 and the bronchiolar Clara cells) through proliferation contribute to metabolic increases. This is corroborated by the fact that the time course of metabolic increases coincides with that of morphological changes (Dungworth et al., 1975; Schwartz et al., 1976). However, the issue is not entirely resolved since protection against oxidants is afforded by the development of metabolic defense mechanisms. Increased activity of antioxidant enzymes evidently prevents further damage against continued exposure as well as high-level exposure (Chow and Tappel, 1972, 1973; Chow et al., 1976; Mustafa and Lee, 1976; Plopper et al., 1978).

8.3. Altered Response

Based on biochemical, morphological, host defense (particle clearance), and other physical and physiological parameters, it has become increasingly evident that pulmonary response to O_3, NO_2, or other oxidants is altered as a result of recurrent or continued exposure. In general, the pulmonary response (effect) is diminished, and the phenomenon is referred to as adaptation. As an altered response, adaptation deserves consideration because it is relevant to the human response to recurrent or long-term exposure to O_3 or NO_2 at ambient or near-ambient levels.

In earlier studies, the phenomenon of *tolerance* was used to describe conditions in which animals previously exposed to a lung irritant (e.g., O_3) could withstand an otherwise lethal dose of that irritant. The phenomenon of *cross-tolerance* was used to describe conditions in which animals previously exposed to a lung irritant (e.g., O_3) could tolerate an otherwise lethal dose of another irritant (e.g., NO_2). The tolerance or cross-tolerance phenomenon has been extensively studied using O_3, NO_2, and other lung irritants at doses capable of causing massive lung injury and/or death (Coffin and Stokinger, 1977, and references cited therein). In subsequent studies, the term *adaptation* has been introduced to denote a similar phenomenon, with lower concentrations of O_3 or NO_2 causing a mild to moderate form of lung injury (Dungworth et al., 1975; Cross et al., 1976; Mustafa and Tierney, 1978; Evans et al., 1985).

In any event, several mechanisms have been proposed to explain tolerance or adaptation. These include (1) hormones or neurohormonal processes, influencing (increasing or decreasing) pulmonary oxidant sensitivity; (2) increased levels of metabolites or activities of metabolic pathways, increasing pulmonary ability to combat oxidant toxicity; and (3) increased capacity to repair injured cells, providing a population of new cells resistant to oxidant toxicity (Fairchild, 1963; Fairchild and Graham, 1963; Coffin and Stokinger, 1977; Mustafa and Tierney, 1978; Frager et al., 1979; Evans et al., 1985; Nikula et al., 1988; Wright et al., 1990).

The opposite of adaptation is hypersensitivity (or hyperresponsiveness), which generally develops among asthmatics or potential asthmatics as they sustain recurrent O_3 or NO_2 exposure. In particular, a previous exposure to O_3 or NO_2 increases sensitivity of the airways to bronchospastic agents such as histamine (Henry et al., 1970; Matsumura et al., 1972; L. Y. Lee et al., 1977). Although both adaptation and hyperresponsiveness are complex, representing an altered response, a simplistic explanation has been offered based on O_3 and NO_2 as both sensory and pulmonary irritants (Lindvall, 1985). As sensory irritants, they stimulate the free nerve endings in the nose, throat, and eyes, causing a burning and itching sensation and tear flow (lacrimation). Recurrent exposure can raise the threshold for both odor and sensory irritation by an adaptation. However, in hypersensitivity, the irritation effects reach deep into the airways and sensitize the smooth muscles by a complex mechanism, resulting in bronchial constriction and increased airway resistance.

Development of tolerance or adaptation should not imply that we become immune to injury from continued O_3 or NO_2 exposure. Adaptation may be apparently beneficial, but animal studies have shown that continuous or intermittent exposure for an extended period of time may bring about unacceptable lung injury (e.g., development of chronic lung disease).

9. SUMMARY

Ozone and NO_2 are ubiquitous oxidizing air pollutants. They occur not only in urban smog but also in certain occupational and indoor environments. In the United States, the national ambient air quality standard (NAAQS) for O_3 is 0.12 ppm ($240 \mu g/m^3$) 1 hr average, not to be exceeded more than once per year. Likewise, NAAQS for NO_2 is 0.05 ppm ($100 \mu g/m^3$) annual arithmetic mean. However, these standards are exceeded in many cities in the United States and other urban regions of the world, despite various measures taken to control emissions of nitrogen oxides and hydrocarbons (precursors of O_3). A large segment of urban dwellers, therefore, receives O_3 and NO_2 exposures that exceed the possible safety margins. Both O_3 and NO_2 are toxic, but O_3 is much more toxic than NO_2. Although many short-term effects of O_3 have been documented, the consequences of long-term exposure to O_3 or a combination of O_3 and NO_2 are yet to be determined.

As judged from biochemical and morphological studies in animals receiving relatively short-term exposures, the mechanisms of O_3 and NO_2 effects in the lung involve direct as well as free radical-mediated oxidation of biomolecules (e.g., lipids, proteins, and various reducing substances), resulting in damage to membrane structure, disruption of cellular metabolism, and ultimately cell damage or death. An acute exposure to O_3 or NO_2 causes lung injury involving the ciliated cells in the airways and the type 1 epithelial cells in the alveolar region. The effects are particularly localized at the junction of terminal bronchioles and alveolar ducts, as evident from loss of cells and accumulation of inflammatory cells. In a typical short-term exposure, the lung tissue response is biphasic: an initial injury phase characterized by cell damage and loss of enzyme activities, followed by a repair phase characterized by increased metabolic activity and proliferation of metabolically active cells, (e.g., the alveolar type 2 cells and the bronchiolar Clara cells). Chronic exposure to O_3 or NO_2 can cause and/or exacerbate lung disease, perhaps including increased lung tumor incidence in susceptible animal models. Ozone exposure also causes extrapulmonary effects involving the blood, spleen, central nervous system, and other organs. A combination of O_3 and NO_2, both of which occur in photochemical smog, can produce effects that may be additive or synergistic. Synergistic lung injury occurs, possibly due to formation of more powerful radicals and chemical intermediates. Dietary antioxidants (e.g, vitamin E, vitamin C, and selenium) can offer protection against O_3 or NO_2 effects.

ACKNOWLEDGMENTS

Special thanks to James J. J. Clark and Nagaraj Vishwanath for their assistance in the manuscript preparation.

REFERENCES

Adkins, B., Jr., Van Stee, E. W., Simmons, J. E., and Eustis, S. L. (1986). Oncogenic response of strains A/J mice to inhaled chemicals. *J. Toxicol. Environ. Health* **17**, 311–322.

Alpert, S. M., Schwartz, B. B., Lee, S. D., and Lewis, T. R. (1971). Alveolar protein accumulation: A sensitive indicator of low level oxidant toxicity. *Arch. Intern. Med.* **128**, 69–73.

Arner, E. C., and Rhoades, R. O. (1973). Long-term nitrogen dioxide exposure. Effects on lung lipids and mechanical properties. *Arch. Environ. Health* **26**, 156–160.

Atkinson, R., and Lloyd, A. C. (1984). Evaluation of kinetic and mechanistic data for modeling of photochemical smog. *J. Phys. Chem. Ref. Data* **13**, 315–444.

Ayaz, K. L., and Csallany, A. S. (1978). Long term NO_2 exposure of mice in the presence and absence of vitamin E. II. Effect of glutathione peroxidase. *Arch. Environ. Health.* **33**, 292–296.

Azoulay, E., Soler, P., Blayo, M. C., and Basset, F. (1977). Nitric oxide effects on lung structure and blood oxygen affinity in rats. *Bull. Eur. Physiopathol. Respir.* **13**, 629–644.

Ballew, M., Calabrese, E. J., and Moore, G. S. (1983). The effect of dietary vitamin C on ozone-induced oxidative changes in guinea pig erythrocytes. *J. Environ. Sci. Health.* **A18**, 597–610.

Barry, B. E., Miller, F. J., and Crapo, J. D. (1983). Alveolar epithelial injury caused by inhalation of 0.25 ppm ozone. In S. D. Lee, M. G. Mustafa, and M. A. Mehlman (Eds.), *International Symposium on the Biomedical Effects of Ozone and Related Photochemical Oxidants*, Princeton Scientific, Princeton, NJ, pp. 299–309.

Bhalla, D. K., and Crocker, T. T. (1986). Tracheal permeability in rats exposed to ozone. *Am. Rev. Respir. Dis.* **134**, 572–579.

Bhalla, D. K., Mannix, R. C., Kleinman, M. T., and Crocker, T. T. (1986). Relative permeability of nasal, tracheal, and bronchoalveolar mucosa to macromolecules in rats exposed ozone. *J. Toxicol. Environ. Health* **17**, 269–283.

Bieri, J. G., Corash, L., and Hubbard, V. S. (1983). Medical uses of vitamin E. *N. Engl. J. Med.* **308**, 1063–1071.

Blair, W. H., Henry, M. C., and Ehrlich, R. (1969). Chronic toxicity of nitrogen dioxide. II. Effect on histopathology of lung tissue. *Arch. Environ. Health.* **19**, 186–192.

Blank, M. L., Dalbey, W., Nettesheim, P., Price, J., Creasia, D., and Snyder, F. (1978). Sequential changes in phospholipid composition and synthesis in lungs exposed to nitrogen dioxide. *Am. Rev. Respir. Dis.* **117**, 273–280.

Borek, C., Ong, A., and Cleaver, J. E. (1988). DNA damage from ozone and radiation in human epithelial cells. *Toxicol. Ind. Health.* **4**, 547–553.

Brummer, M. E. G., Schwartz, L. W., and McQuillen, N. K. (1977). A quantitative study of lung damage by scanning electron microscopy: Inflammatory cell response to high-ambient levels of ozone. *Scanning Electron Microsc.* **2**, 513–518.

Burton, G. W., Joyce, A., and Ingold, K. U. (1983). Is vitamin E the only lipid-soluble, chain-breaking antioxidant in human blood plasma and erythrocyte membranes? *Arch. Biochem. Biophys.* **221**, 281–290.

Cabral-Anderson, L. J., Evans, M. J., and Freeman, G. (1977). Effects of NO_2 on the lungs of rats. I. Morphology. *Exp. Mol. Pathol.* **27**, 353–365.

Carmichael, N. G., Winder, C., Borges, S. H., Backhouse, B. L., and Lewis, P. D. (1982). The health implications of water treatment with ozone. *Life Sci.* **30**, 117–129.

Chao, S. C., and Jaffe, S. (1972). Gas phase reaction of nitrogen dioxide and ethylene at 25 °C. *J. Chem. Phys.* **56**, 1987–1988.

Chaudhari, S., and Dutta, S. (1981). Possible formation of nitrosamine in guinea pigs following exposure to nitrogen dioxide and dimethylamine. *J. Toxicol. Environ. Health.* **7**, 753–763.

Chen, L. H., Lee, M. S., Hsing, W. F., and Chen, S. H. (1980). Effect of vitamin C on tissue antioxidant status of vitamin E deficient rats. *Int. J. Nutr. Res.* **50**, 156–162.

Chiodi, H., Collier, C. R., and Mothler, J. G. (1983). In vitro methemoglobin formation in human blood exposed to NO_2. *Environ. Res.* **30**, 9–15.

Chow, C. K. (1979). Nutritional influence on cellular antioxidant defense systems. *Am. J. Clin. Nutr.* **32**, 1066–1081.

Chow, C. K. (1983). Influence of dietary vitamin E on susceptibility to ozone exposure. In S. D. Lee, M. G. Mustafa, and M. A. Mehlman (Eds.), *International Symposium on the Biomedical Effects of Ozone and Related Photochemical Oxidants*. Princeton Scientific, Princeton, NJ, pp. 75–93.

Chow, C. K., and Tappel, A. L. (1972). An enzymatic protective mechanism against lipid peroxidation damage to lungs of ozone-exposed rats. *Lipids* **7**, 518–524.

Chow, C. K., and Tappel, A. L. (1973). Activities of pentose shunt and glycolytic enzymes in lungs of ozone-exposed rats. *Arch. Environ. Health* **26**, 205–208.

Chow, C. K., Dillard, C. J., and Tappel, A. L. (1974). Glutathione peroxidase system and lysozyme in rats exposed to ozone or nitrogen dioxide. *Environ. Res.* **7**, 311–319.

Chow, C. K., Hussain, M. Z., Cross, C. E., Dungworth, D. L., and Mustafa, M. G. (1976). Effects of low levels of ozone on lung. I. Biochemical responses during recovery and reexposure. *Exp. Mol. Pathol.* **25**, 182–188.

Chow, C. K., Plopper, C. G., and Dungworth, D. L. (1979). Influence of dietary vitamin E on the lungs of ozone-exposed rats: A correlated biochemical and morphological study. *Environ. Res.* **20**, 309–317.

Chow, C. K., Plopper, C. G., Chiu, M., and Dungworth, D. L. (1981). Dietary vitamin E and pulmonary biochemical and morphological alterations of rats exposed to 0.1 ppm ozone. *Environ. Res.* **24**, 315–324.

Clemons, G. K., and Garcia, J. F. (1980). Changes in thyroid function after short-term ozone exposure in rats. *J. Environ. Pathol Toxicol.* **4**, 359–369.

Clemons, G. K., and Wei, D. (1984). Effect of short-term ozone exposure on exogenous thyroxine levels in thyroidectomized and hypophysectomized rats. *Toxicol. Appl. Pharmacol.* **74**, 86–90.

Coffin, D. L., and Stokinger, H. E. (1977). Biologic effects of air pollutants. In A. C. Stern (Ed.), *Air Pollution*. Academic Press, New York, Vol. 2, pp. 231–360.

Coffin, D. L., Gardner, D. E., and Blommer, E. J. (1976). Time-dose response for nitrogen dioxide exposure in an infectivity model system. *Environ. Health Perspect.* **13**, 11–15.

Costa, D. L., Kutzman, R. S., Lehmann, J. R., Popenoe, E. A., and Drew, R. T. (1983). A subchronic multi-dose ozone study in rats. In S. D. Lee, M. G. Mustafa, and M. A. Mehlman (Eds.), *International Symposium on the Biomedical Effects of Ozone and Related Photochemical Oxidants*. Princeton Scientific, Princeton, NJ, pp. 369–393.

Council of European Communities (CEC) (1985). Council directive on air quality standards for nitrogen dioxide. *Off. J. Eur. Commun.* **L87**, 1–7.

Creasia, D. A., Nettesheim, P., and Kim, J. C. S. (1977). Stimulation of DNA synthesis in the lungs of hamsters exposed intermittently to nitrogen dioxide. *J. Toxicol. Environ. Health* **2**, 1173–1181.

Cross, C. E., DeLucia, A. J., Reddy, A. K., Hussain, M. Z., Chow, C. K., and Mustafa, M. G. (1976). Ozone interaction with lung tissue: Biochemical approaches. *Am. J. Med.* **60**, 929–935.

Csallany, A. S., and Ayaz, K. L. (1978a). The effects of intermittent nitrogen dioxide exposure on vitamin E-deficient and sufficient rats. *Toxicol. Lett.* **2**, 97–107.

Csallany, A. S., and Ayaz, K. L. (1978b). Long-term NO_2 exposure of mice in the presence or absence of vitamin E. I. Effect on body weight and lipofusion in pigments. *Arch. Environ. Health* **33**, 285–291.

Dawson, S. V., and Schenker, M. B. (1979). Health effects of inhalation of ambient concentrations of nitrogen dioxide. *Am. Rev. Respir. Dis.* **120**, 281–292.

DeLucia, A. J., Hoque, P. M., Mustafa, M. G., and Cross, C. E. (1972). Ozone interaction with rodent lung. Effect on sulfhydryl and sulfhydryl-containing enzyme activities. *J. Lab. Clin. Med.* **80**, 559–566.

DeLucia, A. J., Mustafa, M. G., Hussain, M. Z., and Cross, C. E. (1975). Ozone interaction with rodent lung. III. Oxidation of reduced glutathione and mixed disulfide formation between protein and nonprotein sulfhydryls. *J. Clin. Invest.* **55**, 794–802.

Demerjian, K. L., Kerr, J. A., and Calvert, J. G. (1974). The mechanism of photochemical smog formation. *Adv. Environ. Sci. Technol.* **4**, 1–262.

DeNicola, D. B., Rebar, A. H., and Henderson, R. H. (1981). Early damage indicators in the lung. V. Biochemical and cytological response to NO_2 inhalation. *Toxicol. Appl. Pharmacol.* **60**, 301–312.

Diggle, W. M., and Gage, J. C. (1955). The toxicity of ozone in the presence of oxides of nitrogen. *Br. J. Ind. Med.* **12**, 60–64.

Dillard, C. J., Urribarri, C. N., Reddy, K., Fletcher, B., Taylor, S., de Lumen, B., and Tappel, A. L. (1972). Increased lysosomal enzymes in lungs of ozone-exposed rats. *Arch. Environ. Health* **25**, 426–431.

Dillard, C. J., Sagai, M., and Tappel, A. L. (1980). Respiratory pentane: Measure of *in vivo* lipid peroxidation applied to rats fed diets varying in polyunsaturated fats, vitamin E., and selenium and exposed to nitrogen dioxide. *Toxicol. Lett.* **6**, 251–256.

Dixon J. R., and Mountain, J. T. (1965). Role of histamine and related substances in the development of tolerance to edemagenic gases. *Toxicol. Appl. Pharmacol.* **7**, 756–766.

Donovan, D. H., Williams, S. J., Charles, J. M., and Menzel, D. B. (1977). Ozone toxicity: Effect of dietary vitamin E and polyunsaturated fatty acids. *Toxicol. Lett.* **1**, 135–139.

Drozdz, M., Kicharz, E., and Szyja, J. (1977). Effect of chronic exposure to nitrogen dioxide on collagen content in lung and skin of guinea pigs. *Environ. Res.* **13**, 369–377.

Dubick, M. A., Heng, H., and Rucker, R. B. (1985). Effects of protein deficiency and food restriction on lung ascorbic acid and glutathione in rats exposed to ozone. *J. Nutr.* **115**, 1050–1056.

Dumelin, E. E., Dillard, C. J., and Tappel, A. L. (1978a). Breath ethane and pentane as measures of vitamin E protection of *Macaca radiata*, against 90 days of exposure to ozone. *Environ. Res.* **15**, 38–43.

Dumelin, E. E., Dillard, C. J., and Tappel, A. L. (1978b). Effect of vitamin E and ozone on pentane and ethane expired by rats. *Arch. Environ. Health* **33**, 129–135.

Dungworth, D. L., Castleman, W. L., Chow, C. K., Mellick, P. W., Mustafa, M. G., Tarkington, B., and Tyler, W. S. (1975). Effects of ambient levels of ozone on monkeys. *Fed. Proc., Fed. Am. Soc. Exp. Biol.* **34**, 1670–1674.

Ehrlich, R. (1966). Effect of nitrogen dioxide on resistance to respiratory infection. *Bacteriol. Rev.* **30**, 604–614.

Ehrlich, R. (1980). Interaction between environmental pollutants and respiratory infections. *Environ. Health Perspect.* **35**, 89–100.

Ehrlich, R., Findlay, J. C., Fenters, J. D., and Gardner, D. E. (1977). Health effects of short-term inhalation of nitrogen dioxide and ozone mixtures. *Environ. Res.* **14**, 223–231.

Ehrlich, R., Findlay, J. C., and Gardner, D. E. (1979). Effects of repeated exposures to peak concentrations of nitrogen dioxide and ozone on resistance to streptococcal pneumonia. *J. Toxicol. Environ. Health* **5**, 631–642.

Elsayed, N. M. (1987). Influence of vitamin E on Polyamine metabolism in ozone-exposed rat lungs. *Arch. Biochem. Biophys.* **255**, 392–399.

Elsayed, N. M., and Mustafa, M. G. (1982). Dietary antioxidants and the biochemical response to oxidant inhalation. I. Influence of dietary vitamin E on the biochemical effects of nitrogen dioxide exposure in rat lung. *Toxicol. Appl. Pharmacol.* **66**, 319–328.

Elsayed, N. M., Hacker, A., Mustafa, M., Kuehn, K., and Schrauzer, K. (1982). Effects of decreased glutathione peroxidase activity on the pentose phosphate cycle in mouse lung. *Biochem. Biophys. Res. Commun.* **104**, 564–569.

Elsayed, N. M., Hacker, A. D., Kuehn, K., Mustafa, M. G., and Schrauzer, G. N. (1983). Dietary antioxidants and the biochemical response to oxidant inhalation. II. Influence of dietary selenium on the biochemical effects of ozone exposure in mouse lung. *Toxicol. Appl. Pharmacol.* **71**, 398–406.

Elsayed, N. M., Mustafa, M. G., Hacker, A. D., Kuehn, K., and Schrauzer, G. N. (1984). Dietary antioxidants and the biochemical response to oxidant inhalation. III. Selenium influence on mouse lung response and tolerance to ozone. *Biol. Trace Elem. Res.* **6**, 249–261.

Elsayed, N. M., Kass, R., Mustafa, M. G., Hacker, A. D., Ospital, J. J., Chow, C. K., and Cross, C. E. (1988). Effect of dietary vitamin E level on the biochemical response of rat lung to ozone. *Drug-Nutr. Interact.* **5**, 373–386.

Elsayed, N. M., Ellingson, A. S., Tierney, D. F., and Mustafa, M. G. (1990). Effects of ozone inhalation on polyamine metabolism and tritiated thymidine incorporation into DNA of rat lungs. *Toxicol. Appl. Pharmacol.* **102**, 1–8.

Environmental Protection Agency (EPA) (1978). *Air Quality Criteria for Ozone and Other Photochemical Oxidants*, EPA-600/8-79-004. Environmental Criteria and Assessment Office, US EPA, Research Triangle Park, NC.

Enivornmental Protection Agency (EPA) (1982). *Review of the National Ambient Air Quality Standards for Nitrogen Oxides: Assessment of Scientific and Technical Information*, EPA-450/5-82-002. Office of Air Quality Planning and Standards, US EPA, Research Triangle Park, NC.

Environmental Protection Agency (EPA) (1986). *Air Quality Criteria for Ozone and Other Photochemical Oxidants*, EPA-600/8-84-020cF, Vol. IV. Environmental Criteria and Assessment Office, US EPA, Research Triangle Park, NC.

Environmental Protection Agency (EPA) (1987). *Ultraviolet Radiation and Melanoma with a Special Focus on Assessing the Risks of Stratospheric Ozone Depletion*. Office of Air and Radiation, US EPA, Washington, DC.

Eustis, S. L., Schwartz, L. W., Kosch, P. C., and Dungworth, D. L. (1981). Chronic bronchiolitis in nonhuman primates after prolonged ozone exposure. *Am. J. Pathol.* **105**, 121–137.

Evans, M. J., Stephens, R. J., Cabral, L. J., and Freeman, G. (1972). Cell renewal in the lungs of rats exposed to low levels of NO_2. *Arch. Environ. Health* **24**, 180–188.

Evans, M. J., Cabral, L. J., Stephens, R. J., and Freeman, G. (1973). Renewal of alveolar epithelium in the rat following exposure to NO_2. *Am. J. Pathol.* **70**, 175–198.

Evans, M. J., Cabral, L. J., Stephens, R. J., and Freeman, G. (1975). Transformation of alveolar Type 2 cells to Type 1 cells following exposure to NO_2. *Exp. Mol. Pathol.* **22**, 142–150.

Evans, M. J., Johnson, L. V., Stephens, R. J., and Freeman, G. (1976a). Renewal of the terminal bronchiolar epithelium in the rat following exposure to NO_2 or O_3. *Lab. Invest.* **35**, 246–257.

Evans, M. J., Johnson, L. V., Stephens, R. J., and Freeman, G. (1976b). Cell renewal in the lungs of rats exposed to low levels of ozone. *Exp. Mol. Pathol.* **24**, 70–83.

Evans, M. J., Johnson, L. V., Stephens, R. J., and Freeman, G. (1977). Effects of NO_2 on the lungs of aging rats. II. Cell proliferation. *Exp. Mol. Pathol.* **27**, 366–376.

Evans, M. J., Dekker, N. P., Cabral-Anderson, L. J., and Shami, S. G. (1985). Morphological basis of tolerance to ozone. *Exp. Mol. Pathol.* **42**, 366–376.

Fairchild, E. J., II (1963). Neurohormonal factors in injury from inhaled irritants. *Arch. Environ. Health* **6**, 85–92.

Fairchild, E. J., II, and Graham, S. L. (1963). Thyroid influence on the toxicity of respiratory irritant gases, ozone and nitrogen dioxide. *J. Pharmacol. Exp. Ther.* **139**, 177–184.

Fairchild, E. J., II, Graham, S. L., Hite, M., Killens, R., and Scheel, L. D. (1964). Changes in thyroid I^{131} activity in ozone-tolerant and ozone-susceptible rats. *Toxicol. Appl. Pharmacol.* **6**, 607–613.

Ferrell, P. M. (1980). Deficiency states, pharmacological effects, and nutrient requirements. In L. J. Machlin (Ed.), *Vitamin E: A Comprehensive Treatise.* Dekker, New York, pp. 520–620.

Filipowicz, C., and McCauley, R. (1986). The effects of chronic ozone exposure on pulmonary collagen content and collagen synthesis in rats. *J. Appl. Toxicol.* **6**, 87–90.

Fletcher, B. L., and Tappel, A. L. (1973). Protective effects of alpha-tocopherol in rats exposed to toxic levels of ozone and nitrogen dioxide. *Environ. Res.* **6**, 165–175.

Folinsbee, L. J., Bedi, J. F., and Horvath, S. M. (1981). Combined effects of ozone and nitrogen dioxide on respiratory function in man. *Am. Ind. Hyg. Assoc. J.* **42**, 534–541.

Ford-Hutichinson, A. W., Bray, M. A., Doig, M. V., Shipley, M. E., and Smith, M. J. H. (1980). Leukotriene b, a potent chemokinetic and aggregating substance released from polymorphonuclear leukocytes. *Nature (London)* **286**, 264–265.

Forman, H. J., and Fisher, A. B. (1981). Antioxidant enzymes of rat granular pneumocytes—constitutive levels and effect of hypoxia. *Lab. Invest.* **45**, 1–6.

Foster, W. M., Costa, D. L., and Langenback, E. G. (1987). Ozone exposure alters tracheobronchial mucociliary function in humans. *J. Appl. Physiol.* **63**, 996–1002.

Frager, N. B., Phalen, R. F., and Kenoyer, J. L. (1979). Adaptations to ozone in reference to mucociliary clearance. *Arch. Environ. Health* **34**, 51–57.

Freeman, B. A., and Mudd, J. B. (1981). Reaction of ozone with sulfhydryls of human erythrocytes. *Arch. Biochem. Biophys.* **208**, 212–220.

Freeman, G., and Haydon, G. B. (1964). Emphysema after low-level exposure to nitrogen dioxide. *Arch. Environ. Health* **8**, 125–128.

Freeman, G., Crane, S. C., Stephens, R. J., and Furiosi, N. J. (1968). Pathologenesis of the nitrogen dioxide-induced lesion in the rat lung. A review and presentation of new observations. *Am. Rev. Respir. Dis.* **98**, 429–443.

Freeman, G., Crane, S. C., Stephens, R. J., and Furiosi, N. J. (1969). Healing in rat lung after subacute exposure to nitrogen dioxide. *Am. Rev. Respir. Dis.* **100**, 662–676.

Freeman, G., Crane, S. C., Stephens, R. J., Furiosi, N. J., Stephens, R. J., Evans, M. J., and Moore, W. D. (1972). Covert reduction in ventilatory surface in rats during prolonged exposure to subacute nitrogen dioxide. *Am. Rev. Respir. Dis.* **106**, 563–577.

Freeman, G., Juhos, L. T., Furiosi, N. J., Mussenden, R., Stephens, R. J., and Evans, M. J. (1974). Pathology of pulmonary disease from exposure to interdependent ambient gases (nitrogen dioxide and ozone). *Arch. Environ. Health* **29**, 203–210.

Friedman, M., Gallo, J. M., Nichols, H. P., and Bromberg, P. A. (1983). Changes in inert gas rebreathing parameters after ozone exposure in dogs. *Am. Rev. Respir. Dis.* **128**, 851–856.

Fukase, O., Watanabe, H., and Ishimura, K. (1978). Effects of exercise on mice exposed to ozone. *Arch. Environ. Health* **33**, 198–200.

Gardner, D. E., Holzman, R. S., and Coffin, D. L. (1969). Effects of nitrogen dioxide on pulmonary cell population. *J. Bacteriol.* **98**, 1041–1043.

Gardner, D. E., Illing, J. W., Miller, F. J., and Coffin, D. L. (1974). The effect of ozone on pentobarbital sleeping time in mice. *Res. Commun. Chem. Pathol. Pharmacol.* **9**, 689–700.

Gardner, D. E., Miller, F. J., Blommer, E. J., and Coffin, D. L. (1979). Influence of exposure mode on the toxicity of NO_2. *Environ. Health Perspect.* **30**, 23–29.

Gardner, M. B. (1966). Biological effects of urban air pollution: lung tumors in mice. *Arch. Environ. Health* **12**, 305–313.

Gelmont, D., Stein, R. A., and Mead, J. F. (1981). The bacterial origin of rat breath pentane. *Biochem. Biophys. Res. Commun.* **102**, 932–936.

Giordano, A. M., and Morrow, P. E. (1972). Chronic low-level nitrogen dioxide exposure and mucociliary clearance. *Arch. Environ. Health* **25**, 443–449.

Glaze, W. H. (1986). Reaction products of ozone: A review. *Environ. Health Perspect.* **69**, 151–157.

Goldstein, B. D. (1973). Hydrogen peroxide in erythrocytes. Detection in rats and mice inhaling ozone. *Arch. Environ. Health* **26**, 279–280.

Goldstein, B. D. (1976). Combined exposure to ozone and nitrogen dioxide. *Environ. Health Perspect.* **13**, 107–110.

Goldstein, B. D. (1979a). The pulmonary and extrapulmonary effects of ozone. *Ciba Found. Symp.* **65**, 295–319.

Goldstein, B. D. (1979b). Combined exposure to ozone and nitrogen dioxide. *Environ. Health Perspect.* **30**, 87–89.

Goldstein, B. D., and Balchum, O. J. (1967). Effect of ozone on lipid peroxidation in the red blood cell. *Proc. Soc. Exp. Biol. Med.* **126**, 356–358.

Goldstein, B. D., Balchum, O. J., Demopoulos, H. D., and Duke, P. S. (1968). Electron paramagnetic resonance spectroscopy. Free radical signals associated with ozonation of linoleic acid. *Arch. Environ. Health* **17**, 46–49.

Goldstein, B. D., Lodi, C., Collinson, C., and Balchum, O. J. (1969). Ozone and lipid peroxidation. *Arch. Environ. Health* **18**, 631–635.

Goldstein, B. D., Buckley, R. D., Cardenas, R., and Balchum, O. J. (1970). Ozone and vitamin E. *Science* **169**, 605–606.

Goldstein, B. D., Solomon, S., Pasternack, B. S., and Bickers, D. R. (1975). Decrease in rabbit lung microsomal cytochrome P-450 levels following ozone exposure. *Res. Commun. Chem. Pathol. Pharmacol.* **10**, 759–762.

Goldstein, E., Eagle, M. C., and Hoeprich, P. D. (1973). Effect of nitrogen dioxide on pulmonary bacterial defense mechanisms. *Arch. Environ. Health* **26**, 203–204.

Goldstein, E., Warshauer, D., Lippert, W., and Tarkington, B. (1974). Ozone and nitrogen dioxide exposure: Murine pulmonary defense mechanisms. *Arch. Environ. Health* **28**, 85–90.

Goldstein, E., Peek, N. F., Parks, N. J., Hines, H., Steffey, E. P., and Tarkington, B. (1977). Fate and distribution of inhaled nitrogen dioxide in rhesus monkeys. *Am. Rev. Respir. Dis.* **115**, 403–412.

Goldstein, E., Goldstein, F., Peek, N. F., and Parks, N. (1980). Absorption and transport of nitrogen oxide. In S. D. Lee (Ed.), *Nitrogen Oxides and Their Effects on Health.* Ann Arbor Science Publishers, Ann Arbor, MI, pp. 143–160.

Gooch, P. C., Creasia, D. A., and Brewen, J. G. (1976). The cytogenetic effect of ozone: Inhalation and *in vitro* exposures. *Environ. Res.* **12**, 188–195.

Gordon, T., Taylor, B. F., and Amdur, M. O. (1981). Ozone inhibition of tissue cholinesterase in guinea-pigs. *Arch. Environ. Health* **36**, 284–288.

Goyer, R. A., and Mehlman, M. A. (1977). *Toxicology of Trace Elements.* Hemisphere Publishing Corp., Washington, DC, pp. 191–240.

Graham, J. A., Menzel, D. B., Miller, F. J., Illing, J. W., and Gardner, D. E. (1981). Influence of ozone on pentobarbital-induced sleeping time in mice, rats, and hamsters. *Toxicol. Appl. Pharmacol.* **61**, 64–73.

Graham, J. A., Miller, F. J., Gardner, D. E., Ward, R., and Menzel, D. B. (1982a). Influence of ozone and nitrogen dioxide on hepatic microsomal enzymes in mice. *J. Toxicol. Environ. Health* **9**, 849–856.

Graham, J. A., Menzel, D. B., Miller, F. J., Illing, J. W., and Gardner, D. E. (1982b). Effect of ozone on drug-induced sleeping time in mice pretreated with mixed-function oxidase inducers and inhibitors. *Toxicol. Appl. Pharmacol.* **62**, 489–497.

Graham, J. A., Gardner, D. E., Blommer, E. J., Housem, D. E., Menache, M. G., and Miller, F. J. (1987). Influence of exposure patterns of nitrogen dioxide and modifications by ozone on susceptibility to bacterial infectious disease in mice. *J. Toxicol. Environ. Health* **21**, 113–125.

Gross, K. B., and White, H. J. (1987). Functional and pathologic consequences of a 52-week exposure to 0.5 ppm ozone followed by a clean air recovery period. *Lung* **165**, 283–295.

Guerrero, R. R., Rounds, D. E., Olson, R. S., and Hackney, J. D. (1979). Mutagenic effects of ozone on human cells exposed *in vivo*, and *in vitro*, based on sister chromatid exchange analysis. *Environ. Res.* **18**, 336–346.

Guidotti, T. L. (1978). The higher oxides of nitrogen: Inhalation toxicology. *Environ. Res.* **15**, 443–473.

Guth, D. J., and Mavis, R. D. (1985). Biochemical assessment of acute nitrogen dioxide toxicity in rat lung. *Toxicol. Appl. Pharmacol.* **81**, 128–138.

Guth, D. J., and Mavis, R. D. (1986). The effect of α-tocopherol content on the acute toxicity of nitrogen dioxide. *Toxicol. Appl. Pharmacol.* **84**, 304–314.

Haagen-Smit, A. J. (1952). Chemistry and physiology of Los Angeles smog. *Ind. Eng. Chem.* **44**, 1342–1346.

Haagen-Smit, A. J. (1963). Photochemistry and smog. *J. Air Pollut. Control Assoc.* **13**, 446–454.

Hacker, A. D., Elsayed, M., Mustafa, M. G., Ospital, J. J., and Lee, S. D. (1976). Effects of short-term nitrogen dioxide exposure on lung collagen synthesis. *Am. Rev. Respir. Dis.* **113**, 107.

Hackney, J. D., Linn, W. S., Mohler, J. G., Pederson, E. E., Breisacher, P., and Russo, A. (1975a). Experimental studies on human health effects of air pollutants. II. Four-hour exposure to ozone alone and in combination with other pollutant gases. *Arch. Environ. Health* **30**, 379–384.

Hackney, J. D., Linn, W. S., Law, D. C., Karuza, S. K., Greenberg, H., Buckley, R. D., and Pederson, E. E. (1975b). Experimental studies on human health effects of air pollutants. III. Two-hour exposure to ozone alone and in combination with other pollutant gases. *Arch. Environ. Health* **30**, 385–390.

Halliwell, B., Wasil, M., and Grootveld, M. (1987). Biologically significant scavenging of the myeloperoxidase-derived oxidant hypochlorous acid by ascorbic acid. *FEBS Lett.* **213**, 15–18.

Hassett, C., Mustafa, M. G., Coulson, W. F., and Elashoff, R. M. (1985a). Murine lung carcinogenesis following exposure to ambient ozone concentrations. *JNCI, J. Natl. Cancer Inst.* **75**, 771–777.

Hassett, C., Mustafa, M. G., Coulson, W. F., and Elashoff, R. M. (1985b). Splenomegaly in mice following exposure to ambient levels of ozone. *Toxicol. Lett.* **26**, 139–144.

Hatton, D. J., Leach, C. S., Nicogossian, A. E., and DiFerrante, N. (1972). Collagen breakdown and nitrogen dioxide inhalation. *Arch. Environ. Health* **32**, 33–36.

Heffner, J. E., and Repine, J. E. (1989). State of the art—pulmonary strategies of antioxidant defense. *Am. Rev. Respir. Dis.* **140**, 531–554.

Henry, M. C., Findlay, J., Spangler, J., and Ehrlich, R. (1970). Chronic toxicity of NO_2 in squirrel monkeys. *Arch. Environ. Health* **20**, 566–570.

Heys, A. D., and Dormandy, T. L. (1981). Lipid peroxidation in iron-overloaded spleens. *Clin. Sci.* **60**, 295–301.

Hu, P. C., Miller, F. J., Daniels, M. J., Hatch, G. E., Graham, J. A., Gardner, D. E., and Selgrade, M. K. (1982). Protein accumulation in lung lavage fluid following ozone exposure. *Environ. Res.* **29**, 377–388.

Hurst, D. J., and Coffin, D. L. (1971). Ozone effect on lysosomal hydrolases of alveolar macrophages in vitro. *Arch. Intern. Med.* **127**, 1059–1063.

Husain, M. M., and Dehnen, W. (1978). Effect of NO_2 and SO_2 inhalation on benzo(a)pyrene metabolism in rat lung. *Arch. Toxicol.* **40**, 207–210.

Hussain, M. Z., Cross, C. E., Mustafa, M. G., and Bhatnagar, R. S. (1976). Hydroxyproline contents and prolyl hydroxylase activities in lungs of rats exposed to low levels of ozone. *Life Sci.* **18**, 897–904.

Ichinose, T., and Sagai, M. (1982). Studies on biochemical effects of nitrogen dioxide. III. Changes of the antioxidative protective systems in rat lungs of lipid peroxidation by chronic exposure. *Toxicol. Appl. Pharmacol.* **66**, 1–8.

Iqbal, Z. M., Dahl, K., and Epstein, S. S. (1980). Role of nitrogen dioxide in the biosynthesis of nitrosamines in mice. *Science* **207**, 1475–1477.

Iqbal, Z. M., Dahl, K., and Epstein, S. S. (1981). Biosynthesis of dimethylnitrosamine in dimethylamine-treated mice and exposure to nitrogen dioxide. *JNCI, J. Natl. Cancer Inst.* **67**, 137–141.

Johnson, D. A. (1980). Ozone inactivation of human alpha-1-antiproteinase inhibitor. *Am. Rev. Respir. Dis.* **121**, 1031–1038.

Joigne, J., and Bader, H. (1979). The role of hydroxyl radical reactions in ozonation process in aqueous solutions. *Water Res.* **10**, 377–386.

Kagawa, J., and Tsuru, K. (1979). Respiratory effects of 2-hour exposure to ozone and nitrogen dioxide alone and in combination in normal subject performing intermittent exercise. *Nippon Kyobu Shikkan Gakkai Zasshi (Jpn. J. Thorac. Dis.)* **17**, 765–774.

Kavlock, R. J., Daston, G., and Grabowski, C. T. (1979). Studies on the development toxicity of ozone. I. Prenatal effects. *Toxicol. Appl. Pharmacol.* **48**, 19–28.

Kavlock, R. J., Meyer, E., and Grabowski, C. T. (1980). Studies on the developmental toxicity of ozone: postnatal effects. *Toxicol. Lett.* **5**, 3–9.

Kehrl, H. R., Vincent, L. M., Kowalsky, R. J., Horstman, D. H., O'Neil, J. J., McCartney, W. H., and Bromberg, P. A. (1987). Ozone exposure increases respiratory epithelial permeability in humans. *Am. Rev. Respir. Dis.* **135**, 1124–1128.

Kenoyer, J. L., Phalen, R. F., and Davis, J. R. (1981). Particle clearance from the respiratory tract as a test of toxicity: Effect of ozone on short and long term clearance. *Exp. Lung. Res.* **2**, 111–120.

Kleinerman, J. (1970). Effects of NO_2 in hamsters: Autoradiographic and electron microspic aspects. *At. Energy Comm. Symp. Ser.* **18**, 271–279.

Kleinerman, J. (1977). Some effects of nitrogen dioxide on the lung. *Fed. Proc., Fed. Am. Soc. Exp. Biol.* **36**, 1714–1718.

Klimmek, R., Krettek, C., and Werner, H. W. (1988). Ferrihemoglobin formation by amyl nitrite and sodium nitrite in different species in vivo and in vitro. *Arch. Toxicol.* **62**, 152–160.

Kobayashi, T., Noguchi, T., Kikuno, M., and Kubota, K. (1980). Effect of acute nitrogen dioxide exposure on the composition of fatty acids in lung and liver phospholipids. *Toxicol. Lett.* **6**, 149–155.

Kobayashi, T., Todoraki, T., and Sato, H. (1987). Enhancement of pulmonary metastasis of murine fibrosarcoma NR-FS by ozone exposure. *J. Toxicol. Environ. Health* **20**, 135–145.

Konisberg, A. S., and Bachman, C. H. (1970). Ozonized atmosphere and gross motor activity of rats. *Int. J. Biometeorol.* **14**, 261–266.

Koppenol, W. H. (1982). The reduction potential of the couple O_3/O_3^-, consequences for mechanisms of ozone toxicity. *FEBS Lett.* **140**, 169–172.

Koren, H. S., Devlin, R. B., Graham, D. E., Mann, R., McGee, M. P., Horstman, D. H., Kozumbo, W. J., Becker, S., House, D. E., McDonnell, W. F., and Bromberg, P. A. (1989). Ozone-induced inflammation in the lower airways of human subjects. *Am. Rev. Respir. Dis.* **139**, 407–415.

Kosaka H., and Tyuma, I. (1987). Mechanism of autocatalytic oxidation of oxyhemoglobin by nitrite. *Environ. Health Perspect.* **73**, 147–151.

Kotin, P., and Falk, H. L. (1956). The experimental induction of pulmonary tumors in strain-A mice after their exposure to an atmosphere of ozonized gasoline. *Cancer (Philadelphia)* **9**, 910–917.

Kotin, P., Falk, H., and McCammon, C. J. (1958). The experimental induction of pulmonary tumors and changes in the respiratory epithelium in C57BL mice following their exposure to an atmosphere of ozonized gasoline. *Cancer (Philadelphia)* **11**, 473–481.

Kratzing, C. C., and Willis, R. J. (1980). Decreased levels of ascorbic acid in lung following exposure to ozone. *Chem.-Biol. Interact.* **30**, 53–56.

Kuraitis, K. V., Richters, A., and Sherwin, R. P. (1981). Spleen changes in animals inhaling ambient levels of nitrogen dioxide. *J. Toxicol. Environ. Health* **7**, 851–859.

Kurz, M. E., and Pryor, W. A. (1978). Radical production from the interaction of closed-shell molecules. 9. Reaction of ozone with *tert*,-butyl hydroperoxide. *J. Am. Chem. Soc.* **100**, 7953–7959.

Lambelet, P., Saucy, F., and Loliger, J. (1985). Chemical evidence for interactions between vitamins E and C. *Experientia* **41**, 1384–1388.

Last, J. A., and Greenberg, D. B. (1980). Ozone-induced alterations in collagen metabolism of rat lungs. II. Long-term exposure. *Toxicol. Appl. Pharmacol.* **55**, 108–114.

Last, J. A., and Kaizu, T. (1980). Mucous glycoprotein secretion by tracheal explants: effects of pollutants. *Environ. Health Perspect.* **35**, 131–138.

Last, J. A., Jennings, M. D., Schwartz, L. W., and Cross, C. E. (1977). Glycoprotein secretion by tracheal explants cultured from rats exposed to ozone. *Am. Rev. Respir. Dis.* **116**, 695–703.

Last, J. A., Greenberg, D. B., and Castleman, W. L. (1979). Ozone-induced alterations in collagen metabolism of rat lungs. *Toxicol. Appl. Pharmacol.* **51**, 247–258.

Last, J. A., Reiser, K. M., Tyler, W. S., and Rucker, R. B. (1984). Long-term consequences of exposure to ozone. I. Lung collagen content. *Toxicol. Appl. pharmacol.* **72**, 111–118.

Last, J. A., Warren, D. L., Pecquet-Goad, E., and Witschi, H. P. (1986). Modification of lung tumor development in mice by ozone. *JNCI, J. Natl. Cancer Inst.* **78**, 149–154.

Lee, J. S., Afifi, A. A., and Mustafa, M. G. (1989). Effects of short-term, single and combined exposure of rats to NO_2 and O_3 on lung tissue enzyme activities. *Inhalation Toxicol.* **1**, 21–35.

Lee, J. S., Mustafa, M. G., and Afifi, A. A. (1990). Effects of short-term, single and combined exposure to low level NO_2 and O_3 on lung tissue enzyme activities in rats. *J. Toxicol. Environ. Health* **29**, 293–305.

Lee, L. Y., Bleecker, E. R., and Nadel, J. A. (1977). Effects of ozone on bronchomotor response to inhaled histamine aerosols in dogs. *J. Appl. Physiol.: Respir., Environ. Exercise Physiol.* **43**, 626–631.

Lee, S. D. (Ed.) (1980). *Nitrogen Oxides and Their Effects on Health.* Ann Arbor Science Publishers, Ann Arbor, MI, pp. 1–382.

Lee, S. D. (Ed.) (1985). *Evaluation of the Scientific Basis for Ozone/Oxidants Standards.* Air Pollution Control Association, Pittsburgh.

Lee, S. D., Mustafa, M. G., Mehlman, M. A. (Eds.) (1983). *International Symposium on the Biomedical Effects of Ozone and Related Photochemical Oxidants.* Princeton Scientific, Princeton, NJ.

Leung, H. W. (1983). Effect of nitrogen dioxide exposure on rat lung lipids. *Res. Commun. Chem. Pathol. Pharmacol.* **40**, 519–552.

Leung, H. W., Wang, M. J., and Mavis, R. D. (1981). The cooperative interaction between vitamin E and vitamin C in suppression of peroxidation of membrane phospholipids. *Biochim. Biophys. Acta* **664**, 266–272.

Lindvall, T. (1985). Health effects of nitrogen dioxide and oxidants. *Scand. J. Work Environ. Health* **11**, Suppl. 3, 3–28.

Lippmann, M. (1989). Health effects of ozone—a critical review, *JAPCA* **39**, 672–695.

Logan, J. A., Prather, M. J., Wofsy, S. C., and McElroy, M. B. (1981). Tropospheric chemistry: A global perspective. *J. Geophys. Res.* **86**, 7210–7254.

Lowry, T., and Schuman, L. M. (1956). Silo fillers disease: A syndrome caused by nitrogen dioxide. *JAMA, J. Am. Med. Assoc.* **162**, 153–160.

Lunan, K. D., Short, P., Negi, D., and Stephens, R. J. (1977). Glucose-6-phosphate dehydrogenase response of postnatal lungs to NO_2 and O_3. In C. L. Sanders, R. P. Schneider, G. E. Dagle, and H. A. Ragan (Eds.), *Pulmonary Macrophages and Epithelial Cells,* Energy Research and Development Administration, Washington, DC, ERDA Symp. Ser. 43, pp. 236–247.

Machlin, L. J., and Bendich, A. (1987). Free radical tissue damage: Protective role of antioxidant nutrients. *FASEB J.* **1**, 441–445.

Madden, M. C., Eling, T. E., and Friedman, M. (1987). Ozone inhibits endothelial cell cyclooxigenase activity through formation of hydrogen peroxide. *Prostaglandins* **34**, 445–463.

Marz, T., Bender, M., Kerr, H., and Kulle, T. (1975). Observations of aberrations in chromosomes of lymphocytes from human subjects exposed to ozone at concentrations of 0.5 ppm for 6 and 10 hours. *Mutat. Res.* **31**, 299–302.

Matsumura, Y., Mizuno, K., Miyamoto, T., Suzuki, T., and Oshima, Y. (1972). The effects of ozone, nitrogen dioxide, and sulfur dioxide on the experimentally induced allergic respiratory disorder in guinea pigs. *Am. Rev. Respir. Dis.* **105**, 262–267.

Maugh, T. H. (1982). New link between ozone and cancer. *Science* **216**, 396–397.

Mautz, W. J., Kleinman, M. T., Phalen, R. F., and Crocker, T. T. (1988). Effects of exercise. I. Ixsosure on toxic interactions between inhaled oxidant and aldehyde air pollutants. *J. Toxicol. Environ. Health* **25**, 165–177.

McCay, P. B. (1985). Vitamin E: Interactions with free radicals and ascorbate. *Annu. Rev. Nutr.* **5**, 323–340.

Mead, J. F., Gan-Elepano, M., and Hirahara, F. (1980). Initiation of peroxidation by nitrogen dioxide in natural and modern membrane systems. In S. D. Lee (Ed.), *Nitrogen Oxides and Their Effects on Health.* Ann Arbor Science Publishers, Ann Arbor, MI, pp. 191–197.

Mehlman, M. A., and Borek, C. (1987). Toxicity and biochemical mechanisms of ozone. *Environ. Res.* **42**, 36–53.

Mellick, P. W., Dungworth, D. L., Schwartz, L. W., and Tyler, W. S. (1977). Short term morphologic effects of high ambient levels of ozone on lungs of rhesus monkeys. *Lab. Invest.* **36**, 82–90.

Melton, C. E. (1982). Effects of long-term exposure to low levels of ozone: A review. *Aviat. Space Environ. Med.* **53**, 105–111.

Menzel, D. B. (1970). Toxicity of ozone, oxygen and radiation. *Annu. Rev. Pharmacol.* **10**, 379–394.

Menzel, D. B. (1971). Oxidation of biologically active reducing substances by ozone. *Arch. Environ. Health* **23**, 149–153.

Menzel, D. B. (1976). The role of free radicals in the toxicity of air pollutants (nitrogen oxides and ozone). In W. A. Pryor (Ed.), *Free Radicals in Biology.* Academic Press, New York, Vol. 2, pp. 181–202.

Menzel, D. B. (1984). Ozone: An overview of its toxicity in man and animals. *J. Toxicol. Environ. Health* **13**, 183–204.

Menzel, D. B., Roehm, J. N., and Lee, S. D. (1972). Vitamin E: The biological and environmental antioxidant. *J. Agric. Food Chem.* **20**, 481–486.

Miller, F. J. (1979). Biomathematical modeling applications in the evaluation of ozone toxicity. In J. B. Mudd and S. D. Lee (Eds.), *Assessing Toxic Effects of Environmental Pollutants.* Ann Arbor Science Publishers, Ann Arbor, MI, pp. 263–286.

Miller, F. J., Graham, J. A., Illing, J. W., and Gardener, D. E. (1980). Extrapulmonary effects of NO_2 as reflected by pentobarbital-induced sleeping time in mice. *Toxicol. Lett.*

Morrow, P. E. (1975). An evaluation of recent NO_x toxicity data and an attempt to derive an ambient air standard for NO_x by established toxicological procedures. *Environ. Res.* **10**, 92–112.

Morrow, P. E. (1984). Toxicological data on NO_x : An overview. *J. Toxicol. Environ. Health* **13**, 205–227.

Mudd, J. B., and Freeman, B. A. (1977). Reaction of ozone with biological membranes. In S. D. Lee (Ed.), *Biochemical Effects of Environmental Pollutants,* Ann Arbor Science Publishers, Ann Arbor, MI, pp. 97–133.

Mustafa, M. G. (1975). Influence of dietary vitamin E on lung cellular sensitivity to ozone in rats. *Nutr. Rep. Int.* **11,** 473–476.

Mustafa, M. G. (1985). General enzymology of the lung. In H. P. Witschi and J. D. Brain (Eds.), *Toxicology of Inhaled Materials.* Springer-Verlag, Berlin, pp. 369–419.

Mustafa, M. G. (1990). Biochemical basis of toxicity. *Free Radical Biol. Med.* **9,** 245–265.

Mustafa, M. G., and Cross, C. E. (1974). Effects of short-term ozone exposure on lung mitochondrial oxidative and energy metabolism. *Arch. Biochem. Biophys.* **162,** 585–594.

Mustafa M. G., and Lee, S. D. (1976). Pulmonary biochemical alterations resulting from ozone exposure. *Ann. Occup. Hyg.* **19,** 17–26.

Mustafa, M. G., and Tierney, D. F. (1978). Biochemical and metabolic changes in the lung with oxygen, ozone, and nitrogen dioxide. *Am. Rev. Respir. Dis.* **118,** 1061–1090.

Mustafa, M. G., DeLucia, A. J., York, G. K., Arth, C., and Cross, C. E. (1973). Ozone interaction with rodent lung. II. Effects on oxygen consumption of mitochondria. *J. Lab. Clin. Med.* **82,** 357–365.

Mustafa, M. G., Hacker, D., Ospital, J. J., Hussain, M. Z., and Lee, S. D. (1977). Biochemical effects of environmental oxidant pollutants in animal lungs. In S. D. Lee (Ed.), *Biochemical Effects of Environmental Pollutants.* Ann Arbor Science Publishers, Ann Arbor, MI, pp. 59–96.

Mustafa, M. G., Elsayed, N., Lim, J. S. T., Postlethwait, E., and Lee, S. D. (1979). Effects of nitrogen dioxide on lung metabolism. In D. Grosjean (Ed.), *Nitrogenous Air Pollutants: Chemical and Biological Implications.* Ann Arbor Science Publishers, Ann Arbor, MI, pp. 165–178.

Mustafa, M. G., Faeder, E. J., and Lee, S. D. (1980a). Biochemical effect of nitrogen dioxide on animal lungs. In S. D. Lee (Ed.), *Nitrogen Dioxides and Their Effects on Health.* Ann Arbor Science Publishers, Ann Arbor, MI, pp. 161–179.

Mustafa, M. G., Faeder, E. J., and Lee, S. D. (1980b). Biochemical basis of pulmonary response to ozone and nitrogen dioxide injury. In R. S. Bhatnagar (Ed.), *Molecular Basis of Environmental Toxicity.* Ann Arbor Science Publishers, Ann Arbor, MI, pp. 151–172.

Mustafa, M. G., Elsayed, N. M., von Dohlen, F. M., Hasset, C. M., Postlethwait, E. M., Quinn, C. L., Graham, J. A., and Gardner, D. E. (1984). A comparison of biochemical effects of nitrogen dioxide, ozone, and their combination in mouse lung. I. Intermittent exposure. *Toxicol. Appl. Pharmacol.* **72,** 82–90.

Mustafa, M. G., Hassett, C. M., Newell, G. W., and Schrauzer, G. N. (1988). Pulmonary carcinogenic effects of ozone. In C. Maltoni and I. J. Selikoff (Eds.), *Living in a Chemical World: Occupational and Environmental Significance of Industrial Carcinogens.* N. Y. Acad. Sci., New York, pp. 714–723.

Myers, A. B., Dubick, M. A., Reiser, K. M., Gerriets, J. E., Last, J. A., and Rucker, R. B. (1984). Ozone exposure, food restriction and protein deficiency: Changes in collagen and elastin in rodent lung. *Toxicol. Lett.* **23,** 43–49.

Nakajima, T., Oda, H., Kusumoto, S., and Nogami, H. (1980). Biological effects of nitrogen dioxide and nitric oxide. In S. D. Lee (Ed.), *Nitrogen Oxides and Their Effects on Health.* Ann Arbor Science Publishers, Ann Arbor, MI, pp. 121–141.

National Academy of Sciences (NAS) (1977). *Nitrogen Oxides.* Committee on Medical and Biological Effects of Environmental Pollutants, NAS, Washington, DC.

National Academy of Sciences (NAS) (1980). *Recommended Dietary Allowances.* National Academy Press, Washington, DC, 9th ed., pp. 63–69.

National Academy of Sciences (NAS) (1984). *Causes and Effects of Changes in Stratospheric Ozone: Update 1983.* National Academy Press, Washington, DC.

National Research Council (NRC) (1977). *Ozone and Other Photochemical Oxidants,* Vol. 19. Committee on Medical and Biologic Effects of Environmental Pollutants, Subcommittee on Ozone and Other Photochemical Oxidants, National Academy of Sciences, Washington, DC.

Nettesheim, P., Hanna, M. G., Jr., Doherty, D. G., Newell, R. F., and Hellman, A. (1970). Effects of chronic exposure to artificial smog and chromium oxide dust on the incidence of lung tumors in mice. In M. G. Hanna, P. Nettesheim, and J. G. Gilbert (Eds.), *Inhalation Carcinogenesis*. USAEC Division of Technical Information Extension, Washington, DC, Symp. Ser. No. 18, pp. 305–317.

Nikula, K. J., Wilson, D. W., Giri, S. N., Plopper, C. G., and Dungworth, D. L. (1988). The response of the rat tracheal epithelium to ozone exposure: Injury adaptation and repair. *Am. J. Pathol.* **131**, 373–384.

Oda, H., Kusomoto, S., and Nakajima, T. (1975). Nitrosyl-hemoglobin formation in the blood of animals exposed to nitric oxide. *Arch. Environ. Health* **30**, 453–456.

Oda, H., Nogami, H., and Nakajima, T. (1980a). Reaction of hemoglobin with nitric oxide and nitrogen dioxide in mice. *J. Toxicol. Environ. Health* **6**, 673–678.

Oda, H., Nogami, H., Kusomoto, S., Nakajima, T., and Kurata, A. (1980b). Lifetime exposure to 2.4 ppm nitric oxide in mice. *Environ. Res.* **22**, 254–263.

Oomichi S., and Kita, H. (1974). Effect of air pollutants on ciliary activity of respiratory tract. *Bull. Tokyo Med. Dent. Univ.* **21**, 327–343.

Ospital, J. J., Hacker, A. D., and Mustafa, M. G. (1981). Biochemical changes in rat lungs after exposure to nitrogen dioxide. *J. Toxicol. Environ. Health* **8**, 47–58.

Packer, J. E., Slater, T. F., and Wilson, R. W. (1979). Direct observation of a free radical interaction between vitamin E and vitamin C. *Nature (London)* **278**, 737–738.

Palmer, M. S., Swanson, D. H., and Coffin, D. L. (1971). Effect of ozone on benzpyrene hydroxylase activity in the Syrian golden hamster. *Cancer Res.* **31**, 730–733.

Pitts, J. N., Jr. (1983). Formation and fate of gaseous and particulate mutagens and carcinogens in real and simulated atmospheres. *Environ. Health Perspect.* **47**, 115–140.

Platt, U. F., Winer, A. M., Blerman, H. W., Atkinson, R., and Pitts, J. N., Jr. (1984). Measurement of nitrate radical concentrations in continental air. *Environ. Sci. Technol.* **16**, 365–369.

Plopper, C. G., Chow, C. K., Dungworth, D. L., Brummer, M., and Nemeth, T. J. (1978). Effect of low level of ozone on rat lungs. II. Morphological responses during recovery and re-exposure. *Exp. Mol. Pathol.* **29**, 400–411.

Plopper, C. G., Chow, C. K., Dungworth, D. L., and Tyler, W. S. (1979). Pulmonary alterations in rats exposed to 0.2 and 0.1 ppm ozone: A correlated morphological and biochemical study. *Arch. Environ. Health* **34**, 390–395.

Postlethwait, E. M., and Mustafa, M. G. (1981). The fate of inhaled nitrogen dioxide in the isolated perfused rat lung. *J. Toxicol. Environ. Health* **7**, 861–872.

Postlethwait, E. M., and Mustafa, M. G. (1983). Formation of N-nitrosomopholine in isolated rat lungs during nitrogen dioxide ventilation. *Carcinogenesis (London)* **4**, 777–778.

Postlethwait, E. M., and Mustafa, M. G. (1989). Effect of altered dose rate on NO_2 uptake and transformation in isolated lungs. *J. Toxicol. Environ. Health* **26**, 497–507.

Pryor, W. A. (1981). Mechanisms and detection of pathology caused by free radicals. Tobacco smoke, nitrogen dioxide and ozone. In J. D. McKinney (Ed.), *Environmental Health Chemistry*. Ann Arbor Science Publishers, Ann Arbor, MI, pp. 445–466.

Pryor, W. A. (1991). Can vitamin E protect humans against the pathological effects of ozone in smog? *Am. J. Clin. Nutr.* **53**, 702–722.

Pryor, W. A., and Lightsey, J. W. (1981). Mechanisms of nitrogen dioxide reactions: Initiation of lipid peroxidation and production of nitrous acid. *Science* **214**, 435–437.

Pryor, W. A., Prier, D. G., and Church, D. F. (1981). Radical production from the interaction of ozone and PUFA as demonstrated by electron spin resonance spin-trapping techniques. *Environ. Res.* **24**, 42–52.

Pryor, W. A., Lightsey, J. W., and Church, D. F. (1982). Reaction of nitrogen dioxide with alkenes and polyunsaturated fatty acids: Addition and hydrogen abstraction mechanisms. *J. Am. Chem. Soc.* **104**, 6685–6692.

Pryor, W. A., Dooley, M. M., and Church, D. F. (1983). Mechanisms for the reaction of ozone with biological molecules, the source of the toxic effects of ozone. In S. D. Lee, M. G. Mustafa, and M. A. Mehlman (Eds.), *International Symposium on the Biomedical Effects of Ozone and Photochemical Oxidants.* Princeton Scientific, Princeton, NJ, pp. 7–19.

Ramazzotto, L., Jones, C. R., and Cornell, F. (1971). Effect of nitrogen dioxide on the activities of cytochrome oxidase and succinic dehydrogenase on homogenates of some organs of the rat. *Life Sci.* **10**, 601–604.

Rawley, D. A., and Halliwell, B. (1983). Formation of hydroxyl radicals from hydrogen peroxide and iron salts by superoxide- and ascorbate-dependent mechanisms: Relevance to the pathology of rheumatoid disease. *Clin. Sci.* **64**, 649–653.

Reasor, M. J., Adams, G. K., Brooks, J. K., and Rubin, R. J. (1979). Enrichment of albumin and IgG in the airway secretions of dogs breathing ozone. *J. Environ. Sci. Health* **C13**, 335–346.

Reed, D., Glaser, S., and Kaldor, J. (1980). Ozone toxicity symptoms among flight attendants. *Am. J. Ind. Med.* **1**, 43–54.

Richters, A., and Kuraitis, K. (1981). Inhalation of NO_2 and blood borne cancer cell spread to the lungs. *Arch. Environ. Health* **36**, 36–39.

Richters, A., and Richters, V. (1983). A new relationship between air pollutant inhalation and cancer. *Arch. Environ. Health* **38**, 69–75.

Roberts, H. J. (1981). Perspective on vitamin E as therapy. *JAMA, J. Am. Med. Assoc.* **246**, 129–131.

Roehm, J. N., Hadley, J. G., and Menzel, D. B. (1972). The influence of vitamin E on the lung fatty acids of rats exposed to ozone. *Arch. Environ. Health* **24**, 237–242.

Rowland, J. R., and Gause, E. M. (1971). Reaction of nitrogen dioxide with blood and lung components. *Arch. Intern. Med.* **128**, 94–100.

Sagai, M., and Ichinose, T. (1987). Lipid peroxidation and antioxidative protection mechanism in rat lungs upon acute and chronic exposure to nitrogen dioxide. *Environ. Health Perspect.* **73**, 179–189.

Sagai, M., Ichinose, T., Oda, H., and Kubota, K. (1981). Studies on biochemical effects of nitrogen dioxide: I. Lipid peroxidation as measured by ethane exhalation of rats exposed to nitrogen dioxide. *Lipids* **16**, 64–67.

Sagai, M., Ichinose, T., Oda, H., and Kubota, K. (1982). Studies on biochemical effects of nitrogen dioxide. II. Changes of the protective systems in rat lungs and of lipid peroxidation by acute exposure. *J. Toxicol. Environ. Health* **9**, 153–164.

Sagai, M., Ichinose, T., Oda, H., and Kubota, K. (1984). Studies on the biochemical effects of nitrogen dioxide. IV. Relation between the change of lipid peroxidation and the antioxidative protective system in rat lungs upon life span exposure to low levels of NO_2. *Toxicol. Appl. Pharmacol.* **73**, 444–456.

Sagai, M., Arakawa, K., Ichinose, T., and Shimojo, N. (1987). Biochemical effects of combined gases of nitrogen dioxide and ozone. I. Species differences of lipid peroxides and phospholipids in lungs. *Toxicology* **46**, 251–265.

Schneider, T., and Grant, L. (Eds.) (1982). *Air Pollution by Nitrogen Oxides.* Elsevier, Amsterdam.

Schwartz, W. L., Dungworth, D. L., Mustafa, M. G., Tarkington, B. K., and Tyler, W. S. (1976). Pulmonary response of rats to ambient levels of ozones: Effects of 7-day intermittent or continuous exposure. *Lab. Invest.* **34**, 565–578.

Selgrade, M. K., Mole, M. L., Miller, F. F., Hatch, G. E., Gardner, D. E., and Hu, P. C. (1981). Effect of NO_2 inhalation and vitamin C deficiency on protein and lipid accumulation in the lung. *Environ. Res.* **26**, 422–437.

Seltzer, J., Bigby, B. G., Stulbarg, M., Holtzman, M. J., Nadel, J. A., Ueki, I. F., Leikauf, G. D., Goetzel, E. J., and Boushey, H. A. (1986). O_3-induced change in bronchial reactivity to methacholine and airway inflammation in humans. *J. Appl. Physiol.* **60**, 1321–1326.

Sevanian, A., Mead, J. F., and Stein, R. A. (1979). Epoxide products of lipid peroxidation in rat lungs. *Lipids* **14**, 634–643.

Sevanian, A., Hacker, A. D., and Elsayed, N. (1982a). Influence of vitamin E and nitrogen dioxide on lipid peroxidation in rat lung and liver microsomes. *Lipids* **17**, 269–277.

Sevanian, A., Elsayed, N., and Hacker, A. D. (1982b). Effect of vitamin E deficiency and nitrogen dioxide exposure on lung lipid peroxidation: Use of lipid epoxides and malonaldehyde as measures of peroxidation. *J. Toxicol. Environ. Health* **10**, 743–756.

Sherwin, R. P., and Carlson, D. A. (1973). Protein content of lung lavage fluid of guinea pigs exposed to 0.4 ppm nitrogen dioxide. *Arch. Environ. Health* **27**, 90–93.

Sherwin, R. P., and Layfield, L. J. (1974). Proteinuria in guinea-pigs exposed to 0.5 ppm nitrogen dioxide. *Arch. Environ. Health* **28**, 336–341.

Sherwin, R. P., and Richters, V. (1971). Lung capillary permeability. Nitrogen dioxide exposure and leakage of tritiated serum. *Arch. Intern. Med.* **128**, 61–68.

Sherwin, R. P., Richters, V., Brooks, M., and Buckley, R. D. (1968). The phenomenon of macrophage congregation *in vitro* and its relationship to *in vivo* NO$_2$ exposure of guinea-pigs. *Lab. Invest.* **18**, 269–277.

Sherwin, R. P., Dibble, J., and Weiner, J. (1972). Alveolar wall cells of the guinea-pig. *Arch. Environ. Health* **24**, 43–47.

Sherwin, R. P., Richters, V., and Okimoto, D. (1983). Type 2 pneumocyte hyperplasia in the lungs of mice exposed to an ambient level (0.3 ppm) of ozone. In S. D. Lee, M. G. Mustafa, and M. A. Mehlman (Eds.), *International Symposium on the Biomedical Effects of Ozone and Related Photochemical Oxidants.* Princeton Scientific, Princeton, NJ, pp. 289–297.

Shimasaki, H. T., Takatori, W. R., Anderson, H. L., Shimasaki, H., Takatori, T., Anderson, W. R., Horten, H. L., and Privett, O. S. (1976). Alteration of lung lipids in ozone exposed rats. *Biochem. Biophys. Res. Commun.* **68**, 1256–1262.

Slater, T. F. (1984). Free radical mechanisms of tissue injury. *Biochem. J.* **222**, 1–15.

South Coast Air Quality Management District (SCAQMD) (1980). *Seasonal and Diurnal Variation in Air Quality in California's South Coast Air Basin.* SCAQMD, El Monte, CA.

South Coast Air Quality Management District (SCAQMD) (1983). *1982 Summary of Air Quality in California's South Coast Air Basin.* SCAQMD, El Monte, CA.

Stephens, R. J., Freeman, G., and Evans, M. J. (1971). Ultrastructural changes in connective tissue in lungs of rats exposed to NO$_2$ *Arch. Intern. Med.* **127**, 873–883.

Stephens, R. J., Freeman, G., and Evans, M. J. (1972). Early response of lungs to low levels of nitrogen dioxide--light and electron microscopy. *Arch. Environ. Health* **24**, 150–179.

Stephens, R. J., Sloan, M. F., Evans, M. J., and Freeman, G. (1974a). Early response of lung to low levels of ozone. *Am. J. Pathol.* **73**, 711–726.

Stephens, R. J., Sloan, M. F., Evans, M. J., and Freeman, G. (1974b). Alveolar type 1 cell response to exposure to 0.5 ppm O$_3$ for short periods. *Exp. Mol. Pathol.* **20**, 11–23.

Tappel, A. L. (1968). Will antioxidant nutrients slow aging process? *Geriatrics* **23**, 97–105.

Tepper, J. L., Weiss, B., and Cox, C. (1982). Microanalysis of ozone depression of motor activity. *Toxicol. Appl. Pharmacol.* **64**, 317–326.

Tepper, J. L., Weiss, B., and Wood, R. W. (1983). Behavioral indices of ozone exposure. In S. D. Lee, M. G. Mustafa, and M. G. Mehlman (Eds.), *International Symposium on the Biomedical Effects of Ozone Related Photochemical Oxidants.* Princeton Scientific, Princeton, NJ, pp. 515–526.

Thomas, H. V., Mueller, P. K., and Lyman, R. L. (1968). Lipoperoxidation of lung lipids in rats exposed to nitrogen dioxide. *Science* **159**, 532–553.

Thomas, T., and Rhoades, R. A. (1970). C-1 palmitate incorporation by rat lung: Effect of nitrogen dioxide. *Proc. Soc. Exp. Biol. Med.* **108**, 1181–1183.

Tice, R. R., Bender, M. A., Ivett, J. L., and Drew, R. T. (1978). Cytogenetic effects of inhaled ozone. *Mutat. Res.* **58**, 293–304.

Tierney, D. F. (1974). Intermediary metabolism of the lung. *Fed. Proc., Fed. Am. Soc. Exp. Biol.* **33**, 2232–2237.

Toothill, C. (1967). The chemistry of the *in vivo* reaction between hemoglobin and various oxides of nitrogen. *Br. J. Anaesth.* **39**, 405–412.

Trams, E. G., Lauter, C. J., Brown, E. A. B., and Young, O. (1972). Cerebral cortical metabolism after chronic exposure to ozone. *Arch. Environ. Health* **24**, 153–159.

Trzeciak, H. I., Kosimder, S., Kryk, K., and Kryk, A. (1977). The effects of nitrogen oxides and their neutralization products with ammonia on the lung phospholipids of guinea pigs. *Environ. Res.* **14**, 87–91.

Tsuda, H., Kushi, A., Yoshida, D., and Goto, F. (1981). Chromosomal aberrations and sister-chromatid exchanges induced by gaseous nitrogen dioxide in cultured Chinese hamster cells. *Mutat. Res.* **89**, 303–309.

Tyson, C. A., Lunan, K. D., and Stephens, R. J. (1982). Age-related differences in GSH-shuttle enzymes in NO_2 or O_3 exposed rat lungs. *Arch. Environ. Health* **37**, 167–176.

van Stee, E. W., Sloane, R. A., Simmons, J. E., and Baunnemann, K. D. (1983). In vivo formation of N-nitrosomorpholine in CD-1 mice exposed by inhalation to nitrogen dioxide and by gavage to morpholine. *JNCI, J. Natl. Cancer Inst.* **70**, 375–379.

von Nieding, G. (1978). Possible mutagenic properties and carcinogenic action of the irritant gaseous pollutant NO_2, O_3 and SO_2. *Environ. Health Perspect.* **22**, 91–92.

von Nieding, G., and Wanger, H. M. (1979). Effects of NO_2 on chronic bronchitis. *Environ. Health Perspect.* **29**: 137–142.

von Nieding, G., Wagner, H. M., Krekeler, H., Lollgen, H., Fries, W., and Beuthan, A. (1979). Controlled studies of human exposure to single and combined action of NO_2, O_3, and SO_2. *Int. Arch. Occup. Environ. Health* **43**, 195–210.

Wagner, W. D., Duncan, B. R., Wright, P. G., and Stokinger, H. E. (1965). Experimental study of threshold limit of NO_2. *Arch. Environ. Health* **10**, 455–466.

Walton, J. R., and Packer, L. (1980). Free radical damage and protection: Relationship to cellular aging and cancer. In L. J. Machlin (Ed.), *Vitamin E: A Comprehensive Treatise*. Dekker, New York, pp. 495–517.

Weiss, B., Ferin, J., Merigan, W., Stern, S., and Cox, C. (1981). Modification of rat operant behavior by ozone exposure. *Toxicol. Appl. Pharmacol.* **58**, 244–251.

Werthamer, S., Schwartz, L. H., and Soskind, L. (1970). Bronchial epithelial alterations and pulmonary neoplasia induced by ozone. *Pathol. Microbiol.* **35**, 224–230.

Whitten, G. Z. (1983). The chemistry of smog formation: A review of current knowledge. *Environ. Int.* **9**, 447–463.

Williams, S. J., Charles J. M., and Menzel, D. B. (1980). Ozone induced alterations in phenol red absorption from the rat lung. *Toxicol. Lett.* **6**, 213–219.

Willis, R. J., and Kratzing, C. C. (1974). Ascorbic acid in rat lung. *Biochem. Biophys. Res. Commun.* **59**, 1250–1253.

Willis, R. J., and Kratzing, C. C. (1976). Extracellular ascorbic acid in lung. *Biochim. Biophys. Acta* **144**, 108–117.

Wilson, R. L. (1983). Free radical protection: Why vitamin E, not vitamin C, β-carotene or gluta-thione? *Ciba Found. Symp.* **101**, 19–44.

Witschi, H. P. (1988). Ozone, nitrogen dioxide and lung cancer: A review of some recent issues and problems. *Toxicology* **48**, 1–20.

Witting, L. A. (1980). Vitamin E and lipid antioxidants in free-radical initiated reactions. In W. A. Pryor (Ed.), *Free Radicals in Biology*. Academic Press, New York, Vol. 4, pp. 295–319.

Witz, G., Amoruso, M. A., and Goldstein, B. D. (1983). Effect of ozone on alveolar macrophage function: Membrane dynamic properties. In S. D. Lee, M. G. Mustafa, and M. A. Mehlman (Eds.), *International Symposium on the Biomedical Effects of Ozone and Related Photochemical Oxidants*. Princeton Scientific, Princeton, NJ, pp. 263–272.

World Health Organization (WHO) (1977). *Environmental Health Criteria 4: Oxides of Nitrogen.* United Nations Environment Programme and WHO, Geneva, Switzerland.

World Health Organization (WHO) (1979). *Photochemical Oxidants: Environmental Health Criteria 7.* WHO, Geneva, Switzerland.

Wright, E. S., Dziedzic, D., and Wheeler, C. S. (1990). Cellular, biochemical and functional effects of ozone: New research and perspectives on ozone health effects. *Toxicol. Lett.* **51**, 125–145.

Yokoyama, E., and Frank, R. (1972). Respiratory uptake of ozone in dogs. *Arch. Environ. Health* **25**, 132–138.

Yokoyama, E., Ichikawa, I., and Kawai, K. (1980). Does nitrogen dioxide modify the respiratory effects of ozone? In S. D. Lee (Ed.), *Nitrogen Oxides and Their Effects on Health.* Ann Arbor Science Publishers, Ann Arbor, MI, pp. 217–229.

Yuen, T. G. H., and Sherwin, R. P. (1971). Hyperplasia of Type 2 pneumocytes and nitrogen dioxide (10 ppm) exposure. *Arch. Environ. Health* **22**, 178–188.

Zelac, R. E., Cromroy, H. L., Bolch, W. E., Jr., Dunavant, B. G., and Bevis, H. A. (1971a). Inhaled ozone as a mutagen. I. Chromosome aberrations induced in Chinese hamster lymphocytes. *Environ. Res.* **4**, 262–282.

Zelac, R. E., Cromroy, H. L., Bolch, W. E., Jr., Dunavant, B. G., and Bevis, H. A. (1971b). Inhaled ozone as a mutagen. II. Effect on the frequency of chromosome aberrations observed in irradiated Chinese hamsters. *Environ. Res.* **4**, 325–342.

12

SOME HEMATOLOGICAL EFFECTS OF OXIDANTS

Oguz K. Baskurt

Department of Physiology, Akdeniz University, Medical Faculty, 07070 Antalya, Turkey

Sema Yavuzer

Department of Physiology, Ankara University, Medical Faculty, Ankara, Turkey

Environmental Oxidants, Edited by Jerome O. Nriagu and Milagros S. Simmons.
ISBN 0–471–57928–9 © 1994 John Wiley & Sons, Inc.

1. INTRODUCTION

Oxygen is the major mediator of energy release from organic molecules, which are the source of energy for most of the organisms living on the earth; for this reason it is essential for life. However, excess molecular oxygen and its metabolic products can be highly toxic. All aeorobic cells generate a certain amount of oxygen free radicals such as superoxide anion and hydroxyl radical. These oxygen free radicals can contribute to cellular damage.

On the other hand, the antioxidant defense mechanisms of most cells prevent these species from causing injury. However, oxidative cell injury may occur if the rate of formation of free radicals is increased and/or the antioxidant defense is impaired. The balance between oxidant attack and antioxidant defense can be disturbed under a variety of pathologic conditions or under the influence of certain chemicals.

Although all living cells can be oxidatively attacked, blood tissue should have a unique place in the discussion of the biological actions of oxidants because it is the oxygen supply system in higher animals. This chapter focuses on the effects of oxidative processes on red blood cells.

2. BASIC MECHANISMS

Formation of free radicals is the key step in oxidative damage. On the one hand generated free radicals attack cell components, while on the other they trigger several reactions resulting in the generation of new radicals in a positive feedback. Free radical reactions can be described in three steps as initiation, propagation, and termination (Pryor, 1976).

2.1. Initiation

A molecule becomes a free radical by gaining or losing electron(s), resulting in unpaired electron(s) in the outer orbitals. In general, this reaction requires high energy transfer and is unlikely to occur except under special conditions. Most biologically important free radical reactions involve the reduction of molecular oxygen, resulting in the generation of reactive oxygen species. Actually, molecular oxygen is itself a biradical, with one unpaired electron in each of its two outer orbitals (Southern and Powis, 1988 a). One might expect that this biradical would be highly reactive in removing electrons from a nonradical molecule, generating a new radical. However, this reactivity is reduced because of the spinning of these electrons in the same direction. An electron spin inversion must occur or the molecule must acquire one electron at a time to take part in an oxidative reaction. Transition metals such as iron and copper facilitate the transfer of single electrons to molecular oxygen (Hill, 1981). Transfer of one electron to oxygen yields the superoxide anion radical.

$$O_2 + X \longrightarrow X^+ + O_2^{\cdot -}$$

The electron source might be a biological molecule such as hemoglobin, cytochrome, quinone, or thiol or a redox metal such as iron (Saltman, 1989). The superoxide anion radical has a mild reactivity (Southern and Powis, 1988a). It is dismutated to hydrogen peroxide

$$2O_2^{\cdot -} + 2H^+ \longrightarrow O_2 + H_2O_2$$

and this reaction is catalyzed by superoxide dismutase (SOD). Hydrogen peroxide also is neither a reactive oxidizing agent nor a free radical. However, in the presence of redox metals such as iron it rapidly generates the highly reactive hydroxyl radical (Fenton reaction):

$$H_2O_2 + Fe^{2+} \longrightarrow OH\cdot + OH^- + Fe^{3+}$$

Hydrogen peroxide may also react with a superoxide anion to yield hydroxyl radical (Haber–Weiss reaction):

$$H_2O_2 + O_2^{\cdot -} \longrightarrow OH\cdot + OH^- + O_2$$

This reaction is also catalyzed by iron.

2.2. Propagation

Free radicals are highly reactive and may attack a great variety of biological materials. When a free radical reacts with a nonradical molecule, other free radicals are formed and chain reactions are induced, such as the chain propagation of lipid peroxidation in biological membranes. Although the initial free radicals may only have local effects, products of the chain reactions can have distant degenerating effects.

2.3. Termination

Free radical chain reactions are terminated when two radicals react and the electrons in the outer orbitals are paired. Alternatively, a special molecule called a free radical scavenger reacts with the radical, transfers to it a hydrogen atom, and itself becomes a stable radical. Although the product of this reaction is a new radical, it is poorly reactive and so the chain reaction is terminated.

3.   OXIDATIVE STRESS IN THE RED BLOOD CELL

Erythrocytes are among the most susceptible cells to oxidative damage. This high susceptibility is the consequence of their unique features, in turn closely related to their basic physiological function. They are exposed to high oxygen tension, they are rich in polyunsaturated fatty acids, and they contain a large amount of iron. This combination provides a very favorable medium for biological oxidation processes mostly based on oxygen free radicals (Chiu et al., 1989).

3.1.   Generation of Free Radicals—Attack

In the highly catalytic medium of the red blood cell, oxygen free radicals are continuously generated; in addition, there are free radicals in the extracellular environment of the red blood cell. Hemoglobin oxidation and lipid peroxidation together play the major role in initiating and propagating the oxidative attack in erythrocytes.

3.1.1.   Hemoglobin Oxidation

Hemoglobin, the major constituent of erythrocytes, has a dual role in oxidative processes in these cells. Besides its catalytical action on lipid peroxidation, hemoglobin is itself among the components most susceptible to oxidation. This susceptibility seems to be closely related to the physiological function of the molecule, since a special mode of oxidation is its basic activity.

In the deoxy form, heme iron is in the high-spin ferrous state, having six electrons in the outer shell of which four are unpaired. When the molecule binds oxygen, one of the unpaired electrons is partially transfered to the oxygen molecule, leaving the iron in the ferric, low-spin state. The oxygen molecule itself becomes a bound superoxide anion

$$Hb(Fe^{2+}) + O_2 \longrightarrow Hb(Fe^{3+})O_2^-$$

(Wintrobe et al., 1981a). When the oxygen is given up by hemoglobin, the shared electron generally remains with the heme iron, transforming it into the high-spin ferrous state again. However, it is also possible for the electron to remain bound to the oxygen, this time yielding a free superoxide anion radical and a heme moiety with a ferric iron:

$$Hb(Fe^{3+})O_2^- + H_2O \longrightarrow Hb(Fe^{3+})H_2O + O_2^{\cdot-}$$

This hemoglobin product is known as methemoglobin (Wintrobe et al., 1981a).

Methemoglobin formation is likely to occur if water or other anions gain access to the heme. This type of methemoglobin formation is important for the cellular oxidant balance, as it introduces free radicals into the erythrocyte internal milieu. Alternatively, methemoglobin may also be formed by the action of superoxide anion on oxyhemoglobin

$$Hb(Fe^{3+})O_2^- + O_2^{\cdot -} + H_2O \longrightarrow Hb(Fe^{3+})H_2O + O_2 + O_2^{\cdot -}$$

and H_2O_2 on deoxyhemoglobin

$$2Hb(Fe^{2+}) + 2H_2O_2 \longrightarrow 2Hb(Fe^{3+})H_2O + O_2$$

thus generating no new free radicals (Wintrobe et al., 1981a).

Methemoglobin is nonfunctional since it cannot bind oxygen. It can be reduced to the functional hemoglobin mostly by enzymatic mechanisms. However, if the oxidative denaturation proceeds, irreversible hemichromes may be formed. Sulfhemoglobin is another oxidative hemoglobin product. It is formed by the incorporation of a sulfur atom into a hemoglobin peroxide (Wintrobe et al., 1981a). Sulfhemoglobin formation is also irreversible.

3.1.2. Lipid Peroxidation

Polyunsaturated fatty acids occupy an important place in the composition of red blood cells, making them highly susceptible to oxidation. Lipid peroxidation is initiated by free radical species attacking and removing a hydrogen atom from a methylene group usually adjacent to a double bond, leaving an unpaired electron on the carbon atom (Saltman, 1989). After a molecular rearrangement, an oxygen molecule is added to this carbon radical and a peroxy radical is formed. This radical may act on other lipid molecules, removing hydrogen atoms and further generating other lipid radicals and lipid hydroperoxides, thus starting the chain reaction. The lipid hydroperoxide is stable unless it comes into contact with transition metals. In the presence of transition metals it produces more radicals by homolytic decomposition. Homolytic decomposition is initiated by the breakage of the O—O bond in *tert*-butyl hydroperoxide; however, the energy needed for this reaction is very high (Chiu et al., 1989). Considering this high energy requirement, it seems clear that the homolytic decomposition of polyunsaturated fatty acids would never start in vivo without a catalyst. Transition metals are known to catalyze this bond breakage. Hemoglobin, the major constituent of the red blood cell cytoplasm, is a powerful catalyst for the initiation and propagation of lipid peroxidation (Chiu et al., 1989). These oxidative lipid products may also yield cyclic peroxides, which are further oxidized to malonyldialdehyde (MDA) (Saltman, 1989). These reactions are propagated until the termination step predominates.

It is obvious from the preceding discussion that these reactions are highly dependent on the catalytic effects of certain species such as transition metals, including iron and copper. These transition metals exert their effect even if they are bound to proteins or to low-molecular weight compounds.

3.1.3.   Overall Oxidative Threat to the Red Blood Cell

In addition to oxygen itself, several compounds exist that are readily active in generating free radicals in erythrocytes. This results in the exposure of red cell components to a continuous free radical attack. This continuous oxidant attack is evidenced by spontaneous formation of methemoglobin at a rate of 0.5–3% per day (Hsieh and Jaffe, 1975) and by the detection of lipid peroxidation products in normal erythrocytes under physiological in vivo conditions (Dikmenoglu et al., 1991). This oxidant threat is limited by the antioxidant defense and specific repair mechanisms.

Free radicals can react with all kinds of biomolecules as a consequence of their extremely high reactivity. In red blood cells, both cytoplasmic and membrane proteins, as well as lipid components, are readily susceptible to free radical damage (Chiu et al., 1989; Pacifici et al., 1988; Tappel, 1973). The protein components of the red cell attacked by free radicals include enzymes and structural proteins. Erythrocyte membrane skeletal proteins are also affected by oxidant stress; of these, spectrin is the most prone to damage as a consequence of lipid peroxidation in the cellular membrane (Chiu et al., 1989). The alterations in lipid and protein structure may have important functional consequences.

3.2.   Antioxidant Defense

It is not surprising that the red cell has powerful antioxidant defense mechanisms considering that it is under a continuous oxidant threat. The red cell contains both enzymatic and nonenzymatic antioxidants. It is also equipped with specific repair mechanisms that act after the oxidant damage occurs.

3.2.1.   Nonenzymatic Defense—Chemical Antioxidants

Chemical antioxidants prevent extensive oxidant damage by either preventing the generation of free radicals or inactivating the formed free radicals (Simic and Taylor, 1988). Antioxidants that prevent free radical generation inhibit the initiation of free radical processes; for example, metal complexing agents prevent the Haber–Weiss reaction. On the other hand, a group of chemicals called free radical scavengers react with free radicals, transforming them into nonradical form and themselves becoming stable radicals

$$R\cdot + S \text{ (scavenger)} \longrightarrow RH + S\cdot \text{ (stable radical)}$$

The new stable radical is poorly reactive compared to the original free radical (mostly an oxy radical), and so the chain reaction in the propagation step of the free radical process is terminated (Chiu et al., 1989).

The most important biological free radical scavenger in the red blood cell is vitamin E. Vitamin C regenerates vitamin E from its radical formed in the reaction with a free radical. In turn, the vitamin C radical formed in the regeneration reaction is reduced back to vitamin C enzymatically, depending on NADH (nicotinamide adenine dinucleotide, reduced).

3.2.2. Enzymatic Defense—Antioxidant Enzymes

The red blood cell contains specific enzymes to inactivate various free radicals or compounds that have the potential to generate free radicals (i.e., hydrogen peroxide).

The superoxide anion radical is dismutated by the enzyme superoxide dismutase. Erythrocyte superoxide dismutase is a copper- and zinc-containing enzyme that increases the dismutation rate of superoxide anion radical by up to 10^4 (McCord and Fridovich, 1969; Southorn and Powis, 1988a). The metal site of the enzyme is first reduced by one superoxide anion radical

$$\text{SOD-Me}^{2+} + O^{\cdot-}_2 \longrightarrow \text{SOD-Me}^+ + O_2$$

then reoxidized by a second superoxide anion

$$\text{SOD-Me}^+ + O^{\cdot-}_2 + 2H^+ \longrightarrow \text{SOD-Me}^{2+} + H_2O_2$$

yielding a molecular oxygen in the first and a hydrogen peroxide molecule in the second step (Southorn and Powis, 1988a).

Although hydrogen peroxide itself is not a free radical, because it has the potential to generate the highly reactive $OH\cdot$ radical it is an active component of the oxidative process in the red blood cell. At low concentrations, hydrogen peroxide is inactivated by the enzyme glutathione peroxidase, using reduced glutathione as the cofactor and yielding oxidized glutathione and water:

$$2GSH + H_2O_2 \longrightarrow GSSG + 2H_2O$$

At high concentrations, catalase is the important enzyme in removing hydrogen peroxide

$$2H_2O_2 \longrightarrow O_2 + 2H_2O$$

(Southorn and Powis, 1988a).

Glutathione peroxidase is also active in the reduction of lipid peroxides by glutathione, thus preventing the propagation of lipid peroxidation reactions.

3.2.3. Repair Mechanisms

Although the red blood cell has efficient protective mechanisms against oxidative attack, some biomolecules can be affected. These include both lipids and proteins, and the proper function of the red blood cell is strongly dependent on the structural and/or biochemical integrity of these biomolecules.

It is important to remember that the mature red blood cell is not capable of synthesizing proteins and lipids. Membrane lipids can be renewed by exchange with the plasma lipid pool. However, the red blood cell does not have mechanisms for protein renewal. It has been proposed that several specific mechanisms exist for the prevention of permanent damage, including glutathione-mediated sulfhydryl group reductions (Chiu et al., 1989). However, oxidized proteins may be removed by proteolysis (Davies, 1988; Pacifici et al., 1988) and the protein content of the cell can be found to be decreased due to oxidant stress (Goldberg and Boches, 1982).

On the other hand, in addition to the direct transfer of several lipid units from plasma, certain other lipid types can be generated by the action of endogenous phospholipases and other enzymes within the erythrocyte membrane. Endogenous phospholipases also act to remove oxidized phospholipids from the membrane (Chiu et al., 1989).

Lipid repair mechanisms may be stimulated under oxidative stress by a poorly understood mechanism (Chiu et al., 1989). However, oxidant stress can also interfere with these mechanisms since the protein enzymes taking part are also prone to oxidant damage.

3.3. Balance

It is obvious that the functional state of the red blood cell is strongly dependent on the balance between attacking oxidants (free radicals) and defense mechanisms. Damage to cellular structure and functional impairment may occur when the production of free radicals is increased or defense mechanisms are deficient.

Glucose-6-phosphate dehydrogenase (G6PD) deficiency is a typical reason for impaired antioxidant defense (Saltman, 1989). The primary pathway for the generation of nicotinamide adenine dinucleotide phosphate (NADPH), which is necessary to keep glutathione (GSH) in the reduced state, is the pentose phosphate pathway initiated by G6PD (Wintrobe et al., 1981a). GSH is used by the glutathione peroxidase enzyme in the breakdown of hydrogen peroxide to water. Additionally, catalase uses NADPH as a cofactor (Kirkman et al., 1987). Therefore, if G6PD is deficient in erythrocytes, decomposition of hydrogen peroxide is reduced and severe biochemical and morphological alterations are observed in addition to anemia. G6PD deficiency provides strong evidence for continuous oxidant attack and the importance of red blood cell antioxidant defense mechanisms.

Except in cases of genetic enzymatic deficiency, oxidant defense attempts to keep pace with oxidant attack despite changes in the extent of oxidant stress, thus maintaining the precise balance. Erythrocyte antioxidant enzyme activities are found to be enhanced if free radical generation is increased for a prolonged period (Baskurt et al., 1994; Medeiros et al., 1983).

3.3.1. Threat to Balance

In addition to the endogenous generation of free radicals in the red blood cell, exposure to oxidants comes from other sources as well. The composition of the microenvironment of the red blood cell is not constant. It changes as the cell travels through different parts of the circulatory system. This microenvironment may contain oxidants and free radicals, especially under certain circumstances. Several chemical species with oxidant activity are generated during metabolic processes. During oxidative phosphorylation, oxygen is reduced to water and oxygen free radical intermediates are formed. As these intermediates are tightly bound to active sites of the enzyme systems, they are not active in inducing oxidant stress (Southorn and Powis, 1988a). However, during certain metabolic processes (e.g., prostaglandin metabolism), unbound free radicals can be formed (Simic and Taylor, 1988).

Granulocytes and macrophages generate superoxide anion and hydrogen peroxide as a part of their physiological function (Nare, this volume). These oxidizing agents can diffuse into the extracellular space and interact with erythrocytes nearby (Claster et al., 1984; Weiss, 1980). Obviously, this interaction between white and red blood cells is enhanced if the white cells are activated.

Endogenous oxidizing species are also generated in several tissues during ischemia and reperfusion periods (Southorn and Powis, 1988b). During an ischemic period, oxygen free radicals may be released from the disrupted mitochondrial electron transport chain or by the oxidation of biomolecules including catecholamines and cellular metabolites. Antioxidant enzyme activities in hypoxic tissues may also be suppressed, favoring an increase in local oxidant stress (Meerson et al., 1982).

However, the most important contribution of ischemia to overall oxidant stress is produced when reperfusion occurs. A burst of free radical production results from the reperfusion of ischemic tissues (Fig. 1). During ischemia, the enzyme xanthine dehydrogenase is converted to xanthine oxidase by a protease activated by increased intracellular free calcium. Xanthine oxidase can transfer electrons from hypoxanthine to oxygen and form superoxide anion radical. Hypoxanthine is produced during ischemia by the catabolism of adenosine triphosphate. The second substrate, oxygen, is supplied to the tissue during reperfusion. Therefore, the result of reperfusion after an ischemic period is production of the superoxide anion (and hydrogen peroxide) in large amounts. This oxidant species can easily affect red blood cells in the blood, reperfusing the ischemic region.

In addition to this endogenous production of oxidant species, a large family of drugs and other chemicals can induce oxidant stress in biological tissues (Ziegler, 1988). A large number of antineoplastic drugs can form free radicals, and it is believed that these free radicals may be involved in the cytotoxic action of the drug (Southorn and Powis, 1988b). Several drugs such as phenylhydrazine induce

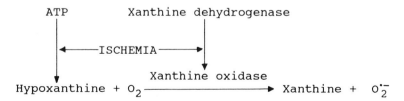

Figure 1. Generation of superoxide anion radical ($O_2^{\cdot-}$) during reperfusion after ischemia.

anemia by oxidative mechanisms (Goldberg et al., 1976). Drugs with quinone moieties also directly react with hemoglobin, causing oxidation and denaturation (Saltman, 1989). Occupational and environmental exposure to certain chemicals (pollutants) may also influence oxidative processes in several tissues, including blood (Constance and Nagel, 1984; Medeiros et al., 1983; Tappel, 1973).

The red blood cell defends itself against oxidant stress using well-developed antioxidant defense mechanisms that allow it to remain functional throughout its lifetime. However, if the capacity of these antioxidant mechanisms is exceeded by increased endogenous and/or exogenous oxidant attack, the cell can no longer be kept in the optimal functional state. Potential damage to the cell covers a wide spectrum from slight morphological and functional alterations to total destruction.

3.4. Oxidative Damage in the Red Blood Cell

Under the influence of oxidant stress, both the morphology and the function of the red blood cell are altered. Oxidatively stressed erythrocytes have a strikingly altered appearance resembling certain types of hemoglobinopathies (Saltman, 1989). Irregular and spiculated, the cells generally do not possess the typical biconcave discoid shape. The molecular mechanisms of these morphological and related functional alterations have begun to be understood in recent years.

3.4.1. Molecular Mechanisms

All lipid and protein components of the red blood cell are affected by oxidant stress. These components may be either related to the membrane or cytoplasmic.

Hemoglobin is the most important cytoplasmic component of the red blood cell and it plays an important role in overall oxidative processes, as discussed above. Other cytoplasmic proteins, especially enzymes, are also subjected to oxidant damage and resulting functional impairment. Protein sulfhydryl groups are extremely susceptible to free radical attack. Proteins and enzymes undergo polymerization, polypeptide chain scission, and chemical changes in amino acids when attacked by free radicals (Tappel, 1973). These changes can also be mediated by lipid peroxidation products.

The alterations in membrane structure due to oxidant attack warrant further discussion. The erythrocyte membrane is an unique structure mainly made up of a liquid bilayer supported by a special protein network, called the erythrocyte membrane skeleton. This protein network is responsible for the structural integrity of the membrane as well as for maintaining the normal biconcave discoid shape of the red cell (Mohandas et al., 1983; Shohet et al., 1981). Additionally, the erythrocyte membrane skeleton is believed to play the major role in regulating the rheologic properties of the red cell (Chasis and Shohet, 1987; Chien and Sung, 1990; Mohandas et al., 1983; Takakuwa et al., 1990). The main component of the membrane skeleton is the protein spectrin, which is made up of alpha and beta subunits and forms a network at the cytoplasmic side of the lipid bilayer (Chasis and Shohet, 1987; Mohandas et al., 1983; Shiga et al., 1990).

Several other proteins are incorporated into this main network, such as actin, tropomyosin, band 4 proteins, and ankyrin. The network is attached to the membrane-integral proteins by means of special interactions.

Oxidative damage in the membrane generally starts in the phospholipids but may readily spread to the membrane proteins, including the erythrocyte membrane skeleton. Spectrin sulfhydryl groups are positioned very close to the polyunsaturated fatty acids in aminophospholipids, rendering them particularly susceptible to damage (Chiu et al., 1989). As a result, spectrin units may become cross-linked by disulfide bonds (Haest et al., 1978; Palek and Liu, 1979; Wali et al., 1987). Products of lipid peroxidation, such as malonyldialdehyde, can also cross-link the membrane components containing amino groups (Chiu et al., 1989; Jain et al., 1983).

Such a cross-linking is not limited to membrane structures. Hemoglobin is also shown to be cross-linked with spectrin as a consequence of oxidant attack (McKenney et al., 1990; Snyder et al., 1985). The formation of hemoglobin–spectrin complexes seems to be independent of lipid peroxidation (Snyder et al., 1985).

Another alteration observed under oxidant stress that is not directly related to lipid peroxidation is increased cation permeability (van der Zee et al., 1985). Increased passive cation (mostly potassium) permeability is likely to be caused by the oxidation of membrane SH groups. Altered water permeability accompanies the potassium leakage and results in red cell dehydration (Jain et al., 1983). On the other hand, lipid peroxidation can damage the membrane structure, forming pores permeable even to hemoglobin (Chiu et al., 1989).

Snyder et al. (1985) reported that hydrogen peroxide treatment altered the lateral organization of membrane phospholipids, although the transbilayer distribution was unaltered. Furthermore, Jain (1985) observed that lipid peroxidation caused externalization of phosphatidylserine, exposing it on the outer surface of the membrane.

3.4.2. *Functional Aspects*

Obviously, these biochemical and structural alterations have important functional consequences. Hemoglobin, the protein on which the physiological mission of the red blood cell depends, can easily be rendered nonfunctional by oxidant attack. However, this alteration is mostly reversible by special mechanisms, and the nonfunctional hemoglobin does not accumulate in concentrations high enough to totally disrupt oxygen carrying in normal cells. On the other hand, the alterations related to the membrane structures are more important from a functional aspect. These alterations may even lead to destruction of erythrocytes.

Hydrogen peroxide treatment of red cell suspensions is a widely used model for investigating functional alterations due to oxidant stress in erythrocytes (Mino et al., 1978; Stocks and Dormandy, 1971). Incubation of erythrocytes with hydrogen peroxide results in hemolysis, and the degree of hemolysis in hydrogen peroxide-containing suspensions has been widely used to determine the susceptibility of erythrocytes to oxidant stress (Kuypers et al., 1990; Mino et al., 1978).

However, functional alterations that begin far before the cell lyses are more effective in determining the behavior of the red blood cell in vivo.

The polymerization of membrane components caused by free radical attack can easily affect the mechanical properties of the erythrocyte membrane. It has been reported that phospholipid bilayer rigidity increased after lipid peroxidation (Dobretsov et al., 1977). Moreover, the cross-linking within the erythrocyte membrane skeleton and between the membrane skeletal proteins and hemoglobin significantly increases the rigidity of the red blood cell membrane. Several groups reported that red blood cell deformability was impaired in suspensions treated with hydrogen peroxide or lipid peroxidation products (malonyldialdehyde) (Jain et al., 1983; Kuypers et al., 1990; Pfafferott et al., 1982; Snyder et al., 1985). Kuypers et al. (1990) reported that the deformability of erythrocytes can be completely lost through hydrogen peroxide treatment at concentrations causing no significant hemolysis. Therefore, red cell deformability is far more sensitive to oxidant stress.

Deformability is an important property of the red blood cell, significantly contributing to the flow behavior of blood both in bulk flow conditions and in the microcirculation (Chien, 1987). Passage of a normal erythrocyte through a capillary with a diameter less than half its own diameter is highly dependent on this unique mechanical property. Impaired erythrocyte deformability may disturb tissue perfusion, especially if the vascular geometry is also altered due to disease.

Furthermore, it is believed that red cell deformability is among the determinants of red cell survival. Aged red blood cells were found to be more rigid (Linderkamp and Meiselman, 1982; Nash and Meiselman, 1983; Tillmann et al., 1980) and it is possible that these rigid cells are trapped in the reticuloendothelial system, principally the spleen. Senescent red blood cells possess several properties similar to those of oxidatively stressed cells, such as increased hemoglobin–spectrin association (McKenney et al., 1990). These observations led to the hypothesis that oxidative damage to the red cell membrane may play an important role in red cell aging (Winterbourne, 1985). McKenney et al. (1990) reported that the in vivo survival of hydrogen peroxide-damaged red blood cells was decreased. Jain et al. (1983) observed similar changes with malonyldialdehyde-treated red cells.

Apart from alterations in deformability, changes in surface characteristics may also play a role in decreasing in vivo survival of red blood cells. Phagocytosis by monocytes is enhanced after exposure of phosphatidylserin on the cell surface due to lipid peroxidation (Schwartz et al., 1985). Hydrogen peroxide increased the adherence and phagocytosis of erythrocytes by monocytes in a dose-dependent manner (Snyder et al., 1985). The lipid peroxidation product malonyldialdehyde was also shown to be effective in enhancing the phagocytosis of normal erythrocytes (Hebbel and Miller, 1988).

Wali et al. (1987) reported that the adherence of oxidant-treated erythrocytes to the cultured endothelial cells was significantly increased. They proposed that this increased adherence might be the result of membrane-surface alterations due to spectrin cross-linking and the organization of phospholipids.

Alterations in the red blood cell under the influence of oxidants are especially important when they interact with the behavior of the cell in the circulatory system. The rheologic behavior of the red blood cell is important in determining tissue perfusion at all levels. Therefore, impaired erythrocyte deformability, together with the increased adherence of these cells, may result in tissue perfusion problems.

4. ENVIRONMENTAL INFLUENCES

Blood cells circulate in the plasma throughout the body, and plasma is the unique transport medium for all material entering the organism. These materials include many chemicals capable of inducing biochemical changes in tissues, as well as respiratory gases and nutrients. Some of these chemicals are ingested in the diet or are used as drugs for their biochemical activity.

Toxic methemoglobinemia is a well-known clinical condition, reflecting the oxidative stress induced by exogenous chemicals (Wintrobe et al., 1981b). A large number of chemicals and drugs are capable of inducing methemoglobin formation. Some of them directly oxidize hemoglobin, such as nitrites, nitrates, chlorates, and quinones. Others induce oxidative damage to hemoglobin, possibly by generating active intermediates in vivo. These indirect oxidizing agents include sulfonamides, aniline dyes, and a number of drugs. These chemicals may enter the blood through the gastrointestinal tract or by absorption through the skin.

Living organisms also have a insensible exposure to a wide variety of chemicals that exist in their environment. Although other routes are also possible (e.g., absorption through the skin), most of these chemicals reach the internal environment of higher animals through the respiratory tract and come into contact with the blood. Therefore, chemicals carried in the respired air in the gaseous or particulate form are of significant importance to changes in blood composition and function.

Some chemicals to which humans are exposed via the respiratory tract, through both occupational and environmental exposure, can be identified as atmospheric pollutants. Atmospheric pollutants cover wide spectrum, including a large number of chemicals with an oxidant potential. Important atmospheric oxidants are ozone and nitrogen oxides. Additionally, certain other pollutants that are not accepted as classical oxidants, such as sulfur dioxide, may affect oxidation processes in a wide variety of tissues (Haider, 1985). A very important source of oxidants is cigarette smoke (Pryor et al., 1983). It contains a large number of oxidant species and these oxidants might be responsible for the well-established effects of smoking. Priority is usually given to influences on the respiratory apparatus when studying the health effects of these exposures. However, it is certainly clear that the effects of the respired pollutants are not limited to this system.

Kapalin (1963) reported altered hematological parameters in children living in an area with high dust fallout and high sulfur dioxide concentrations. Furthermore, he suggested that these alterations were the manifestation of the adaptation of red cells to changes in oxidation–reduction conditions, which are disturbed in the presence of sulfur dioxide. This appears to be one of the earliest reports attributing to air pollutants a potential for influencing hematological parameters through oxidative mechanisms, although it was not supported by clear biochemical evidence.

Goldstein and Balchum (1967) demonstrated that in vitro exposure of erythrocytes to ozone resulted in increased lipid peroxidation accompanied by spherocyte formation and increased osmotic fragility. This study provided evidence for the ozone effects mediated by free radical reactions, supporting the hypothesis proposed after the study by Brinkman et al. (1964). Goldstein et al. (1968) also reported a decrease in blood acetylcholinesterase activity, which was previously shown to be influenced by lipid peroxides (O'Malley et al., 1966). Ozone is the first air pollutant to induce hematological alterations by experimentally confirmed oxidative mechanisms.

However, ozone is not the only atmospheric pollutant having oxidative potential. The polluted atmosphere contains a large number of chemical species, a considerable number of which have such potential; however, a description of atmospheric composition and chemistry is beyond the scope of this chapter. The cumulative health effects of these pollutants can be investigated by epidemiological studies on populations exposed to air pollution. Hematological parameters were included in a very limited number of studies, in contrast to the effects of air pollution on other systems, especially the respiratory system.

Medeiros et al. (1983) measured hemoglobin oxidative denaturation products and erythrocyte antioxidant enzymes in a population exposed to high levels of air pollution. Methemoglobin and sulfhemoglobin levels were higher in this population than in a control population. Red blood cell superoxide dismutase and glutathione peroxidase activities were also higher in the population living in the polluted area, and this change was attributed to an adaptive response against increased concentrations of superoxide radical and hydrogen peroxide. Baskurt et al. (1990a) also reported increased methemoglobinemia and sulfhemoglobinemia after a heavily polluted period, indicated by about a fivefold increase in atmospheric sulfur dioxide concentrations compared with preexposure levels. This change was accompanied by a decrease in red blood cell deformability (Baskurt et al., 1990a).

The exact nature of the pollutants inducing these changes in the red blood cells is not clear in either of the studies. Medeiros et al. (1983) indicated that their polluted atmosphere contained powerful redox species such as polyaromatic hydrocarbons, pentachlorophenol, nitrobenzene, and anilines, in addition to a large amount of carbon monoxide, sulfur dioxide, and nitrogen oxides. The main pollutant in the city in which Baskurt et al. (1990a) worked was sulfur dioxide resulting from heating, but a detailed analysis of the pollutants was not presented.

Sulfur dioxide has also been shown to cause oxidative alterations in biological systems. Haider et al. (1981, 1982) first demonstrated that sulfur dioxide may alter lipid composition, lipase activity, and lipid peroxidation in rat and guinea pig neural tissues. Haider (1985) also studied the effect of sulfur dioxide inhalation on several different tissues including liver, heart, lung, and kidney. He reported that lipid peroxidation increased in heart and lung, whereas it decreased in liver and kidney after exposure to sulfur dioxide at a concentration of 10 ppm for 30 days, one hour daily. These results indicated that sulfur dioxide is capable of inducing biochemical alterations in tissues not directly exposed to the gas, besides having the high irritant potential to which most of its health effects were previously attributed. This effect should include alterations in oxidative mechanisms since the lipid peroxidation in different tissues changed.

Further evidence for sulfur dioxide-mediated oxidative alterations was provided by Baskurt (1988). It was reported that sulfhemoglobinemia increased in rats exposed to sulfur dioxide at 0.87 ppm concentration for 24 hr, compared with a control group. Methemoglobin values remained unchanged while osmotic hemolysis increased slightly. Further studies (Baskurt et al., 1990b) revealed that after a similar sulfur dioxide exposure, hemoglobin content of rat erythrocyte membrane ghosts increased, and this was accepted as evidence for increased association between hemoglobin and membrane structures. Furthermore, erythrocyte deformability in rats exposed to sulfur dioxide was impaired (Baskurt et al., 1990b).

Prolonged sulfur dioxide inhalation resulted in increased lipid peroxidation, in addition to increased sulfhemoglobin ratios, in guinea pig red blood cells (Dikmenoglu et al., 1991). On the other hand, lipid peroxidation induced by incubation with hydrogen peroxide decreased, indicating that the effectiveness of the antioxidant defense mechanisms was enhanced (unpublished data). The lipid peroxidation response to hydrogen peroxide treatment of red blood cells from sulfur dioxide inhalation by guinea pigs was normalized after the inhibition of catalase by sodium azide, supporting this suggestion (unpublished data). In further studies, catalase activity was directly shown to be enhanced in the red blood cells of guinea pigs similarly exposed to sulfur dioxide (Baskurt et al., 1994). In all cases, these alterations were accompanied by impairment of red blood cell deformability.

This collection of experimental data provides considerable evidence of oxidative damage in red blood cells, which may result in deterioration of their mechanical properties, under the influence of sulfur dioxide inhalation. However, the basic chemical mechanisms of this relation are not clear. Sulfur dioxide is not as powerful an oxidant as ozone or nitrogen oxides. This aspect needs further evaluation. Some indirect influences should also be considered. White blood cells may play some role in the oxidative damage to erythrocytes under the influence of sulfur dioxide inhalation, as white cell counts in prolonged exposures were found to be increased significantly (Baskurt et al., 1994).

5. CONCLUSION

It is obvious that oxidants, including atmospheric pollutants with oxidative potential, may induce structural and functional alterations in the red blood cell.

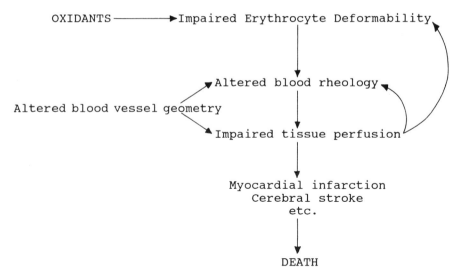

Figure 2. Role of oxidants in tissue perfusion problems.

As a consequence of these alterations, the flow properties of blood might be disturbed under the influence of oxidants. The resulting hemorheological changes would be expected to interfere with normal hemodynamics and may result in impaired tissue perfusion in general (Fig. 2). This should be considered in evaluating the health effects of oxidants, because it may provide some insight into certain physiopathological mechanisms related to environmental epidemiology.

Air pollution episodes have generally been characterized by increased cardiovascular mortality (Ministry of Health, 1954; Schrenk, 1949). Deaths were common among those in the population who had cardiovascular problems, although no specific pathological changes could be attributed to air pollutant effects. Maintenance of "normal" hemorheology is particularly important in individuals with cardiovascular problems (Baskurt et al., 1991). Disturbance of normal hemorheology under the influence of air pollutants—oxidants—may easily disturb the critical balance between the demand of and the supply to vital tissues. Therefore, the hemorheological consequences of air pollution effects may provide a possible physiopathological link between cardiac mortality and air pollution.

REFERENCES

Baskurt, O. K. (1988). Acute hematologic and hemorheologic effects of sulfur dioxide inhalation. *Arch. Environ. Health* **43**, 344–348.

Baskurt, O. K., Levi, E., Caglayan, S., Dikmenoglu, N., and Kutman, M. N. (1990a). Hematological and hemorheological effects of air pollution. *Arch. Environ. Health* **45**, 224–228.

Baskurt, O. K., Levi, E., and Caglayan, S. (1990b). Effect of sulfur dioxide inhalation on erythrocyte deformability. *Clin. Hemorheol.* **10**, 485–489.

Baskurt, O. K. Levi, E., Caglayan, S., Dikmenoglu, N., Ucer, O., Guner, R., and Yorukan, S. (1991). The role of hemorheologic factors in the coronary circulation. *Clin. Hemorheol.* **11**, 121–127.

Baskurt, O. K., Dikmenoglu, N., Ficicilar, H., Celebi, M., and Yavuzer, S. (1994). Oxidant stress and deformability in the red blood cell under the influence of sulfur dioxide inhalation. *DOGA Turkish J. Med. Sci.* **19** (in press).

Brinkman, R., Lamberts, H. B., and Veninga, T. S. (1964). Radiomimetic toxicity of ozonised air. *Lancet* **1**, 133–136.

Chasis, J. A., and Shohet, S. B. (1987). Red cell biochemical anatomy and membrane properties. *Annu. Rev. Physiol.* **49**, 237–248.

Chien, S. (1987). Red cell deformability and its relevance to blood flow. *Annu. Rev. Physiol.* **49**, 177–192.

Chien, S., and Sung, L. A. (1990). Molecular basis of red cell membrane rheology. *Biorheology* **27**, 327–344.

Chiu, D., Kuypers, F., and Lubin, B. (1989). Lipid peroxidation in human red cells. *Semin. Hematol.* **26**, 257–276.

Claster, S., Chiu, D. T. Y., Quintanilha, A., and Lubin, B. (1984). Neutrophils mediate lipid peroxidation in human red cells. *Blood* **64**, 1079–1084.

Constance, M. P., and Nagel, R. L. (1984). Sulfhemoglobinemia, clinical and molecular aspects. *N. Engl. J. Med.* **310**, 1579–1584.

Davies, K. J. A. (1988). A secondary antioxidant defense role for proteolytic systems. In M. G. Surgenor, K. A. Taylor, J. F. Ward, and C. von Sonntag (Eds.), *Oxygen Radicals in Biology and Medicine.* Plenum, New York and London, pp. 531–535.

Dikmenoglu, N., Baskurt, O. K., Levi, E., Caglayan, S., and Guler, S. (1991). How does sulphur dioxide affect erythrocyte deformability? *Clin. Hemorheol.* **11**, 497–499.

Dobretsov, G. E., Borschevskaya, T. A., Petrov, V. A., and Vladimir, Y. A. (1977). The increase of phospholipid bilayer rigidity after lipid peroxidation. *FEBS Lett.* **84**, 125–128.

Goldberg, A. L., and Boches, F. S. (1982). Oxidized proteins in erythrocytes are rapidly degraded by the adenosine triphosphate-dependent proteolytic system. *Science* **215**, 1107–1109.

Goldberg, B., Stern, A., and Peisach, J. (1976). The mechanism of superoxide anion generation by its interaction of phenylhydrazine with hemoglobin. *J. Biol. Chem.* **251**, 3045–3051.

Goldstein, B. D., and Balchum, O. J. (1967). Effect of ozone on lipid peroxidation in the red blood cell. *Proc. Soc. Exp. Biol. Med.* **126**, 356–358.

Goldstein, B. D., Pearson, B., Lodi, C., Buckley, R. D., and Balchum, O. J. (1968). The effect of ozone on mouse blood in vivo. *Arch. Environ. Health* **16**, 648–650.

Haest, C. W. M., Plasa, G., Kamp, D., and Deuticke, B. (1978). Spectrin as a stabilizer of the phospholipid asymmetry in the human erythrocyte membrane. *Biochim. Biophys. Acta.* **509**, 21–32.

Haider, S. S. (1985). Effect of exhaust pollutant sulfur dioxide on lipid metabolism of guinea pig organs. *Ind. Health* **23**, 81–87.

Haider, S. S., Hasan, M., Hasan, S., Khan, S. R., and Ali, S. F. (1981). Regional effects of sulfur dioxide exposure on guinea pig brain lipids, lipid peroxidation and lipase activity. *Neurotoxicology* **2**, 443–450.

Haider, S. S., Hasan, M., and Khan, N. H. (1982). Air pollutant sulfur dioxide-induced alterations on the levels of lipids, lipid peroxidation and lipase activity in various regions of the rat brain. *Acta Pharmacol. Toxicol.* **41**, 45–50.

Hebbel, R. P., and Miller, W. J. (1988). Unique promotion of erythrophagocytosis by malondialdehyde. *Am. J. Hematol.* **29**, 222–225.

Hill, H. A. O. (1981). Oxygen, oxidases, and the essential trace metals. *Philos. Trans. R. Soc. London* **B294**, 119–128.

Hsieh, H. S., and Jaffe, E. R. (1975). The metabolism of methemoglobin in human erythrocytes. In D. M. Surgenor (Ed.), *The Red Blood Cell.* Academic Press, New York, 2nd ed., p. 799.

Jain, S. K. (1985). In vivo externalization of phosphatidylserine and phosphatidylethanolamine in the membrane bilayer and hypercoagulability by the lipid peroxidation of erythrocytes in rats. *J. Clin. Invest.* **76**, 281–286.

Jain, S. K., Mohandas, N., Clark, M., and Shohet, S. B. (1983). The effect of malonyldialdehyde, a product of lipid peroxidation, on the deformability, dehydration and ^{51}Cr-survival of erythrocytes. *Br. J. Haematol.* **53**, 247–255.

Kapalin, V. L. (1963). The red blood picture in children from different environments. *Rev. Czech. Med.* **9**, 65–81.

Kirkman, H. N., Galiano, S., and Gaetani, G. F. (1987). The function of catalase-bound NADPH. *J. Biol. Chem.* **262**, 660–666.

Kuypers, F. A., Scott, M. D., Schott, M. A., Lubin, B., and Chiu, D. T. Y. (1990). Use of ektacytometry to determine red cell susceptibility to oxidative stress. *J. Lab. Clin. Med.* **116**, 535–545.

Linderkamp, O., and Meiselman, H. J. (1982). Geometric, osmotic and membrane mechanical properties of density seperated human red cells. *Blood* **59**, 1121–1127.

McCord, J. M., and Fridovich, I. (1969). Superoxide dismutase: An enzymatic function for erythrocuprein (hemocuprein). *J. Biol. Chem.* **244**, 6049–6055.

McKenney, J., Valeri, C. R., Mohandas, N., Fortier, N., Giorgio, A., and Snyder, L. M. (1990). Decreased in vivo survival of hydrogen peroxide-damaged baboon red blood cells. *Blood* **76**, 206–211.

Medeiros, M. H. G., Bechara, E. J. H., Naoum, P. C., and Mourao, C. A. (1983). Oxygen toxicity and hemoglobinemia in subjects from a highly polluted town. *Arch. Environ. Health* **38**, 11–16.

Meerson, F. Z., Kagan, V. E., Kozlov, Y. P., Belkina, L. M., and Arkhipenko, Y. V. (1982), The role of lipid peroxidation in pathogenesis of ischemic damage and the antioxidant protection of the heart. *Basic Res. Cardiol.* **77**, 465–485.

Ministry of Health (1954). *Mortality and Morbidity During the London Fog of December 1952,* Report No. 95. Her Majesty's Stationery Office, London.

Mino, M., Nishida, Y., Murata, K., Takegawa, M., Katsui, G., and Guguchi, Y. (1978). Studies on the factors influencing the hydrogen peroxide hemolysis test. *Nutr. Sci. Vitaminol.* **24**, 383–395.

Mohandas, N., Chasis, J. A., and Shohet, S. B. (1983). The influence of membrane skeleton on red cell deformability, membrane material properties and shape. *Semin. Hematol.* **20**, 225–242.

Nash, G. B., and Meiselman, H. J. (1983). Red cell and ghost viscoelasticity-effects of hemoglobin concentration and in vivo aging. *Biophys. J.* **43**, 63–73.

O'Malley, B. W., Mengel, C. E., Meriweth, W. D., and Zirkle, L. G. (1966). Inhibition of erythrocyte acetylcholinesterase by peroxides. *Biochemistry* **5**, 40–45.

Pacifici, R. E., Lin, S. W., and Davies, K. J. A. (1988). The measurement of protein degradation in response to oxidative stress. In M. G. Surgenor, K. A. Taylor, J. F. Ward, and C. von Sonntag (Eds.), *Oxygen Radicals in Biology and Medicine.* Plenum, New York and London, pp. 531–535.

Palek, J., and Liu, S. C. (1979). Dependence of spectrin organization in red blood cell membrane on cell metabolism: Implication for control red cell shape, deformability and surface area. *Semin. Hematol.* **14**, 75–93.

Pfafferott, C., Meiselman, H. J., and Hochstein, P. (1982). The effect of malonyldialdehyde on erythrocyte deformability. *Blood* **59**, 12–15.

Pryor, W. A. (1976). The role of free radical reactions in biological systems. In W. A. Pryor (Ed.) *Free Radicals in Biology.* Academic Press, New York, pp. 1–49.

Pryor, W. A., Prier, D. G., and Church, D. F. (1983). Electron-spin resonance study of mainstream and sidestream cigarette smoke: Nature of the free radicals in gas-phase smoke and in cigarette tar. *Environ. Health Perspect.* **47**, 345–355.

Saltman, P. (1989). Oxidative stress: A radical view. *Semin. Hematol.* **26**, 249–256.

Schrenk, H. H. (1949). Air pollution in Donora, epidemiology of the unusual smog episode of October 1948: Preliminary report. *Public Health Bull.* **306**.

Schwartz, R. S., Tanaka, Y., Fidler, I. J., Chiu, D. T. Y., Lubin, B., and Schroit, A. J. (1985). Increased adherence of sickled and phosphatidylserine-enriched human erythrocytes to cultured human peripheral blood monocytes. *J. Clin. Invest.* **75**, 1965–1972.

Shiga, T., Maeda, N., and Kon, K. (1990). Erythrocyte rheology. *CRC Crit. Rev. Oncol. Hematol.* **10**, 9–48.

Shohet, S. B., Card, R. T., Clark, M., Greenquist, A. C., Mohandas, N., Shelton, D., and Wyatt, J. (1981). The erythrocyte cytoskeleton and its apparent role in cellular functions. In *The Function of Red Blood Cells: Erythrocyte Pathobiology*. Alan R. Liss, New York, pp. 35–58.

Simic, M. G., and Taylor, K. A. (1988). Introduction to peroxidation and antioxidation mechanisms. In M. G. Surgenor, K. A. Taylor, J. F. Ward, and C. von Sonntag (Eds.), *Oxygen Radicals in Biology and Medicine*. Plenum, New York and London, pp. 1–10.

Snyder, L. M., Fortier, N. L., Trainor, J., Jacobs, J., Leb, L., Lubin, S., Chiu, D., Shohet, S., and Mohandas, N. (1985). Effect of hydrogen peroxide exposure on normal human erythrocyte deformability, morphology, surface characteristics, and spectrin-hemoglobin cross-linking. *J. Clin. Invest.* **76**, 1971–1977.

Southorn, P. A., and Powis, G. (1988a). Free radicals in medicine. I. Chemical nature and biologic reactions. *Mayo Clin. Proc.* **63**, 381–389.

Southorn, P. A., and Powis, G. (1988b). Free radicals in medicine. II. Involvement in human disease. *Mayo Clin. Proc.* **63**, 390–408.

Stocks, J., and Dormandy, T. L. (1971). The autoxidation of human red cell lipids induced by hydrogen peroxide. *Br. J. Haematol.* **20**, 95–111.

Takakuwa, Y., Ishibashi, T., and Mohandas, N. (1990). Regulation of red cell membrane deformability and stability by skeletal protein network. *Biorheology* **27**, 357–365.

Tappel, A. L. (1973). Lipid peroxidation damage to cell components. *Fed. Proc., Fed. Am. Soc. Exp. Biol.* **32**, 1870–1874.

Tillmann, W., Levin, C., Prindull, G., and Schroter, W. (1980). Rheological properties of young and aged human erythrocytes. *Klin. Wochenschr.* **58**, 469–574.

van der Zee, J., Dubbelman, T. M. A. R., and van Steveninck, J. (1985). Peroxide-induced membrane damage in human erythrocytes. *Biochim. Biophys. Acta* **818**, 38–44.

Wali, R. K., Jaffe, S., Kumar, D., Sorgente, N., and Kalra, V. K. (1987). Increased adherence of oxidant-treated human and bovine erythrocytes to cultured endothelial cells. *J. Cell. Physiol.* **133**, 25–36.

Weiss, S. J. (1980). The role of superoxide in the destruction of erythrocyte targets by human neutrophils. *J. Biol. Chem.* **255**, 9912–9917.

Winterbourne, C. C. (1985). Hemoglobin oxidation and interrelationship with lipid peroxidation in the red cell. In J. W. Eaton, D. K. Konzen, and J. G. White (Eds.), *Cellular and Molecular Aspects of Aging: The Red Cell as a Model*. Alan R. Liss, New York, p. 173.

Wintrobe, M. M., Lee, G. R., Boggs, D. R., Bithell, T. C., Foerster, J., Athens, J. W., and Lukens, J. N. (1981a). The mature erythrocyte. In *Clinical Hematology*. Lea & Febiger, Philadelphia, pp. 75–107.

Wintrobe, M. M., Lee, G. R., Boggs, D. R., Bithell, T. C., Foerster, J., Athens, J. W., and Lukens, J. N. (1981b). Methemoglobin and other disorders usually accompanied by cyanosis. In *Clinical Hematology*. Lea & Febiger, Philadelphia, pp. 1011–1020.

Ziegler, D. M. (1988). Mechanisms for the generation of oxygen radicals by drugs. In M. G. Surgenor, K. A. Taylor, J. F. Ward, and C. von Sonntag (Eds.), *Oxygen Radicals in Biology and Medicine*. Plenum, New York and London, pp. 531–535.

13

EFFECTS OF FREE RADICALS FROM HYPOXIC CELL RADIOSENSITIZERS, HYPOXIC CELL CYTOTOXINS, AND BIOREDUCTIVE ANTICANCER DRUGS ON THE BIOLOGICAL ENVIRONMENT

Hitoshi Hori

Department of Biological Science and Technology, Faculty of Engineering, University of Tokushima, Tokushima 770, Japan

Hideko Nagasawa

Pharmaceutical Institute, School of Medicine, Keio University, Shinjuku, Tokyo 160, Japan

Hiroshi Terada

Faculty of Pharmaceutical Sciences, University of Tokushima, Tokushima 770, Japan

Environmental Oxidants, Edited by Jerome O. Nriagu and Milagros S. Simmons.
ISBN 0–471–57928–9 © 1994 John Wiley & Sons, Inc.

---

1. INTRODUCTION

The dependence of cell death by ionizing radiation on the presence of oxygen has been known from the beginning of this century. This fact is particularly relevant to the treatment of malignant tumors by radiotherapy. Often, as such tumor cells grow, the parts within tissues and organs no longer receive an adequate blood supply and thus become hypoxic. While radiation treatment may destroy most of the tumor cells, the hypoxic cells are more resistant and can serve as "nuclei" for subsequent regrowth. There has, therefore, been considerable interest in various therapies combining increased oxygen exposure with radiation, especially using various oxygen-mimetic agents that make hypoxic cells more sensitive to ionizing radiation. These agents are known as hypoxic cell radiosensitizers, or sometimes simply radiosensitizers. Optimal radiosensitizers would make hypoxic cells indistinguishable from aerated cells in their response to radiation.

However, recent results of extensive studies have suggested that, rather than being a problem, hypoxic cells may be exploited to advantage (Hall, 1991). An alternative therapeutic strategy is the preferential killing of hypoxic cells by the use of cytotoxic agents. This suggests that drugs that are more effective under hypoxic than under aerobic conditions should be very useful for therapy of tumors. These drugs are called hypoxic cell cytotoxins. It has been shown that molecular oxygen is required for some chemotherapeutic agents to kill cells, particularly those agents such as bleomycin that exert their effects via the production of free radicals. We review here the actions of hypoxic cell radiosensitizers, hypoxic cell cytotoxins, and bioreductive anticancer drugs in the biological matrix (i.e., the biological environment), taking into consideration the effect of oxygen.

2. HYPOXIC CELL RADIOSENSITIZERS

Oxygen is the simplest agent that dramatically modifies the biological effect of ionizing radiation. It has been known since the beginning of this century, from radiation experiments by Schwarz (1909) using his own forearm, that cells irradiated under nitrogen (anoxic) or in the presence of very limited amounts of oxygen (hypoxic) are less sensitive to ionizing radiation than those irradiated in the presence of air (aerobic) or oxygen (oxic). The possibility that this causes a problem in radiotherapy was soon appreciated (Crabtree and Cramer, 1933), but it was not until the pioneering work of Gray and colleagues (Gray et al., 1953; Thomlinson and Gray, 1955) that it was shown to impose serious limitations on radiotherapy. They suggested that viable hypoxic cells might be present in a thin rim around necrotic areas in the tumor tissue. The value of the oxygen enhancement ratio (OER) for sparsely ionizing radiation, such as by X-rays and γ-rays, usually centers at around 3. Thus, the dose required to achieve the same biological effect in the absence of oxygen is about three times higher than that in its presence. The mechanism of the oxygen effect is still not fully understood. The most popular model to explain it is the repair/fixation competition model (Alper and Howard-Flanders, 1956). It has been postulated that intracellular thiols (RSH) repair some radiation damage (Alexander and Charlesby, 1955) that otherwise could be fixed by oxygen (Ormerod and Alexander, 1963). In experiments using various strains of *Escherichia coli* K12, some strains deficient in enzymes synthesizing glutathione (GSH) showed increased resistance to oxygen-dependent damage (Michael et al., 1981). This has been taken as good evidence for the repair/fixation competition model of the oxygen effect.

As 80% of a cell is water, most of the radiation-related energy during X-ray irradiation is absorbed by water molecules, from which various ions ($H_2O^{\cdot+}$, e^-) and electrically excited water molecules (H_2O^*) are produced within an extremely short period (10^{-16} sec).

$$H_2O \xrightarrow{\text{ionizing radiation}} H_2O^{\cdot+} + e^-$$

$$H_2O \xrightarrow{\text{ionizing radiation}} H_2O^*$$

These forms decay to free radicals in 10^{-14} to 10^{-13} sec.

$$H_2O^{\cdot+} + H_2O \longrightarrow H_3O^+ + \cdot OH$$

$$H_2O^* \longrightarrow \cdot H + \cdot OH$$

The electrons become hydrated (e_{aq}^-) within 10^{-12} to 10^{-11} sec. Hydroxyl radicals ($\cdot OH$) are considered to be responsible for a large part of the damage done to cellular DNA and to membranes by ionizing radiation (von Sonntag, 1987). Single- and double-strand breaks in DNA are very important damaging events. Oxygen, present in most biological environments, aggravates the damage.

An organic radical ($R\cdot$) in a major target molecule (RH) such as DNA is produced either by direct ionization or by reaction with $\cdot OH$ produced by radiolysis of a neighboring water molecule (indirect effect) (Hall, 1988, pp. 10–12). The direct action of radiation is the dominant process in radiation with high linear energy transfer (LET), such as by neutrons or α-particles, while an indirect action is dominant in sparsely ionizing radiation by X- and γ-rays.

$$RH \longrightarrow R\cdot + H\cdot \qquad \text{(direct effect)}$$

$$RH + \cdot OH \longrightarrow R\cdot + H_2O \qquad \text{(indirect effect)}$$

These organic radicals (R·) can be "repaired" by reaction with reductants, such as GSH.

$$R\cdot + GSH \longrightarrow RH + GS\cdot$$

$$2GS\cdot \longrightarrow GSSG$$

However, the presence of oxygen may "fix" the damage by forming peroxyl radicals ($RO_2\cdot$), followed by irreversible damage.

$$R\cdot + O_2 \longrightarrow RO_2\cdot$$

Furthermore, the hydrated electrons e_{aq}^- and H· can react with oxygen to produce superoxide ($O_2^{\cdot-}$) and its conjugated acid, the hydroperoxyl radical (·OOH), respectively.

$$e_{aq}^- + O_2 \longrightarrow O_2^{\cdot-}$$

$$H\cdot + O_2 \longrightarrow \cdot OOH$$

$$O_2^{\cdot-} + H^+ \longrightarrow \cdot OOH$$

Because the reactivities of these radicals with organic compounds are lower than that of ·OH, their effects in damage fixation are not clear at present. Under low concentrations of oxygen (10^{-6}–10^{-4} M), addition of superoxide dismutase (SOD) to the incubation medium partially protected E. coli B/r cells from damage by ionizing radiation (Ewing and Jones, 1987).

As described above, the absence of oxygen decreases the sensitivity of cells to ionizing radiation. Hypoxic cells within solid tumors are resistant to ionizing radiation and some chemotherapeutic agents, and their survival may subsequently result in tumor regrowth. Of various available methods to overcome this problem, chemical sensitizers are expected to be the most promising. One class of radiosensitizers which mimics oxygen is hypoxic cell radiosensitizers. They make hypoxic cells more sensitive than normally aerated cells to ionizing radiation. Thus, a hypoxic cell radiosensitizer, which is an "oxygen mimic," will be specific for hypoxic cells.

In 1973, it was proposed that the 5-nitroimidazole derivative metronidazole (Fig. 1), which is a well-known trichomonacide, is useful as a hypoxic cell radiosensitizer (Foster and Wilson, 1973). Accordingly, over the past 20 years, nitroheterocyclic compounds such as metronidazole have been developed as hypoxic cell radiosensitizers for clinical use (Coleman et al., 1988; Adams et al., 1991; Fielden et al., 1992). Nitroimidazoles, electron-deficient heterocyclic compounds, have received particular attention as chemical modifiers because they increase the radiation sensitivity of hypoxic cells, being selectively cytotoxic to them, and can increase the effectiveness of chemotherapeutic agents (Hall, 1988, pp. 182–191). Misonidazole (MISO in Figure 1) is the best known of these

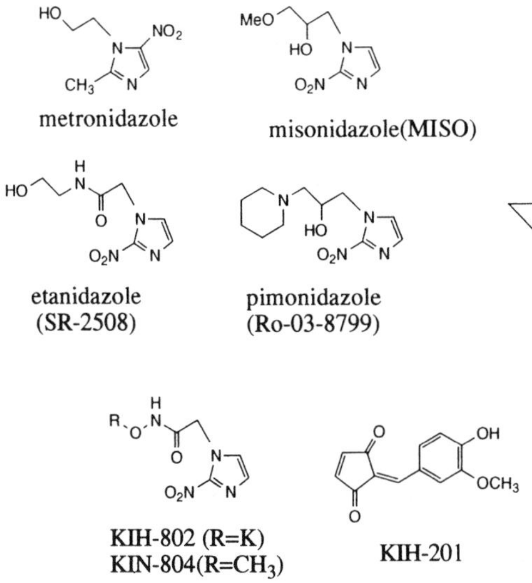

Figure 1. Representative hypoxic cell radiosensitizers.

radiosensitizers (Adams et al., 1976a, b), and the first 2-nitroimidazole compound tested in clinical trials (Dische et al., 1977; Dische, 1985).

However, its neurotoxicity prevents its use at an effective dose to produce optimum sensitization. Therefore, many atempts have been made to find more effective and less toxic sensitizers than MISO for clinical use (Stratford, 1992). Etanidazole (SR-2508 in Fig. 1) is less neurotoxic owing to its hydrophilicity. Pimonidazole (Ro-03-8799 in Fig. 1), which is more electron affinic than MISO, is a more effective radiosensitizer, and owing to the presence of a weakly basic piperazine group it is concentrated in tumor cells, which have a lower intracellular pH than normal cells. Since the successful development of the potent compound RSU 1069, which has an aziridine moiety as an alkylating moiety in its side-chain (Adams et al., 1984), much effort has been devoted to developing effective nitroheterocyclic compounds by introducing additional active functional groups. Recently, the nitroheterocyclic radiosensitizers KIH-802 and KIN-804 (Fig. 1), with a hydroxamic acid moiety possessing additional biochemical function in their side-chains, have been developed (Hori et al., 1989; Sasai et al., 1990; Nagasawa et al., 1992).

In general, the oxygen-mimicking hypoxic cell radiosensitizers are designed to take the place of oxygen and produce irreversible damage by forming a complex between DNA and the sensitizer radical (DNA–S·).

$$DNA· + S \longrightarrow DNA–S· \qquad \text{(damage fixation)}$$

Most of them are nitroaromatic compounds ($ArNO_2$), as described above, and thus are electron affinic. Their radiosensitization is quantitatively related to the one-electron redox potential of their first redox couple $ArNO_2/ArNO_2^{\cdot-}$ (Adams et al., 1976b, 1979). However, recent kinetic studies on nitro radical formation and decay showed that the disproportionate reaction rate of the nitro radical is a more discriminatory parameter of radiosensitizing efficiency than the redox potential (Guissani et al., 1990).

Despite considerable study, the underlying molecular mechanisms of the radiosensitization process remain poorly understood (Wardman, 1987). It has generally been believed that nitroaromatic compounds act as radiosensitizers primarily by causing one-electron oxidation of DNA bases. Details of the molecular mechanisms of this reaction have been studied mainly in model systems involving the pyrimidine base radicals generated by ionizing radiation (von Sonntag, 1987). It is known that nitroaromatic compounds oxidize α-hydroxyalkyl radicals (Fig. 2). An intermediate adduct has been detected with nitrobenzenes (Jagannadham and Steenken, 1984) and nitroimidazoles (Wardman, 1984). An intermediate nitroxide radical adduct is formed between the sensitizer and the radical center on DNA bases (Fig. 3) (Steenken and Jagannadham, 1985). Nitroaromatic compounds increase the yield of thymine glycol generated in irradiated solution (Nishimoto et al., 1983). Furthermore, these compounds have been found to sensitize phosphate elimination from 5'-nucleotides (Raleigh et al., 1973a,b). A possible mechanism of DNA strand breaks is shown in Figure 4. The balance between pure electron transfer and radical-addition followed by electron transfer and α elimination at C-5' is dependent on the solvent polarity (Jagannadham and Steenken, 1984). A recent study on strand breaks formed under anaerobic conditions by the radiomimetic antitumor antibiotic neocarzinostatin (NCS) in the presence of MISO showed that the 3'-(formyl phosphate)-ended DNA and nitroso-MISO were derived from β fragmentation of MISO-DNA adduct similar to that shown in Figure 4 (Kappen et al., 1989). The ability of MISO to increase DNA strand breakage by

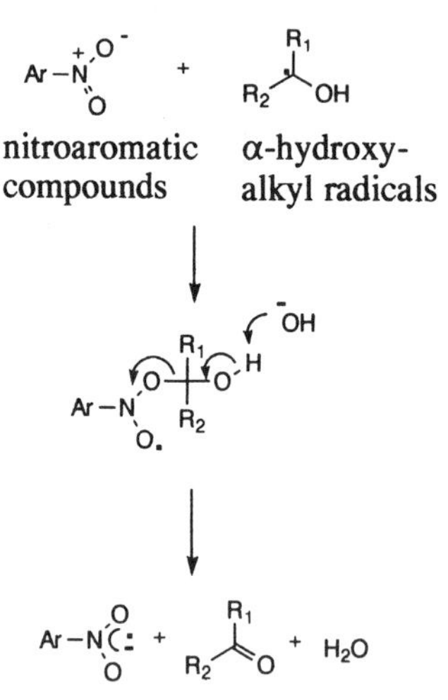

Figure 2. Reaction of nitroaromatic compounds with α-hydroxyalkyl radicals.

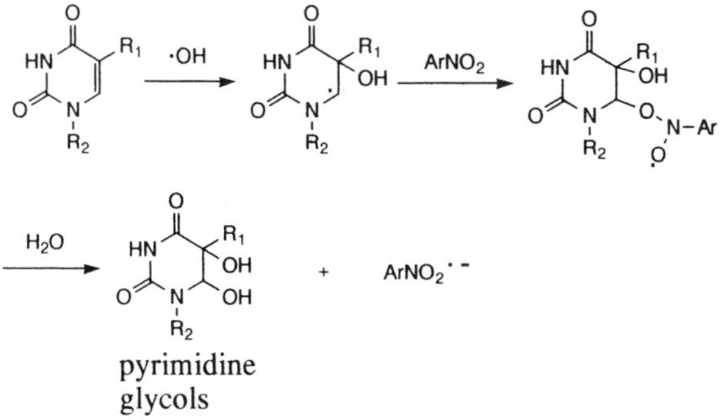

Figure 3. Formation of pyrimidine glycols by oxidation with a nitroaromatic radiosensitizer.

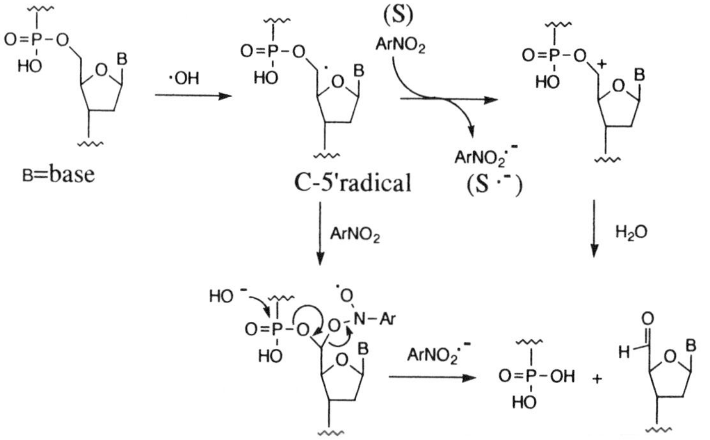

Figure 4. Oxidation of a C-5' radical in a DNA fragment by a nitroaromatic radiosensitizer.

NCS used as a radiomimetic drug instead of radiation suggests that a similar mechanism, involving oxygen transfer from the nitro group of MISO, can occur in the radiosensitization reaction.

Radiosensitizer 2-nitromidazoles are reported to have various effects on cellular metabolism, such as causing depletion of nonprotein thiols (NPSH) and protein thiols (Biaglow, 1982), stimulation of oxygen consumption (Anisworth et al., 1978; Greenstock et al., 1978), inhibition of glycolysis (Varnes and Biaglow, 1982), and perturbation of purine metabolism (Henderson and Zomber, 1980), especially at higher concentrations than their effective concentrations as radiosensitizers. If a compound that affects cellular metabolism at concentrations similar to its radiosensitizing concentrations can be developed, this bi- or multifunctional compound should be useful as a potent and specific hypoxic cell radiosensitizer.

In this connection, it is noteworthy that the effects of ionizing radiation and hyperthermia are reported to be increased in the presence of inhibitors of oxidative phosphorylation in mitochondria (Tannock and Rotin, 1989). In fact, weakly acidic uncouplers of energy transduction in mitochondria, such as 2,4-dinitrophenol (DNP), N-phenylanthranilic acid, and carbonyl cyanide m-chlorophenylhydrazone (CCCP), showed significant effects on cell survival after irradiation (Anderson et al., 1985; Hodgkiss, 1987). Slight radioprotection was observed with aerobic cells depleted of ATP by post-irradiation treatment with DNP (Nishizawa et al., 1979). In contrast, uncouplers including DNP radiosensitize aerobic bacteria (Anderson et al., 1985)

As the efficiency of hypoxic cell radiosensitizers seems to depend on the energy status, it is important to take into consideration the energy status of tumor cells in the development of selective and potent radiosensitizers. In this connection, it is noteworthy that 2-arylidene-4-cyclopentene-1,3-diones were developed as hypoxic cell radiosensitizers and hypoxic cell cytotoxins (Hori et al., 1987), and their derivative KIH-201 [2-(4'-hydroxy-3'-methoxybenzylidene)-4-cyclopentene-1,3-dione] (Fig. 1) was found to be a very potent inhibitor of the P_i-transporter in mitochondria (Koike et al., 1988).

3. HYPOXIC CELL CYTOTOXINS

In studies on the action of hypoxic radiosensitizers, many electron affinic radiosensitizers were found to be preferentially cytotoxic to hypoxic cells, even in the absence of radiation. This suggested that these agents, defined by hypoxic cell cytotoxins, could be used as effective cytotoxins specific to hypoxic tumor cells without irradiation. As hypoxic cell cytotoxins were originally detected by the secondary effects of some radiosensitizers, most possess some degree of radiosensitizing activity, and thus their effects on hypoxic cells have sometimes been examined in combination with radiation. As the principle of the action of hypoxic cell cytotoxins is very important in cancer therapy, there have been extensive recent attempts to develop effective compounds. The modes of action of these compounds are reviewed in this section.

Lin et al. (1972) proposed that selective cytotoxicity can be achieved by taking into consideration the fact that hypoxic cells can perform reductive metabolism more extensively than aerobic cells. Quinone-reductive alkylating agents such as mitomycin C and the nitroimidazole hypoxic cell radiosensitizers have this property. In fact, mitomycin C exhibits preferential cytotoxicity on cells under hypoxic conditions in vitro but not in vivo. The nitroimidazole MISO and other hypoxic cell radiosensitizers are preferentially cytotoxic to hypoxic cells in vivo, as described in Section 2, but they are effective only above their clinically available concentrations. Thus, these two types of compounds are not ideal in cancer therapy.

Accordingly, further attempts have been made to develop potent and specific hypoxic cell cytotoxins. The most promising compound is 3-amino-1,2,4-ben-

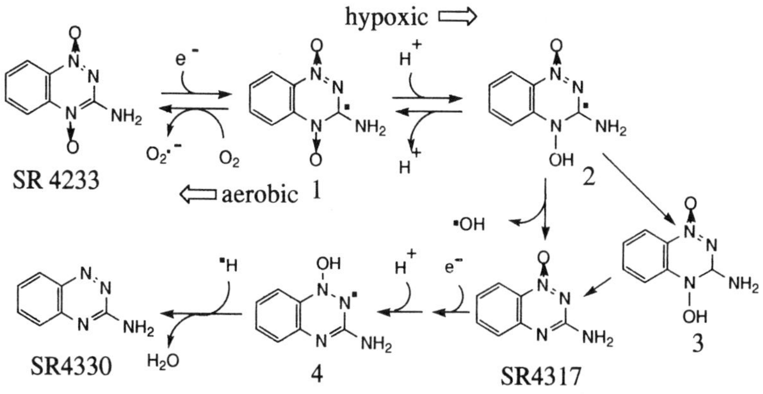

Figure 5. Metabolic pathway of SR 4233 under hypoxic and aerobic conditions.

zotriazine-1,4-dioxide (SR 4233), which has the chemical structure shown in Figure 5 (Zeman et al., 1986). Studies on various cells in culture showed that SR 4233 is about 100-fold more potent on hypoxic cells than on aerobic cells. Thus, it is a strong and selective cytotoxin to hypoxic cells (Zeman et al., 1986). Moreover, in combination with irradiation, SR 4233 specifically induced hypoxia in tumor cells in mice. A nontoxic dose of SR 4233 enhanced the cytotoxic action of irradiation on tumor cells (Zeman et al., 1986, 1988; Brown and Lemmon, 1990), and its activity was increased in the presence of the hypoxic inducer flavone acetic acid (Brown, 1987; Sun and Brown, 1989).

SR 4233 is metabolized more rapidly under hypoxic conditions than under aerobic conditions both in vitro and in vivo (Zeman et al., 1986). As shown in Figure 5, SR 4233 is reduced to SR 4317 and SR 4330 in hypoxic cells (Baker et al., 1988; Laderoute and Rauth, 1986; Walton et al., 1989; Costa et al., 1989). As both these products are nontoxic to cells under hypoxic and aerobic conditions (Zeman et al., 1988), the intermediate(s) (either **1** or **2** or both) should be effective. In the absence of oxygen, the reduction of SR 4233 proceeds finally to SR 4330, and the SR 4233 radical(s) (**1** or **2** or both) abstracts a hydrogen radical ·H from DNA, causing strand breakage leading ultimately to cell death. In contrast, under aerobic conditions, the SR 4233 radical (**1**) reacts with oxygen by one-electron reduction, generating superoxide $O_2^{·-}$ (Laderoute et al., 1988). It is noteworthy that in the presence of O_2, the reduction reaction does not proceed further than the first step (from SR 4233 to **1**), and intermediate **1** is oxidized to the original compound SR 4233. This would explain why SR 4233 is metabolized under hypoxic conditions more rapidly than under aerobic conditions. The major enzyme responsible for the bioreduction of SR 4233 appears to be P-450 in liver microsomes under both aerobic and hypoxic conditions (Walton et al., 1989).

Superoxide radicals produced during SR 4233 metabolism under aerobic conditions are suggested to be responsible for aerobic cytotoxicity. Thus, the toxicity of SR 4233 under aerobic conditions may be dependent not only on the amount of reductase, but also on the concentration of oxygen radical scavengers, or perhaps on the catalase or GSH-associated enzymes, present in these cells (Biedermann et al., 1991). The cytotoxic effect of SR 4233 on hypoxic cells appears to result from two factors: the rate of drug metabolism and the ability of DNA to repair double-strand breaks.

Recently, the phenazine di-N-oxide compound 2-[(3'-aminopropyl)amino] phenazine 5,10-di-N-oxide (PDNO in Figure 6) was designed as a new bioreductive agent to produce diffusible oxygen radicals, such as the hydroxy radical $\cdot$OH and superoxide $O_2^{\cdot-}$, and concomitant DNA strand scission under physiological conditions (Nagai et al., 1991). The PDNO radical produced by one-electron reduction of PDNO is suggested to be involved in the cytotoxic action of PDNO. This radical produces $\cdot$OH under hypoxic conditions and $O_2^{\cdot-}$ under aerobic conditions by a mechanism similar to that of SR 4233. Conceivably, the PDNO radical and these diffusible oxygen radicals are directly associated with DNA strand scission.

When PDNO was incubated aerobically with $\phi\chi174$ replicative DNA either in the presence or absence of the reductive compound dithiothreitol (DTT), relaxation of supercoiled DNA was observed only in the presence of DTT (Nagai et al., 1991). Furthermore, substantial conversion of PDNO to the parent phenazine compound was observed when phenazine di-N-oxide was incubated with DTT and DNA under hypoxic conditions, but not under aerobic conditions.

In the case of hypoxic cell radiosensitizers that possess an NO_2 group, such as the 2-nitroimidazole derivative MISO, hypoxic cytotoxicity was found to be caused by highly toxic substances formed from the parent compound via

PDNO

NCS-Chrom   **Figure 6.** Chemical structure of NCS–Chrom and PDNO.

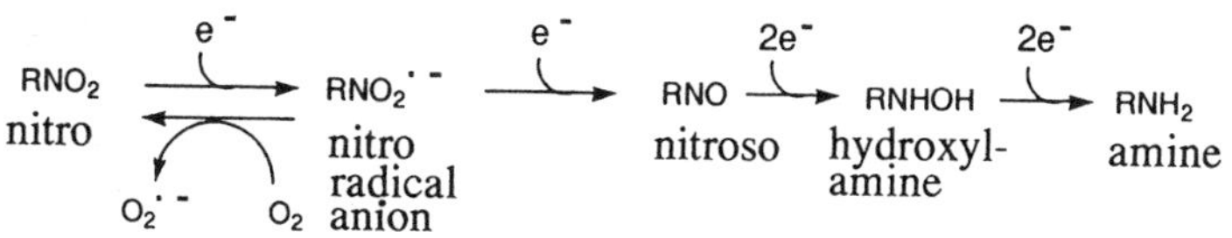

Figure 7. Nitro-reduction of 2-nitroimidazole radiosensitizers under anaerobic and aerobic conditions.

metabolic nitroreduction in hypoxic cells, according to the reaction (Flockhart et al., 1978; Alexander, 1986) summarized in Figure 7. In the presence of oxygen, the nitro radical anion is back-oxidized to the parent nitro compound, resulting in a futile cycle without generation of an active intermediate. This back-oxidation reaction results in lack of cell death under aerobic conditions. The $O_2^{\cdot-}$ formed by this reaction might account for some of the side effects of the hypoxic cell radiosensitizers on normal aerobic tissues, although there is no direct evidence for this. On the other hand, under hypoxic conditions, the bioreduction proceeds to form some cytotoxic species (Whitmore and Varghese, 1986). In addition, Bolton and McClelland (1989) proposed the following model for the hypoxic cytotoxic reaction. Reduction of 2-nitroimidazoles (**1**) to 2-(hydroxyamino)imidazoles (**2** or **3**) causes formation of the nitrenium ion (**4** or **5**), which reacts with various cellular nucleophiles such as GSH and proteins containing OH and SH groups (Fig. 8). The adducts thus formed are cytotoxic.

Electrochemical studies by Tocher and Edwards (1990) suggested that the nitro radical anion ($RNO_2^{\cdot-}$) is responsible for the DNA-damaging action of nitroheterocyclic drugs. In a recent report on the interaction of $RNO_2^{\cdot-}$ with DNA, they also proposed that the effect of the nitro radical anion, and the interaction of the 2-electron addition product nitroso (RNO) with target substances, were directly associated with the action of these bioreductive compounds

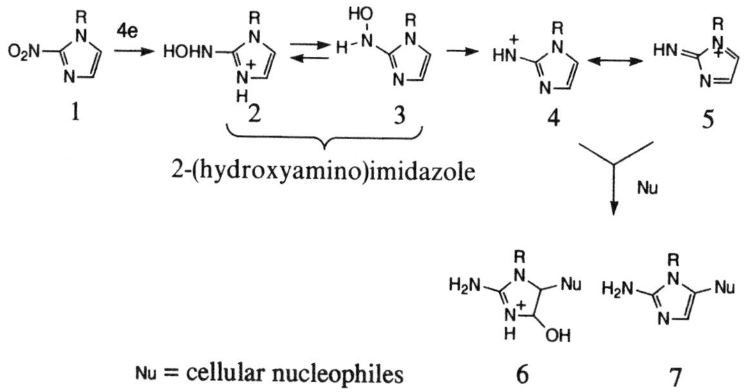

Figure 8. Proposed mechanisms for reaction of 2-(hydroxyamino)imidazoles with cellular nucleophiles.

(Tocher and Edwards, 1992). The nitroso-reduction products were suggested to be active intermediates in the expression of chemosensitization by 2-nitroimidazoles (Mulcahy et al., 1990).

In summary, the selective toxicity of hypoxic cell cytotoxins toward hypoxic mammalian cells is mediated by a reactive oxidizing radical, which is formed directly from these compounds to cause DNA strand scission. This mode of action appears to be different from those of other classes of bioreductively activated anticancer drugs such as the nitroimidazoles, mitomycins, and NCS (see Section 4). Ultimately, hypoxic cell cytotoxins must be developed as effective anticancer drugs targeted to hypoxic cells in solid tumors. Heterocyclic di-N-oxides appear to be promising initial compounds in the design of anticancers of this type.

4. BIOREDUCTIVE ANTICANCER DRUGS

In general, anticancer drugs are used for cancer therapy with no consideration the effect of O_2. They directly attack DNA, splitting its double strand. They seem to be effective only in the presence of oxygen, unlike hypoxic cell radiosensitizers and hypoxic cell cytotoxins, but the effect of oxygen has not been studied extensively.

There are two types of anticancer drugs. One type, represented by cyclophosamide and cisplatin, attacks DNA bases by alkylation. The major action of these alkylating agents is interaction with guanine residues in runs of guanines of DNA to produce a guanine-alkylating agent adduct under physiological conditions. Although free radicals are formed in degradation of DNA, there is no reasonable evidence that they are directly responsible for the anticancer effect of these drugs. Moreover, it is not apparent whether oxygen participates in the action of alkylating agents.

Drugs of the second class, such as mitomycin, NCS, esperamicin, and adriamycin, attack the sugar moiety of DNA. They can be classified as bioreductive anticancer drugs because their radicals responsible for attacking DNA are produced by reductive activation. Thiols or NADPH are necessary for their reductive activation. NCS is a representative anticancer drug of this type that has been studied extensively in relation to the effect of oxygen.

NCS is a naturally occurring protein antitumor antibiotic isolated from *Streptomyces carzinostaticus* variant F-41 (Ishida et al., 1965) that induces base release and DNA strand breaks (Kappen and Goldberg, 1978; Burger et al., 1978). It is a complex, consisting of a 1:1 mixture of an acidic single-chain polypeptide with 113 amino acid residues (NCS apoprotein) (Kim et al., 1993; Meinhofer et al., 1972) and a chromophore (NCS–Chrom, MW: 659 in Fig. 6) (Edo et al., 1985), of 10,717 molecular weight, but no carbohydrate.

There is strong circumstantial evidence that damage of DNA is induced by free radicals. Chin and Goldberg (1986) showed that superoxide is produced during the spontaneous degradation of NCS. Production of superoxide from NCS is not affected by the presence of thiols, although earlier work showed that DNA damage is stimulated at least 1000-fold by thiols (Kappen and Goldberg, 1978). Chin and Goldberg (1986) concluded that the generation of superoxide by NCS is not correlated with its ability to cause DNA damage, regardless of the presence of thiols. Furthermore, the failure of hydroxyl radical scavengers to inhibit drug-induced single-strand breaks in supercoiled DNA in the absence of thiol also suggests that a diffusible hydroxyl radical is not involved in this reaction.

On the other hand, NCS–Chrom possesses a novel bicyclo[7.3.0]dodeca-diynene system (1 in Fig. 9), which is responsible for the full activity of NCS. Recently, a series of novel anticancer antibiotics possessing a bicyclodiyne moiety have been discovered. Esperamicins, calicheamicins, and dynemicins contain a

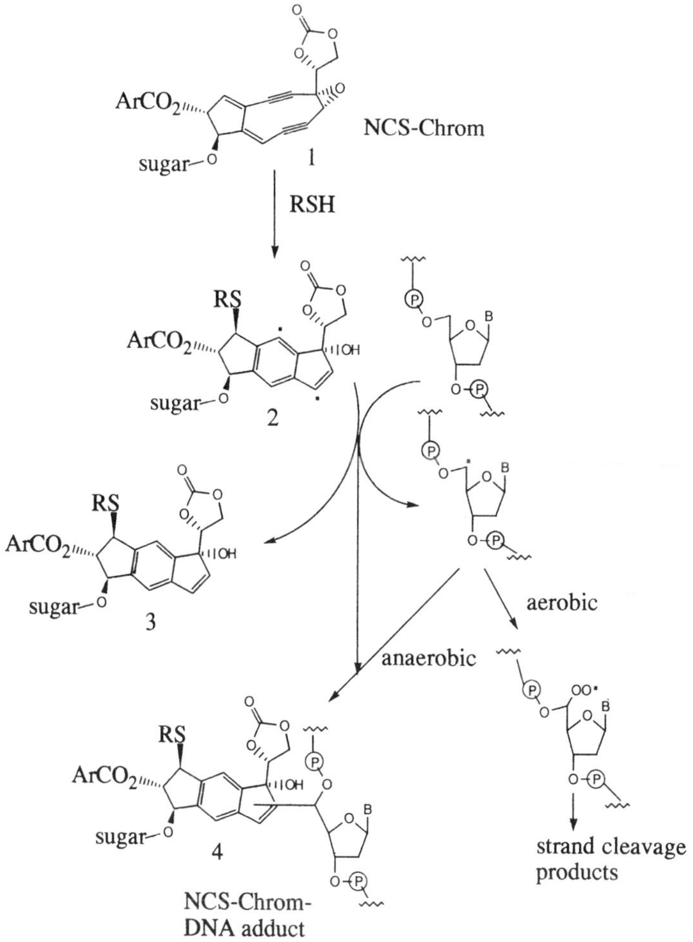

Figure 9. Proposed mechanisms for reductive activation of NCS–Chrom-induced damage under anaerobic and aerobic conditions.

bicyclo-[7.3.1]-1, 5-diyne-3-ene moiety substituted for an oligosaccharide or an anthraquinone, and have been reported to form free radicals responsible for their action (for reviews, see Goldberg, 1991; Nicolaou and Dai, 1991). NCS–Chrom is also similar to enediyne anticancer antibiotic having a bicyclo[7.3.0]diyne chromophore.

The naphthoate group of NCS–Chrom intercalates into DNA and the rest of the molecule becomes located in the minor groove (Lee and Goldberg, 1989). The reaction of NCS–Chrom with thiol produces a cumulene intermediate (Myers, 1987; Myers et al., 1988; 1992; Myers and Proteau, 1989). This thiol activation of NCS–Chrom is facilitated dramatically through participation of the carbohydrate amino group as an internal base. On decay of the intermediate, the putative biradical (**2** in Figure 9) is formed, which attacks DNA by hydrogen atom abstraction resulting in the stable indene derivative (**3**). Thus, the activation and the consequent hydrogen abstraction reaction of NCS–Chrom to generate a DNA radical, presumably a carbon-dependent radical at deoxyribose, does not require O_2 but has a dose-dependent relation with thiol. As shown in Figure 9, under aerobic conditions, base release and DNA strand breaks are produced (Kappen and Goldberg, 1985). Under anaerobic conditions there is little, if any, DNA strand breakage; instead, the radical at C-5′ of deoxyribose interacts with the intercalated NCS–Chrom to form a covalent drug–DNA adduct (**4**) (Povirk and Goldberg, 1984), as shown in Figure 9. Thus, NCS acts as an alkylating agent under anaerobic conditions. NCS–Chrom has been shown to exert its anticancer activity via a free radical mechanism regardless of the presence of O_2. It is noteworthy that mitomycin C, like NCS, exerts its effect as a bioreductive alkylating anticancer agent and a hypoxic cell cytotoxin (Holden et al., 1992).

In summary, the most widely accepted mechanism of the cytotoxic effects of bioreductive anticancer drugs is as follows. First the drugs induce a reductive product by reaction with reductants such as NADPH and thiols. Then the reductive product forms an adduct DNA, inhibiting replication of DNA and subsequently inhibiting cell division. The reductive action product also forms oxygen radicals such as $\cdot OH$ and $O_2^{\cdot-}$ in the presence of O_2 and Fe^{3+}. These radicals may be responsible for damage of DNA. However, further studies are necessary to understand more exactly the mechanisms of action of anticancer drugs of this class (Goldberg, 1991).

5. CONCLUSION

Considerable progress has been made in understanding the role of reactive free radicals, including oxygen radicals, in the action of hypoxic cell-specific anticancer drugs such as hypoxic cell radiosensitizers, hypoxic cell cytotoxins, and bioreductive anticancer agents in the biological environment. Solid tumor tissues exist in a heterogeneous biological (biochemical and physiological) environment (Coleman et al., 1988). This inherent heterogeneity is a major factor that makes development of effective therapies for solid tumors very difficult. Recent progress in biochemical research (Argiles and Azcón-Bieto, 1988; Michell and Coleman, 1992) must be taken into consideration in development of effective therapy specific to tumor cells.

Physiological differences between tumor tissues and normal tissues, such as in oxygen concentration, pH, temperature, and glucose concentration, are now being exploited to therapeutic advantage (Coleman et al., 1988; Sartorelli, 1988). Hypoxic cells in solid tumors are an obstacle to effective cancer treatment because radiotherapy and most anticancer drugs are more cytotoxic toward normally oxic tumor cells than toward hypoxic tumor cells. Even the most potent enediyne anticancer antibiotics such as NCS–Chrom, esperamicins, calicheamicins, and dynemicins require oxygen molecules to cause effective damage of DNA. Residual malignant cells, surviving by hypoxia after irradiation and attack by anticancer drugs, can subsequently proliferate.

Few data are available on the exact role of oxidants in the action of anticancer drugs. Moreover, the effect of limited oxygen concentration in the critical biological environment of tumor tissues is sometimes overlooked. Therefore, it is very important to take into consideration the participation of free radicals in the presence of oxygen and bioreductive enzymes in the action of hypoxic cell anticancer drugs such as hypoxic cell radiosensitizers, hypoxic cell cytotoxins, and bioreductive anticancer drugs (Workman, 1992). It is also important to study the action of anticancer drugs on hypoxic cells in relation to the energy status of the tumor cells (Karczmar et al., 1992) and mitochondria present in these cells (Pedersen, 1978), because the action of these drugs is dependent on the energy status of the tumor cells.

REFERENCES

Adams, G. E., Denekamp, J., and Fowler, J. F. (1976a). Biological basis of radiosensitization by hypoxic-cell radiosensitizers. *Chemotherapy* 7, 187–206.

Adams, G. E., Flockhart, I. R., Smithen, C. E., Stratford, I. J., Wardmen, P., and Watts, M. E. (1976b). Electron affinic sensitization. VII. A correlation between structures, one-electron reduction potentials, and efficiencies of nitroimidazoles as hypoxic cell radiosensitizers. *Radiat. Res.* 67, 9–20

Adams, G. E., Clarke, E. D., Flockhart, I. R., Jacobs, R. S., Sehmi, D. S., Stratford, I. J., Wardman, P., Watts, M. E., Parrick, J., Wallace, R. G., and Smithen, C. E. (1979). Structure-activity relationships in the development of hypoxic cell radiosensitizers. I. Sensitization efficiency. *Int. J. Radiat. Biol. Relat. Stud. Phys. Chem. Med.* 35, 133–150

Adams, G. E., Ahmed, I., Sheldon, P. W., and Stratford, I. J. (1984). Radiation sensitization and chemopotentiation: RSU 1069, a compound more efficient than misonidazole *in vitro* and *in vivo*. *Br. J. Cancer* 49, 571–578.

Adams, G. E., Stratford, I. J., Bremner, J. C. M., Cole, S., Edwards, H. S., and Fielden, E. M. (1991). Nitroheterocyclic compounds as radiation sensitizers and bioreductive drugs. *Radiother, Oncol.* 20, Suppl. 1, 85–91.

Alexander, P. (Ed.) (1986). Bioreduction in the activation of drugs. *Biochem. Pharmacol.* 35, 1–122.

Alexander, P., and Charlesby, A. (1955). Physico-chemical methods of protection against ionizing radiations. In Z. M. Bacq and P. Alexander (Eds.), *Radiobiology Symposium 1954.* Butterworth, London, pp. 49–60.

Alper, T., and Howard-Flanders, P. (1956). Role of oxygen in modifying the radiosensitivity of *E. coli* B. *Nature (London)* **178**, 978–979.

Anderson, R. F., Patel, K. B., and Evans, M. D. (1985). Changes in the survival curve shape of *E. coli* cells following irradiation in the presence of uncouplers of oxidative phosphorylation. *Int. J. Radiat. Biol.* **48**, 495–504.

Anisworth, P. L., Channon, M., Sridhar, R., Gushulak, B., and Tustanoff, E. R. (1978). Effect of radiosensitizing agents on electron transport systems. *Can. J. Biochem.* **56**, 457–461.

Argiles, J. M., and Azcón-Bieto, J. (1988). The metabolic environment of cancer. *Mol. Cell. Biochem.* **81**, 3–17.

Baker, M. A., Zeman, E. M., Hirst, V. K., and Brown, J.M. (1988). Metabolism of SR 4233 by Chinese hamster ovary cells: Basis of selective hypoxic cytotoxicity. *Cancer Res.* **48**, 5947–5952.

Biaglow, J. E. (1982). Oxygen, hydrogen donors and radiation response. *Adv. Exp. Med. Biol.: Hyperthermia* **57**, 147–175.

Biedermann, K. A., Wang, J., Graham, R. P., and Brown, J. M. (1991). SR 4233 cytotoxicity and metabolism in DNA repair-competent and repair-deficient cell cultures. *Br. J. Cancer* **63**, 358–362.

Bolton, J. L., and McClelland, A. (1989). Kinetics and Mechanism of decomposition in aqueous solution of 2-(hydroxyamino)imidazoles. *J. Am. Chem. Soc.* **111**, 8172–8181.

Brown, J. M. (1987). Exploitation of bioreductive agents with vasoactive drugs. *Radiat. Res.: Proc. Int. Cong., 8th,* Vol. 2, p. 719.

Brown, J. M., and Lemmon, M. J. (1990). SR 4233: A tumor specific radiosensitizer active in fractionated radiation regimes. *Radiother. Oncol.* **20**, Suppl. 1, 151–156.

Burger, R. M., Peisach, J., and Horwitz, S. B. (1978). Effect of light and oxygen on neocarzinostatin stability and DNA-cleaving activity. *J. Biol. Chem.* **253**, 4830–4832.

Chin, D.-H., and Goldberg, I. H. (1986). Generation of superoxide free radical by neocarzinostatin and its possible role in DNA damage. *Biochemistry* **25**, 1009–1015.

Coleman, C. N., Bump, E. A., and Kramer, R. A. (1988). Chemical modifiers of cancer treatment. *J. Clin. Oncol.* **6**, 709–733.

Costa, A. K., Baker, M. A., Brown, J. M., and Trudell, J. R. (1989). *In vitro* hepatotoxicity of SR 4233 (3-amino-1,2,4-benzotriazine-1,4-dioxide), a hypoxic cytotoxin and potential antitumor agent. *Cancer Res.* **49**, 925–929.

Crabtree, H. G., and Cramer, W. (1933). The action of radium on cancer cells. II. Some factors determining the susceptibility of cancer cells to radium. *Proc. R. Soc. London, Ser. B* **113**, 238–250.

Dische, S. (1985). Chemical sensitizers for hypoxic cells: A decade of experience in clinical radiotherapy. *Radiother. Oncol.* **3**, 97–115.

Dische, S., Saunders, M.I., Lee, M. E., Adams, G. E., and Flockhart, I. R. (1977). Misonidazole-A drug for trial in radiotherapy and oncology. *Br. J. Cancer* **35**, 567–579.

Edo, K., Mizugaki, M., Koide, Y., Seto, H., Furikata, K., Otake, N., and Ishida, N. (1985). The structure of neocarzinostatin chromophore possessing a novel bicyclo[7.3.0] dodecadiyne system. *Tetrahedron Lett.* **26**, 331–334.

Ewing, D., and Jones, S. R. (1987). Superoxide removal and radiation protection in bacteria. *Arch. Biochem. Biophys.* **254**, 53–62.

Fielden, E. M., Adams, G. E., Cole, S., Naylor, M. A., O'Neill, P., Stephens, M. A., and Stratford, I. J. (1992). Assessment of a range of novel nitroaromatic radiosensitizers and bioreductive drugs. *Int. J. Radiat. Oncol. Biol. Phys.* **22**, 707–711.

Flockhart, I. R., Large, P., Malcolm, J. L., Martin, T. R., and Troup, D. (1978). Pharmacokinetic and metabolic studies of the hypoxic cell radiosensitizer misonidazole. *Xenobiotics* **8**, 97–105.

Foster, J. L., and Wilson, R. L. (1973). Radiosensitization of anoxic cells by metronidazole. *Br. J. Radiol.* **46**, 234–235.

Goldberg, I. H. (1991). Mechanism of neocarzinostatin action: Role of DNA microstructure in determination of chemistry of bistranded oxidative damage. *Acc. Chem. Res.* **24**, 191–198.

Gray, L. H., Conger, A. D., Ebert, M., Hornsey, S., and Scott, O. C. A. (1953). The concentration of oxygen dissolved in tissues at the time of irradiation as a factor in radiotherapy. *Br. J. Radiol.* **26**, 638–648.

Greenstock, C. L., Biaglow, J. E., and Durand, R. E. (1978). Effects of sensitizers on cell respiration: II. The effects of hypoxic cell sensitizers on oxygen utilization in cellular and chemical models. *Br. J. Cancer* **37**, Suppl. III, 11–15.

Guissani, A., Henry, Y., Lougmani, N., and Hickel, B. (1990). Kinetic studies of four types of nitroheterocyclic radicals by pulse radiolysis correlation of pharmacological properties to decay rates. *Free Radical Biol. Med.* **8**, 173–189.

Hall, E. J. (1988). *Radiobiology for the Radiologist*, 3rd ed. Lippincott, Philadelphia.

Hall, E.J. (1991). Hypoxia revisited. *J. Natl. Cancer Inst.* **83**, 56.

Henderson, J.F., and Zomber, G. (1980). Effects of misonidazole on purine metabolism in Ehrlich ascites tumor cells *in vitro*. *Biochem. Pharmacol.* **29**, 2533–2536.

Hodgkiss, R. J. (1987). The effect of electron-affinic radiosensitizers on ATP levels in V79–397A Chinese hamster cells. *Biochem. Pharmacol.* **36**, 393–396.

Holden, S. A., Teicher, B. A., Ara, G., Herman, T. S., and Coleman, C. N. (1992). Enhancement of alkylating agent activity by SR-4233 in the FSaIIC murine fibrosarcoma. *J. Natl. Cancer Inst.* **84**, 187–193.

Hori, H., Maezawa, H., Iitaka, Y., Ohsaka, T., Shibata, T., Mori, T., and Inayama, S. (1987). 2-Arylidene-4-cyclopentene-1,3-diones designed as non-nitro radiosensitizers and hypoxic cytotoxins. *Jpn. J. Cancer Res. (Gann)* **78**, 1128–1133.

Hori, H., Murayama, C., Mori, T., Shibamoto, Y., Abe, M., Onoyama, Y., and Inayama, S. (1989). KIH-802: 2-nitroimidazole-1-acetohydroxamate as a hypoxic cell radiosensitizer. *Int. J. Radiat. Oncol. Biol. Phys.* **16**, 1029–1032.

Ishida, N., Miyazaki, K., Kumagai, K., and Rikimaru, M. (1965). Clinical studies of carzinostatin, an antitumor substance. *J. Antibiot.* **8**, 68–76.

Jagannadham, V., and Steenken, S. (1984). One-electron reduction of nitrobenzenes by α-hydroxy-alkyl radicals via addition/elimination. An example of an organic inner-sphere electron-transfer reaction. *J. Am. Chem. Soc.* **106**, 6542–6551.

Kappen, L.S., and Goldberg, I. H. (1978). Activation and inactivation of neocarzinostatin-induced cleavage of DNA. *Nucleic Acids Res.* **5**, 2959–2967.

Kappen, L.S., and Goldberg, I. H. (1985). Activation of neocarzinostatin chromophore and formation of nascent DNA damage do not require molecular oxygen. *Nucleic Acids Res.* **13**, 1637–1648.

Kappen, L. S., Lee, T. R., Yang, C.-C., and Goldberg, I. H. (1989). Oxygen transfer from the nitro group of a nitroaromatic radiosensitizer to a DNA sugar damage product. *Biochemistry* **28**, 4540–4542.

Karczmar, G. S., Arbeit, J. M., Toy, B. J., Speder, A., and Weiner, M. W. (1992). Selective depletion of tumor ATP by 2-deoxyglucose by ^{31}P magnetic resonance spectroscopy. *Cancer Res.* **52**, 71–76.

Kim, K.-H., Kwon, B.-M., Myers, A. G., and Rees, D. C. (1993). Crystal structure of neocarzinostatin, an antitumor protein-chromophore complex. *Science* **262**, 1042–1045.

Koike, H., Hori, H., Inayama, S., and Terada, H. (1988). Effect of arylidenecyclopenetendione radiosensitizers on ATP synthesis in mitochondria: Action as potent inhibitors of phosphate transport. *Biochem. Biophys. Res. Commun.* **155**, 1066–1074.

Laderoute, K. R., and Rauth, A. M. (1986). Identification of two major reduction products of the hypoxic cell toxin 3-amino-1,2,4-benzotriazine-1,4-dioxide. *Biochem. Pharmacol.* **35**, 3417–3420.

Laderoute, K. R., Workman, P., and Rauth, A. M. (1988). Molecular mechanisms for the hypoxia-dependent activation of 3-amino-1, 2, 4-benzotriazine-1, 4-dioxide (SR 4233). *Biochem. Pharmacol.* **37**, 1487–1495.

Lee, S. H., and Goldberg, I. H. (1989). Sequence specific, strand-selective, and directional binding of neocarzinostatin chromophore to oligodeoxyribonucleotides. *Biochemistry* **28**, 1019–1026.

Lin, A. J., Cosby, L. A., Shansky, C. W., and Sartorelli, A. C. (1972). Bioreductive alkylating agents: 1. Benzoquinone derivatives. *J. Med. Chem.* **15**, 1247–1252.

Meinhofer, J., Maeda, H., Glaser, C. B., Crombos, J., and Kuromizu, M. (1972). Primary structure of neocarzinostatin: Antitumor protein. *Science* **178**, 875–876.

Michael, B. D., Harrop, H. A., and Held, K. D. (1981). Time scale and mechanisms of the oxygen effect in irradiated bacteria. In M. A. J. Rodgers and E. L. Powers (Eds.), *Oxygen and Oxy-radicals in Chemistry and Biology*. Academic Press, New York, pp. 285–296.

Michell, J. B., and Coleman, C. N. (1992). Keynote address: Biochemical modification of therapeutic response. *Int. J. Radiat. Oncol. Biol. Phys.* **22**, 483–484.

Mulcahy, R. T., Gipp, J. J., Ublacker, G. A., and McClelland, R. A. (1990). Enhancement of melphalan (L-PAM) toxicity by reductive metabolites of 1-methyl-2-nitroimidazole, a model nitroimidazole chemosensitizing agent. *Biochem. Pharmacol.* **40**, 2671–2676.

Myers, A. G. (1987). Proposed structure of the neocarzinostatin chromophore-methyl thioglycolate adduct; A mechanism for the nucleophilic activation of neocarzinostatin. *Tetrahedron Lett.* **28**, 4493–4496.

Myers, A. G., and Proteau, P. J. (1989). Evidence for spontaneous, low-temperature biradical formation from a highly reactive neocarzinostatin chromophore-thiol conjugate. *J. Am. Chem. Soc.* **111**, 1146–1147.

Myers, A. G., Proteau, P. J., and Handel, T. M. (1988). Stereochemical assignment of neocarzinostatin chromophore. Structure of neocarzinostatin chromophore-methyl thioglycolate adducts. *J. Am. Chem. Soc.* **110**, 7212–7213.

Myers, A. G., Harrington, P. M., and Kwon, B. (1992). Evidence for aminoglycoside participation in thiol activation of neocarzinostatin chromophore. Synthesis and reactivity of the epoxy-dienediyne core. *J. Am. Chem. Soc.* **114**, 1086–1087.

Nagai, K., Carter, B. J., Xu, J., and Hecht, S. M. (1991). DNA cleavage by oxygen radicals produced in the absence of metal ions or light. *J. Am. Chem. Soc.* **113**, 5099–5100.

Nagasawa, H., Bando, M., Hori, H., Satoh, T., Tada, T., Onoyama, Y., and Inayama, S. (1992). Radiosensitizing, toxicological, and pharmacokinetic properties of hydroxamate analogues of nitroimidazoles as bifunctional radiosensitizers/chemical modifiers. *Int. Radiat. Oncol. Biol. Phys.* **22**, 561–564.

Nicolaou, K. C., and Dai, W.-M. (1991). Chemistry and biology of the enediyne anticancer antibiotics. *Angew. Chem., Int. Ed. Engl.* **30**, 1387–1416.

Nishimoto, S., Ide, H., Wada, T., and Kagiya, T. (1983). Radiation-induced by hydroxylation of thymine promoted by electron-affinic compounds. *Int. J. Radiat. Biol. Relat. Stud. Phys. Chem. Med.* **44**, 585–600.

Nishizawa, K., Sato, C., and Morita, T. (1979). Alterations in the survival of X-irradiated cells by 2, 4-dinitrophenol depending on ATP deprivation. *Int. J. Radiat. Biol. Relat. Stud. Phys. Chem. Med.* **35**, 15–22.

Ormerod, M. G., and Alexaner, P. (1963). On the mechanism of radiation protection by cysteamine: An investigation by means of electron spin resonance. *Radiat. Res.* **18**, 495–509.

Pedersen, P. L. (1978). Tumor mitochondria and bioenergetics of cancer cells. *Prog. Exp. Tumor Res.* **22**, 190–274.

Povirk, L. F., and Goldberg, I. H. (1984). Competition between anaerobic covalent linkage of neocarzinostatin chromophore to deoxyribose in DNA and oxygen-dependent strand breakage and base release. *Biochemistry* **23**, 6304–6311.

Raleigh, J. A., Greenstock, C. L., and Kremers, W. (1973a). Chemical radiosensitization of phosphate ester cleavage *Int. J. Radiat. Biol. Relat. Stud. Phys. Chem. Med.* **23**, 457–467.

Raleigh, J. A., Greenstock, C. L., Whitehouse, J., and Kremers, W. (1973b). Radiosensitization of phosphate release from 3′- and 5′-nucleotides: Correlations between chemical change and biological inactivation. *Int. J. Radiat. Biol. Relat. Stud. Phys. Chem. Med.* **24**, 595–603.

Sartorelli, A. C. (1988). Therapeutic attack of hypoxic cells of solid tumors: Presidential address. *Cancer Res.* **48**, 775–778.

Sasai, K., Shibamoto, Y., Takahashi, M., Zhou, L., Hori, H., Nagasawa, H., Shibata, T., Inayama, S., and Abe, M. (1990). KIH-802, and acetohydroxamic acid derivative of 2-nitroimidazole, as a new potent hypoxic cell radiosensitizer: Radiosensitizing activity, acute toxicity, and pharmacokinetics. *Cancer Chemother. Pharmacol.* **26**, 112–116.

Schwarz, G. (1909). Über Desensibilisierung gegen Rontogen- und Radiumstrahlen. *Muench. Med. Wochenschr.* **56**, 1217–1218.

Steenken, S., and Jagannadham, V. (1985). Reaction of 6-yl radicals of uracil, thymine, and cytosine and their nucleosides and nucleotides with nitrobenzenes via addition to give nitroxide radicals. OH^- catalyzed nitroxide heterolysis. *J. Am. Chem. Soc.* **107**, 6818–6826.

Stratford, I. J. (1992). Keynote address: Concepts and development in radiosensitization of mammalian cells. *Int. J. Radiat. Oncol. Biol. Phys.* **22**, 529–532.

Sun, J. R., and Brown, J. M. (1989). Enhancement of the antitumor effect of flavone acetic acid by the bioreductive cytotoxic drug SR 4233 in a murine carcinoma. *Cancer Res.* **49**, 5664–5670.

Tannock, I. F., and Rotin, D. (1989). Acid pH in tumors and its potential for therapeutic exploitation. *Cancer Res.* **49**, 4373–4384.

Thomlinson, R. H., and Gray, L. H. (1955). The histological structure of some human lung cancers and the possible implications for radiotherapy. *Br. J. Cancer* **9**, 539–549.

Tocher, J. H., and Edwards, D. I. (1990). Electrochemical characteristics of nitroheterocyclic compounds of biological interest. V. Measurement and comparison of nitro radical lifetimes. *Int. J. Radiat. Biol.* **57**, 45–53.

Tocher, J. H., and Edwards, D. I. (1992). The interaction of reduced metronidazole with DNA bases and nucleosides. *Int. J. Radiat. Oncol. Biol. Phys.* **22**, 661–663.

Varnes, M. E., and Biaglow, J. E. (1982). Inhibition of glycolysis of mammalian cells by misonidazole and other radiosensitizing drugs. Prevention of thiols. *Biochem. Pharmacol.* **31**, 2345–2351.

von Sonntag, C. (1987). *The Chemical Basis of Radiation Biology.* Taylor & Francis, London, pp. 116–275.

Walton, M. I., Wolf, C. R., and Workman, P. (1989). Molecular enzymology of the reductive bioactivation of hypoxic cell cytotoxins. *Int. J. Radiat. Oncol. Biol. Phys.* **16**, 983–986.

Wardman, P. (1984). Radiation chemistry in the clinic: Hypoxic cell radiosensitizers for radiotherapy. *Radiat. Phys. Chem.* **24**, 293–305.

Wardman, P. (1987). The mechanism of radiosensitization by electron-affinic compounds. *Radiat. Phys. Chem.* **30**, 423–432.

Whitmore, G. F., and Varghese, A. J. (1986). The biological properties of reduced nitroheterocycles and possible underlying biochemical mechanisms. *Biochem. Pharmacol.* **35**, 97–103.

Workman, P. (1992). Keynote address: Bioreductive mechanisms. *Int. J. Radiat. Oncol. Biol. Phys.* **22**, 631–637.

Zeman, E. M. Brown, J. M., Lemmon, M. J., Hirst, V. K., and Lee, W. W. (1986). SR-4233: A new bioreductive agent with high selective toxicity for hypoxic mammalian cells. *Int. J. Radiat. Oncol. Biol. Phys.* **12**, 1239–1242.

Zeman, E. M., Hirst, V. K., Lemmon, M. J., and Brown, J. M. (1988). Enhancement of radiation-induced tumor cell killing by hypoxic cell toxin SR 4233. *Int. J. Radiat. Oncol. Biol. Phys.* **12**, 209–218.

14

THE ROLE OF ACTIVE OXYGEN SPECIES IN LUNG TOXICITY INDUCED BY MINERAL FIBERS AND PARTICULATES

Yvonne M. W. Janssen and Paul J. A. Borm

Department of Health Risk Analysis and Toxicology, University of Limburg, 6200 MD, Maastricht, The Netherlands

Joanne P. Marsh and Brooke T. Mossman

Department of Pathology, University of Vermont, Burlington, Vermont 05405

Environmental Oxidants, Edited by Jerome O. Nriagu and Milagros S. Simmons.
ISBN 0–471–57928–9 © 1994 John Wiley & Sons, Inc.

1. INTRODUCTION

Inhalation of mineral fibers or particulates results in their deposition in lung tissue. The degree of penetration into the lung and the persistence of the minerals depend in part on the physicochemical characteristics of the particulates. For example, several types of asbestos fibers in the amphibole family (crocidolite) persist for a number of years in the lung, whereas others, such as chrysotile asbestos fibers, are more degradable and disappear faster (Mossman et al., 1990a). The differences in persistence in the lung may be due to elements such as magnesium or silica, which leach out of the fiber and result in dissociation.

Another important determinant affecting the persistence of mineral fibers or particulates within the lung relates to the ability of macrophages to engulf these minerals during phagocytosis and transport them out of the lung via the bronchial tree or the lymphatic system (Kamp et al., 1992). Due to the fibrous structure of some minerals, phagocytosis is incomplete, causing the phenomenon of "frustrated phagocytosis." Figure 1 shows a macrophage attempting to engulf an asbestos fiber. At high fiber concentrations, this results in death of the macrophage and renewed release of fibers into the lung. Moreover, at nonlethal fiber concentrations, the macrophage can be chronically activated, thus releasing a number of inflammatory mediators and active oxygen species (AOS) (Farber et al., 1990; Mossman and Marsh, 1989; Freeman and Crapo, 1982). The persistent accumulation of fibers in the lung, chronic activation of macrophages, and

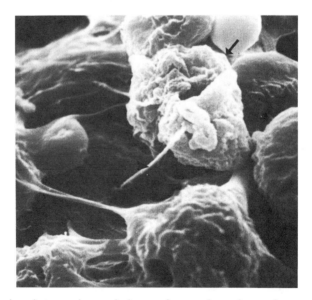

Figure 1. Scanning electron microscopic image of macrophages in an airspace phagocytizing chrysotile asbestos fibers. Macrophages are indicated by arrows; arrowheads represent chrysotile asbestos fibers.

accumulation of inflammatory cells cause local tissue damage. Aberrant repair of injury then results in collagen deposition leading to pulmonary fibrosis. The type and extent of fibrosis depend on the type of mineral inhaled and the site of its deposition within the lung (Mossman et al., 1990a). For example, nodular areas of fibrosis are observed following inhalation of silica particulates, whereas a more diffuse pattern of fibrosis is caused by inhalation of asbestos fibers.

In addition to fibrosis, two forms of malignant disease arising from different cell types are observed following exposure to asbestos. Bronchogenic carcinomas or lung cancer have been documented following occupational exposure to asbestos. Additive or synergistic relationships between cigarette smoking and asbestos exposure have been observed in workers (Mossman et al., 1990a). On the other hand, malignant mesothelioma arising from the pleura of the lung is not associated with cigarette smoking.

Evidence for the development of lung cancer following exposure to other mineral dusts is more controversial. Experimental studies have demonstrated development of pulmonary malignancies in laboratory animals following exposure to silica dusts (Craighead, 1992). However, epidemiological studies have not strenghtened a cause and effect relationship between silica exposure and lung cancer in humans (Craighead, 1992; McDonald, 1989; Pairon et al., 1991). Studies examining the association between exposure to mineral dusts and the development of lung cancer have been complicated by the frequent finding that a particular mineral appears to be contaminated with other minerals. This increases the difficulty of relating exposure to a specific mineral to the etiology of a specific disease. For example, amphibole asbestos fibers are sometimes found as contaminants of commercially mined minerals, including chrysotile, talc, and vermiculite (Mossman et al., 1990a).

2. PHYSICOCHEMICAL CHARACTERISTICS OF MINERALS RELATED TO AOS FORMATION

Various investigators have shown that cell-free mineral preparations have the capability to generate AOS (Eberhardt et al., 1985; Weitzman and Graceffa, 1984; Dalal et al., 1989a; Fubini et al., 1989). A number of techniques are available to measure AOS formation, including electron spin resonance (ESR). The fact that many AOS are very reactive and consequently unstable necessitates the use of chemical spin traps to generate a longer lived intermediate that can be identified by its specific resonance pattern. These techniques have shown that different types of asbestos, including chrysotile, crocidolite, and amosite asbestos, can generate hydroxyl radical ($OH\cdot$) in the presence of hydrogen peroxide (Eberhardt et al., 1985; Weitzman and Graceffa, 1984; Zalma et al., 1987). Transition metals present on many minerals, but most notably on crocidolite asbestos, drive a reaction known as the Fenton (modified Haber–Weiss) reaction and results in $OH\cdot$ formation.

The importance of iron in this process is demonstrated by studies using the iron chelator desferoxamine, which stabilizes iron and consequently inhibits the formation of OH· (Kamp et al., 1992). The ability to generate AOS in an aqueous solution is a property that is not specific to asbestiform fibers, for some other mineral fibers including glass have this ability as well (Gulumian and van Wyk, 1987). AOS production by minerals appears to be directly related to the amount of available iron present on the fibers. Fresh grinding or crushing of the fiber makes more iron available and consequently enhances OH· formation. In addition to mineral fibers, a number of nonfibrous mineral dusts generate AOS in aqueous solutions. For instance, ESR spectroscopy revealed that freshly ground anthracite coal generates AOS (Dalal et al., 1989b). Nickel or copper arsenides, present in metal mines or industrial sites, form AOS (Costa et al., 1989a,b). Grinding of silica cleaves bonds in a silica tetrahydral lattice, generating free radical sites on the surface of silica particles. ESR spectroscopy has shown formation of SiO· and Si· radicals from freshly ground silica, and has indicated OH· formation upon reaction of silicon-based surface radicals with water (Dalal et al., 1989a; Fubini et al., 1989; Castranova et al., 1989).

3. FORMATION AND DETOXIFICATION OF AOS BY CELLS

In addition to the intrinsic ability of mineral dusts to generate AOS, these species are also formed as a consequence of the encounter of target cells with mineral dusts. As mentioned above, during phagocytosis of fibers by pulmonary macrophages or polymorphonuclear leukocytes (PMN), a membrane-bound NADPH-oxidase is activated and causes formation of a realm of AOS (Mossman et al., 1987; Mossman and Marsh, 1991). The pathway of AOS formation is shown in Figure 2. However, other processes activated within the cells can also cause AOS formation. For instance, release of the cytokine tumor necrosis factor (TNF) after exposure to mineral dusts, including asbestos and silica (Dubois et al., 1989), is also thought to cause oxidative stress (Larrick and Wright, 1990; Vilcek and Lee, 1991). Activation of the enzyme xanthine oxidase, observed after a number of

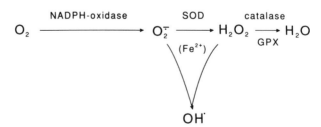

Figure 2. Formation and removal of active oxygen species. Activation of a membrane-bound NADPH-oxidase in phagocytes by mineral dusts causes formation of $O_2^{\cdot-}$, which is converted to H_2O_2 by SOD. Subsequent action of catalase or GPX detoxifies H_2O_2, leading to the formation of H_2O. $O_2^{\cdot-}$, superoxide anion; OH·, hydroxyl radical; H_2O_2, hydrogen peroxide; SOD, superoxide dismutase; GPX, glutathione peroxidase.

stresses, is another source of AOS formation within cells (Friedl et al., 1989; Till et al., 1991).

Mammalian cells have an elaborate system of enzymatic and nonenzymatic antioxidants that scavenge AOS (Farber et al., 1990; Sun, 1990; Heffner and Repine, 1991). The functions of some antioxidant enzymes are outlined in Figure 2. Mammalian cells contain three different superoxide dismutases (SOD), which convert superoxide anion ($O_2^{\cdot-}$) into H_2O_2, a nonradical intermediate. Two SODs contain copper and zinc: CuZnSOD, localized within peroxisomes (Keller et al., 1991), and extracellular superoxide dismutase, present in the extracellular milieu of cells (Marklund, 1984). The third form of SOD contains manganese (MnSOD) and is situated in mitochondria (Wispé et al., 1989). Both catalase and glutathione peroxidase (GPX) are enzymes that convert H_2O_2 into H_2O. These enzymatic antioxidants are complemented by a number of nonenzymatic factors including glutathione, vitamins C and E, albumin, ceruloplasmin, and uric acid (Heffner and Repine, 1989). Normally, a balance exists between formation of AOS and antioxidant defenses. However, oxidant-induced damage might ensue after excessive production of oxidants or deficient functioning of the antioxidant system (Farber et al., 1990).

Our laboratory has been investigating responses of cells after exposure to various mineral dusts. We have focused on (1) production of AOS following exposure to mineral dusts, (2) alteration of antioxidant defenses following mineral exposure, and (3) implications of these processes in cell or tissue damage, proliferation, and carcinogenesis.

4. FORMATION OF AOS AFTER EXPOSURE TO MINERAL DUSTS

A number of investigators have focused on generation of AOS following exposure of cells or tissues to mineral dusts. These studies have used reduced cytochrome c to measure $O_2^{\cdot-}$, phenol-red horseradish peroxidase to measure H_2O_2, and chemiluminescence techniques.

Results of experiments measuring extracellular AOS formation by cells following exposure to mineral dusts have been summarized in a recent review (Kamp et al., 1992). A number of different cell types from humans and animals have been used, including alveolar macrophages, peritoneal macrophages, and PMN, to measure AOS production following in vitro exposure to mineral dusts. Some investigators have measured AOS formation in lavaged cells in healthy individuals exposed to mineral dusts, or in workers with occupational lung disease caused by exposure to minerals (Kamp et al., 1992). In some studies, AOS formation following mineral dust exposure could be correlated with cytotoxicity, whereas in other cases this correlation was not evident.

The fibrous nature of asbestos is important in triggering $O_2^{\cdot-}$ release from inflammatory cells. We previously demonstrated that the nonfibrous particles riebeckite, mordenite, and glass are significantly less active in $O_2^{\cdot-}$ formation by

AM than the chemically identical fibers crocidolite, erionite, and code-100 fiberglass, respectively (Hansen and Mossman, 1987).

Measurement of products of lipid peroxidation is another approach used by several laboratories to demonstrate involvement of AOS in mineral dust-induced toxicity. Lipid peroxidation has been demonstrated following mineral exposure in inflammatory cells and whole lungs of laboratory animals (Kamp et al., 1992; Petruska et al., 1990b). Measurement of the products of lipid peroxidation in the urine of workers exposed to asbestos or silica showed increased levels of thiobarbituric-acid-reactive material, indicating the involvement of AOS following occupational exposure to these minerals (Kamal et al., 1989).

Other studies addressing the involvement of AOS in mineral dust-induced cell or tissue damage have used scavengers of AOS. For example, our laboratory has demonstrated that cytotoxicity in rat lung fibroblasts, hamster tracheal epithelial (HTE) cells, or AM induced by asbestos could be ameliorated or prevented by concomitant exposure to antioxidants (Mossman et al., 1986; Mossman and Marsh, 1989, 1991; Shatos et al., 1987). Similarly, silica-induced cell death was decreased by glutatione or glutathione precursors (Voisin et al., 1987). More important, systemic administration of polyethylene glycol-conjugated catalase in rats during inhalation of crocidolite asbestos reduced the pulmonary inflammation, injury, and fibrosis normally observed following asbestos inhalation (Mossman et al., 1990b).

5. ALTERATION OF ANTIOXIDANT DEFENSES AFTER MINERAL DUST EXPOSURE

Following exposure to AOS, cells can alter their antioxidant defenses in order to prevent AOS-mediated damage. We hypothesized that exposure to mineral dusts causing formation of AOS gives rise to alterations in antioxidant enzymes. We examined antioxidant enzymes in vitro using different target cells of mineral dust-induced disease, in rat lungs, and in workers exposed to mineral dusts.

One technique used in our laboratory to study regulation of antioxidant enzymes following exposure to AOS or mineral dusts is Northern blot analysis to determine the gene expression of different antioxidant enzymes. In this technique, total RNA is extracted from cells or lung tissue, electrophoresed and transferred onto nitrocellulose filters, and hybridized with [^{32}P] cDNA probes encoding antioxidant enzyme genes (Sambrook et al., 1989). With this technique we demonstrated differential gene regulation of antioxidant enzymes in HTE cells following exposure to H_2O_2 or the AOS-generating system xanthine plus xanthine oxidase (X/XO). X/XO treatment of cells led to increased gene expression of MnSOD, whereas exposure of cells to H_2O_2 caused increased mRNA levels of catalase and increased expression of glutathione peroxidase and, to a lesser extent, MnSOD (Shull et al., 1991).

In recent studies, we examined oxidant stress responses in human adult lung fibroblasts (HAL) and human pleural mesothelial cells (HMC) following exposure to asbestos or X/XO (Janssen et al., 1992a). Results of these studies show that HMC respond to asbestos or X/XO with increases in mRNA levels of MnSOD. In addition, X/XO or asbestos caused increased gene expression of heme oxygenase (HO) (Janssen et al., 1992a), an enzyme induced in mammalian cells following oxidative stress and implicated in cell defense against AOS (Applegate et al., 1991; Keyse and Tyrrell, 1989). In HAL cells, crocidolite asbestos caused increases in mRNA levels of HO, whereas MnSOD gene expression remained unaltered (Janssen et al., 1992a). Figure 3 shows mRNA levels of MnSOD and HO in HMC cells following exposure to crocidolite asbestos or X/XO. Crocidolite or X/XO cause increased gene expression of MnSOD and HO in HMC cells, whereas addition of polystyrene beads, a negative particulate control, failed to alter mRNA levels of these enzymes. HMC cells are more sensitive than HAL cells to the cytotoxic effects of asbestos or X/XO. These initial studies suggest that HMC cells are not compromised in their ability to increase antioxidant defense following exposure to AOS or mineral dusts.

In addition, we investigated antioxidant enzymes in rat lung after inhalation of crocidolite asbestos or cristobalite silica (Janssen et al., 1992b). In crocidolite-exposed rat lungs we observed increased gene expression of MnSOD, GPX, and catalase, increased MnSOD protein levels, and increased enzyme activities of total SOD, catalase, and GPX. In contrast, cristobalite caused striking increases in MnSOD mRNA and protein levels, whereas gene expression of other anti-

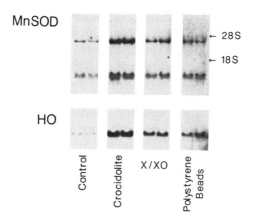

Figure 3. Northern blot of MnSOD and heme oxygenase (HO) in HMC cells following exposure to crocidolite asbestos or xanthine plus xanthine oxidase (X/XO). Confluent HMC cells were treated with Union Internationale Contre Le Cancer (UICC) reference crocidolite asbestos ($1.9\ \mu g/cm^2$), X/XO, a generating system of AOS (0.1 U/mL, Calbiochem, La Jolla, CA), or polystyrene beads ($10\ \mu g/cm^2$, Polysciences, Warrington, PA). After 24 hr of exposure, RNA was extracted, electrophoresed, and transferred onto nitrocellulose filters (Sambrook et al., 1989). Blots were hybridized with cDNA probes encoding MnSOD or HO, as described elsewhere (Applegate et al., 1991; Janssen et al., 1992b). Ribosomal RNA species (28 S and 18 S) are indicated by arrows. Note the multiple mRNA species for MnSOD (Shull et al., 1991).

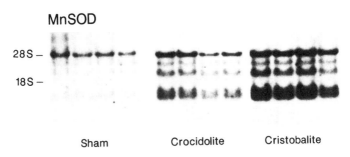

Figure 4. Northern blot of MnSOD in rat lung after nine days of inhalation of crocidolite asbestos or cristobalite silica. Rats were exposed to $7-10 \ mg/m^3$ air of crocidolite or cristobalite. RNA was processed as described in Figure 3. Increases in MnSOD mRNA levels are apparent after inhalation of either mineral.

oxidant enzymes remained unaltered. In addition, no alterations in activities of SOD, catalase, or GPX were observed. These results indicate distinct mechanisms of regulation of antioxidant enzymes following, inhalation of crocidolite or cristobalite (Janssen et al., 1992b). Figure 4 shows MnSOD mRNA levels in rat lung after nine days of inhalation of crocidolite asbestos or cristobalite silica, at which point we observed increased gene expression of MnSOD. Both minerals caused increases in MnSOD mRNA levels in lung, although more striking increases were observed following cristobalite exposure. Using immunocytochemistry, we showed that MnSOD protein was localized predominantly in mitochondria of type II pneumocytes. Quantitative increases in MnSOD protein were apparent in type II cells following inhalation of crocidolite or cristobalite, which correlated with overall increases in MnSOD immunoreactive protein in whole lung (Holley et al., 1992).

Several investigators have demonstrated induction of MnSOD gene expression after exposure to bacterial endotoxin (Lipopolysaccharide, LPS), TNF, or interleukin 1 (IL-1) (Wong and Goedel, 1988; Masuda et al., 1988; Shaffer et al., 1990). TNF and IL-1 are released by inflammatory cells after in vitro exposure to asbestos or silica (Dubois et al., 1989; Driscoll et al., 1989). Therefore, TNF could mediate the increases in gene expression and protein levels of MnSOD that we observed in rats following exposure to asbestos or silica (Driscoll et al., 1992).

Studies examining workers exposed to silica-containing dusts, including coal mine dust, have revealed alterations of antioxidant defenses in peripheral blood. For example, activities of some antioxidant enzymes were correlated with radiological abnormalities in red blood cells of coal workers but not in control workers. Furthermore glutathione levels were decreased in coal workers in the early stages of pneumoconiosis, but were increased in patients with progressive massive fibrosis (Borm et al., 1986; Engelen et al., 1990).

In summary, these studies reveal specific alterations of some components of the antioxidant defense system following exposure to mineral dusts in vitro in isolated cell systems, in rat lungs, and in workers exposed to mineral dusts. Increases in antioxidants could reflect an adaptation to mineral-dust-induced oxidant stress. However, in our rapid onset inhalation models of asbestosis, rats develop acute pulmonary damage and subsequently a diffuse fibrosis, despite increases in antioxidant defenses. This might indicate that increases in antioxidant defenses are insufficient to protect the rats from disease at the high airborne concentrations of minerals applied in our model ($7-10\,\text{mg/m}^3$ air) and generally used by others.

6. INVOLVEMENT OF AOS IN PROLIFERATION INDUCED BY ASBESTOS

In addition to altering antioxidant defenses, mineral dusts cause proliferative alterations in pulmonary target cells of asbestos-induced disease. Our laboratory has focused on ornithine decarboxylase (ODC), a rate-limiting enzyme in polyamine synthesis that is essential for cell division (Gilmour et al., 1987; O' Brien, 1976). Tumor promotors will induce ODC enzyme activity, and therefore ODC enzyme induction may be linked to tumor promotion (Marsh and Mossman, 1991). Initial studies showed that the fibrous nature of asbestos fibers is critical in augmenting ODC activity in HTE cells (Marsh and Mossman, 1988). In addition to asbestos, glass fibers were also potent inducers of ODC activity. Exposure of HTE cells to X/XO also caused increases in ODC mRNA and enzyme activity, showing that AOS have the ability to induce ODC (Marsh and Mossman, 1991). Recent studies showed that antioxidant enzymes could diminish asbestos-induced ODC mRNA levels and enzyme activity, indicating that AOS mediate ODC induction by asbestos (Marsh and Mossman, 1991). Figure 5 shows a Northern blot of ODC mRNA levels in rat lung following nine days of inhalation of asbestos or silica. Inhalation of either mineral results in increased mRNA levels of ODC in lung. These findings indicate that proliferative changes occur in lung following inhalation of these mineral dusts, presumably by oxidant-dependent mechanisms.

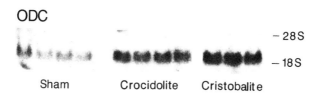

Figure 5. Northern blot of ODC in rat lung after nine days of inhalation of crocidolite asbestos or cristobalite silica, both at $7-10\,\text{mg/m}^3$ air. Inhalation of crocidolite or cristobalite causes increases in ODC mRNA levels in rat lung.

7. SUMMARY AND CONCLUSIONS

A causal role for AOS in the etiology of asbestos-induced lung disease has been demonstrated by our laboratory (Mossman and Marsh, 1989; Shatos et al., 1987; Mossman et al., 1990b). AOS are generated by acellular mechanisms, driven by transition metals on the fibers and by inflammatory cells following phagocytosis of fibers within the lung. AOS cause a multitude of effects. They damage a number of macromolecules, including proteins, lipids, and DNA. In addition, they alter genetic processes affecting the expression of various genes. We have shown that the gene expression of antioxidant enzymes is augmented following exposure to mineral dusts, including asbestos and silica in rat lung, and following exposure to asbestos in HMC cells in vitro. These increases in antioxidant defenses can be regarded as a reaction to mineral dust-induced oxidative stress that might lead to adaptation to subsequent oxidative injury.

Aside from their ability to induce antioxidant defenses, mineral dusts at low doses also have proliferative effects in cells in vitro and in whole lung. We have shown that ODC, an enzyme intrinsic to cell proliferation, is increased in rat lung following inhalation of asbestos or silica and in vitro in HTE cells after exposure to asbestos. Evidence for the involvement of AOS in these responses was obtained from studies in our laboratory showing that antioxidants could ameliorate increases in ODC gene expression and activity induced by asbestos in HTE cells (Marsh and Mossman, 1991). The balance between increases in antioxidant defenses and increases in proliferation could be an important determinant of the effects of certain mineral dusts on the lung and the development of disease.

Clearly, AOS are not the only factors mediating pulmonary effects of mineral dusts, but are part of a cascade of processes triggered within the lung following inhalation of minerals. A number of cytokines or growth factors are generated in the lung along with AOS following mineral deposition, some of which have profound effects on disease processes elicited in the lung (Borm et al., 1988; Piquet et al., 1990). However, our laboratory has focused on AOS as "second messengers" of asbestos-induced toxicity and has demonstrated both in vitro and in vivo that these species are critical to the development of asbestos-induced pulmonary damage, fibrosis, and cell proliferation driven by asbestos fibers (Mossman and Marsh, 1991). More studies are needed to acertain the role of AOS in pulmonary effects caused by other mineral dusts.

ACKNOWLEDGMENTS

The authors thank Judith Kessler for providing the illustrations. This work was supported by grants from NIH (RO1) and the EPA to B. T. Mossman and CEC to P. J. A. Borm.

REFERENCES

Applegate, L. A., Luscher, P., and Tyrrell, R. M. (1991). Induction of heme oxygenase: A general response to oxidant stress in cultured mammalian cells. *Cancer Res.* **51**, 974–978.

Borm, P. J. A., Bast, A., Wouters, E. F. M., Slangen, J. J. M., Swaen, G. M. H., and de Boorder, Tj. (1986). Red blood cell antioxidant parameters in silicosis. *Int. Arch. Occup. Environ. Health* **58**, 235–244.

Borm, P. J. A., Palmen, N., Engelen, J. J. M., and Buurman, W. (1988). Spontaneous and stimulated release of tumor necrosis factor (TNF-α) from blood monocytes of miners with coal workers' pneumoconiosis. *Am. Rev. Respir. Dis.* **138**, 1589–1594.

Castranova, V., Vallyathan, V., Van Dyke, K., and Dalal, N. S. (1989). Use of chemiluminescence assays to monitor the surface characteristics and biological reactivity of freshly fractured vs aged silica. In B. T. Mossman and R. O. Bégin (Eds.), *Effects of Mineral Dusts on Cells.* Springer-Verlag, Berlin, pp. 181–188.

Costa, D., Guignard, J., and Pezerat, H. (1989a). Production of free radicals arising from the surface activity of minerals and oxygen. Part II. Arsenides, sulfides, and sulfoarsenides of iron, nickel and copper. *Toxicol. Ind. Health* **5**, 1079–1097.

Costa, D., Guignard, J., and Pezerat, H. (1989b). Production of free radicals by non-fibrous materials in a cell-free buffer medium. In B. T. Mossman and R. O. Bégin (Eds.), *Effects of Mineral Dusts on Cells.* Springer-Verlag, Berlin, pp. 189–196.

Craighead, J. E. (1992). Do silica and asbestos cause lung cancer. *Arch. Pathol. Lab. Med.* **116**, 16–20.

Dalal, N. S., Shi, X., and Vallyathan, V. (1989a). Potential role of silicon-oxygen radicals in acute lung injury. In B. T. Mossman and R. O. Bégin (Eds.), *Effects of Mineral Dusts on Cells.* Springer-Verlag, Berlin, pp. 265–272.

Dalal, N. S., Suryan, M. M., Vallyathan, V., Green, F. H. Y., Jafari, B., and Wheeler, R. (1989b). Detection of reactive free radicals in fresh coal mine dust and their implication for pulmonary injury. *Ann. Occup. Hyg.* **33**, 79–84.

Driscoll, K. E., Lindenschmidt, R. C., Maurer, J. K., and Higgins, J. M. (1989). Release of interleukin-1 and tumor necrosis factor by rat alveolar macrophages after in vivo and in vitro exposure to mineral dusts. In B. T. Mossman and R. O. Bégin (Eds.), *Effects of Mineral Dusts on Cells.* Springer-Verlag, Berlin, pp. 101–108.

Driscoll, K. E., Strzelecki, J., Hassenbein, D., Janssen, Y. M. W., Marsh, J., Oberdorster, G., and Mossman, B. T. (1992). Tumor necrosis factor (TNF): Evidence for the role of TNF in increased expression of manganese superoxide dismutase after inhalation of mineral dusts. *Ann. Occup. Hyg.* (in press).

Dubois, C. M., Bissonnette, E., and Rola-Pleszczynski, M. (1989). Asbestos fibers and silica particles stimulate rat alveolar macrophages to release tumor necrosis factor: Autoregulatory role of leukotriene B4. *Am. Rev. Respir. Dis.* **139**, 1257–1264

Eberhardt, M. K., Roman-Franco, A. A., and Quiles, M. R. (1985). Asbestos-induced decomposition of hydrogen peroxide. *Environ. Res.* **37**, 287–292.

Engelen, J. J., Borm, P. J., van Sprundel, M., and Leenaerts, L. (1990). Blood anti-oxidant parameters at different stages of pneumoconiosis in coal workers. *Environ. Health Perspect.* **84**, 165–172.

Farber, J. L., Kyle, M. E., and Coleman, J. B. (1990). Biology of disease. Mechanisms of cell injury by activated oxygen species. *Lab. Invest.* **62**, 670–679.

Freeman, B. A., and Crapo, J. D. (1982). Biology of disease. Free radicals and tissue injury. *Lab. Invest.* **47**, 412–426.

Friedl, H. P., Till, G. O., Ryan, U. S., and Ward, P. A. (1989). Mediator-induced activation of xanthine oxidase in endothelial cells. *FASEB J.* **3**, 2512–2518.

Fubini, B., Bolis, V., Giamello, E., Pugliese, L., and Volante, M. (1989). The formation of oxygen reactive radicals at the surface of the crushed quartz dusts as a possible cause of silica pathogenicity. In B. T. Mossman and R. O. Bégin (Eds.), *Effects of Mineral Dusts on Cells.* Springer-Verlag, Berlin, pp. 205–214.

Gilmour, S. K., Verma, A. K., Madara, T., and O'Brien, T. G. (1987). Regulation of ornithine decarboxylase gene expression in mouse epidermis and epidermal tumors during two-stage tumorigenesis. *Cancer Res.* **47**, 1221–1225.

Gulumian, M., and van Wyk, J. A. (1987). Hydroxyl radical production in the presence of fibres by a Fenton-type reaction. *Chem-Biol. Interact.* **62**, 89–97.

Hansen, K., and Mossman, B. T. (1987). The generation of superoxide from alveolar macrophages exposed to asbestiform and non-fibrous particles. *Cancer Res.* **47**, 1681–1686.

Heffner, J. E., and Repine, J. E. (1989). Pulmonary strategies of antioxidant defense. *Am. Rev. Respir. Dis.* **140**, 531–554.

Heffner, J. E., and Repine, J. E. (1991). Antioxidants and the lung. In R. G. Chrystal, J. B. West, P. J. Barnes, N. S. Chernlack, and E. R. Weibel (Eds.), *The Lung: Scientific Foundations.* Raven Press, New York, p. 1811.

Holley, J. A., Janssen, Y. M. W., Mossman, B. T., and Taatjes, D. (1992). Increased manganese superoxide dismutase protein in Type II epithelial cells of rat lungs after inhalation of crocidolite asbestos or cristobalite silica. *Am. J. Pathol.* **141**, 475–485.

Janssen, Y. M. W., Marsh, J. P., Absher, M. P., Gabrielson, E., Borm, P. J. A., and Mossman, B. T. (1992a). Gene expression of heme oxygenase and antioxidant enzymes in human mesothelial cells and human adult lung fibroblasts after exposure to asbestos or active oxygen species. *Am. Rev. Respir. Dis.* **145**, 713 (abstr).

Janssen, Y. M. W., Marsh, J. P., Absher, M. P., Hemenway, D., Vacek, P. M., Leslie, K. O., Borm, P. J. A., and Mossman, B. T. (1992b). Expression of antioxidant enzymes in rat lungs after inhalation of asbestos or silica. *J. Biol. Chem.* **267**, 10625–10630.

Kamal, A. A. M., Gomaa, A., Khafif, M. E., and Hammad, A. S. (1989). Plasma lipid peroxides among workers exposed to silica or asbestos dusts. *Environ. Res.* **49**, 173–180.

Kamp, D. W., Graceffa, B., Pryor, W. A., and Weitzmann, S. A. (1992). The role of free radicals in asbestos induced diseases. *Free Radical Biol. Med.* **12**, 293–315.

Keller, G. A., Warner, T. G., Steimer, K. S., and Hallewell, R. A. (1991). Cu, Zn superoxide dismutase is a peroxisomal enzyme in human fibroblasts and hepatoma cells. *Proc. Natl. Acad. Sci. U.S.A.* **88**, 7381–7385.

Keyse, S. M., and Tyrrell, R. M. (1989). Heme oxygenase is the major 32-kDa stress protein induced in human skin fibroblasts by UVA radiation, hydrogen peroxide, and sodium arsenite. *Proc. Natl. Acad. Sci. U.S.A.* **86**, 99–103.

Larrick, J. W., and Wright, S. C. (1990). Cytotoxic mechanism of tumor necrosis factor-alpha. *FASEB J.* **4**, 3215–3223.

Marklund, S. L. (1984). Extracellular superoxide dismutase and other superoxide dismutase isoenzymes in tissues from nine mammalian species. *Biochem. J.* **222**, 649–655.

Marsh, J. P., and Mossman, B. T. (1988). Mechanisms of induction of ornithine decarboxylase activity in tracheal epithelial cells by asebstiform fibers. *Cancer Res.* **48**, 709–714.

Marsh, J. P., and Mossman, B. T. (1991). Role of asbestos and active oxygen species in activation and expression of ornithine decarboxylase in hamster tracheal epithelial cells. *Cancer Res.* **51**, 167–173.

Masuda, A., Longo, D., Kobayashi, Y., Appella, E., Oppenheim, J. J., and Matsushima, K. (1988). Induction of manganese superoxide dismutase by interleukin 1. *FASEB J.* **2**, 3087–3091.

McDonald, J. C. (1989). Silica, silicosis and lung Cancer. *Br. J. Ind. Med.* **46**, 289–291.

Mossman, B. T., and Marsh, J. P. (1989). Evidence supporting a role for active oxygen species in asbestos-induced toxicity and lung disease. *Environ. Health Perspect.* **81**, 91–94.

Mossman, B. T., and Marsh, J. P. (1991). Role of active oxygen species in asbestos-induced cytotoxicity, cell proliferation, and carcinogenesis. In C. C. Harris, J. F. Lechner, and B. R. Brinkley (Eds.), *Cellular and Molecular Aspects of Fiber Carcinogenesis.* Cold Spring Harbor Lab. Press, Cold Spring Harbor, NY, pp. 159–168.

Mossman, B. T., Marsh, J. P., and Shatos, M. A. (1986). Alteration of superoxide dismutase activity in tracheal epithelial cells by asbestos and inhibition of cytotoxicity by antioxidants. *Lab. Invest.* **54**, 204–212.

Mossman, B. T., Marsh, J. P., Shatos, M. A., Doherty, J., Gilbert, R., and Hill, S. (1987). Implication of active oxygen species as second messengers of asbestos toxicity. *Drug Chem. Toxicol.* **10**, 157–180.

Mossman, B. T., Bignon, J., Corn, M., Seaton, A., and Gee, J. B. L. (1990a). Asbestos: Scientific developments and implications for public policy. *Science* **247**, 294–301.

Mossman, B. T., Marsh, J. P., Sesko, A., Hill, S., Shatos, M. A., Doherty, J., Petruska, J., Adler, K. B., Hemenway, D., Mickey, R., Vacek, P., and Kagan, E. (1990b). Inhibition of lung injury, inflammation, and interstitial pulmonary fibrosis by polyethylene glycol-conjugated catalase in a rapid inhalation model of asbestosis. *Am. Rev. Respir. Dis.* **141**, 1266–1271.

O'Brien, T. G. (1976). The induction of ornithine decarboxylase as an early, possibly obligatory, event in mouse skin carcinogenesis. *Cancer Res.* **36**, 2644–2653.

Pairon, J. C., Brochard, P., Jaurand, M. C., and Bignon, J. (1991). Silica and lung cancer: A controversial issue. *Eur. Respir. J.* **4**, 730–744.

Petruska, J. M., Marsh, J., Bergeron, M., and Mossman, B. T. (1990a). Brief inhalation of asbestos compromises superoxide production in cells from bronchoalveolar lavage. *Am. J. Respir. Cell Mol. Biol.* **2**, 129–136.

Petruska, J. M., Wong, S. H. Y., Sunderman, F. W., Jr., and Mossman, B. T. (1990b). Detection of lipid peroxidation in lung and in bronchoalveolar lavage cells and fluid. *Free Radical Biol. Med.* **9**, 51–58.

Piguet, P. F., Collart, M. A., Grau, G. E., Sappino, A. P., and Vassalli, P. (1990). Requirement of tumour necrosis factor for the development of silica-induced pulmonary fibrosis. *Nature (London)* **344**, 245–247.

Sambrook, J., Fritsch, E. F., and Maniatis, T. (Eds.) (1989). *Molecular Cloning: A Laboratory Manual.* Cold Spring Harbor Lab. Press, Cold Spring Harbor, NY.

Shaffer, J. B., Treanor, C. P., and Del Vecchio, P. L. (1990). Expression of bovine and endothelial cell antioxidant enzymes following TNF-α exposure. *Free Radical Biol. Med.* **8**, 497–502.

Shatos, M. A., Doherty, J. M., Marsh, J. P., and Mossman, B. T. (1987). Prevention of asbestos-induced cell death in rat lung fibroblasts and alveolar macrophages by scavengers of active oxygen species. *Environ. Res.* **44**, 103–116.

Shull, S., Heintz, N. H., Periasamy, M., Manohar, M., Janssen, Y. M. W., Marsh, J. P., and Mossman, B. T. (1991). Differential regulation of antioxidant enzymes in response to oxidants. *J. Biol. Chem.* **266**, 24398–24403.

Sun, Y. (1990). Free radicals, antioxidant enzymes, and carcinogenesis. *Free Radical Biol. Med.* **8**, 583–599.

Till, G. O., Friedl, H. P., and Ward, P. A. (1991). Lung injury and complement activation: Role of neutrophils and xanthine oxidase. *Free Radical Biol. Med.* **10**, 379–386.

Vilcek, J., and Lee, T. H. (1991). Tumor necrosis factor. New insights into the molecular mechanisms of its multiple actions. *J. Biol. Chem.* **266**, 7313–7316.

Voisin, C., Aerts, C., and Wallaert, B. (1987). Prevention of in vitro oxidant-mediated alveolar macrophage injury by cellular glutathione and precursors. *Bull. Eur. Physiopathol. Respir.* **23**, 309–313.

Weitzman, S. A., and Graceffa, P. (1984). Asbestos catalyzes hydroxyl and superoxide radical generation from hydrogen peroxide. *Arch. Biochem. Biophys.* **228**, 373–376.

Wispé, J. R., Clark, J. C., Burhans, M. S., Kropp, P. E., Korfhagen, T. R., and Whitsett, J. A. (1989). Synthesis and processing of the precursor for human mangano-superoxide dismutase. *Biochim. Biophys. Acta.* **994**, 30–36.

Wong, G. H. W., and Goedel, D. V. (1988). Induction of manganous superoxide dismutase by tumor necrosis factor: Possible protective mechanism. *Science* **242**, 941–943.

Zalma, R., Bonneau, L., Jaurand, M. C., Guignard, J., and Pezerat, H. (1987). Formation of oxy-radicals by oxygen reduction arising from the surface activity of asbestos. *Can. J. Chem.* **65**, 2338–2341.

15

OXIDATIVE STRESS AND ASBESTOS

Janusz Z. Byczkowski

ManTech Environmental Technology, Inc., Dayton, Ohio 45437–0009

Arun P. Kulkarni

Toxicology Program, College of Public Health MDC-56, University of South Florida, Tampa, Florida 33612

Environmental Oxidants, Edited by Jerome O. Nriagu and Milagros S. Simmons.
ISBN 0–471–57928–9 © 1994 John Wiley & Sons, Inc.

1. INTRODUCTION

Asbestos comprises a group of morphologically heterogeneous mineral fibers, composed mainly of hydrated silicates, containing several transition metals either as structural components or as contaminants (Timbrell, 1970; Platek et al., 1985). The health effects and mode of action of asbestos, and the pathogenesis of asbestos-related diseases have been reviewed recently in several excellent publications (Vallyathan and Green, 1985; Murphy, 1988; Mossman et al., 1990; Selikoff, 1991). A link between asbestos exposure, lipid peroxidation (Goodglick et al., 1989), and pulmonary fibrosis was confirmed experimentally by Petruska et al. (1991). Iron, present on the surface of asbestos fibers, is apparently responsible for the pro-oxidant properties and tissue injury caused by asbestos (Weitzman and Weitberg, 1985; Weitzman et al., 1988).

It is well known that iron in the presence of hydrogen peroxide, lipid hydroperoxides, or superoxide anion radical can generate a whole family of reactive oxygen species. These species promote damage to the tissues and cause toxicity by means of free radical reactions with lipids, DNA, and protein. This deleterious condition caused by reactive oxygen species is known as an oxidative stress. It has been reported that asbestos fibers can catalyze generation of hydroxyl radicals from hydrogen peroxide (Weitzman and Graceffa, 1984; Eberhardt et al., 1985; Gulumian and VanWyk, 1987), presumably because of the iron present on the surface of the asbestos fibers (Weitzman and Weitberg, 1985; Gulumian and Kilroe-Smith, 1987; Weitzman et al., 1988; Kandaswami et al., 1988; Goodglick et al., 1989). While Igushi and Kojo (1989) questioned the role of iron in asbestos-mediated lipid peroxidation, their investigation showed an inhibition of lipid peroxidation by free radical scavengers and antioxidants, and hence indirectly confirmed the involvement of free radicals. The investigations of Gulumian and Kilroe-Smith (1987) documented that asbestos fibers mediate peroxidation of microsomal lipid in rat lung. This peroxidation depended on iron but was not supported by β-nicotinamide-adenine dinucleotide phosphate reduced (NADPH). It therefore seems that, in the case of asbestos, the ferrous/ferric ions two-electron redox cycling, usually supported by NADPH, does not initiate lipid peroxidation in lung tissue. Rather, the lipid peroxidation is caused by one-electron transfer reactions and subsequent production of free radicals.

Several health effects of asbestos have been linked to reactive oxygen species and/or lipid peroxidation, and the available data on this subject have been adequately reviewed by Mossman et al. (1990) and recently by Kamp et al. (1992). Therefore, in this chapter we have focused on the possible mechanism linking a non-specific, but most common and deadly, effect of asbestos—bronchogenic carcinoma—with the oxidative metabolism of the most ubiquitous precarcinogenic component of tobacco smoke—benzo[a]pyrene.

1.1. Benzo[a]pyrene Metabolism and Bioactivation

In general, polycyclic aromatic hydrocarbons are the most widespread precarcinogens in the environment, but expression of their carcinogenic activity requires metabolic activation to both "proximate" and "ultimate" carcinogenic species. The net carcinogenic potential of benzo[a]pyrene depends upon the balance between the activation and inactivation processes (Fig. 1). According to generally accepted terminology, metabolic oxidation (in contrast to one-electron oxygenation) and hydrolysis by epoxide hydrolase result in the formation of intermediate compounds called "proximate carcinogens." These proximate carcinogens undergo neutralization or further oxygenation to form the "ultimate carcinogen." For example, 7,8-dihydrodiol-9,10-epoxide is considered to be the ultimate carcinogenic metabolic derivative of benzo[a]pyrene (Thakker et al., 1978; Kapitulnik et al., 1978). This metabolite is produced by several processes: cytochrome P-450-dependent mixed-function oxidation, co-oxidation with prostaglandin synthetase or lipoxygenase, and possibly direct catalytic epoxidation with several inorganic and organic peroxyl radicals (Dix and Marnett, 1983; Cavalieri

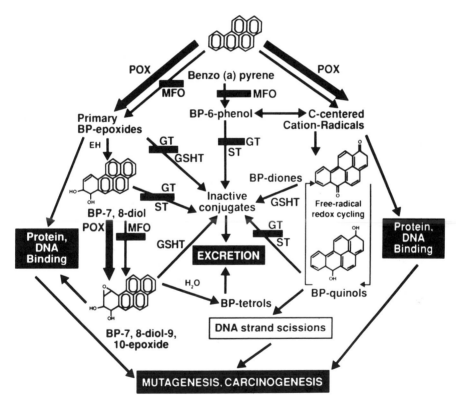

Figure 1. A simplified scheme of benzo[a]pyrene metabolism with indication of the effects of pro-oxidant conditions triggered by asbestos and/or Fe.FeEDTA in the presence of superoxide (heavy arrows and bars across the arrows), according to Gower (1988) and Byczkowski and Gessner (1987a).

    Key: MFO, mixed function oxidase; POX, pro-oxidant conditions; EH, epoxide hydrolyase; GT, UDP-glucuronyl-transferase; ST, sulfotransferase; GSHT, glutathione *S*-transferase; arrows, normal metabolic reaction; thick arrows, peroxidative reactions; black bar, decreased enzymatic acitivity by peroxidation.

and Rogan, 1984; Byczkowski and Kulkarni, 1989, 1990a, b, c). The proximate carcinogenic intermediates may be inactivated by further metabolism, such as conjugation with glutathione, glucuronic acid, and/or sulfate (Thakker et al., 1985; Ball et al., 1979).

The bioactivation of benzo[a]pyrene by iron-containing particles has been demonstrated in several models, including the isolated, perfused lung (Warshawsky et al., 1984). In the latter system, Fe_2O_3 treatment resulted in decreased metabolism of benzo[a]pyrene, but with a concomitant enhancement in the formation of a proximate carcinogen, benzo[a]pyrene-7,8-dihydrodiol. The same investigation also demonstrated the importance of the pulmonary alveolar macrophages in benzo[a]pyrene bioactivation and the defense of the lung against particles.

1.2. Lipid Peroxidation and the Role of Iron

In the lung, both bioactivation and inactivation of carcinogens may be influenced by lipid peroxidation, triggered by reactive oxygen species (DelMaestro et al., 1980; Hogberg et al., 1973; Bentley et al., 1979). The formation of reactive oxygen species can result from either macrophage activation (Doelman and Bast, 1990) or the redox cycling of certain transition metals, such as iron (Johnston, 1984; Gower and Wills, 1984; Weitzman and Weitberg, 1985). Transition metals are contained in asbestos fibers, either as structural constituents (e.g., crocidolite contains up to 27% iron) or as contaminants adsorbed on their surface (e.g., chrysotile contains 1–2.6% iron) (Timbrell, 1970).

Iron can contribute significantly to the toxicity and carcinogenicity of asbestos fibers, because of both its significant concentrations at the surface and its known reactivity. Iron participates in oxygen activation and has been shown to influence benzo[a]pyrene metabolism in several mammalian systems (Warshawsky et al., 1984; Dixon et al., 1970; Thomson et al., 1974).

Dix et al. (1985) presented evidence that heme-bound iron may stimulate lipid peroxidation and enhance the formation of ultimate carcinogenic metabolite from benzo[a]pyrene-7,8-dihydrodiol. Although different types of asbestos contain transition metals other than iron, the spectrographic analyses showed that the relative concentrations of these metals are one to three orders of magnitude lower than that of iron (Timbrell, 1970).

2. IN VIVO AND IN VITRO STUDIES

2.1. Epidemiological Studies

In 1968, Selikoff et al. reported that workers employed as insulators and exposed to asbestos fibers had markedly increased risk for bronchogenic carcinoma if they smoked cigarettes (Table 1). The demonstrated synergistic effect between exposure to asbestos fibers and smoking has been verified by numerous epidemiologi-

Table 1   Relationship between Exposure to Asbestos and/or Cigarette Smoking and Death Rate from Lung Cancer[a]

Group	Death Rate/100,000 Man-Years from Lung Cancer (Excluding Mesothelioma)
No asbestos work and no smoking	11.3
No asbestos work and smoking	122.6
Asbestos work and no smoking	80.2
Asbestos work and smoking	693.8

[a] Data from Frank (1979).

cal and experimental studies (Saracci, 1977; Hammond et al., 1979), and the relevant literature was reviewed extensively by Mossman and Craighead (1986), Mossman and Gee (1989), Mossman et al. (1983, 1990). Table 1 shows data presented at the New York Academy of Sciences conference *Health Hazards of Asbestos Exposure*, which established the general acceptance of the synergism in asbestos-exposed cigarette smokers by the scientific community worldwide, as summarized by Frank (1979).

2.2.   Enhancement of Benzo[a]pyrene Carcinogenicity by Asbestos

One of the putative precarcinogenic components of tobacco smoke that may cause bronchogenic carcinoma is benzo[a]pyrene. Several investigations have addressed possible mechanisms of interaction between asbestos fibers and benzo[a]pyrene. For example, asbestos fibers can enhance the oncogenic transformation of cultured cells in the presence of benzo[a]pyrene (DiPaolo et al., 1983). Kandaswami and O'Brien (1980, 1981) and Kandaswami et al. (1982, 1985) suggested that asbestos fibers impair the cytochrome P-450-dependent two-electron oxidation of benzo[a]pyrene, but simultaneously they enhance its macromolecular binding via one-electron activation (O'Brien et al., 1984). In agreement with this proposal, chrysotile asbestos was found to enhance DNA binding of benzo[a]pyrene in human fibroblasts (Chang et al., 1983; Hart et al., 1980).

A possible rationale to explain the asbestos–benzo[a]pyrene interaction has been suggested by Lakowicz and Bevan (1979), who postulated an increased bioavailability of benzo[a]pyrene as a result of adsorbtion on asbestos fibers. However, Hubbard et al. (1986) criticized this hypothesis. Subsequently, Byczkowski and Gessner (1987c,d) suggested that asbestos might supply the catalytically active iron capable of activating benzo[a]pyrene under pro-oxidant conditions. Furthermore, it was shown that the reactive oxygen species involved in the peroxidation of hepatic microsomal lipids (see review by Byczkowski and Gessner, 1988) can shift the balance between bioactivation and glucuronidation of benzo[a]pyrene metabolites, favoring the accumulation of epoxides and other products capable of protein binding (Fig. 1).

2.3. Asbestos-Enhanced Carcinogenesis

The synergistic effect of asbestos on benzo[a]pyrene-mediated carcinogenesis in the respiratory tract was first reported by Miller et al. (1965). Subsequent investigations, using Syrian hamster embryo cells and mouse embryo fibroblast cultures, confirmed the synergism of oncogenic transformation in vitro (DiPaolo et al., 1983; Brown et al., 1983). However, Paterour et al. (1985) found no synergism in a two-stage model that employed rat pleural mesothelial cells with benzo [a]pyrene as an initiator and chrysotile as a promoter. This apparent contradiction with the previous studies suggested that the synergism between benzo[a] pyrene and asbestos always requires the simultaneous presence of both agents and does not occur in the sequential exposure protocol (Brown et al., 1983; Paterour et al., 1985). It seems, therefore, that asbestos fibers affect the activation of benzo[a]pyrene to carcinogenic metabolites (proximate and/or ultimate) rather than promote the oncogenic transformation initiated by benzo[a]pyrene.

Most investigators have employed Canadian chrysotile asbestos in their biochemical studies. This type of asbestos fiber has been shown to cause extensive lipid peroxidation in vitro. It is the most active initiator of lipid peroxidation among the three types of UICC (Union Internationale Contre Le Cancer) Asbestos Reference Standard samples tested (Weitzman and Weitberg, 1985) and, in fact, it was the most extensively used type of asbestos in the United States. Chrysotile asbestos does not contain structurally bound iron (Mossman et al., 1990) and thus, its pro-oxidant potential can be drastically reduced by washing the traces of contaminating iron from the fiber surface with solutions containing iron chelator such as desferrioxamine B (Weitzman and Weitberg, 1985). On the other hand, the iron content can be significantly increased by loading with the exogenous iron (Byczkowski and Gessner, 1987c). Moreover, the chrysotile fibers, when administered concurrently with benzo[a]pyrene by intratracheal instillation, were shown to reach the apex of the lung and were capable of increasing benzo[a]pyrene-induced tumor formation (Miller et al., 1965).

2.4. Phagocytic Cells

Hesterberg et al. (1987) showed that the cytotoxicity of asbestos fibers depends on phagocytosis and that chrysotile asbestos was phagocytized more readily than crocidolite. Phagocytic cells release superoxide anion radicals and other reactive oxygen species to the surrounding medium (see review by Byczkowski and Gessner, 1988). Chrysotile fibers can activate exocytosis in polymorphonuclear leukocytes and cause membrane damage in a calcium-dependent manner (Elferink et al., 1989).

Since pulmonary subcellular membranes contain substantial amounts of naturally occurring unsaturated fatty acids (Holtzman et al., 1986), they are very likely to serve as the substrates for lipid peroxidation triggered by iron and reactive oxygen species released by the phagocytic cells. Another potential source of peroxidizable lipids is lung surfactant (Lachmann, 1989; Haagsman and van

Golde, 1991). It was proposed that quantitative and functional abnormalities of lung surfactant may participate in bleomycin-induced fibrosis (Osanai et al., 1991). Moreover, it was shown that the composition of lung surfactant may be affected by exposure to smoke (Oulton et al., 1991).

3. A SEARCH FOR THE MOLECULAR MECHANISM

Prior to the studies by Byczkowski and Gessner (1987c, d), a few experiments were designed to explore the molecular interaction between asbestos and benzo[a]pyrene. Dixon et al. (1970) examined the effects of trace metals associated with asbestos fibers on the lung microsomal hydroxylation of benzo[a]pyrene. Various metal cations present in the asbestos fibers either stimulated or inhibited benzo[a]pyrene hydroxylase activity when applied in a reagent form. The same cations, however, only inhibited hydroxy-benzo[a]pyrene formation when they were extracted from chrysotile asbestos fibers and applied as a solution. The authors did not examine the formation of other metabolites of benzo[a]pyrene. Later, Lakowicz and Hylden (1978) suggested that the physical sorptive properties of the fiber surface enhance the uptake of benzo[a]pyrene by biomembranes. Both studies did not explain, however, why more macromolecular-bound and fewer polar metabolites of benzo[a]pyrene are produced in the presence of asbestos.

Several years later, DiPaolo et al. (1983) suggested that asbestos fibers may modify detoxification of reactive benzo[a]pyrene metabolites. Recently, it was shown that asbestos may actually stimulate the production of ultimately carcinogenic benzo[a]pyrene-7,8-dihydrodiol-9,10-epoxide (Byczkowski and Kulkarni, 1990a, b).

4. INVOLVEMENT OF TRANSITION METALS AND FREE RADICALS

The role of iron in superoxide-mediated damage and benzo[a]pyrene metabolism was studied using mouse liver microsomes (Byczkowski and Gessner, 1986, 1987a, b, e). In microsomes incubated with a superoxide-generating system (without NADPH), the addition of Fe.FeEDTA (equimolar mixture of Fe^{3+} *plus* $Fe^{2+} \cdot$ ethylene diamine tetraacetic acid) increased total metabolism of benzo[a]-pyrene by 79% (Table 2). In this system, the reactive oxygen species were supplied in the form of superoxide generated by xanthine-xanthine oxidase, and microsomal lipid peroxidation was triggered by the addition of catalytically active iron (Byczkowski and Gessner, 1987b).

More reactive metabolites of benzo[a]pyrene were detected in the presence of NADPH in incubations containing liver microsomes preincubated with xanthine-xanthine oxidase plus Fe.FeEDTA, washed and then probed with benzo[a]pyrene and UDPGA (uridine-5'-diphosphoglucuronic acid). In this

Table 2   Effect of Superoxide Generation by Xanthine-Xanthine Oxidase plus Fe.FeEDTA on One-Electron Oxidation of Benzo[a]pyrene (B[a]P) in the Presence of Mouse Liver Microsomes without NADPH

| Metabolites | B[a]P Product Formation (pmol/mL) | |
	Control	Xanthine-Xanthine Oxidase + Fe.FeEDTA
Diones	71 ± 7	120 ± 8*
4, 5-Epoxy	0 ± 0	32 ± 4*
Protein-bound	151 ± 11	247 ± 12*
Total products	222 ± 18	399 ± 30*

* Significant difference from the corresponding control $P < 0.05$ by paired t-test. For details, see Byczkowski and Gessner (1987 b).

system, the product(s) of lipid peroxidation generated by reactive oxygen species decreased the total metabolism of benzo[a]pyrene (especially the amount of monohydroxyphenols) but, at the same time, diones, 4,5-epoxide, and protein-bound metabolites of benzo[a]pyrene were substantially increased (Table 3). These changes, along with a decrease in UDP-glucuronyltransferase activity (Byczkowski and Gessner, 1987a, e), were only partially prevented by superoxide dismutase. Therefore, concurrent damage to the microsomal mixed-function oxidase system and to the UDP-glucuronyltransferase precluded further inactivation of reactive benzo[a]pyrene metabolites (Fig. 1). These effects were apparently dependent upon catalytically active iron and were presumably caused by reactive oxygen species, lipid hydroperoxides, and/or peroxyl radicals.

Table 3   Effect of in Vitro Pretreatment of Mouse Liver Microsomes with Xanthine-Xanthine Oxidase plus Fe.FeEDTA on Benzo[a]pyrene (B[a]P) Metabolism via Oxidation and Glucuronidation in the Presence of NADPH

| Metabolites | B[a] P Product Formation | |
	Control (pmol/mg protein)	Pretreated with Xanthine-Xanthine Oxidase + Fe.FeEDTA (% of control)
Polar	100 ± 4	156 ± 5*
Gluncuronides	1222 ± 183	75 ± 5
Tetrols/triols	144 ± 27	53 ± 25
Diols	125 ± 61	25 ± 12
Diones	242 ± 52	120 ± 15*
4,5-Epoxy	3 ± 10	2876 ± 619*
Monohydroxy	706 ± 72	4 ± 0*
Protein-bound	74 ± 8	307 ± 50*
Total metabolites	2616 ± 388	66 ± 6*

* Significant difference from the corresponding control $P < 0.05$ by paired t-statistics. For details, see Byczkowski and Gessner (1987 b).

It is well known that iron ions, their chelates, and other transition metal compounds can catalyze one-electron transfer reactions (Halliwell and Gutteridge, 1984). Because the particle-activated microphages generate superoxide anion radical and other reactive oxygen species (DelMaestro et al., 1980; DelMaestro, 1984; Johnston, 1984; Byczkowski and Gessner, 1988), it is possible that the previously reported enhancement of benzo[a]pyrene activation by Fe_2O_3 particles in the isolated, perfused lung (Warshawsky et al., 1984) was the result of reactive oxygen species produced by Fe^{2+}/Fe^{3+} cycling. Perhaps these reactions represent the underlying biochemical mechanism of asbestos action on the pulmonary metabolism of benzo[a]pyrene.

In the nonenzymatic model studies conducted by Byczkowski and Kulkarni (1990a, b, 1992) as well, tetravalent vanadium (vanadyl) catalyzed co-oxygenation of benzo[a]pyrene and benzo[a]pyrene-7,8-dihydrodiol during the peroxidation of unsaturated lipids (described in detail in Chapter 16).

Other studies with mouse liver microsomes suggested that asbestos fibers impair detoxification of benzo[a]pyrene by initiating microsomal lipid peroxidation and enhancing its activation by co-oxygenation (Table 4), supporting the conclusions of Kandaswami and O'Brien (1980, 1981), Kandaswami et al. (1982, 1983, 1985), and O'Brien et al. (1984). These results clearly demonstrated that asbestos (Canadian chrysotile) can supply catalytically active iron capable of activating benzo[a]pyrene in peroxidized microsomal preparations (Byczkowski and Gessner, 1987c, d). This observation, as well as other findings, suggest that reactive oxygen species decrease the microsomal hydroxylation of benzo[a]

Table 4   Effects of Asbestos on Oxidation of Benzo[a]pyrene (B[a]P) by Superoxide-Peroxidized Mouse Liver Microsomes without NADPH

Metabolites	B[a]P Product Formation by Peroxidized Microsomes [pmol/mL]		
	Control	Asbestos	Asbestos + Fe.FeEDTA
Polar	ND[a]	ND	18 + 5*
Tetrols/triols	18 + 3	103 + 8*	185 + 13**
Diols	ND	18 + 3*	60 + 8**
1,6-Dione	50 + 15	427 + 75*	447 + 78
3,6-Dione	42 + 10	362 + 58*	545 + 87
6,12-Dione	53 + 13	463 + 33*	643 + 45**
4,5-Epoxy	ND	ND	20 + 5*
Monohydroxy	ND	ND	ND
Nonextractable	53 + 10	553 + 58*	490 + 53
Total products	216 + 20	1927 + 190*	2408 + 193**

[a] ND, below detectable amount. For details, see Byczkowski and Gessner (1987c).
* Significant difference from the corresponding control $P < 0.05$ by paired t-test.
** Significant difference from samples containing peroxidized microsomes plus asbestos $P < 0.05$ by paired t-test.

pyrene (Byczkowski and Gessner, 1987b) and reduce the subsequent conjugation of its metabolites (Byczkowski and Gessner, 1987a, e), but increase the production of free epoxides and macromolecular-binding metabolites (Byczkowski and Gessner, 1986).

Benzo[a]pyrene does not react directly with the superoxide (Lee-Ruff et al., 1986), but requires the presence of a catalyst capable of activating the hydrocarbon and displacing the hydrogen atom (usually in position C-6). The intermediate radical cation generated in this reaction is prone to nucleophilic attack by oxygen. A probable mechanism of such activation is shown in Figure 2 (left side). From experiments with peroxidized and re-isolated microsomes (Byczkowski and Gessner, 1987a, c, d), it was clear that some product(s) of lipid peroxidation

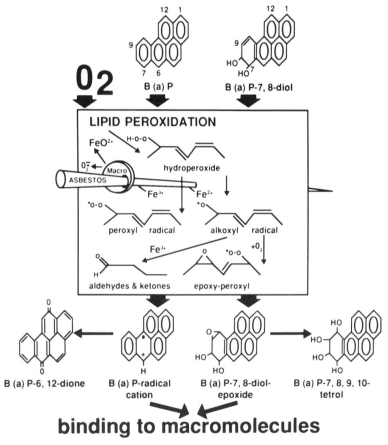

Figure 2. A probable mechanism of activation of benzo[a]pyrene and benzo[a]pyrene-7, 8-dihydro-diol by oxygen-containing free radicals in the presence of catalytically active iron and/or asbestos (macro-alveolar macrophages), according to Byczkowski and Gessner (1987b, c, d, 1989) and Byczkowski and Kulkarni (1990a, b, c).

interact with benzo[a]pyrene in the presence of catalytically active asbestos fibers. The findings of Cavalieri et al. (1987) suggest that hydroperoxides or peroxyl radicals may displace hydrogen at the C-6 position and produce the benzo[a]pyrene radical cation. According to Graceffa and Weitzman (1987), asbestos may serve as a catalyst, yielding benzo[a]pyrene 6-oxo-radicals from the 6-substituted derivative of benzo[a]pyrene.

4.1. Free-Radical Co-oxygenation

The studies reported by Byczkowski and Gessner (1987c, d) and Byczkowski and Kulkarni (1990a,b) indicate that, in a system containing either the peroxidized microsomes (Table 4) or autoperoxidized linoleic acid, asbestos fibers catalyze co-oxygenation of benzo[a]pyrene to its dione derivatives in the presence of lipid peroxidation product(s). The levels of 6,12-dione and total metabolites increased dramatically when iron-loaded asbestos fibers were used. No increase in benzo[a]pyrene dione production was evident when a fresh linoleic acid, containing no hydroperoxides, was used.

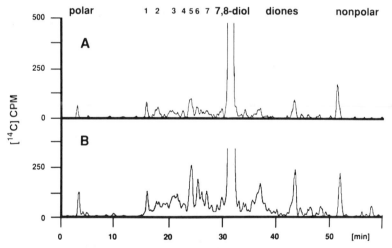

Figure 3. Results of the reversed-phase HPLC (high-performance liquid chromatography) analysis of the products of [^{14}C] (+)-benzo[a]pyrene-*trans*-7,8-dihydrodiol co-oxygenation by Canadian chrysotile asbestos in the presence of (A) fresh and (B) partially autoperoxidized linoleic acid, according to Byczkowski and Kulkarni (1990a, b).

Key: 1 = benzo[a]pyrene-*trans-anti*-7,8,9,10-tetrahydrotetrol; 4 = benzo[a]pyrene-*cis-syn*-7,8,9,10-tetrahydrotetrol; 2,3,5,6,7 = unidentified metabolites.

Incubation conditions: 12 min at 30° C in 50 mM tris-HCl (pH 7.3), final volume of 0.2 mL containing 50 μM benzo[a]pyrene-7, 8-dihydrodiol, 1 mg asbestos (UICC Asbestos Reference Standard), and (A) 1 mM linoleic acid or (B) autoperoxidized linoleic acid containing about 2% hydroperoxides (by spectrophotometry at 234 nm).

In other experiments, when Canadian chrysotile asbestos was incubated with (+)-benzo[a]pyrene-trans-7,8-dihydrodiol, little trans-anti-7,8,9,10-tetrahydrotetrol was produced upon addition of fresh linoleic acid (Fig. 3A). The addition to this system of partially autoperoxidized linoleic acid containing about 2% hydroperoxides caused an increase in the production of diones and benzo[a]pyrene-trans-anti-tetrahydrotetrol (Fig 3B). The production of the latter metabolite is diagnostic for peroxyl radical-mediated co-oxygenation (Dix and Marnett, 1984). A similar phenomenon has been described for the hematin-arachidonate hydroperoxide system by Dix et al. (1985). Furthermore, in the mechanism proposed by these authors, benzo[a]pyrene-7,8-dihydrodiol epoxide (the ultimate carcinogen produced stereospecifically during co-oxygenation) undergoes subsequent hydrolysis in an aqueous environment and yields trans-anti-benzo[a]pyrene-7,8,9,10-tetrahydrotetrol. We also observed this metabolite as a major product in the asbestos–linoleate hydroperoxide system (Fig. 2, right side).

These results suggest that asbestos fibers can actually catalyze the generation of peroxyl radicals from linoleate hydroperoxide (Byczkowski and Kulkarni, 1990a,b) in a manner analogous to the production of hydroxyl radicals from hydrogen peroxide (Weitzman and Weitberg, 1985). One cannot, however, exclude the possibility of direct benzo[a]pyrene-7,8-dihydrodiol epoxidation by singlet oxygen (e.g., the Russel mechanism) generated during transition metal-catalyzed linoleate peroxidation on the fiber surface (Gardner, 1989).

5. CONCLUSIONS

Over the past 20 years, numerous investigations have focused on asbestos–benzo[a]pyrene synergism, including epidemiological studies, whole animal cocarcinogenesis investigations, perfused organ experiments, and studies of the molecular biology of cultured cells. The biochemical toxicologic approach seems to provide the most probable explanation for the interaction. From the experimental evidence reviewed in this chapter, it is now possible to propose a mechanism that explains the synergistic interaction.

In the respiratory tract, asbestos fibers activate alveolar macrophages to release reactive oxygen species, causing local pro-oxidant conditions. This milieu is further affected by asbestos-triggered lipid peroxidation and the generation of peroxyl radicals. Lipid peroxidation has a deleterious effect on microsomal phase I and II detoxification pathways. The net effect of these concurrent reactions is the liberation of more pre- and proximate carcinogenic derivatives of benzo[a]-pyrene for the reaction with peroxyl radicals. In situ co-oxygenation generates benzo[a]pyrene quinones and results in one-electron redox cycling. The proximate carcinogen is subjected to epoxidation, yielding benzo[a]pyrene dihydrodiol epoxide. Both dione and diol epoxide metabolites of benzo[a]pyrene can interact with cellular macromolecules, and presumably cause an increased frequency of DNA mutations that lead to cancer.

ACKNOWLEDGMENTS

Some of the work reported here was supported in part by a grant to A.P.K. from the Council for Tobacco Research USA, Inc.

REFERENCES

Ball, L. M., Plummer, J. L., Smith, B. R., and Bend, J. R. (1979). Benzo(a)pyrene oxidation, conjugation and disposition in the isolated perfused rabbit lung: Role of the glutathione S-transferases. *Med. Biol.* **57**, 298–305.

Bentley, D. N., Wood, G. C., and Graham, A. B. (1979). Effect of lipid peroxidation on the activity of microsomal UDP-glucuronyltransferase. *Med. Biol.* **57**, 274–280.

Brown, R. C., Poole, A., and Fleming, G. T. A. (1983). The influence of asbestos dust on the oncogenic transformation of C3H10T1/2 cells. *Cancer Lett.* **18**, 221–227.

Byczkowski, J. Z., and Gessner, T. (1986). Modification of microsomal metabolism of benzo(a)pyrene [B(a)P] by superoxide generated in vitro. *Fed. Proc., Fed. Am. Soc. Exp. Biol.* **45**, 3146.

Byczkowski, J. Z., and Gessner, T. (1987a). Effects of superoxide generated in vitro on glucuronidation of benzo(a)pyrene metabolites by mouse liver microsomes. *Int. J. Biochem.* **19**, 531–537.

Byczkowski, J. Z., and Gessner, T. (1987b). Action of xanthine-xanthine oxidase system on microsomal benzo(a)pyrene metabolism in vitro. *Gen. Pharmacol.* **18**, 385–395.

Byczkowski, J. Z., and Gessner, T. (1987c). Asbestos-catalyzed oxidation of benzo(a)pyrene by superoxide-peroxidized microsomes. *Bull. Environ. Contam. Toxicol.* **39**, 312–317.

Byczkowski, J. Z., and Gessner, T. (1987d). Superoxide-initiated lipid peroxidation and benzo(a)pyrene metabolism in vitro. *Int. Congr. Oxygen Radicals, Contri. Abstr., 4th*, UCSD, La Jolla, pp. 47–50.

Byczkowski, J. Z., and Gessner, T. (1987e). Interaction between vitamin K_3 and benzo(a)pyrene metabolism in uninduced microsomes. *Int. J. Biochem.* **19**, 1173–1179.

Byczkowski, J. Z., and Gessner, T. (1988). Biological role of superoxide ion-radical. Minireview. *Int. J. Biochem.* **20**, 569–580.

Byczkowski, J. Z., and Gessner, T. (1989). Effects of inhibition of NADPH: Cytochrome P-450 reductase on benzo(a)pyrene metabolism in mouse liver microsomes. *Int. J. Biochem.* **21**, 525–529.

Byczkowski, J. Z., and Kulkarni, A. P. (1989). Lipoxygenase-catalyzed epoxidation of benzo(a) pyrene-7,8-dihydrodiol. *Biochem. Biophys. Res. Commun.* **159**, 1199–1205.

Byczkowski, J. Z., and Kulkarni, A. P. (1990a). Peroxidation of linoleic acid by environmental pollutants: Hydrated SO_2, reduced vanadium, and asbestos. *Toxicologist* **10**, 172(685).

Byczkowski, J. Z., and Kulkarni, A. P. (1990b). Lipid peroxidation and benzo(a)pyrene derivative co-oxygenation by environmental pollutants. *Bull. Environ. Contam. Toxicol.* **45**, 633–640.

Byczkowski, J. Z., and Kulkarni, A. P. (1990c). Lipid peroxidation-coupled co-oxygenation of benzo(a)pyrene and benzo(a)pyrene-7,8-dihydrodiol in human term placental microsomes. *Placenta* **11**, 17–26.

Byczkowski, J. Z., and Kulkarni, A. P. (1992). Vanadium redox cycling, lipid peroxidation and co-oxygenation of benzo(a)pyrene-7,8-dihydrodiol. *Biochim. Biophys. Acta* **1125**, 134–141.

Cavalieri, E. L., and Rogan, E. G. (1984). One-electron and two-electron oxidation in aromatic hydrocarbon carcinogenesis. In W. A. Pryor (Ed.), *Free Radicals in Biology*. Academic Press, Orlando, FL, Vol. 6, pp. 323–369.

Cavalieri, E. L., Wong, A., and Rogan, E. G. (1987). Evidence for distinct binding sites in the cumene hydroperoxide-dependent metabolism of benzo(a)pyrene catalyzed by cytochrome P-450. *Biochem. Pharmacol.* **36**, 435–440.

Chang, M. J., Singh, N. P., and Hart, R. W. (1983). Effects of chrysotile co-exposure on BaP binding in normal human fibroblasts. *Environ. Health Perspect.* **51**, 241–244.

DelMaestro, R. F. (1984). Systemic consequences of $O_2^{\cdot-}$ — production. In L. Packer (Ed.), *Methods in Enzymology.* Academic Press, Orlando, FL, Vol. 105, pp. 386–389.

DelMaestro, R. F., Thaw, H. H., Bjork, J., Planker, M., and Arfors, K. E. (1980). Free radicals as mediators of tissue injury. *Acta Physiol. Scand., Suppl.* **492**, 43–57.

DiPaolo, J. A., DeMarinis, A. J., and Doniger, J. (1983). Asbestos and benzo(a)pyrene synergism in the transformation of Syrian hamster embryo cells. *Pharmacology* **27**, 65–73.

Dix, T. A., and Marnett, L. J. (1983). Metabolism of polycyclic aromatic hydrocarbon derivatives to ultimate carcinogens during lipid peroxidation. *Science* **221**, 77–79.

Dix, T. A., and Marnett, L. J. (1984). Detection of the metabolism of polycyclic aromatic hydrocarbon derivatives to ultimate carcinogens during lipid peroxidation. In Le Packer (Ed.), *Methods in Enzymology.* Academic Press, Orlando, FL, Vol. **105**, pp. 347–352.

Dix, T. A., Fontana, R., Panthani, A., and Marnett, L. J. (1985). Hematin-catalyzed epoxidation of 7,8-dihydroxy-7,8-dihydrobenzo(a)pyrene by polyunsaturated fatty acid hydroperoxides. *J. Biol. Chem.* **260**, 5358–5365.

Dixon, J. R., Lowe, D. B., Richards, D. E., Cralley, L. J., and Stokinger, H. E. (1970). The role of trace metals in chemical carcinogenesis: Asbestos cancers. *Cancer Res.* **30**, 1068–1074.

Doelman, C. J., and Bast, A. (1990). Oxygen radicals in lung pathology. *Free Radical Biol. Med.* **9**, 381–400.

Eberhardt, M. K., Roman-Franco, A. A., and Quiles, M. R. (1985). Asbestos-induced decomposition of hydrogen peroxide. *Environ. Res.* **37**, 287–292.

Elferink, J. G. R., Deierkauf, M., Kramps, J. A., and Koerten, H. K. (1989). An activating and cytotoxic effect of asbestos on polymorphonuclear leukocytes. *Agents Actions* **26**, 213–215.

Frank, A. L. (1979). Public health significance of smoking–asbestos interactions. *Ann. N.Y. Acad. Sci.* **330**, 791–794.

Gardner, H. W. (1989). Oxygen radical chemistry of polyunsaturated fatty acids. *Free Radical Biol. Med.* **7**, 65–86.

Goodglick, L. A., Pietras, L. A., and Kane, A. B. (1989). Evaluation of the casual relationship between crocidolite asbestos-induced lipid peroxidation and toxicity to macrophages. *Am. Rev. Respir. Dis.* **139**, 1265–1273.

Gower, J. D. (1988). A role of dietary lipids and antioxidants in the activation of carcinogens. *Free Radical Biol. Med.* **5**, 95–111.

Gower, J. D., and Wills, E. D. (1984). The generation of oxidation products of benzo(a)pyrene by lipid peroxidation: A study using gamma-irradiation. *Carcinogenesis (London)* **5**, 1183–1189.

Graceffa, P., and Weitzman, S. A. (1987). Asbestos catalyzes the formation of the 6-oxobenzo(a)pyrene radical from 6-hydroxybenzo(a)pyrene. *Arch. Biochem. Biophys.* **257**, 481–484.

Gulumian, M., and Kilroe-Smith, T. A. (1987). Crocidolite-induced lipid peroxidation in rat lung microsomes. I. Role of different ions. *Environ. Res.* **43**, 267–273.

Gulumian, M., and VanWyk, J. A. (1987). Hydroxyl radical production in the presence of fibers by a Fenton-type reaction. *Chem.-Biol. Interact.* **62**, 89–97.

Haagsman, H., and van Golde, L. M. G. (1991). Synthesis and assembly of lung surfactant. *Annu. Rev. Physiol.* **53**, 441–464.

Halliwell, B., and Gutteridge, J. M. C. (1984). Role of iron in oxygen radical reactions. In L. Packer (Ed.), *Methods in Enzymology.* Academic Press, Orlando, FL, Vol. 105, pp. 47–53.

Hammond, E. C., Selikoff, I. J., and Seidman, H. (1979). Asbestos exposure, cigarette smoking and death rates. *Ann. N. Y. Acad. Sci.* **330**, 473–490.

Hart, R. W., Daniel, F. B., Kindig, O. R., and Beach, C. A. (1980). Elemental modifications and polycyclic aromatic hydrocarbon metabolism in human fibroblasts. *Environ. Health Perspect.* **34**, 59–68.

Hesterberg, T. W., Ririe, D. G., Barrett, J. C., and Nettesheim, P. (1987). Mechanisms of cytotoxicity of asbestos fibers in rat tracheal epithelial cells in culture. *Toxicol. In Vitro* 1, 59–65.

Hogberg, J., Bergstrand, A., and Jakobsson, S. V. (1973). Lipid peroxidation of rat-liver microsomes: Its effect on the microsomal membranes and some membrane-bound microsomal enzymes. *Eur. J. Biochem.* 37, 51–59.

Holtzman, M. J., Grunberger, D., and Hunter, J. A. (1986). Phospholipid fatty acid composition of pulmonary airway epithelial cells: Potential substrates for oxygenation. *Biochim. Biophys. Acta* 877, 459–464.

Hubbard, S. A., Davis, P. J. B., Hunt, C. M., and McDonald, T. (1986). Effects of benzo(a)pyrene-coated particles in a bacterial mutagenicity test and on macrophages in culture. *Food Chem. Toxicol.* 24, 697–698.

Igushi, H., and Kojo, S. (1989). Possible generation of hydrogen peroxide and lipid peroxidation of erythrocyte membrane by asbestos: Cytotoxic mechanism of asbestos. *Biochem. Int.* 18, 981–990.

Johnston, R. B., Jr. (1984). Measurement of $O_2^{\cdot -}$-secreted by monocytes and macrophages. In L. Packer (Ed.), *Methods in Enzymology*. Academic Press, Orlando, FL, Vol. 105, pp. 365–370.

Kamp, D. W., Graceffa, P., Pryor, W. A., and Weitzman, S. A. (1992). The role of free radicals in asbestos-induced diseases. *Free Radical Biol. Med.* 12, 293–315.

Kandaswami, C., and O'Brien, P. J. (1980). Effects of asbestos on membrane transport and metabolism of benzo(a)pyrene. *Biochem. Biophys. Res. Commun.* 97, 794–801.

Kandaswami, C., and O'Brien, P. J. (1981). Pulmonary metabolism of benzo(a)pyrene: Effect of asbestos. *Biochem. Pharmacol.* 30, 811–814.

Kandaswami, C., Subrahmanyam, V. V., and O'Brien, P. J. (1982) The effect of asbestos on the bioactivation of benzo(a)pyrene. *Polynucl. Aromat. Hydrocarbons: Phys. Biol. Chemi., Int. Symp. 6th*, Columbus OH, 1981, pp. 389–403.

Kandaswami, C., Rahimtula, M., and O'Brien, P. J. (1983). Effect of chrysotile asbestos pretreatment of rats on the hepatic microsomal metabolism of benzo(a)pyrene. *Polynucl. Aromat. Hydrocarbons: Formation, Metab. Meas., Int. Symp., 7th*, Columbus OH, 1982, pp. 649–661.

Kandaswami, C., Rhimtula, M., and O'Brien, P. J. (1985). Effect of asbestos on the microsomal metabolism and bioactivation of polynuclear aromatic hydrocarbon. *NATO ASI Ser.* G3, 221–228.

Kandaswami, C., Morin, G., and Sirois, P. (1988). Lipid peroxidation in rat alveolar macrophages exposed to chrysotile fibers. *Toxicol. In Vitro* 2, 117–120.

Kapitulnik, J., Wislocki, P. G., Levin, W., Yagi, H., Jerina, D., and Conney, A. H. (1978). Tumorigenicity studies with diol-epoxides of benzo(a)pyrene which indicate that ($\pm$) -*trans*-7β, 8α-dihydroxyl-9α-epoxy-7,8,9, 10-tetrahydrobenzo (a)pyrene is an ultimate carcinogen in newborn mice. *Cancer Res.* 38, 354–358.

Lachmann, B. (1989). Animal studies of surfactant replacement therapy. *Dev. Pharmacol. Ther.* 13, 164–172.

Lakowicz, J. R., and Bevan, D. R. (1979). Effects of asbestos, iron oxide, silica, and carbon black on the microsomal availability of benzo(a)pyrene. *Biochemistry* 18, 5176–5179.

Lakowicz, J. R., and Hylden, J. L. (1978). Asbestos-mediated membrane uptake of benzo(a)pyrene observed by fluorescence spectroscopy. *Nature (London)* 275, 446–448.

Lee-Ruff, E., Kazarians-Moghaddam, H., and Katz, M. (1986). Controlled oxidations of benzo(a)pyrene. *Can. J. Chem.* 64, 1297–1303.

Miller, L., Smith, W. E., and Berliner, S. W. (1965). Tests for effect of asbestos on benzo(a)pyrene carcinogenesis in the respiratory tract. *Ann. N. Y. Acad. Sci.* 132, 489–500.

Mossman, B. T., and Craighead, J. E. (1986). Mechanisms of asbestos associated bronchogenic carcinoma. In *Asbestos-Related Malignancy*. Grune & Stratton, New York, pp. 137–150.

Mossman, B. T., and Gee, J. B. L. (1989). Asbestos-related diseases. *N. Engl. J. Med.* 320, 1721–1730.

Mossman, B. T., Light, W., and Wei, E. (1983). Asbestos: Mechanisms of toxicity and carcinogenicity in the respiratory tract. *Annu. Rev. Pharmacol. Toxicol.* **23**, 595–615.

Mossman, B. T., Bignon, J., Corn, M., Seaton, A., and Gee, J. B. L. (1990). Asbestos: Scientific developments and implications for public policy. *Science* **247**, 294–301.

Murphy, R. L. H., Jr. (1988). Update on asbestos. In R. B. George (Ed.), *Pulmonary and Critical Care Update.* Continuing Professional Education Center, Princeton, NJ, Vol. 3, lesson 25, pp. 2–8.

O'Brien, P. J., Rahimtula, M., and Kandaswami, C. (1984). Asbestos enhanced metabolic activation of carcinogenic polycyclic aromatic hydrocarbons and arylamines. *Am. Assoc. Cancer Res. Abstr.* **25**, (43), 11.

Osanai, K., Takahashi, K., Sato, S., Iwabuchi, K., Ohtake, K., Sata, M., and Yasui, S. (1991). Changes of lung surfactant and pressure-volume curve in bleomycin-induced pulmonary fibrosis. *J. Appl. Physiol.* **70**, 1300–1308.

Oulton, M., Moores, H. K., Scott, J. E., Janigan, D. T., and Hajela, R. (1991). Effects of smoke inhalation on surfactant phospholipids and phospholipase A2 activity in the mouse lung. *Am. J. Pathol.* **138**, 195–202.

Paterour, M. J., Bignon, J., and Jaurand, M. C. (1985). In vitro transformation of rat pleural mesotheliomal cells by chrysotile fibers and/or benzo(a)pyrene. *Carciongenesis (London)* **6**, 523–529.

Petruska, J. M., Leslie, K. O., and Mossman, B. T. (1991). Enhanced lipid peroxidation in lung lavage of rats after inhalation of asbestos. *Free Radical Biol. Med.* **11**, 425–432.

Platek, S. F., Groth, D. H., Ulrich, C. E., Stettler, L. E., Finnell, M. S., and Stoll, M. (1985). Chronic inhalation of short asbestos fibers. *Fundam. Appl. Toxicol.* **5**, 327–340.

Saracci, R. (1977). Asbestos and lung cancer: An analysis of the epidemiological evidence on the asbestos-smoking interaction. *Int. J. Cancer* **20**, 323–331.

Selikoff, I. J. (1991). Asbestos disease—1990–2020: The risk of asbestos risk assessment. *Toxicol. Ind. Health* **7**, 117–127.

Selikoff, I. J., Hammond, H. C., and Churg, J. C. (1968). Asbestos exposure, smoking, and neoplasia. *JAMA, J. Am. Med. Assoc.* **204**, 104–110.

Thakker, D. R., Yagi, H., and Jerina, D. M. (1978). Analysis of polycyclic aromatic hydrocarbons and their metabolites by high-pressure liquid chromatography. In S. Fleischer and L. Packer (Eds.), *Methods in Enzymology.* Academic Press, New York, Vol. 52, pp. 279–296.

Thakker, D. R., Yagi, H., Levin, W., Wood, A. W., Cooney, A. H., and Jerina, D. M. (1985). Polycyclic aromatic hydrocarbons: Metabolic activation to ultimate carcinogens. In M. Anders (Ed.), *Bioactivation of Foreign Compounds.* Academic Press, Orlando, FL, pp. 177–242.

Thomson, R., Webster, I., and Kilroe-Smith, T. A. (1974). The metabolism of benzo(a)pyrene in rat liver microsomes: The effect of asbestos-associated metal ions and pH. *Environ. Res.* **7**, 149–157.

Timbrell, V. (1970). Characteristics of the International Union Against Cancer standard reference samples of asbestos. *Pneumoconiosis, Proc. Int. Conf. 3rd*, Johannesburg, *1969*, pp. 28–36.

Vallyathan, V., and Green, F. H. Y. (1985). The role of analytical techniques in the diagnosis of asbestos-associated disease. *CRC Crit. Rev. Clin. Lab. Sci.* **22**, 1–42.

Warshawsky, D., Bingham, E., and Niemeier, R. W. (1984). The effects of co-carcinogen, ferric oxide, on the metabolism of benzo(a)pyrene in the isolated perfused lung. *J. Toxicol. Environ. Health* **14**, 191–209.

Weitzman, S. A., and Graceffa, P. (1984). Asbestos catalyzes hydroxyl and superoxide radical generation from hydrogen peroxide. *Arch. Biochem. Biophys.* **228**, 373–376.

Weitzman, S. A., and Weitberg, A. B. (1985). Asbestos-catalysed lipid peroxidation and its inhibition by desferroxamine. *Biochem. J.* **225**, 259–262.

Weitzman, S. A., Chester, J. F., and Graceffa, P. (1988). Binding of deferroxamine to asbestos fibers in vitro and in vivo. *Carcinogenesis (London)* **9**, 1643–1645.

16

EFFECTS OF TRANSITION METALS ON BIOLOGICAL OXIDATIONS

Arun P. Kulkarni

Toxicology Program, College of Public Health MDC-56, University of South Florida, Tampa, Florida 33612

Janusz Z. Byczkowski

ManTech Environmental Technology, Inc., Dayton, Ohio 45437-0009

Environmental Oxidants, Edited by Jerome O. Nriagu and Milagros S. Simmons.
ISBN 0-471-57928-9 © 1994 John Wiley & Sons, Inc.

1. INTRODUCTION

Transition metals are ubiquitous in the environment as both building elements and pollutants. Due to industrial activity, huge amounts of these elements are released into the atmosphere or redistributed into different environmental compartments (Nriagu and Pacyna, 1988). In addition, each year, thousands of tons of oxidized transition metals are released in the fly ash generated during fossil fuel burning. They are fouling the air and polluting drinking water. Unlike organic pollutants, these metals are not biodegradable and they tend to build up in the ecosystem.

The atoms of transition metals have valence electron shells incompletely filled in at least one of their oxidation states. Thus, they can accept or donate one electron at a time. The one-electron mechanism can overcome spin restrictions on accepting electrons by ground-state molecular oxygen. This implies that transition metals, in general, are good catalysts for oxidation reactions. Several transition metals are found at the catalytic centers of many enzymes such as oxidases and oxygenases. Therefore, at low concentrations, several metals are required by living organisms as essential trace elements, and they must be supplied in the diet. At higher concentrations, however, they may became detrimental and toxic. Transition metals, such as iron, copper, and vanadium, can initiate lipid peroxidation in several tissues, including brain and lung. Lipid peroxidation is a deleterious, free radical process that damages biomembrane integrity. Moreover, some free radical intermediates can mediate further tissue damage and deplete cellular antioxidants. Also, products of lipid peroxidation can bioactivate precarcinogens to yield reactive chemical species that bind DNA, in addition to the direct DNA damage caused by the metal itself (e.g., nickel or chromium) or its redox cycling products (Kasprzak, 1991; Klein et al., 1991; Standeven and Wetterhahn, 1991). Exposure to transition metals may initiate fibrotic changes in the lung as well as behavioral changes and impairment of brain function (White et al., 1990; Floyd, 1991). In patients on chronic hemodialysis therapy, accumulation of transition metals was linked to neurodegenerative brain disorders (Tsukamoto et al., 1990).

These recent discoveries point to oxidative stress as a causal mechanism of toxicity and have opened a new avenue for the mechanistic understanding of biochemical changes in organisms exposed to transition metals.

The purpose of this chapter is to provide a review of the effects of some transition metals on oxidation processes in a few selected biological systems, and to discuss some of the attempts to elucidate their specific modes of action.

1.1. Main Source of Transition Metal Pollution

Fossil fuels contain significant amounts of metallic elements. During combustion, these metals are oxidized and dispersed with the fly ash. Once released into the biosphere, they enter the human body following ingestion of polluted water or inhalation of air. The easily oxidizable transition metals, in particular, pose environmental and health hazards due to their pro-oxidant properties. For example, vanadium and nickel are the main metallic elements found in natural oil. They are primarily associated with the heavy fractions (Semple et al., 1990) and can account for about three-quarters of the fly ash mass.

Similarly, different kinds of coal contain significant amounts of iron, copper, and manganese, in addition to vanadium. Manganese may be intentionally added to petroleum fuels as an antiknocking agent.

1.2. Transition Metal Toxicity

Some of these pollutants have received considerable attention due to their inherent toxicity and carcinogenicity (Kasprzak, 1991; Klein et al., 1991; Standeven and Wetterhahn, 1991). For example, nickel is generally recognized as a toxic and carcinogenic element both in experimental animals and in man. Extensive research by Misra et al. (1990) has clearly demonstrated the pro-oxidant properties of nickel. Similar toxic properties were found to be associated with chromium [Alcedo and Wetterhahn, 1990; Agency for Toxic Substance and Disease Regulation (ATSDR), 1989]. Compared to those of nickel or chromium, the adverse effects of short-term exposure to vanadium (Younes and Strubelt, 1991), iron (Aisen et al., 1990), and copper (Halliwell and Gutteridge, 1984), as well as manganese (Hammond and Beliles, 1980), are less obvious. Usually, the physiological concentrations of these transition metal ions in animal tissues are extremely low (e.g., 10^{-23} M Fe, 10^{-18} M Cu, 10^{-12} M Mn), and generally all the measurable amounts of the metals exist as complexes or chelates with biological ligands (Byczkowski and Sorenson, 1984). Therefore, under conditions of environmental or occupational exposure, the target tissue must cope with a local concentration of transition metals often exceeding a thousand or even several million times the normal physiological levels.

1.3.   Target Organs for Transition Metals

Besides the liver (Zychlinski and Byczkowski, 1989) and kidney, two other main organs are considered to be primary targets for transition metals; the lung and brain, depending on the route of administration.

The lung is a very obvious target organ for transition metals inhaled with air. For instance, vanadium pentoxide caused pronounced damage to the lung after chronic intratracheal administration (Zychlinski and Byczkowski, 1990; Zychlinski et al., 1991). However, it was also demonstrated that after chronic treatment *per os*, vanadium may cause a significant increase in the lung's insoluble collagen content (Kowalska, 1989). Several studies have indicated that transition metals such as iron and vanadium may participate in the pulmonary bioactivation of precarcinogens (Warshawsky et al., 1984; Byczkowski and Kulkarni, 1990).

A much less well understood target organ for transition metal intoxication is the brain. Nevertheless, chronic exposure to manganese, copper, and vanadium has been linked with neurotoxicity and some depressive disorders (Hammond and Beliles, 1980; Naylor, 1983). Increased concentrations of vanadium, along with elevated aluminum levels, were found in some patients on chronic hemodialysis treatment (Tsukamoto et al., 1990).

At present, only fragmentary knowledge exists regarding the actions and interactions of transitional metals with biological systems. Similarly, our inadequate mechanistic understanding of changes in the target organs does not allow us to predict or reasonably estimate the risk associated with exposure to these agents. Unresolved questions of particular importance are the mechanisms of interaction between noncarcinogenic transition metals and carcinogenic agents, the link between long-term exposure to transition metals and lung fibrosis, and the cause-and-effect relationship between transition metal-induced oxidative stress in the brain and neurotoxicity or neurodegeneration.

2.   IRON

Iron is known to cause both acute poisoning (e.g., from overdoses of sugar-coated "iron pills" or iron-fortified vitamin preparations) and chronic overload (Aisen et al., 1990). Chronic iron overload may result from diet, disease (Berger et al., 1990), or environmental and/or occupational exposure.

Inhalation exposure to iron oxides leading to diffuse pulmonary fibrosis is most common in welders, steel workers, miners, and jewelry makers (Hammond and Baliles, 1980). The most pronounced effect of the treatment of lung preparations with iron in vitro and in vivo is the initiation of lipid peroxidation. The peroxidation of biomembrane lipids by iron usually requires the simultaneous presence of ferrous and ferric cations (Aust et al., 1990). It seems, therefore, that the β-nicotinamide-adenine dinucleotide phosphate reduced (NADPH)-supported, two-electron redox cycling of ferrous/ferric ions is not necessarily responsible for lipid peroxidation. Rather, one-electron transfer reactions generating free radicals may be involved. Reports indicate that increased iron concentrations in some discrete regions of brain can cause lipid peroxidation (Subbarao et al., 1990), which has been linked to the pathogenesis of Parkinson's disease (Aisen et al., 1990).

2.1. Free Radicals and Reactive Oxygen Species

In the presence of either hydrogen peroxide or superoxide anion radical, the addition of iron generates a whole family of reactive oxygen species due to Fenton chemistry and the Haber–Weiss reactions (Byczkowski and Gessner, 1988). Similarly, other transition metals can also produce highly reactive free radicals by cleaving the hydrogen peroxide or lipid (hydro)peroxides (Goldstein and Czapski, 1990; Sawyer, 1990). Excessive generation of hydroxyl radicals, superoxide radicals, hydrogen peroxide, singlet oxygen, peroxyl radicals, and alkoxyl radicals is harmful and has been implicated in many diseases (Marks, 1987; Yagi, 1987). These reactive oxygen species, in addition to initiating lipid peroxidation, cause damage to biomembranes, nucleic acids (Halliwell and Aruoma, 1991), and proteins. They may hydroxylate proline (Zemlan et al., 1989) and initiate a fibrotic response in the lung (Fantone and Phan, 1988). They are involved in Parkinson's as well as Alzheimer's diseases (Adams and Odunze, 1991; Volicer and Crino, 1990), and can affect the metabolic fate of benzo[a]pyrene and possibly other precarcinogens in the biosystem (Byczkowski and Gessner, 1988).

3. VANADIUM

Vanadium occurs naturally as a contaminant of fossil fuels; in the past few years, as much as 66,000 tons of vanadium were released into the atmosphere each year and further redistributed in different environmental compartments (Nriagu and Pacyna, 1988). In addition to burning, the massive incidental and/or intentional spilling of vanadium-containing crude oil into sea or coastal waters has a particularly dramatic effect on the environment (Sadiq and Zaidi, 1984). Venezuelan crude oil is especially rich in vanadium and the fly ash resulting from its combustion may contain up to 80% vanadium compounds (Hudson, 1964). This points to the importance of vanadium as a major toxic pollutant. In line with this reasoning, Japan recently added vanadium to the list of standard indices of urban air pollution. Unfortunately, the threat from vanadium is not yet generally recognized and our understanding of vanadium toxicity is still far from complete.

3.1. Exposure to Vanadium

Currently, occupational exposure to vanadium is quite common in the petro-chemical, mining, and steel industries. It also occurs among operators of fuel-oil-powered generators (Bowden et al., 1953), but acute intoxication is very rare.

In addition to fossil fuels, some ores contain a significant amount of this metal. For example, vanadium is often associated with uranium ores and may increase the occupational risk to workers employed in uranium mines (Paschoa et al., 1987).

Huge amounts of vanadium are usually deposited in the exhaust systems of engines and generators powered by fuel oil as well as diesel fuels. Vanadium toxicity is a concern for many industrial workers and army, air force, navy, and marine personnel exposed to its compounds on land and sea. In addition to limited vanadium exposure in the workplace (Zychlinski, 1980), the general population is increasingly exposed to this metal as result of increased utilization of vanadium-containing natural oil (Schiff and Graham, 1984) and coal.

Vanadium-bearing particles may persist in the lungs for many years (Paschoa et al., 1987). For example, in the inhabitants of U.S. cities, vanadium deposits in the lungs are markedly elevated in the older population (Tipton and Shaffer, 1964). Increased concentrations of vanadium were also found in chronic hemodialysis patients (Hosokawa and Yoshida, 1990).

3.2.   Health Effects of Vanadium

The data on the environmental and biological distribution of vanadium as well as its action on biological systems were reviewed by Nechay (1984). In oxygenated blood, absorbed vanadium circulates as polyvanadate $\{[V(V)]$, isopolyanions containing vanadium in the $+5$ oxidation state$\}$. In tissue, in the presence of reducing endogenous compounds (such as glutathione-SH) (Bruech et al., 1984), vanadium exists as vanadyl $\{[V(IV)]$, the cationic form of vanadium in the $+4$ oxidation state$\}$ (Erdmann et al., 1984).

Vanadium(V) inhibits (Na^+, K^+) ATPase (adenosine triphosphatase), and at high concentrations may exert a digitalis-like action. Vanadium interferes with the physiological functions of the intestinal, vascular, and *vans deferens* smooth muscles; heart muscle; kidney; eye, ear, brain, and nervous tissues; the lungs and respiratory tract; the liver; and the immunological system (Nechay, 1984, and references therein), as well as the skin and reproductive system (Jandhyala and Hom, 1983, and references therein). The reported animal studies have established vanadium as a teratogen that accumulates in the placenta and fetal skeleton. Vanadium reaches significant levels in human placenta, especially in the full-thickness mid-disc tissues (Ganong et al., 1988).

Co-occurrence of vanadium and polycyclic aromatic hydrocarbons in the exhaust of combustion engines is increasing the possibility of synergistic mutagenesis and carcinogenesis due to co-oxygenation of hydrocarbons and their metabolites (Byczkowski and Kulkarni, 1990, 1992b).

3.2.1.   Pulmonary Effects of Vanadium

Experiments performed on young male rats showed a statistically significant, negative correlation of -0.94 between concentrations of vanadium pentoxide and LD_{50} for seven different natural oil-generated, industrial fly ashes (Zychlinski,

1980). The LD_{50} established for pure vanadium pentoxide, applied intratra-cheally, was 5.6 mg/kg of body weight. Animals that received $0.1 LD_{50}$ of vanadium pentoxide once a month for one year showed a significant increase in lung hydroxyproline contents along with increased total lung weight and hyper-trophy of the connective tissue. The treated animals after 10 months had a significantly lower total body weight, and the weight loss continued through the eleventh and twelfth months. In contrast to this decreased body weight, the average lung weight measured at the end of the twelfth month was significantly increased in the exposed animals. No significant changes were noted in the weight of kidneys or livers (Zychlinski et al., 1991). Average hydroxyproline content per animal was 9.62 mg in the lungs of exposed rats, compared to 5.62 mg in the lungs of control rats ($p \leqslant 0.001$; cf. Czarnowski et al., 1978). Histologically, lungs of the vanadium pentoxide-exposed animals showed marked inflammatory changes (interstitial and sometimes bronchial pneumonia), marked lymphocytic infiltra-tion, frequent blood extravasations into the lung parenchyma, connective tissue hypertrophy with reduced alveolar air capacity, and significantly increased hydroxyproline content.

These changes may indicate an early, initial stage of pulmonary fibrosis that is frequently associated with lung injury by reactive oxygen species and their metabolites (Autor and Schmitt, 1977; Chvapil and Peng, 1975; Johnson et al., 1981). Inability of the lung to inactivate reactive oxygen species was considered a causative factor in the initiation of the fibrotic response, for example, by bleomycin (Fantone and Phan, 1988). In a recent study, Kowalska (1989) demonstrated an increased collagen cross-linking and thus, elevated insoluble collagen content in the lungs of rats treated during their lifetime with vanadium(V) in drinking water. These combined results suggested that, during chronic exposure to vanadium, the lung is the main target organ for toxicity (Zychlinski et al., 1991).

The question arises of the possible mechanism(s) of vanadium's toxic action. Although, plenty of experimental work has been done at the cellular and subcellular levels (Simons, 1979), no single clear-cut mechanism has been identi-fied for its pulmonary toxicity in vivo. It seems, therefore, that several diverse modes of action may be involved. It also seems that most of the effects can be accounted for by vanadium redox cycling and generation of reactive free radicals. Quite similar mechanisms may mediate pulmonary toxicity of other transition metals, such as iron, copper, and manganese.

4.   COPPER, MANGANESE, AND OTHER TRANSITION METALS

Despite several physiological homeostatic mechanisms that regulate the level of copper in the organism (Winge and Mehra, 1990), overexposure resulting in copper toxicosis is common. Inhalation exposure to copper is known to cause a metal fume fever, or "brass chills" disease. Moreover, increased incidence of lung cancer was reported in coppersmiths (Hammond and Belilies, 1980).

In the case of manganese, pulmonary exposure (in most cases to MnO_2) is quite common in mining and in the battery manufacturing industry. The acute symptoms of intoxication include pneumonitis, epithelial necrosis, and mononuclear proliferation (Hammond and Belilies, 1980). However, more serious symptoms of chronic manganese intoxication appear in the central nervous system. The damage is manifested by Parkinson's-like encephalopathy as well as neurodegeneration (briefly reviewed by Zychlinski et al., 1976).

In contrast to these reports on toxicity, it was also suggested that manganese may protect cells against oxidative stress (Stadtman et al., 1990). Thus, manganese can catalyze a superoxide dismutase-like dismutation of $O_2^{\cdot -}$ as well as disproportionation of H_2O_2. On the other hand, manganese complexes can generate oxygen-containing free radical intermediates and facilitate Fenton-type reactions (Berlett et al., 1990).

5. COCARCINOGENESIS AND CO-OXYGENATION

There is no consensus regarding possible mechanism of carcinogenic or cocarcinogenic action of transition metals such as iron, copper, vanadium, and manganese in the lung and other tissues (Kasprzak, 1991; Klein et al., 1991). Recent experiments have suggested, however, that transition metals during redox cycling can promote free radical damage of DNA and/or proteins in tissue (Prutz, 1989; Cramp et al., 1989; Stadtman, 1990). They can oxidize some amino acids (Yim et al., 1990) and may actually catalyze co-oxygenation of some precarcinogens.

The metabolic activation of benzo[a]pyrene by iron-containing particles has been well established in several models and is described in detail in Chapter 15.

Experiments conducted at the College of Public Health, University of South Florida, pointed to lipoxygenase-catalyzed reactions as a potential pathway for the oxidative metabolism of xenobiotics (Kulkarni and Cook, 1988a,b; Byczkowski and Kulkarni, 1989; Kulkarni et al., 1992). For example, rat lung lipoxygenase in the presence of linoleate can catalyze the co-oxygenation of benzo[a]pyrene-7,8-dihydrodiol to its reactive epoxide, a putative ultimate carcinogen (Byczkowski and Kulkarni, 1992a). Hydroperoxides, generated under the pro-oxidant conditions evoked by transition metals, apparently can activate this pathway.

5.1. Free Radical Co-oxygenation by Vanadium

The mechanisms of co-oxygenation was investigated extensively with vanadium as a generator of free radicals. In the model system containing linoleic acid, under aerobic conditions, vanadyl(IV) initiated lipid peroxidation. The kinetics of the process was sigmoidal, as measured by oxygen uptake. Vanadate(V) was ineffective in the linoleate model; however, it was efficiently reduced by NAD(P)H in the lung preparations. Thus, under aerobic conditions, in a highly oxygenated tissue such as lung, it can trigger lipid peroxidation in the presence of NADPH (Zychlinski et al., 1991).

Vanadium redox cycling, fueled by NADPH and oxygen, was also demonstrated in human-term placental microsomes (Byczkowski et al., 1988). It has been shown that vanadyl(IV) and vanadate(V) may trigger lipid peroxidation due to the formation of a reactive peroxy-vanadyl complex with superoxide. It is well known that some free radical intermediates of lipid peroxidation, such as lipid peroxyl radicals, are able to co-oxygenate xenobiotics. For example, recent experiments have shown that the products of linoleic acid oxidation by lipoxygenase can co-oxygenate benzo[a]-pyrene-7,8-dihydrodiol (Byczkowski and Kulkarni, 1989). In vitro, one can employ (+)-stereoisomer of this compound as an indicator of the peroxidative process. The co-oxygenation reaction yields a reactive epoxide that undergoes spontaneous hydrolysis, and *anti-trans*-tetrol stereoisomer accumulates in the medium as the major stable metabolite.

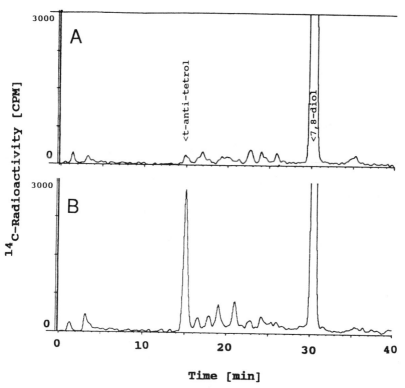

Figure 1. High performance liquid chromatographic (HPLC) profile of $50\,\mu M[^{14}C](+)$-benzo[a]-pyrene-7,8-dihydrodiol co-oxygenation by peroxyl radicals generated during peroxidation of 1 mM linoleic acid with 0.5 mM vanadyl sulfate (B) compared to the control without linoleic acid (A). (For details, see Byczkowski and Kulkarni, 1990.)

This metabolite of (+)-benzo[a]pyrene-7,8-dihydrodiol co-oxygenation by lipid peroxyl derivatives can easily be detected by HPLC (Byczkowski and Kulkarni, 1989). As shown in Figure 1, a nonenzymatic co-oxygenation model system consisting of vanadyl(IV) plus linoleic acid mainly resulted in the production of the *trans-anti*-tetrahydrotetrol of [^{14}C] (+)-benzo[a]pyrene-7,8-dihydrodiol (the first peak in Figure 1B). This result clearly suggests that peroxyl radicals formed during vanadium-initiated lipid peroxidation can be trapped by benzo[a]pyrene-7,8-diol to produce the ultimate carcinogenic, benzo[a]pyrene-7,8-dihydrodiol-9,10-epoxide (Byczkowski and Kulkarni, 1990, 1992b).

6. LUNG

Several studies suggest a causative role for free radicals (Doelman et al., 1990) and unsaturated fatty acid-derived metabolites in toxic lung injury (Cortesi and Privett, 1972). Toxic oil syndrome, a case of massive intoxication that occurred in Spain, manifests itself as a severe bronchopneumonia after ingestion of adulterated rapeseed oil containing high concentrations of partially peroxidized linoleic acid. The toxicity was ascribed to the linoleic acid-derived free radicals (Toxic Epidemic Syndrome Study Group, 1982). Normally, linoleic acid, a major essential fatty acid, is converted in the body into arachidonate, but it can "auto-oxidize" to conjugated diene hydroperoxides through free radical (one-electron) chain reactions (Gardner, 1989b).

6.1. Effects of "Auto-oxidation" Catalyzed by Transition Metals

Transition metals, even in trace amounts, serve as excellent catalysts for the peroxidation of polyunsaturated fatty acids. Thus, auto-oxidation reactions in biological systems are in fact catalyzed by traces of transition metals (Miller et al., 1990). Reactive, oxygen-containing intermediates generated during auto-oxidation of linoleate may cause cell injury by attacking vital cellular molecules, including other lipids, proteins, and DNA. Moreover, it has been demonstrated that peroxidized linoleic acid uncouples oxidative phosphorylation and causes mitochondrial swelling (Brown et al., 1988).

6.2. Pulmonary Lipoxygenases

Peroxidation of polyunsaturated fatty acids can be initiated and propagated by both enzymatic and nonenzymatic processes. In a reaction mimicking the process of auto-oxidation, lipoxygenases [EC 1.13.1.13] catalyze oxidation of unsaturated fatty acids to the hydroperoxy fatty acid (Reinaud et al., 1989); however, the stereochemical fidelity maintained by lipoxygenase demonstrates that the enzyme is not simply an auto-oxidation catalyst (Gardner, 1989 a). Lipoxygenases contain a non-heme bound iron. Some lipoxygenases can peroxidize esterified fatty acids, phospholipids, and even fatty acids embedded in biomembranes (Kuhn et al., 1988). Under physiological conditions, lipoxygenases supply the lung tissue with leukotrienes, lipoxins, and other biologically active metabolites of unsaturated fatty acids (Dahlen et al., 1987; Cuss and Barnes, 1987). In pathological states, the same fatty acid-derived metabolites can probably mediate several allergic reactions in the respiratory tract, as well as bronchoconstriction, lysosomal membrane leakiness, and endothelium-dependent vasorelaxation (Stahl et al., 1989). They are also suspected of playing some role in the symptoms of lung inflammation, edema, acute asthmatic state anaphylaxis, and fibrosis, as well as in free radical lung damage (Dahlen et al., 1987, and references therein). Epoxy-derivative "leukotoxin" was shown to impair the function of mitochondria (Hayakawa et al., 1986).

Linoleate-derived free radicals can also initiate and further propagate membrane lipid peroxidation. However, lipoxygenase that was isolated and partially purified from rat lung cytosol exhibited relatively low activity toward linoleic acid, compared to arachidonate (Yokoyama et al., 1983). It seems that lipid peroxidation in the lung may instead be initiated by transition metals, and that the released hydroperoxides may further stimulate lipoxygenase activity (Kulkarni et al., 1990) and thus enhance lipid peroxidation.

6.3. Reactive Oxygen Species and Macrophages

The lung is particularly sensitive to environmental pollutants, since it is directly exposed to air. High concentrations of oxygen induce lung damage as a consequence of increased generation of superoxide, hydrogen peroxide, and other reactive oxygen species by the lung organelles, followed by migration of polymorphonuclear cells that in turn became activated, inducing even more damage (Freeman and Tanswell, 1985; Repine and Tate, 1983; Crapo, 1986). The structural and functional integrity of pulmonary endothelial cells and their plasma membranes is of major importance for normal cellular and organ function. The pulmonary endothelium is particularly susceptible to injury because it is vulnerable to noxious agents that are inhaled as well as to those delivered to the lung by way of pulmonary circulation.

Recent evidence indicates that a variety of endogenous and exogenous agents can cause structural derangement and loss of normal endothelial cell function. In animals fed a linoleic acid-rich diet, the content of linoleate in lung biomembranes was markedly increased (Gower, 1988). Since linoleate-derived free radicals can cause peroxidative cleavage of membrane lipids, it is plausible that exposure to linoleate and its subsequent peroxidation alter the physical state of lipids in the biomembranes of pulmonary endothelial cells, and that these alterations lead to derangements in biomembrane function, similar to those seen in oxygen toxicity (Patel and Block, 1988). Peroxidative cleavage of membrane lipids can lead to alterations in the cholesterol to phospholipid ratio, the unsaturation index, fatty acyl chain length, and the percentage distribution of fatty acids.

Peroxidation of biomembrane lipids has been implicated in cell injury due to free radicals of oxygen, ionizing radiation, aging, and a host of chemical compounds and drugs. Release of reactive oxygen species appears to contribute to the development of acute edematous lung injury (the so-called adult respiratory distress syndrome). The liberation of H_2O_2 from neutrophils has been hypothesized to cause lung endothelial cell damage and lung injury (Tate and Repine, 1983; Shasby et al., 1982; Tate et al., 1982, 1984).

6.4. Pulmonary Fibrosis

Interstitial pulmonary fibrosis is also a frequent result of many forms of lung injury associated with the generation of oxygen-derived free radicals and their reactive metabolites (Chvapil and Peng, 1975; Weiss and Muggia, 1980; Hunninghake and Fauci, 1979; Johnson et al., 1981; Phan et al., 1980). This suggests that a loss in the ability of the lung to inactivate reactive free radicals and their metabolites may play a role in the initiation of the fibrotic response. In this connection, the recent observation that lipoxygenase can hydroxylate proline may be of significance (Byczkowski et al., 1991).

7. BRAIN

Early reports on lipid peroxidation in the brain came from animal models of hyperbaric oxygen injury (Haugaard, 1968). It was shown that oxygen injury damages normal respiration in brain preparations (Elliot and Libet, 1942; Mann and Quastel, 1946) and inhibits many enzymatic activities, including those that produce acetylcholine (Stadie et al., 1945). Further studies provided evidence that increased ferrous/ferric levels play a key role in brain lipid peroxidation (Jamieson, 1989). It was also demonstrated that superoxide anion radicals, released by xanthine-xanthine oxidase, stimulate lipid peroxidation and inhibit γ-aminobutyric acid uptake by synaptosomes (Braughler, 1985).

It has been suggested that some degenerative disorders of the central nervous system (CNS) and death of neural tissue may be caused by free radicals and lipid peroxidation (Braughler and Hall, 1990; Ciuffi et al., 1991). It was even speculated that some mineral catalysts could play a role in the activation of "scrapie agent" and the pathogenesis of Alzheimer's disease (Jeandel et al., 1989; Evans et al., 1991). Transition metals are known to catalyze free radical reactions and initiate lipid peroxidation. Free radicals were implicated in an Alzheimer-type senile dementia (Harman, 1988). Indeed, autopsy samples of an Alzheimer's patient's brain cortex showed an increased vulnerability to lipid peroxidation in vitro (Subbarao et al., 1990). Despite these recent results, the cause(s) of Alzheimer's disease are still unknown (Volicer and Crino, 1990). Although pretreatment of experimental animals with lipid-soluble antioxidants decreased the extent of brain damage caused by lipid peroxidation (Hall and Braughler, 1990), no treatment is currently available for humans that stops or decreases the process of lipid peroxidation in this tissue.

CNS is very sensitive to peroxidative damage, because of the high level of polyunsaturated lipids in the neuronal cell membranes and poor antioxidant protection (Braughler and Hall, 1990; Floyd, 1990; Phillis and Clough-Helfman, 1990). On the other hand, one kind of brain lipid—lecithins—serves as an important storage site for choline, which is an essential precursor of the neurotransmitter acetylcholine. Biosynthesis of acetylcholine also requires the "high-enery" intermediate acetyl-CoA, and therefore is sensitive to any decrease in energy conservation and ATP biosynthesis. It has repeatedly been reported that in the brains of Alzheimer's patients, the level of acetylcholine is markedly decreased (Katzman, 1986).

In comparison to other tissues, such as liver or heart, the brain is poorly protected against free radical insult by its antioxidant enzymes. For instance, catalase activity in the brain is much lower than in the liver (about 1/160). This is also true for glutathione peroxidase (1/20), glutathione levels (about 1/3), and superoxide dismutase (about 1/2)(Sohal et al., 1990). On the other hand, cerebral arteries can peroxidize arachidonic acid and generate 5- and 15-hydroxy derivatives (Schulz et al., 1990). Despite this extensive background, no information is currently available on the role of brain lipid peroxidation in long-term exposure to transition metals.

Moreover, a large body of information has been gathered on brain lipid peroxidation using models of ischemia reperfusion and traumatic brain injury (Braughler and Hall, 1990; Hall and Braughler, 1990). There is, however, a surprising lack of information on the role of lipid peroxidation and mitochondrial/synaptosomal membrane damage in chronic intoxication by transition metals (e.g., iron overload, copper or vanadium exposure, manganese intoxication). In particular, there is no available information on whether brain lipid peroxidation, impaired synaptosomal/mitochondrial function, and induction of degenerative disorders of the CNS are caused directly by the environmental toxicant or rather by endogenous factors (e.g., stimulated lipoxygenase activities). Also, it is unknown whether antioxidants and some vitamins can in fact protect brain biomembranes against toxic injury.

Review of the relevant literature indicates that the following known factors can initiate lipid peroxidation in the brain in vivo: hemoglobin, lipoxygenases, non-heme iron, and other transition metals. Whereas hemoglobin plays an important pathogenic role in trauma and stroke, non-heme iron and some other transition metals, as well as lipoxygenases, seem to be likely initiators of lipid peroxidation in chronic intoxication. They may increase biomembrane fragility, as in Alzheimer's disease (Hajimohammadreza et al., 1990).

7.1. Hemoglobin

Large amounts of hemoglobin are released during microstrokes, extravasations, and brain trauma. Its deleterious action is prevented and sometimes reversed by haptoglobin (Braughler and Hall, 1990).

7.2.   Lipoxygenases in the Brain

Brain tissue stimulated with calcium ionophore (A23187) generates mono-hydroxytetraenoic acids and leukotrienes. The isolation of 9-monohydroxy-eicosatetraenoic acid (9-HETE) suggested the presence of unusual lipoxygenase activity in the brain, and it was even postulated that some lipoxygenase products may serve as neuromodulators (Samuelsson et al., 1987). Arachidonic acid caused a vasogenic edema that was blocked by the 5-lipoxygenase inhibitor BW755C. It was postulated that brain 5-lipoxygenase may generate leukotrienes and deleterious lipid hydroperoxides.

Hydroperoxides can propagate lipid peroxidation and may enhance leuko-triene generation by stimulating the 5-lipoxygenase. Accordingly, exogenous glutathione peroxidase, which destroys lipid hydroperoxides, inhibited 5-lipoxy-genase activity in brain (Black and Hoff, 1985). Moreover, it has been shown that hydroperoxides produced by lipid peroxidation may further stimulate the enzy-matic activity of lipoxygenase in the brain (Fig. 2). This effect was especially marked in aged-brain cytosol preparations (Kulkarni et al., 1991). It seems that even small amounts of hydroperoxides generated in transition metal-initiated lipid peroxidation may stimulate latent dioxygenase activity. This can lead to exaggerated lipid peroxidation and marked biomembrane destruction.

These results suggest that both lipoxygenase and transition metals can initiate lipid peroxidation in brain preparations and may lead to free radical damage of biomembranes and enzymes in the brain. Hydroperoxides and some neurotrans-mitters can modulate this process.

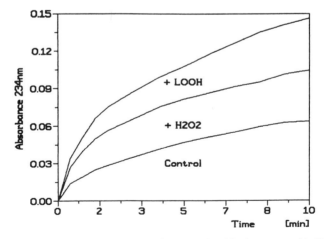

Figure 2. Effects of linoleate hydroperoxide (LOOH, 30 μM) and hydrogen peroxide (H_2O_2, 3.3 nM) on rat brain lipoxygenase activity. The dioxygenation reaction catalyzed by lipoxygenase was measured as an increase in optical density at 234 nm (for conjugated dienes) evoked by aged rat brain cytosol (7 days at 4 °C) in the presence of freshly prepared linoleic acid (1 mM). The result is representative for several experiments with different rat brain aged cytosol preparations. (According to Kulkarni et al., 1991.)

7.3. Non-heme Iron

Microinjections of ferrous iron produced focal edema in rat brain. The damage was correlated with tissue levels of malondialdehyde (index of lipid peroxidation). The effect was blocked by ferritin and other iron-binding proteins. Pretreatment with vitamin E (600 mg/kg) and selenium reduced the iron-induced edema and lipid peroxidation (Willmore and Rubin, 1984).

7.4. Other Transition Metals

Several transition metals (e.g., copper, vanadium, and manganese) known for their neurotoxicity (Hammond and Beliles, 1980) or linked to the etiology of depressive disorders (Naylor, 1983), can form complexes with catecholamines and initiate phospholipid peroxidation (Sotomasu et al., 1990). Similarly, in the presence of iron, catecholamines (dopa, dopamine) caused death of cultured neurons (Tanaka et al., 1991). It seems that, in addition to initiating lipid peroxidation, damage to biomembranes, and increased synaptosomal membrane permeability, copper, vanadium, and possibly also manganese can impair the bioenergetic functions of brain mitochondria.

7.5. Brain Mitochondria

It was shown that transition metals accumulating in brain mitochondria can affect their function and impair biomembrane selective permeability (Byczkowski and Sorenson, 1984). Actually, rat brain mitochondrial oxidative phosphorylation was more sensitive to a single, massive dose (200 mg/kg) of MnO_2 given intratracheally than was the nonmitochondrial enzyme acetylcholinesterase (AChE) (Zychlinski et al., 1976). Brain AChE activity was not changed up to two weeks after the MnO_2 exposure. On the fourteenth day, AChE activity decreased to about 70% of the control but returned to 100% within the next two weeks (Fig. 3). In contrast, oxidative phosphorylation efficiency decreased significantly in brain mitochondria four days after exposure, and was hardly measurable on day 14 (32% of control; Fig. 3).

Figure 4 shows the effects of manganese on the oxidation of succinate (+ glutamate) by rat brain mitochondria at state 4 [without adenosine diphosphate (ADP)], state 3 (in the presence of ADP), and after uncoupling with 2,4-dinitrophenol (DNP). It was noted that in the manganese-treated mitochondria, the state 3 and uncoupled respiratory rates as well as ADP:O ratios were markedly decreased (Byczkowski et al., 1976). A similar pattern of changes was reported as a characteristic result of treatment with peroxidized linoleate and linolenate on mitochondrial membranes in vitro, and it was suggested that these changes may be responsible for Reye's syndrome in infants (Brown et al., 1988).

In the brain, mitochondria are particularly sensitive to peroxidation, because they contain cytochrome P-450 (Iscan et al., 1990), which can promote lipid peroxidation. Impaired oxidative phosphorylation and energy conservation by

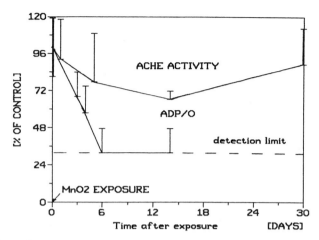

Figure 3. Effects of intratracheal exposure to MnO_2 (200 mg/kg) on acetylcholinesterase activity (AChE) and mitochondrial oxidative phosphorylation effeciency (ADP/O) in rat brain. The values are mean + SD, expressed as % of control activity (for AChE, n = 3 and for ADP/O, $n = 5$) (According to Byczkowski et al., 1976 and Zychlinski et al., 1976.)

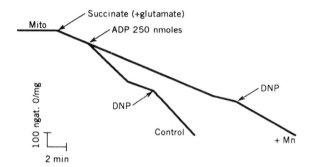

Figure 4. Effects of Mn^{2+} (365 ng atoms/mg of mitochondrial protein) on succinate (+ glutamate, 5 mM) oxidation by rat brain mitochondria (+ Mn) in comparison to control. Uncoupled respiration was induced by addition of 2, 4-dinitrophenol (DNP, 50 µM); state 3 respiration was induced by ADP (250 nmol). The results are representative for several in vitro experiments with different rat brain mitochondrial preparations. (For details, see Byczkowski et al., 1976.)

mitochondria are very likely to limit acetyl-CoA production and decrease brain acetylcholine levels. It was noted that copper causes quite similar impairment of the energy conservation and bioenergetic functions in mitochondria (Byczkowski and Borysewicz, 1976; Byczkowski et al., 1977). Similar deleterious effects were caused by vanadium (Byczkowski et al., 1979). Subsequent experiments with vanadium-treated mitochondria in vitro and in vivo linked its deleterious action on biomembranes to one-electron redox cycling and lipid peroxidation

(Zychlinski and Byczkowski, 1990; Zychlinski et al., 1991; Byczkowski and Kulkarni, 1990). According to the mechanism postulated for the toxic action of vanadium in human tissue, lipid peroxidation is initiated by a free radical complex of the transition metal with superoxide, leading to NADPH depletion, free radical release, and biomembrane destruction (Byczkowski et al., 1988; Byczkowski and Kulkarni, 1992b).

REFERENCES

Adams, J. D., and Odunze, I. N. (1991). Oxygen free radicals and Parkinson's disease. *Free Radical Biol. Med.* **10**, 161–169.

Agency for Toxic Substance and Disease Regulation (ATSDR) (1989). *Toxicological Profile for Chromium*, ATSDR/TP-88/10. U.S. Public Health Service, Washington, DC.

Aisen, P., Cohen, G., and Kang, J. O. (1990). Iron toxicosis. *Int. Rev. Exp. Pathol.* **31**, 1–46.

Alcedo, J. A., and Wetterhahn, K. E. (1990). Chromium toxicity and carcinogenesis. *Int. Rev. Exp. Pathol.* **31**, 85–108.

Aust, S. D., Miller, D. M., and Samokyszyn, V. M. (1990). Role of iron in model lipid peroxidation systems. In P. C. Beaumont, D. J. Deeble, B. J. Parsons, and C. Rice-Evans (Eds.), *Free Radicals, Metals and Biopolymers*. Richelieu Press, London, pp. 251–282.

Autor, A. P., and Schmitt, S. L. (1977). Pulmonary fibrosis and paraquat toxicity. In A. P. Autour (Ed.), *Biochemical Mechanisms of Paraquat Toxicity*. Academic Press, New York, pp. 175–186.

Berger, H. M., Lindeman, J. H. N., van Zoern-Grobben, D., Houdkamp, E., Schrijver, J., and Kanhai, H. H. (1990). Iron overload, free radical damage, and rhesus haemolytic disease. *Lancet* **335**, 933–936.

Berlett, B. S., Chock, P. B., Yim, M. B., and Stadtman, E. R. (1990). Manganese(II) catalyzes the bicarbonate-dependent oxidation of amino acids by hydrogen peroxide and the amino acid-facilitated dismutation of hydrogen peroxide. *Proc. Natl. Acad. Sci. U.S.A.* **87**, 389–394.

Black, K. L., and Hoff, J. T. (1985). Leukotrienes increase blood-brain barrier permeability following intraparenchymal injections in rats. *Ann. Neurol.* **18**, 349–351.

Bowden, A. T., Draper, P., and Rawling, H. (1953). The problem of fuel oil depoisition in open-cycle gas turbines. *Proc.— Inst. Mech. Eng.* **167**, 291–313.

Braughler, J. M. (1985). Lipid peroxidation-induced inhibition of γ-aminobutyric acid uptake in rat brain synaptosomes. Protection by glucocorticoids. *J. Neurochem.* **44**, 1282–1288.

Braughler, J. M., and Hall, E. D. (1990). Central nervous system trauma and stroke. I. Biochemical considerations for oxygen radical formation and lipid peroxidation. *Free Radical Biol. Med.* **6**, 289–301.

Brown, R. E., Bhuvaneswaran, C., and Brewster, M. (1988). Effects of peroxidized polyunsaturated fatty acids on mitochondrial function and structure: Pathogenetic implications for Reye's syndrome. *Ann. Clin. Lab. Sci.* **18**, 337–343.

Bruech, M., Quintanilla, M. E., Legrum, W. Koch, J., Netter, K. J., and Fuhrmann, G. F. (1984). Effects of vanadate on intracellular reduction equivalents in mouse liver and the fate of vanadium in plasma, erythrocytes and liver. *Toxicology* **31**, 283–289.

Byczkowski, J. Z., and Borysewicz, R. (1976). Action of some neurotropic drugs and Cu^{2+} cations on mitochondrial membrane permeability. *Gen. Pharmacol.* **7**, 365–369.

Byczkowski, J. Z., and Gessner, T. (1987). Effects of superoxide generated in vitro on glucuronidation of benzo(a)pyrene metabolites by mouse liver microsomes. *Int. J. Biochem.* **19**, 531–537.

Byczkowski, J. Z., and Gessner, T. (1988). Biological role of superoxide ion-radical. Minireview. *Int. J. Biochem.* **20**, 569–580.

Byczkowski, J. Z., and Kulkarni, A. P. (1989). Lipoxygenase-catalyzed epoxidation of benzo(a)pyrene-7, 8-dihydrodiol.*Biochem. Biophys. Res. Commun.* **159**, 1190–1205.

Byczkowski, J. Z., and Kulkarni, A. P. (1990). Lipid peroxidation and benzo(a)pyrene derivative co-oxygenation by environmental pollutants. *Bull. Environ. Contam. Toxicol.* **45**, 633–640.

Byczkowski, J. Z., and Kulkarni, A. P. (1992a). Linoleate-dependent co-oxygenation of benzo(a)pyrene and benzo(a)pyrene-7,8-dihydrodiol by rat cytosolic lipoxygenase. *Xenobiotica* **22**, 609–618.

Byczkowski, J. Z., and Kulkarni, A. P. (1992b). Vanadium redox cycling, lipid peroxidation and co-oxygenation of benzo(a)pyrene-7,8-dihydrodiol. *Biochim. Biophys. Acta* **1125**, 134–141.

Byczkowski, J. Z., and Sorenson, J. R. J. (1984). Effects of metal compounds on mitochondrial function: A review. *Sci. Total Environ.* **37**, 133–162.

Byczkowski, J. Z., Zychlinski, L., Stachowiak, M., and Byczkowski, S. (1976). Effects of manganese on substrate oxidation and oxidative phosphorylation in rat liver mitochondria. *Pol. J. Pharmacol. Pharm.* **28**, 323–327.

Byczkowski, J. Z., Hac, E. E. M., and Korolkiewicz, Z. K. (1977). Cu^{2+} and salicylate interaction with the energy transformation in mitochondria. *Stud. Biophys.* **66**, 189–199.

Byczkowski J. Z., Zychlinski, L., and Tluczkiewicz, J. (1979). Interaction of vanadate with respiratory chain of rat liver and wheat seedling mitochondria. *Int. J. Biochem.* **10**, 1007–1011.

Byczkowski, J. Z., Wan, B., and Kulkarni, A. P. (1988). Vanadium-mediated lipid peroxidation in microsomes from human term placenta. *Bull. Environ. Contam. Toxicol.* **41**, 696–703.

Byczkowski, J. Z., Ramgoolie, P. J., and Kulkarni, A. P. (1991). Proline hydroxylation by soybean lipoxygenase. *Biochem. Int.* **25**, 639–646.

Chvapil, M., and Peng, Y. M. (1975). Oxygen and lung fibrosis. *Arch. Environ. Health* **30**, 528–532.

Ciuffi, M., Gentilini, G., Franchi-Micheli, S., and Zilletti, L. (1991). Lipid peroxidation induced "in vivo" by iron–carbohydrate complex in the rat brain cortex. *Neurochem. Res.* **16**, 43–49.

Cortesi, R., Privett, O. S. (1972). Toxicity of fatty ozonides and peroxides. *Lipids* **7**, 715–721.

Cramp, W. A., George, A. M., Khan, H., and Yatvin, M. B. (1989). The role of copper in radiation and drug induced changes in the quarternary structure of DNA. In P. C. Beaumont, D. J. Deeble, B. J. Parsons, and C. Rice-Evans (Eds.), *Free Radicals, Metal Ions and Biopolymers*. Richelieu Press, London pp. 127–141.

Crapo, J. D. (1986). Morphologic changes in pulmonary oxygen toxicity. *Annu. Rev. Physiol.* **48**, 721–731.

Cuss, F. M., and Barnes, P. J. (1987). Epithelial mediators. *Am. Rev. Respir. Dis.* **136**, S32–S35.

Czarnowski, W., Zychlinski, L., Zawistowski, S., and Byczkowski, S. (1978). Effect of intratracheal administration of vanadium pentoxide on some organs and lung hydroxyproline level in rat. *Bromatol. Chem. Toksykol.* **11**, 191–196.

Dahlen, S.-E., Kumlin, M., Bjorck, T., Raud, J., and Hedquist, P. (1987). Leukotrienes and related eicosanoids. *Am. Rev. Respir. Dis.* **136**, S24–S28.

Doelman, C. J. A., Leurs, R., Oosterom, W. C., and Bast, A. (1990). Mineral dust exposure and free radical-mediated lung damage. *Exp. Lung Res.* **16**, 41–55.

Elliot, K. A. C., and Libet, B. (1942). Studies of the metabolism of brain suspensions. I. Oxygen uptake. *J. Biol. Chem.* **143**, 227–246.

Erdmann, E., Werdan, K., Krawietz, W., Schmitz, W., and Scholtz, H. (1984). Vanadate and its significance in biochemistry and pharmacology. *Biochem. Pharmacol.* **33**, 945–950.

Evans, P. H., Klinowski, J., and Yano, E. (1991). Cephaloconiosis: A free radical perspective on the proposed particulate-induced etiopathogenesis of Alzheimer's dementia and related disorders. *Med. Hypotheses* **34**, 209–219.

Fantone, J. C., and Phan, S. H. (1988). Oxygen metabolite detoxifying enzyme levels in bleomycin-induced fibrotic lungs. *Free Radical Biol. Med.* **4**, 399–402.

Floyd, R. A. (1990). Role of oxygen free radicals in carcinogenesis and brain ischemia. *FASEB J.* **4,** 2587–2597.

Floyd, R. A. (1991). Oxidative damage to behavior during aging. *Science* **254,** 1597.

Freeman, B. A., and Tanswell, A. K. (1985). Biochemical and cellular aspects of pulmonary oxygen toxicity. *Adv. Free Radical Biol. Med.* **1,** 133–164.

Ganong, C. A., Coffin, C. M., and Manci, E. A. (1988). Vanadium in human placentae. *FASEB J.* **2,** A647 (2035).

Gardner, H. W. (1989a). Soybean lipoxygenase-1 enzymatically forms both (9S)- and (13S)-hydroperoxides from linoleic acid by a pH-dependent mechanism. *Biochim. Biophys. Acta* **1001,** 274–281.

Gardner, H. W. (1989b). Oxygen radical chemistry of polyunsaturated fatty acids. *Free Radical Biol. Med.* **7,** 65–86.

Goldstein, S., and Czapski, G. (1990). Transition metal ions and oxygen radicals. *Int. Rev. Exp. Pathol.* **31,** 133–164.

Gower, J. D. (1988). A role for dietary lipids and antioxidants in the activation of carcinogens. *Free Radical Biol. Med.* **5,** 95–111.

Hajimohammadreza, I., Bremer, M. J., Eagger, S., Burns, A., and Levy, R. (1990). Platelet and erythrocyte membrane changes in Alzheimer's disease. *Biochim. Biophys. Acta* **1025,** 208–214.

Hall, E. D., and Braughler, J. M. (1990). Central nervous system trauma and stroke. II. Physiological and pharmacological evidence for involvement of oxygen radicals and lipid peroxidation. *Free Radical Biol. Med.* **6,** 303–313.

Halliwell, B., and Aruoma, O. I. (1991). DNA damage by oxygen-derived species: Its mechanism and measurement in mammalian system. *FEBS Lett.* **281,** 9–19.

Halliwell, B., and Gutteridge, J. M. C. (1984). Role of iron in oxygen radical reactions. In L. Packer (Ed.), *Methods in Enzymology.* Academic Press, Orlando, FL, Vol. **105,** 47–53.

Hammond, P. B., and Beliles, R. P. (1980). Metals. In J. Doull, C. D. Klaassen, and M. O. Amdur (Eds.), *Toxicology, The Basic Science of Poisons.* Macmillan, New York, pp. 409–467.

Harman, D. (1988). The aging process. In M. G. Simic, K. A. Taylor, J. F. Ward, and C. von Sonntag, (Eds.), *Oxygen Radicals in Biology and Medicine.* Plenum, New York, pp. 1057–1065.

Haugaard, N. (1968). Cellular mechanisms of oxygen toxicity. *Physiol. Rev.* **48,** 311–373.

Hayakawa, M., Sugiyama, S., Takamura, T., Yokoo, K., Iwata, M., Suzuki, K., Taki, F., Takahashi, S., and Ozawa, T. (1986). Neutrophils biosynthesize leukotoxin, 9,10-epoxy-12-octadecenoate. *Biochem. Biophys. Res. Commun.* **137,** 424–430.

Hosokawa, S., and Yoshida, O. (1990). Vanadium in chronic hemodialysis patients. *Int. J. Artif. Organs* **13,** 197–199.

Hudson, T. G. F. (1964). *Vanadium Toxicology and Biological Significance.* Elsevier, Amsterdam.

Hunninghake, G. W., and Fauci, A. S. (1979). Pulmonary involvement in the collagen vascular diseases. *Am. Rev. Respir. Dis.* **119,** 471–503.

Iscan, M., Reuhl, K., Weiss, R., and Maines, M. D. (1990). Regional and subcellular distribution of cytochrome P-450-dependent drug metabolism in monkey brain: The olfactory bulb and the mitochondrial fraction have high levels of activity. *Biochem. Biophys. Res. Commun.* **169,** 858–863.

Jamieson, D. (1989). Oxygen toxicity and reactive oxygen metabolites in mammals. *Free Radical Biol. Med.* **7,** 87–108.

Jandhyala, B. S., and Hom, G. J. (1983). Physiological and pharmacological properties of vanadium. *Life Sci.* **33,** 1325–1340.

Jeandel, C., Nicolas, M. B., Dubois, F., Nabet-Belleville, F., Penin,F., and Cuny, G. (1989). Lipid peroxidation and free radical scavengers in Alzheimer's disease. *Gerontology,* **35,** 275–282.

Johnson, K.J., Fantone, J. C., Kaplan, J., and Ward, P. A. (1981). In vivo damage of rat lungs by oxygen metabolites. *J. Clin. Invest.* **67,** 983–993.

Kasprzak, K.S. (1991). The role of oxidative damage in metal carcinogenicity. *Chem. Res. Toxicol.* **4,** 604–615.

Katzman, R. (1986). Alzheimer's disease. *N. Engl. J. Med.* **314,** 964–973.

Klein, C. B., Frenkel, K., and Costa, M. (1991). The role of oxidative processes in metal carcinogenesis. *Chem. Res. Toxicol.* **4,** 592–604.

Kowalska, M. (1989). Changes in rat lung collagen after life-time treatment with vanadium. *Toxicol. Lett.* **47,** 185–190.

Kuhn, H., Schewe, T., Rapoport, S. M., and Brash, A. R. (1988). Enzymatic lipid peroxidation: A step in the breakdown of mitochondria during the maturation of red blood cells. In M. G. Simic, K. A. Taylor, J. F. Ward, and C. von Sonntag (Eds.), *Oxygen Radicals in Biology and Medicine*. Plenum, New York, pp. 945–948.

Kulkarni, A. P., and Cook, D. (1988a). Hydroperoxidase activity of lipoxygenase: A potential pathway for xenobiotic metabolism in the presence of linoleic acid. *Res. Commun. Chem. Pathol. Pharmacol.* **61,** 305–314.

Kulkarni, A. P., and Cook, D. (1988b). Hydroperoxidase activity of lipoxygense: Hydrogen peroxide-dependent oxidation of xenobiotics. *Biochem. Biophys. Res. Commun.* **155,** 1075–1081.

Kulkarni, A. P., Mitra, A., Chaudhuri, J., Byczkowski, J. Z., and Richards, I. (1990). Hydrogen peroxide: A potent activator of dioxygenase activity of soybean lipoxygenase. *Biochem. Biophys. Res. Commun.* **166,**417–423.

Kulkarni, A. P., Naidu, A., Naidu, A., Ramgoolie, P. J., and Byczkowski, J. Z. (1991). Rat brain lipoxygenase: Hydroperoxidase activity and effects of hydroperoxides and monoamines on dioxygenase. *Toxicologist* **11,** 46(86).

Kulkarni, A. P., Cai, Y., and Richards, I. (1992) Rat pulmonary lipoxygenase: Dioxygenase activity and role in xenobiotic metabolism. *Int. J. Biochem.* **24,** 255–261.

Mann, P. J. G., and Quastel, J. H. (1946). Toxic effects of oxygen and of hydrogen peroxide on brain metabolism. *Biochem. J.* **40,** 139–144.

Marks, J. L. (1987). Oxygen radicals linked to many diseases. *Science* **235,** 529–531.

Miller, D. M., Buettner, G. R., and Aust, S. D. (1990). Transition metals as catalyst of "autooxidation" reactions. *Free Radical Biol. Med.* **8,** 95–108.

Misra, M., Rodriguez, R. E., and Kasprzak, K. S. (1990). Nickel induced lipid peroxidation in the rat: Correlation with nickel effect on antioxidant defense system. *Toxicology* **64,** 1–17.

Naylor, G. J. (1983). Vanadium and affective disorders. *Biol. Psychiatry* **18,** 103–113.

Nechay, B. R. (1984). Mechanisms of action of vanadium. *Annu. Rev. Pharmacol. Toxicol.* **24,** 501–524.

Nriagu, J. O., and Pacyna, J. M. (1988). Quantitative assessment of world-wide contamination of air, water and soil by trace metals. *Nature (London)* **333,** 134–139.

Paschoa, A. S., Warenn, M. E., Singh, N. P., Bruenger, F. W., Miller, S. C., Cholewa, M., and Jones, K. W. (1987). Localization of vanadium-containing particles in the lungs of uranium/vanadium miners. *Biol. Trace Elem. Res.* **13,** 275–282.

Patel, J. M., and Block, E. R. (1988). The effect of oxidant gases on membrane fluidity and function in pulmonary epithelial cell. *Free Radical Biol. Med.* **4,** 121–134.

Phan, S. H., Thrall, R. S., and Ward, P. A. (1980). Bleomycin-induced pulmonary fibrosis in rats: Demonstration of increased rate of collagen synthesis. *Am. Rev. Respir. Dis.* **121,** 501–506.

Phillis, J. W., and Clough-Helfman, C. (1990). Free radicals and ischaemic brain injury: Protection by the spin trap agent PBN. *Med. Sci. Res.* **18,** 403–404.

Prutz, W. (1989). Copper ions as redox catalysts in DNA model systems. In P. C. Beaumont, D. J. Deeble, B. J. Parsons, and C. Rice-Evans (Eds.), *Free Radicals, Metal Ions and Biopolymers*. Richelieu Press, London, pp. 117–126.

Reinaud, O., Delaforge, M., Boucher, J. L., Rocchiccioli, F., and Mansuy, D. (1989). Oxidative metabolism of linoleic acid by human leukocytes. *Biochem. Biophys. Res. Commun.* **161,** 883–891.

Repine, J. E., and Tate, R. M. (1983). Oxygen radicals and lung edema. *Physiologist* **26**, 177–181.

Sadiq, M., and Zaidi, T. H. (1984). Vanadium and nickel content of Nowruz spill tar flakes on the Saudi Arabian coastline and their probable environmental impact. *Bull. Environ. Contam. Toxicol.* **32**, 635–639.

Samuelsson, B., Dahlen, S.-E., Lindgren, J. A., Rouzer, C. A., and Serhan, C. N. (1987). Leukotrienes and lipoxins: Structures, biosynthesis, and biological effects. *Science* **237**, 1171–1176.

Sawyer, D. T. (1990). The chemistry of dioxygen species (O_2, $O_2^{\cdot-}$, HOO·, HOOH) and their activation by transition metals. *Int. Rev. Exp. Pathol.* **31**, 109–131.

Schiff, L. J., and Graham, J. A. (1984). Cytotoxic effect of vanadium and oil-fired fly ash on hamster tracheal epithelium. *Environ. Res.* **34**, 390–402.

Schulz, R., Jancar, S., and Cook, D. A. (1990). Cerebral arteries can generate 5- and 15-hydroxyeicosatetraenoic acid from arachidonic acid. *Can. J. Physiol. Pharmacol.* **68**, 807–813.

Semple, K. M., Cyr, N., Fedorak, P. M., and Westlake, D. W. S. (1990). Characterization of asphaltenes from Cold Lake heavy oil: Variations in chemical structure and composition with molecular size. *Can. J. Chem.* **68**, 1092–1099.

Shasby, D. M., VanBenthuysen, K. M., Tate, R. M., Shasby, S. S., McMurty, I. F., and Repine, J. E. (1982). Granulocytes mediated acute ematodus lung injury in rabbits and isolated rabbit lungs perfused with phorbol myristate acatate. Role of oxygen radicals. *Am. Rev. Respir. Dis.* **125**, 443–447.

Simons, T. J. B. (1979). Vanadate—a new tool for biologists. *Nature (London)* **281**, 337–338.

Sohal, R. S., Sohal, B. H., and Brunk, U. T. (1990). Relationship between antioxidant defenses and longevity in different mammalian species. *Mech. Ageing Deve.* **217**, 227–235.

Sotomatsu, A., Nakano, M., and Hirai, S. (1990). Phospholipid peroxidation induced by the catechol–Fe^{3+} (Cu^{2+}) complex: A possible mechanism of nigrostriatal cell damage. *Arch. Biochem. Biophys.* **283**, 334–341.

Stadie, W. C., Riggs, B. C., and Haugaard, N. (1945). Oxygen poisoning. VIII. The effect of high oxygen pressure on enzymes: The system synthesizing acetyl choline. *J. Biol. Chem.* **161**, 189–196.

Stadtman, E. R. (1990). Metal ion-catalyzed oxidation of proteins: Biochemical mechanisms and biological consequences. *Free Radical Biol. Med.* **9**, 315–325.

Stadtman, E. R., Berlett, B. S., and Chock, P. B. (1990). Manganese-dependent disproportionation of hydrogen peroxide in bicarbonate buffer. *Proc. Natl. Acad. Sci. U.S.A.* **87**, 384–389.

Stahl, G. L., Tsao, P., Lefer, A. M., Ramphal, J. Y., and Nicolaou, K. C. (1989). Pharmacological profile of lipoxins A5 and B5: New biologically active eicosanoids. *Eur. J. Pharmacol.* **163**, 55–60.

Standeven, A. M., and Wetterhahn, K. E. (1991). Is there a role for reactive oxygen species in the mechanism of chromium(VI) carcinogenesis? *Chem. Res. Toxicol.* **4**, 616–625.

Subbarao, K. V., Richardson, J. S., and Ang, L. C. (1990). Autopsy samples of Alzheimer's cortex show increased peroxidation *in vitro*. *J. Neurochem.* **55**, 342–345.

Tanaka, M., Sotomatsu, A., Kanai, H., and Hirai, S. (1991). Dopa and dopamine cause cultured neuronal death in the presence of iron. *J. Neurol. Sci.* **101**, 198–203.

Tate, R.M., and Repine, J. E. (1983). Neutrophils and the adult respiratory distress syndrome. *Am. Rev. Respir. Dis.* **128**, 552–559.

Tate, R. M., VanBenthuysen, K. M., Shasby, D. M., McMurtry, I. F., and Repine, J. E. (1982). Oxygen radical mediated permeability edema and vasoconstriction in isolated perfused rabbit lungs. *Am. Rev. Respir. Dis.* **126**, 802–806.

Tate, R. M., Morris, H. G., Shroeder, W. R., and Repine, J. E. (1984). Oxygen metabolites stimulate thromboxane production and vasoconstriction in isolated saline perfused rabbit lungs. *J. Clin. Invest.* **78**, 608–613.

Tipton, I. H., and Shaffer, J. J. (1964). Statistical analysis of lung trace element levels. *Arch. Environ. Health* **8**, 56–62.

Toxic Epidemic Syndrome Study Group (1982). Toxic epidemic syndrome, Spain 1981. *Lancet* **2**, 697–702.

Tsukamoto, Y., Saka, S., Kumano, K., Iwanami, S., Ishida, O., and Marumo, F. (1990). Abnormal accumulation of vanadium in patients on chronic hemodialysis therapy. *Nephron* **56**, 368–373.

Volicer, L., and Crino, P. B. (1990). Involvement of free radicals in dementia of the Alzheimer type: A hypothesis. *Neurobiol. Ageing* **11**, 567–571.

Warshawsky, D., Bingham, E., and Niemeier, R. W. (1984). The effects of co-carcinogen, ferric oxide, on the metabolism of benzo(a)pyrene in the isolated perfused lung. *J. Toxicol. Environ. Health* **14**, 191–209.

Weiss, R. B., and Muggia, F. M. (1980). Update cytotoxic drug-induced pulmonary disease. *Am. J. Med.* **68**, 259–266.

White, R. F., Feldman, R. G., and Travers, P. H. (1990). Neurobehavioral effects of toxicity due to metals, solvents, and insecticides. *Clin. Neuropharmacol.* **13**, 392–412.

Willmore, L. J., and Rubin, J. J. (1984). Effects of anti-peroxidants on $FeCl_2$-induced lipid peroxidation and focal edema in rat brain. *Exp. Neurol.* **83**, 62–70.

Winge, D. R., and Mehra, R. K. (1990). Host defense against copper toxicity. *Int. Rev. Exp. Pathol.* **31**, 47–83.

Yagi, K. (1987). Lipid peroxides and human diseases. *Chem. Phys. Lipids* **45**, 337–351.

Yim, M. B., Berlett, B. S., Chock, P. B., and Stadtman, E. R. (1990). Manganese(II)-mediated catalytic activity for hydrogen peroxide dismutation and amino acid oxidation: Detection of free radical intermediates. *Proc. Natl. Acad. Sci. U.S.A.* **87**, 349–352.

Yokoyama, C., Mizuno, K., Mitachi, H., Yoshimoto, T., Yamamoto, S., and Pace-Asciak, C. R. (1983). Partial purification and characterization of arachidonate 12-lipoxygenase from rat lung. *Biochim. Biophys. Acta* **750**, 237–243.

Younes, M., and Strubelt, O. (1991). Vanadate-induced toxicity towards isolated perfused rat livers: The role of lipid peroxidation. *Toxicology* **66**, 63–74.

Zemlan, F. P., Thienhaus, O. J., and Bosmann, H. B. (1989). Superoxide dismutase activity in Alzheimer's disease. Possible mechanism for paired helical filament formation. *Brain Res.* **476**, 160–162.

Zychlinski, L. (1980). Toxicological appraisal of workplaces exposed to the dust containing vanadium pentoxide. *Bromatol. Chem. Toksykol.* **8**, 195–199.

Zychlinski, L., and Byczkowski, J. Z. (1989). Effects of intratracheal V_2O_5 administration on rat liver mitochondria. *Toxicologist* **9**, 132 (528).

Zychlinski, L., and Byczkowski, J. Z. (1990). Inhibitory effects of vanadium pentoxide on respiration of rat liver mitochondria. *Arch. Environ. Contam. Toxicol.* **19**, 138–142.

Zychlinski, L., Stachowiak, M., Byczkowski, J. Z., and Cempel, M. (1976). Effects of intratracheal administration of manganese dioxide on liver and brain enzyme activities in rat. *Bromatol. Chem. Toksykol.* **9**, 307–313.

Zychlinski, L., Byczkowski, J. Z., and Kulkarni, A. P. (1991). Toxic effects of long-term intratracheal administration of vanadium pentoxide in rats. *Arch. Environ. Contam. Toxicol.* **20**, 295–298.

17

IMPACT OF OXIDANTS ON A FOREST AREA ASSESSED WITH DYNAMIC MAPS

Wolf Dieter Grossmann

Department for Applied Landscape Ecology, Environmental Research Center Leipzig, D-04318 Leipzig, Germany

Environmental Oxidants, Edited by Jerome O. Nriagu and Milagros S. Simmons.
ISBN 0–471–57928–9   © 1994 John Wiley & Sons, Inc.

---

1. INTRODUCTION

Research on forest dieback was done in several locations in Germany and Austria to assess the overall contribution to damage by individual factors. The method used here combines terrestrial data collection, dynamic models, risk maps concerning forest damage produced with geographical information systems, aerial photography, remote sensing data, and monitoring of the actual development supported by "dynamic maps" depicting predicted development. One result is a simple dynamic model that predicts forest damage. This model seems to reproduce the past development of forest damage quite accurately. It was used to assess the silvicultural effects of different policies to curb emissions of precursors to ozone. The findings indicate: (1) ozone must be the cause of much of the damage, but it cannot explain all symptoms; (2) other contributing factors were nonozone secondary pollutants. It must still be determined how applicable these results are in the other areas.

2. SITUATION

Since 1900, the emissions of many pollutants have dramatically and more or less simultaneously increased in Austria (Fig. 1 a–d). Many of these pollutants contribute directly or indirectly to forest damage. The vegetation also emits considerable amounts (200,000 tonnes/year) of volatile organic compounds (VOCs), in addition to the 500,000 tonnes/year produced by anthropogenic sources (Orthofer and Urban, 1989). Austria has a large share of forest area (44% of the total area). Development of emissions was similar in Western Germany; emissions were about 10 times as high, which corresponds to the 10-fold greater size of its population and GNP. Emissions of VOC's are relatively lower in Germany due to its lower percentage of forest area.

This increase has leveled off since the 1970s for carbon monoxide and since the 1980s for most other pollutants. Pollutants differ in their effects on different tree species. Their local impact depends on their local concentrations modified not only directly by orography (slope, aspect, altitude) and prevailing wind directions, but also indirectly by soil (less damage in good soils), climate (more damage during hot, dry years), and biological factors (insects and diseases).

Rapidly increasing forest damage has been reported from Western Germany since about 1980 (Table 1). The numbers from 1980 to 1983 express more the increasing awareness of the problem than the actual increase in damage. It is now

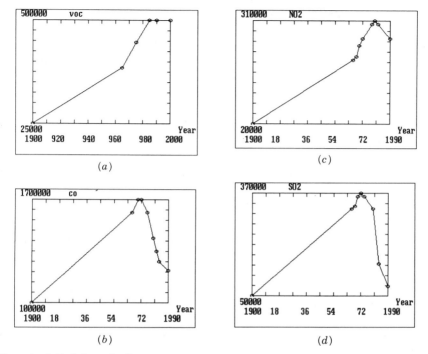

Figure 1a–d. Emissions of pollutants during 1900 to 1990 in tonnes/year; numbers based on Ulrich (1989) and Austrian Energy Reports (1980–1984).

Table 1 Percentage of Forest Area Affected by Forest Decline in Germany

Year:	1980	1981	1982	1983	1984	1985	1990
Damage (%):	1?	2?	8	34	49	~50	~63

Note: The values for Austria are different as the Austrian criteria for the assessment of damage are also very different. The percentage of damaged area is far less in Austria even in areas immediately adjacent to Germany. The actual development of damage in Austria seems to be similar if comparable criteria are selected. Damage is reported to have developed comparably in other European countries.

being discussed how suitable the criteria are that were used in the early 1980s to determine forest damage. Hence, there is uncertainty with respect to the size of the problem. A slight recovery from overall damage in the years 1986 to 1989, combined with a shift in damage from coniferous to deciduous trees, led to the question of whether the problem of forest decline even exists (e.g., Prinz, 1987). This is scientifically a difficult problem because of the interaction of many factors and their spatial and temporal variability.

3. GOAL OF THE RESEARCH

The goal was to assess the overall contribution to damage by individual factors. Therefore, a method had to be developed to deal with complexity, spatial and temporal variability, and uncertainty. It was applied in several locations over the last 10 years.

4. BRIEF DESCRIPTION OF THE METHOD

The method is multifaceted or multilayered, combining:

Strategic evaluations of the problem (e.g., regarding diversity of affected sites, of affected species, and of pollutants)

Dynamic evaluations with models on the interaction of factors (e.g., deposition or removal of pollutants by foliage), possibly simultaneously with the process of damage

Detailed evaluations of local spatial characteristics (e.g., orography, microclimate, soils, species and age of trees).

The evaluations of these three layers are combined to produce a time series of geographical maps, so-called "dynamic maps," that depict the spatiotemporal development of forest decline and recovery. These maps were used for forecasts that were then compared with the actual and ongoing development. The method combines spatial and dynamic modeling, terrestrial data collection, false-color infrared photography, and remote sensing data. It is iterative because the comparisons of forecasts and observed development were used to improve the scientific hypotheses, the resulting models, and the data. It turned out that the different evaluations within this method had different strengths and weaknesses. For example, the perception with terrestrial data differs from the perception with false-color infrared, which, in turn, differs from the forecasts, and so on.

5. RESULTS AND EXPERIENCES

5.1. The Damaging Factors Allow a Quick Recovery of Forests

Modeling results in 1983/1984 for the area of Pfaffenhofen in Bavaria (Germany) predicted a fast recovery of forests if precursors of ozone (VOCs, carbon monoxide, and catalytically acting nitrogen oxides) were decreased (Fig. 2). In the official statistics from 1983, the contribution by cars to VOC emissions was overestimated by a factor of 2 (65% instead of about 30%). Thorough inventories of emissions were done by Meisterhofer et al. (1986) and Orthofer and Urban (1989). The actual effect of catalyzers should be correspondingly lower than depicted in Figure 2.

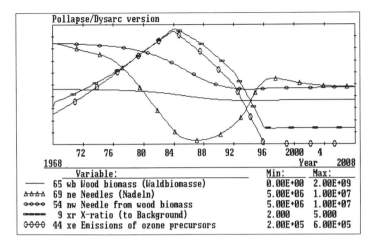

Figure 2. Prediction of forest recovery in 1984. (Adapted from Grossmann et al., 1984).

The model in Figure 2 depicts the development of needle biomass and concentrations of ozone for the years 1968 to 2008. The minimum and maximum columns give the minimum value and maximum value for each curve. Ozone is given as peak values in multiples of 30 ppb during the summer, as this is the threshold value of damage for sensitive species (curve xr, Fig. 2).

The possibility of quick recovery was disputed until about 1986; it was subsequently confirmed by the recovery of coniferous forests after 1984 (Fig. 3). But this

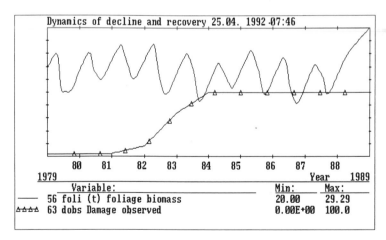

Figure 3. Dynamics of forest damage and recovery, observed (line with open triangles) and calculated (solid line). Higher values of foliage biomass indicate a healthier state of forests. There was increasing damage in 1982 and 1983 and recovery in 1984 and, in particular, in 1985 and 1989. Part of the change in calculated foliage biomass is due to the annual shedding of one needle set.

pattern of increasing damage and consecutive recovery was due to higher than average sun irradiation during the summers of 1981 to 1984 and lower than average sun irradiation during the summers of 1985 to 1988, which resulted in corresponding changes in the monthly mean values of ozone concentration. This recovery was also another indication of the potential importance of ozone in forest decline.

5.2. A Synergistic Damage Function

The predicted dynamics were also combined with spatial assessment depicting how the risk of forest damage is spatially distributed. This combination shows the spatiotemporal development of damage. The model and the spatial assessment were based on a synergistic combination, D, of the impacts of primary pollutants, p, and secondary pollutants, s (see Figs. 4 and 5)

$$D = w_1(p) \cdot w_2(s)$$

where p is concentrations of primary pollutants (SO_2 and NO_x in $\mu g/m^3$, contribution by NO_x weighted $1/3$ compared to SO_2), and s is concentration of ozone in ppb. The functions w_1 and w_2 are Weybul functions, as depicted in Figures 4 and 5. These functions were derived from Jacobson (1977) and SRU (1983) and afterwards adapted to give the observed pattern of damage. The map with the prediction for 1984 was compared to the actual damage. The fit was satisfactory, indicating the potential relevance of the damage function specified above.

5.3. Results in the Rosalia Forest Area

Results could be improved in the Rosalia Demonstration Forest of the Viennese Agricultural University because of the availability of detailed data [in particular,

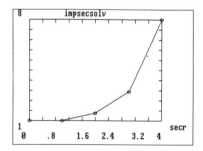
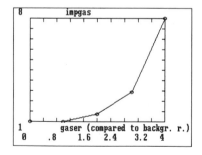

Figures 4 and 5. Weybul functions between concentration of pollutants and resulting damage (dose–response functions) $w_1(p)$ right and $w_2(s)$ left side. The value 1 up to concentrations of secr = 1 denotes that no damage occurs. Secr is in multiples of 30 ppb. In addition to the Weybul functions, locally varying factors are used in the risk maps: prevailing wind directions, slope, aspect, species of trees, and age of trees. In the dynamic model, the use of functions w_1 and w_2 depicts a dose–response relationship.

Table 2   Forest Damage in the Rosalia

Degree of damage	Spruce	Fir	Larch	Pine	Beech
1 (healthy)	22	15	82	17	55
2 (slightly ill)	49	37	14	56	35
3 (ill)	25	32	1	25	5
4 (severely ill)	3	11	1	1	1
5 (dead or dying)	0	5	0	0	0

From Sagl (1986).

Sagl (1986) and Krapfenbauer (1986 a, b)]. Rosalia is close to Wiener Neustadt in Austria. It covers an area of about 600 ha. Forest damage (Table 2) was above the Austrian average in 1985.

5.3.1.   Data Describing the Rosalia Forest Area

The following data were available:

1. Time series of SO_2, NO, NO_2, ozone, and temperature since 1985 (time series of emissions were extended backwards by German data for the years 1900 to 1970; numbers were adapted for Austria according to the relative size and industrial production of these two countries)
2. Assessment of forest damage in 1985 (Table 2)
3. Map depicting the spatial distribution of forest damage (this map was made available to the computer team only after they had produced their prediction)
4. Soil map with pH values
5. Nutrient content of soils, leaves, and branches of about 60 trees and tree-ring analyses of these trees
6. Map of age and species (composition) of stands
7. Orography (slope, aspect, altitude)
8. Microclimate (seasonal wind directions and seasonal precipitation).

Data were also available on depositions of ions from six nearby locations.

5.3.2.   Damaging Factors in the Rosalia

Important damaging factors mentioned in the summary of the very thorough report by Forschungsbeirat Waldschäden (FBW) (1986) are impact of ozone, acidification of soils with too-low buffering capacity by deposition of pollutants and excessive deposition of nitrogen compounds. Rosalia is a clean-air area with very low depositions of ions (Table 3). There seems to be no damaging factor, with the one exception of ozone (Fig. 6).

Almost none of the aforementioned damaging factors could be responsible for the observed damage, according to this overview, with the exception of ozone, which should cause damage at these concentrations according to Heck and

Table 3 Damaging Factors in the Rosalia Area Compared to International Limits (+) in Comparable Areas

Substance	Measurement ($\mu g/m^3$)	Limits ($\mu g/m^3$)
SO_2	7 (mean annual value)	15, mean annual value (Wentzel, 1983) 25, mean annual value (IUFRO) 30, mean annual value (Swiss Federal Parliament 1987, WHO),
NO_x	12 (mean annual value)	30, value set this low to prevent synergisms between NO_x and SO_2 (Swiss Federal Parliament, 1987),
Ozone	> 160 during several days (80 ppb). Episodes up to 340 ($2 \mu g/m^3 = 1$ ppb)	100, 98% of 0.5 hr mean values 120, 1 hr mean value, to be surpassed, at most, once per year (Swiss Federal Parliament 1987) 60, mean value during vegetation period (WHO, VDI, 1987),
Total deposition of nitrogen (+)	10 kg N/ha/yr	Unknown. Present state: 20–60 kg/ha/yr in wide areas of Western Europe. Extreme values: 4–85 (Sartorius, 1986; Masuch et al., 1986)
pH of rain (+)	Rarely < pH 4 (3% of precipitation), pH 3.62 to 7.3	Damage (deficiencies of Mg) due to extensive precipitation with pH < 3.6 if soils are low in nutrients [see Ulrich (1984, 1987) for more precise description]

Notes: (1) Peak values in the Rosalia area: Max. monthly mean of SO_2 was 27, max. daily mean was 108, NO_x was generally below 1 (Kolb and Scheifinger, 1986). Data were measured and additionally calculated.

(2) Other primary pollutants: Not measured. According to the area-related inventory of emissions, there was no likely damaging factor. No unusual concentrations of potentially damaging factors were measured in the soil or in the leaves and branches [base: analysis of chemical elements, e.g., Pb, S, and N (Krapfenbauer et al., 1986b)].

(3) State of the soil: Sufficient supply of nutrients; in particular, no deficiency of Mg, in no areas were pH values very low [e.g., no pH value less than 3.8 (Krapfenbauer et al., 1986b)]. No remarkable concentration of heavy metals. Nutrient supply of trees: good, only Mg with 0.8 µg/g was bordering on the upper limit of deficiency (according to Zöttl 1985) or above the area of deficiency (according to Forschungsbeirat Waldschäden, 1986).

(4) Damaging insects, diseases: Not significant.

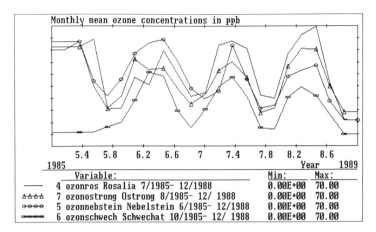

Figure 6. Time series of monthly mean values of ozone (in ppb) in the Rosalia forest and other locations within 100 km. The lower ozone values in Schwechat are most likely due to emissions of primary pollutants from a local refinery (which destroy ozone).

Brandt (1977), Smith (1981, 1984), SRU (1983), Guderian et al. (1984, 1985), FBW (1986), Heck et al. (1986), or Woodman and Cowling (1987).

However, the impact of ozone does not explain all of the observed damage. For example, fir (*Abies alba*) withstood six to eight weeks of ozone concentrations of 300 ppb without damage (Prinz et al., 1984). In the demonstration forest, 48% of all fir trees were injured, including 8% that were severely ill. Damage in beech was already observed at concentrations of 75 ppb ozone after eight weeks in the experiments described by Prinz et al. (1984). The observed symptoms are not typical for ozone damage but correspond to the normal symptoms observed in similar orographic locations (as described in FBW, 1986). See also Hanisch and Kilz (1990) for an extensive description, including 225 color plates, of symptoms of forest damage in spruce and pine.

5.3.3. *Synergisms between Damaging Factors*

Prinz and colleagues (1984) and Guderian and colleagues (1985) have shown the existence of several synergisms between damaging factors, in particular, ozone and SO_2 and ozone and NO_x. However, such synergisms used in the earlier model could not explain the observed damage, simply because the concentrations of SO_2 or NO_x observed and calculated in the Rosalia area were too low to significantly increase any impact of the ozone.

Moreover, wet deposition of SO_2, NO_x, and their compounds (Table 3) will most likely not contribute significantly to damage caused by ozone. Deposition of H^+ ions should be correspondingly low. The mean annual pH value of rain in the background station Exelberg (10 km to the north of Vienna, and 60 km to the north of the Rosalia area) was 4.68 in 1985/1986 (summer 4.98, winter 4.33). The extreme values occurred during two very brief periods of rain, with pH values of 3.62 and 7.38 (Puxbaum and Ober, 1987). Smidt (1986, 1987) reported similar pH values of rain (between 4.3 and 6.09 in the years 1983–1985) from six other stations in the state of Lower Austria where Rosalia is located. Neither should the observed low deposition rates of nitrogen contribute much to the damage, given the good state of the soil. Not even a deficiency of magnesium exists in the Rosalia area.

Hence, all known synergistic explanations have failed in the Rosalia. If damage was above average in spite of low concentrations of primary pollutants, then the role of primary pollutants in the Rosalia has been drastically overestimated.

5.3.4. Evaluation of Nonozone Secondary Pollutants

As ozone alone could not explain the observed damage, a search for an additional, unknown pollutant was undertaken. This pollutant must be windborne according to the spatial pattern of damage. Additionally, the unknown pollutant must cause symptoms different from those of ozone.

An evaluation of nonozone secondary pollutants, here called "products," was done with respect to their concentrations and phytotoxicity (see also Table 4) (Altshuller, 1983). Rain and fog may deposit products, many of which are highly soluble, such as hydrogen peroxide, nitric acid, sulfuric acid, formic acid, and acetic acid. Ozone is not very soluble. Hydrogen peroxide causes symptoms similar to those of SO_2 but different from those of ozone (Masuch et al., 1985, 1986).

According to this evaluation, nonozone products could contribute considerably to the observed damage. Moreover, some cause symptoms different from those caused by ozone. Therefore, factor p (primary pollutants) in the synergistic damage function D specified above was replaced by a weighted sum of nonozone oxidants and other secondary pollutants (e.g., aerosols of SO_4 and NO_3), with a

Table 4   Concentrations of Secondary Pollutants Compared to Ozone

Substance	Percentage	Concentration ($\mu g/m^3$)	Relative Phytotoxicity (%)	Relative Effect
Ozone	100[a]	40–340[c,d]	1 (1)[a]	100
Formaldehyde	20[b]	8–68[f]	3.3(2)[b]	66
Formic acid	4[b]	1–21[c]	1 (3)[c]	4
Peroxyacetylnitrate (PAN)	5[b]	10[e]	1[e]	5
Nitric acid	6[b]	0.2–14[c]	1[c] (3)	6

Note: Ratio of relative phytotoxicity of ozone to other secondary substances: 100/81.

[a] The respective concentrations of ozone are here set equal to 100%; measurements are from Hann (1986), Krapfenbauer (1986b), Puxbaum and Ober (1987).

[b] Average values vary considerably (Altshuller, 1983).

[c] Measurement at the Exelberg (Puxbaum and Ober, 1987).

[d] Measurement at the Heuberg Station, Demonstration Forest Rosalia.

[e] Swiss measurement (Landolt et al., 1985).

[f] Recalculation of the measured ozone concentrations according to (b), here feasible due to similar ratio of ozone/formic acid (Ehhalt, 1986; Gäb and Hellpointner, 1987).

slight residual contribution by primary pollutants. The resulting synergistic damage function is

$$D = w_1 (so) \cdot w_2 (o)$$

with the functions w_1 and w_2 as defined in Section 5.2 but with different arguments; o being ozone concentration in ppb with added (negligible) concentrations of SO_2 and concentrations of NO_x weighted with the factor 1/3 (all arguments here designate gaseous, not readily soluble substances); and so being an indicator of nonozone products weighted as specified in Table 4, based on ozone concentrations. Aerosols of SO_4 and NO_3 would also be included here. (This argument mainly applies to substances that are readily soluble.)

5.3.5. *Prediction of Spatial Aspects of Damage*

This function was then used for a spatial prediction of damage in the Rosalia (*Calc*); the actual spatial distribution of damage was not known. The calculations were the basis for a risk map, which was based on damage assessments from measured and calculated concentrations of ozone, other products, other pollutants (SO_2, NO_x), orography, prevailing wind directions and their seasonal distributions, age and species of trees, and soil types. The function for the spatial assessment of damage is reported in Kopcsa and Grossmann (1991). The risk map covers about 600 ha of Rosalia in about 6000 polygons.

5.3.6. *Comparisons of Prediction and Observations*

Actual damage was observed by terrestrial data collection (*Terr*) and was also independently derived from aerial false-color infrared photography (*Aer*).

Terr and *Aer* had more differences than *Terr* and *Calc* or *Aer* and *Calc* combined. Damage in lower altitudes in *Terr* was higher than in *Aer*, and damage in higher areas was lower in *Terr* than in *Aer*. *Calc* was halfway between *Terr* and *Aer*. Deviations in *Calc* from what is probably the actual situation (that is, those polygons where *Aer* and *Terr* are in agreement but *Calc* deviates) can be attributed to damage in *Calc* being overestimated in protected, elevated areas in the southern part. There are two reasons for this overestimation. First, ozone values were measured in one site in the eastern part of Rosalia and were extrapolated to other locations based on relative elevation. Ozone profiles measured in the Alps were used for this extrapolation (Kolb and Scheifinger, 1986). However, when additional data from two more stations in the Rosalia became available after 1986, it was shown that the ozone profile in the Rosalia is independent of elevation and different from that in the Alps (Krapfenbauer, 1987). Second, protection of forest areas by hills located upwind seems to decrease damage. Such protection exists in the southern parts of Rosalia, where calculated damage was too high. Protection was not made part of the equation to assess the risk of forest damage.

In all, the spatial fit between the predicted and observed patterns of damage supports the assumptions made here; the prediction of damage agreed with the observations in 80% of the area. The new synergistic function outlined above should also have caused considerable damage in Pfaffenhofen, a test site in Bavaria, Germany, as the information in Table 4 generally should be fairly applicable.

5.3.7. *Comparison of the Spatiotemporal Prediction with Actual Further Developments*

Another test of the assumptions was performed. The spatiotemporal prediction of damage was given in 1985 as dynamic maps for all years from 1985 to 2000. The prediction for 1988 was compared in 1988/1989 with "reality," based on 1988 terrestrial data collection and evaluation of 1988 FF-IC aerial photography.

Prediction of damage was, on average, too high by one damage class. The overall health of forests had improved by an average one-half damage class, whereas the dynamic model had on average predicted deterioration by one-half damage class, although the dynamic map was highly heterogeneous in changes.

Analysis of this deviation between predicted and actual damage suggests that it had five causes, each of which *increased* the overestimation; contrary to normal expectations, errors did not partially offset one another. The five reasons for the deviation are as follows:

1. A further increase in emissions of VOCs was assumed in the model, but emissions actually remained constant and may even have decreased slightly since 1985.

2. Carbon monoxide was not included as a precursor to ozone, although it seemed to contribute about 20% to prevailing ozone concentrations (Orthofer and Urban, 1989). Emissions of carbon monoxide were reduced after 1972 (Fig. 1). Therefore, one precursor that also decreased from 1985 to 1989 was left out. As a consequence of (1) and (2), too-high concentrations of ozone were used in the model.

3. Climate was assumed to be constant, whereas the actual sun irradiation was above average between 1981 and 1984 and below average between 1985 and 1988. This further increased the discrepancy between assumed and actual concentrations of ozone.

4. The historical data on the development of forest damage (Table 1) were incorrect according to almost all experts. The model could not produce the reported steep increase between the years 1981 and 1984. Considerable forest damage most likely already existed in the 1970s but was observed only in the 1980s, indicating the fast unfolding of a drama that did not exist. (The new historical data on forest damage in Germany no longer include the years 1981 to 1983. They now start with 1984.) As a consequence of the calibration with false data, the model predicted too much damage.

5. Several severely damaged stands have been cut since 1985; the reforestations are producing healthy trees. No cuts were included in the prediction.

In conclusion, reasons (1), (2), (4), and (5) together should explain the overestimation of damage by one-half damage class, and reason (3) should be responsible for another deviation by one-half damage class due to recovery.

5.3.8.   Effects of Climate Variations and Buildup of Secondary Pollutants

The 1985 model was revised to include actual time series of sun irradiation and to calculate concentrations of ozone from emissions of VOCs and CO. (These values had to be calculated, as only a few years of measurements of ozone were available.) At the same time, the model was simplified as much as possible, so that it can still reproduce the known or supposed past dynamics of forest decline and recovery. The resulting model seems to produce a nice fit to the past dynamics of forest decline and recovery with a minimal amount of data. The main input data in the model include:

1. Time series of emissions of SO_2, NO_x, CO, and VOCs for a state or nation from 1900 to 1990
2. Time series of temperature and sun irradiation (hours)—monthly mean values (here from 1953 to the end of 1988); no other time series are used
3. Forest area, nonforest area, above-ground standing biomass, and biomass of foliage.

In Figure 7, the model portrays an even-aged stand of spruce about 60 years old. The development of foliage and standing biomass over the years from 1960 to 2000 is depicted. There is only the normal decrease of biomass due to harvest, but beginning in the 1970s, a pattern of damage and recovery appears. The recovery in the late 1980s is due to lower sun irradiation in the summers from 1985 onward. In Figure 8, this evaluation is repeated, but the 1980–1984 climate is repeated from 1985 to 1989. Ozone measurements were available for only a few sites for a very limited period of time. Hence, a very simple submodel was included to

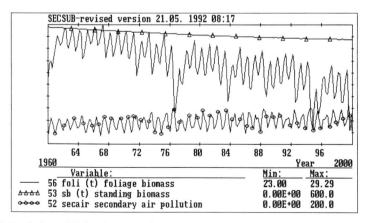

Figure 7. Revised model that includes climate and calculates ozone concentrations. The slight recovery of foliage between 1985 and 1988 should be noted.

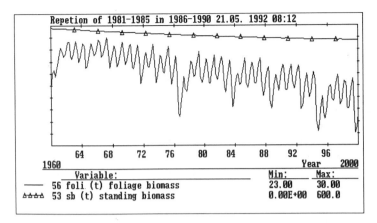

Figure 8. Consequences of higher sun irradiation from 1986 to 1989; recovery of foliage is less compared to Figure 7.

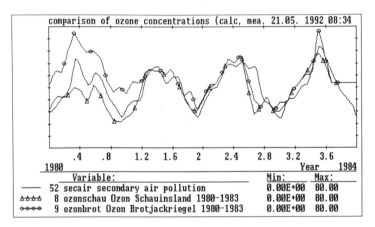

Figure 9. Predicted and calculated concentrations of ozone for two locations in Germany.

calculate ozone concentrations. In Figures 9 and 10, calculated and actual ozone concentrations are shown for two areas in both Germany and Austria. In Figure 11, the long-term development of monthly mean values is compared to measured annual values. This submodel seems to work well.

The model output already shows damage in 1976, with subsequent recovery and new increasing damage beginning in 1989. Most forest scientists agree that there had been damage in 1976. Data do not exist, but tree-ring analysis seems to confirm stress of trees, in particular in the very sunny year 1976. Recovery indeed occurred for most coniferous stands, whereas the health of deciduous trees continued to deteriorate. According to Prinz (1987) most deciduous species are more sensitive to ozone than are coniferous species. The dynamics of this model

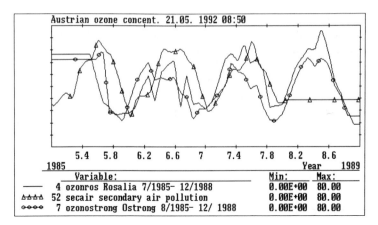

Figure 10. Predicted and calculated concentrations of ozone for two locations in Austria.

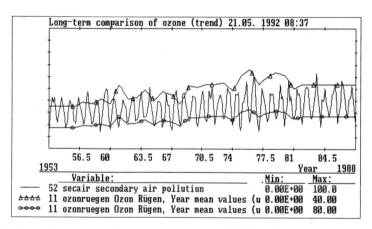

Figure 11. Comparison of monthly mean values of ozone (calculated) and measured annual mean values from a background station (Rügen).

version now seem more accurate than the observed development of damage and recovery.

The translation into spatiotemporal patterns also turned out to be very satisfactory. This model seems to explain the damage sufficiently well for those sites that were tested. But there is far too little knowledge on the phytotoxic effects of the products. Also, too few data on their concentration and dynamics are available.

The method outlined here can be used to develop an acceptable model that can depict the likely dynamics of forest damage. It was also possible to anticipate the pattern of damage. A prediction in 1985 was less successful due to wrong data and too little knowledge about decisive causes of the process of forest damage. But the method also made it possible to identify key factors. It is effective in support of research on ecological and environmental problems. The possibility of comparing predictions or expectations is of utmost importance as (1) observations of reality tend to be seriously questionable and (2) complex systems are often evolving, so that predictions and expectations are often wrong. The combination of predictions and observation and their mutual agreement and disagreement are therefore essential to further scientific research.

6. CONCLUSIONS

The basic requirements for the simplest possible model on the dynamics of forest decline and recovery are as follows:

Such a model needs a simple module on the generation of ozone, since sufficient time series on ozone are not available for the past 20 years.

The model needs time series of emissions of precursors for products, because product concentrations very much depend on the availability of precursors.

The model needs monthly mean values of sun irradiation to allow a sufficiently precise prediction of ozone concentrations.

The model needs to include seasonal changes of growth and senescence of vegetation to prevent recovery of forests during the winter season of low ozone concentrations.

The model also needs emissions of biogenic VOCs, which can constitute about 20% of all emissions in Germany and 30% in Austria. Otherwise, the evaluation of emission control policies may be misleading.

The dynamics of damage caused by products or by acidification tend to be very similar in both spatial and temporal respects: both increase during sunny, hot years. However, forest recovery based on decrease of products tends to be fast. Therefore, the biological model portion must be able to portray this feature of actual forests.

A model should depict appropriate biomass of foliage as a fixed ratio of standing biomass, although leaf biomass decreases very rapidly with age of stands. A model that has to portray realistic amounts of foliage biomass must do bookkeeping on the fraction of the different age classes in the forest area that is modeled. This requires a fairly sophisticated submodel.

ACKNOWLEDGMENT

The research for this paper was funded by the Austrian Federal Ministry of Science and Research.

REFERENCES

Altshuller, A. P. (1983). Measurements of the products of atmospheric photochemical reactions in laboratory studies and in ambient air-relationships between ozone and other products. *Atmos. Environ.* **17**, 2383–2427.

Bruckmann, P., Buck, M., and Eynck, P. (1980). Modelluntersuchungen über den Zusammenhang zwischen Vorläufer- und Photooxidantienkonzentrationen. *Staub—Reinhalt. Luft* **40**, 412–417.

Bufalini, J. J., Walter, T. A., and Bufalini, M. M. (1976). Ozone formation potential of organic compounds. *Environ. Sci. Technol.* **10**(9), 908–912.

Bundesamt für Umweltschutz (1985). *Luftschadstoff- Emissionen in der Schweiz: Ausmass, Entwicklung und bisher getroffene Massnahmen.* Bundesamt für Umweltschutz, Bern.

Bundesamt für Umweltschutz (1987). *Luftbelastung 1986. Messresultate des Nationalen Beobachtungsnetzes für Luftfremdstoffe (NABEL),* Schriftenreihe Umweltschutz No. 67. Bundesamt für Umweltschutz, Bern.

Chapman, E. G., and Sklarew, D. S. (1986). Organic acids in springtime Wisconsin precipitation samples. *Atmos. Environ.*

Davis, D. D., and Wilhour, R. G. (1976). *Susceptibility of Woody Plants to Sulfur Dioxide and Photochemical Oxidants. A Literature Review,* EPA-660/3-76-102. Corvallis Environmental Research Laboratory, U. S. Environmental Protection Agency, Washington, DC.

Derwent, T. G., and Hov, O. (1982). The potential for secondary pollution formation in the atmospheric boundary layer in a high pressure situation over England. *Atmos. Environ.* **16**, 655–665.

Deutsches MAB Nationalkomitee (1983). *Ziele, Fragestellungen und Methoden. Ökosystemforschung Berchtesgaden,* MAB- Mitt. No. 16. Deutsches Nationalkomitee MAB, Bonn.

Ehhalt (1986). Personal communication.

Energiebericht (1986). *Energiebericht der österreichischen Bundesregierung.* Bundesministerium für Handel, Gewerbe und Industrie.

Forschungsbeirat Waldschäden (FBW) (1986). *2. Bericht. Forschungsbeirat Waldschäden/Luftverunreinigungen der Bundesregierung und der Länder. Literaturabteilung des Kernforschungszentrums Karlsruhe.* Karlsruhe, Germany.

Gäb, S., and Hellpointner, E. (1987). Persönliche Mitteilung. (beide: GSF, D-8050 Freising.)

Grossmann, W. D. (1986). *Systemprojekt Lehrforst Rosalia: Untersuchungen zum Problem der neuartigen Waldschäden,* Abschlussbericht, 214 + 6 Seiten. Bundesministerium für Wissenschaft und Forschung, Wien.

Grossmann, W. D. (1987). Products of photooxidation as the main factors of the new forest decline—results and considerations. In L. Kairiukstis, S. Nilsson, and A. Straszak (Eds.), *Forest Decline and Reproduction: Regional and Global Consequences.* International Institute for Applied Systems Analysis (IIASA), Laxenburg, Austria, WP-85–75, pp. 79–100.

Grossmann, W. D. (1989). *Das Modell SECSUB: Waldschäden und Sekundärschadstoffe.* Lehrstuhl für Landschaftsökologie Haber TU München/Weihenstephan.

Grossmann, W. D. (1991a). *Einsatz von Risikokarten in der Waldschadensproblematik. Konzept. Probleme und Ergebnisse im Lehrforst Rosalia.* Centralblatt für das Forstwesen, Wien.

Grossmann, W. D. (1991b). *Einsatz von dynamischen Modellen in der Waldschadensproblematik. Konzept, Probleme und Ergebnisse im Lehrforst Rosalia.* Centralblatt für das Forstwesen, Wien.

Grossmann, W. D., and Grossmann, B. (1985). *Schlussbericht; Beratung des Bundesminsteriums für Wissenschaft und Forschung in der Waldschadensproblematik.* BMWF, Wien.

Grossmann, W. D., and Orthofer, R. (1987). *Grobbeurteilung von Kohlenwasserstoffemissionen im Hinblick auf ihre möglichen Beiträge zum Waldsterben,* ÖFZS-Bericht A-1121, Seibersdorf.

Grossmann, W. D., Schaller, J., and Sittard, M. (1984). "Zeitkarten"—Eine neue Methodik zum Test von Hypothesen und Gegenmassnahmen bei Waldschäden. *Allg. Forst Z.* **33/34**, 837–843.

Guderian, R. (1984). Impact of photochemical oxidants on vegetation in the Federal Republic of Germany. In *The Evaluation and Assessment of the Effects of Photochemical Oxidants on Human Health, Agricultural Crops, Forestry, Materials and Visibility*. Swedish Environmental Institute, pp. 76–91.

Guderian, R., Tingey, D. T., and Rabe, R. (1984). Wirkungen von Photooxidantien auf Pflanzen. In Umweltbundesamt (Ed.)., *Luftqualitätskriterien für photochemische Oxidantien*. Erich Schmidt Verlag, Berlin, pp. 205–427.

Guderian, R., Küppers, K., and Six, R. (1985). *Wirkungen von Ozon, Schwefeldioxid und Stickstoffdioxid auf Fichte und Pappel bei unterschiedlicher Versorgung mit Magnesium und Kalzium sowie auf die Blattflechte Hypogymnia physodes*. VDI-Kommission Reinhaltung der Luft, VDI-Verlag, Waldschäden, Düsseldorf. pp. 657–702.

Haber, W., Lenz, R., Bachhuber, R., Schall, P., Grossmann, W. D., Tobias, K., and Kerner, H. F. (1990). *Prüfung von Hypothesen zum Waldsterben mit Einsatz dynamischer Feedback-Modelle und flächenbezogener Bilanzierungsrechnung für 4 Schwerpunktforscungsräume der Bundesrepublik Deutschland*. TU München/Weihenstephan, Lehrstuhl für Landschaftsökologie.

Hanisch, B., and Kilz, E. (1990). *Monitoring of Forest Damage. Spruce and Pine*. Ulmer, London.

Hann, W. (1986). Private Communication.

Heck, W. W., and Brandt, C. S. (1977). Effects on vegetation: Native, crops, forests. In A. C. Stern (Ed.), *Air Pollution*. Academic Press, New York, pp. 158–229.

Heck, W. W., Heagle, A. S., and Shriner, D. S. (1986). Effects on vegetation: Native, crops, forests. In A. C. Stern (Ed.) *Air Pollution*, 3rd Edn. Academic Press, Orlando, FL.

Jacobson, J. S. (1977). The effects of photochemical oxidants on vegetation. *VDI-Beri.* **270**, 163–173.

Kelly, N. A. (1985). Ozone/precursor relationships in the Detroit metropolitan area derived from captive-air irradiations and an empirical photochemical model. *J. Air Pollut. Control Assoc.* **35**, 27–34.

Kok, G. L. (1980). Measurements of hydrogen peroxide in rainwater. *Atmos. Environ.* **14**, 653–656.

Kolb, H., and Scheifinger, H. (1986). *Immissionsklimatologische Studie zum Lehrforst "Rosalia."* Institut für Meteorologie und Geophysik der Universität, Wien.

Kopcsa, A., and Grossmann, W. D. (1991). *Erstellen der Risikokarten für den Lehrforst Rosalia*. In special issue of "Centralblatt für das gesamte Forstwesen," Vienna.

Krapfenbauer, A. (1986a). *Luftverschmutzung Photooxidantien. Welche Hoffnung bleibt dem Wald?* Universität für Bodenkultur, Wien.

Krapfenbauer, A. (1986b). Ergebnisse des Teilprojektes I der Universität für Bodenkultur im Rahmen des Projektes Lehrforst Rosalia. In Sagl (1986, pp. 147–165).

Krapfenbauer, A. (1987). *Messergebnisse von der NÖ Luftgütemesstelle ROSALIA, den Messtürmen und dem MLU-Luftgütemesswagen*. Universität für Bodenkultur, Wien.

Krause, G. H. M., and Prinz, B. (1989). *Experimentelle Untersuchungen der LIS zur Aufklärung möglicher Ursachen der neuartigen Waldschäden*, No. 80. Landesanstalt für Immissionschutz, Essen.

Landolt, W., Joos, F., and Mächler, H. (1985). Erste Messungen des PAN- Gehaltes der Luft im Raume Birmensdorf (ZH). *Schwei. Z. Forstwes.* **136**, 421–426.

Marenco, A. (1986). Variations of CO and O_3 in the troposphere: Evidence of O_3 photochemistry. *Atmos. Environ.* **20**, 911–918.

Masuch, G., Kettrup, A., Mallant, R., and Slanina, J. (1985). Wirkungen von wasserstoffperoxidhaltigem saurem Nebel auf die Laubblätter junger Buchen (*Fagus sylvatica* L.). In *VDI-Kommission Reinhaltung der Luft. 1985. Waldschäden. Einflussfaktoren und ihre Bewertung*. VDI Verlag, Berlin.

Masuch, G., Kettrup, A., Mallant, R., and Slanina, J. (1986). Effects of H_2O_2-containing acidic fog on young trees. *Int. J. Environ. Chem.* **27**, 183–213.

Meisterhofer, H., Grossmann, W. D., Wurst, F., Loibl, W., and Orthofer, R. (1986). *Studie zu den Emissionen organischer Luftschadstoffe (Kohlenwasserstoffe)*. Umweltfond, Wien.

Mulawa, P. A., Cadle, S. H., Lipari, F., Ang, C. C., and Vandervennet, R. T. (1986). Urban dew: Its composition and influence on dry deposition rates. *Atmos. Environ.* **20**, 1389–1396.

Müller, K. P. (1986). Organische Säuren im Niederschlag. In K. D. Höfgen and H. Bauer (Eds.), *Bestimmung von organischen Stoffen im Niederschlag*. GSF, München, pp. 39–45.

National Research Council of Canada (NRCC) (1975). *Photochemical Air Pollution: Formation, Transports and Effects*. NRCC, Ottawa.

Orthofer, R. (1987). *Abschätzung des Beitrages der Kohlenwasserstoffe zu den Waldschäden*. ÖFZS-Bericht, Österreichisches Forschungszentrum Seibersdorf.

Orthofer, R., and Urban, G. (1989). *Abschätzung der Emissionen von flüchtigen organischen Verbindungen in Österreich*, OEFZS-4492. Österreichisches Forschungszentrum Seibersdorf, Seibersdorf, 163 pp.

Prinz, B. (1987). Zusammenfassung des Themenbereiches E: Oberirdischer Wirkungspfad, Stress- und Stoffwechselphysiologie. In *Projektleitung Biologie, Ökologie, Energie, zusammengestellt von E. Stüttgen 1987. Statusseminar zum BMFT-Förderschwerpunkt Ursachenforschung zu Waldschäden*. Kernforschungsanlage, Zentralbibliothek, Jülich.

Prinz, B., Krause, G. H. M., and Jung, K.-D. (1984). Neuere Untersuchungen der LIS zu den neuartigen Waldschäden. *Düsseldorfer Geobot. Kolloq.* **1**, 25–36.

Puxbaum, H., and Ober, E. (1987). *Backgroundstation Exelberg*. TU Wien. Institut für Analytische Chemie, Umweltbundesamt, Wien.

Sagl, W. (1986). *Endbericht zum Forschungsauftrag "Anwendung der Methode der Zeitkarten auf Österreich"*. Universität für Bodenkultur, Wien.

Sartorius, R. (1986). *Stoffliche Belastung des Bodens über die Atmosphäre*, Bodenschutz, No. 51. Dt. Rat für Landespflege, pp. 39–42.

Schulze, E.-D. (1989). Air pollution and forest decline in a spruce (*Picea abies*) forest. *Science* **244**, 776–783.

Schweizerischer Bundesrat. (1987). *Bericht Luftreinhalte-Konzept*. Schweizerischer Bundesrat, Bern.

Smidt, S. (1986). Bulk-Niederschlagsmessungen in Waldgebieten Österreichs. *Allg. Forstztg.* **236**, 339–341.

Smidt, S. (1987). *Zusammenstellung der Stickstoffdepositionen in Österreich in Vorbereitung*. Forstliche Bundesversuchsanstalt, Wien.

Smith, W. H. (1981). *Air Pollution and Forests*. Springer-Verlag, New York.

Smith, W. H. (1984). Auswirkungen von regionalen Luftschadstoffen auf die Wälder in den USA. *Forstwiss. Centralbl.* **103**, 48–61.

SRU (1983). *Waldschäden und Luftverunreinigungen*. Deutscher Bundestag, Bonn.

Statusseminar (1987). *Statusseminar zum BMFT-Förderschwerpunkt Ursachenforschung zu Waldschäden. Tagungsbericht, Projektleitung Biologie, Ökologie, Energie, zusammengestellt von E. Stüttgen 1987*. Kernforschungsanlage, Zentralbibliothek, Jülich.

Stern, A. C. (Ed.) (1976–1986). *Air Pollution*, 3rd ed., Vols. 1–4, 6–8. Academic Press, Orlando; FL.

Ulrich, B. (1981). Eine ökosystemare Hypothese über die Ursachen des Tannensterbens (*Abies alba* Mill.). *Forstwiss. Centralbl.* **100**, 228–236.

Ulrich, B. (1983). Gefahren für das Waldökosystem durch Saure Niederschläge. In *Immissionsbelastungen von Waldökosystemen. Landesanstalt für Ökologie, Landschaftsentwicklung und Forstplanung Nordrhein-Westfalen*. Recklinghausen, 9–25.

Ulrich, B. (1984). Langzeitwirkungen von Luftverunreinigungen auf Waldökosysteme. *Düsseldorfer Geobot. Kolloq.* **1**, 11–23.

Ulrich, B. (1987). Einführung und Überblick über den Themenbereich F "Bodenchemie, Nährstoffhaushalt (incl. Meliorationsmassnahmen). In *Statusseminar zum BMFT-Förderschwerpunkt Ursachenforschung zu Waldschäden. Tagungsbericht, Projektleitung Biologie, Ökologie, Energie,*

516 Assessment of Oxidative Damage in Forest Areas

zusammengestellt von E. Stüttgen 1987. Kernforschungsanlage, Zentralbibliothek, Jülich, pp. 296–305.

Ulrich, B. (1989). Stabilitätsbedingungen von Waldökosystemen. Forschungsantrag an das Bundesministerium für Forschung und Technologie Bonn im Rahmen der Errichtung von Ökosystemforschungszentren. In *Berichte des Forschungszentrums Waldökosysteme*, Reihe B, Bd. 14. Selbstverlag, Universität Göttingen.

Umweltbundesamt (1986). *Daten zur Umwelt 1986/87*. Erich Schmidt Verlag, Berlin.

van Haut, H. and Prinz, B. (1979). Beurteilung der relativen Pflanzenschädlichkeit organischer Luftverunreinigungen im LIS—Kurzzeittest. *Staub—Reinhalt. Luft* **39**, 408–414.

VDI Kommission Reinhaltung der Luft (1987). *Maximale Immissions-Konzentrationen für Ozon.*, VDI, Düsseldorf.

Wentzel, K. F. (1983). Ursachen des Waldsterbens in Mitteleuropa. *Allg. Forst Z.* 1365–1368.

Woodman, J. N., and Cowling, E. B. (1987). Airborne chemicals and forest health. *Environ. Sci. Technol.* **21**, 120–126.

Zöttl, H. W. (1985). Waldschäden und Nährelementversorgung. Düsseldorf: *Düsseldorfer Geobot. Kolloq.* **2**, 31–41.

18

USE OF OZONE AND OTHER STRONG OXIDANTS FOR HAZARDOUS WASTE MANAGEMENT

Susan J. Masten and Simon H. R. Davies

Department of Civil and Environmental Engineering, Michigan State University, East Lansing, Michigan 48824

Environmental Oxidants, Edited by Jerome O. Nriagu and Milagros S. Simmons.
ISBN 0–471–57928–9 © 1994 John Wiley & Sons, Inc.

1.   INTRODUCTION

Chemical oxidation processes have been used to oxidize pollutants to terminal end products (CO_2 and H_2O) or to intermediate products that are more readily biodegradable or less toxic. Ozone, hydrogen peroxide, chlorine, chlorine dioxide, persulfate, and potassium permanganate are oxidants used in water and wastewater treatment. In this paper we will focus on the processes that are most commonly used for hazardous waste treatment. These are ozone alone, ozone in combination with UV irradiation, ozone with added hydrogen peroxide, hydrogen peroxide alone, hydrogen peroxide in combination with Fe(II) (Fenton's reagent), and hydrogen peroxide with UV irradiation.* With the exception of ozone and hydrogen peroxide, these processes are grouped together and referred to as advanced oxidation processes (AOPs) because they all result in the formation of OH radicals in sufficient quantity to affect water treatment (Glaze, 1987). Other AOPs, such as titanium dioxide-sensitized photocatalysis, will not be discussed in this paper.

2.   OZONE

2.1.   Ozone as an Oxidant

Ozone is a very strong oxidant, more powerful than other oxidants commonly used in water and wastewater treatment. The use of ozonation as a treatment process began in Nice, France, in 1907 with the application of ozone to disinfect drinking water. In Europe, ozone technology flourished, so that by 1936 there were more than 100 ozone installations in France and 30 to 40 elsewhere. Meanwhile, in the United States, chlorine had become the disinfectant of choice, and ozone was used in a limited number of facilities mainly for taste and odor control. By the 1960s, ozone was also being used for odor control in wastewater

---

* For the sake of brevity, we will use shortened forms of the process descriptions in this paper. Ozone in combination with UV radiation will be referred to as ozone/UV, ozone with hydrogen peroxide as ozone/H_2O_2, ozone at high pH as ozone/high pH. Hydrogen peroxide in combination with UV radiation will be referred to as H_2O_2/UV.

treatment (Griffin, 1965; Miller, 1966). In more recent years, as environmental engineers have become more concerned with the presence of recalcitrant organics that are not removed by conventional treatment processes, interest in ozonation has soared (Evans, 1972). Studies have been conducted to determine the potential of using ozone to oxidize such organic compounds as surfactants, chlorinated hydrocarbons, phenols, pesticides, and benzene. While a comprehensive review of the literature on the ozonation of specific organic chemicals is beyond the scope of this paper, we will attempt to provide an overview of the applications of ozonation in the treatment of hazardous wastes.

2.2. Reaction Mechanisms Involving Ozone

The mechanism by which ozone reacts with organic chemicals has been extensively studied in nonparticipating solvents (i.e., solvents that do not participate in ozonolysis reactions, such as that of carbon tetrachloride and *n*-pentane) and, to a lesser degree, in water (a participating solvent). Bailey (1978, 1982) provides an exhaustive review of the literature through the 1970s.

The reaction of ozone with alkenes occurs via a 1,3-dipolar cyclic addition of ozone across the carbon-carbon double bond and involves the formation of an unstable intermediate known as a molozonide (Bailey, 1958, 1971; Huisgen, 1963a, b; Bailey and Thompson, 1966; Bailey and Lane, 1967; Bauld et al., 1968; Criegee, 1968; Durham and Greenwood, 1968). This mechanism is known as the Criegee mechanism. In water, the molozonide rapidly decomposes to form aldehydes, ketones, and/or organic acids (Pryde et al., 1968). Gilles and Kuszkowski (1983) studied the ozonolysis of haloalkenes. They found that while the Criegee mechanism could explain the formation of ozonolysis products, other chemicals were produced via decomposition reactions, hydrolysis reactions, and halogen rearrangements of the parent compound or of the various reaction products formed.

Ozone does not appear to react with aromatic compounds via the Criegee mechanism. Electrophilic reactions occur on molecular sites having a strong electronic density. Aromatic compounds with electron-donating groups (e.g., —OH, —NH$_2$) tend to have high electron densities at the carbons at the ortho and para positions, resulting in a tendency for the ozone to attack at these positions (Langlais et al., 1991). These reactions generally result in the formation of carboxylic acids. Aromatic compounds with electron-withdrawing groups (e.g., —COOH, —NO$_2$) are more likely to react via nucleophilic substitution (Doré and Legube, 1983). The initial attack of the ozone molecule is usually at the destabilized meta position (Langlais et al., 1991). Legube (1983) reported that aromatic compounds react with ozone to yield polyhydroxy aromatics, unsaturated aliphatics (including alcohols, dicarboxylic acids, and esters), saturated aliphatics, quinoids, and finally CO$_2$ and H$_2$O.

The reaction of phenolic compounds has been extensively studied; the mechanism appears to be dependent on pH. Under acidic or neutral conditions, electrophilic attack by ozone is thought to occur at the ortho or para position

(Eisenhauer, 1968, 1971; Doré and Legube, 1983; Decoret et al., 1984; Gurol and Vatistas, 1987). Singer and Gurol (1983) report that electrophilic attack on phenolate ions occurs at neutral or basic pH. Gurol and Vatistas (1987) suggest that the OH radical reaction predominates. They found that the oxidation of phenol resulted in the formation of diphenols and quinones from aromatic nuclear hydroxylation. The breakdown of the ring structure resulted in the formation of muconic acid and similar derivatives. More complete oxidation of the compounds mentioned above resulted in the formation of glyoxylic, formic acid, and glyoxal. Similar products from the ozonation of phenol were observed by Jarret et al. (1983).

As early as 1904, research was conducted to identify the products of the ozonation of biphenyl (Harris and Weiss, 1904). It was found that ozonation yielded a tetraozonide instead of the expected hexaozonide. Subsequent work conducted by Copeland et al. (1960, 1961) showed that when biphenyl was oxidized by ozone in chloroform (a nonparticipating solvent), benzoic acid was obtained in 86% yield. As water is a participating solvent, the oxidation products, reaction rates, and reaction mechanisms would be expected to differ from those observed in organic solvents. More recently, Yokoyama et al. (1974) found that, in water, polychlorinated biphenyls (PCBs) were degraded slowly by ozone. Berdin et al. (1989) discovered that the oxidation of trichlorobiphenyl with ozone is initiated by the addition of potassium hydroxide. In 2.5×10^{-3} M KOH at $20\,°C$, 99.4% of the trichlorobiphenyl reacted.

The reaction of polycyclic aromatic hydrocarbons is believed to occur via a 1,3-dipolar cyclic addition (Legube et al., 1986). Legube and colleagues found that the reaction of ozone and naphthalene in water yielded numerous products including oxalic acid, formic acid, phthalaldehydic acid, and 1,4-naphthoquinone. Van Cauwenberghe et al. (1979) studied the gas-phase reaction of benzo[a]pyrene and ozone. They found that substitution reactions resulted in the formation of quinones and epoxides, whereas ring cleavage reactions resulted in the formation of dialdehydes, benzanthronedicarboxylic acid, and other dicarboxylic acid compounds. Corless et al. (1990) investigated the use of ozone in the oxidation of pyrene. Almost complete (85%) disappearance of pyrene occurred at an ozone concentration of $5\,mg\,L^{-1}$ and a reaction time of 1 min. (The initial concentration of pyrene was $10\,\mu g\,L^{-1}$.) The two products isolated were thought to be short-chain aliphatics.

2.3.   Kinetics of the Reactions of Ozone with Organic Chemicals

The kinetics of the ozonation of organic chemicals have been extensively studied [see Langlais et al. (1991) for a review of the literature]. Unsaturated aliphatic compounds react faster than saturated hydrocarbons (Hoigné and Bader, 1983a). Aromatic compounds with electron-donating groups react faster than aromatic chemicals with electron-withdrawing groups. For example, Hoigné and Bader (1983a) determined that, in aqueous solution, the reaction of ozone with phenol > toluene > benzene > chlorobenzene > nitrobenzene. In general, aldehydes react faster than alcohols, and carboxylic acids are essentially unreactive with ozone. Aliphatic amines are reactive with ozone (Langlais et al., 1991). However, as the protonated form of the amine is much less reactive than the ionized form of the amine, the rate of reaction decreases with decreasing pH. In general, the more chlorinated a compound, the less easily it is oxidized (Hoigné and Bader, 1983a).

In carbon tetrachloride (Williamson and Cvetanović, 1968; Pryor et al., 1983) and water (Hoigné and Bader, 1983a; Hoigné, 1988), it has been found that the decay of ozone in the presence of an alkene (at constant pH) is first-order both in ozone and reactant. Selected rate constants obtained by these and other researchers for the reaction of ozone and chlorinated alkenes are compared in Table 1. Except for tetra- and trichloroethylene, the rate constants obtained in water appear to be much larger (by a factor of 5 to 38) than those obtained in carbon tetrachloride. Nevertheless, in both solvents, several trends hold: the rate constants decrease with increasing chlorine substitution, and cis-substituted compounds react more slowly than those that are trans-substituted. As the addition of ozone across a double bond is an electrophilic substitution, the addition of chlorine to a molecule would slow the reaction kinetics. Steric hindrance by the chlorine group could also be a factor, as appears to be the case with *cis*- and *trans*-chloroethylene.

Chlorine substitution also slows the rate at which aromatic compounds are oxidized. Hoigné and Bader (1983a) found that the rate constants for the reaction of ozone with benzene, chlorobenzene, 1,4-dichlorobenzene, and 1,2,4-trichlorobenzene all differed by less than a factor of 4, with chlorobenzene being the least reactive and 1,4-dichlorobenzene being the most reactive.

Gould (1987) studied the kinetics of ozonation of phenolic compounds at constant pH and found that the first-order rate constants for a series of substituted phenols were well correlated with the Hammett polar constant. A strong cor-

Table 1   Rates of Reaction of Chloroalkenes with Ozone in Carbon Tetrachloride and Water

	k_{O_3} (L mol^{-1} sec^{-1})			
	CCl$_4$		H$_2$O (pH2)	
	Ref. a	Ref. b	Ref. c	Ref. d
Tetrachloroethylene	1.0	0.007	—	< 0.1
Trichloroethylene	3.6	1.8	17 ± 4	—
cis-Dichloroethylene	35.7	21.3	800	—
trans-Dichloroethylene	591	590	5700 ± 100	—
1,1-Dichloroethylene	22.1	25.5	—	110 ± 20
Chloroethylene	1,180	—	—	—
3-Chloro-1-propene	11,000	11,700	—	—
1,1-Dichloro-1-propene	—	178	—	940

Note: Temperature = $25 \pm 1\,°C$.

References: (a) Williamson and Cvetanović (1968), $[M]_0 = 5.0$ mM; (b) Pryor et al. (1983), $[M]_0 = 3.0$ mM; (c) Hoigné and Bader (1983a), $[M]_0 = 0.1$ mM; (d) Masten (1986); $[M]_0 = 0.4$ mM.

relation was also observed for the kinetics of ozonation of a homologous series of 2-alkyl-4,6-dinitrophenols and the steric parameter. As electron density in the ring increased, the reaction rate increased. The reaction rate was observed to decrease as the size of the substituted group increased. The rate at which phenolic compounds react is dependent upon pH. Hoigné and Bader (1983b) found that undissociated phenolic species have rate constants that range from 10^3 to 10^4 M^{-1} sec^{-1}, while the rate constants of the dissociated forms are on the order of 10^9 M^{-1} sec^{-1}. In contrast, Anderson (1977) and Joshi and Shambaugh (1982) observed that the rate constants for phenol and its anion varied by no more than a factor of 27, not six orders of magnitude. Gilbert (1982) found similar results with 2-nitro-p-cresol in that an increase in reactivity occurred with increasing pH, but only on the order of a factor of 3. While the reason for the differences in these results is unclear, it may lie in the techniques used by Hoigné and Bader to calculate their rate constants. It appears that they measured the rate of ozone decomposition (not phenol decomposition) in the pH range from 2 to 6 and then extrapolated, assuming a slope of 1, to a pH equal to the pK for phenol dissociation. From the rate constant for the decomposition of ozone at that pH they then calculated the rate constant for the phenolate ion. The other authors measured the rate of disappearance of phenol.

The kinetics of the aqueous reaction of ozone with polycyclic aromatic hydrocarbons (PAHs) have been studied by a number of researchers. Radding et al. (1976) determined the rate constants for the reaction of ozone and several PAHs in water at 25 °C. The rate constants and half-lives are given in Table 2. Lee and

Table 2   Rate Constants for the Reaction of Ozone and Various PAHs in Water

Compound	Rate Constant $k(M^{-1} sec^{-1})$	Ref.
Naphthalene	3,000	(c)
Pyrene	170	(a)
Benzo[a]pyrene	110	(a)
Benz[a]anthracene	260	(a)
Dimethylbenzanthracene	> 680	(a)
Dibenzanthracene	280	(a)
Fluorene	Fast, completely destroyed	(d)
Phenanthrene	Rapid oxidation of center ring, then very slow oxidation	(d)
	15,000	(b)
Anthracene	Fast	(d)
Acenaphthene	Fast	(d)
Acenaphthylene	Fast	(d)
3,4-Benzofluoranthene	Intermediate	(d)
11,12-Benzofluoranthene	Intermediate	(d)

References: (a) Radding et al. (1976); (b) Butković et al. (1983); (c) Hoigné and Bader (1983a); (d) Water Pollution Control Federation (1986).

Hunter (1985) showed that greater than 99% decomposition of naphthalene, fluoranthrene, benzo[*b*]fluoranthene, benzo[*k*]fluoranthene, and benzo[*a*]pyrene could be achieved when aqueous solutions of these compounds were oxidized using an ozone dosage of 21.3 mg L^{-1} and 1-hr contact time. The most recalcitrant of the compounds studied by Lee and Hunter was benzo[*a*]anthracene. Under the conditions given above, 54% of this compound was degraded.

2.4.   Decomposition of Ozone

Ozone decomposes rapidly in water to form more reactive secondary oxidants, the most important being the hydroxyl radical. The formation of the hydroxyl radical is important in that ·OH is a more powerful and less selective oxidant than ozone. As a result, many compounds that are unreactive with ozone can be oxidized by ·OH.

The mechanism by which ozone decomposes in water has been extensively studied (Weiss, 1935; Alder and Hill, 1950; Staehelin and Hoigné, 1982, 1985; Bühler et al., 1984; for reviews, see Peleg, 1976; Forni et al., 1982; Staehelin, 1984). As shown in Figure 1, ozone reacts with hydroxide ions to form an ozonide ion that reacts via a cyclic mechanism to form a hydroxyl radical (Staehelin and Hoigné, 1985). Although Tomiyasu et al. (1985) proposed a different ozone degradation pathway that did not involve the formation of the two intermediates, HO_3 and HO_4, they, too, proposed the formation of an ozonide ion that reacts to form ·OH. As the hydroxide ion is a good promotor of ozone decomposition, the half-life of ozone is very short under alkaline conditions.

In waters containing species such as humic acid, carbonate species, iron, carboxylic acids, and primary alcohols, the mechanisms involved in ozone decomposition are more complex than in pure water (see Fig. 2). The species mentioned above can act as initiators, promoters, or inhibitors of free radical reactions. Initiation reactions involve the formation of superoxide ion (O_2^-), while promoters are capable of regenerating this ion. Inhibitors decrease the rate at which the superoxide ion is formed, resulting in a decrease in the rate of hydroxyl radical formation. A more detailed discussion of these processes is presented by Langlais et al. (1991).

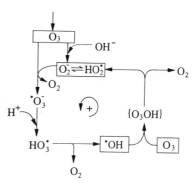

Figure 1. The decomposition of ozone in pure waters occurs via a cyclic mechanism. The oxidants that can oxidize organic chemicals are indicated in boxes. [Reprinted with permission from Staehelin and Hoigne, 1985. Copyright (1985) American Chemical Society.]

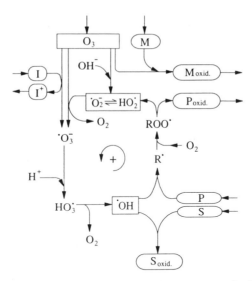

Figure 2. The decomposition of ozone in waters containing reactant species. M designates a micro-pollutant that reacts directly with ozone; I, an initiator that reacts with ozone to initiate the chain reaction mechanism; S, a scavenger that reacts with ·OH to terminate the chain reaction; and P, a promoter that reacts with ·OH to form a radical species, resulting in chain propagation. Oxidant species are indicated in boxes. [Reprinted with permission from Staehelin and Hoigne, 1985. Copyright (1985) American Chemical Society.]

   Both hydrogen peroxide and UV irradiation can act as free radical initiators when they react with ozone. Thus, the combination of either hydrogen peroxide or UV irradiation with ozone is useful for the generation of hydroxyl radicals.

   The mechanism of ozone photolysis has been discussed by various researchers (Taube, 1957; Fortin et al., 1972; Prengle, 1983; Cobos et al., 1983; Peyton and Glaze, 1987, 1988; Glaze and Kang, 1989a, b). Ozone absorbs ultraviolet light (< 334 nm) to produce both singlet-D and triplet-P oxygen atoms. As the quantum yields of triplet-P oxygen atoms are several orders of magnitude less than that of singlet oxygen, only singlet oxygen is important in the reactions that result in the formation of ·OH. Taube (1957) and others (Hoigné and Bader, 1987; Peyton and Glaze, 1987) showed that the photolysis of ozone results in the production of hydrogen peroxide. The hydrogen peroxide can undergo direct photolysis or its conjugate base can react with ozone. Both reactions result in the formation of hydroxyl radicals. As the absorption coefficient of hydrogen peroxide at approximately 254 nm is small compared to that of ozone,* the photolysis of hydrogen peroxide (when using a low-pressure mercury lamp) is important only after ozone is depleted.

---

* The molar absorptivity for hydrogen peroxide at 260 nm has been measured as $13\,M^{-1}\,cm^{-1}$ (Morgan et al., 1988); ozone, in contrast, has a molar absorptivity of $3000\,M^{-1}\,cm^{-1}$ at 258 nm (Nowell and Hoigné, 1987).

The decomposition of ozone can be enhanced by hydrogen peroxide (Taube and Bray, 1940; Forni et al., 1982; Staehelin and Hoigné, 1982). Staehelin and Hoigné found that at pH 2, hydrogen peroxide had little effect on ozone decomposition. At pH values above 5, hydrogen peroxide was found to greatly enhance the rate of ozone decomposition. A second-order rate expression was used to describe the rate at which ozone decomposes:

$$-\frac{d[O_3]}{dt} = k[O_3][HO_2^-] \qquad k = (5.5 \pm 1.0) \times 10^6 \, M^{-1} \, sec^{-1}$$

Glaze and Kang (1989a, b) extended this basic equation to describe the kinetics of ozone decomposition in a sparged, semibatch reactor. Included in the model are the initiation reactions involving ozone and hydrogen peroxide (Staehelin and Hoigné, 1982); the propagation reactions involving superoxide, ozonide, and HO_3 (Bühler et al., 1984); and numerous radical scavenging reactions. The mass transfer characteristics of the reactor are also considered in the model. The model accurately predicts the rate of reaction of tetrachloroethylene with O_3/H_2O_2 in the mass transfer-limited region. The effect of bicarbonate and other radical scavengers on the oxidation of trichlorethylene and tetrachloroethylene was also accurately predicted by the model (Glaze and Kang, 1989b).

3. HYDROGEN PEROXIDE

Hydrogen peroxide is a powerful oxidant with a redox potential of $+1.77$ V for the two-electron reduction to water (Eilbeck and Mattock, 1987). Nevertheless, due to the slow rate at which most organics are oxidized by hydrogen peroxide alone, hydrogen peroxide in the treatment of organic hazardous wastes should be used in combination with Fe(II) (Fenton's reagent) or low-wavelength UV irradiation.

3.1. Fenton's Reagent

Fenton's reagent [Fe(II) + H_2O_2] is powerful oxidant. In the presence of Fe^{2+}, hydrogen peroxide decomposes to form the hydroxyl radical, $\cdot OH$ (Walling, 1975):

$$Fe^{2+} + H_2O_2 \longrightarrow OH^- + \cdot OH + Fe^{3+}$$

The $\cdot OH$ radical can react with an organic compound, C, or it can be scavenged by Fe^{2+} or hydrogen peroxide:

$$C + \cdot OH \longrightarrow products$$
$$Fe^{2+} + \cdot OH \longrightarrow OH^- + Fe^{3+}$$
$$H_2O_2 + \cdot OH \longrightarrow H_2O + HO_2^{\cdot}$$

It has been proposed that the regeneration of Fe^{2+} occurs via a series of steps (Pignatello, 1992):

$$Fe^{3+} + H_2O_2 \longrightarrow FeOOH^{2+} + H^+$$

$$FeOOH^{2+} \longrightarrow HO_2^{\bullet} + Fe^{2+}$$

$$Fe^{3+} + HO_2^{\bullet} \longrightarrow Fe^{2+} + O_2 + H^+$$

The addition of Fe^{2+} will tend to be more effective in promoting the degradation of organic compounds than the addition of Fe^{3+}. However, the presence of Fe^{3+} may enhance the production of OH radicals, particularly if Fe^{2+} is also present (Barbeni et al., 1987).

Fenton's reagent is known to oxidize many organic compounds. The reaction rate has been shown to increase with increasing pH (Walling, 1975). However, at pH greater than 4 to 5, ferric iron precipitates and consequently the reaction slows down. This decrease in the reaction rate is presumed to be due, in part, to the fact that iron atoms buried in the precipitate are not accessible to the hydrogen peroxide (Mill and Haag, 1989).

Numerous studies have been conducted on the applicability of Fenton's reagent or Fenton-like systems to the degradation of various compounds, including chlorophenols (Barbeni et al., 1987; Bowers et al., 1989; Ravikumar and Gurol, 1992), phenol (Al-Hayek and Doré, 1990), polychlorinated biphenyls (Sedlak and Andren, 1991), trichloroethylene (Ravikumar and Gurol, 1992), azo dyes (Kitao et al., 1982), nitrophenol (Murphy et al., 1989), and 2,4-D (Pignatello, 1992).

In addition to the study of the degradation of organic compounds by the classical Fenton's reagent, studies have been conducted using various "Fenton-like" reagents. In these systems, other forms of iron and other metal ions have also been employed as catalysts. Ferric iron (Murphy et al., 1989) and iron oxide (Al-Hayek et al., 1985; Al-Hayek and Doré, 1990; Mill and Haag, 1989) have been used as a catalyst. The addition of complexing agents can increase the effectiveness of the catalyst. Ethylene diaminetetraacetic acid (EDTA) is particularly effective in this regard (Walling, 1975; Mill and Haag, 1989). For example, Mill and Haag found that in presence of 7.5 mM EDTA, the rate of oxidation of O, S-diethyl methylphosphonothionate (DEMP) was greatest at pH 8.3. At this pH, the reaction proceeded very rapidly ($t_{1/2}$ = ca. 1 min). At pH 6.7, the half-life for the disappearance of DEMP is about 100 min. Sun and Pignatello (1992) studied the effect of the addition of 50 chelating agents on the Fe(III)-catalyzed oxidation of 2,4-dichlorophenoxyacetic acid (2,4-D). They found that 20 of the compounds studied formed complexes capable of oxidizing 2,4-D. Pignatello (1992) found that the Fe^{3+}-photo-assisted catalysis of 2,4-D degradation by hydrogen peroxide was markedly greater when the sample was irradiated with visible light containing a small UV component.

3.2. Hydrogen Peroxide/UV Irradiation

The photolysis of hydrogen peroxide results in the formation of hydroxyl radicals in a direct process with a quantum yield of two OH radicals formed per quantum of radiation absorbed (Baxendale and Wilson, 1957). While the quantum yield for this process is high, the molar absorptivity of hydrogen peroxide at wavelengths > 200 nm is low (Morgan et al., 1988). Hence, the concentration of hydrogen peroxide or the light intensity used must be very high in order to appreciably increase the rate at which the target compound is oxidized.

Sundström et al. (1986) used H_2O_2/UV to oxidize halogenated alkanes and alkenes. At pH 6.8, greater than 98% removal of trichloroethylene (TCE) could be achieved in 45 min using an initial molar ratio of H_2O_2: TCE of 4.5 and a dosage of 58 mg $L^{-1}H_2O_2$. The order of reactivity observed was trichloroethylene > tetrachloroethylene > chloroform > dichloromethane > ethylene   dibromide > tetrachloroethane > carbon tetrachloride. The first-order rate constants ranged from 0.09 (for trichloroethylene) to 0.003 min^{-1} (for carbon tetrachloride). Symons et al. (1989) also investigated the use of H_2O_2/UV for the oxidation of solvents. Using a H_2O_2 concentration of approximately 2 mg L^{-1} and a UV intensity of 5.8 W L^{-1}, an order of reactivity similar to that noted above was observed. Benzene and 1,4-dichlorobenzene were removed at much faster rates than were the aliphatic solvents. Glaze et al. (1987) treated a solution containing 500 μg L^{-1} of trichloroethylene with hydrogen peroxide (added at a rate of 0.14 mg L^{-1} min^{-1}). Trichloroethylene degraded at a rate of approximately 0.7 min^{-1}. Hydrogen peroxide accumulated in the reactor; it reached a concentration of 2 mg L^{-1} after 20 min. The relative reactivities obtained by Glaze et al. for the oxidation of the compounds by H_2O_2/UV tended to follow the relative reactivities for the reaction of the target chemical and the OH radical (Buxton et al., 1988; Lal et al., 1988; Haag and Yao, 1992), thereby suggesting a radical reaction mechanism involving ·OH.

Guittonneau et al. (1988) investigated the photooxidation of chloro- and nitroaromatic compounds in the presence and absence of hydrogen peroxide. The chloroaromatic compounds were more rapidly photodecomposed than the nitroaromatic compounds. The addition of hydrogen peroxide (1 mM) greatly enhanced the rates of reaction. Hydroxylated derivatives were formed as intermediates. Complete oxidation yielded nitrate, chloride, and carbon dioxide. Malaiyandi et al. (1980) used high concentrations of hydrogen peroxide (approximately, 0.3 M) in their studies of the effect of H_2O_2/UV oxidation of dissolved organic impurities in water. Reductions in total organic carbon (TOC) levels of 88% and 98% were obtained for distilled and tap water, respectively. No accumulation of hydrogen peroxide was observed. While this study indicates that H_2O_2/UV would be useful for the reduction of organic chemical concentrations in water, the concentrations of hydrogen peroxide used are probably much higher than what would be economically feasible in engineered systems.

Moza et al. (1988) investigated the oxidation of chlorinated phenols by hydroxyl radicals generated by the photolysis of hydrogen peroxide. Using light of wavelength > 290 nm, the degradation of 2-chloro-, 2,4-dichloro-, and 2,4,6-trichlorophenol could be enhanced in the presence of 55 mg L^{-1} H$_2$O$_2$. Trichlorophenol (TCP) was the most reactive; 100% removal of TCP was obtained in 3 hr. Dechlorination of these compounds resulted, with the formation of quinones, polyhydroxy, and carbonyl compounds. The effects of pH, wavelength, and background substrates on the oxidation of several aromatic compounds were investigated by Mansour (1985). The extent to which the target compound degraded in a given reaction time was greater when the hydrogen peroxide was irradiated with light of wavelength < 290 nm than with light of wavelength > 290 nm. The order of reactivity for the compounds studied was p-cresol $>$ benzene $>$ phenol $>$ toluene $>$ allyl alcohol $>$ methanol $>$ dimethyl phthalate. For a given hydrogen peroxide concentration, the extent of degradation of benzene was greater at pH 2.3 than at pH 4.6. When benzene in tap water was irradiated in the presence of hydrogen peroxide (0.53 M), approximately 40% of the benzene reacted after 2 hr. In distilled water, only approximately 25% had reacted after the same amount of time. It can be speculated that the irradiation of the tap water may result in a greater concentration of OH radicals than is produced in the distilled water.

4. ADVANCED OXIDATION PROCESSES INVOLVING OZONE

Hoigné and Bader (1987) compared the different mechanisms by which OH radicals are generated in ozone/UV, ozone/high pH, and ozone/H$_2$O$_2$ systems. They concluded that at high pH, hydrogen peroxide more effectively catalyzes the decomposition of ozone. In ozone/UV systems, hydrogen peroxide is produced as a result of photolysis. At low light intensities where the photolytic degradation of ozone is slow, hydrogen peroxide appears to control the rate of ozone decomposition. At high light intensities, the presence of hydrogen peroxide is unimportant.

Yao et al. (1992) recently reported the development of models to describe the kinetics of butyrate oxidation for H$_2$O$_2$/UV, O$_3$, O$_3$/UV, ozone/H$_2$O$_2$, and ozone/H$_2$O$_2$/UV systems. Comparisons between the experimental data and the predictions from the model show good agreement for losses of H$_2$O$_2$, ozone, and butyrate in these systems.

4.1. Ozone/UV

Peyton and Glaze (1988) investigated the mechanisms by which refractory compounds are oxidized in irradiated solutions containing ozone. They found that in photolytic ozonation systems, hydrogen peroxide was produced at a rate nearly equivalent to the mass transfer rate of ozone into solution. Hydroxyl radicals were formed by the reaction of hydrogen peroxide with ozone via a pathway involving the formation of superoxide. As the optimum conditions for photolytic oxidation vary with substrate and concentration, they proposed a kinetic model that could be used to predict what these optimum conditions might be.

Gurol and Vatistas (1987) compared the removal of phenolic compounds (phenol, p-cresol, 3,4-xylenol, and catechol) using ozone, ozone/UV, and UV irradiation alone. It was found that at neutral and high pH, the free radical reaction appeared to be the predominant mechanism by which these compounds were removed. This conclusion was based upon the observation that the compounds of interest were nonselectively removed at neutral and high pH. In ozone and ozone/UV systems, the higher removal efficiencies obtained with increasing pH are thought to be due to an increase in the mass transfer of ozone. The increase in the ozone mass transfer rate at high pH is due to an increase in the rate at which ozone decomposes. The use of UV irradiation in combination with ozone resulted in more complete oxidation of reaction products, as evidenced by an increased reduction in TOC (as compared to ozone alone).

The oxidation of 2-chlorophenol was studied by Khan et al. (1985) in a semibatch-type continuous flow-stirred tank reactor. Ozone/UV was found to be significantly more effective at oxidizing chlorophenol than either ozone or UV irradiation alone. Within 50 min, the concentration of the target chemical was reduced from 200 ppm to less than 1 ppm. Likewise, ozone/UV was found to be more effective than either ozone or UV irradiation alone for the oxidation of chlorendic acid (Stowell et al., 1990). Ozone/UV greatly increased the rate at which chlorendic acid was oxidized. The effects of pH, bicarbonate concentration, and UV intensity on the ozonolysis of chlorendic acid were consistent with a hydroxyl radical mechanism.

4.2. Ozone/γ-Irradiation and Ozone/Electron Beam Irradiation

Two processes that have been virtually overlooked despite their demonstrated promise are ozone in combination with γ-irradiation (Gehringer et al., 1988) and with electron beam irradiation (Gehringer et al., 1991) Gehringer et al., (1988, 1991) have shown that trichloroethylene and tetrachloroethylene can be efficiently oxidized by either of these processes. Using the ozone/electron beam treatment, the organic chlorine content in the water was reduced by more than 95%. No mutagenic activity of the product water was observed using the Ames test.

4.3. Ozone/H_2O_2

Ozone/H_2O_2 has been promoted as the most practical of the AOPs because of its simplicity and ease of operation (Namba and Nakayama, 1982). Like the other AOPs, this process results in the formation of hydroxyl radicals that then oxidize refractory organics. Namba and Nakayama's research with low-molecular-weight alcohols by supported the hypothesis that hydroxyl radicals are the most important oxidants in this process.

Paillard et al. (1988) used oxalic acid to determine the optimum conditions for operating an ozone/H_2O_2 system. It was found that with oxalic acid, the rates of oxidation were highest with a pH of 7.5, an initial hydrogen peroxide concentration of 60 to 70 μM. The consumption of hydrogen peroxide was 0.5 mole H_2O_2 per mole of ozone introduced. Adding bicarbonate as an OH radical scavenger reduced the rate at which the target compounds were oxidized. Paillard's results confirm the findings of Brunet et al. (1984), who found that when O_3/H_2O_2 was used, oxalic acid degraded faster at pH 7 than at pH 10 or 5.3. Both research groups found that at high initial concentrations of hydrogen peroxide, the rate at which oxalic acid was oxidized was reduced. Brunet et al. proposed that this is due to the consumption of OH radicals by excess hydrogen peroxide.

The effect of contact time and hydrogen peroxide dosage on the ozonation of o-chloronitrobenzene was investigated by Duguet et al. (1990). It was found that a H_2O_2:ozone mass ratio of 0.4 offered the best removal. With an ozone and hydrogen peroxide dosage of 8 and 3 mg L^{-1}, respectively, and a 20-min retention time, greater than 99% reduction in chloronitrobenzene could be achieved.

Ozone/H_2O_2 was found to be more efficient for the oxidation of formic acid, acetic acid, and p-nitrophenol than either ozone or hydrogen peroxide alone (Slater et al., 1985). Similar results were obtained by Duguet et al. (1985) with chloroform.

4.4. Comparison of AOPs

Masten and Hoigné (1992) compared the efficiency of ozonation and AOPs for chlorinated hydrocarbons using a closed batch system. The direct reaction between substrates and ozone predominated at lower pH, which resulted in the efficient oxidation of the olefin, 1,1-dichloropropene. At high pH, ozonation resulted in a more efficient oxidation of the chlorinated alkanes (compared to that observed at the lower pH), with a corresponding decrease in the efficiency of the oxidation of 1,1-dichloropropene. Ozone/H_2O_2 and ozone/UV also resulted in a greater increase in the efficiency of oxidation of the alkanes and trichloroethylene, with a corresponding decrease in that of 1,1,-dichloropropene. None of the chlorinated hydrocarbons was effectively oxidized using H_2O_2/UV (254 nm). Masten and Butler (1986) showed that 1,2-dichloroethane (DCA) is essentially unreactive with ozone at low pH. Trichloroethylene (TCE) reacted slowly. Both compounds were oxidized under conditions in which OH radicals became an important secondary oxidant. Similar conclusions have been reported for DCA, TCE, and tetrachloroethylene (PCE) (Hoigné and Bader, 1983a). Glaze et al. (1980) observed an increase in the oxidation rate for such recalcitrant compounds as PCBs and trihalomethanes (THMs) when UV light was used in combination with ozone (as compared to that observed for ozone alone).

The effectiveness of AOPs in the removal of chlorinated hydrocarbons has been studied in continuous flow systems (Bourbigot et al., 1985; Guittonneau et al., 1987; Glaze et al., 1987; Galbraith et al., 1992). Galbraith et al. (1992) investigated the effect of pH, humic acid concentration, and bicarbonate concentration on the efficiency of removal of trichlorobenzene using AOPs. The effect of pH on the oxidation of trichlorobenzene using ozone/UV was assessed. As shown in Figure 3, it was found that in the pH range from 2 to 7.5, pH had no significant effect on the removal efficiency; > 95% removal of trichlorobenzene was ob-

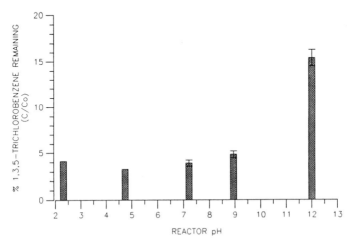

Figure 3. Effect of pH on the efficiency of oxidation of 1,3,5-trichlorobenzene using ozone/UV treatment. Initial reactor TCB concentration was $1280 \pm 89\,\mu g\,L^{-1}$. Ozone dosage was $0.15\,mg\,min^{-1}$. Ozone solution is diluted by half in the reactor.

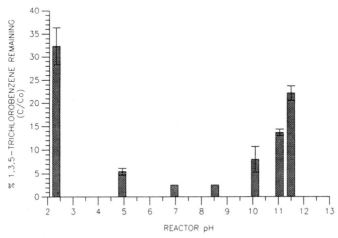

Figure 4. Effect of pH on the efficiency of oxidation of 1,3,5-trichlorobenzene using ozone/H_2O_2 treatment. Initial reactor TCB concentration was $1280 \pm 89\,\mu g\,L^{-1}$. Hydrogen peroxide was dosed at a concentration of $72\,\mu M$. Ozone dosage was $0.15\,mg/min$. Ozone solution is diluted by half in the reactor.

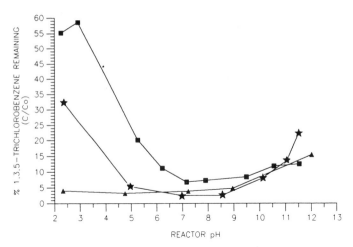

Figure 5. Comparison of the effect of pH on the efficiency of oxidation of 1,3,5-trichlorobenzene using ozone, ozone/UV, and ozone/H_2O_2 treatment. Curve with star represents ozone/H_2O_2 treatment (initial TCB concentration = 1.43 ± 0.09 mg L^{-1}; hydrogen peroxide dosage = $60 \mu M$). Curve with square represents ozone treatment (initial TCB concentration = 1.50 ± 0.08 mg L^{-1}). Curve with triangle represents ozone/UV treatment (initial TCB concentration = 1.28 ± 0.09 mg L^{-1}; UV intensity 1.41 W for reactor volume). Ozone dosage = 0.15 mg min^{-1} for all experiments.

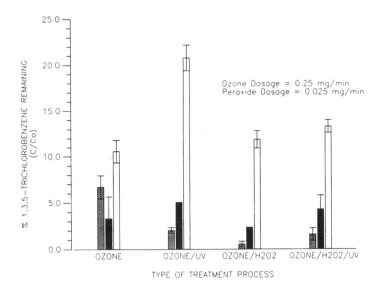

Figure 6. Effect of Aldrich humic acid on the efficiency of oxidizing 1,3,5-trichlorobenzene using ozone and three AOPs. Shaded bar, no humic acid; solid bar, humic acid concentration = 1.6 mg L^{-1}; open bar, humic acid concentration = 10 mg L^{-1}. Hydrogen peroxide dosage = $60 \mu M$, UV intensity = 1.41 W for the reactor volume, ozone dosage = 0.15 mg min^{-1} for all experiments.

tained in all cases. The removal efficiency decreased slightly at pH 9 to approximately 90%. At pH 12, a further decrease in efficiency (to 85%) was observed. For ozone/H_2O_2 treatment, the efficiency of the process was affected to a much greater extent by pH. As shown in Figure 4, the extent to which trichlorobenzene is oxidized is greatest at pH 7 to 8.5. In this pH range, >97% removal is obtained. Figure 5 compares the effect of pH on ozone, ozone/UV, and ozone/H_2O_2 processes. The greatest pH effect occurs with ozone alone. At high pH, it is apparent that for the treatment of trichlorobenzene, there is no benefit in combining ozone with either UV irradiation or hydrogen peroxide addition, because ozone is as efficient as the other processes used. The effect of humic acid can be observed in Figure 6. Humic acid at a concentration of $2 \, mg \, L^{-1}$ (typical of the concentration of organic matter in natural waters) only slightly decreased the efficiency of oxidation. With humic acid $(2 \, mg \, L^{-1}) > 95\%$, oxidation of trichlorobenzene was achieved at pH 7.0. Interestingly, it appears that the presence of humic acid at $2 \, mg \, L^{-1}$ enhances the oxidation of trichlorobenzene when ozone is used alone. At a concentration of $10 \, mg \, L^{-1}$, humic acid significantly reduced the extent to which trichlorobenzene was oxidized. This is consistent with the findings of Masten (1991) with chlorinated aliphatic compounds.

4.5. Commercial Processes

A number of treatment system using AOPs are commercially available; four examples are discussed below. The ULTROX® process was the first commercially available AOP. It was patented in 1988 (Ultrox International, 1988). This process uses low-intensity UV irradiation in combination with ozone and hydrogen peroxide. Early work indicated that the Ultrox process could reduce PCB levels in water by 97 to 99% (Arisman and Musick, 1980). Zeff (1990) explains that PCBs are dechlorinated and hydroxylated by this process. He also hypothesizes that the effectiveness of the Ultrox process is due to the transformation of ozone or hydrogen peroxide into hydroxyl radicals, the excitation of target compounds to a higher energy level, and/or initial attack on the target chemical by the UV light.

The ability of the Ultrox process to oxidize a wide variety of chemicals in different matrices has been studied. Methanol, methylene chloride, vinyl chloride, methylene chloride, 1,1-dichloroethane, trichloroethylene, tetrachloroethylene, benzene, toluene, and xylene were all removed from groundwater that had been heavily contaminated by a chemical plant. Nondetectable levels were achieved for these compounds after treatment. The process was less successful in the removal of 1,1-dichloroethane, 1,2-dichloroethane (Ultrox International 1988). Phenols were shown to be oxidized to form catechol, muconic acid, muconaldehyde, glyoxylic acid, oxalic acid, and formic acid (Zeff, 1990). While not applicable to all wastewaters, Barich et al. (1992) indicates that the Ultrox is useful for treating wastes containing trichlorethylene, benzene, phenol, PCBs, TNT, pesticides, and cyanide.

The perox-pure™ process developed by Peroxidation Systems, Inc., uses hydrogen peroxide in combination with high-intensity UV light (Peroxidation Systems, Inc., 1992). While the manufacturer's literature provides no specific information on the process, Roy (1990) notes that up to 98% removal of trichloroethylene, dichloroethylene, and vinyl chloride were achieved with this process.

Ozone/H_2O_2 treatment is commercially available from Carus Chemical Co. (1989) under the trade name HYDROZON®. Ozone is bubbled into a (column-type) contactor containing stainless steel packing for additional contact. Ozonation is followed by dual-media filtration and, if necessary, these two processes are next followed by an additional ozonation step and then granular-activated carbon (GAC). Treatment efficiencies, as measured by the removal of UV absorptivity at 254 nm, of 42 to 70% were obtained after filtration. Organics removal from Lake Constance water of 72% was achieved after GAC.

The CATAZONE process is a heterogeneous catalytic process in which water is ozonated in the presence of a solid catalyst composed of TiO_2. Hydrogen peroxide may be added, if necessary. Initial studies (Paillard et al., 1991) have shown almost complete oxidation of oxalic acid to CO_2 and water. TOC removal of 94% was obtained with O_3/TiO_2, compared to 50% with O_3/H_2O_2 and 30% with ozone alone. Bicarbonate did not seem to effect the efficiency of TOC removal, as it does with other AOPs. It is thought that this is because oxidation occurs at or near the catalyst surface. Later work (Paillard, 1992) on pesticides has shown that aldrin, dieldrin, and hexachlorobenzene are effectively removed by the CATAZONE process, whereas atrazine is best removed by ozone/H_2O_2.

5.   TREATMENT OF LIQUID WASTES AND SOILS

A comprehensive review of all the applications of ozone and AOPs is beyond the scope of this chapter. Nevertheless, we have given several examples of the uses of ozone in industrial and hazardous waste treatment.

5.1.   Paper and Pulp Wastes

Paper and pulp wastes typically have high BOD (biological oxygen demand) and COD (chemical oxygen demand) levels, contain high concentrations of chlorophenolic compounds, chloroacetones, and chloroform (Gergov et al., 1988; Heimburger et al., 1988), and are toxic (Heimburger et al., 1988). These effluents are also highly colored and odorous (Gergov et al., 1988). Ozone and AOPs have offered an alternative to conventional technologies for reducing the problems associated with paper and pulp effluents.

In 1970, Huriet and Gelly patented a process using ozone for the decolorization of kraft black liquors. Furgason et al. (1973) found that primary effluents from a kraft bleach plant could be decolorized from dark brown to slightly yellow. A 15% reduction in COD also resulted. Nebel et al. (1974) discovered that treating secondary effluents from pulp and paper plants resulted in the decolorization of the effluent along with the elimination of COD, BOD, and turbidity. Ng et al. (1978) showed that when kraft mill wastes are ozonated, color is most affected, followed by BOD. The TOC levels of the effluents were only slightly reduced, and no changes in toxicity resulted.

In more recent years, various researchers have investigated the use of ozone in combination with AOPs and other processes. Prat et al. (1989) investigated using ozonation for the treatment of kraft bleach plant effluents. They determined the rates of reaction for the two classes of colored compounds found in these effluents. The rate constant for the more rapidly oxidizable colored fraction was $81 \pm 7 \, L \, mg^{-1} \, min^{-1}$, while that for the more slowly oxidable colored fraction was $3.1 \pm 0.45 \, L \, mg^{-1} \, min^{-1}$. The rate constant for the noncolored fraction was determined to be $2.3 \pm 0.20 \, L \, mg^{-1} \, min^{-1}$. The more rapidly oxidizable fraction made up approximately 90% of the color. Treatment with ozone was more effective than with either hydrogen peroxide alone or H_2O_2/UV (Prat et al., 1987). Sozańska and Sozańska (1991) treated secondary effluents from a biological treatment plant treating wastewater from paper and pulp manufacturing. They found that when the effluents had been treated with alum or lime coagulation prior to ozonation, better removals of COD and color were obtained than when the effluents were either ozonated (without any subsequent treatment) or chlorinated prior to ozonation. Alum coagulation followed by ozonation resulted in approximately 50% reduction in COD, complete odor control, and a slight increase in BOD_5. Chlorination prior to ozonation had a detrimental effect on the ability of the ozone to remove COD, produced THMs, and resulted in foaming problems upon ozonation.

The effectiveness of ozone and ozone/UV in the decolorization of kraft bleach plant effluent was examined by Prat et al. (1990) using a continuous flow system in which ozone was bubbled into a column reactor and the effluent was pumped at a constant flow rate. The effect of pH (2.2 vs. 11.2) was insignificant. Complete color removal could be obtained with a 13-min retention time and an ozone concentration of $32.4 \, mg \, L^{-1}$. At $15 \, mg$ ozone L^{-1} and pH 2.2, the use of UV had no effect on color removal. However, at $32 \, mg$ ozone L^{-1} and pH 2.2, ozone/UV was less effective at color removal than ozone alone.

5.2. Treatment of Pesticide Wastewaters

An extensive review of the literature on the ozonation of pesticides has been prepared by Reynolds et al. (1989). Five classes of pesticides were reviewed: chlorinated hydrocarbons, organophosphorus insecticides, phenoxyalkyl acid derivatives, organonitrogen compounds, and phenolic derivatives. Where available, they also documented product identification. Organophosphorus insecticides were found to be more readily oxidized by ozone than were the organochlorinated compounds. Three phenoxyalkyl acid pesticides: 2,4-D [(2,4-dichlorophenoxy)acetic acid], MCPA (4-chloro-*o*-tolyloxyacetic acid), and MCPB [4-(4-chloro-*o*-tolyloxy)butyric acid] were readily and nearly completely oxidized by ozone. On the contrary, the concentration of 2,4,5-T [(2,4,5-trichlorophenoxy)acetic acid] was reduced by only 34% under typical drinking water treatment conditions. The organonitrogen compounds were readily oxidized and ring cleavage occurred. Complete removal of the phenol derivatives studied was also demonstrated.

The use of ozone in combination with biological treatment is appealing in that the economics of treatment could be improved by using ozone to render recalcitrant chemicals more biodegradable. The ozonation products could then be treated using bioremediation which is more cost-efficient. Somich et al. (1988) investigated the use of ozonation in combination with UV irradiation for the treatment of agricultural pesticide wastewaters. It was found that the ozonation of alachlor resulted in the oxidation of the alkyl side-chain and the opening of the aromatic ring. Dechlorination did not occur. In soil, the untreated alachlor degraded very slowly; however, the products of both ozonation and photolysis were rapidly metabolized in soil.

Kearney et al. (1987) treated 11 pesticides with ozone/UV. The UV irradiation unit consisted of 66 low-pressure mercury lamps with a maximum energy output of 455 W (at 254 nm). The degradation rates were concentration-dependent. In solutions containing 100 mg L^{-1} or less of the formulated pesticide, removal of all 11 pesticides was essentially complete in less than 3 hr. The solutions containing 1000 mg L^{-1} of the pesticide required much longer treatment times; in several cases, even after 5 hr of treatment, 20 to 40% of the chemical still remained. Oxidation by ozone alone was compared to ozone/UV treatment for bentazon and metribuzin only. At initial pesticide concentrations of 1000 mg L^{-1}, ozone was more effective. This may be due to the presence of OH radical scavengers in the formulation. Photodehalogenation appeared to be an important detoxification mechanism.

Kearney et al. (1987) found that with the O_3/UV treatment system described above, the concentration of atrazine could be reduced from 10 mg L^{-1} to less than 1 mg L^{-1} in approximately 0.5 hr. Duguet et al. (1990) found that with 6.2 mg L^{-1} ozone and 3.5 mg L^{-1} hydrogen peroxide, 91% removal of atrazine could be obtained with a 10-min retention time in a continuous flow reactor. Based upon the demonstrated efficiency of AOPs for the oxidation of atrazine in laboratory-scale experiments, O_3/UV has been placed on-line at the Le Mont-Valérien water treatment plant to oxidize atrazine present in raw water obtained from the Seine River (Duguet et al., 1991).

5.3. Soil Treatment Using Ozone

The application of ozone and AOPs to soil treatment has not been as extensively studied as their application in liquid waste treatment. Only a few studies have been conducted to study ozone reactions in soil systems (Willms et al., 1988; Masten, 1991; Yao and Masten, 1992; Alperin et al., 1992). Alperin et al. (1992) investigated using high-intensity UV irradiation to treat PCB-contaminated soils. With the use of surfactant solutions to enhance photodegradation, 99.9% and

99.6% removal of Aroclor 1260 and 1242, respectively, could be attained after 16 hr of irradiation. (450 W medium-pressure mercury vapor lamps were used.) Some of the losses may have been due to volatilization, as the author states elsewhere in the article that when the lamps were placed 4 in. from the soil, the temperature of the soil rose to 150 °C. In the experiments discussed above, the lamps were placed 0.25 in. from the soil. Using similar conditions as those in the experiment mentioned above, except that the lamps were placed much farther from the soil so that the soil temperature was only 30 °C, only 30% reduction in PCB concentration was obtained.

Yao and Masten (1992) investigated the applicability of ozone to the oxidation of polycyclic aromatic hydrocarbons (PAHs) in soils. The soil studies were conducted by passing gaseous ozone through a dry soil column contaminated with the target PAH. The experimental results are presented in Tables 3 through 5. A control experiment, in which oxygen was passed through phenanthrene-contaminated soil at a flow rate of 2.64 L hr^{-1} for 7 hr, showed that no volatilization of phenanthrene occurred.

After passing ozone through the soil at a rate of 253 mg O_3 hr^{-1} for 2.3 hr, greater than 95% of the phenanthrene had disappeared. The final concentration of phenanthrene in the soil was below detection limits (< 5 ppm). When the ozone dosage was reduced to a rate of 150 mg O_3 hr^{-1} and the soil was treated for 1 hr, approximately 70% removal of phenanthrene was still obtained.

After passing ozone through the soil at a flux of 500 mg O_3 hr^{-1} for 1 hr, 40% removal of chrysene can be achieved, compared to greater than 95% removal of the phenanthrene by the same ozone dosage. When the ozone dosage was increased to a flux of 594 mg O_3 hr^{-1} and the soil was treated for 4 hr, 50% removal of chrysene was achieved.

Table 3 Treatment of Phenanthrene in Soil Columns (initial concentration = 100 µg PH g^{-1} soil)

Ozone Flux (mg hr^{-1})	Gas Flow Rate (L hr^{-1})	Run Time (hr)	% Removal
0.00	2.64	7.0	—
940	13.6	8.0	> 95
882	12.8	6.0	> 95
253	3.66	2.3	> 95
170	2.46	2.5	87
151	2.35	1.0	60[a]
			84[b]

Note: Phenanthrene recovery rate from soil: 78.8%.

[a] Twelve extracts analyzed.

[b] Six extracts analyzed.

Table 4   Treatment of Chrysene in Soil Columns (initial concentration = 100 µg CH g^{-1} soil)

Ozone Flux (mg hr^{-1})	Gas Flow Rate (L hr^{-1})	Run Time (hr)	% Removal
594	8.65	4.0	50
558	8.68	2.0	43
611	8.97	1.0	33
501	7.45	1.0	40
412	6.00	1.0	39
339	4.94	1.0	39[a]
			65[b]

Note: CH recovery rate from soil: 82.7%.

[a] Twelve extracts analyzed.

[b] Six extracts analyzed.

Table 5   Treatment of Pyrene In Soil Columns (initial concentration = 100 µg PY g^{-1} soil)

Ozone Flux (mg hr^{-1})	Gas Flow Rate (L hr^{-1})	Run Time (hr)	% Removal
598	8.65	4.0	91
581	8.40	1.0	83
315	4.55	1.0	79
218	3.22	1.0	> 95
119	1.72	1.0	71
26.3	0.38	1.0	53

Note: PY recovery rate from soil: 105%.

When considering the reactivity of PAHs with ozone, the bond-localization energy is considered to be the parameter of greatest interest (Bailey, 1982). If one compound has lower bond-localization energy than the other, it is more easily attacked by ozone. Thus, the reaction rate with ozone would be expected to be highest for the compound with the lowest bond-localization energy. Based upon bond-localization energies, the order of reactivity for the compounds studied should be pyrene > phenanthrene > chrysene > naphthalene (Bailey, 1982).

The order of reactivity observed in soils was phenanthrene > pyrene > chrysene. The order of reactivity for pyrene and chrysene are consistent with the order of reactivity that would be expected if the direct ozone reaction predominated in dry soil. However, the order of reactivity of pyrene and phenanthrene are not consistent with the order expected based upon bond-localization energies.

Table 6 Partition Coefficients for Selected PAHs

PAH	Log K_{ow}	Log K_{oc}	Log S_w
Naphthalene	3.36	5.00	−5.35
Phenanthrene	4.57	6.12	−6.89
Chrysene	5.61	6.27	−9.80
Pyrene	5.18	6.51	−7.92

A possible explanation for this inconsistency may be the sorption of the hydrophobic PAHs onto the soil, which protects the PAHs from being oxidized by ozone. The sediment-to-water partition coefficients (K_{oc}) of the PAHs studied, together with octanol-to-water and aqueous solubility (Kayal and Connell, 1990), are presented in Table 6. The log K_{oc} values for phenanthrene and pyrene are 6.12 and 6.51, respectively. Thus, the soil could be expected to sorb pyrene more strongly than it sorbed phenanthrene. This might explain why a higher ozone dosage was required for the removal of pyrene than for the removal of phenanthrene.

5.4. Degradation of VOCs in Soil Slurries

Masten (1991) has shown that in the presence of soils volatile organic compounds (VOCs) can be oxidized by ozone. The effect of Eustis soil (0.66% organic matter) and a simulated soil containing 0.7% organic matter (prepared by coating activated alumina with Aldrich humic acid) on the decomposition of cis-dichloroethylene (c-DCE) and perchloroethylene (PCE) was studied. Almost complete oxidation (> 99%) of c-DCE occurred in soil slurries containing 1.0 g of soil per 10.0 mL water. The soils were treated with 22 ± 1.6 mg L^{-1} ozone. The extent of the reaction of tetrachloroethylene in the presence of these two soils was much less than that of c-DCE; this would be expected, since PCE is more difficult to oxidize than c-DCE. Nonetheless, at pH 6.6, the application of 0.17 mg of ozone resulted in approximately 40% decomposition of PCE in the presence of 1.0 g of Eustis soil per 10 mL of water.

5.5. Degradation of Soil Contaminants Using Fenton's Reagent

Watts et al. (1990, 1991) studied the degradation of pentachlorophenol (PCP) in soils using Fenton's reagent. At pH 3, in soils containing 0.2% and 0.5% organic carbon, the levels of PCP were below detection limits three days after the application of the Fenton's reagent. The optimal pH for treatment in soil systems was around pH 2 to 3. Both the natural iron and the iron minerals present in the soil effectively catalyzed the reaction. It was found that the addition of iron adversely affected the efficiency of the reaction, since the peroxide consumption rate increased when iron was added.

The degradation of PAHs in soils from manufactured-gas plant sites using Fenton's reagent was investigated by Gauger et al. (1991) and Kelley et al. (1991). Gauger and colleagues found that, in soil containing 173 μg g^{-1} total aromatics, over 90% of the total aromatics were removed by the addition of Fenton's reagent. More than 97% of 2-to 6-ring compounds were removed. In a more heavily contaminated soil, containing 8586 μg g^{-1} total aromatics, more than 70% of the aromatics were removed by the addition of Fenton's reagent. The use of Fenton's reagent as a pre- or posttreatment in combination with biological treatment showed promise, as the chemical treatment was particularly effective in removing the 4-,5-, and 6-aromatic ring compounds that were not readily biodegradable (Kelley et al., 1991).

Ravikumar and Gurol (1992) studied the oxidation of PCP and trichloroethylene (TCE) in soil using either Fenton's reagent or hydrogen peroxide. They found that TCE was more readily oxidized than PCP and that both contaminants could be effectively degraded using either Fenton's reagent or hydrogen peroxide. The addition of ferrous iron increased the effectiveness of the oxidation reaction.

6. CONCLUDING REMARKS

As is evident from this review, much of the work that has been conducted on the treatment of hazardous chemicals has been completed in aqueous solutions containing a single organic chemical (or, in some cases, several compounds). The effects of background electrolytes, other reactive species, pH, and light intensity and wavelength need to be much more extensively studied. In addition, few studies have been conducted to determine the oxidation products or the toxicity of these products. Another area that should be investigated is the application of ozone, hydrogen peroxide, and AOPs in the remediation of soils, aquifer soils, and sediments.

The coupling of biological and chemical processes is potentially very useful for the treatment of recalcitrant chemicals in such complex matrices as leachates, soils, and sediments. While bioremediation is useful for the treatment of chemicals that can be either metabolized or cometabolized by microorganisms, many chemicals are degraded by microorganisms either very slowly or not at all. Chemical treatment using strong oxidants has much potential, but complete mineralization of the target compounds is often prohibitively expensive. The use of combined chemical/biological treatment seems to have the advantages of both treatment systems, in that even hydrophobic recalcitrant compounds, such as the PAHs, can be oxidized to more polar and bioavailable products using chemical treatment, and these oxidation products are potentially more biodegradable. Since biological degradation could follow chemical pretreatment, the use of chemical oxidation for complete mineralization is no longer necessary, thus reducing the cost of remediation. This area should be explored more extensively.

ACKNOWLEDGMENTS

Preparation of this publication was, in part, supported by funds from Office of Research and Development, U.S. Environmental Protection Agency, under Grant R815750 to the Great Lakes and Mid-Atlantic Hazardous Substance Research Center; the Office of Exploratory Research, U.S. Environmental Protection Agency, under Grant R-816922–01–0; and the National Science Foundation under Grant BCS 9016994. The content of this publication does not necessarily represent the views of the agencies mentioned above.
The authors wish to thank Michael Galbraith for providing the figures.

REFERENCES

Alder, M. G., and Hill, G. R. (1950). The kinetics and mechanism of hydroxide ion catalyzed ozone decomposition in aqueous solution. *J. Am. Chem. Soc.* **72**, 1884–1886.

Al-Hayek, N., and Doré, M. (1990). Oxidation of phenols in water by hydrogen peroxide on alumine supported iron. *Water Res.* **24**, 973–982.

Al-Hayek, N., Eymery, J. P., and Doré, M. (1985). Catalytic oxidation of phenols with hydrogen peroxide. *Water Res.* **19**, 675–666.

Alperin, E. S., Fox, R. D., Groen, A., and Miller, R. A. (1992). UV photolysis of PCB contaminated soils. Presented at the 85th Annual Meeting and Exhibition of the Air and Waste Management Association, Kansas City, MO, Paper No. 92–26.06.

Anderson, G. L. (1977). Ozonation of high levels of phenol in water. *AIChE Symp. Seri.* **166** (73), 265–271.

Arisman, R. K., and Musick, R. C. (1980). Experience in operation of a UV-ozone (ULTROX®) pilot plant for destroying PCB's in industrial waste effluent. *Proc. 35th Annu. Purdue Ind. Waste Conf.*, Lafayette, IN, pp. 802–808.

Bailey, P. S. (1958). The reactions of ozone with organic compounds. *Chem. Rev.* **58**, 926–1009.

Bailey, P. S. (1971). Complexes of ozone with carbon pi systems. *J. Am. Chem. Soc.* **93**, 3552–3554.

Bailey, P. S. (1978). *Ozonation in Organic Chemistry.* Academic Press, New York, Vol. 1.

Bailey, P. S. (1982). *Ozonation in Organic Chemistry.* Academic Press, New York, Vol. 2, pp. 43–76.

Bailey, P. S., and Lane, A. G. (1967). Competition between complete and partial cleavage during ozonation of olefins. *J. Am. Chem. Soc.* **89**, 4473–4479.

Bailey, P. S., and Thompson, J. S. (1966). Structure of the initial ozone-olefin adduct. *J. Am. Chem. Soc.* **88**, 4098–4099.

Barbeni, M., Minero, C., Pellizetti, E., Borgarello, E., and Serpone, N. (1987). Chemical degradation of chlorophenols with Fenton's reagent. *Chemosphere*, **16**, 2225–2237.

Barich, J. T., Zeff, J. D., and Leitis, E. (1992). New applications of UV technology at municipal drinking water, Superfund and industrial facilities. Presented at the Environmental Technology Expo Conference, Chicago.

Bauld, N. L., Thompson, J. A., Hudson, C. E., and Bailey, P. S. (1968). Stereospecificity in ozonide and cross-ozonide formation. *J. Am. Chem. Soc.* **90**, 1822–1830.

Baxendale, J. H., and Wilson, J. A. (1957). The photolysis of hydrogen peroxide at high light intensities. *Trans. Faraday Soc.* **53**, 344–356.

Berdin, Y. S., Zubarev, S. V., Galutkina, K. A., Karazeeva, L. N., and Proskuryakov, V. A. (1989). Oxidation of trichlorobiphenyl with ozone in an aqueous medium. *Zh. Prikl. Khim.* **62**, 2160–2162; *Chem. Abstr.* **112**, 83763f (1990).

Bourbigot, M. M., Brunet, R., Zeana, Z., and Doré, M. (1985). L'utilisation de l' ozone et des rayons ultraviolets en traitement d'eau. *Proc. Wasser Berlin* (Zürich, Switzerland, International Ozone Association), pp. 287–314.

Bowers, A. R., Gaddipati, P., Eckenfelder, W. W., Jr., and Monsen, R. M. (1989). Treatment of toxic or refractory wastewater with hydrogen peroxide. *Water Sci. Technol.* **21**, 477–486.

Brunet R., Bourbigot, M. M., and Doré, M. (1984). Oxidation of organic compounds through the combination ozone-hydrogen peroxide. *Ozone: Sci. Eng.* **6**, 163–183.

Bühler, R., Staehelin, J., and Hoigné, J. (1984). Ozone decomposition in water studied by flash photolysis. I. HO_2/O_2^- and HO_3/O_3^- as intermediates. *J. Phys. Chem.* **88**, 2560–2564.

Butković, V., Klainc, L., Orhanović, M., Turk, J., and Gusten, H. (1983). Reaction rates of polycyclic aromatic hydrocarbons with ozone in water. *Environ. Sci. Technol.* **17**, 546–548.

Buxton, G. V., Greenstock, C. L., Heiman, W. P., and Ross, A. B. (1988). Critical review of rate constants for reactions of hydrated electrons, hydrogen atoms and hydroxyl radicals in aqueous solution. *J. Phys. Chem. Ref. Data* **17**, 513–886.

Carus Chemical Co. (1989). HYDROZON®. *The Natural Way to Treat Water.* Ottawa, IL, Form No. 280, p. 23.

Cobos, C., Castellano, E., and Schumacher, H. J. (1983). The kinetics and mechanism of ozone photolysis at 253.7 nm. *J. Photochem.* **21**, 291–312.

Copeland, P. G., Dean, R. E., and McNeil, D. (1960). The ozonolysis of polycyclic aromatic hydrocarbons. Part I. *J. Chem. Soc.*, pp. 3230–3234.

Copeland, P. G., Dean, R. E., and McNeil, D. (1961). The ozonolysis of polycyclic aromatic hydrocarbons. Part II. *J. Chem. Soc.*, pp. 1232–1238.

Corless, C. E., Reynolds, G. L., Graham, N. J. D., and Perry, R. (1990). Ozonation of pyrene in aqueous solution. *Water Res.* **24**, 1119–1123.

Criegee, R. (1968). Neues aus der chemie der ozonide. *Chemia* **22**, 392–396.

Decoret, C., Royer, J., Legube, B., and Doré, M. (1984). Experimental and theoretical studies of the mechanism of the initial attack of ozone on some aromatics in aqueous medium. *Environ. Technol. Lett.* **5**, 207–218.

Doré, M., and Legube, B. (1983). Mechanism of the ozonation of monocyclic aromatic compounds. *J. Fr. Hydrol.* **14**, 11–30.

Duguet, J. P., Brodard, E., Dussert, B., and Mallevialle, J. (1985). Improvement in the effectiveness of ozonation of drinking water through the use of hydrogen peroxide. *Ozone: Sci. Eng.* **7**, 241–258.

Duguet, J. P., Anselme, C., Mazounie, P., and Mallevialle, J. (1990). Application of combined ozone-hydrogen peroxide for the removal of aromatic compounds from a groundwater. *Ozone: Sci. Eng.* **12**, 281–294.

Duguet, J. P., Wable, O., Bernazeau, F., and Mallevialle, J. (1991). Removal of atrazine from the Seine River by the ozone-hydrogen peroxide combination in a full scale plant. *IOA Ozone News* **19**, 20.

Durham, L. J., and Greenwood, F. L. (1968). Ozonolysis. X. The molozonide as an intermediate in ozonolysis of *cis* and *trans* alkenes. *J. Org. Chem.* **33**, 1629–1632.

Eilbeck, W. J., and Mattock, G. (1987). *Chemical Processes in Waste Water Treatment.* Ellis Horwood, Chichester, England, p. 140.

Eisenhauer, H. R. (1968). The ozonation of phenolic wastes. *J. Water Pollut. Control Fed.* **40**, 1887–1899.

Eisenhauer, H. R. (1971). Dephenolization by ozonolysis. *Water Res.* **5**, 467–472.

Evans, F. L., III (1972). Ozone technology—current status. In F. L. Evans, III (Ed.), *Ozone in Water and Wastewater Treatment.* Ann Arbor Science Publishers, Ann Arbor, MI, pp. 1–14.

Forni, L., Bahnemann, D., and Hart, E. J. (1982). Mechanism of the hydroxide ion initiated decomposition of ozone in aqueous solution. *J. Phys. Chem.* **86**, 255–259.

Fortin, C. J., Snelling, D. R., and Tardif, C. (1972). The ultraviolet flash photolysis of ozone and reaction with H_2O. *Can. J. Chem.* **50**, 2747–2760.

Furgason, R. R., Harding, H. L., and Smith, M. A. (1973). Ozone treatment of waste effluent. *NTIS Rep.* **PB-200, 008**.

Galbraith, M., Shu, M.-M., Davies, S. H., and Masten, S. J. (1992). Use of ozone, ozone, peroxide, and ozone/UV for the generation of OH radicals to oxidize chlorinated organics. Presented at the 24th Mid-Atlantic Industrial Waste Conference, Morgantown, WV, 1992, pp. 411–430.

Gauger, W. K., Srivastava, V. J., Hayes, T. D., and Linz, D. G. (1991). Enhanced biodegradation of polycyclic aromatic hydrocarbons in manufactured gas plant wastes. In C. Akin and J. Smith (Eds.) *Gas, Oil, Coal, and Environmental Biotechnology III.* Institute of Gas Technology, Chicago, pp. 75–92.

Gehringer, P., Proksch, E., Szinovatz, W., and Eschweiler, H. (1988). Decomposition of trichloroethylene and tetrachloroethylene in drinking water by a combined radiation/ozone treatment. *Water Res.* **22**, 645–646.

Gehringer, P., Proksch, E., Szinovatz, W., and Eschweiler, H. (1991). Oxidation of volatile chlorinated contaminants in drinking water by a combined ozone/electron beam treatment. *IOA Ozone News* **19**, 25.

Gergov, M., Priha, M., Talka, E., Valttila, O., Kangas, A., and Kukkonen, K. (1988). Chlorinated organic compounds in effluent treatment at kraft mills. *Tappi J.* **71**(12), 175–184.

Gilbert, E. (1982). Ozonation of aromatic compounds: pH dependence. *Water Sci. Technol.* **14**, 849–861.

Gilles, C. W., and Kuczkowski, R. L. (1983). The ozonolysis of haloalkenes in solution. *Isr. J. Chem.* **23**, 446–450.

Glaze, W. H. (1987). Drinking water treatment with ozone. *Environ. Sci. Technol.* **21**, 224–230.

Glaze, W. H., and Kang, J.-W. (1989a). Advanced oxidation processes. Description of a kinetic model for oxidation of hazardous materials in aqueous media with ozone and hydrogen peroxide in a semibatch reactor. *Ind. Eng. Chem. Res.* **28**, 1573–1580.

Glaze, W. H., and Kang, J.-W. (1989b). Advanced oxidation processes. Test of a kinetic model for the oxidation of organic compounds with ozone and hydrogen peroxide in a semibatch reactor. *Ind. Eng. Chem. Res.* **28**, 1580–1587.

Glaze, W. H., Peyton, G. R., Huang, F. Y., Burleson, J. L., and Jones, P. C. (1980). *Oxidation of Water Supply Refractory Species by Ozone with Ultraviolet Radiation,* EPA-600/2–80–110. U.S. Environmental Protection Agency, Washington, DC.

Glaze, W. H., Kang, J.-W., and Chapin, D. H. (1987). The chemistry of water treatment processes involving ozone, hydrogen peroxide, and ultraviolet radiation. *Ozone: Sci. Eng.* **9**, 335–352.

Gould, J. P. (1987). Correlation between chemical structure and ozonation kinetics: Preliminary observations. *Ozone: Sci. Eng.* **9**, 207–216.

Griffin, G. E. (1965). Good neighbor plant. *Am. City* **89**, 99.

Guittonneau, S., de Laat, J., Doré, M., Duguet, J. P., and Bonnel, C. (1987). Etude de la dégradation de quelques composes organochlores volatils par photolyse du peroxyde d'hydrogene en milieux aqueux. *Rev. Sci. Eau* **1**, 35–54.

Guittonneau, S., de Laat, J., Doré, M., Duguet, J. P., and Bonnel, C. (1988). Comparative study of the photodegradation of aromatic compounds in water by UV and H_2O_2/UV. *Environ. Technol. Lett.* **9**, 1115–1128.

Gurol, M. D., and Vatistas, R. (1987). Oxidation of phenolic compounds by ozone and ozone/UV radiation. *Water Res.* **8**, 895–900.

Haag, W. R., and Yao, C. C. D. (1992). Rate constants for reaction of hydroxyl radicals with several drinking water contaminants. *Environ. Sci. Technol.* **26**, 1005–1013.

Huriet, B., and Gelly, P. (1970). Treatment of pulp factory wastewater. French Pat. 1,599,588.

Harris, C. D., and Weiss, V. (1904). *Ber. Dtsch. Chem. Ges.* **37**, 3431 (as cited in Bailey, 1982, p. 39).

Heimburger, S. A., Blevins, D. A., Bostwick, J. H., and Donnini, G. P. (1988). Kraft mill bleach effluents: Recent developments aimed at decreasing their environmental impact. Part 1. *Tappi J.* **71**(10), 51–60.

Hoigné, J. (1988). The chemistry of ozone in water. In S. Stucki, (Ed.), *Process Technologies for Water Treatment*. Plenum, New York, pp. 121–143.

Hoigné, J., and Bader, H. (1983a). Rate constants of direct reactions of ozone with organic and inorganic compounds in water. I. Non-dissociating compounds, *Water Res.* **17**, 173–183.

Hoigné, J., and Bader, H. (1983b). Rate constants of direct reactions of ozone with organic and inorganic compounds in water. II. Dissociating compounds. *Water Res.* **17**, 185–194.

Hoigné, J., and Bader, H. (1987). Combination of ozone/UV and ozone/hydrogen peroxide: Formation of secondary oxidants. In H. R. Naef, (Ed.), *Proceedings of the Eighth Ozone World Congress*. International Ozone Association, Zürich, Switzerland, pp. K83–K98.

Huisgen, R. (1963a). 1,3-Dipolar cycloadditions, Past and future. *Angew. Chem.* **2**, 565–598.

Huisgen, R. (1963b). Kinetics and mechanisms of 1,3-dipolar cycloadditions. *Angew. Chem.* **2**, 633–645.

Jarret, M., Bermond, A., and Ducauze, C. (1983). Application de la chromatographie en phase liquide à la détermination des produits formés par ozonation de solutions aqueuses de phénol. *Analusis* **11**, 185–189.

Joshi, M. G., and Shambaugh, R. L. (1982). The kinetics of ozone-phenol reaction in aqueous solutions. *Water Res.* **16**, 933–938.

Kayal, S. I., and Connell, D. W. (1990). Partitioning of unsubstituted polycyclic aromatic hydrocarbons between surface sediments and the water column in the Brisbane River estuary. *Aust. J. Mar. Freshwater Res.* **41**, 443–456.

Kearney, P. C., Muldoon, M. T., and Somich, C. J. (1987). UV-Ozonation of eleven major pesticides as a waste disposal pretreatment. *Chemosphere* **16**, 2321–2330.

Kelley, R. L., Gauger, W. K., and Srivastava, V. J. (1991). Application of Fenton's reagent as a pre-treatment step in biological degradation of polycyclic aromatic hydrocarbons. In C. Akin and J. Smith (Eds.), *Gas, Oil, Coal, and Environmental Biotechnology III*. Institute of Gas Technology, Chicago, pp. 105–120.

Khan, S. R., Huang, C. R., and Bozzelli, J. W. (1985). Oxidation of 2-chlorophenol using ozone and ultraviolet radiation. *Environ. Prog.* **4**, 229–238.

Kitao, T., Kiso, Y., and Yahashi, R. (1982). Studies on the mechanism of decolorization with Fenton's reagent. *Mizii Shori Gijutsu* **23**, 1019–1026.

Lal, M., Schöneich, C., Mönig, J., and Asmus, K. D. (1988). Rate constants for the reactions of halogenated organic radicals. *Int. J. Radiat. Biol.* **54**, 773–785.

Langlais, B., Reckhow, D. A., and Brink, D. R. (1991). *Ozone in Water Treatment: Application and Engineering*. Lewis Publishers, Chelsea, MI.

Lee, Y. S., and Hunter, J. V. (1985). Effect of ozonation and chlorination on Environmental Protection Agency Priority Pollutants. In R. L. Jolley et al. (Eds.), *Water Chlorination: Chemistry, Environmental Impact and Health Effects*. Lewis Publishers, Chelsea, MI, Vol. 5, pp. 1515–1526.

Legube, B. (1983). Contribution à l' étude de l' ozonation de composés aromatiques en solution aqueuse. Ph.D. Thesis, Université de Poitiers, France (as cited in Langlais et al., 1991).

Legube, B., Sugimitsu, H., Guyon, S., and Doré, M. (1986). Ozonation of naphthalene in aqueous solution. II. *Water Res.* **20**, 209–214.

Malaiyandi, M., Sadar, M. H., Lee, P., and O'Grady, R. (1980). Removal of organics in water using hydrogen peroxide in presence of ultraviolet light. *Water Res.* **14**, 1131–1135.

Mansour, M. (1985). Photolysis of aromatic compounds in water in the presence of hydrogen peroxide. *Bull. Environ. Contam. Toxicol.* **34**, 89–95.

Masten, S. J. (1986). Mechanisms and kinetics of ozone and hydroxyl radical reactions with model aliphatic and olefinic compounds. Ph.D. Thesis, Harvard University, Cambridge, MA.

Masten, S. J. (1991). Ozonation of VOCs in the presence of humic acid and soils. *Ozone: Sci. Eng.* **13**, 287–313.

Masten, S. J., and Butler, J. N. (1986). Ultraviolet-enhanced ozonation of organic compounds: 1,2-Dichloroethane and trichloroethylene as model substrates. *Ozone: Sci. Eng.* **8**, 339–353.

Masten, S. J., and Hoigné, J. (1992). Comparison of ozone and hydroxyl radical-induced oxidation of chlorinated hydrocarbons in water. *Ozone: Sci. Eng.* **14**, 197–214.

Mill, T., and Haag, W. R. (1989). Novel metal/peroxide systems for the treatment of organic compounds in drinking water. Extended Abstract presented before the meeting of the Division of Environmental Chemistry, American Chemical Society, Miami Beach, FL, pp. 342–345.

Miller, F. J. (1966). Upline sewage treatment. *Water Wastes Eng.* **3**, 52.

Morgan, M. S., Van Trieste, P. F., Garlick, S. M., Mahon, M. J., and Smith, A. L. (1988). Ultraviolet molar absorptivities of aqueous hydrogen peroxide and hydroperoxyl ion. *Anal. Chim. Acta* **215**, 325–329.

Moza, P. N., Fytianos, K., Samanidou, V., and Korte, F. (1988). Photodecomposition of chlorophenols in aqueous medium in the presence of hydrogen peroxide. *Bull. Environ. Contam. Toxicol.* **41**, 678–682.

Murphy, P., Murphy, W. J., Boegli, M., Price, K., and Moody, C. D. (1989). Fenton-like reaction to neutralize formaldehyde waste solutions. *Environ. Sci. Technol.* **23**, 166–169.

Namba, K., and Nakayama, S. (1982). Hydrogen peroxide-catalyzed ozonation of refractory organics. 1. Hydroxyl radical formation. *Bull. Chem. Soc. Jpn.* **55**, 3339–3340.

Nebel, C., Gottschling, R. D., and O'Neil, H. J. (1974). *Ozone Decolorization of Pulp and Paper Secondary Effluents.* Welsbach Ozone Systems, Philadelphia.

Ng, K. S., Mueller, J. C., and Walden, C. C. (1978). Ozone treatment of kraft mill wastes. *J. Water Pollut. Control Fed.* **50**, 1742–1749.

Nowell, L. H., and Hoigné, J. (1987). Interactions of iron(II) and other transition metals with aqueous ozone. In H. R. Naef (Ed.), *Proceedings of the Eighth Ozone World Congress.* International Ozone Association, Zürich, Switzerland.

Paillard, H. (1992). Paper presented at the IWSA/AWWA sponsored workshop (April 13–14, 1992, Coral Gables, FL) on Advanced Oxidation for Drinking Water. Reviewed in *IOA Ozone News* **20**, 24–25.

Paillard, H., Brunet, R., and Doré, M. (1988). Optimal conditions for applying an ozone-hydrogen peroxide oxidizing system. *Water Res.* **22**, 91–103.

Paillard, H., Doré, M., and Bourbigot, M. M. (1991). Prospects concerning applications of catalytic ozonation in drinking water treatment. *IOA Ozone News* **19**, 21.

Peleg, M. (1976). The chemistry of ozone in the treatment of water. *Water Res.* **10**, 361–365.

Peroxidation Systems, Inc. (1992). *Perox-pure™ process description.* Tucson, AZ.

Peyton, G. R., and Glaze, W. H. (1987). Mechanism of photolytic ozonation. *ACS Symp. Ser.* **327**, 76–88.

Peyton, G. R., and Glaze, W. H. (1988). Destruction of pollutants in water with ozone in combination with ultraviolet radiation. 3. Photolysis of aqueous ozone. *Environ. Sci. Technol.* **22**, 761–767.

Pignatello, J. J. (1992). Dark and photoassisted Fe^{3+}-catalyzed degradation of chlorophenoxy herbicides by hydrogen peroxide. *Environ. Sci. Technol.* **26**, 944–951.

Prat, C., Vicente, M., and Esplugas, S. (1987). Treatment of bleaching waters in the paper industry by hydrogen peroxide and ultraviolet radiation. *Water Res.* **22**, 663–668.

Prat, C., Vicente, M., and Esplugas, S. (1989). Ozonization of bleaching waters of the paper industry. *Water Res.* **23**, 51–55.

Prat, C., Vicente, M., and Esplugas, S. (1990). Ozone and ozone/UV decolorization of bleaching waters of the paper industry. *Ind. Eng. Chem. Res.* **29**, 349–355.

Prengle, H. W., Jr. (1983). Experimental rate constants and reactor considerations for the destruction of micropollutants and trihalomethane precursors by ozone with ultraviolet radiation. *Environ. Sci. Technol.* **17**, 743–747.

Pryde, E. H., Moore, D. J., and Cowan, J. C. (1968). Hydrolytic, reductive and pyrolytic decomposition of selected ozonolysis products. Water as an ozonization medium. *J. Am. Oil Chem. Technol. Soc.* **45**, 888–894.

Pryor, W. A., Giamalva, D., and Church, D. F. (1983). Kinetics of ozonation. I. Electron-deficient alkenes. *J. Am. Chem. Soc.* **105**, 6858–6881.

Radding, S. B., Mill, T., Gould, C. W., Liu, D. H., Johnson, H. L., Bomberger, D. C., and Fojo, C. V. (1976). *The Environmental Fate of Selected Polynuclear Aromatic Hydrocarbons*, EPA 560/5-75-009. U.S. Environmental Protection Agency, Washington, DC.

Ravikumar, J. X., and Gurol, M. D. (1992). Fenton's reagent as a chemical oxidant for soil contaminants. In W. W. Eckenfelder, A. R. Bowers, and J. A. Roth (Ed.), *Chemical Oxidation Technologies for the Nineties*, Technomic Publishing Co., Lancaster, PA, Vol. 2, pp. 206–229.

Reynolds, G., Graham, N., Perry, R., and Rice, R. G. (1989). Aqueous ozonation of pesticides: A review. *Ozone: Sci. Eng.* **11**, 339–382.

Roy, K. A. (1990). UV-oxidation technology: Shining star or flash in the pan? *Hazmat World* June, pp. 35–39.

Sedlak, D. L., and Andren, A. W. (1991). Aqueous-phase oxidation of polychlorinated biphenyls by hydroxyl radicals. *Environ. Sci. Technol.* **25**, 1419–1427.

Singer, P. C., and Gurol, M. D. (1983). Dynamics of the ozonation of phenol. I. Experimental observations. *Water Res.* **17**, 1163–1171.

Slater, D., McNeillie, A., and Redmayne, W. H. (1985). The use of a combination of ozone and hydrogen peroxide in the oxidation of some organic compounds. In R. Perry and A. E. McIntyre (Eds.), *The Role of Ozone in Water and Wastewater Treatment*. Selper Ltd., London.

Somich C. J., Kearney, P. C., Muldoon, M. T., and Elsasser, S. (1988). Enhanced soil degradation of alachlor by treatment with ultraviolet light and ozone. *J. Agric. Food Chem.* **36**, 1322–1326.

Sozańska, Z., and Sozańska, M. M. (1991). Efficiency of ozonation as a unit process in the treatment of secondary effluents from the pulp and paper industry. *Ozone: Sci. Eng.* **13**, 521–534.

Staehelin, J. (1984). Ozonzerfall in Wasser: Kinetik der Initiierung durch OH^--Ionen und H_2O_2 sowie der Folgereaktionen der OH^- und O_2^--Radikale. Ph.D. Thesis No. 7342. Swiss Federal Institute of Technology, Zürich.

Staehelin, J., and Hoigné, J. (1982). Decomposition of ozone in water: Rate of initiation by hydroxide ions and hydrogen peroxide. *Environ. Sci. Technol.* **16**, 676–681.

Staehelin, J., and Hoigné, J. (1985). Decomposition of ozone in water in the presence of organic solutes acting as promotors and inhibitors of radical chain reactions. *Environ. Sci. Technol.* **19**, 1206–1213.

Stowell, J. P., Masten, S. J., and Jensen, J. N. (1990). Dechlorination of chlorendic acid with ozone. *Proceedings of the Ninth Ozone World Congress*. New York, *1989*. International Ozone Association, Zürich, Switzerland, pp. 482–495.

Sun, Y., and Pignatello, J. J. (1992). Chemical treatment of pesticide wastes. Evaluation of Fe(III) chelates for catalytic hydrogen peroxide oxidation of 2, 4-D at circumneutral pH. *J. Agric. Food Chem.* **40**, 322–327.

Sundstrom, D. W., Klei, H. E., Nalette, T. A., Reidy, D. J., and Weir, B. A. (1986). Destruction of halogenated aliphatics by ultraviolet catalyzed oxidation with hydrogen peroxide. *Hazard. Waste Hazard. Mater.* **3**, 101–110.

Symons, J. M., Prengle, H. W., and Belhateche, D. (1989). Use of ultraviolet irradiation and hydrogen peroxide for the control of solvent contamination in small water utilities. *Proc. Annu. Conf. Am. Water Works Assoc.*, Part 2, pp. 1403–1418.

Taube, H. (1957). Photochemical reactions of ozone in solution. *Trans. Faraday Soc.* **53**, 656–665.

Taube, H., and Bray, W. C. (1940). Chain reactions in aqueous solutions containing ozone, hydrogen peroxide and acid. *J. Am. Chem. Soc.* **62**, 3357–3373.

Tomiyasu, H., Fukutomi, H., and Gordon, G. (1985). Kinetics and mechanisms of ozone decomposition in basic aqueous solution. *Inorg. Chem.* **24**, 2962–2966.

Ultrox International (1988). Oxidation of organic compounds in water. U.S. Pat. 4,792,407.

Van Cauwenberghe, K., Van Vaeck, L., and Pitts, J. N., Jr. (1979). Chemical transformations of organic pollutants during aerosol sampling. *Adv. Mass Spectrom.* **8B**, 1499–1507.

Walling, C. (1975). Fenton's chemistry revisited. *Acc. Chem. Res.* **8**, 125–131.

Water Pollution Control Federation (1986). *Removal of Hazardous Wastes in Wastewater Facilities—Halogenated Organics*. Facilities Development, Water Pollution Control Federation, Alexandria, VA, Manual of Practice No. FD-11, pp. 68–71.

Watts, R. J., Udell, M. D., Rauch, P. A., and Leung, S. W. (1990). Treatment of pentachlorophenol-contaminated soils using Fenton's reagent. *Hazard. Waste Hazard. Mater.* **7**, 335–345.

Watts, R. J., Udell, M. D., and Leung, S. W. (1991). Treatment of contaminated soils using catalyzed hydrogen peroxide. In W. W. Eckenfelder, A. R. Bowers, and J. A. Roth (Eds.), *Chemical Oxidation—Technology for the Nineties*. Technomic Publishing Co., Lancaster, PA, pp. 37–50.

Weiss, J. (1935). The radical HO_2 in solution. *Trans. Faraday Soc.* **31**, 668–681.

Williamson, D. G., and Cvetanović, R. J. (1968). Rates of reaction of ozone with chlorinated and conjugated olefins. *J. Am. Chem. Soc.* **90**, 4248–4252.

Willms, R. S., Wetzel, D. M., Reible, D. D., and Harrison, D. P. (1988). Aqueous phase oxidation: The effect of soil on oxidation kinetics. *Hazard. Waste Hazard. Mater.* **5**, 65–71.

Yao, C. C. D., Haag, W. R., and Mill, T. (1992). Kinetic features of advanced oxidation processes for treating aqueous chemical mixtures. In W. W. Eckenfelder, A. R. Bowers, and J. A. Roth (Eds.), *Chemical Oxidation—Technology for the Nineties*. Technomic Publishing Co., Lancaster, PA, Vol. 2, pp. 112–139.

Yao, J.-J., and Masten, S. J. (1992). The use of *in-situ* ozonation for the removal of PAHs from unsaturated soils. In B. R. Reed and W. A. Sack (Eds.), *Hazardous and Industrial Wastes, Proc. 24th Industrial Waste Conference*. West Virginia University, Morgantown, pp. 642–651.

Yokoyama, K., Sato, T., Yoshiyasu, I., and Imamura, T. (1974). Degradation of organic substances in water by ozone. *Mitsubishi Denki Giho* **48**, 1233–1238; *Chem. Abstr.* **82**, 89803c (1975).

Zeff, J. D. (1990). On-site destruction of water phase PCBs utilizing ULTROX® ultraviolet oxidation technology. Presented at the International Conference for the Remediation of PCB Contamination, Houston, TX.

19

ALTERNATIVE ATTAINMENT CRITERIA FOR A SECONDARY FEDERAL STANDARD FOR OZONE

E. Henry Lee

ManTech Environmental Technologies, Inc., Corvallis, Oregon 97333

William E. Hogsett and David T. Tingey

U.S. Environmental Protection Agency, Corvallis, Oregon 97333

Environmental Oxidants, Edited by Jerome O. Nriagu and Milagros S. Simmons.
ISBN 0–471–57928–9 © 1994 John Wiley & Sons, Inc.

1. INTRODUCTION

The 1970 Clean Air Act and its amendments mandated the U.S. Environmental Protection Agency (EPA) to establish, periodically review, and (as necessary) update the primary and secondary national ambient air quality standards (NAAQSs). The primary and secondary NAAQSs were designated to protect public health and public welfare, respectively, from known or anticipated adverse effects of criteria air pollutants. Drawing from scientific information on health studies, the EPA determined the lowest concentrations of air pollutants at which adverse health effects occurred and (to provide a margin of safety) established the primary NAAQSs below these levels. Similarly, the EPA determined the lowest concentrations of air pollutants at which adverse environmental effects were detectable and established the secondary NAAQSs at these levels.

In the case of ozone (O_3), the EPA determined that the primary NAAQS criterion (established in 1971) was also sufficient for the secondary NAAQS; thus, both standards originally specified that the average hourly O_3 concentration must not exceed 0.08 ppm more than once per year. The primary and secondary O_3 NAAQSs were both revised (to their present design) in 1979; the revision increased the maximum allowable hourly concentration from 0.08 to 0.12 ppm, and this level was not to be exceeded more than once a year on average (Federal Register, 1979). Attainment status was determined using a three-year moving average. These changes to the original O_3NAAQS criteria reflected a shift from an effects-based NAAQS to an attainment-based NAAQS (U.S. General Accounting Office, 1988).

The EPA's purpose in changing to a three-year averaging period was to minimize yearly fluctuations in attainment status due to short-term weather conditions [U.S. Environmental Protection Agency (USEPA), 1979]. Rather than reducing meteorological effects on attainment status, however, the three-year moving average actually increased the probability of nonattainment for areas at or near the NAAQS level (Javitz, 1980; Chock, 1989). Given a 1-in-365 probability of a daily exceedance (i.e., one exceedance per year on average), there is a 35% chance of four or more exceedances in three years, versus a 26% chance of two or more exceedances in one year by applying the binomial distribution. That is, of the roughly 22% of the metropolitan statistical areas (MSAs) that have a design value (i.e., a fourth-highest concentration in three years) of between 0.12 and 0.13 ppm (USEPA, 1991b), slightly more than a third will violate the current NAAQS, whereas only a quarter would violate a one-year attainment strategy. In general, the three-year averaging period increases the probability of violations for areas that barely meet the NAAQS, and it can cause historically compliant areas to violate the NAAQS for one or more years following an unusual meteorological event.

In addition to the weather-related stochastic problems with the current NAAQS criteria, data from health and vegetation exposure studies suggest that the current primary and secondary O_3 NAAQS requirements may be insufficiently stringent to protect public health and welfare (USEPA, 1986, 1987a, 1988,

1992; Lee et al., 1988, 1989a, b, 1991a; Tingey et al., 1991). For example, following its 1988 review of health data for possible updating of the primary O_3 NAAQS, the EPA recommended that the public-health-related O_3 level of concern should fall somewhere between 0.08 and 0.12 ppm (USEPA, 1988). Also, retrospective studies using field data from the National Crop Loss Assessment Network (NCLAN) have indicated that in order to protect against a 10% or greater yield reduction for 50% of the crops, the secondary NAAQS level should be lowered from 0.12 to 0.094 ppm (Tingey et al., 1991).

Results of work involving O_3 effects on crops indicate that sufficient federal protection of public welfare requires modifications not only to the level but also to the mathematical form of the current secondary O_3 NAAQS. The mathematical form of the current O_3 NAAQS ignores key features of exposure that are important for minimizing the risk of crop damage, and hence is inappropriate for protecting against known or anticipated adverse effects to crops (Lee et al., 1988, 1989 a, b, 1991 a). The form of the current O_3 NAAQS may also be problematic in terms of attainment concerns. For example, the probabilities of crossover and nonattainment during a three-year (vs. one-year) period, discussed earlier, have been attributed more to the inherently high variability (i.e., non-robustness) of the NAAQS's mathematical form than to its three-year averaging period (Chock, 1989, 1991; Fairley and Blanchard, 1991). Tingey et al. (1991) compared an attainment strategy* based on the sum of hourly concentrations at or above 0.06 ppm over the growing season (denoted SUM06), which relates well to agricultural production, with the current attainment strategy (second-highest daily maximum, or 2ndHDM), which is based on an extreme value statistic and relates poorly to agricultural production. They found that, at the tested 10% level of protection for half the crops, fewer sites violated the SUM06 strategy than the 2ndHDM strategy.

Changes to the mathematical form of the primary O_3 NAAQS have been proposed by Chock (1989, 1991) and Fairley and Blanchard (1991). Because of the limited amount of available health-effects data (USEPA, 1988), these proposals were based primarily on attainment concerns. To minimize the effects of meteorological fluctuations on attainment status, Chock (1991) recommended a two-component compliance test (a strict criterion for achieving attainment and a looser criterion for moving into nonattainment) and the use of a robust indicator, such as the 95th percentile. Fairley and Blanchard (1991) proposed a similar two-test scheme, based on number of exceedances, to minimize the frequency of crossovers from attainment to violation and vice versa. However, the attainment levels of these compliance tests were selected to produce the same number of violations as the current primary O_3 NAAQS and were unrelated to known or potential health effects. Thus, the extent to which either of these

---

*The reader should substitute the term "air quality indicator" for the terms "attainment strategy" and "strategy" throughout the text. Air quality indicators are rules designed to protect public welfare. Attributes of an air quality indicator include an index for characterizing ambient condition over a season, an evaluation level of concentration, and a violation rate over some averaging period.

proposed strategies would protect public health (or welfare) against O_3 effects is unknown.

The objective of this study is to develop a new attainment strategy, as an alternative to the current secondary O_3 NAAQS, that adequately addresses both protection of public welfare and minimization of NAAQS violations and crossovers. Specifically, we compare the zero-exceedance annual SUM06-based strategy, which is linked to biological response, with the annual number of exceedances strategy (which is similar in form to the current NAAQS strategy) in terms of several attainment properties. The rationale for using an exposure index that relates well to biological response in the NAAQS is to allow specification of the attainment level based on the distribution of effects. We test various smoothing techniques (i.e., use of an averaging period and/or two-component compliance test) that mitigate the effects of meteorological fluctuations, in an attempt to define an effects-based SUM06 strategy that is comparable to the current attainment-based secondary NAAQS in terms of violations and crossovers. Our purpose in comparing the two types of strategies and discussing their relative merits is to provide guidance to decision makers involved in the O_3 standard-setting process.

2. METHODS

2.1. Attainment Properties Compared Between Strategies

For the development of a protective strategy, we defined *stringency* in terms of risk of plant damage due to O_3 and explored the effects on nonattainment status by modifying the form and/or level of the current secondary NAAQS. We used the term *sufficient stringency* to describe protection against 10% or greater crop reduction for at least 50% of the crops, as determined from vegetation effects studies conducted by the Agency's NCLAN program. Sufficient stringency must be maintained to protect public welfare and, on this basis, we compared competing strategies based on expected number of exceedances (i.e., current NAAQS) and SUM06 indicators in terms of attainment properties that are desirable in a standard. We will clarify the nature of the SUM06 indicator and specified protection target level in subsequent sections. We compared the expected number of exceedances (Ex Ex) and SUM06 strategies, operating at the two specified stringency levels, in terms of the following attainment properties:

1. *Efficiency.* The efficiency of an attainment strategy refers to the number/percentage of area violations; the fewer violations, the greater the efficiency. Efficiency depends upon the form and stringency level of the strategy as well as on the averaging period. Reducing the level of the current secondary NAAQS for O_3 will result in an increase in the number of violations and thus an efficiency loss. Our objective in proposing modifications to the form of the current O_3 NAAQS was to achieve sufficient stringency with minimum loss in efficiency.

2. *Stability.* The stability of an attainment strategy is expressed in terms of the

number/percentage of crossovers from attainment to violation or vice versa that are due to short-term changes in weather patterns. From a regulator's viewpoint, changes in attainment status should be directly linked to changes in emission levels and be insensitive to meteorological fluctuations (Fairley and Blanchard, 1991). That is, the attainment strategy should not cause a historically compliant MSA to bounce out of attainment if emissions in that MSA have not increased. Stability against meteorological fluctuations is an attainment property that, like efficiency, conflicts with effects concerns to protect welfare against damaging levels of O_3 independent of meteorology or emission levels. Stability is inversely related to the variability of the mathematical form of the strategy (Chock, 1989; McCurdy and Atherton, 1990). As mentioned earlier, Chock (1989) proposed using a more robust air quality index (i.e., 95th percentile) to improve stability. However, because the strategies being compared were tested at different stringency levels, it was difficult to ascertain whether the observed differences in stability between the strategies were due to differences in form or stringency level. Our comparisons of the two attainment strategies in terms of stability were always done at the same stringency level.

3. *Consistency.* The consistency of two attainment strategies is defined in terms of the number/percentage of areas that are in joint compliance or in violation of both strategies. In this study, comparisons in terms of consistency were made at equal and unequal stringency levels. The desirability of identifying an alternative attainment strategy that is 100% consistent with the current NAAQS is undeniable from a regulatory perspective. Given the concern that the current secondary NAAQS for O_3 may be insufficiently stringent to protect public welfare, however, developing an alternative attainment strategy that is both sufficiently stringent and 100% consistent with the current NAAQS is unlikely. A more realistic approach is to seek a strategy that minimizes inconsistencies in compliance status.

2.2. Areas Used in the Analyses

Comparisons of the Ex Ex and SUM06 strategies were conducted using ambient air quality data for the years 1982 to 1989 from the USEPA National Air Data Branch's Aerometric Information Retrieval System (AIRS) database. Violations were counted on an area basis, consistent with the federal approach, except that we focused on metropolitan statistical areas (MSAs) and not consolidated MSAs, since the latter could change from year to year. Typically, about 86% of the monitoring sites were located within one of the U.S. urban areas designated as MSAs according to EPA guidelines (USEPA, 1987b); the remaining 14% were located in non-MSA areas. (There were far fewer non-MSA areas with sufficient data for our analyses.) Of the 331 US MSAs, only 222 had ambient monitors with adequate data for one or more years from 1982 to 1989, and only about 200 could provide adequate data for any given year. Monitoring sites in Nevada were excluded from the analyses because some of the monitoring data did not meet the EPA's quality assurance criteria. In addition, only those MSAs with at least 70% valid daily maxima during the EPA-defined O_3 season for three or more consecutive years met EPA requirements for checking compliance with the current NAAQS. The Agency designates the O_3 season on a state-by-state basis; the season ranges from January to December in the southernmost states, from April to October in midlatitude states, and from May to September in some northern states. Thus, because few non-MSAs possessed three or more consecutive years of required data, this analysis focused particularly on the MSAs.

2.3. Methods of Analysis

Analysis of the Ex Ex strategy upon which the current O_3 NAAQS is based was straightforward when there were no invalid daily maximum concentrations during the O_3 season, but became complicated when missing-value adjustments were required to determine compliance. Invalid days immediately preceding and following days with one or more hourly concentrations greater than 0.09 ppm were assumed to be as likely as valid days to have an exceedance of 0.125 ppm (USEPA, 1979). The remaining invalid days were assumed to have daily maxima less than 0.125 ppm. Due to rounding, the level of the NAAQS was actually 0.125 ppm rather than 0.120 ppm, and *exceedance* was defined as a daily maximum O_3 concentration that is at or above 0.125 ppm. In the simplest case in which daily maximum concentrations were assumed to be independent and the probability of an exceedance was estimated by the ratio of observed exceedances to number of valid days, the number of annual expected exceedances was

$$e = v + v/n \cdot (N - n - z)$$

where $e = $ Ex Ex, $v = $ observed exceedances, $n = $ number of valid days, $N = $ number of days in the O_3 season, and $z = $ number of invalid days with daily maxima less than 0.125 ppm. [Thus, $(N - n - z) = $ number of invalid days amidst daily maxima greater than 0.09 ppm.] For the three-year moving average comparisons, the annual Ex Ex for individual monitoring sites in an MSA were compared and the maximum readings for each of three consecutive years were averaged to determine the MSA attainment status for that period. Attainment was considered to be achieved if the three-year average Ex Ex was less than 1.05 (rather than less than 1) to allow for the potential adjustment for invalid days.

In contrast to analysis of the current O_3 NAAQS indicator (Ex Ex), we defined the annual SUM06 indicator for the maximum three-month period within the O_3 season because (1) the maximum three-month period encompasses the growing season for most economically important crops and is the period of greatest potential O_3 impact; (2) three months (about 91 days) corresponds to the average fumigation (78 days) for the 31 NCLAN experiments used to determine the form and level of the proposed attainment strategy; and (3) the shortest EPA-defined O_3 season is four months (June to September for Montana and South Dakota). The three-month SUM06 values for all monitoring sites within an MSA (adjusted for missing values as described in the next paragraph) were compared, and the maximum value averaged over the designated period of years was used to determine compliance. The proposed strategy was attained when the average three-month SUM06 was less than 24.4 ppm-hr (or 34.5 ppm-hr depending upon specified stringency level).

Because the SUM06 indicator was calculated on an hourly rather than a daily basis, attainment analysis using SUM06 required a different statistical method than was used for the Ex Ex analysis to make allowances for incomplete data. To avoid understating the ambient condition, the monthly SUM06 value calculated from available hours was adjusted upwards by a factor of M/m, where M = number of hours in a month and m = number of available hours in a month. The simple scaling adjustment assumes that missing hourly concentrations occur at random according to a uniform distribution for each day, and is a method that has compared favorably with other adjustment procedures and has been recommended for dealing with invalid and missing days (Lee et al., 1991b). To avoid potential biases, scaling was used only in months that had at least 50% or more valid daily maxima. The SUM06 data for months with less than 50% valid days were imputed using the larger of the SUM06 values for adjustment months with 90% or more available hours and the unadjusted monthly SUM06 value.

We restricted comparison of the Ex Ex and SUM06 strategies to the same averaging period (one, three, or four years) so that differences in attainment

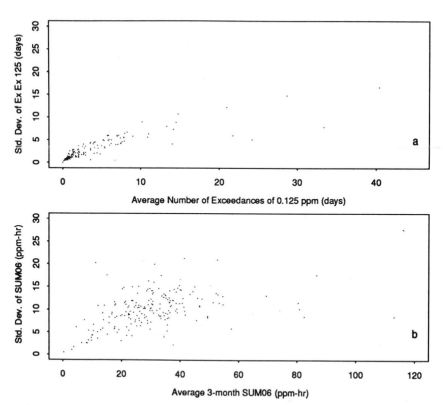

Figure 1. Standard deviations versus the means of annual (a) number of exceedances of 0.125 ppm (Ex Ex 125) and (b) maximum 3-month SUM06 for 199 metropolitan statistical areas (MSAs), for 1982–1989.

properties could definitively be attributed to differences in form or level. A one-year averaging period corresponds to the ecological period for interrelating crop effects data with ambient air quality data and is optimal for establishing a specified target protection level. An averaging period of three years, which is the current NAAQS criterion, mitigates the effects of short-term weather changes. While a three-year averaging may be adequate for smoothing the Ex Ex indicator, it may be inadequate for smoothing the SUM06 indicator due to greater variation for the latter, as shown in Figure 1. We therefore included in our study two additional modifications to the SUM06 and Ex Ex compliance tests for the purpose of minimizing meteorological effects. First, we evaluated a longer (four-year) averaging period. Second, we evaluated a two-component compliance test similar to Chock's (1991) that preserved three-year averaging but did not require 100% compliance.

For the two-component compliance test, we took Ex Ex and SUM06 values for three consecutive years and performed a one-sided test based on the sample mean $\bar{x}$ and range w to determine whether the population mean was greater than the parameter μ_1 (e.g., > 24.4 ppm-hr for SUM06 and > 1.05 days for the Ex Ex). Rather than using the popular t test proposed by Chock (1989), which is based on the sample mean and standard deviation, we used Lord's range test, in which t_w is defined as $(\bar{x} - \mu_1)/w$ with an α level of significance (Lord, 1947). It was easier to calculate t_w than t, and the efficiency of the range test relative to the t test was above 95% for sample sizes up to 20. An MSA must first pass a strict t_w test with a significance level of α_1 (say, 0.30) to achieve compliance. When the MSA has passed this strict test for at least two consecutive three-year periods, it must pass a looser t_w test, with a significance level of α_2 ($< \alpha_1$, say, 0.16), to maintain compliance. These levels were chosen to correspond roughly to the $\alpha = 0.30$ and 0.10 levels suggested by Chock (1991) for achieving efficiency equal to the current NAAQS. It is not clear how to select the significance levels α_1 and α_2 to achieve a specified protection target level based on effects, because crop effects are determined from a single growing season and not over multiple seasons. The t_w values for $n = 3$ were 0.1867 when $\alpha = 0.30$ and 0.4000 when $\alpha = 0.16$.

3. SELECTION OF AN ALTERNATIVE ATTAINMENT STRATEGY

These proposed modifications to the current secondary NAAQS for O_3 were intended to improve the protection level and attainment properties as determined from crop effects and ambient air quality data, respectively. Of the proposed changes, the optimization of effects and attainment properties was dictated primarily by the form of the attainment strategy. Because the CAA mandated the establishment of a secondary O_3 NAAQS to protect public welfare, it was necessary to define the functional correspondence between exposure index and the known or anticipated adverse effects as determined from vegetation effects studies. Effects concerns dictate that the index of the NAAQS be related to biological response such that it functionally corresponds to protection of public welfare. In choosing an alternative attainment strategy, therefore, our foremost objective was to identify an appropriate exposure index. This index is the most important component of the NAAQS and dictates the attainment level and, ultimately, the biological and statistical properties of the NAAQS.

Evidence from air pollution research was thoroughly reviewed in the U.S. EPA air quality criteria documents and their supplements (USEPA, 1978, 1986, 1988, 1992), and formed the scientific basis for determining the present form of the secondary O_3 NAAQS. A comprehensive review of published O_3 exposure research, in particular those studies conducted by the U.S. EPA's NCLAN program, had detailed the biological, environmental, and exposure-kinetic factors (e.g., concentration, duration, temporal flux, threshold, respite time) that influence the magnitude of vegetation responses (Hogsett et al., 1988). In particular, to establish the form of the secondary NAAQS, the key features of exposure relevant to eliciting a plant response were identified. The scientific evidence from individual air pollution studies varied in conclusiveness but together suggested that short-term high concentrations were more important than long-term low concentrations in eliciting plant response for agricultural crops (Heck et al., 1966; Heck and Tingey, 1971; Henderson and Reinert, 1979; Nouchi and Aoki, 1979; Reinert and Nelson, 1979; Stan et al., 1981; Musselman et al., 1983; Ashmore, 1984; Tonneijck, 1984; Hogsett et al., 1985). The O_3 effect, indicated by plant damage and/or visible foliar injury, increased with increasing concentration, but approached an asymptotic value at high concentrations. Evidence existed for a number of crops that equal doses of O_3 did not produce equal effects when applied at different exposure durations and concentrations. The effect of O_3 was influenced more by concentration than exposure duration (Guderian et al., 1985) and was postulated to be the result of the cumulative impact of repeated peak concentrations, rather than low concentrations (USEPA, 1986).

In addition to concentration and duration of exposure, research showed that dynamics (e.g., frequency, respite time, and episode duration) can also influence the biological response of plants to O_3. Musselman et al. (1983) reported greater O_3 effects on yield and growth in red kidney beans exposed to a simulated ambient regime than in beans exposed to a repeated daily uniform distribution, when both exposures were at the same dose and duration. Also, in long-term exposure studies of alfalfa, timothy, and orchard grass in field open-top chambers, an episodic exposure with peak concentration near 0.20 ppm was found to have a significantly greater negative impact on growth than did an equal dose of a daily peak exposure with high concentrations near 0.12 ppm (Hogsett et al., 1985, and unpublished data). Other research also indicated that uniform distributions of concentrations over a long exposure duration are of less concern than variable or episodic distributions. For example, a constant O_3 exposure caused less ethylene production and foliar necrosis in bean plants than did a daily peak exposure of the same concentration and duration (Stan and Schicker, 1982).

Based on the scientific evidence cited above, the form of the current NAAQS, which is based on an extreme value statistic, does not differentiate between temporal patterns of exposure and therefore does not relate to biological response. Recent retrospective analyses of existing plant-response data (largely from the NCLAN program) conducted by Lee et al. (1987, 1988, 1989a, b, 1991a), Lefohn et al. (1988), and Musselman et al. (1988) concurred with these observations and were in agreement that (1) seasonal mean indices are not among the best indices and (2) exposure indices that cumulated the hourly O_3 concentrations over the season, as well as preferentially weighted the higher concentrations and concentrations near anthesis, best relate crop yield to O_3 exposure. Several NCLAN studies, replicated through time with varying exposure durations, showed greater yield reductions during the longer exposure periods (Lee et al. 1991a; Lefohn et al., 1988); this suggests that the effects of O_3 are cumulative and increase with exposure duration. These retrospective studies illustrated the shortcomings of commonly used exposure indices, including seasonal means, dose (or total exposure), number of daily exceedances, and single-peak indices that weighted all concentrations equally, did not cumulate hourly concentrations, or gave undue weight to an extreme concentration (Tingey et al., 1989, 1991; Lefohn et al., 1988; Lee et al., 1988). Other factors, including environmental, edaphic, and biological (e.g., phenological development) factors, contributed to the within- and between-year variations in plant response and were considered in developing exposure indices for use in mathematical models of the exposure–response relationship (Lee et al., 1988).

Concentration and, to a lesser extent, duration now appear to be the key factors in exposure that determine plant response. Exposure indices that differentiate and order the O_3 exposures according to the magnitude of biological response should both emphasize the peak concentrations and cumulate the hourly O_3 concentrations. The SUM06 index cumulates all hourly concentrations of 0.06 ppm or greater. It is considered simpler to understand and implement as the index of a secondary NAAQS than the mathematically similar sigmoid-weighted indices that sum all hourly O_3 concentrations and weight concentrations by an increasing function of the O_3 value (Lee et al., 1988; Lefohn and Runeckles, 1987). Air quality indicators, including expected exceedances of 0.10 ppm, second-highest daily maximum, and 95th percentile, are inadequate for relating exposure and plant response based on predictive ability (Lee et al., 1988; Tingey et al., 1989). We therefore selected SUM06 as the index of the alternate attainment strategy. Our selection was based primarily on functional models developed from the crop response studies conducted by the NCLAN program. The SUM06 index did reasonably well in ordering the experimental O_3 exposures according to biological response, and its use is consistent with the general consensus that the effects of O_3 are caused by intermittent high O_3 episodes rather than low concentrations.

3.1. Attainment Levels Used in the Comparisons

As mentioned earlier, the stringency of an attainment strategy is defined as the degree to which the strategy protects against plant damage due to O_3. Once the NAAQS index is selected, stringency is determined primarily by selection of numerical level. In setting the numerical level of secondary NAAQS, the EPA administrator must determine the acceptable magnitude of crop loss and the proportion of the species/cultivar population to be protected. In this paper, we concluded that a strategy is sufficiently stringent if it protects against 10% or greater yield reduction for at least 50% of the crops, as proposed by Tingey et al. (1991). To achieve this degree of stringency, the levels of the SUM06 and 2ndHDM strategies cannot exceed 24.4 ppm-hr and 0.094 ppm, respectively (Tingey et al., 1991). Compliance with the 2ndHDM < 0.094 ppm criterion is mathematically equivalent to attainment of less than two expected exceedances of the 0.094 ppm criterion (Ex Ex 094) for complete data. McCurdy and Atherton (1990) found negligible differences in attainment properties between the 2ndHDM and Ex Ex strategies when missing values were present. Hence, we do not distinguish between the 2ndHDM and Ex Ex strategies.

We also compared the attainment properties of the SUM06 and Ex Ex strategies at the level of the current NAAQS, which, based on the predicted relative yield loss curves previously generated from 31 NCLAN experiments and an 0.125 ppm attainment level, would protect against 18.5% crop loss for at least 50% of the species/cultivars (Figure 2a). To be equally stringent as the annual

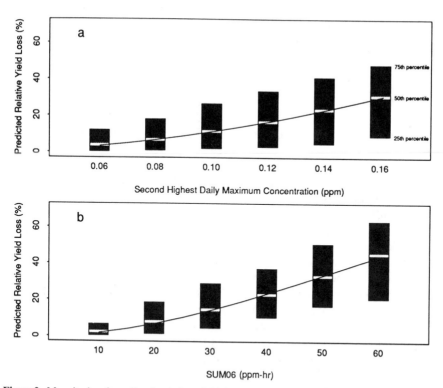

Figure 2. Magnitude of predicted relative yield losses for 49 National Crop Loss Assessment Network program cases as a function of (*a*) the second-highest daily maximum concentration and (*b*) SUM06 (Lee et al., 1989b).

expected exceedances of the 0.125 ppm criterion (Ex Ex 125), the SUM06 level would need to be 34.5 ppm-hr (Figure 2b).

In generalization and extrapolation of the NCLAN results, the Ex Ex and SUM06 exposure indices, defined over the individual growing seasons of 21 to 119 days, were calculated from either a 7-hr or 12-hr fumigation period rather than from all hours. Hourly O_3 concentrations outside the fumigation period were not recorded for individual chambers but could be inferred from supplemental readings of ambient and charcoal-filtered (CF) monitors when available. The additional 17 hr (or 12 hr) of ambient O_3 concentrations had little or no consequence on the 2ndHDM values but did supplement the 7-hr (or 12-hr) SUM06 nonfiltered (NF) chamber values by as much as 38%. For 22 of the 31 NCLAN studies, supplemental values of SUM06 outside the fumigation period ranged from 0.1 to 11.6 ppm-hr with a 3.3 ppm-hr average (Lefohn and Foley, 1991). By understanding the SUM06 treatment values in the individual exposure-response equations developed by Lee et al. (1991a), the 7-hr SUM06 values of 24.4 and 34.5 ppm-hr underestimated the level associated with 10% and 18.5% yield loss, respectively, in half the cases. The bias in the 7-hr SUM06 levels was expected to be less than 2 ppm-hr while the bias in the 2ndHDM levels was expected to be negligible based on cursory calculations. We considered the 7-hr (or 12-hr) SUM06 values of 24.4 ppm-hr and 34.5 ppm-hr associated with 10% and 18.5% losses, respectively, for half the cases to be reasonable approximations to the 24-hr SUM06 values.

4. RESULTS

Annual trends in ambient O_3 concentrations were presented using the 2ndHDM, Ex Ex 125, Ex Ex 094, and maximum three-month SUM06 for 222 MSAs with at least 70% valid daily maximum values for 1982 to 1989 (Figure 3). Due to their extremely large air quality indicator values (e.g., Ex Ex 125 > 120 days), data for Los Angeles–Long Beach and San Bernardino–Riverside, CA, MSAs were not included in Figure 3. All four of these annual indicators ranked 1988, 1987, and 1983 as the three highest O_3 years, but these peaks (particularly in 1988) were most evident with the SUM06 and Ex Ex 094 indicators—a ranking that is consistent with previous trends reports (USEPA, 1987a, 1990, 1991a; Curran and Frank, 1991). The peaks during these years (again, particularly in 1988) have been attributed primarily to unusual meteorological conditions that were conducive to O_3 formation, rather than to changes in emissions.

The variation in ambient conditions for the years 1982 to 1989 provides a suitable range of conditions for revealing the impact of meteorological fluctuation on efficiency, stability, and consistency with the current NAAQS of equally stringent SUM06- and Ex Ex-based strategies. Tables 1 through 3 show the joint distribution of MSAs in compliance or violation of the following pairs of equally stringent strategies during 1982 to 1989: (1) the annual Ex Ex 125 $\geqslant 2$ days and annual SUM06 $\geqslant 34.5$ ppm-hr tests; (2) the annual Ex Ex 094 $\geqslant 2$ days and an-

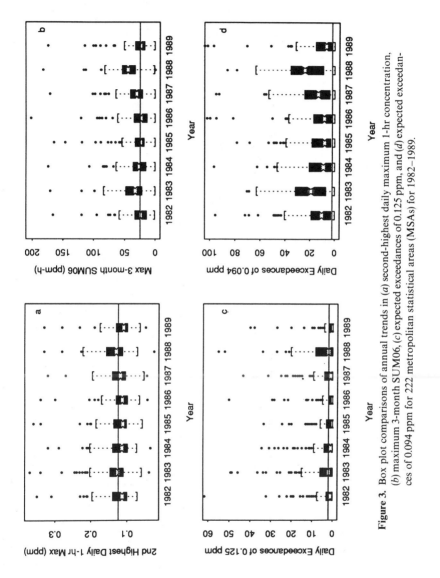

Figure 3. Box plot comparisons of annual trends in (*a*) second-highest daily maximum 1-hr concentration, (*b*) maximum 3-month SUM06, (*c*) expected exceedances of 0.125 ppm, and (*d*) expected exceedances of 0.094 ppm for 222 metropolitan statistical areas (MSAs) for 1982–1989.

561

Table 1 Distribution of Metropolitan Statistical Areas (MSAs) that Violate or Comply with Two Equally Stringent Strategies Based on Annual Expected Exceedances of 0.125 ppm ≥ 2 Days and the Maximum 3-Month SUM06 Value ≥ 34.5 ppm-hr for 1982–1989

Year	Annual Exceedance Attainment Strategy	Annual SUM06 Attainment Strategy[a]		
		SUM06 < 34.5 ppm-hr	SUM06 ≥ 34.5 ppm-hr	Marginal Distribution
1982	Expected exceedances of 0.125 ppm < 2 days	111(58)	5(3)	116(61)
	Expected exceedances of 0.125 ppm ≥ 2 days	35(18)	39(21)	74(39)
	Marginal distribution	146(77)	44(23)	190
1983	Expected exceedances of 0.125 ppm < 2 days	68(35)	23(12)	91(47)
	Expected exceedances of 0.125 ppm ≥ 2 days	36(18)	68(35)	104(53)
	Marginal distribution	104(53)	91(47)	195
1984	Expected exceedances of 0.125 ppm < 2 days	111(57)	12(6)	123(63)
	Expected exceedances of 0.125 ppm ≥ 2 days	30(15)	41(21)	71(37)
	Marginal distribution	141(73)	53(27)	194

Year		Column 1	Column 2	Column 3
1985	Expected exceedances of 0.125 ppm < 2 days	121(61)	11(6)	132(67)
	Expected exceedances of 0.125 ppm ≥ 2 days	34(17)	31(16)	65(33)
	Marginal distribution	155(79)	42(21)	197
1986	Expected exceedances of 0.125 ppm < 2 days	120(60)	18(9)	138(69)
	Expected exceedances of 0.125 ppm ≥ 2 days	31(16)	30(15)	61(31)
	Marginal distribution	151(76)	48(24)	199
1987	Expected exceedances of 0.125 ppm < 2 days	96(48)	23(12)	119(60)
	Expected exceedances of 0.125 ppm ≥ 2 days	25(13)	54(27)	79(40)
	Marginal distribution	121(61)	77(39)	198
1988	Expected exceedances of 0.125 ppm < 2 days	44(22)	41(20)	85(42)
	Expected exceedances of 0.125 ppm ≥ 2 days	12(6)	104(52)	116(58)
	Marginal distribution	56(28)	145(72)	201
1989	Expected exceedances of 0.125 ppm < 2 days	134(66)	13(6)	147(73)
	Expected exceedances of 0.125 ppm ≥ 2 days	29(14)	26(13)	55(27)
	Marginal distribution	163(81)	39(19)	202

[a] Cell entries in the two-way tables are the frequency (%) of the MSAs classified according to two strategies.

Table 2 Distribution of Metropolitan Statistical Areas (MSAs) that Violate or Comply with Two Equally Stringent Strategies Based on Annual Expected Exceedances of 0.094 ppm ≥ 2 days and the Maximum 3-month SUM06 Value ≥ 24.4 ppm-hr for 1982–1989

Year	Annual Exceedance Attainment Strategy	Annual SUM06 Attainment Strategy[a]		
		SUM06 < 24.4 ppm-hr	SUM06 ≥ 24.4 ppm-hr	Marginal Distribution
1982	Expected exceedances of 0.094 ppm < 2 days	40(21)	0(0)	40(21)
	Expected exceedances of 0.094 ppm ≥ 2 days	57(30)	93(49)	150(79)
	Marginal distribution	97(51)	93(49)	190
1983	Expected exceedances of 0.094 ppm < 2 days	15(8)	0(0)	15(8)
	Expected exceedances of 0.094 ppm ≥ 2 days	40(21)	140(72)	180(92)
	Marginal distribution	55(28)	140(72)	195
1984	Expected exceedances of 0.094 ppm < 2 days	35(18)	1(1)	36(19)
	Expected exceedances of 0.094 ppm ≥ 2 days	54(28)	104(54)	158(81)
	Marginal distribution	89(46)	105(54)	194

Year		Column 1	Column 2	Marginal distribution
1985	Expected exceedances of 0.094 ppm < 2 days	26(13)	0(0)	26(13)
	Expected exceedances of 0.094 ppm ≥ 2 days	76(39)	95(48)	171(87)
	Marginal distribution	102(52)	95(48)	197
1986	Expected exceedances of 0.094 ppm < 2 days	31(16)	1(1)	32(16)
	Expected exceedances of 0.094 ppm ≥ 2 days	76(38)	91(46)	167(84)
	Marginal distribution	107(54)	92(46)	199
1987	Expected exceedances of 0.094 ppm < 2 days	22(11)	2(1)	24(12)
	Expected exceedances of 0.094 ppm ≥ 2 days	45(23)	129(65)	174(88)
	Marginal distribution	67(34)	131(66)	198
1988	Expected exceedances of 0.094 ppm < 2 days	11(5)	2(1)	13(6)
	Expected exceedances of 0.094 ppm ≥ 2 days	23(11)	165(82)	188(94)
	Marginal distribution	34(17)	167(83)	201
1989	Expected exceedances of 0.094 ppm < 2 days	33(16)	2(1)	35(17)
	Expected exceedances of 0.094 ppm ≥ 2 days	79(39)	88(44)	167(83)
	Marginal distribution	112(55)	90(45)	202

[a] Cell entries in the two-way tables are the frequency (%) of the MSAs classified according to two strategies.

Table 3 Distribution of Metropolitan Statistical Areas (MSAs) that Violate or Comply with the Current Standard (or Its Modification that Lowers the Nominal Level from 0.125 to 0.094 ppm) and an Alternative Attainment Strategy Based on the 3-Year Average Maximum 3-Month SUM06 $\geq$ 24.4 ppm-hr for 3-Year Averaging Periods between 1982 and 1989

| Year | Current or Modified Standard (i.e., nominal level is $x = 0.125$ or 0.094 ppm, respectively) | Current or Modified Standard (i.e., $x = 0.125$ or 0.094 ppm, resp.) vs. SUM06 Strategy[a] | | | | | |
| | | Nominal level $x = 0.125$ ppm | | | Nominal level $x = 0.094$ ppm | | |
		SUM06 < 24.4	SUM06 $\geq$ 24.4	Marginal Distribution	SUM06 < 24.4	SUM06 $\geq$ 24.4	Marginal Distribution
1984	Expected exceedances of x < 1.05 days	48(28)	31(18)	79(46)	10(6)	0(0)	10(6)
	Expected exceedances of $x \geq$ 1.05 days	10(6)	84(49)	94(54)	48(27)	117(67)	165(94)
	Marginal distribution	58(34)	115(66)	173	58(33)	117(67)	175
1985	Expected exceedances of x < 1.05 days	54(31)	33(19)	87(49)	7(4)	0(0)	7(4)
	Expected exceedances of $x \geq$ 1.05 days	12(7)	77(44)	89(51)	59(33)	112(63)	171(96)
	Marginal distribution	66(38)	110(63)	176	66(37)	112(63)	178

Year		Obs 1	Obs 2	Marginal	Obs 1	Obs 2	Marginal
1986	Expected exceedances of $x < 1.05$ days	71(39)	35(19)	106(58)	17(9)	0(0)	17(9)
	Expected exceedances of $x \geq 1.05$ days	15(8)	61(34)	76(42)	69(38)	97(53)	166(91)
	Marginal distribution	86(47)	96(53)	182	86(47)	97(53)	183
1987	Expected exceedances of $x < 1.05$ days	67(35)	49(26)	116(61)	15(8)	0(0)	15(8)
	Expected exceedances of $x \geq 1.05$ days	14(7)	60(32)	74(39)	66(35)	109(57)	175(92)
	Marginal distribution	81(43)	109(57)	190	81(43)	109(57)	190
1988	Expected exceedances of $x < 1.05$ days	44(23)	41(22)	85(45)	11(6)	0(0)	11(6)
	Expected exceedances of $x \geq 1.05$ days	6(3)	99(52)	105(55)	39(20)	142(74)	181(94)
	Marginal distribution	50(26)	140(74)	190	50(26)	142(74)	192
1989	Expected exceedances of $x < 1.05$ days	34(18)	53(28)	87(46)	11(6)	0(0)	11(6)
	Expected exceedances of $x \geq 1.05$ days	8(4)	95(50)	103(54)	31(16)	148(78)	179(94)
	Marginal distribution	42(22)	148(78)	190	42(22)	148(78)	190

[a] Cell entries in the two-way tables are the frequency (%) of the MSAs classified according to two strategies

Table 4 Distribution of Metropolitan Statistical Areas (MSAs) that Violate or Comply with Current Standard and a Pair of Two-Component Compliance Tests that Test the Hypotheses that the Average Expected Exceedance of 0.125 ppm (Ex Ex 125) is Greater than 1.05 and the Average Maximum 3-month SUM06 is Greater than 24.4 ppm-hr, respectively, for 3-Year Averaging Periods Between 1982 and 1989[a]

| | | Current Standard Vs. Two-stage Compliance t_w–Tests of Ex Ex 125 and SUM06 Indicators[a] | | | | | |
| | | H_o: Ex Ex 125 = 1.05 vs H_a: Ex Ex 125 > 1.05 | | | H_o: SUM06 = 24.4 ppm vs H_a: SUM06 > 24.4 ppm | | |
Year	Current Standard	Accept H_o: Compliance	Reject H_o: Violation	Marginal Distribution	Accept H_o: Compliance	Reject H_o: Violation	Marginal Distribution
1984	Expected exceedances of 0.125 < 1.05 days	79(46)	0(0)	79(46)	63(37)	16(9)	79(46)
	Expected exceedances of 0.125 ≥ 1.05 days	6(4)	85(50)	91(54)	16(9)	75(44)	91(54)
	Marginal distribution	85(50)	85(50)	170	79(46)	91(54)	170
1985	Expected exceedances of 0.125 < 1.05 days	87(49)	0(0)	87(49)	62(35)	25(14)	87(49)
	Expected exceedances of 0.125 ≥ 1.05 days	8(5)	81(46)	89(51)	21(12)	68(39)	89(51)
	Marginal distribution	95(54)	81(46)	176	83(47)	93(53)	176

1986	Expected exceedances of 0.125 < 1.05 days	106(59)	0(0)	106(59)	79(44)	27(15)	106(59)
	Expected exceedances of 0.125 ≥ 1.05 days	11(6)	64(35)	75(41)	24(13)	51(28)	75(41)
	Marginal distribution	117(65)	64(35)	181	103(57)	78(43)	181
1987	Expected exceedances of 0.125 < 1.05 days	116(62)	0(0)	116(62)	80(43)	36(19)	116(62)
	Expected exceedances of 0.125 ≥ 1.05 days	6(3)	66(35)	72(38)	21(11)	51(27)	72(38)
	Marginal distribution	122(65)	66(35)	188	101(54)	87(46)	188
1988	Expected exceedances of 0.125 < 1.05 days	85(45)	0(0)	85(45)	62(33)	23(12)	85(45)
	Expected exceedances of 0.125 ≥ 1.05 days	36(19)	69(36)	105(55)	23(12)	82(43)	105(55)
	Marginal distribution	121(64)	69(36)	190	85(45)	105(55)	190
1989	Expected exceedances of 0.125 < 1.05 days	87(46)	0(0)	87(46)	56(29)	31(16)	87(46)
	Expected exceedances of 0.125 ≥ 1.05 days	32(17)	71(37)	103(54)	22(12)	81(43)	103(54)
	Marginal distribution	119(63)	71(37)	190	78(41)	112(59)	190

[a] Cell entries in the two-way tables are the frequency (%) of the MSAs jointly classified according to two attainment strategies.

nual SUM06 $\geq$ 24.4 ppm-hr tests; and (3) the three-year Ex Ex 094 $\geq$ 1.05 days and three-year SUM06 $\geq$ 24.4 ppm-hr tests. In addition, Table 3 compares the unequally stringent three-year Ex Ex 125 $\geq$ 1.05 days (current NAAQS) and the three-year SUM06 $\geq$ 24.4 ppm-hr tests to show the effects of tightening the current NAAQS to a sufficiently stringent level of protection. Table 4 shows the joint distribution of MSAs in compliance with or in violation of the following pairs of equally efficient but unequally stringent attainment strategies: (1) the current NAAQS and the two-component Ex Ex 125 $\geq$ 1.05 days statistical test (denoted t_w Ex Ex 125) and (2) the current NAAQS and the two-component SUM06 $\geq$ 24.4 ppm-hr statistical test (t_w SUM06 24.4).

4.1. Comparison in Terms of Efficiency

The relative efficiency of competing attainment strategies was determined by comparing the proportion of MSAs in violation of the strategies. For example, referring to Table 1, 44 of the 190 MSAs (23%) violated the annual SUM06 $\geq$ 34.5 ppm-hr test in 1982 versus 74 of 190 MSA violations (39%) of the annual Ex Ex 125 test, meaning that in 1982 the SUM06 test was more efficient than the Ex Ex test. At both tested levels of protection and regardless of the use of averaging periods or statistical tests, the SUM06-based strategy was more efficient than the Ex Ex-based test. Moreover, the relative efficiency of SUM06 increased with an increase in stringency level. We attribute efficiency differences between the SUM06 and Ex Ex indicators to the better functional correspondence with biological response of the SUM06 indicator and the higher precision in estimating the numerical level of the SUM06 index.

Except for 1988 (a high O_3 season), the annual SUM06-based test produced fewer violations than the equally stringent annual Ex Ex-based test at the stringency level that protects against 18.5% loss for 50% of the crops (Table 1). When the stringency of the SUM06 and Ex Ex tests was increased to protect against 10% loss for 50% of the crops, there were again fewer MSA violations of the SUM06 test than of the Ex Ex test for all years including 1988 (Table 2). At this higher level of stringency, the percentage of MSAs that violated the SUM06 strategy was about 77% during high O_3 years (1983, 1987, 1988) and 48% during normal years, an average violation increase of 11% from the less stringent level. In comparison, at the higher level of stringency, about 92% and 81% of the MSAs violated the Ex Ex strategy during high and normal O_3 years, respectively, a 39% average increase in violations from the lower stringency level.

For the MSAs with three consecutive years of adequate air quality data, increasing the averaging period from one to three years reduced violations of the SUM06 and Ex Ex tests proportionately, such that, as with the one-year tests, fewer MSAs violated the three-year SUM06 $\geq$ 24.4 ppm-hr test than the Ex Ex 094 test (Table 3). At this stringency level, the percentage of MSA violations of the three-year SUM06 and Ex Ex strategies fluctuated randomly about horizontal lines at 65% and 94%, respectively. Similarly, the percentage of MSA violations of the three-year SUM06 $\geq$ 34.5 ppm-hr and Ex Ex 125 (i.e., current

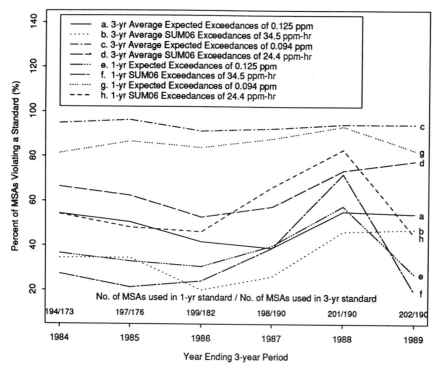

Figure 4. Percentages of metropolitan statistical areas (MSAs) violating (*a*) the present standard; (*b*) the 3-year average SUM06 exceeding 34.5 ppm-hr; (*c*) the 3-year average number of exceedances of 0.094 ppm; (*d*) the 3-year average SUM06 exceeding 24.4 ppm-hr; (*e*) the annual number of exceedances of 0.125 ppm; (*f*) the annual maximum 3-month SUM06 exceeding 34.5 ppm-hr; (*g*) the annual number of exceedances of 0.094 ppm; and (*h*) the annual maximum 3-month SUM06 exceeding 24.4 ppm-hr.

NAAQS) strategies varied about horizontal lines at 35% and 50%, respectively (Fig. 4). Similar results were also obtained for a four-year averaging period; the SUM06 $\geq$ 24.4 ppm-hr strategy was more efficient than the Ex Ex 094 strategy in terms of percentage of MSA violations (Fig. 5).

Figure 5 shows incremental differences in percentage of violations between strategies based on three- and four-year averaging periods and moderate decreases in percent violations when a two-stage compliance test is introduced. With respect to the two-stage compliance test, the comparison of the SUM06 and Ex Ex indicators was done on the basis of unequally stringent levels because the significance levels used for the compliance test, $\alpha_1 = 0.30$ and $\alpha_2 = 0.16$, were chosen to achieve equal efficiency for the two-stage SUM06 $\geq$ 24.4 ppm-hr compliance test and the current NAAQS. Still, because of the stochastic nature of exposure, the current NAAQS and the two-stage SUM06 test produced an equal number of violations only in the periods 1982 to 1984 (91 violations) and 1986 to 1988 (105 violations) (Table 4).

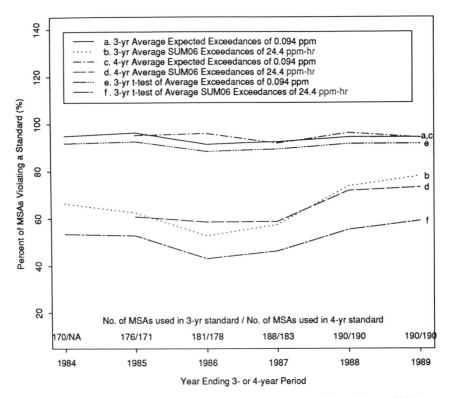

Figure 5. Percentages of MSAs violating (*a*) the 3-year average number of exceedances of 0.094 ppm; (*b*) the 3-year average SUM06 exceeding 24.4 ppm-hr; (*c*) the 4-year average number of exceedances of 0.094 ppm; (*d*) the 4-year average maximum 3-month SUM06 exceeding 24.4 ppm-hr; (*e*) the 3-year t_w-test of exceedances of 0.094 ppm; and (*f*) the 3-year t_w-test of SUM06 exceeding 24.4 ppm-hr.

Our results show that the relative efficiency of the SUM06-based strategy compared with the Ex Ex-based strategy increases with increasing levels of protection. Thus, the SUM06 indicator is better than the Ex Ex indicator in terms of minimizing violation increases to achieve a target protection level that is higher than that of the current NAAQS.

4.2. Comparison in Terms of Stability

Figure 4 illustrates how areas can bounce in and out of attainment from year to year due to meteorological fluctuations, most notably when compliance is determined from annual indicator values (lines e–h in Figure 4). MSA violations of annual attainment strategies rose sharply in 1988 from previous levels (due to an unusually hot summer) and dropped sharply in 1989 when summer temperatures returned to normal. The use of three-year moving averages (lines a–d in Figure 4) enhances stability somewhat.

Of the annual attainment strategies, only the annual Ex Ex 094 test displayed satisfactory stability against meteorological fluctuations, due primarily to its inefficiency in achieving sufficient stringency. Based on the annual Ex Ex 094 test, no crossovers occurred during the years 1984 to 1989 for about 70% of the MSAs, whereas only 37% of the MSAs had zero crossovers based on the equally stringent annual SUM06 $\geqslant$ 24.4 ppm-hr test (Fig. 6a). At the less stringent level, about 50% of the MSAs had no crossovers based on the annual Ex Ex 125 test versus 36% with zero crossovers based on the annual SUM06 $\geqslant$ 34.5 ppm-hr test. In addition, 27 and 40% of the MSAs had exactly two crossovers based on annual attainment strategies (excluding the Ex Ex 094 test) in the years 1984 to 1989. In most cases, new violations between 1987 and 1988 and subsequent return to attainment in 1989 accounted for the two crossovers (Table 5).

These data suggest that stability depends strongly on meteorological stability as well as on the variability inherent in an attainment strategy's form. Because the SUM06 indicator was more sensitive to weather changes than the Ex Ex indicator, the annual SUM06 strategy had more MSA crossovers than the Ex Ex strategy at both levels of stringency.

The three-year moving average was effective in increasing stability of both the SUM06- and Ex Ex-based strategies. MSAs with less than two crossovers increased to $> 90\%$ based on the three-year Ex Ex strategy and to about 83% based on the corresponding (equally stringent) SUM06 strategy at both levels of stringency (Figure 6b). Still, 15% of the MSAs bounced out and in of attainment based on the three-year SUM06 strategy, slightly more than the 11% with two crossovers based on the current NAAQS.

Increasing the averaging period from three to four years resulted in marginal improvement in the stability of the Ex Ex strategy but, at both stringency levels, increased the stability of the SUM06 strategy to equal or exceed the stability level of the current NAAQS (Fig. 7). The two-stage compliance test was nearly as effective as the four-year moving average in improving the stability of the SUM06 tests to the stability level of the current NAAQS (Fig. 7). Note that the stability of the SUM06 $\geqslant$ 34.5 ppm-hr strategy was nearly identical to the stability of the SUM06 $\geqslant$ 24.4 ppm-hr strategy when smoothing techniques were used, and so data for the less stringent SUM06 strategy was omitted from Figure 7 and Table 5. Also, the Ex Ex 094 strategy showed little or no change in stability when the averaging period was changed from three to four years or when the two-stage compliance test was applied, and so data for this strategy also was omitted from Figure 7 and Table 5.

The use of averaging and/or a statistical compliance test proves effective in increasing the stability of attainment strategies, but this increase in instability comes at the expense of stringency and efficiency. Typically, increases in stability through the use of moving averages and/or statistical tests, in conjunction with a robust indicator, change the trend in MSA violations toward a horizontal target line, thereby modulating the proportion of area violations toward a historic average of normal and high O_3 years. Because stringency is based on a single season as determined from crop effects studies, the effects of smoothing

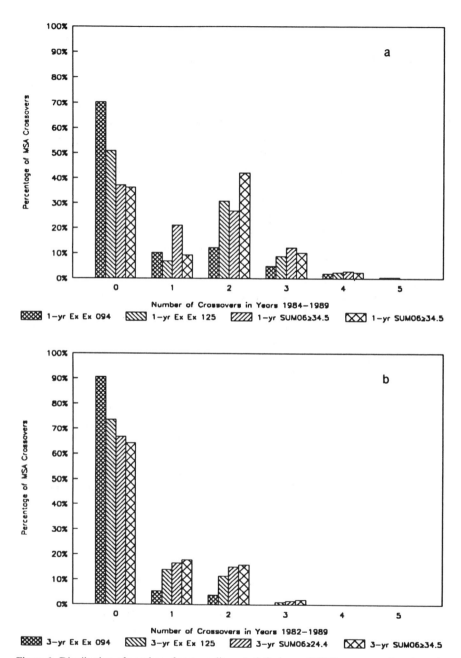

Figure 6. Distribution of number of metropolitan statistical area (MSA) crossovers based on the expected number of daily exceedances of 0.094 and 0.125 ppm (Ex Ex 094 and Ex Ex 125, respectively) and the maximum 3-month SUM06 values exceeding 24.4 and 34.5 ppm-hr for averaging periods of (*a*) one year and (*b*) three years, for the years 1982–1989.

Table 5 Number of Metropolitan Statistical Areas (MSAs) Experiencing Crossovers from Attainment to Nonattainment or from Nonattainment to Attainment Based on Strategies Using the Expected Exceedances of 0.125 ppm and maximum 3-Month SUM06 ≥ 24.4 ppm-hr, Alone or in Combination with a Two-Component t-Test strategy, for Averaging Periods of One, Three, and Four Years Between 1982 and 1989

Annual Expected Exceedances of 0.125 ppm

Year	#MSAs	Total# Crossovers	#Crossovers from Nonatt. to Attainment	#Crossovers from Attainment to Nonatt.
1985	186	26(14.0)	16(8.6)	10(5.4)
1986	192	33(17.2)	18(9.4)	15(7.8)
1987	193	47(24.4)	14(7.3)	33(17.1)
1988	194	48(24.7)	6(3.1)	42(21.6)
1989	191	65(34.0)	63(33.0)	2(1.0)

3-Yr Average Expected Exceedances of 0.125 ppm

Year	#MSAs	Total# Crossovers	#Crossovers from Nonatt. to Attainment	#Crossovers from Attainment to Nonatt.
1985	171	10(5.8)	7(4.1)	3(1.8)
1986	178	16(9.0)	15(8.4)	1(0.6)
1987	183	12(6.6)	7(3.8)	5(2.7)
1988	190	34(17.9)	1(0.5)	33(17.4)
1989	189	8(4.2)	5(2.6)	3(1.6)

4-Yr Average Expected Exceedances of 0.125 ppm

Year	#MSAs	Total# Crossovers	#Crossovers from Nonatt. to Attainment	#Crossovers from Attainment to Nonatt.
1985	NA	NA	NA	NA
1986	170	10(5.9)	7(4.1)	3(1.8)
1987	176	11(6.3)	10(5.7)	1(0.6)
1988	180	31(17.2)	3(1.7)	28(15.6)
1989	187	4(2.1)	3(1.6)	1(0.5)

3-Yr t_{tw}-Test of Expected Exceedances of 0.125 ppm

Year	#MSAs	Total# Crossovers	#Crossovers from Nonatt. to Attainment	#Crossovers from Attainment to Nonatt.
1985	165	8(4.8)	7(4.2)	1(0.6)
1986	172	18(10.5)	17(9.9)	1(0.6)
1987	179	8(4.5)	3(1.7)	5(2.8)
1988	186	9(4.8)	3(1.6)	6(3.2)
1989	187	8(4.3)	3(1.6)	5(2.7)

Annual 3-month SUM06 Exceeding 24.4 ppm-hr

Year	#MSAs	Total# Crossovers	#Crossovers from Nonatt. to Attainment	#Crossovers from Attainment to Nonatt.
1985	186	38(20.4)	26(14.0)	12(6.5)
1986	192	41(21.4)	22(11.5)	19(9.9)
1987	193	50(25.9)	5(2.6)	45(23.3)
1988	194	40(20.6)	4(2.1)	36(18.6)
1989	191	83(43.5)	78(40.8)	5(2.6)

3-Yr Average SUM06 Exceeding 24.4 ppm-hr

Year	#MSAs	Total# Crossovers	#Crossovers from nonatt. to Attainment	#Crossovers from Attainment to Nonatt.
1985	170	12(7.1)	8(4.7)	4(2.4)
1986	172	25(14.5)	20(11.6)	5(2.9)
1987	181	18(9.9)	6(3.3)	12(6.6)
1988	187	35(18.7)	2(1.1)	33(17.6)
1989	189	12(6.3)	3(1.6)	9(4.8)

4-Yr Average SUM06 Exceeding 24.4 ppm-hr

Year	#MSAs	Total# Crossovers	#Crossovers from Nonatt. to Attainment	#Crossovers from Attainment to Nonatt.
1985	NA	NA	NA	NA
1986	168	11(6.5)	6(3.6)	5(3.0)
1987	174	10(5.7)	4(2.3)	6(3.4)
1988	181	24(13.3)	1(0.6)	23(12.7)
1989	185	7(3.8)	3(1.6)	4(2.2)

3-Yr t_{tw}-Test of SUM06 Exceeding 24.4 ppm-hr

Year	#MSAs	Total# Crossovers	#Crossovers from Nonatt. to Attainment	#Crossovers from Attainment to Nonatt.
1985	165	15(9.1)	8(4.8)	7(4.2)
1986	172	17(9.9)	16(9.3)	1(0.6)
1987	179	15(8.4)	6(3.4)	9(5.0)
1988	186	18(9.7)	1(0.5)	17(9.1)
1989	187	18(9.6)	7(3.7)	11(5.9)

Note: Column entries are the frequency (%) of MSAs in transition.

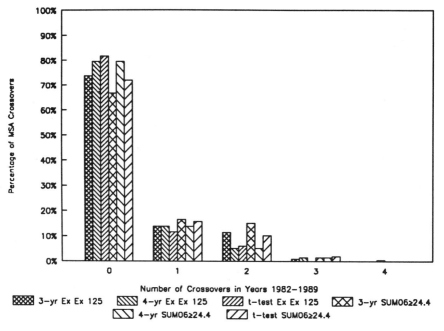

Figure 7. Distribution of number of metropolitan statistical area (MSA) crossovers for the daily expected exceedances of 0.125 ppm (Ex Ex 125) and the three-month SUM06 $\geqslant$ 24.4 ppm-hr indicators, in conjunction with averaging periods of three or four years and/or a two-component compliance test, for years 1982–1989.

techniques on the stringency level of attainment strategies are unknown but would appear to decrease the protection level.

4.3. Comparison in Terms of Consistency

As already discussed, the index of the strategy dictates the efficiency and stability properties of individual attainment tests; at both levels of stringency, the SUM06 indicator had greater efficiency but less stability than the Ex Ex indicator. Differences in efficiency between the SUM06- and Ex Ex-based strategies imply less than 100% consistency between the two tests. We calculated the consistency between the SUM06- and Ex Ex-based strategies (i.e., the percentage of MSAs in compliance with or in violation of both strategies) as the sum of diagonal cell proportions from the joint distributions given in Tables 1 to 3. For example, referring to the year 1982 in Table 1, a total of 150 (111 + 39) out of 190 MSAs (79%) were in joint compliance with or in violation of the annual Ex Ex 125 and SUM06 $\geqslant$ 34.5 ppm-hr tests, designed to protect against 18.5% loss for 50% of the crops. For the entire 1982 to 1989 period, joint compliance or violation of the annual Ex Ex 125 and SUM06 $\geqslant$ 34.5 ppm-hr tests ranged between 70 and 79%. During the same period, between 60 and 87% of the MSAs were in joint compliance with or in violation of the Ex Ex 094 and SUM06 $\geqslant$ 24.4 ppm-hr tests (Table 2). At this protection level, the two types of strategies were more consistent during the high O_3 years (1983, 1987, and 1988) than during other years. Also, for nearly all of the MSAs that violated only one strategy at the sufficiently stringent level, the violation occurred for the Ex Ex 094 strategy, as indicated by the off-diagonal cell proportions in Table 2.

The effect on consistency of using a three-year moving average was marginal. At the sufficiently stringent protection level, between 62 and 84% of the MSAs were classified as being in joint compliance with or in violation of the three-year Ex Ex 094 and SUM06 $\geqslant$ 24.4 ppm-hr strategies during 1982 to 1989 (Table 3), results similar to those for the annual strategies. However, because the moving averages increased the transition time from violation to resumed compliance following a high O_3 year, the two strategies were most consistent during or following a high O_3 year. Because of the greater efficiency of the SUM06 strategy, all of the MSAs that violated only one strategy were in violation of the three-year Ex Ex 094 strategy (Table 3). Use of a four-year moving average gave virtually identical results as the three-year moving average (data not presented).

Among the three-year attainment strategies, consistency with the current NAAQS was highest for the t_w-based version of the current NAAQS. At least 94% of the MSAs were in joint compliance with or in violation of the current NAAQS and the t_w Ex Ex 125 tests for all periods except 1986 to 1988 and 1987 to 1989 (Table 4). The lower level of the t_w Ex Ex 125 test's consistency with the current NAAQS in the periods 1986 to 1988 and 1987 to 1989 (81% and 83%, respectively) occurred because the t_w compliance test was designed to prevent historically compliant areas from bouncing in and out of attainment during and following peak O_3 seasons such as the one that occurred in 1988. In contrast, among all of the three-year attainment strategies, consistency with the current NAAQS was lowest for the three-year Ex Ex 094 strategy (or the t_w Ex Ex 094 test), ranging between 48 and 63% (Table 6).

Between 73 and 80% of the MSAs were in joint compliance with or in violation of the current NAAQS and the three-year SUM06 $\geqslant$ 34.5 ppm-hr strategy for the years 1982 to 1989 (Table 6); consistency was largely unaffected by the use of the two-stage compliance test for all three-year periods including 1986 to 1988. Finally, between 67 and 77% of the MSAs were in joint compliance with or in violation of the current NAAQS and the three-year SUM06 $\geqslant$ 24.4 ppm-hr strategy for the years 1982 to 1989 (Table 6); again, consistency did not change appreciably with the use of the two-stage compliance test.

Consistency with the current NAAQS was largely dependent upon the index of the strategy and favored the use of the more efficient indicator, namely SUM06, in order to achieve a sufficiently stringent level of protection. Consistency was marginally affected by the use of the t_w statistical test and, for the SUM06 indicator, decreased marginally as stringency increased from the level attained by the current NAAQS to a sufficiently stringent level. Because the SUM06 indicator emphasized different characteristics of exposure than the extreme value Ex Ex indicator, the attainment status of as many as one-third of the MSAs

Table 6 Percent of Metropolitan Statistical Areas (MSAs) that Jointly Violate or Comply with the Current Standard and Alternative Attainment Strategy Based on the 3-Year Moving Averages of Expected Exceedances of 0.094 and 0.125 ppm (Ex Ex 094 and Ex Ex 125, respectively) and SUM06 Indices (SUM06 $\geq$ 24.4 ppm-hr and SUM06 $\geq$ 34.5 ppm-hr), Alone or in Conjunction with a Two-Component Compliance Test, for 3-Year Averaging Periods between 1982 and 1989

3-Yr Avg. Period Ending	3-Yr Expected Exceedances Strategies				3-Yr SUM06 Attainment Strategies			
	Ex Ex 125 Strategy		Ex Ex 094 Strategy		SUM06 $\geq$ 34.5 ppm-hr		SUM06 $\geq$ 24.4 ppm-hr	
	w/o t_w-test (%)	w/Test (%)	w/o t_w-test (%)	w/Test (%)	w/o t_w-test (%)	w/Test (%)	w/o t_w-test (%)	w/Test (%)
1984	100	96	60	62	75	67	77	71
1985	100	95	56	58	73	72	75	74
1986	100	94	59	53	75	74	73	72
1987	100	97	48	49	73	74	67	70
1988	100	81	62	63	80	76	75	76
1989	100	83	61	62	76	75	68	72

differed according to whether the SUM06 strategy or the current NAAQS was used for testing.

Unlike Lefohn and Foley (1991), we did not compare the consistency of the current NAAQS with the annual SUM06 strategy because use of different averaging periods complicates attribution of differences in attainment status to either the form or level of the strategies. Because of differences in designation of MSA violations of the current NAAQS for three-year averaging periods between 1986 and 1989 and the annual SUM06 ≥ 24.4 ppm-hr strategy for years 1987, 1988, and 1989, Lefohn and Foley (1991) concluded that the SUM06 indicator, as the form of a secondary standard, did not correlate well with peak concentrations, and therefore resulted in apparent inconsistent protection of vegetation. The authors' analysis did not corroborate their findings because (1) the comparison was not done on an equally stringent basis; (2) the authors did not control for differences in averaging periods of one and three years; and (3) the authors did not compare the current NAAQS with the Ex Ex 094 strategy, annual or otherwise. Had the authors compared the current NAAQS with the Ex Ex 094 strategy, the Ex Ex indicator, being inconsistent with the current NAAQS, would also be discounted for use as the form of a secondary standard.

5. DISCUSSION

In our study, we emphasized the need to implement a federal O_3 standard that provides adequate protection (based on vegetation effects data) against crop damage due to O_3. We proposed the use of the SUM06 indicator as the form of the secondary standard because of its ability to address effects concerns, and we set the target protection level to protect against 10% crop damage for 50% of the crops (i.e., sufficient stringency) based on evidence from individual NCLAN studies that collectively supported a 10% yield reduction target protection level (Tingey et al., 1991). It was determined that, to achieve this degree of stringency, the attainment level of the SUM06 strategy would need to be set at 24.4 ppm-hr. The protection level attained by the current secondary NAAQS is unclear, because crop effects are determined from a single growing season, not for seasons spanning three years. However, using the Ex Ex indicator on which the current NAAQS is based, the 0.125 ppm attainment level corresponds to an 18.5% yield reduction threshold for 50% of the crops and, for an annual attainment strategy, would need to be lowered to 0.094 ppm to realize a 10% protection level.

The effects of using a three-year moving average to monitor target protection level have not been studied but would result in some degree of decreased protection. An attainment strategy that mandates 100% (or less) compliance with the nominal level over multiple growing seasons, say three years, can experience O_3 levels thrice the nominal level in one growing season and stay within attainment of the test. For example, an area with SUM06 values of 0, 60, and 0 ppm-hr for three consecutive years would comply with the three-year average SUM06 < 24.4 ppm-hr test but, at a SUM06 level of 60 ppm-hr in the second year, half or more of the crops would experience yield losses of 42% or greater (Tingey et al., 1991). Hence, for the establishment of a protective standard, it is important that the averaging period match the ecological period in which effects are determined.

However, because past and current decisions on the secondary NAAQS have focused more on regulatory concerns than effects concerns, the use of a three-year averaging period to mitigate the effects of short-term meteorological fluctuations has been introduced. We therefore also tested variations of the SUM06 $\geqslant$ 24.4 ppm-hr using (1) averaging periods of three and four years and (2) a two-component compliance test that is strict in identifying compliance and, for areas that have been in attainment for several periods, is less strict in identifying a move to nonattainment (Chock, 1991). As noted earlier, multiple-year attainment strategies offer less protection than an annual attainment strategy, but how much less is not known. We used significance levels of 0.30 and 0.16 for the statistical test to designate attainment and subsequent nonattainment, respectively; these levels were designed to achieve equal efficiency with the current NAAQS rather than equal stringency, similar to Chock (1991). Because it is a multiple-year strategy, the two-component compliance test suffers from the same disadvantages as the moving average strategies. As noted by Chock (1989), the power of the compliance test may be poor for an averaging period of three years, depending upon the standard deviation (or range, in the case of t_w). For example, an area with SUM06 values of 0, 0, and 100 ppm-hr for three consecutive years would comply with the three-year period because of high imprecision in estimation of the population mean (i.e., $t_w = 0.089 < t_{w,0.3} = 0.187$). In addition, the test's precision cannot be specified a priori and varies through time and space.

We compared the SUM06 and Ex Ex strategies on an equally stringent basis in terms of efficiency, stability, and consistency with each other and with the current NAAQS. The attainment properties were most influenced by the form of the strategy and, except for stability, were optimized using the SUM06 indicator. At the level sufficiently stringent to protect against 10% yield loss for 50% of the crops, the SUM06 $\geqslant$ 24.4 ppm-hr strategy was considerably more efficient and more consistent with the current NAAQS than the equally stringent Ex Ex 094 strategy, regardless of averaging period and statistical test. While lower in stability than the Ex Ex strategy, the SUM06 strategy using either a four-year moving average or the statistical test was as stable as the current NAAQS.

Stability is clearly a desirable attainment property from a regulatory viewpoint when the intent of designating an area in violation of a standard is to identify the need for emission control strategies, but its importance in a protective standard is debatable. If the intent of designating an area in violation of a standard is to identify conditions associated with potential adverse effects to crops, then stability is unnecessary. Because of our interest in establishing an effects-based attainment strategy, we considered stability to be less important than stringency, efficiency, or even consistency with the current NAAQS. We would prefer not to smooth the annual SUM06 $\geqslant$ 24.4 ppm-hr attainment strategy. However, because attainment concerns need to be addressed, we propose using the SUM06 $\geqslant 24.4$ ppm-hr strategy with a three-year averaging period as the secondary O_3 NAAQS, although, as discussed earlier, the nominal numerical level may need to be lowered slightly to achieve sufficient stringency. We need to explore the distribution of effects and the distribution of attainment as determined from crop effects and air quality data, respectively, to evaluate the impact of a moving average on the specified target protection level.

6. CONCLUSION

Effects and regulatory concerns can both be addressed by the selection of an air quality indicator whose form is related to crop response. The advantage of selecting this type of attainment strategy is that the attainment level can be based on the varying magnitudes of vegetation response. By basing the attainment level on the distribution of effects rather than the distribution of attainment, the strategy has direct links to known or anticipated effects and is not site- or year-specific. The approach can be generalized to different levels of protection and allows additional crop information to be included in the formulation of an alternative air quality indicator and associated level.

The question of whether attainment or effects concerns are more important in O_3 regulation is debatable. Regardless of one's viewpoint, however, tightening of the current NAAQS with minimum sacrifice to the number of violations, changes in attainment status, and number of crossovers from year to year, requires modifications to the form, level, and possibly averaging period of the NAAQS. Based on effects and attainment concerns, we propose the SUM06-based strategies as guidelines from which a secondary O_3 NAAQS can be recommended.

ACKNOWLEDGMENTS

The authors acknowledge contribution of data from the U.S. Environmental Protection Agency's National Air Data Branch (NADB) and the cooperation of Thomas Link of NADB in providing the O_3 data. The research described in this paper was conducted at the EPA's Research Laboratory in Corvallis, OR. The work upon which this publication is based was performed pursuant to Contract No. 68-C8-0006, Option V, with the U.S. Environmental Protection Agency.

REFERENCES

Ashmore, M. R. (1984). Effects of ozone on vegetation in the United Kingdom. In P. Greenfelt (Ed.), *Evaluation of the Effects of Photochemical Oxidants on Human Health, Agricultural Crops, Forestry, Materials and Visibility* (Proceedings of the International Workshop). Swedish Environmental Research Institute, Goteburg, pp. 92–104.

Chock, D. (1989). The need for a more robust ozone air quality standard. *JAPCA* **39**, 1063–1072.

Chock, D. (1991). Issues regarding the ozone air quality standards. *J. Air Waste Manage. Assoc.* **41**, 148–152.

Curran, T. C., and Frank, N. H. (1991). Ambient ozone trends using alternative indicators. In R. L. Berglund, D. R. Lawson, and D. J. McKee (Eds.), *Transactions of the Tropospheric Ozone and the Environment Specialty Conference.* Air & Waste Management Assoc., Pittsburgh, pp. 749–759.

Fairley, D., and Blanchard, C. L. (1991). Rethinking the ozone standard. *J. Air Waste Manage. Assoc.* **41**, 928–936.

Federal Register (1979). Revisions to the national primary and secondary ambient air quality standards for photochemical oxidants. *Fed. Regist.* **44**(28), 8202–8221.

Guderian, R., Tingey, D. T., and Rabe, R. (1985). Effects of photochemical oxidants on plants. In R. Guderian (Ed.), *Air Pollution by Photochemical Oxidants: Formation, Transport Control, and Effects on Plants.* Springer-Verlag, Berlin, pp. 129–333.

Heck, W. W., and Tingey, D. T. (1971). Ozone time-concentration model to predict foliar injury. In H. M. Englund and W. T. Beery (Eds.), *Proceedings of the Second International Clean Air Congress, 1970.* Academic Press, New York, pp. 249–255.

Heck, W. W., Dunning, J. A., and Hindawi, I. J. (1966). Ozone: Nonlinear relation of dose and injury to plants. *Science* **151**, 511–515.

Henderson, W. R., and Reinert, R. A. (1979). Yield response of four fresh market tomato cultivars after acute ozone exposure in the seedling stage. *J. Am. Soc. Hortic. Sci.* **104**, 754–759.

Hogsett, W. E., Tingey, D. T., and Holman, S. R. (1985). A programmable exposure control system for determination of the effects of pollutant exposure regimes on plant growth. *Atmos. Environ.* **19**, 1135–1145.

Hogsett, W. E., Tingey, D. T., and Lee, E. H. (1988). Exposure indices: Concepts for development and evaluation of their use. In W. W. Heck, O. C. Taylor, and D. T. Tingey (Eds.), *Assessment of Crop Loss from Air Pollutants.* Elsevier, London, pp. 107–138.

Javitz, H. S. (1980). Statistical interdependencies in the ozone national ambient air quality standard. *J. Air Pollut. Control Assoc.* **30**, 58–59.

Lee, E. H., Tingey, D. T., and Hogsett, W. E. (1987). *Selection of the Best Exposure-Response Model Using Various 7-hour Ozone Exposure Indices.* U.S. Environmental Protection Agency, Office of Air Quality Planning and Standards, Research Triangle Park, NC.

Lee, E. H., Tingey, D. T., and Hogsett, W. E. (1988). Evaluation of ozone exposure indices in exposure-response modeling. *Environ. Pollut.* **53**, 43–62.

Lee, E. H., Tingey, D. T., and Hogsett, W. E. (1989a). *Evaluation of Ozone Exposure Indices for relating exposure to Plant Production and for Estimating Agricultural Losses,* EPA/600/3-89/039. U.S. Environmental Protection Agency, Office of Air Quality Planning and Standards, Research Triangle Park, NC.

Lee, E. H., Tingey, D. T., and Hogsett, W. E. (1989b). *Interrelation of Experimental Exposure and Ambient Air Quality Data for Comparison of Ozone Exposure Indices and Estimating Agricultural Losses,* EPA/600/3-89/047. U.S. Environmental Protection Agency, Office of Air Quality Planning and Standards, Research Triangle Park, NC.

Lee, E. H., Tingey, D. T., and Hogsett, W. E. (1991a). Efficacy of ozone exposure indices in the standard setting. In R. L. Berglund, D. R. Lawson, and D. J. McKee (Eds.), *Transactions of the Tropospheric Ozone and the Environment Specialty Conference.* Air & Waste Management Assoc., Pittsburgh, pp. 255–271.

Lee, E. H., Tingey, D. T., and Hogsett, W. E. (1991b). Adjusting ambient ozone air quality indicators for missing values. In *1991 Proceedings of the Business and Economics Section, American Statistical Association.* American Statistical Association, Alexandria, VA, pp. 198–203.

Lefohn, A. S., and Foley, J. K. (1991). *Protecting Agricultural Crops from Ozone Exposures—Key Issues and Future Research Directions,* API Publ. No. 305. Health and Environmental Affairs Department, American Petroleum Institute, Washington, DC.

Lefohn, A. S., and Runeckles, V. C. (1987). Establishing standards to protect vegetation—Ozone exposure/dose considerations. *Atmos. Environ.* **21**, 561–568.

Lefohn, A. S., Laurence, J. A., and Kohut, R. J. (1988). A comparison of indices that describe the relationship between exposure to ozone and reduction in the yield of agricultural crops. *Atmos. Environ.* **22**, 1229–1240.

Lord, E. (1947). The use of range in place of standard deviation in the *t*-test. *Biometrika* **34**, 41–67.

McCurdy, T., and Atherton, R. (1990). Variability of ozone air quality indicators in selected metropolitan statistical areas. *J. Air Waste Manage. Assoc.* **40**, 477–486.

Musselman, R. C., Oshima, R. J., and Gallavan, R. E. (1983). Significance of pollutant concentration distribution in the response of "red kidney" beans to ozone. *J. Am. Soc. Hortic. Soc.* **108**, 347–351.

Musselman, R. C., McCool, P. M., and Younglove, T. (1988). Selecting ozone exposure statistics for determining crop yield loss from air pollutants. *Environ. Pollut.* **53**, 63–78.

Nouchi, I., and Aoki, K. (1979). Morning glory as a photochemical oxidant indicator. *Environ. Pollut.* **18**, 289–303.

Reinert, R. A., and Nelson, P. V. (1979). Sensitivity and growth of twelve elatior begonia cultivars to ozone. *Hort. Science* **14**, 747–748.

Stan, H. J., and Schicker, S. (1982). Effects of repetitive ozone treatment on bean plants—stress ethylene production and leaf necrosis. *Atmos. Environ.* **16**, 2267–2270.

Stan, H. J., Schicker, S., and Kassner, H. (1981). Stress ethylene evolution of bean plants—a parameter indicating ozone pollution. *Atmos. Environ.* **15**, 391–395.

Tingey, D. T., Hogsett, W. E., and Lee, E. H. (1989). Analysis of crop loss for alternative ozone exposure indices. In T. Schneider, S. D. Lee, G. J. R. Wolters, and L. D. Grant (Eds.), *Atmospheric Ozone Research and Its Policy Implications*. Elsevier, Amsterdam, pp. 219–227.

Tingey, D. T., Hogsett, W. E., Lee, E. H., Herstrom, A. A., and Azevedo, S. H. (1991). An evaluation of various alternative ambient ozone standards based on crop yield loss data. In R. L. Berglund, D. R. Lawson, and D. J. McKee (Eds.), *Transactions of the Tropospheric Ozone and the Environment Specialty Conference*. Air & Waste Management Assoc., Pittsburgh, pp. 272–288.

Tonneijck, A. E. G. (1984). Effects of peroxyacetyl nitrate (PAN) and ozone on some plant species. In P. Greenfelt (Ed.), *Evaluation of the Effects of Photochemical Oxidants on Human Health, Agricultural Crops, Forestry, Materials and Visibility* (Proceedings of the International Workshop). Swedish Environmental Research Institute, Goteburg, pp. 118–127.

U.S. Environmental Protection Agency (USEPA) (1978). *Air Quality Criteria for Ozone and Other Photochemical Oxidants*, EPA-600/8-78-004. U.S. Environmental Protection Agency, Environmental Criteria and Assessment Office, Research Triangle Park, NC. Available from NTIS, Springfield, VA, PB80-124753.

U.S. Environmental Protection Agency (USEPA) (1979). *Guidelines for the Interpretation of Ozone Air Quality Standards*, EPA-450/4-79-003. U.S. Environmental Protection Agency, Monitoring and Data Analysis Division, Research Triangle Park, NC.

U.S. Environmental Protection Agency (USEPA) (1986). *Air Quality Criteria for Ozone and Other Photochemical Oxidants*, EPA-600/8-84-020. U.S. Environmental Protection Agency, Environmental Criteria and Assessment Office, Research Triangle Park, NC.

U.S. Environmental Protection Agency (USEPA) (1987a). *Report of the Clean Air Scientific Advisory Committee (CASAC)—Recommendations for Future Research on National Ambient Air Quality Standards for Ozone and Lead.*, GAO-CASAC-87-036. U.S. Environmental Protection Agency, Office of the Administrator, Science Advisory Board, Washington, DC.

U.S. Environmental Protection Agency (USEPA) (1987b). *Aerometric Information Retrieval System User's Guide*, Vol. VII. U.S. Environmental Protection Agency, Office of Air Quality Planning and Standards, Research Triangle Park, NC.

U.S. Environmental Protection Agency (USEPA) (1988). *Review of the National Ambient Air Quality Standards for Ozone Assessment of Scientific and Technical Information*, OAQPS Draft Staff Paper. U.S. Environmental Protection Agency, Office of Air Quality Planning and Standards, Research Triangle Park, NC.

U.S. Environmental Protection Agency (USEPA) (1990). *National Air Quality and Emissions Trends Report, 1988,* EPA-450/4-90-002. U.S. Environmental Protection Agency, Office of Air Quality Planning and Standards, Research Triangle Park, NC.

U.S. Environmental Protection Agency (USEPA) (1991a). *National Air Quality and Emissions Trends Report, 1989,* EPA-450/4-91-003. U.S. Environmental Protection Agency, Office of Air Quality Planning and Standards, Research Triangle Park, NC.

U.S. Environmental Protection Agency (USEPA) (1991b). *Ozone and Carbon Monoxide Areas Designated Nonattainment.* U.S. Environmental Protection Agency, Office of the Air Quality Planning and Standards, Research Triangle Park, NC.

U.S. Environmental Protection Agency (USEPA) (1992). Summary of Selected New Information on Effects of Ozone on Health and Vegetation-Supplement to 1986 Air Quality Criteria for Ozone and Other Photochemical Oxidants, EPA-600/8-88/105F. U.S. Environmental Protection Agency, Office of Research and Development, Washington, D.C.

U.S. General Accounting Office (1988). *Air Pollution: Ozone Attainment Requires Long-term Solutions to Solve Complex Problems,* GAO/RCED-88-40. U.S. General Accounting Office (Resources, Community, and Economic Development Division), Washington, DC.

20

THE SELECTIVE CATALYTIC REDUCTION OF NO$_x$ EMISSIONS FROM UTILITY BOILERS

Arun B. Mukherjee

Department of Limnology and Environmental Protection, University of Helsinki, SF-00710 Helsinki, Finland

Environmental Oxidants, Edited by Jerome O. Nriagu and Milagros S. Simmons.
ISBN 0–471–57928–9 © 1994 John Wiley & Sons, Inc.

1. INTRODUCTION

Large amounts of fossil fuels such as coal, oil, natural gas, and to some extent biofuels are used in power production, industrial processes, and transportation. Due to the increase in the use of these fuels, release of sulfur dioxides, nitrogen oxides, and carbon dioxide gases has also increased. In the atmosphere, these gases disperse widely and contribute to tropospheric photochemical smog, ozone formation, wet and dry acid deposition, stratospheric ozone depletion, and the greenhouse effect [Singh et al., 1976; Curtzen et al., 1978; National Academy of Science (NAS), 1979; World Meteorological Organization (WMO), 1986; Tripac, 1987]. Recently, great interest has been shown in oxides of nitrogen, that is, in nitric oxide (NO) and nitrogen dioxide (NO$_2$), which are jointly termed NO$_x$ because they are pathogenic in humans, animals, and plants and initiate noxious photochemical processes in the atmosphere.

Most NO$_x$ is generated by the combustion of fossil fuels. The global contribution of NO$_x$ from technosystems, calculated as nitrogen, is estimated to be $35-58 \times 10^{12}$ g yr^{-1} (Bosch and Janssen, 1987). In recent decades, both combustion and postcombustion methods have been used to control NO$_x$ from power plants. However, stricter regulations on NO$_x$ emission have forced power plants in many countries to select postcombustion methods such as the ammonia-based selective catalytic reduction (SCR) process.

The SCR process was developed in Japan in the 1970s for the reduction of NO$_x$ emissions from coal-, oil-, and gas-fired boilers (Lowe et al., 1989). The process was put into commercial use in Europe and the United States in the 1980s, and can result in 60 to 90% reduction of NO$_x$ from stationary sources (Sengoku et al., 1980; Slack, 1980; Jones, 1981; Lowe, 1984; Ando, 1985b; Hein et al., 1985; Andreasen and Morsing, 1990). Recent study (Hjalmarsson and Soud, 1991) indicates that as of the end of 1990, SCR systems have been installed on 171 coal-fired power plants having a total capacity of approximately 39.0 GW [gigawatts (electrical)].

This study focuses on SCR NO$_x$ control technology applied to coal-, oil-, and gas-fired boilers and on SCR catalysts and their dimensioning, geometry, maintenance, and poisoning.

2.   NITROGEN OXIDE EMISSIONS

2.1.   NO$_x$ Formation Mechanisms

There are seven oxides of nitrogen: NO, NO$_2$, NO$_3$, N$_2$O, N$_2$O$_3$, N$_2$O$_4$, and N$_2$O$_5$. Nitric oxide (NO) and nitrogen dioxide (NO$_2$) known together as NO$_x$, are emitted mostly from fuel combustion in mobile and stationary sources (Amann, 1989), but the burning of biomass may also be an important source (Logan, 1983).

The mechanisms of NO$_x$ formation during the combustion of fossil fuels are as follows:

- Formation of thermal NO$_x$ by fixation of atmospheric nitrogen with oxygen

Table 1   NO$_x$ Emissions from Both Stationary and Mobile Sources in Europe in 1980

Country	RAINS model	ECE	OECD
		(1000 tonnes as NO$_2$)	
Albania	28	9	—
Austria	239	216	216
Belgium	439	442	336
Bulgaria	357	150	—
Czechoslovakia	796	1204	—
Denmark	250	247	245
Finland	234	280	278
France	1944	2560	1962
Federal Republic of Germany	2891	2950	2936
German Democratic Republic	850	965	—
Greece	239	127	217
Hungary	305	270	—
Ireland	89	67	75
Italy	1458	1480	1599
Luxembourg	31	23	23
Netherlands	577	548	551
Norway	169	181	185
Poland	1597	1500	—
Portugal	149	166	165
Romania	661	390	—
Spain	950	950	937
Sweden	333	318	318
Switzerland	186	196	205
Turkey	356	175	—
UK	2394	1916	1924
USSR	9454	2790	—
Yugoslavia	394	190	—

Source: ECE (1988); OECD (1989); Springmann (1989).

- Formation of fuel NO_x
- Formation of prompt NO_x.

2.2. Emission Inventories

Recently estimated global NO_x emissions as well as those for several European countries have been summarized [Organization for Economic Cooperation and Development (OECD), 1983; Economic Commission for Europe (ECE), 1986; Amann, 1989; Sloss, 1990; Ministry of the Environment, U.K., 1990]. About 60% of NO_x emissions are anthropogenic; the rest come from natural sources, mainly the denitrification of bacteria in soils and waters, lightning, and transport from the stratosphere. Hameed and Dignon (1988) have estimated a 4% increase in global NO_x emissions per year since 1950.

Due to the wide range of fuels and combustion techniques used, there are uncertainties in estimating emissions of NO_x from energy production. Singh (1987) estimated total global emissions of NO_x (as $NO + NO_2$) from this source to be between 25 and 75×10^6 tonne yr^{-1}, whereas the Ministry of the Environment, U.K. (1990), estimated it (expressed as NO_2 from anthropogenic and natural sources) to be 153×10^6 tonne yr^{-1}.

Springmann (1989) estimated NO_x emissions based on the RAINS model for 27 countries in Europe. These values are mostly within the quite reasonable range presented by the European Commission for Europe (ECE) and the OECD (Table 1).

3. THE SELECTIVE CATALYTIC REDUCTION PROCESS

3.1. Introduction and Background

In recent decades, the control of NO_x emissions from utility boilers has been both studied and implemented (Ando, 1985b; Shimoto and Muzio, 1986). In the 1970s, the SCR technique for limiting NO_x emissions from stationary sources was developed solely in Japan. This technique claims a higher reduction of NO_x emissions than is presently available from combustion technologies. Due to its high reliability, more than 200 commerical plants in Japan have already adopted this process (Ando, 1987), and the technology has also been exported to Europe and the United States since the end of the 1970s. The first SCR installation in Europe was commissioned in Germany in 1985 (Rentz and Leibfritz, 1987).

3.1.1. Process Description

In the SCR process, NO_x is selectively reduced by NH_3 in the presence of oxygen over any one of a number of catalysts. As described by Kolar (1990), the predominant reactions are as follows:

$$4NO + 4NH_3 + O_2 \longrightarrow 4N_2 + 6H_2O \tag{1}$$

$$4NO + 4NH_3 + O_2 \rightarrow 4N_2 + 6H_2O$$
$$6NO_2 + 8NH_3 \rightarrow 7N_2 + 12H_2O$$

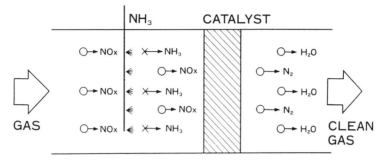

Figure 1. The reduction of NO_x with ammonia by the SCR technique at a temperature between 300 and 400 °C. (From Babcock-Hitachi K.K, with permission.)

$$6NO + 4NH_3 \longrightarrow 5N_2 + 6H_2O \qquad (2)$$
$$6NO_2 + 8NH_3 \longrightarrow 7N_2 + 12H_2O \qquad (3)$$
$$2NO_2 + 4NH_3 + O_2 \longrightarrow 3N_2 + 6H_2O \qquad (4)$$

In the SCR process, the most important component is of course the catalyst. The optimum temperature is between 300 and 400 °C and the main by-products are molecular N_2 and water.

The catalyst is generally mounted in the vertical position with the flue gases flowing from top to bottom. Occasionally, it is placed horizontally. The basic principle of the SCR catalyst reactor is shown in Figure 1 and the whole complex may be divided into three sections: the SCR catalyst, the ammonia storage and injection section, and the flue gas section.

It should be noted that the efficiency of the SCR process depends very much on the catalyst composition and flue gas temperature. In addition, the following parameters also influence the NO_x reduction efficiency:

- Reaction temperature
- Molar ratio of NH_3 to NO_x
- Space velocity of the flue gas passing through the reactor and the area velocity
- Type of catalyst

3.2. The SCR Catalyst

A number of catalysts for reducing NO_x have been studied but mostly parallel-flow-type catalysts of a honeycomb or plate shape are used for utility boilers

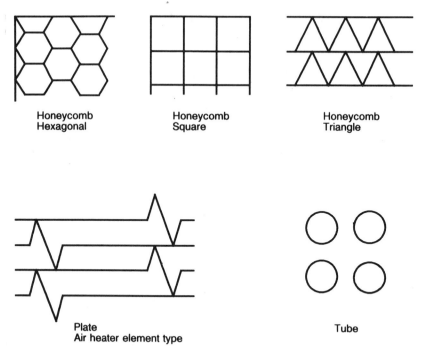

Honeycomb
Hexagonal

Honeycomb
Square

Honeycomb
Triangle

Plate
Air heater element type

Tube

Figure 2. Cross section of a parallel-flow catalyst. (Redrawn from Ando, 1985a.)

(Klovsky et al., 1980; Sengoku et al., 1980; Lachman, 1986; Bosch and Janssen, 1987).

Occasionally, packed-bed catalysts of activated coke or zeolite pellets are used. The extruded-type catalyst is generally known as honeycomb-shaped. This shape of catalyst is made of a mixture of carrier material, such as TiO$_2$ or TiO$_2$/SiO$_2$, and active components, such as WO$_3$, V$_2$O$_5$, or MoO$_3$. Besides these compounds, many other types of materials have been extruded into honeycomb-shaped catalysts (Bosch and Janssen, 1987). In this shape of catalyst, pitch or channel diameter and wall thickness are very important. The pitch diameter varies from 4 to 10 mm and wall thickness is generally about 1 mm (Ando, 1985a). Figure 2 shows the major types of parallel-flow catalysts used in the SCR technique; the dimensions of extruded-type catalysts are depicted in Figure 3.

Pitch diameter should be large enough that no blocking occurs, even with a heavy load of fly ash. According to Ando (1985b), in order to minimize dust deposition, a hexagonal honeycomb-type catalyst (angle 120°) offers better service than a triangular (angle 60°) honeycomb type. However, the square (grid) honeycomb-type catalyst is still the most popular. The relative popularity of the different types of catalysts used in Europe is shown in Figure 4. It can be seen that about 96% of catalysts are of the honeycomb type. This large share of the marketing volume is due to large pitch size, which prevents blocking of the catalyst by flue gas laden with fly ash (ECE, 1992).

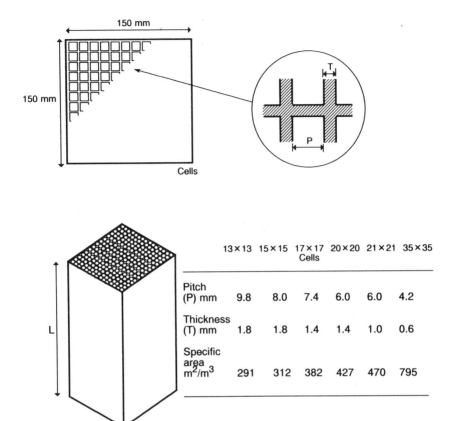

Figure 3. Grid-type catalyst dimensions. (Redrawn from Robie and Ireland, 1990.)

Plate-type catalysts consist of a metal (iron) net on which the active substance is applied. The plates can be corrugated or a mixture of plain and corrugated (Fig. 2). Plate types have higher resistance to deposition and blockage than honeycomb types. The packaging of the plates is such that, due to the difference in physical and mechanical properties of the active and carrier materials, the layers may detach under thermal or mechanical stress (ECE, 1992). On the other hand, Babcock Hitachi (Kuroda, 1985) has developed a plate-type catalyst for coal fired-boiler flue gas that has a lower pressure drop (below 100 mm H_2O) than that in the moving bed pellet-type catalyst (pressure drop 150–200 mm H_2O). Siemens AG (1990) claims that their plate-type catalyst has high thermal and mechanical stability. Their plates also have high erosion resistance. It is further reported that there is less problem of disposal with a plate-type catalyst which can be treated as scrap in a furnace. But before charging into the furnace, the active materials should be removed from the surface of the plates.

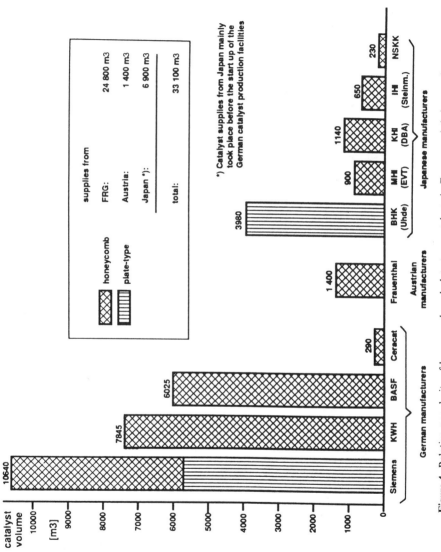

Figure 4. Relative popularity of honeycomb and plate-type catalysts in Europe and their manufacturers. (Siemens AG, 1990, with permission.)

3.2.1. Catalyst Composition

Many catalyst compositions are cited for use in the reduction of NO_x and their application range has been widened by the development of new catalyst systems such as low- and high-temperature catalysts (Jónes, 1981; Jung et al., 1987; Bosch and Janssen, 1987; Williams et al., 1988). The transition metal oxides of group B in the periodic table mainly act as materials for the catalyst. The metals include: Cu, Ag; Zn; Ti, Zr; V, Ni, Ta; Cr, Mo, W, Mn, Te, Re; and all elements in group VIII (Chen et al., 1990). However, the composition of a catalyst should be specified on the basis of the following: required NO_x reduction, flue gas temperature, dust load, conversion of SO_2 to SO_3, permissible NH_3 slip, presence of alkaline metals, and presence of arsenic.

The following types of catalysts are generally in use in SCR reactors in power plants throughout Japan and Europe (Jones, 1981; Jung et al., 1987; Williams

Table 2   An Example of a Typical Catalyst for the SCR Process

Compensation	
TiO_2	> 90 wt.%
Active Compounds	
V_2O_5	$1-5$ wt.%
MoO_3	
+	$5-10$ wt.%
WO_2	
Operating Temperature:	$270-400\,°C$
Dimensions	
Plate Catalyst	
Plate distance	$6-10$ mm
Plate thickness	$1.5-2$ mm
Plate length	$450-650$ mm
Cross-sectional area	464×464 mm^2
Specific surface area	$250-500$ m^2 m^{-3}
Extruded Catalyst	
Pitch size	$3.7-7.4$ mm
Wall thickness	$1.0-2.4$ mm
Length	$350-1000$ mm
Cross-sectional area	150×150 mm^2
Specific surface area	$427-860$ m^2 m^{-3}
Space velocity	
High-dust	$1000-3000$ m hr^{-1}
Tail gas	$5000-10000$ m hr^{-1}
Oil	$5000-10000$ m hr^{-1}
Natural gas	$5000-10000$ m hr^{-1}
Area velocity	$6-12$ m hr^{-1}

Source: ECE (1992).

et al., 1988; Hjalmarsson, 1990; ECE, 1992): titanium oxide, iron oxide, zeolite, and activated coke. Details of the active compounds in catalysts are not readily available, but the ECE (1992) has compiled information on the composition and dimensions of a typical TiO$_2$-based catalyst, as shown in Table 2.

3.2.2. Catalyst Deactivation

Catalyst life and activity are adversely affected mainly by poisoning, sintering and dust deposition, and erosion. Coal contains many harmful elements. During the combustion process, the compounds of these elements exist in the flue gas and fly ash and are adsorbed by the active sites of the catalyst, resulting in "catalyst poisoning." These harmful compounds may be subdivided into two groups: alkali metal oxides and heavy metals, especially arsenic.

Alkali Metal Oxides. Deactivation of V$_2$O$_5$/TiO$_2$-based catalysts by alkali metal oxides has been studied in detail (Shikada and Fujimoto, 1983; Yoshida et al., 1984; Kasaoka et al., 1984). When vanadium is used as an active material in the catalyst, it may be deactivated due to the formation of alkali-vanadium compounds (Martin et al., 1988). Chen et al. (1990) observed that the poisoning effect of CaO is less than that of other alkaline metal oxides. Hjalmarsson (1990) pointed out that when fly ash contains less than 10 to 15% (wt %) total alkaline metal oxides (CaO, MgO, Na$_2$O, and K$_2$O), its poisonous effect on the catalyst is less. This phenomenon is generally observed with brown coal, in which the CaO in the ash is quite high. In the United States, this problem is also observed with eastern bituminous coal, which contains large amounts of alkali metals. In Finnish coal-fired power plants, the fly ash normally contains about 6.5% CaO, 4.5% MgO, 0.7% Na$_2$O, and 2.2% K$_2$O.

Heavy Metals, especially Arsenic. Arsenic is poisonous to SCR catalysts but is less poisonous than the alkali metals, except for lithium (Yang et al., 1989). Arsenic poisoning was reported in wet bottom boilers with ash recycling in Germany (Necker, 1989), as well as in pilot plant tests conducted by Balling and Hein (1989).

Coal contains arsenic, which, when subjected to temperatures above 1300 °C in the boiler, is converted to gaseous arsenic trioxide (As$_2$O$_3$) (Schönbucher, 1987).

Arsenic concentration in the flue gas also depends upon the presence of CaO in the fly ash. As$_2$O$_3$ reacts with CaO above 500 °C to form calcium arsenite in the following reaction:

$$3CaO + As_2O_3 + O_2 \longrightarrow Ca_3(AsO_4)_2 \qquad (5)$$

Experiments conducted in the Datteln K1 power plant indicate that CaO can suppress arsenic poisoning. A 1% addition of CaO to coal increased the CaO in the granulate by 3.8%; at the same time, the arsenic content of the flue gas dropped from 0.55 to 0.1 mg m^{-3} (Fig. 5) (Gutberlet, 1988). Hence, the CaO

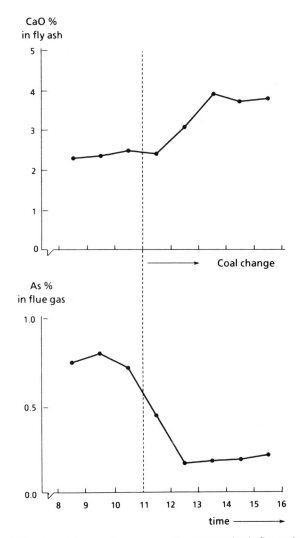

Figure 5. CaO addition to coal decreased gaseous arsenic concentration in flue gas in an SCR reactor at the Datteln K1 power plant in Germany. (Redrawn from Gutberlet, 1988.)

content in coal should be between 3 and 5% to suppress the arsenic content in the flue gas (ECE, 1992).

Siemens AG has also effectively studied the deactivation of catalysts by arsenic poisoning (Balling and Hein, 1989). In dry-bottom boilers, the only source of arsenic is the fuel, whereas in wet-bottom boilers, the fuel, recirculation ash, and melting of ash in the combustion chamber are arsenic sources. Here it was stressed that it is the gaseous arsenic that deactivates the catalyst, and not the

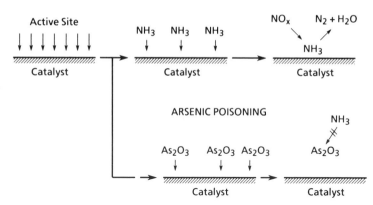

Figure 6. The mechanism of catalyst poisoning by arsenic. (Redrawn from Kuroda et al., 1989.)

arsenic in the ash. It has already been mentioned that it is possible to suppress this problem by adding lime, but pilot plant studies carried out by Siemens AG (1990) indicate that dosing with lime is not a permanent solution to this problem because the activity of the catalyst is eventually affected.

The arsenic content of the coal, the operation mode, and the reactivity of the fly ash may all affect arsenic poisoning. The mechanism of catalyst poisoning by arsenic and how the NO$_x$ efficiency decreases with the increase of arsenic on the active sites of the catalyst are depicted in Figures 6 and 7, respectively. Gaseous arsenic apparently has the strongest poisonous effect on the active surface of the catalyst. But during periods of shutdown, if the moisture (i.e., humidity) cannot be controlled in the SCR reactor, then the arsenic in the fly ash may also have a poisonous effect on the catalyst. However, all these problems can be reduced by the following: less fly ash recycling in the wet-bottom boiler, low ash content in the coal, presence of the active compound in the catalyst, additives in the fly ash, control of humidity in the reactor during shutdown.

Chen et al. (1990) observed the poisoning effect of lead on the SCR catalyst, but it was not as strong as that of other alkali metals. The poisonous effect of selenium is the same as that of arsenic (Gutberlet, 1988), but there is no detailed study of selenium poisoning in fuel-fired power plants.

Dust Deposition and Sintering. The pores of the catalyst may be blocked by the deposition of dusts through capillary condensation and of fine fly ash. Hence, the choice of catalyst geometry and the position of the catalyst in the gas flow line are very important in avoiding pore blockage. On the other hand, condensation of water drops on the surface of the catalyst occurs at lower temperatures and the water may dissolve species of fly ash, thus reducing the activity of the catalyst. However, during SCR processes, it is believed that fine dust deposits on the surface of the catalyst cause a greater deactivation problem than arsenic poison-

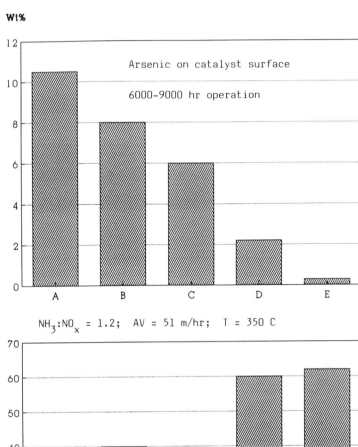

Wt%

Arsenic on catalyst surface

6000-9000 hr operation

$NH_3:NO_x = 1.2$;   AV = 51 m/hr;   T = 350 C

Plant

Fuel firing system	Hard coal	
	Wet bottom	
Recirculation-rate 100%		0%

Figure 7. De-NO$_x$ efficiency decreases with increase of arsenic on the catalyst surface. (Redrawn from Kuroda et al., 1989.)

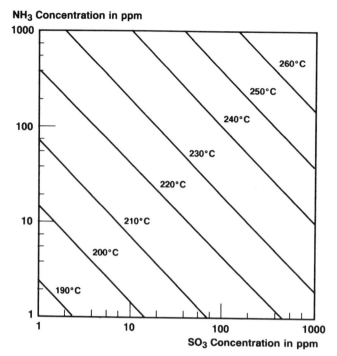

NH$_3$ Concentration in ppm

SO$_3$ Concentration in ppm

Figure 8. The dew point of ammonium bisulfate. (Redrawn from Ando et al., 1977.)

ing (personal communication with P. A. Lowe, 1991). This is probably due to masking of the surface of the catalyst.

There are many parameters, such as the presence of V_2O_5 in catalysts, process temperature, the fuel firing system, the boiler load, and oxygen concentration, that influence the oxidation of SO_2 in the flue gases. The SO_3 produced reacts with NH_3, forming ammonium sulfate and bisulfate as follows:

$$NH_3 + SO_3 + H_2O \longrightarrow (NH_4)HSO_4 \tag{6}$$

$$NH_3 + (NH_4)HSO_4 \longrightarrow (NH_4)_2SO_4 \tag{7}$$

However, condensation and deposition of ammonium bisulfate generally occur only when the temperature falls below the recommended operation temperature or dew point of this compound (Robie and Ireland, 1990). The dew point of ammonium bisulfate, which depends upon the concentration of NH_3 and SO_3 in the flue gas, is shown in Figure 8.

Sintering of the catalyst, which is a change of crystalline structure, reduces the number of pores and active sites. It generally occurs at higher temperatures. Robie and Ireland (1990) pointed out that under normal operating conditions, catalysts are damaged only slowly by the sintering mechanism. This type of damage is not recoverable. At higher temperatures, the sintering of dusts and deposition of fly ash may totally destroy the catalyst.

Erosion. Erosion is mainly a function of the gas velocity in the catalyst (Cichanowicz and Offen, 1987). Pilot plant experiments at Rheinisch-Westfälischen Electrizitätswerke indicated that even abrasive fly ash did not cause any erosion problems in catalysts. It must be said, however, that this pilot plant is 25% oversized, which may explain why the catalyst is able to resist erosion. Hjalmarsson (1990) pointed out that the presence of large amounts of fly ash and the uneven concentration of particulates and size distribution may be the main causes of erosion in catalysts.

3.2.3. Catalyst Lifetime

Catalyst lifetime strongly affects the cost of the SCR process. The lifetime guarantees vary from one vendor to another, depending especially on a limited number of fuel properties. Catalyst deactivation modes such as surface masking by bisulfate compounds, arsenic poisoning, sulfur content, fly ash, and alkaline metal compounds should also be kept in mind. Experience in Japan suggests that the lifetime of an SCR catalyst is six to seven years for gas-fired boilers, five years for oil-fired boilers, and three years for coal-fired boilers (Ando, 1983). Some recent data on guaranteed and typical expected lifetimes of SCR reactors in western Europe are shown in Table 3. In reality, there will be exceptions. In Germany for example, a wet-bottom boiler with 100% dust recirculation experienced shorter catalyst lifetime due to arsenic poisoning.

3.2.4. Catalyst Maintenance

Correct maintenance and repair work will increase plant efficiency and prolong catalyst life. Successful maintenance depends to a great extent on careful study of the daily operation of the plant. The following process data are key to the success

Table 3 Typical Lifetime of SCR Catalyst in Western Europe

Application	Guaranteed Lifetime (Year)	Expected Lifetime (Year)
Coal		
High dust	2	4
Low dust	2	6
Tail end	2	6
Oil	3	6
Natural gas	6	8

Source: Robie and Ireland (1990).

of maintenance: NO$_x$ concentration, inlet and outlet temperature of the flue gas in the reactor, O$_2$ and SO$_2$ content in the flue gas, NH$_3$ injection, mixing air volume, running hours, pressure loss across the catalyst, NH$_3$ slip, temperature in the catalyst.

3.2.5. *Disposal of Deactivated Catalyst*

There is no information on the correct disposal of spent catalysts, since nearly all the original catalysts are still in use. As for fuel used, catalyst lifetime increases in the following order: coal < oil < natural gas.

When a catalyst is deactivated, the catalyst manufacturer should collect all the spent catalyst, since its composition is proprietary (Ando, 1985a). Recently, Japanese suppliers have observed that the recovery of metals from the catalyst is quite costly, so they simply crush the spent catalysts and bury them in landfills (personal communication with P. A. Lowe, 1991). It is also reported that in Japan spent catalyst is used as a raw material for portland cement (personal communication with M. Yanai, 1990). In Germany, export of catalyst wastes is prohibited (Cichanowicz and Offen, 1987). In Finland, if the compounds in a catalyst are classified as hazardous, they are treated in the hazardous waste disposal plant in Riihimäki. This means that disposal costs are quite high. However, disposal costs and mode of disposal vary considerably from one country to another.

In the United States, the EPA has classified spent catalyst as a hazardous waste that must be disposed of in a class-one landfill. An independent company in the United States has estimated the disposal costs of spent catalyst to be US$315 per ton (Robie and Ireland, 1990). Research is continuing in Japan and Germany on catalyst recycling techniques (Cichanowicz and Offen, 1987).

3.3. SCR Systems

The arrangement of the catalyst in the flue gas flow depends on the gas temperature and its composition. On this basis, the SCR reactor is placed in one of three different positions known as the high-dust, low-dust, and tail-end configurations. The system arrangements are shown in Figure 9 and the common arrangements in Europe and Japan in Figure 10.

3.3.1. *High-dust Configuration*

In the high-dust configuration, the NO$_x$ reduction plant is placed just behind the steam generator and upstream of the air preheater, as shown in Figure 9. The main advantage of this arrangement is that preheating of the flue gas is not necessary. But clogging due to dust load and the presence of SO$_2$ gas from the combustion of coal and oil may reduce the NO$_x$ reduction rate. If the flue dust is recirculated, arsenic poisoning may also occur. If the dust load is abnormal, soot blowers become necessary, adding to capital and operating expenses. The flue gas temperature in the reactor normally varies between 300 and 400 °C but possibly can be reduced to 300 to 350 °C (Lowe et al., 1989). More reactors are needed in this arrangement than in the other two.

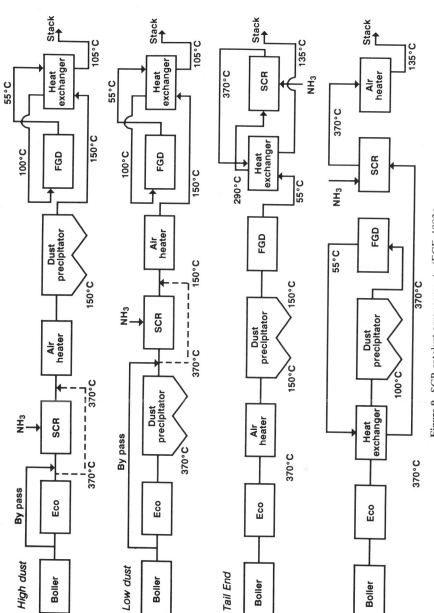

Figure 9. SCR catalyst arrangements. (ECE, 1992.)

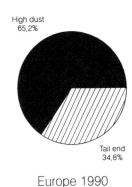
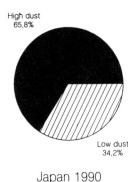

High dust
65,2%

Tail end
34,8%

High dust
65,8%

Low dust
34,2%

Europe 1990

Japan 1990

Figure 10. Trends in SCR catalyst arrangements in Europe and Japan. (Redrawn from Hjalmarsson, 1990.)

The principal advantages of the high-dust system are low capital investment and operating costs, provided it is installed in a new plant or integrated into an existing boiler system in such a way that no special problems are encountered (Siemens AG, 1990).

3.3.2. Low-dust Configuration

In this arrangement, the catalyst is situated between the hot dust precipitator and the air preheater. The flue gases entering the SCR reactor therefore contain very little or no dust. But the gases are richer in SO$_2$. This arrangement is quite common in Japan, where there is no stringent residual dust content requirement. Sooth blowers are necessary in this system because fine particles 0.1 to 3.0 μm in size have a tendency to be deposited or to agglomerate on the surface of the catalyst (Ando, 1985b). In addition, there is a tendency for ammonium bisulfate compounds to be deposited in the air preheater. The induced draft fan may be affected by the ammonium bisulfate compounds, and a special design may therefore be required. Fly ash caught by the cold precipitator may create a disposal problem, as some unreacted ammonia may pass through with the fly ash. To minimize this, NH$_3$ slip should be kept below 3 ppm (Ando, 1985b).

3.3.3. Tail-end Arrangement

This variant is almost exclusively employed in western Europe due to space limitations and catalyst deactivation, which would result from fly ash recirculation. In this arrangement, the SCR unit is placed at the end of the chain of the flue gas cleaning equipment. [i.e., after the flue gas desulfurization (FGD) unit]. This means that the flue gas contains very little dust and SO$_2$ when entering the catalyst. After the FGD unit, the flue gas temperature is quite low (55–70 °C), which is not suitable for the catalyst. For this reason, reheating of the flue gas is necessary. This is done partly by utilizing waste heat from the process, but there is still a temperature difference (20–40 °C) that must be compensated for by fuel-fired burners or by a heat exchanger.

3.4. Impact of Ammonia Slip

Ammonia slip is the unreacted ammonia that leaves the catalyst reactor with the flue gases. This unreacted NH_3 can react with SO_3 in the gases, forming ammonium compounds according to Reactions (6) and (7). These compounds have a damaging effect on various parts of the system, including the air preheater, FGD system, induced draft fan, wastewater system, and the management and disposal of the fly ash.

It is noted that problems with ammonium bisulfate are quite serious when using medium or high sulfur-content coals. If the SO_3 concentration exceeds the NH_3 concentration in the flue gas, its deposition on the air preheater tubes accelerates corrosion and increases pressure drop, causing damage or shutdown of the air preheater. In the case of low-sulfur coal or oil, the NH_3 concentration may exceed the SO_3 concentration in the outlet flue gas of the reactor, which will lead to the formation of dry ammonium sulfate powder that will not foul the air preheater (Robie and Ireland, 1990). However, to achieve better reduction of NO_x and a longer lifetime for equipment, NH_3 slip should be kept to a minimum (0.5–2 vppm) (ECE, 1992).

4. ENVIRONMENTAL IMPACT OF THE SCR PROCESS

It is possible to reduce NO_x emissions from utility boilers by 80 to 90% using pollution control technology based on the SCR process, in which NH_3 injection plays an important role. Due to the incomplete reaction of NH_3 with NO_x, some NH_3 will be emitted to the environment from the process. The amount of unreacted NH_3 depends upon the NH_3/NO_x molar ratio and the activity of the catalyst. The reaction of SO_3 with NH_3 and its effects on the downstream equipment have been discussed earlier. It is estimated that 80% of the unreacted NH_3 salts are discharged with the fly ash, as can be seen from the materials balance of NH_3 depicted in Figure 11. In Germany, the removal efficiency depends on the form of the NH_3 in the flue gas, the pH of the scrubbing liquor, and the type of conductor. Here, the NH_3 salt is in the submicron size, for which a high-pressure conductor is more effective.

If the fly ash contains a large amount of NH_3, there is a risk of NH_3 emissions from fly ash to the environment. The process may also have an impact on groundwater if NH_3 leaches from ash that has been used as landfill. The NH_3-bearing wash water from the air preheater and FGD system may also have an effect on the groundwater if it is not denitrified in a wastewater treatment plant before discharge.

It is reported (Jones, 1981; Jones et al., 1982) that NH_3 salts or gaseous NH_3 may form a visible stack plume. The chance of a visible plume forming is minimized when the flue gas passes through a cold electrostatic precipitator, which removes nearly all the NH_3 salts. However, systems using a hot electrostatic precipitator may have the problem of $(NH_4)_2SO_4/NH_4HSO_4$ plume formation. This can be avoided if a high-pressure drop mist eliminator is used in

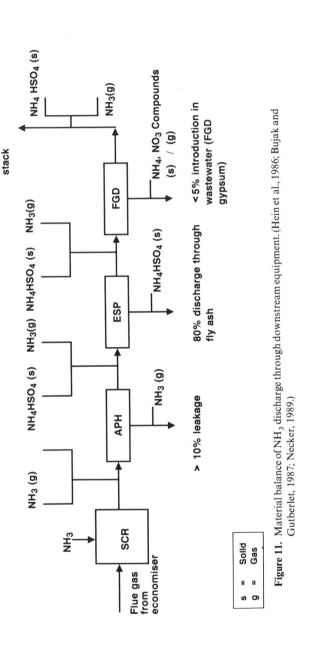

Figure 11. Material balance of NH_3 discharge through downstream equipment. (Hein et al., 1986; Bujak and Gutberlet, 1987; Necker, 1989.)

the FGD equipment. A visible plume may also form due to the reaction of gaseous NH_3 and SO_2 (g). This happens only when NH_3 discharge from the stack exceeds 10 ppm (Jones et al., 1982). No reports on plume formation from the SCR process have been cited in the literature, however. Under normal operating conditions, NH_3 emission from the stack is negligible and as a result, the risk plume formation is non-existent or negligible.

ACKNOWLEDGMENTS

In 1990, a feasibility study funded by Imatran Voima Oy was carried out on the SCR process. This paper is taken from that study and the author wishes to thank the management of Imatran Voima Oy for their permission to publish the paper. The author also wishes to extend his gratitude to many companies and educational establishments for providing valuable documents and information.

It is impossible to mention the names of all the contributors, but I am particularly indebted to Professor O. Rentz, University of Karlsruhe, Germany; Ms A-K. Hjalmarsson, IEA Coal Research, England; Dr. P. A. Lowe, INTECH Inc., USA; Mr. W. Ellison, Ellison Consultants, USA; Mr. C. B. Sedmann, U.S. Environmental Protection Agency; Mr. J. E. Cichanowicz, EPRI, USA; Mr. S. Oikarinen, Ministry of Trade & Industry; Mr. H. Hahkala, Imatran Voima Oy, Finland; and Mr. Beer, Siemens AG, Germany, and Babcock Hitachi K.K.

Finally, I am also indebted to Mr. J. Lepikko, Imatran Voima Oy, for his comments on the original draft, and the staff of the Department of Limnology and Environmental Protection at Helsinki University for their assistance and cooperation during the course of the study.

REFERENCES

Amann, M. (1989). *Potential and Costs for Control of NO_x Emissions in Europe,* SR-89-1. International Institute for Applied Systems Analysis (IIASA), Laxenburg, Austria.

Ando, J. (1983). *NO_x Abatement for Stationary Sources in Japan,* EPA-600/7-83-027. U.S. Environmental Protection Agency, Industrial Environmental Research Lab., NC.

Ando, J. (1985a). *Recent Development in SO_2 and NO_x Abatement Technology for Stationary Sources in Japan,* EPA-600/7-85-040. U.S. Environmental Protection Agency, Office of Research and Development, Washington DC.

Ando, J. (1985b). Review of Japanese NO_x abatement technology for stationary sources. Paper presented at NO_x Symposium Karlsruhe 1985, University of Karlsruhe, Karlsruhe, Germany, pp. A1–42.

Ando, J. (1987). *SO_2 and NO_x Control Technology Development in Japan* (draft report to U.S. Environmental Protection Agency).

Ando, J., Tohata, H., Nagata, K., and Laseke, B. A. (1977). *NO_x abatement for stationary sources in Japan,* EPA-600/7-77-103b. U.S. Environmental Protection Agency, Office of Research and Development, Washington, D.C. 20460.

Andreasen, J., and Morsing, P. (1990). *A Look at Selective Reduction of Nitrogen Oxides Produced in Power Generation.* Int. Power Generation Magazine, United Kingdom.

Balling, L., and Hein, D. (1989). De-NO_x catalytic for various types of furnaces and fuels—development, testing, operation. In *Proceedings of the Joint EPRI/EPA 1989 Symposium on Stationary Combustion NO_x Control,* EPRI GS-6433. Electric Power Research Institute, Palo Alto, CA, Vol. 2, pp. 7A-27–7A-40.

Bosch, H., and Janssen, F. (1987). Catalytic reduction of nitrogen oxides—A review on the fundamentals and technology. *Catal. Today* **2**(4), 369–521.

Bujak, W., and Gutberlet, H. (1987). Untersuchungen zum Eintrag von NH$_3$ aus deNO$_x$-Anlagen in REA-Abwasser. *VGB Kraftwerkstechnik* **67**, 876–883.

Chen, J. P., Buzanowiski, M. A., Yang, R. T., and Cichanowicz, J. E. (1990). Deactivation of the vanadia catalyst in the selective reduction process. *J. Air, Water, Manage. Assoc.* **40**, 1403–1409.

Cichanowicz, J. C., and Offen, G. R. (1987). Applicability of European SCR experience to U.S. utility operation. In *Proceedings of the Joint EPRI/EPA 1987 Symposium on Stationary Combustion NO$_x$ Control,* EPRI. Electric Power Research Institute, Palo Alto, CA, pp. 28-1–28-18.

Curtzen, P. J., Isaksen, I. S., and McAfee, J. R. (1978). The impact of chlorcarbon industry on the ozone layer. *J. Geophys. Res.* **83**, 345–363.

Economic Commission for Europe (ECE) (1986). *Technologies for Controlling NO$_x$ Emissions from Stationary Sources,* Report of the NO$_x$ Task Force. Published by Institute for Industrial Production, University of Karlsruhe, Karlsruhe, Germany.

European Commission for Europe (ECE) (1988). *Annual Review of Strategies and Policies for Air Pollution Abatement,* Report EB.AIR/R.32, Paris.

European Commission for Europe (ECE) (1992). In O. Rentz (Ed.), *Operating Experiences with NO$_x$ Abatement at Stationary Sources,* Report of the NO$_x$ Task Force. Institut für industriebetrieblehre und industrielle produktion, University of Karlsruhe, Karlsruhe, Germany (in draft).

Gutberlet, H. (1988). Einfluss der Feuerungsart auf die Vergiftung von DeNO$_x$ Katalysatoren durch Arsen [Influence of furnace type poisoning of DeNO$_x$ catalysts by arsenic]. *VGB Kraftwerkstechnik* **68**(3), 297–293.

Hameed, S., and Dignon, J. (1988). Changes in the geographical distributions of global emissions of NO$_x$ and SO$_x$ from fossil-fuel combustion between 1966 and 1980. *Atmos. Environ.* **22**(3), 441–449.

Hein, K. R. G., König, J., and Hoppe, V. (1985). *1985 EPA/EPRI Joint Symposium on Stationary Combustion NO$_x$ Control,* Boston, MA, Paper No. 8a-2.

Hein, K. R. G., König, J., and Bals, M. (1986). Problematik des ammoniakschlupfes bei Anlagen zur NO$_x$-Minderung. *VGB Kraftwerkstechnik* **66**(9), 861–866.

Hjalmarsson, A.-K. (1990). *NO$_x$ Control Technologies for Coal Combustion,* IEACR/24. IEA Coal Research, London.

Hjalmarsson, A.-K., and Soud, H. N. (1991). *NO$_x$ Control Installations on Coal-fired Plants,* IEACR/34. IEA Coal Research, London, pp. 1–23.

Jones, G. D. (1981). *Selective Catalytic Reduction and NO$_x$ Control in Japan,* USEPA 600/S 7-81-030. U.S. Environmental Protection Agency, Research Triangle Park, NC, pp. 1–5.

Jones, G. D., Glover, R. L., Behrens, G. P., and Shirley, T. E. (1982). *Impact on NO$_x$ Selective Catalytic Reduction Processes on Flue Gas Cleaning Systems,* EPA-600/S7-82-025b. U.S. Environmental Protection Agency, Research Triangle Park, NC, pp. 1–7.

Jung, H. J., Becker, E. R., Lis, R. E., and Keck, L. (1987). *Proc. 80th Annu. Meet. and Exhib. Air Pollut. Control Assoc.,* New York, Paper No. 87–52.5.

Kasaoka, S., Sasaoka, E., and Nanba, H. (1984). Deactivation mechanism of vanadium pentaoxide-titanium dioxide catalyst by deposited alkali salts and regeneration method of deactivated catalyst in reduction of nitric oxide with ammonia (in Japanese). *Nippon Kagaku Kaishi,* p. 486.

Klovsky, J. R., Kordia, P. B., and Lim, C. T. (1980). Evaluation of a new zeolite catalyst for NO$_x$ reduction with NH$_3$. *Ind. Eng. Chem. Prod. Res. Dev.* **19**, 218.

Kolar, J. (1990). *Stickstoffoxide und luftreinhaltung.* Springer-Verlag, Berlin.

Kuroda, H. (1985). Babcock Hitachi NO$_x$ abatement technology. In *NO$_x$ Symposium Karlsruhe 1985.* University of Karlsruhe, Karlsruhe, Germany, pp. L1–L39.

Kuroda, H., Morita, I., Murataka, T., Nakajima, F., Kato, Y. and Kato, A. (1989). Recent developments in the SCR system and its operational experiences. In *Proceedings of the joint EPRI/ EPA*

1989 Symposium on Stationary Combustion NO_x Control, EPRI GS-6423. Electric Power Research Institute, Palo Alto, CA, Vol. 2, pp. 6A.39–6A.55.

Lachman, I. M. (1986). Ceramic honeycombs for catalysts and industrial applications. *Fachber. Sprechsaal* **119**(12), 1116.

Logan, J. A. (1983). Nitrogen oxides in the tropospheric, global and regional budgets of NO_x. *J. Geophys. Res.* **90**, 10463–10482.

Lowe, P. A. (1984). Review of Japanese NO_x control technology retrofitted to coal-fired boilers. NUS 4465, January, USA.

Lowe, P. A., Ellison, W., and Radak, L. (1989). Assessment of Japanese SCR technology for oil-fired boilers and its applicability in the U.S.A. In *Proceedings of the joint EPRI/EPA 1989 Symposium on Stationary Combustion NO_x Control*, EPRI GS-6423. Electric Power Research Institute, Palo Alto, CA, Vol. 2, pp. 7A.41–7A.52.

Martin, C., Rivers, V., and Gonzalez-Elipe, A. R. (1988). Effect of sodium on the reducibility of V ions during propene adsorption on V_2O_5/TiO_2 catalyst. *J. Catal.* **114**, 473.

Ministry of the Environment, UK (1990). *Oxides of Nitrogen in the United Kingdom*. Building 1, Victoria Road, South Ruislip, Middlesex HA4 ONZ, England, pp. 1–16.

National Academy of Science (NAS) (1979). *Stratospheric Ozone Depletion by Halocarbons: Chemistry and Transport*. NAS, Washington, DC.

Necker, P. (1989). Experience gained by Neckwerke from operation of SCR $DeNO_x$ units. In *Proceedings of the Joint EPRI/EPA 1989 Symposium on Stationary Combustion Nitrogen Oxide Control*, EPRI GS-6423. Electric Power Research Institute, Palo Alto, CA, Vol. 2, pp. 6A19–6A38.

Organization for Economic Co-operation and Development (OECD) (1983). *Control Technology for Nitrogen Oxide Emissions from Stationary Sources*. OECD, Paris, pp. 51–56.

Organization for Economic Cooperation and Development (OECD) (1989). *Control of Major Air Pollutants—Emission Inventory for OECD Europe*, ENV/AIR/87.7. Air Management Policy Group, OECD, Paris.

Rentz, O., and Leibfritz, R. (1987). Overview of recent developments in NO_x control in Europe. In *Proceedings of the Joint EPRI/EPA Symposium on Stationary Combustion Nitrogen Oxide Control*, EPRI CS-5361. Electric Power Research Institute, Palo Alto, CA, Vol. 1, pp. 5.1–5.27.

Robie, C. P., and Ireland, P. A. (1990). *Technical Feasibility and Cost of SCR NO_x Control Utility Applications*. United Engineers and Constructors, Denver, CO.

Schönbucher, B. (1987). Costs of $DeNO_x$ plant on the basis of the SCR process. Unit 7 of Energie-Versorgung Schwaben AG at Heilbronn power station. In O. Rentz, J. Remmers, and E. Plinke (Eds.), *Workshop on Emission Control Costs: Methodology and Example cases*. Institute for Industrial Production, University of Karlsruhe, Karlsruhe, Germany.

Sengoku, T., Todo, Y., Yokoyama, N., and Howell, B. M. (1980). The development of a catalytic No_x reduction system for coal-fired steam generators. Paper presented at the EPA/EPRI Joint Symposium on Stationary Combustion NO_x Control, Denver, CO.

Shikada, T. and Fujimoto, K. (1983). Effect of added alkali salts on the activities of supported vanadium oxide catalysts for nitric oxide reduction. *Chem. Lett.* p. 77.

Shimoto, G. H., and Muzio, L. J. (1986). *Selective Catalytic Reduction for Coal-fired Power Plants–Pilot Plant Results*, Research Project No. 1256. Electric Power Research Institute, Palo Alto, CA, CS-4386.

Siemens, A. G. (1990). *Survey about Siemens de–NO_x Technology*. Erlangen, Germany.

Singh, H. B. (1987). Reactive nitrogen in the troposphere. *Environ. Sci. Technol.* **21**(4), 320–326.

Singh, H. B., Fowler, D. P., and Peyton, T. O. (1976). Atmospheric carbon tetrachloride: Another man-made pollutant. *Science* **192**, 1231–1234.

Slack, A. V. (1980). *Applicability of Japanese NO$_x$ Control in the U.S.*, DOE/MC/15091–1612.

Sloss, L. (1990). *Nitrogen Oxides from Coal Combustion.* IEA Coal Research, London.

Springmann, F. (1989). *Emissions in Europe,* Working paper. International Institute for Applied Systems Analysis (IIASA), Laxenburg, Austria (in manuscript).

Tripac, D. A. (1987). The role of nitrous oxides (N$_2$O) in global climate change and stratospheric ozone depletion. In *Proceedings of the Joint EPRI/EPA 1987 Symposium on Stationary Combustion Nitrogen Oxide Control,* EPRI CR-5361. Electric Power Research Institute, Palo Alto, CA, Vol. 1, pp. 4.1–4.10.

Williams, J. L., Lachman, I. M., and Rosenbusch, T. F. (1988). Mordenite-titania-silica honeycombs catalyzed for deNO$_x$. *Proc. 81st Annu. Meet. Exhib. Air Pollut. Control Assoc.,* Dallas, Texas, Paper No. 88–83.3.

World Meteorological Organization (WMO) (1986). *Atmospheric Ozone 1985: Assessment of our Understanding of the Processes Controlling its Present Distribution and Change,* Global Ozone Research and Monitoring Project Report No. 16, 3 vols. WMO, Geneva.

Yang, R. T., Chen, J. P., Buzanowski, M. A., and Cichanowicz, J. E. (1989). Catalyst poisoning in the selective catalytic reduction reaction. In *Proceedings of the Joint EPRI/EPA 1989 Symposium on Stationary Combustion Nitrogen Oxide Control,* EPRI GS-6423. Electric Power Research Institute, Palo Alto, CA, Vol. 2, pp. 8.1–8.7.

Yoshida, H., Takahashi, K., Sekiya, Y., Morikawa, S., and Kurita, S. (1984). Study of deterioration behavior of catalyst for reduction of NO$_x$ with ammonia. *Proc. Int. Conf. Catal., 8th, 1983,* Berlin, pp. 111–649.

INDEX

609